One-Stop In

Log on to **il.msmath2.net**

Illinois Online Book
- Complete Student Edition
- Links to Online Study Tools

Illinois Online Study Tools
- Chapter Readiness
- Extra Examples
- Self-Check Quizzes
- Vocabulary Review
- Chapter Test Practice
- Standardized Test Practice

Online Activities
- WebQuest Projects
- Game Zone Activities
- Career Links
- Data Updates

Online Workbooks
- Practice: Skills
- Spanish Practice: Skills
- Practice: Word Problems
- Spanish Practice: Word Problems
- Math Skills Maintenance
- Parent and Student Study Guide
- Spanish Parent and Student Study Guide

Graphing Calculator Keystrokes
- Calculator Keystrokes for other calculators

Illinois Edition

Mathematics

Applications and Concepts

Course 2

Cahokia Mounds near Collinsville

Contents

Illinois Teacher Advisory Board	IL2
Illinois Mathematics Assessment Framework, Grade 7	IL4
How to Prepare for the ISAT	IL6
Countdown to ISAT	IL8

 Glencoe

New York, New York · Columbus, Ohio · Chicago, Illinois · Peoria, Illinois · Woodland Hills, California

ISBN: 0-07-869342-X *(Illinois Student Edition)*

Illinois Teacher Advisory Board

Tina E. Beer
Mathematics Teacher
Algonquin Middle School
Des Plaines, Illinois

Lisa Doig-West
Instructional Coordinator
Frankfort School District 157-C
Frankfort, Illinois

Lesley A. Ebel
Mathematics and Science
 Teacher
Ira Jones Middle School
Plainfield, Illinois

Nancy O. Hogan
Mathematics Teacher
Bourbonnais Upper Grade
 Center
Bourbonnais, Illinois

Linda M. Jeeninga
Mathematics Teacher
East Middle School, Alton
 School District
Alton, Illinois

Katherine A. Kendall
Science and Math Resource
 Specialist
Joliet Public Schools District 86
Joliet, Illinois

Dr. Lavern W. Little
Middle School Manager
Chicago Public Schools
Chicago, Illinois

Gary F. Oscko
Mathematics Department
 Chair and Teacher
Lincoln Middle School
East St. Louis, Illinois

Vanessa Reese-Clark
Mathematics Teacher
Douglass Junior High School
Chicago, Illinois

Don Reinhard
Mathematics Teacher
Kaneland Middle School
Maple Park, Illinois

Mary M. Sliga
Mathematics Teacher
Herrick Middle School
Downers Grove, Illinois

Karen Vandewiel
Mathematics Teacher
Central Middle School
Tinley Park, Illinois

Photo Credits: IL1 Michael S. Lewis/CORBIS, **IL2** Jason Lindsey/Alamy, **IL5** CORBIS

Shedd Aquarium in Chicago

Illinois Mathematics Assessment Framework, Grade 7

Assessment Objective

STATE GOAL 6 NUMBER SENSE

Standard 6A–Representations and Ordering

6.7.01	Read, write, and recognize equivalent representations of positive powers of 10.
6.7.02	Read, write, recognize, model, and interpret integers, including translating numerical expressions.
6.7.03	Recognize, translate between, and apply multiple representations of rational numbers (decimals, fractions, mixed numbers, and percents less than 100%).
6.7.04	Represent repeated factors using exponents.
6.7.05	Order and compare integers, terminating decimals, fractions, and mixed numbers.
6.7.06	Identify and locate integers, decimals, and fractions/mixed numbers on a number line, and estimate the locations of square roots.
6.7.07	Solve problems involving descriptions of numbers, including characteristics and relationships (e.g., square numbers, prime/composite, prime factorization, greatest common factor, least common multiple).

Standards 6B, 6C–Computation, Operations, Estimation, and Properties

6.7.08	Solve problems and number sentences involving addition, subtraction, multiplication, and division using integers, fractions, and decimals.
6.7.09	Identify and apply order of operations to simplify numeric expressions involving whole numbers (including exponents), fractions, and decimals.
6.7.10	Identify and apply the following properties of operations with rational numbers: • the commutative and associative properties for addition and multiplication; • the distributive property; • the additive and multiplicative identity properties; • the additive and multiplicative inverse properties; and • the multiplicative property of zero.
6.7.11	Demonstrate and apply the relationships between addition/subtraction and multiplication/division with rational numbers.
6.7.12	Make estimates appropriate to a given situation, and analyze what effect the estimation method used has on the accuracy of results.
6.7.13	Estimate the square root of a number less than 1,000 between two whole numbers (e.g., $\sqrt{41}$ is between 6 and 7)

Standard 6D–Ratios, Proportions, and Percents

6.7.14	Create and explain ratios that represent a given situation.
6.7.15	Use proportional reasoning to model and solve problems.
6.7.16	Read, write, recognize, model, and interpret percents from 0% to 100%.
6.7.17	Solve number sentences and problems involving fractions, decimals, and percents (e.g., 50% of 10 is the same as 1/2 of 10 is the same as 0.5×10, sales tax, tips, interest, discounts).

STATE GOAL 7 MEASUREMENT

Standards 7A, 7B, 7C–Units, Tools, Estimation, and Applications

7.7.01	Select and use appropriate standard units and tools to measure length, mass/weight, capacity, and angles. Sketch, with given specifications, line segments, angles, triangles, and quadrilaterals.
7.7.02	Solve problems involving the perimeter and area of polygons and composite figures using diagrams, models, and grids or by measuring or using given formulas (may include sketching a figure from its description).
7.7.03	Compare and estimate length (including perimeter), area, volume, weight/mass, and angles (0° to 180°) using referents.

Illinois Mathematics: Applications and Concepts, Course 2   IL3

Illinois

	Assessment Objective
7.7.04	Determine the volume and surface area of a right rectangular prism using an appropriate formula or strategy.
7.7.05	Solve problems involving unit conversions <u>within the same measurement system</u> for length, weight/mass, capacity, and square units (e.g., 1 ft^2 = 144 in^2).
7.7.06	Solve problems involving scale drawings and maps.

STATE GOAL 8   ALGEBRA

Standard 8A–Representations, Patterns, and Expressions

8.7.01	Determine a missing term in a sequence, extend a sequence, and construct and identify a rule that can generate the terms of an arithmetic or geometric sequence.
8.7.02	Write an expression using variables to represent unknown quantities.
8.7.03	Simplify algebraic expressions by identifying and combining like terms.
8.7.04	Recognize equivalent forms of algebraic expressions.
8.7.05	Evaluate or simplify algebraic expressions with one or more integer variable values (e.g., $a^2 + b$ for $a = 3$ and $b = -4$).

Standard 8B–Connections Using Tables, Graphs, and Symbols

8.7.06	Determine how a change in one variable relates to a change in a second variable.
8.7.07	Represent linear equations and quantitative relationships on a rectangular coordinate system, and interpret the meaning of a specific part of a graph.
8.7.08	Translate between different representations (table, written, graphical, or pictorial) of whole number relationships and linear expressions.
8.7.09	Identify, graph, and interpret inequalities on a number line.

Standards 8C, 8D–Writing, Interpreting, and Solving Equations

8.7.10	Represent and analyze problems with linear equations and inequalities.
8.7.11	Solve linear equations in one variable (e.g., $2x + 3 = 13$) and inequalities involving $<$ or $>$ (e.g., $2x < 6$, $x + 7 > 10$).
8.7.12	Solve word problems involving unknown quantities.

STATE GOAL 9   GEOMETRY

Standard 9A–Properties of Single Figures and Coordinate Geometry

9.7.01	Classify, describe, and sketch regular and irregular two-dimensional shapes according to the number of sides, length of sides, number of vertices, and interior angles.
9.7.02	Solve problems involving two- and three-dimensional shapes.
9.7.03	Solve problems using properties of triangles and quadrilaterals (e.g., opposite sides of a parallelogram are congruent).
9.7.04	Identify, describe, and determine the radius and diameter of a circle.
9.7.05	Graph points and identify coordinates of points on the Cartesian coordinate plane (all four quadrants).
9.7.06	Represent and identify geometric figures using coordinate geometry.
9.7.07	Analyze the results of a combination of transformations.
9.7.08	Identify or analyze relationships of angles formed by intersecting lines.
9.7.09	Identify and sketch acute, right, and obtuse angles.
9.7.10	Solve problems involving complementary and supplementary angles.

Standard 9B–Relationships Between and Among Multiple Figures

9.7.11	Identify a three-dimensional object from its net.
9.7.12	Recognize which attributes (such as shape, perimeter, and area) change or don't change when plane figures are composed, decomposed, or rearranged.
9.7.13	Describe the difference between congruence and similarity.

Assessment Objective

9.7.14	Determine if figures are similar, and identify relationships between corresponding parts of similar figures.
9.7.15	Determine the distance between two points on a horizontal or vertical number line.

Standard 9C–Justifications of Conjectures and Conclusions

This standard is not assessed in isolation. Rather, its essence is assessed indirectly through problems that require this type of thinking.

Standard 9D–Trigonometry

This standard is not assessed on the state assessment until grade 11.

STATE GOAL 10 DATA ANALYSIS, STATISTICS, AND PROBABILITY

Standards 10A, 10B–Data Analysis and Statistics

10.7.01	Read, interpret, and make predictions from data represented in a bar graph, line (dot) plot, Venn diagram (with two circles), chart/table, line graph, scatter plot, circle graph, or histogram.
10.7.02	Compare different representations of the same data.
10.7.03	Create a bar graph, chart/table, line graph, or circle graph for a given set of data.
10.7.04	Identify a reasonable approximation of the line of best fit from a set of data or a scatter plot.
10.7.05	Determine and use the mode, range, median, and mean to interpret data.

Standard 10C–Probability

10.7.06	Solve problems involving the probability of a simple or compound event, including representing the probability as a fraction, decimal, or percent.
10.7.07	Represent all possible outcomes for simple events.
10.7.08	Solve simple problems involving the number of ways objects can be arranged (permutations and combinations).

State Animal
White-tailed Deer

How To...

Prepare for the ISAT

Countdown to ISAT

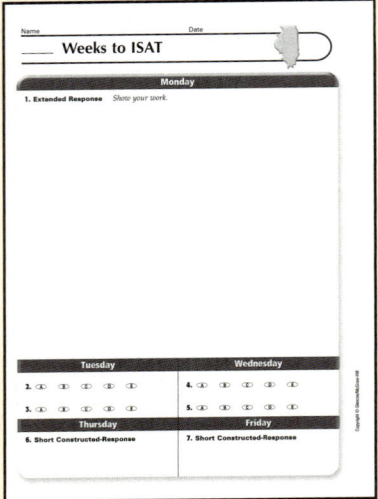

Pages IL8–IL32 of this text include a section called **Countdown to ISAT.** Each page contains 7 problems that are similar to those on the ISAT. You should plan to complete one page each week to help you prepare for the test.

Plan to spend a few minutes each day working on the ISAT problem(s) for that day unless your teacher asks you to do otherwise. If you have difficulty with any problem, you can refer to the lesson that is referenced in parentheses after the problem.

Each week is comprised of an extended response problem, 4 multiple-choice problems, and 2 short constructed response problems as listed below.

Monday	1 extended response
Tuesday	2 multiple choice
Wednesday	2 multiple choice
Thursday	1 short constructed response
Friday	1 short constructed response

Your teacher can provide you with an answer sheet to record your work and your answers for each week. A printable worksheet is also available at il.msmath2.net. At the end of each week, your teacher may want you to turn in the answer sheet.

ISAT Workbook, Grade 7

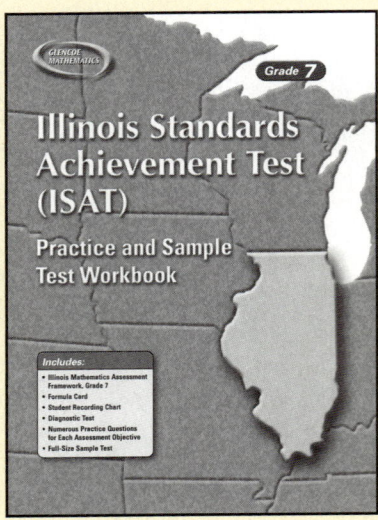

The **Illinois Standards Achievement Test (ISAT) Sample and Practice Test Workbook, Grade 7,** contains a diagnostic test, practice for each Illinois Mathematics Assessment Framework objective, and a sample test.

As you practice and master each objective, you can record your progress in the Student Recording Chart in your workbook.

Your teacher may also ask you to take the sample test at various points throughout the year to see if you're ready to take the real ISAT.

Illinois Mathematics: Applications and Concepts, Course 2

Your Textbook

Your textbook contains many opportunities for you to get ready for the ISAT every day. Take advantage of these so you don't need to cram before the test.

- **Each lesson** contains at least two Standardized Test Practice problems. You can use these problems every day to keep your ISAT skills sharp. The **Mid-Chapter Practice Test** and the **Chapter Practice Test** also include Standardized Test Practice problems.

- **Worked-out examples** in each chapter show you step-by-step solutions of Standardized Test Practice problems. Just like the practice problems, these problems model those that appear on the ISAT. **Test-Taking Tips** are also included.

- Two pages of **Standardized Test Practice** are included at the end of each chapter. These problems may cover any of the content up to and including the chapter they follow.

- Pages 608–613 and 618–625 of the **Preparing for Standardized Tests** section of your textbook discusses various strategies for attacking questions like those that appear on the ISAT. Additional practice problems are also provided.

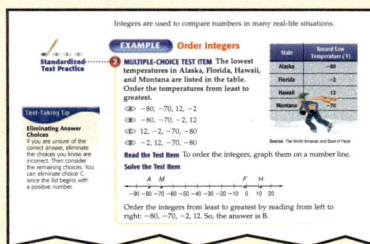

Test-Taking Tips

- ✓ Pace yourself. Don't spend too much time on one problem. You can come back if there is time.

- ✓ Listen to and read the directions carefully.

- ✓ Read each question carefully. It is important that you understand what each question asks.

- ✓ Answer the easy questions first. Then go back and try to answer the more difficult questions.

- ✓ Use logic in more difficult questions. Eliminate as many incorrect answers as you can, then make an educated guess from the remaining answers.

- ✓ Answer every question.

- ✓ Review your work. Check to be sure that you marked only one answer to each question.

- ✓ Be precise in marking your answer document. Check to be sure that the number for the line of circles on your answer document is the same as the number for the question you are answering.

- ✓ Erase completely. If you want to change an answer on your answer document, be sure to erase the unintended mark completely.

Illinois Mathematics: Applications and Concepts, Course 2

Countdown to ISAT

25 Weeks to ISAT

Monday

1. **Extended Response** Show your work and clearly explain your answer. These are some items from the menu of Downtown Pizza and Ice Cream. (Lesson 1-1)

   A. What is the total cost of two slices of pizza, one cheese and one pepperoni, and two scoops of ice cream?

   B. Create an order that costs exactly $4.75.

   C. Which order costs more?
      1) 2 slices of cheese and 1 scoop of ice cream with 3 toppings
      2) 1 slice of salami and olives and 2 scoops of ice cream with no toppings

MENU	
Item	Price
Cheese Slice	$2.00
+ 1 topping	$2.50
+ 2 toppings	$3.00
Ice Cream Scoop	$1.25
2 Scoops	$1.75
Ice Cream Toppings	Free!

Tuesday

2. **Multiple-Choice** Which is the result of evaluating the expression $4.5(7.1 - 2^2)^3 - 59.5 \div 1{,}000$? (Lesson 1-3)
   - Ⓐ 17.1
   - Ⓑ 36.5
   - Ⓒ 43.1855
   - Ⓓ 134
   - Ⓔ 134.0595

3. **Multiple-Choice** Jessica's class went on a field trip to JFK Health World in Barrington. If there were 22 students and one teacher on the trip and the total cost of admission was $115, which was the price of admission? (Children and adults pay the same price.) (Lesson 1-1)
   - Ⓐ $5.00
   - Ⓑ $5.50
   - Ⓒ $5.75
   - Ⓓ $6.00
   - Ⓔ $6.50

Wednesday

4. **Multiple-Choice** Fifteen liters of punch divided equally among 30 people is which number of milliliters per person? (Lesson 1-8)
   - Ⓐ 375
   - Ⓑ 450
   - Ⓒ 500
   - Ⓓ 550
   - Ⓔ 600

5. **Multiple-Choice** Which are the next three terms in the sequence 100, 94, 88, 82, …? (Lesson 1-7)
   - Ⓐ 78, 72, 66
   - Ⓑ 76, 70, 64
   - Ⓒ 80, 76, 72
   - Ⓓ 76, 71, 66
   - Ⓔ 74, 68, 62

Thursday

6. **Short Constructed-Response** Write the product $11 \cdot 11 \cdot 11 \cdot 11 \cdot 11 \cdot 11 \cdot 11$ in exponential form. (Lesson 1-2)

Friday

7. **Short Constructed-Response** Write the number ten million two hundred twenty-six thousand forty. (Prerequisite Skill Lesson 2)

24 Weeks to ISAT

Monday

1. **Extended Response** Show your work and clearly explain your answer. The table shows the population of Boone county for selected years. (Lesson 2-4)

   A. To the nearest person, what is the mean population of Boone County for this 40-year time period?

   B. What is the median population? What is the mode?

   C. In which decade did the population increase the most?

   Boone County Population

Year	Population
1960	20,326
1970	25,440
1980	28,630
1990	30,806
2000	41,786

   Source: http://www.igpa.uiuc.edu

Tuesday

2. **Multiple-Choice** Which is 0.2, 0.46, 0.2001, 0.199, and 0.5 ordered from greatest to least? (Prerequisite Skill Lesson 3)
   - A 0.199, 0.2, 0.2001, 0.46, 0.5
   - B 0.199, 0.2001, 0.2, 0.46, 0.5
   - C 0.5, 0.46, 0.2, 0.2001, 0.199
   - D 0.46, 0.5, 0.199, 0.2, 0.2001
   - E 0.5, 0.46, 0.2001, 0.2, 0.199

3. **Multiple-Choice** Which expression represents the number of blue marbles in a bag containing b blue, 8 red, and 14 green marbles? There are T marbles in all. (Lesson 1-4)
   - A $b - 8 - 14$
   - B $T - 8 - 14$
   - C $T + 8 + 14 - b$
   - D $bT - 8 - 14$
   - E $8b - 14T$

Wednesday

4. **Multiple-Choice** Mr. Tamaguchi had a barbecue. He cooked 3 packages of 8 hot dogs each, 22 hamburger patties, 1 package of 8 soy dogs, 8 chicken breasts, and 12 garlic sausages. If his guests ate 13 hot dogs, 17 hamburgers, half of the soy dogs, 5 chicken breasts, and half of the garlic sausages, which expression gives the number of food items he has left? (Lesson 1-3)
   - A $(3 \cdot 8 - 13) + (22 - 17) + 0.5(5) + 8 + 0.5(12)$
   - B $(3 \cdot 8 - 13) + 22 + 0.5(8) + 3 + 0.5(12)$
   - C $(3 \cdot 8 + 13) + (22 - 17) + 0.5(8) + 3 + 0.5(12)$
   - D $(3 \cdot 8 - 13) + (22 - 17) + 0.5(8) + 3 + 0.5(12)$
   - E $(3 \cdot 8 - 13) + (22 - 17) + 0.5(8) + 0.5(12)$

5. **Multiple-Choice** Evaluate $a^3 + (b - 2)^2$ if $a = -2$ and $b = 3$. (Lesson 1-4)
   - A 7   B 9   C -7 D -3 E 6

Thursday

6. **Short Constructed-Response** Create a bar graph from the data in the table. (Lesson 2-7)

   Weights of Dogs in Kenny's Kennel

Pounds	0–10	11–20	21–30	Over 30
Number of Dogs	5	9	7	9

Friday

7. **Short Constructed-Response** Are the expressions $\dfrac{x}{x - \frac{1}{4}}$ and $\dfrac{4x}{4x - 1}$ equivalent? (Lesson 1-4)

Illinois Mathematics: Applications and Concepts, Course 2

Countdown to ISAT

23 Weeks to ISAT

Monday

1. **Extended Response** Show your work and clearly explain your answer. (Lesson 1-7)

   1, 3, 9, ■, 81, …

   A. Find the missing term in the sequence.

   B. What are the next two terms in the sequence?

   C. What is the rule that generates the terms of this sequence?

Tuesday

2. **Multiple-Choice** Which is the mode of the following data? (Lesson 2-4)

   99.9, 98.9, 97.8, 99.8, 99.9, 98.9, 97.5, 98.9

   - A) 99.9
   - B) 99.8
   - C) 98.9
   - D) 97.8
   - E) 97.5

3. **Multiple-Choice** Which is the best prediction of the cross-country runner's week 5 time to complete the course? (Lesson 2-2)

Lap Times	
Week	Time (min:sec)
1	6:23
2	6:17
3	6:11
4	6:05

   - A) 5:57
   - B) 5:58
   - C) 5:59
   - D) 6:00
   - E) 6:01

Wednesday

4. **Multiple-Choice** Solve $14.3 + t = 17.7$ for t. Which point is the value of t? (Lesson 1-5)

   - A) A
   - B) B
   - C) C
   - D) D
   - E) E

5. **Multiple-Choice** Which is the median of the following data? (Lesson 2-4)

   356, 358, 399, 368, 333, 355, 347

   - A) 363.5
   - B) 357
   - C) 356.5
   - D) 356
   - E) 355

Thursday

6. **Short Constructed-Response** The First United Methodist Church in Chicago is one of the tallest in the world. Solve $2h = 1{,}136$ for h to find the height of the church in feet. (Lesson 1-5)

Friday

7. **Short Constructed-Response** Write the product $\frac{1}{5} \cdot \frac{1}{5} \cdot \frac{1}{5} \cdot \frac{1}{5} \cdot \frac{1}{5} \cdot \frac{1}{5} \cdot \frac{1}{5} \cdot \frac{1}{5} \cdot \frac{1}{5}$ in exponential form. (Lesson 1-2)

22 Weeks to ISAT

Monday

1. **Extended Response** Show your work and clearly explain your answer.

   A. Graph $A = -2$ and $B = 3$ on a number line. (Lesson 3-2)

   B. Evaluate $|A - B|$ to find the distance along the number line between A and B. (Lesson 3-5)

   C. Replace ● with < or > to make the sentence A ● B true. (Lesson 3-2)

Tuesday

2. **Multiple-Choice** The Egyptian Theater in DeKalb was renovated in 1983 after falling into disrepair. Add -55 to 1983 to find which year the theater was built. (Lesson 3-4)
   - A) 1926
   - B) 1927
   - C) 1928
   - D) 1929
   - E) 1930

3. **Multiple-Choice** Which are the coordinates of A? (Lesson 3-3)

   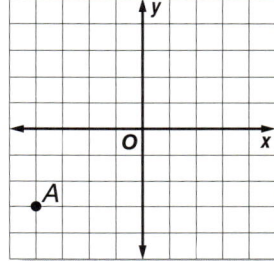

   - A) $(3, -4)$
   - B) $(-3, -4)$
   - C) $(-4, 3)$
   - D) $(-4, -3)$
   - E) $(4, -3)$

Wednesday

4. **Multiple-Choice** A scuba diver begins to dive below the surface of the water at a rate of 3 feet per second. Which integer represents the position of the diver after 15 seconds? (Lesson 3-6)
   - A) 45 feet
   - B) 15 feet
   - C) -15 feet
   - D) -30 feet
   - E) -45 feet

5. **Multiple-Choice** Evaluate the expression $(-7 + 3)^2 \div 2 - (-13) + 55 \div (-11)$. (Lesson 3-7)
   - A) 16
   - B) 12
   - C) 0
   - D) -12
   - E) -16

Thursday

6. **Short Constructed-Response** Evaluate $\dfrac{q^2 - (-p)}{p}$ if $q = -3$ and $p = -3$. (Lesson 3-7)

Friday

7. **Short Constructed-Response** Find the next two numbers in the pattern $-256, 64, -16, 4, \ldots$. (Lesson 3-6)

Illinois Mathematics: Applications and Concepts, Course 2

Countdown to ISAT

21 Weeks to ISAT

Monday

1. **Extended Response** Show your work and clearly explain your answer. Dewayne is selling T-shirts, soft drinks, and candy bars to raise money for his school's Spanish club. (Lesson 4-2)

   A. Write an equation that gives the total T dollars he collects selling t T-shirts at $15.00 each, s soft drinks at $1.00 each, and c candy bars at $0.75 each.

   B. If Dewayne sells 5 T-shirts, 10 candy bars, and s soft drinks for $97.50, how many soft drinks did he sell?

   C. If Dewayne sells $175.00 worth of candy and soft drinks, how many shirts must he sell to collect $400.00?

Tuesday

2. **Multiple-Choice** Solve $10g - 2 = 18$ for g. (Lesson 4-4)
   - A) 0
   - B) 1
   - C) 2
   - D) 3
   - E) 4

3. **Multiple-Choice** A train is traveling at a speed of 112.5 feet per second. Which distance does the train travel in 30 seconds? (Lesson 4-3)
   - A) 1,669 feet
   - B) 2,875 feet
   - C) 2,980 feet
   - D) 3,250 feet
   - E) 3,375 feet

Wednesday

4. **Multiple-Choice** Aaliyah spent 5.75 hours painting her bedroom. How many minutes did it take her to paint the bedroom? (Lesson 1-8)
   - A) 345
   - B) 355
   - C) 396
   - D) 420
   - E) 438

5. **Multiple-Choice** Which expression is the result of simplifying $-7 - 3x + 9$? (Lesson 3-4)
   - A) $3 - 2x$
   - B) $2 + 3x$
   - C) $-2 - 3x$
   - D) $2 - 3x$
   - E) $-2 + 3x$

Thursday

6. **Short Constructed-Response** In 2001, the oldest father in Illinois with a newborn was 83 years old. The oldest mother with a newborn was 29 years younger. Write and solve an equation that gives the age a of the oldest mother. (Lesson 4-1)

Friday

7. **Short Constructed-Response** Amata ordered p posters from a mail-order company. The posters cost $8.95 each and shipping and handling is $5.95. Write an expression that represents the total cost of the order. (Lesson 1-4)

IL12 Illinois Mathematics: Applications and Concepts, Course 2

20 Weeks to ISAT

Monday

1. **Extended Response** Show your work and clearly explain your answer. In 2001, the heaviest baby born in Illinois weighed 13 pounds 12 ounces.

   A. How many ounces did the baby weigh? (Lesson 3-6)

   B. How many pounds did the baby weigh? (Lesson 3-7)

   C. In the same year, the median weight of all babies born in Illinois was 7 pounds 6 ounces. How many ounces more than the median was the heaviest baby? (Lesson 3-5)

Tuesday

2. **Multiple-Choice** Which expression is modeled by the following? (Lesson 3-5a)

   A. $9 - 5$
   B. $9 - (-5)$
   C. $5 - 9$
   D. $-5 - 9$
   E. $5 + 9$

3. **Multiple-Choice** Which is the solution of $2n - 17 > 3$? (Lesson 4-5)

   A. $n > 0$
   B. $n < 7$
   C. $n > 9$
   D. $n > 10$
   E. $n < 10$

Wednesday

4. **Multiple-Choice** Which inequality is graphed on the number line? (Lesson 4-5)

   A. $x < 20$
   B. $x > 20$
   C. $x \leq 20$
   D. $0 \leq x \leq 20$
   E. $x \geq 20$

5. **Multiple-Choice** Which is $-11, 10, 0, 5,$ and -1 ordered from greatest to least? (Lesson 3-2)

   A. $10, 5, 0, -1, -11$
   B. $10, 5, 0, -11, -1$
   C. $-11, 10, 5, -1, 0$
   D. $-11, -1, 0, 5, 10$
   E. $-11, 10, 0, 5, -1$

Thursday

6. **Short Constructed-Response** Graph the points $A(-2, 1)$ and $B(3, -4)$ on a coordinate plane. (Lesson 3-3)

Friday

7. **Short Constructed-Response** Lawanda is three more than m inches tall. Lawanda is taller than Judy. Judy is 58 inches tall. Write and solve an inequality that represents this situation. (Lesson 4-5)

Illinois Mathematics: Applications and Concepts, Course 2

Countdown to ISAT

19 Weeks to ISAT

Monday

1. **Extended Response** Show your work and clearly explain your answer. While water flowed down the drain of his bathtub, Esteban used a ruler and stopwatch to take the data in the table. (Lesson 4-6)

   A. Graph the data and draw a straight line through the points.

   B. Write a linear equation to represent the data.

   C. Use your equation to predict the water level in the tub at 45 seconds.

Draining the Bathtub

Time (seconds)	Water Level (inches)
0	4.0
10	3.5
20	3.0
30	2.5

Tuesday

2. **Multiple-Choice** The largest of Cahokia's mounds is Monk's Mound. It stands 100 feet tall and covers 14 acres. How many square feet are in 14 acres? (There are 43,560 square feet in an acre.) (Lesson 3-6)
   - A 3,111 sq ft
   - B 40,526 sq ft
   - C 566,280 sq ft
   - D 609,840 sq ft
   - E 780,000 sq ft

3. **Multiple-Choice** Which property is shown by the statement $x(1 - 3y) = x(1) - x(3y)$? (Lesson 1-6)
   - A additive inverse property
   - B identity property of multiplication
   - C commutative property of multiplication
   - D associative property of addition
   - E distributive property

Wednesday

4. **Multiple-Choice** Which ordered pair is a solution of the equation $y = 0.5x - 8$? (Lesson 4-6)
   - A $(-2, 7)$
   - B $(5, -9)$
   - C $(2, -7)$
   - D $(1.5, -7.5)$
   - E $(3, 7)$

5. **Multiple-Choice** Which inequality is equivalent to $7z + 3 - 11z - 15 > 0$? (Lesson 4-5)
   - A $z > 4$
   - B $z < -3$
   - C $z \leq 0$
   - D $z > -4$
   - E $z < 1$

Thursday

6. **Short Constructed-Response** In words, write a rule for the variables in the table. (Lesson 4-6)

x	y
0	-5
1	0
2	5

Friday

7. **Short Constructed-Response** Graph the solution of $x - 14 > -19$ on a number line. (Lesson 4-5)

IL14 Illinois Mathematics: Applications and Concepts, Course 2

18 Weeks to ISAT

Monday

1. **Extended Response** Show your work and clearly explain your answer. Use the Venn diagram to answer the questions. (Lesson 5-2)

   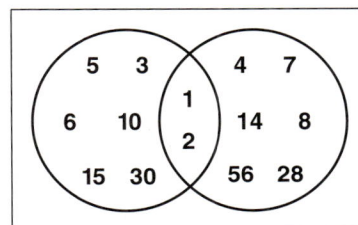

   A. The circle on the left contains the factors of what number?

   B. The circle on the right contains the factors of what number? What factors do these two numbers have in common?

   C. Find the greatest common factor of the two numbers.

Tuesday

2. **Multiple-Choice** The world's heaviest man, Robert Earl Hughes, was born in Fish Hook. At birth he weighed 11.5 pounds. His greatest weight during his life was 1,069 pounds. Which is about the same fraction as his birth weight to his greatest weight? (Lesson 5-4)

   - A $\frac{1}{101}$
   - B $\frac{1}{99}$
   - C $\frac{1}{96}$
   - D $\frac{1}{93}$
   - E $\frac{1}{89}$

3. **Multiple-Choice** Which is the prime factorization of 172? (Lesson 5-1)
   - A $2 \times 3 \times 41$
   - B $2 \times 3 \times 13$
   - C $2 \times 2 \times 43$
   - D $2 \times 5 \times 17$
   - E $2 \times 7 \times 23$

Wednesday

4. **Multiple-Choice** It took 6.5 hours to paint Kevin's room. 2.5 hours of the time was spent painting the trim. What fraction of the total painting time was spent painting trim? (Lesson 5-3)
   - A $\frac{3}{13}$
   - B $\frac{7}{26}$
   - C $\frac{4}{13}$
   - D $\frac{9}{26}$
   - E $\frac{5}{13}$

5. **Multiple-Choice** Twelve minutes is what part of one hour? (Lesson 5-3)
   - A $\frac{1}{5}$
   - B $\frac{1}{6}$
   - C $\frac{1}{7}$
   - D $\frac{1}{8}$
   - E $\frac{1}{9}$

Thursday

6. **Short Constructed-Response** Which fraction, $\frac{7}{15}$ or $\frac{9}{22}$, is closer to $\frac{5}{11}$? (Lesson 5-4)

Friday

7. **Short Constructed-Response** Reynaldo studied math $\frac{4}{5}$ of an hour on Monday and $\frac{11}{15}$ of an hour on Tuesday. Which fraction is greater? (Lesson 5-8)

Countdown to ISAT

17 Weeks to ISAT

Monday

1. **Extended Response** Show your work and clearly explain your answer.
   $\frac{36}{64}, \frac{24}{51}, \frac{66}{126}, \frac{50}{95}$

   A. Simplify all four fractions. (Lesson 5-3)

   B. Order the simplified fractions from least to greatest. (Lesson 5-8)

   C. Write each fraction as a decimal rounded to the nearest hundredth. (Lesson 5-4)

Tuesday

2. **Multiple-Choice** Tawana needs to buy socks and underwear for her son. She wants to buy him the same number of pairs of socks as pairs of underwear. If her son's favorite socks and underwear come in packages of eight pairs and three pairs, respectively, which is the least number of pairs of each she will have to buy? (Lesson 5-7)
   - Ⓐ 16
   - Ⓑ 24
   - Ⓒ 25
   - Ⓓ 28
   - Ⓔ 42

3. **Multiple-Choice** Which is $3\frac{3}{5}$ expressed as a percent? (Lesson 5-5)
   - Ⓐ 3.6%
   - Ⓑ 36%
   - Ⓒ 360%
   - Ⓓ 3,600%
   - Ⓔ 36,000%

Wednesday

4. **Multiple-Choice** Which is the greatest common factor of $72x^4$ and $18x$? (Lesson 5-2)
   - Ⓐ 4
   - Ⓑ $9x$
   - Ⓒ $36x^2$
   - Ⓓ $18x$
   - Ⓔ 24

5. **Multiple-Choice** In 2001, DuQuoin State Fair's attendance was 447,955. In 2000, it was 393,088. To the nearest percent, what percent of the 2000 attendance was the 2001 attendance? (Lesson 5-6)
   - Ⓐ 88%
   - Ⓑ 98%
   - Ⓒ 104%
   - Ⓓ 110%
   - Ⓔ 114%

Thursday

6. **Short Constructed-Response** What is the percent represented by the model? (Lesson 5-5)

Friday

7. **Short Constructed-Response** Mark is having a house-warming party. So far, 12 of his expected 30 guests have arrived. What percent of his expected guests have not yet arrived? (Lesson 5-5)

Illinois Mathematics: Applications and Concepts, Course 2

Countdown to ISAT

16 Weeks to ISAT

Monday

1. **Extended Response** Show your work and clearly explain your answer. Emily sold $\frac{3}{13}$ of her doll collection on Saturday and $\frac{6}{13}$ of it on Sunday.

   A. Estimate the total amount of her collection Emily sold to the nearest $\frac{1}{4}$. (Lesson 6-1)

   B. Is this estimate high or low? (Lesson 6-1)

   C. Find the actual fraction sold. If Emily originally had 26 dolls, how many did she sell? (Lesson 6-2)

Tuesday

2. **Multiple-Choice** In 2001, there were about 89,500 weddings in Illinois. Of these, about $\frac{3}{26}$ occurred in June. Which is about the number of June weddings? (Lesson 6-3b)
   - A 10,200
   - B 10,300
   - C 10,400
   - D 10,500
   - E 10,600

3. **Multiple-Choice** Which is the result of subtracting $7\frac{3}{8}$ from $8\frac{7}{8}$? (Lesson 6-3)
   - A $\frac{3}{8}$
   - B $\frac{7}{8}$
   - C $1\frac{1}{8}$
   - D $1\frac{1}{2}$
   - E $1\frac{5}{8}$

Wednesday

4. **Multiple-Choice** Shelly needs to divide $2\frac{1}{2}$ pounds of sugar into three equal portions. Which is the weight of each portion? (Lesson 6-6)
   - A $\frac{3}{4}$ pound
   - B $\frac{4}{5}$ pound
   - C $\frac{5}{6}$ pound
   - D $\frac{6}{7}$ pound
   - E $\frac{7}{8}$ pound

5. **Multiple-Choice** Which property is shown by the statement $\frac{17}{3} \cdot \frac{3}{17} = 1$? (Lesson 6-5)
   - A additive inverse property
   - B identity property of multiplication
   - C commutative property of multiplication
   - D associative property of addition
   - E multiplicative inverse property

Thursday

6. **Short Constructed-Response** Divide $5\frac{2}{7} \div \frac{4}{7}$. (Lesson 6-6)

Friday

7. **Short Constructed-Response** Multiply $\frac{11}{9} \times 7\frac{5}{22}$. (Lesson 6-4)

Illinois Mathematics: Applications and Concepts, Course 2

Countdown to ISAT

15 Weeks to ISAT

Monday

1. **Extended Response** Show your work and clearly explain your answer.
   (Prerequisite Skill)

   A. Name the points graphed at A and B. Your answers should be fractions.
   B. Copy the number line and graph the point $C = 2\frac{1}{4}$.
   C. Graph a point D, which is $1\frac{1}{2}$ units to the left of B.

Tuesday

2. **Multiple-Choice** One thousand square feet of wall was painted using $3\frac{1}{3}$ gallons of paint. Which equation represents the average area A covered per gallon of paint? (Lesson 4-1)
   - Ⓐ $3\frac{1}{3} = \frac{1,000}{A}$
   - Ⓑ $3\frac{1}{3} = \frac{A}{1,000}$
   - Ⓒ $1,000A = 3\frac{1}{3}$
   - Ⓓ $A = 3\frac{1}{3} \div 1,000$
   - Ⓔ $A + 3\frac{1}{3} = 1,000$

3. **Multiple-Choice** Which is the result of evaluating $u^2 - v \div u + 1$ if $u = \frac{1}{3}$ and $v = \frac{1}{6}$? (Lesson 6-6)
   - Ⓐ $\frac{5}{6}$
   - Ⓑ $\frac{11}{18}$
   - Ⓒ $\frac{5}{9}$
   - Ⓓ $\frac{1}{2}$
   - Ⓔ $\frac{4}{9}$

Wednesday

4. **Multiple-Choice** The table shows three Illinois' State Symbols and their dates of adoption. Which list is ordered from latest to earliest adoption? (Prerequisite Skill)

Illinois' State Symbols	
Symbol	Date
State dance: Square dance	1990
State flower: Violet	1908
State insect: Monarch Butterfly	1975

   - Ⓐ Violet, Monarch, Square dance
   - Ⓑ Square dance, Violet, Monarch
   - Ⓒ Square dance, Monarch, Violet
   - Ⓓ Monarch, Square dance, Violet
   - Ⓔ Violet, Square dance, Monarch

5. **Multiple-Choice** Which is the radius of a circle with diameter 10 feet? (Lesson 6-9)
   - Ⓐ 20 feet
   - Ⓑ 15 feet
   - Ⓒ 12 feet
   - Ⓓ 6 feet
   - Ⓔ 5 feet

Thursday

6. **Short Constructed-Response** If a square with perimeter 12 inches is divided into four smaller squares of equal area, what is the perimeter of a new square? (Lesson 6-8)

Friday

7. **Short Constructed-Response** Use your centimeter ruler to sketch a rectangle with length 3 cm and width 2 cm. (Lesson 6-8)

Illinois Mathematics: Applications and Concepts, Course 2

Countdown to ISAT

14 Weeks to ISAT

Monday

1. **Extended Response** Show your work and clearly explain your answer. The table shows prices of salmon Jose is considering purchasing. (Lesson 7-2)
   A. Which purchase will cost the least per ounce and what is that price?
   B. Which is the best deal among the two wild salmon choices?
   C. If Jose decides to buy three orders of the wild-60 ounce salmon, how much will he have to pay?

Wild & Farmed Salmon	
Type-Weight	Price
Wild-16 ounces	$14.95
Farmed-24 ounces	$16.49
Wild-60 ounces	$49.95
Farmed-40 ounces	$25.95

Tuesday

2. **Multiple-Choice** Tavon's model airplane travels at a maximum speed of 40 miles per hour. If the plane's minimum speed is $\frac{1}{5}$ of its maximum speed, which is the minimum speed? (Lesson 7-2)
   A. 8.5 mph
   B. 8 mph
   C. 7.8 mph
   D. 6 mph
   E. 5 mph

3. **Multiple-Choice** Which property is shown by the statement $S \cdot 0 = 0$? (Prerequisite Skill)
   A. additive inverse property
   B. identity property of multiplication
   C. commutative property of multiplication
   D. associative property of addition
   E. multiplicative property of zero

Wednesday

4. **Multiple-Choice** In Lowden State Park stands a giant statue of a Native American. This statue is the largest reinforced concrete statue in the world. It weighs 268 tons and its foundation goes 30 feet into the ground. To the nearest tenth, which is the ratio of the statue's weight to depth of foundation? (Lesson 7-1)
   A. 5.2 tons per foot
   B. 6.4 tons per foot
   C. 7.8 tons per foot
   D. 8.9 tons per foot
   E. 9.4 tons per foot

5. **Multiple-Choice** In a class of 26 students, 14 have brown eyes. Which is the ratio of brown-eyed to all students in the class? (Lesson 7-1)
   A. 4:7
   B. 6:12
   C. 7:13
   D. 8:11
   E. 5:8

Thursday

6. **Short Constructed-Response** A model car is $\frac{1}{50}$ the size of the actual car it models. If the model car is 4 inches long, what is the length in feet of the actual car? (Lesson 7-3)

Friday

7. **Short Constructed-Response** A 3,000-acre parcel of land is to be donated to a community under the condition that 85% will be used as a nature preserve. What fraction of the land must be used for the preserve, and how many acres? (Lesson 7-5)

Illinois Mathematics: Applications and Concepts, Course 2

Countdown to ISAT

13 Weeks to ISAT

Monday

1. **Extended Response** Show your work and clearly explain your answer. Selected data for 1,000 cars crossing a toll bridge is shown in the table. (Lesson 7-5)

Item	Percent of Cars
Electronic Toll Pay	12
1 passenger	18
2 or more passengers	21
Driver only	61

   A. What fraction of cars had electronic toll pay?
   B. If eight percent of cars had three or more passengers, what fraction of cars had two passengers?
   C. If electronic toll pay reduces the toll by fifty cents per car, how much more money was taken in tolls than would have been the case had all 1,000 cars had electronic toll pay?

Tuesday

2. **Multiple-Choice** Dena drove 336 miles on 14 gallons of gas. Which was her car's gas mileage for the trip? (Lesson 7-2)
   - Ⓐ 28 miles per gallon
   - Ⓑ 24 miles per gallon
   - Ⓒ 22 miles per gallon
   - Ⓓ 20 miles per gallon
   - Ⓔ 18 miles per gallon

3. **Multiple-Choice** A water tower supplies a small town with enough water to last for one week without refilling it. If there are 20 houses in the town and each uses an average of 300 gallons per day, which is the approximate capacity of the tower when it is full? (Lesson 6-7)
   - Ⓐ 32,000 gallons
   - Ⓑ 37,000 gallons
   - Ⓒ 40,000 gallons
   - Ⓓ 42,000 gallons
   - Ⓔ 44,000 gallons

Wednesday

4. **Multiple-Choice** In front of a Wilmington café stands a 30-foot, 500-pound statue of a cartoon-like astronaut, The Gemini Giant. Which is the statue's average weight per foot of height? (Lesson 7-2)
   - Ⓐ $16\frac{2}{3}$ pounds per foot
   - Ⓑ $16\frac{3}{4}$ pounds per foot
   - Ⓒ 17 pounds per foot
   - Ⓓ $17\frac{2}{3}$ pounds per foot
   - Ⓔ 18 pounds per foot

5. **Multiple-Choice** Which property is shown by the statement $14 \times ab = a \times 14b$? (Lesson 1-6)
   - Ⓐ additive inverse property
   - Ⓑ identity property of multiplication
   - Ⓒ commutative property of multiplication
   - Ⓓ associative property of addition
   - Ⓔ multiplicative property of zero

Thursday

6. **Short Constructed-Response** Alejandro hit 2 out of 5 times at bat during one game and 1 out of 4 times at bat during a second game. Determine the ratio of hits to at bats for each game. (Lesson 7-1)

Friday

7. **Short Constructed-Response** Camille sold 97.5% of her allotment of tickets to her school play. If she sold 195 tickets, how many did she originally have? (Lesson 7-8)

12 Weeks to ISAT

Monday

1. **Extended Response** Show your work and clearly explain your answer. One thousand of 120,000 tax-paying residents of a city were asked what they would do with a proposed $1,000 tax credit. Their responses are shown in the circle graph. (Lesson 8-3)

   A. Predict the number of tax-paying residents in the city that would use the tax credit for a vacation.

   B. If 42% of the tax-paying residents saved the tax credit, how much money would be saved?

   C. Predict the maximum number of tax-paying residents that would use the credit for home repair.

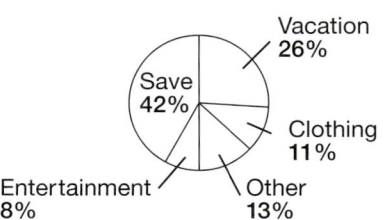

How to Spend $1,000
- Vacation 26%
- Save 42%
- Clothing 11%
- Other 13%
- Entertainment 8%

Tuesday

2. **Multiple-Choice** Which property is shown by the statement $5x^3 \times 1 = 5x^3$? (Lesson 1-6)
   - A additive inverse property
   - B identity property of multiplication
   - C commutative property of multiplication
   - D associative property of addition
   - E multiplicative property of zero

3. **Multiple-Choice** Illinois has a sales tax rate of 6.25%. How much would a $500 television set cost with tax included? (Lesson 8-5)
   - A $526.25
   - B $528.75
   - C $530.00
   - D $531.00
   - E $531.25

Wednesday

4. **Multiple-Choice** Seventy-five percent of employees of a company participate in the company's health plan. If 450 employees participate, how many employees does the company have? (Lesson 8-2)
   - A 475
   - B 500
   - C 550
   - D 600
   - E 650

5. **Multiple-Choice** Jeremy deposited $500 in an account that earns 4.5% simple interest annually. If Jeremy neither deposits nor withdraws money from the account, which is the amount in the account after 4 years? (Lesson 8-6)
   - A $546.01
   - B $570.58
   - C $590.00
   - D $623.09
   - E $651.13

Thursday

6. **Short Constructed-Response** A young cat weighed 9 pounds one year ago. Today, the cat weighs 13.5 pounds. What is the percent increase in weight of the cat? (Lesson 8-4)

Friday

7. **Short Constructed-Response** Which is the least, $\frac{17}{7}$, $2\frac{2}{5}$, or 2.41? (Lesson 5-8)

Countdown to ISAT

11 Weeks to ISAT

Monday

1. **Extended Response** Show your work and clearly explain your answer. The retail price of four items is shown in the table. (Lesson 8-5)

Midtown Bodega	
Item	Price
Leather Jacket	$350.00
Sport Socks	$4.90
Purse	$44.80
Wool Sweater	$89.95

   A. Each item has been marked up 40% from the wholesale price. What was the wholesale price of each item?

   B. If the sale price for the leather jacket is $280.00, what is the percent of discount?

   C. What is the total for all four items with no discount and 7.5% sales tax?

Tuesday

2. **Multiple-Choice** Which is equivalent to $2 \times 10^4 + 8 \times 10^3 + 7 \times 10 + 3$? (Prerequisite Skill Lesson 2)
   - Ⓐ 280,730
   - Ⓑ 28,073
   - Ⓒ 2,807.3
   - Ⓓ 280.73
   - Ⓔ 28.073

3. **Multiple-Choice** Which number is not prime? (Lesson 5-1)
   - Ⓐ 31
   - Ⓑ 41
   - Ⓒ 43
   - Ⓓ 47
   - Ⓔ 51

Wednesday

4. **Multiple-Choice** The Illinois and Michigan Canal is 97 miles long. How long is it in yards? (Lesson 6-7)
   - Ⓐ 100,876 yards
   - Ⓑ 125,625 yards
   - Ⓒ 150,020 yards
   - Ⓓ 170,720 yards
   - Ⓔ 512,160 yards

5. **Multiple-Choice** An amount of money was deposited with a bank at 2.7% simple interest for 8 years. At the end of the 8 years, the amount of interest was $2,160. Which is the amount that was originally deposited? (Lesson 8-6)
   - Ⓐ $8,000
   - Ⓑ $9,000
   - Ⓒ $10,000
   - Ⓓ $11,000
   - Ⓔ $12,000

Thursday

6. **Short Constructed-Response** If Kristin walks three-fourths of a mile each day, how many days will it take her to walk ten and one-half miles? (Lesson 6-6)

Friday

7. **Short Constructed-Response** Simplify the product $5 \times 5 \times y \times y \times y \times x \times x$ using exponents. (Lesson 1-2)

10 Weeks to ISAT

Monday

1. **Extended Response** Show your work and clearly explain your answer.
   Two number cubes, one with sides marked with A, B, C, D, E, and F and the other with U, V, W, X, Y, and Z, are tossed.

   A. How many different outcomes are possible? (Lesson 9-3)

   B. What is the probability that A and Y is the outcome? (Lesson 9-1)

   C. What is the probability that A and Y is not the outcome? (Lesson 9-1)

Tuesday

2. **Multiple-Choice** In 2000, the population of Chicago was 2,896,016, where 1,490,909 were female. The percent of the population that was female is approximately which of the following? (Lesson 8-2)
   - Ⓐ 51.48%
   - Ⓑ 51.41%
   - Ⓒ 50.98%
   - Ⓓ 50.27%
   - Ⓔ 49.98%

3. **Multiple-Choice** Juana is trying to decide between two of three scarves and two of four sweaters to take with her on a weekend trip. How many different scarf and sweater combinations can she choose? (Lesson 9-5)
   - Ⓐ 8   Ⓑ 10   Ⓒ 14   Ⓓ 18   Ⓔ 20

Wednesday

4. **Multiple-Choice** In a spelling bee, there are six finalists. In how many ways can three of the six finalists finish first, second, and third? (Lesson 9-4)
   - Ⓐ 102
   - Ⓑ 120
   - Ⓒ 136
   - Ⓓ 144
   - Ⓔ 169

5. **Multiple-Choice** Which property is shown by the statement $A + (B + 3) = (A + B) + 3$? (Lesson 1-6)
   - Ⓐ additive inverse property
   - Ⓑ identity property of multiplication
   - Ⓒ commutative property of multiplication
   - Ⓓ multiplicative property of zero
   - Ⓔ associative property of addition

Thursday

6. **Short Constructed-Response** Does the line of best fit on the scatter plot have a positive or negative slope? (Lesson 4-7)

Friday

7. **Short Constructed-Response** Is the bar graph representation of the data in the table correct? (Lesson 2-7)

Monthly Sales	
Month	Sales
March	$10,950
April	$12,623
May	$9,547

Illinois Mathematics: Applications and Concepts, Course 2

Countdown to ISAT

9 Weeks to ISAT

Monday

1. **Extended Response** Show your work and clearly explain your answer. Five new products are to be arranged from left to right on a display shelf.
   A. In how many ways can the five products be arranged? (Lesson 9-4)
   B. If the store manager decides that only three of the five should be displayed on the shelf, in how many ways can three of the five be chosen and arranged on the shelf? (Lesson 9-4)
   C. It is almost closing time and there is no time to arrange the products, but enough time to select three of the five products to be arranged on the shelf later. In how many ways can three of the five products be chosen? (Lesson 9-5)

Tuesday

2. **Multiple-Choice** Riverview Amusement Park in Chicago began selling foot-long hot dogs in the 1930s. Placed end-to-end, how many hot dogs would it take to cover one-quarter mile? (Lesson 6-7)
   - A 1,320
   - B 1,289
   - C 1,266
   - D 1,188
   - E 1,049

3. **Multiple-Choice** The ratio of men to women in a remote mining town is 8 to 1. If there are 567 people in the town, how many are women? (Lesson 7-1)
   - A 66 B 65 C 64 D 63 E 62

Wednesday

4. **Multiple-Choice** It took 718 tiles to cover a patio. If each tile is 0.25 square foot in area, what is the area of the patio? (Lesson 6-8)
   - A 179.0 sq ft
   - B 179.5 sq ft
   - C 180.0 sq ft
   - D 180.5 sq ft
   - E 181.0 sq ft

5. **Multiple-Choice** Which expression is equivalent to $24b + 12 \times 4a - 12 \times 0.02$? (Lesson 1-6)
   - A $12(b + 4a) - 0.24$
   - B $24(b + 4a) - 0.24$
   - C $24b + 12(4a - 0.02)$
   - D $24b - 12(4a - 0.02)$
   - E $24b + 12(4a + 0.02)$

Thursday

6. **Short Constructed-Response** Use the scale drawing to find the area of the kitchen. (Lesson 7-4)

Friday

7. **Short Constructed-Response** Evaluate $\frac{a!}{b} \div \frac{c}{7}$ if $a = 4$, $b = 6$, and $c = 2$. (Lesson 9-4)

IL24 Illinois Mathematics: Applications and Concepts, Course 2

8 Weeks to ISAT

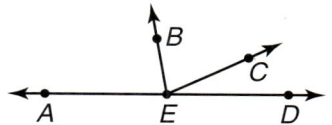
Countdown to ISAT

Monday

1. **Extended Response** Show your work and clearly explain your answer.

   A. Use your protractor to measure ∠AEB and ∠BEC. (Lesson 10-1)

   B. Use your results from part A to find the measure of ∠CED. (Lesson 10-1)

   C. Are ∠AEC and ∠CED complementary, supplementary, or neither? (Lesson 10-3)

Tuesday

2. **Multiple-Choice** Which is the degree measure of the angle? (Lesson 10-1)
   - A 40°
   - B 45°
   - C 50°
   - D 55°
   - E 60°

3. **Multiple-Choice** Solve for x. (Lesson 10-3)
   - A 110
   - B 118
   - C 120
   - D 125
   - E 132

Wednesday

4. **Multiple-Choice** Which is a name that describes the figure? (Lesson 10-5)
   - A parallelogram
   - B trapezoid
   - C rhombus
   - D rectangle
   - E square

5. **Multiple-Choice** If two angles are complementary and the measure of one is 43°, which is the measure of the other angle? (Lesson 10-3)
   - A 147°
   - B 137°
   - C 57°
   - D 53°
   - E 47°

Thursday

6. **Short Constructed-Response** Sketch a triangle with sides 3 cm, 4 cm, and 5 cm. Note that the 5-cm side must be opposite a right angle. (Lesson 10-4)

Friday

7. **Short Constructed-Response** A side of each square of the grid on the map represents six miles in length. What is the perimeter of Knox County? (Lesson 7-4)

TOWNSHIPS OF KNOX COUNTY

	Rio	Ontario	Walnut Grove	Lynn
Henderson →		Sparta	Copley	Victoria
Galesburg →		Knox	Persifer	Truro
	Cedar	Orange	Haw Creek	Elba
	Indian Point	Chestnut	Maquon	Salem

Illinois Mathematics: Applications and Concepts, Course 2   IL25

Countdown to ISAT

7 Weeks to ISAT

Monday

1. **Extended Response** Show your work and clearly explain your answer. The figure is a scale drawing of an L-shaped swimming pool. (Lesson 7-4)

   A. Find the scale factor of the drawing.

   B. What is the perimeter of the pool?

   C. What is the area of the pool?

Tuesday

2. **Multiple-Choice** Which is the measure of the missing angle? (Lesson 10-4)
   - Ⓐ 48°
   - Ⓑ 49°
   - Ⓒ 50°
   - Ⓓ 51°
   - Ⓔ 52°

3. **Multiple-Choice** Tamarr sprinted 100 yards in 11.2 seconds. Which is his approximate average speed in miles per hour? (Lesson 6-7)
   - Ⓐ 18.3
   - Ⓑ 18.5
   - Ⓒ 19.3
   - Ⓓ 19.6
   - Ⓔ 20.2

Wednesday

4. **Multiple-Choice** Solve for x in the parallelogram. (Lesson 10-5)
   - Ⓐ 14
   - Ⓑ 17
   - Ⓒ 21
   - Ⓓ 23
   - Ⓔ 24

5. **Multiple-Choice** Which describes the angle? (Lesson 10-1)
   - Ⓐ straight
   - Ⓑ right
   - Ⓒ open
   - Ⓓ obtuse
   - Ⓔ acute

Thursday

6. **Short Constructed-Response** The statue is a model of a much larger statue of Ramses II that stands in front of the Gold Pyramid House in Wadsworth. The scale factor is $\frac{1}{48}$. What is the height of the actual statue in feet? (Lesson 7-4)

Friday

7. **Short Constructed-Response** Sketch an acute angle. (Lesson 10-1)

Countdown to ISAT

6 Weeks to ISAT

Monday

1. **Extended Response** Show your work and clearly explain your answer. Use the parallelogram to answer the questions.
   A. If the area of the parallelogram is 620 cm², what is the value of b? (Lesson 11-4)
   B. What is the value of a? (Lesson 10-5)
   C. What is the measure of the missing angle? (Lesson 10-5)

Tuesday

2. **Multiple-Choice** Which is the area of the triangle? (Lesson 11-5)

   Ⓐ 180 cm²
   Ⓑ 136 cm²
   Ⓒ 104 cm²
   Ⓓ 96 cm²
   Ⓔ 90 cm²

3. **Multiple-Choice** Which is a perfect square? (Lesson 11-1)
   Ⓐ 96
   Ⓑ 120
   Ⓒ 142
   Ⓓ 169
   Ⓔ 198

Wednesday

4. **Multiple-Choice** Which is the distance between F and G? (Lesson 3-1)

   Ⓐ $1\frac{3}{8}$
   Ⓑ $1\frac{1}{4}$
   Ⓒ $1\frac{1}{8}$
   Ⓓ $1\frac{1}{16}$
   Ⓔ $1\frac{1}{32}$

5. **Multiple-Choice** In 2000, the population density of Ford County was 29.3 persons per square mile. For all of Illinois, it was 223.4. To the nearest hundredth, which was the ratio of the population density of Ford County to that of the entire state of Illinois? (Lesson 7-1)
   Ⓐ 0.09 : 1
   Ⓑ 0.12 : 1
   Ⓒ 0.13 : 1
   Ⓓ 3.81 : 1
   Ⓔ 7.62 : 1

Thursday

6. **Short Constructed-Response** Are the figures similar? (Lesson 10-6)

Friday

7. **Short Constructed-Response** The area of a square concrete patio is 580 square feet. Estimate the length of the sides of the patio to the nearest whole foot. (Lesson 11-2)

Countdown to ISAT

5 Weeks to ISAT

Monday

1. **Extended Response** Show your work and clearly explain your answer. The figure is composed of a triangle and a trapezoid. (Lesson 11-7)

   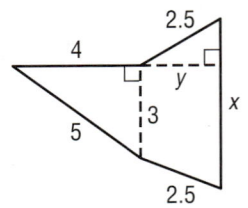

   A. If the perimeter of the figure is 20 units, what is the value of x?
   B. If the area of the figure is 15 square units, what is the value of y?
   C. If the triangle and trapezoid were separated, would the sum of the areas be equal to the area of the original figure? Would the sum of the perimeters be equal to the original perimeter?

Tuesday

2. **Multiple-Choice** Tower Park in Peoria Heights has a water fountain based upon the shape shown. Which is that shape? (Lesson 10-7)

   A nonagon
   B octagon
   C heptagon
   D hexagon
   E pentagon

3. **Multiple-Choice** Which square root lies between 8 and 9? (Lesson 11-2)

   A $\sqrt{48}$
   B $\sqrt{56}$
   C $\sqrt{78}$
   D $\sqrt{82}$
   E $\sqrt{85}$

Wednesday

4. **Multiple-Choice** Which point graphed on the number line is closest to $\sqrt{6}$? (Lesson 11-2)

   A A
   B B
   C C
   D D
   E E

5. **Multiple-Choice** The square of the radius of a circle is 81 m². Which is the diameter of the circle? (Lesson 6-9)

   A 18 m
   B 19 m
   C 22 m
   D 24 m
   E 25 m

Thursday

6. **Short Constructed-Response** What is the area of the parallelogram if its perimeter is 60 inches? (Lesson 11-4)

Friday

7. **Short Constructed-Response** What is the area of the shaded region? (Lesson 11-7)

IL28 Illinois Mathematics: Applications and Concepts, Course 2

Countdown to ISAT

4 Weeks to ISAT

Monday

1. **Extended Response** Show your work and clearly explain your answer. A right rectangular prism is shown.

   A. Find the volume of the figure. (Lesson 12-2)
   B. Find the surface area of the figure. (Lesson 12-4)
   C. If the length of the prism is doubled to 28 cm, what are the new volume and surface area? (Lesson 12-4b)

Tuesday

2. **Multiple-Choice** Which object is represented by the net? (Lesson 12-4a)
   - A. cone
   - B. square pyramid
   - C. cube
   - D. cylinder
   - E. triangular pyramid

3. **Multiple-Choice** Which is the probability that a coin tossed four times will come up heads all four times? (Lesson 9-7)
   - A. 0.03125
   - B. 0.0625
   - C. 0.125
   - D. 0.25
   - E. 0.5

Wednesday

4. **Multiple-Choice** Donley's Wild West Town in Union offers group rates. Groups of 20 or more are $11.50 per person. Which is the price of admission for a group of 31? (Prerequisite Skill Lesson 7)
   - A. $356.50
   - B. $365.50
   - C. $376.50
   - D. $389.50
   - E. $393.50

5. **Multiple-Choice** When x is 2, y is 8, and when x is 5, y is 20. Which is x when y is 220? (Lesson 4-6)
   - A. 27.5
   - B. 36
   - C. 44
   - D. 55
   - E. 110

Thursday

6. **Short Constructed-Response** What type of polygon is ABCDE? (Lesson 10-7)

Friday

7. **Short Constructed-Response** Two triangular pyramids with sides numbered from 1 to 4 are tossed. List all possible outcomes. (Lesson 9-1)

Countdown to ISAT

3 Weeks to ISAT

Monday

1. **Extended Response** Show your work and clearly explain your answer.
   A. Graph quadrilateral $ABCD$ with vertices $A(-5, -5)$, $B(-4, -2)$, $C(-1, -2)$, and $D(0, -5)$. (Lesson 3-3)
   B. Reflect $ABCD$ over the y-axis. (Lesson 10-9)
   C. Translate $A'B'C'D'$ 5 units up and 5 units left. If you disregard the letters, what translation of the first figure would have given the same result as the reflection and the translation? (Lesson 10-8)

Tuesday

2. **Multiple-Choice** Approximately 12 million waste automobile tires are generated in Illinois every year. For 2003, this is approximately 1.05 Illinois residents per waste tire. Which was the approximate population of Illinois? (Lesson 7-3)
   A. 11.8 million
   B. 12.1 million
   C. 12.4 million
   D. 12.6 million
   E. 13.2 million

3. **Multiple-Choice** Which is the same as $j \cdot k \cdot 7 \cdot 7 \cdot k \cdot j$? (Lesson 1-2)
   A. $7^2 jk^2$
   B. $14j^2 k$
   C. $49j^2 k^2$
   D. $28jk$
   E. $49j^3 k^2$

Wednesday

4. **Multiple-Choice** Which object is represented by the net? (Lesson 12-4a)
   A. cone
   B. square pyramid
   C. cube
   D. cylinder
   E. triangular pyramid

5. **Multiple-Choice** In how many ways can a family of five stand in a row for a painting? (Lesson 9-4)
   A. 120
   B. 60
   C. 20
   D. 5
   E. 1

Thursday

6. **Short Constructed-Response** Graph the intersection of $x > -2$ and $x < 4$ on a number line. (Lesson 4-5)

Friday

7. **Short Constructed-Response** State the measure of the eraser to the nearest tenth. (Lesson 12-6)

IL30 *Illinois Mathematics: Applications and Concepts,* Course 2

Countdown to ISAT

2 Weeks to ISAT

Monday

1. **Extended Response** Show your work and clearly explain your answer. Use the figure to answer the questions. (Lesson 10-6)

    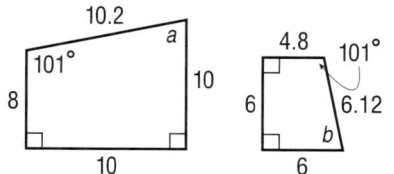

    A. Are angles a and b congruent?

    B. Are the corresponding sides of the two figures proportional? Are the figures similar?

    C. Describe the difference between congruent figures and similar figures.

Tuesday

2. **Multiple-Choice** A bag of multi-colored candy holds 12 red, 13 yellow, 18 green, and 11 blue candies. If one candy is removed from the bag at random, what is the probability that it is neither red nor green? (Lesson 9-1)

    Ⓐ $\frac{5}{9}$ Ⓑ $\frac{4}{9}$ Ⓒ $\frac{7}{18}$ Ⓓ $\frac{1}{3}$ Ⓔ $\frac{5}{18}$

3. **Multiple-Choice** A concert hall has the floor plan shown. The perimeter of the room is 732 feet. Which is the value of d? (Lesson 6-8)

    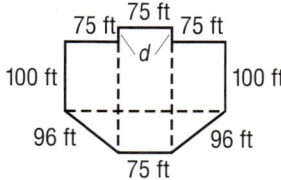

    Ⓐ 40 feet
    Ⓑ 36 feet
    Ⓒ 28 feet
    Ⓓ 20 feet
    Ⓔ 16 feet

Wednesday

4. **Multiple-Choice** Purdue University has a research program that hopes to increase the ethanol production from corn by $\frac{1}{9}$. If the current production of ethanol is 2.7 gallons per bushel of corn, which is the target production for a bushel of corn? (Lesson 7-1)

    Ⓐ 2.8 gallons
    Ⓑ 2.9 gallons
    Ⓒ 3.0 gallons
    Ⓓ 3.1 gallons
    Ⓔ 3.2 gallons

5. **Multiple-Choice** Which is the surface area of a rectangular prism with length 2.5 yards, width 1.5 yards, and height 3.5 yards? (Lesson 12-4)

    Ⓐ 35.5 sq yd
    Ⓑ 32.25 sq yd
    Ⓒ 25.5 sq yd
    Ⓓ 17.75 sq yd
    Ⓔ 13.125 sq yd

Thursday

6. **Short Constructed-Response** Find the missing term in the sequence. (Lesson 1-7)

    $100, 20, \blacksquare, \frac{4}{5}, \frac{4}{25}, \ldots$

Friday

7. **Short Constructed-Response** The sides of a triangle have lengths 11 m, 12 m, and $\sqrt{265}$ m. Estimate the perimeter of the triangle to the nearest whole meter. (Lesson 11-2)

Illinois Mathematics: Applications and Concepts, Course 2

Countdown to ISAT

1 Week to ISAT

Monday

1. **Extended Response** Show your work and clearly explain your answer. Answer the following questions about the numbers 99 and 75.
   A. Create factor trees for the numbers. (Lesson 5-1)
   B. Find the greatest common factor of the numbers. (Lesson 5-2)
   C. Find the least common multiple of the numbers. (Lesson 5-7)

Tuesday

2. **Multiple-Choice** Which inequality is graphed on the number line? (Lesson 4-5)

   Ⓐ $x \leq -4$ or $x > 4$
   Ⓑ $x < -4$ and $x \geq 4$
   Ⓒ $x > -4$ or $x \leq 4$
   Ⓓ $x \leq -4$ and $x > 4$
   Ⓔ $x < -4$ or $x \geq 4$

3. **Multiple-Choice** The Illinois Energy Efficient Affordable Housing Program grants money for the building of affordable and energy efficient homes. If a family moves from a home with $1,700 in annual energy costs to a home built by the program that is 63% more energy efficient, which do you expect to be their annual energy costs? (Lesson 8-4)
   Ⓐ $595
   Ⓑ $629
   Ⓒ $799
   Ⓓ $969
   Ⓔ $1,071

Wednesday

4. **Multiple-Choice** Harold timed himself as he filled five one-gallon jugs with water from a slow-flowing spout. He found that it took 20 minutes to fill one gallon jug with water. If y is the number of filled jugs and t is the time in hours, which equation gives the number of jugs filled for a particular amount of time? (Lesson 4-6)
   Ⓐ $y = 3t$
   Ⓑ $y = \frac{1}{3}t$
   Ⓒ $y = t + 3$
   Ⓓ $y = t - 3$
   Ⓔ $y = t^3$

5. **Multiple-Choice** Which is the volume of a rectangular prism with length 19 m, width 17 m, and height 21 m? (Lesson 12-2)
   Ⓐ $6{,}983 \text{ m}^3$
   Ⓑ $6{,}883 \text{ m}^3$
   Ⓒ $6{,}783 \text{ m}^3$
   Ⓓ $6{,}683 \text{ m}^3$
   Ⓔ $6{,}583 \text{ m}^3$

Thursday

6. **Short Constructed-Response** Cheyenne has six favorite TV shows, but she is allowed to watch only four shows per week. In how many ways can she choose four shows out of the six? (Lesson 9-5)

Friday

7. **Short Constructed-Response** Evaluate $0.78 + \frac{1}{2} \times 7.8^2 - 78 \div 1{,}000$. (Lesson 1-3)

Glencoe Mathematics

Mathematics
Applications and Concepts

Course 2

Illinois Edition

Bailey
Day
Frey
Howard
Hutchens
McClain

Moore-Harris
Ott
Pelfrey
Price
Vielhaber
Willard

McGraw Hill Glencoe

New York, New York · Columbus, Ohio · Chicago, Illinois · Peoria, Illinois · Woodland Hills, California

About the Cover

On the cover of this book, you will find the formula $6a = 720°$. This is the formula for the measure of an interior angle of a regular hexagon. The basic shape for snowflakes is a regular hexagon.

You will also notice the word *symmetry* and the picture of a snowboarder on the cover. Snowboards often are symmetrical. They can be symmetrical front to back and/or side to side. Most snowboards used for freestyle and halfpipe snowboarding have front to back symmetry. You will learn more about symmetry in Chapter 10.

Copyright © 2006 by The McGraw-Hill Companies, Inc.
All rights reserved. Except as permitted under the United States Copyright Act, no part of this publication may be reproduced or distributed in any form or by any means, or stored in a database or retrieval system, without prior permission of the publisher.

The USA TODAY® service mark, USA TODAY Snapshots® trademark and other content from USA TODAY® has been licensed by USA TODAY® for use for certain purposes by Glencoe/McGraw Hill, a Division of The McGraw-Hill Companies, Inc. The USA TODAY Snapshots® and the USA TODAY® articles, charts, and photographs incorporated herein are solely for private, personal, and noncommercial use.

Microsoft® Excel® is a registered trademark of Microsoft Corporation in the United States and other countries.

Send all inquiries to:
Glencoe/McGraw-Hill
8787 Orion Place
Columbus, OH 43240-4027

ISBN: 0-07-869342-X (*Illinois Student Edition*)

Printed in the United States of America.

1 2 3 4 5 6 7 8 9 10 027/043 12 11 10 09 08 07 06 05

Contents in Brief

UNIT 1 — Decimals, Algebra, and Statistics
1. Decimal Patterns and Algebra
2. Statistics: Analyzing Data

UNIT 2 — Integers and Algebra
3. Algebra: Integers
4. Algebra: Linear Equations and Functions

UNIT 3 — Fractions
5. Fractions, Decimals, and Percents
6. Applying Fractions

UNIT 4 — Proportional Reasoning
7. Ratios and Proportions
8. Applying Percent
9. Probability

UNIT 5 — Geometry and Measurement
10. Geometry
11. Geometry: Measuring Two-Dimensional Figures
12. Geometry: Measuring Three-Dimensional Figures

Authors

Rhonda Bailey
Mathematics Consultant
Mathematics by Design
DeSoto, Texas

Roger Day, Ph.D.
Associate Professor
Illinois State University
Normal, Illinois

Patricia Frey
Director of Staffing and Retention
Buffalo City Schools
Buffalo, New York

Arthur C. Howard
Mathematics Teacher
Houston Christian High School
Houston, Texas

Deborah T. Hutchens, Ed.D.
Assistant Principal
Great Bridge Middle School
Chesapeake, Virginia

Kay McClain, Ed.D.
Assistant Professor
Vanderbilt University
Nashville, Tennessee

Beatrice Moore-Harris
Mathematics Consultant
League City, Texas

Jack M. Ott, Ph.D.
Distinguished Professor of
 Secondary Education
 Emeritus
University of South Carolina
Columbia, South Carolina

Ronald Pelfrey, Ed.D.
Mathematics Specialist
Appalachian Rural Systemic
 Initiative
Lexington, Kentucky

Jack Price, Ed.D.
Professor Emeritus
California State Polytechnic
 University
Pomona, California

Kathleen Vielhaber
Mathematics Specialist
Parkway School District
St. Louis, Missouri

Teri Willard, Ed.D.
Assistant Professor of
 Mathematics Education
Central Washington
 University
Ellensburg, Washington

Contributing Authors

USA TODAY
The USA TODAY Snapshots®, created by USA TODAY®, help students make the connection between real life and mathematics.

FOLDABLES Dinah Zike
Educational Consultant
Dinah-Might Activities, Inc.
San Antonio, Texas

Content Consultants

Each of the Content Consultants reviewed every chapter and gave suggestions for improving the effectiveness of the mathematics instruction.

Mathematics Consultants

L. Harvey Almarode
Curriculum Supervisor, Mathematics K–12
Augusta County Public Schools
Fishersville, VA

Claudia Carter, MA, NBCT
Mathematics Teacher
Mississippi School for Mathematics and Science
Columbus, MS

Carol E. Malloy, Ph.D.
Associate Professor, Curriculum Instruction,
    Secondary Mathematics
The University of North Carolina at Chapel Hill
Chapel Hill, NC

Melissa McClure, Ph.D.
Mathematics Instructor
University of Phoenix On-Line
Fort Worth, TX

Robyn R. Silbey
School-Based Mathematics Specialist
Montgomery County Public Schools
Rockville, MD

Leon L. "Butch" Sloan, Ed.D.
Secondary Mathematics Coordinator
Garland ISD
Garland, TX

Barbara Smith
Mathematics Instructor
Delaware County Community College
Media, PA

Reading Consultant

Lynn T. Havens
Director
Project CRISS
Kalispell, MT

ELL Consultants

Idania Dorta
Mathematics Educational Specialist
Miami–Dade County Public Schools
Miami, FL

Frank de Varona, Ed.S.
Visiting Associate Professor
Florida International University
    College of Education
Miami, FL

Teacher Reviewers

Each Teacher Reviewer reviewed at least two chapters of the Student Edition, giving feedback and suggestions for improving the effectiveness of the mathematics instruction.

Royallee Allen
Teacher, Math Department Head
Eisenhower Middle School
San Antonio, TX

Dennis Baker
Mathematics Department Chair
Desert Shadows Middle School
Scottsdale, AZ

Rosie L. Barnes
Teacher
Fairway Middle School–KISD
Killeen, TX

Charlie Bialowas
Math Curriculum Specialist
Anaheim Union High School District
Anaheim, CA

Stephanie R. Boudreaux
Teacher
Fontainebleau Jr. High School
Mandeville, LA

Dianne G. Bounds
Teacher
Nettleton Junior High School
Jonesboro, AR

Susan Peavy Brooks
Math Teacher
Louis Pizitz Middle School
Vestavia Hills, AL

Karen Sykes Brown
Mathematics Educator
Riverview Middle School
Grundy, VA

Kay E. Brown
Teacher, 7th Grade
North Johnston Middle School
Micro, NC

Renee Burgdorf
Middle Grades Math Teacher
Morgan Co. Middle
Madison, GA

Kelley Summers Calloway
Teacher
Baldwin Middle School
Montgomery, AL

Carolyn M. Catto
Teacher
Harney Middle School
Las Vegas, NV

Claudia M. Cazanas
Math Department Chair
Fairmont Junior High
Pasadena, TX

David J. Chamberlain
Secondary Math Resource Teacher
Capistrano Unified School District
San Juan Capistrano, CA

David M. Chioda
Supervisor Math/Science
Marlboro Township Public Schools
Marlboro, NJ

Carrie Coate
7th Grade Math Teacher
Spanish Fort School
Spanish Fort, AL

Toinette Thomas Coleman
Secondary Mathematics Teacher
Caddo Middle Career & Technology School
Shreveport, LA

Linda M. Cordes
Math Department Chairperson
Paul Robeson Middle School
Kansas City, MO

Polly Crabtree
Teacher
Hendersonville Middle School
Hendersonville, NC

Dr. Michael T. Crane
Chairman Mathematics
B.M.C. Durfee High School
Fall River, MA

Tricia Creech, Ph.D.
Curriculum Facilitator
Southeast Guilford Middle School
Greensboro, NC

Lyn Crowell
Math Department Chair
Chisholm Trail Middle School
Round Rock, TX

B. Cummins
Teacher
Crestdale Middle School
Matthews, NC

Debbie Davis
8th Grade Math Teacher
Max Bruner, Jr. Middle School
Ft. Walton Beach, FL

Diane Yendell Day
Math Teacher
Moore Square Museums Magnet Middle School
Raleigh, NC

Wendysue Dodrill
Teacher
Barboursville Middle School
Barboursville, WV

Judith F. Duke
Math Teacher
Cranford Burns Middle School
Mobile, AL

Carol Fatta
Math/Computer Instructor
Chester Jr. Sr. M.S.
Chester, NY

Cynthia Fielder
Mathematics Consultant
Atlanta, GA

Georganne Fitzgerald
Mathematics Chair
Crittenden Middle School
Mt. View, CA

Jason M. Fountain
7th Grade Mathematics Teacher
Bay Minette Middle School
Bay Minette, AL

Sandra Gavin
Teacher
Highland Junior High School
Cowiche, WA

Ronald Gohn
8th Grade Mathematics
Dover Intermediate School
Dover, PA

Larry J. Gonzales
Math Department Chairperson
Desert Ridge Middle School
Albuquerque, NM

Shirley Gonzales
Math Teacher
Desert Ridge Middle School
Albuquerque, NM

Paul N. Hartley, Jr.
Mathematics Instructor
Loudoun County Public Schools
Leesburg, VA

Deborah L. Hewitt
Math Teacher
Chester High School
Chester, NY

Steven J. Huesch
Mathematics Teacher/Department Chair
Cortney Jr. High
Las Vegas, NV

Sherry Jarvis
8th Grade Math/Algebra 1 Teacher
Flat Rock Middle School
East Flat Rock, NC

Teacher Reviewers continued

Mary H. Jones
Math Curriculum Coordinator
Grand Rapids Public Schools
Grand Rapids, MI

Vincent D.R. Kole
Math Teacher
Eisenhower Middle School
Albuquerque, NM

Ladine Kunnanz
Middle School Math Teacher
Sequoyah Middle School
Edmond, OK

Barbara B. Larson
Math Teacher/Department Head
Andersen Middle School
Omaha, NE

Judith Lecocq
7th Grade Teacher
Murphysboro Middle School
Murphysboro, IL

Paula C. Lichiello
7th Grade Math and Pre-Algebra Teacher
Forest Middle School
Forest, VA

Michelle Mercier Maher
Teacher
Glasgow Middle School
Baton Rouge, LA

Jeri Manthei
Math Teacher
Millard North Middle School
Omaha, NE

Albert H. Mauthe, Ed.D.
Supervisor of Mathematics (Retired)
Norristown Area School District
Norristown, PA

Karen M. McClellan
Teacher & Math Department Chair
Harper Park Middle
Leesburg, VA

Ken Montgomery
Mathematics Teacher
Tri-Cities High School
East Point, GA

Helen M. O'Connor
Secondary Math Specialist
Harrison School District Two
Colorado Springs, CO

Cindy Ostrander
8th Grade Math Teacher
Edwardsville Middle School
Edwardsville, IL

Michael H. Perlin
8th Grade Mathematics Teacher
John Jay Middle School
Cross River, NY

Denise Pico
Mathematics Teacher
Jack Lund Schofield Middle School
Las Vegas, NV

Ann C. Raymond
Teacher
Oak Ave. Intermediate School
Temple City, CA

M.J. Richards
Middle School Math Teacher
Davis Middle School
Dublin, OH

Linda Lou Rohleder
Math Teacher, Grades 7 & 8
Jasper Middle School
Jasper, IN

Dana Schaefer
Pre-Algebra & Algebra I Teacher
Coachman Fundamental Middle School
Clearwater, FL

Donald W. Scheuer, Jr.
Coordinator of Mathematics
Abington School District
Abington, PA

Angela Hardee Slate
Teacher, 7th Grade Math/Algebra
Martin Middle School
Raleigh, NC

Mary Ferrington Soto
7th Grade Math
Calhoun Middle School-Ouachita Parish Schools
Calhoun, LA

Diane Stilwell
Mathematics Teacher/Technology Coordinator
South Middle School
Morgantown, WV

Pamela Ann Summers
K–12 Mathematics Coordinator
Lubbock ISD–Central Office
Lubbock, TX

Marnita L. Taylor
Mathematics Teacher/Department Chairperson
Tolleston Middle School
Gary, IN

Susan Troutman
Teacher
Dulles Middle School
Sugar Land, TX

Barbara C. VanDenBerg
Math Coordinator, K–8
Clifton Board of Education
Clifton, NJ

Mollie VanVeckhoven-Boeving
7th Grade Math and Algebra Teacher
White Hall Jr. High School
White Hall, AR

Mary A. Voss
7th Grade Math Teacher
Andersen Middle School
Omaha, NE

Christine Waddell
Teacher Specialist
Jordan School District
Sandy, UT

E. Jean Ware
Supervisor
Caddo Parish School Board
Shreveport, LA

Karen Y. Watts
9th Grade Math Teacher
Douglas High School
Douglas, AL

Lu Wiggs
Supervisor
I.S. 195
New York, NY

Teacher Advisory Board and Field Test Schools

Teacher Advisory Board

Glencoe/McGraw-Hill wishes to thank the following teachers for their feedback on *Mathematics: Applications and Concepts*. They were instrumental in providing valuable input toward the development of this program.

Katie Davidson
Legg Middle School
Coldwater, MI

Lynanne Gabriel
Bradley Middle School
Huntersville, NC

Kathleen M. Johnson
New Albany-Plain Local Middle School
New Albany, OH

Ronald C. Myer
Indian Springs Middle School
Columbia City, IN

Mike Perlin
John Jay Middle School
Cross River, NY

Reema Rahaman
Brentwood Middle School
Brentwood, MO

Diane T. Scheuber
Elizabeth Middle School
Elizabeth, CO

Deborah Sykora
Hubert H. Humphrey Middle School
Bolingbrook, IL

DeLynn Woodside
Roosevelt Middle School,
    Oklahoma City Public Schools
Oklahoma City, OK

Field Test Schools

Glencoe/McGraw-Hill wishes to thank the following schools that field-tested pre-publication manuscript during the 2002–2003 school year. They were instrumental in providing feedback and verifying the effectiveness of this program.

Knox Community Middle School
Knox, IN

Roosevelt Middle School
Oklahoma City, OK

Brentwood Middle School
Brentwood, MO

Elizabeth Middle School
Elizabeth, CO

Legg Middle School
Coldwater, MI

Great Hollow Middle School
Nesconset, NY

Student Advisory Board

The Student Advisory Board gave the authors, editorial staff, and design team feedback on the lesson design, content, and covers of the Student Editions. We thank these students for their hard work and creative suggestions in making *Mathematics: Applications and Concepts* more student friendly.

Front Row: Joey Snyder, Tiffany Pickenpaugh, Craig Hammerstein

Back Row: Brittany Yokum, Alex Johnson, Cimeone Starling, Kristina Smith, Kate Holt, Ben Ball

UNIT 1
Decimals, Algebra, and Statistics

CHAPTER 1

Decimal Patterns and Algebra

Student Toolbox

Prerequisite Skills
- Diagnose Readiness 5
- Getting Ready for the Next Lesson
  9, 13, 17, 21, 27, 33, 36, 41

Reading and Writing Mathematics
- Link to Reading 30, 34
- Reading in the Content Area 14
- Reading Math 10, 18, 39
- Writing Math 8, 11, 26, 35, 37, 40, 44

Standardized Test Practice
- Multiple Choice 9, 13, 17, 21, 23, 27, 28, 33, 41, 45, 49, 50
- Short Response/Grid In 9, 13, 21, 27, 28, 33, 36, 41, 45, 51
- Extended Response 51
- Worked-Out Example 25

Interdisciplinary Connections
- WebQuest Math and Science 3
- Earth Science 45
- Geography 45
- Health 19, 21
- History 31
- Life Science 33
- Music 23
- Physical Science 13, 23
- Technology 10

HANDS-ON Mini Lab 18, 34, 38

The Game Zone
Evaluating Expressions 29

 Snapshots 6, 17

Getting Started 5
1-1 A Plan for Problem Solving 6
1-2 Powers and Exponents 10
1-3 Order of Operations 14
1-4 **Algebra:** Variables and Expressions 18
1-5a **Problem-Solving Strategy:** Guess and Check .. 22
1-5 **Algebra:** Equations 24
 Mid-Chapter Practice Test 28
1-6 **Algebra:** Properties 30
1-7 Sequences 34
1-7b **HANDS-ON Lab:** Exploring Sequences 37
1-8 **Measurement:** The Metric System 38
 Study Skill: Use Power Notes 42
1-9 Scientific Notation 43

ASSESSMENT
 Study Guide and Review 46
 Practice Test 49
 Standardized Test Practice 50

Lesson 1-2, p. 13

UNIT 1

CHAPTER 2: Statistics: Analyzing Data

Getting Started		53
2-1	Frequency Tables	54
2-2a	**Problem-Solving Strategy:** Use a Graph	58
2-2	Making Predictions	60
2-3	Line Plots	64
2-4	Mean, Median, and Mode	69
2-4b	**HANDS-ON Lab:** Are You Average?	73
	Mid-Chapter Practice Test	74
2-5	Stem-and-Leaf Plots	76
2-6	Box-and-Whisker Plots	80
2-6b	**Graphing Calculator Investigation:** Box-and-Whisker Plots	84
2-7	Bar Graphs and Histograms	85
2-7b	**Spreadsheet Investigation:** Multiple-Line and Multiple-Bar Graphs	90
2-8	Misleading Statistics	92
ASSESSMENT		
	Study Guide and Review	96
	Practice Test	99
	Standardized Test Practice	100

Lesson 2-3, p. 67

Student Toolbox

Prerequisite Skills
- Diagnose Readiness 53
- Getting Ready for the Next Lesson 57, 63, 68, 72, 79, 83, 89

Reading and Writing Mathematics
- Link to Reading 64, 69
- Reading in the Content Area 64
- Reading Math 61, 81
- Writing Math 56, 61, 71, 73, 78, 82, 87, 93

Standardized Test Practice
- Multiple Choice 57, 59, 63, 68, 72, 74, 79, 83, 89, 95, 99, 100
- Short Response/Grid In 63, 72, 74, 79, 83, 89, 95, 101
- Extended Response 101
- Worked-Out Example 87

Interdisciplinary Connections
- WebQuest Math and Science 95
- Civics 88
- Earth Science 59
- Geography 94
- History 55, 76
- Music 72
- Reading 59
- Technology 88

HANDS-ON Mini Lab 60, 69

The Game Zone
Finding the Mean 75

 Snapshots 62

UNIT 2
Integers and Algebra

CHAPTER 3 Algebra: Integers

Student Toolbox

Prerequisite Skills
- Diagnose Readiness 105
- Getting Ready for the Next Lesson 108, 111, 115, 124, 131, 137

Reading and Writing Mathematics
- Link to Reading 120
- Reading in the Content Area 109
- Reading Math 107, 113, 118, 127, 139
- Writing Math 107, 114, 119, 127, 130

Standardized Test Practice
- Multiple Choice 108, 111, 115, 116, 124, 131, 133, 137, 141, 145, 146
- Short Response/Grid In 108, 111, 115, 116, 124, 131, 137, 141, 147
- Extended Response 147
- Worked-Out Example 110

Interdisciplinary Connections
- WebQuest Math and Geography 103
- Astronomy 139
- Earth Science 120, 129, 133, 137, 140, 143
- Geography 113, 114, 130
- History 131
- Life Science 133
- Physical Science 141

HANDS-ON Mini Lab 128, 134, 138

The Game Zone
Graphing Points on a Coordinate Plane 117

Getting Started 105
- 3-1 Integers and Absolute Value 106
- 3-2 Comparing and Ordering Integers 109
- 3-3 **Geometry:** The Coordinate Plane 112
- **Mid-Chapter Practice Test** 116
- 3-4a **HANDS-ON Lab:** Adding Integers 118
- 3-4 Adding Integers 120
- **Study Skill:** Use a Flowchart 125
- 3-5a **HANDS-ON Lab:** Subtracting Integers .. 126
- 3-5 Subtracting Integers 128
- 3-6a **Problem-Solving Strategy:** Look for a Pattern .. 132
- 3-6 Multiplying Integers 134
- 3-7 Dividing Integers 138

ASSESSMENT
- **Study Guide and Review** 142
- **Practice Test** 145
- **Standardized Test Practice** 146

Lesson 3-7, p. 139

UNIT 2

CHAPTER 4 — Algebra: Linear Equations and Functions

Getting Started	149
4-1 Writing Expressions and Equations	150
Study Skill: Simplify the Problem	153
4-2a **HANDS-ON Lab:** Solving Equations Using Models	154
4-2 Solving Addition and Subtraction Equations	156
4-3 Solving Multiplication Equations	160
4-4a **Problem-Solving Strategy:** Work Backward	164
4-4 Solving Two-Step Equations	166
Mid-Chapter Practice Test	170
4-5 Inequalities	172
4-6a **HANDS-ON Lab:** Functions and Graphs	176
4-6 Functions and Linear Equations	177
4-7 Lines and Slope	182
ASSESSMENT	
Study Guide and Review	186
Practice Test	189
Standardized Test Practice	190

Lesson 4-3, p. 163

Student Toolbox

Prerequisite Skills
- Diagnose Readiness 149
- Getting Ready for the Next Lesson 152, 159, 163, 169, 175, 181

Reading and Writing Mathematics
- Link to Reading 182
- Reading in the Content Area 156
- Writing Math 155, 174, 176, 179, 184

Standardized Test Practice
- Multiple Choice 152, 159, 163, 165, 169, 170, 175, 181, 185, 189, 190
- Short Response/Grid In 159, 169, 170, 185, 191
- Extended Response 191
- Worked-Out Example 183

Interdisciplinary Connections
- WebQuest Math and Geography 185
- Civics 181
- Geography 165
- Health 173
- Science 156

HANDS-ON Mini Lab 160, 166

Solving Equations 171

 Snapshots 175, 180

UNIT 3
Fractions

CHAPTER 5: Fractions, Decimals, and Percents

Student Toolbox

Prerequisite Skills
- Diagnose Readiness 195
- Getting Ready for the Next Lesson 200, 206, 209, 213, 219, 223, 226

Reading and Writing Mathematics
- Link to Reading 210
- Reading in the Content Area 203
- Reading Math 217
- Writing Math 196, 199, 208, 212, 225

Standardized Test Practice
- Multiple Choice 200, 202, 206, 209, 213, 214, 219, 223, 226, 231, 235, 236
- Short Response/Grid In 200, 206, 209, 213, 214, 219, 223, 226, 231, 237
- Extended Response 237
- Worked-Out Example 229

Interdisciplinary Connections
- WebQuest Math and Health 193
- Earth Science 202
- Geography 220, 221, 222
- History 213
- Life Science 233
- Music 208, 209
- Reading 220

HANDS-ON Mini Lab 197, 207, 224

The Game Zone Finding Factors 215

 Snapshots 209, 222

Getting Started	195
5-1a **HANDS-ON** Lab: Exploring Factors	196
5-1 Prime Factorization	197
5-2a **Problem-Solving Strategy:** Make an Organized List	201
5-2 Greatest Common Factor	203
5-3 Simplifying Fractions	207
5-4 Fractions and Decimals	210
Mid-Chapter Practice Test	214
5-5 Fractions and Percents	216
5-6 Percents and Decimals	220
5-7 Least Common Multiple	224
5-8 Comparing and Ordering Rational Numbers	227
ASSESSMENT	
Study Guide and Review	232
Practice Test	235
Standardized Test Practice	236

Lesson 5-4, p. 210

UNIT 3

CHAPTER 6 Applying Fractions

Getting Started		**239**
6-1	Estimating with Fractions	**240**
6-2	Adding and Subtracting Fractions	**244**
6-3	Adding and Subtracting Mixed Numbers	**248**
6-3b	**Problem-Solving Strategy:** Eliminate Possibilities	**252**
6-4	Multiplying Fractions and Mixed Numbers	**254**
6-5	**Algebra:** Solving Equations	**258**
	Mid-Chapter Practice Test	**262**
6-6	Dividing Fractions and Mixed Numbers	**264**
6-7	**Measurement:** Changing Customary Units	**267**
6-8	**Geometry:** Perimeter and Area	**270**
6-9a	HANDS-ON **Lab:** Circumference	**274**
6-9	**Geometry:** Circles and Circumference	**275**

ASSESSMENT
Study Guide and Review	**278**
Practice Test	**281**
Standardized Test Practice	**282**

Lesson 6-2, p. 245

Student Toolbox

Prerequisite Skills
- Diagnose Readiness 239
- Getting Ready for the Next Lesson
  243, 247, 251, 257, 261, 266, 269, 273

Reading and Writing Mathematics
- Link to Reading 258
- Reading in the Content Area 244
- Writing Math 242, 260, 272, 274, 276

Standardized Test Practice
- Multiple Choice 243, 247, 251, 253, 257, 261, 262, 269, 273, 277, 281, 282
- Short Response/Grid In 257, 261, 266, 269, 273, 277, 283
- Extended Response 283
- Worked-Out Example 259

Interdisciplinary Connections
- WebQuest Math and Health 277
- Astronomy 248
- Earth Science 254, 279
- History 268
- Life Science 257
- Music 276

HANDS-ON **Mini Lab** 244, 258, 264

The Game Zone
Multiplying Fractions 263

 Snapshots 261

UNIT 4
Proportional Reasoning

CHAPTER 7 Ratios and Proportions

Student Toolbox

Prerequisite Skills
- Diagnose Readiness 287
- Getting Ready for the Next Lesson 291, 295, 300, 308, 315, 318, 321

Reading and Writing Mathematics
- Reading in the Content Area 288
- Reading Math 317
- Writing Math 296, 299, 301, 314, 322

Standardized Test Practice
- Multiple Choice 291, 295, 300, 303, 308, 315, 318, 321, 325, 329, 330
- Short Response/Grid In 300, 308, 310, 315, 321, 325, 331
- Extended Response 331
- Worked-Out Example 293

Interdisciplinary Connections
- WebQuest Math and Recreation 285
- Art 308
- Geography 308
- Life Science 295, 298
- Music 294
- Physical Science 303

HANDS-ON **Mini Lab** 292, 304, 316

Game Zone Solving Proportions 311

 Snapshots 320

Getting Started	287
7-1 Ratios	288
7-2 Rates	292
7-2b HANDS-ON **Lab:** Rate of Change	296
7-3 **Algebra:** Solving Proportions	297
7-3b HANDS-ON **Lab:** Wildlife Sampling	301
7-4a **Problem-Solving Strategy:** Draw a Diagram	302
7-4 **Geometry:** Scale Drawings	304
7-4b **Spreadsheet Investigation:** Scale Drawings	309
Mid-Chapter Practice Test	310
7-5 Fractions, Decimals, and Percents	312
7-6 Percents Greater Than 100% and Percents Less Than 1%	316
7-7 Percent of a Number	319
7-8a HANDS-ON **Lab:** Using a Percent Model	322
7-8 The Percent Proportion	323
ASSESSMENT	
Study Guide and Review	326
Practice Test	329
Standardized Test Practice	330

Lesson 7-8, p. 323

UNIT 4

CHAPTER 8 Applying Percent

Getting Started	333
8-1 Percent and Estimation	334
8-1b **Problem-Solving Strategy:** Reasonable Answers	338
8-2 **Algebra:** The Percent Equation	340
8-3a **HANDS-ON Lab:** Sampling	344
8-3 **Statistics:** Using Statistics to Predict	345
Mid-Chapter Practice Test	348
8-4 Percent of Change	350
8-5 Sales Tax and Discount	354
8-6 Simple Interest	358
8-6b **Spreadsheet Investigation:** Simple Interest	361
ASSESSMENT	
Study Guide and Review	362
Practice Test	365
Standardized Test Practice	366

Lesson 8-1, p. 334

Student Toolbox

Prerequisite Skills
- Diagnose Readiness 333
- Getting Ready for the Next Lesson 337, 343, 347, 353, 357

Reading and Writing Mathematics
- Link to Reading 340
- Reading in the Content Area 345
- Reading Math 359
- Writing Math 346, 356

Standardized Test Practice
- Multiple Choice 337, 339, 343, 347, 348, 353, 357, 360, 365, 366
- Short Response/Grid In 347, 353, 357, 367
- Extended Response 367
- Worked-Out Example 355

Interdisciplinary Connections
- Art 353
- Geology 337
- History 359
- Life Science 336
- Music 339, 355
- Physical Science 340
- Technology 341, 354

HANDS-ON Mini Lab 350

The Game Zone
Percent of a Number 349

 Snapshots 343, 345

CHAPTER 9 Probability

UNIT 4

Student Toolbox

Prerequisite Skills
- Diagnose Readiness 369
- Getting Ready for the Next Lesson 373, 377, 380, 383, 390, 396

Reading and Writing Mathematics
- Link to Reading 387
- Reading in the Content Area 378
- Reading Math 398
- Writing Math 376, 382, 386, 389, 395, 397, 399

Standardized Test Practice
- Multiple Choice 373, 377, 380, 390, 392, 396, 401, 406
- Short Response/Grid In 377, 380, 383, 384, 390, 396, 401, 405, 407
- Extended Response 407
- Worked-Out Example 379

Interdisciplinary Connections
- WebQuest Math and Recreation 401
- Civics 389, 400
- Earth Science 392
- Music 383
- Space Science 392
- Technology 373

HANDS-ON **Mini Lab** 374, 381, 393

The Game Zone Probability 385

 Snapshots 396

Getting Started		369
9-1	Simple Events	370
9-2	Tree Diagrams	374
9-3	The Fundamental Counting Principle	378
9-4	Permutations	381
	Mid-Chapter Practice Test	384
9-5a	HANDS-ON Lab: Exploring Combinations	386
9-5	Combinations	387
9-6a	**Problem-Solving Strategy:** Act It Out	391
9-6	Theoretical and Experimental Probability	393
9-6b	HANDS-ON Lab: Experimental Probability	397
9-7	Independent and Dependent Events	398

ASSESSMENT

Study Guide and Review 402
Practice Test 405
Standardized Test Practice 406

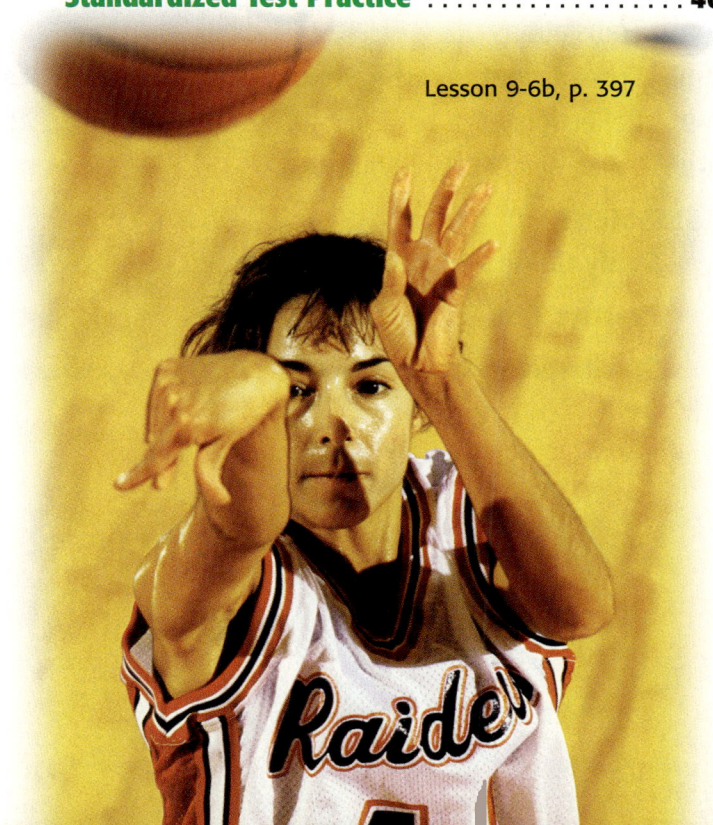

Lesson 9-6b, p. 397

xix

UNIT 5
Geometry and Measurement

CHAPTER 10 Geometry

Getting Started		411
10-1a	**HANDS-ON** Lab: Measuring Angles	412
10-1	Angles	413
10-1b	**HANDS-ON** Lab: Constructing and Bisecting Angles	416
10-2	**Statistics:** Making Circle Graphs	418
10-3	Angle Relationships	422
10-3b	**HANDS-ON** Lab: Constructing Parallel Lines	426
10-4	Triangles	428
10-4b	**HANDS-ON** Lab: Constructing Triangles	432
10-5	Quadrilaterals	434
	Mid-Chapter Practice Test	438
10-6	Similar Figures	440
10-7a	**Problem-Solving Strategy:** Use Logical Reasoning	444
10-7	Polygons and Tessellations	446
10-8	Translations	451
10-8b	**Spreadsheet Investigation:** Dilations	455
10-9	Reflections	456
10-9b	**HANDS-ON** Lab: Rotations	460

ASSESSMENT

Study Guide and Review	462
Practice Test	465
Standardized Test Practice	466

Lesson 10-9, p. 458

Student Toolbox

Prerequisite Skills
- Diagnose Readiness 411
- Getting Ready for the Next Lesson
  415, 421, 425, 431, 437, 443, 450, 454

Reading and Writing Mathematics
- Link to Reading 413, 451
- Reading in the Content Area 422
- Reading Math 434, 441, 447, 460
- Writing Math 412, 414, 417, 424, 427, 430, 433, 435, 442, 448, 461

Standardized Test Practice
- Multiple Choice 415, 421, 425, 431, 437, 438, 443, 445, 450, 454, 459, 465, 466
- Short Response/Grid In 415, 421, 425, 431, 438, 443, 450, 454, 459, 467
- Extended Response 467
- Worked-Out Example 441

Interdisciplinary Connections
- WebQuest Math and History 409
- Art 436, 449, 454, 459
- Earth Science 415
- Geography 431, 446, 465
- Health 425

HANDS-ON Mini Lab 422, 428, 440, 451, 456

The Game Zone
Finding Squares 439

 Snapshots 421, 425

UNIT 5

CHAPTER 11
Geometry: Measuring Two-Dimensional Figures

Student Toolbox

Prerequisite Skills
- Diagnose Readiness 469
- Getting Ready for the Next Lesson
  473, 477, 482, 485, 492, 495, 500

Reading and Writing Mathematics
- Link to Reading 498
- Reading in the Content Area 475
- Reading Math 470, 471, 480, 490
- Writing Math 472, 476, 478, 481, 484, 488, 491, 499, 502

Standardized Test Practice
- Multiple Choice 473, 477, 482, 485, 486, 492, 495, 497, 500, 503, 507, 508
- Short Response/Grid In 473, 477, 482, 485, 486, 492, 495, 500, 503, 509
- Extended Response 509
- Worked-Out Example 499

Interdisciplinary Connections
- Archaeology 480
- Earth Science 497
- Geography 472, 490, 492
- History 494
- Music 492
- Physical Science 471, 472

HANDS-ON Mini Lab 470, 475, 483, 489, 493, 501

The Game Zone
Finding Square Roots 487

 Snapshots 492

Getting Started 469
- 11-1 Squares and Square Roots 470
  - **Study Skill:** Use a Web 474
- 11-2 Estimating Square Roots 475
- 11-3a **HANDS-ON** Lab: The Pythagorean Theorem ... 478
- 11-3 The Pythagorean Theorem 479
- 11-4 Area of Parallelograms 483
  - **Mid-Chapter Practice Test** 486
- 11-5a **HANDS-ON** Lab: Triangles and Trapezoids 488
- 11-5 Area of Triangles and Trapezoids 489
- 11-6 Area of Circles 493
- 11-7a **Problem-Solving Strategy:** Solve a Simpler Problem 496
- 11-7 Area of Complex Figures 498
- 11-8 **Probability:** Area Models 501

ASSESSMENT
- **Study Guide and Review** 504
- **Practice Test** 507
- **Standardized Test Practice** 508

Lesson 11-5, p. 490

UNIT 5

CHAPTER 12 Geometry: Measuring Three-Dimensional Figures

	Getting Started	**511**
12-1a	**HANDS-ON Lab:** Building Three-Dimensional Figures	**512**
12-1	Drawing Three-Dimensional Figures	**514**
12-1b	**Problem-Solving Strategy:** Make a Model	**518**
12-2	Volume of Rectangular Prisms	**520**
12-2b	**Spreadsheet Investigation:** Similar Solids	**523**
12-3	Volume of Cylinders	**524**
	Mid-Chapter Practice Test	**528**
12-4a	**HANDS-ON Lab:** Nets and Surface Area	**530**
12-4	Surface Area of Rectangular Prisms	**532**
12-4b	**HANDS-ON Lab:** Changes in Volume and Surface Area	**536**
12-5	Surface Area of Cylinders	**538**
12-6	**Measurement:** Precision	**542**

ASSESSMENT
Study Guide and Review **546**
Practice Test **549**
Standardized Test Practice **550**

Student Handbook

Built-In Workbooks
Prerequisite Skills **554**
Extra Practice **564**
Mixed Problem Solving **596**
Preparing for Standardized Tests **608**

Reference
English-Spanish Glossary **626**
Selected Answers **645**
Photo Credits **663**
Index **664**

Student Toolbox

Prerequisite Skills
- Diagnose Readiness 511
- Getting Ready for the Next Lesson 517, 522, 527, 535, 541

Reading and Writing Mathematics
- Link to Reading 520
- Reading in the Content Area 520
- Writing Math 513, 525, 531, 533, 537, 543

Standardized Test Practice
- Multiple Choice 517, 519, 522, 527, 535, 541, 545, 549, 550
- Short Response/Grid In 522, 528, 535, 551
- Extended Response 551
- Worked-Out Example 525

Interdisciplinary Connections
- Math and History 545
- Art 519
- History 516
- Physical Science 534
- Space Science 526

HANDS-ON Mini Lab 520, 524, 532, 538

Three-Dimensional Figures 529

USA TODAY Snapshots 544

xxii

HOW TO...
Use Your Math Book

Why do I need my math book?

Have you ever been in class and not understood all of what was presented? Or, you understood everything in class, but at home, got stuck on how to solve a couple of problems? Maybe you just wondered when you were ever going to use this stuff?

These next few pages are designed to help you understand everything your math book can be used for . . . besides homework problems!

BEFORE YOU READ

Have a Goal

- What information are you trying to find?
- Why is this information important to you?
- How will you use the information?

Have a Plan

- Read *What You'll Learn* at the beginning of the lesson.
- Look over photos, tables, graphs, and opening activities.
- Locate boldfaced words and read their definitions.
- Find Key Concept and Concept Summary boxes for a preview of what's important.
- Skim the example problems.

Have an Opinion

- Is this information what you were looking for?
- Do you understand what you have read?
- How does this information fit with what you already know?

xxiii

IN CLASS

During class is the opportunity to learn as much as possible about that day's lesson. Ask questions about things that you don't understand, and take notes to help you remember important information.

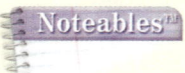 Each time you find this logo throughout your book, use your *Noteables™: Interactive Study Notebook with Foldables™* or your own notebook to take notes.

To help keep your notes in order, try making a Foldables Study Organizer. It's as easy as 1-2-3! Here's a Foldable you can use to keep track of the rules for addition, subtraction, multiplication, and division.

FOLDABLES™ Study Organizer

Operations Make this Foldable to help you organize your notes. Begin with a sheet of 11" × 17" paper.

STEP 1
Fold
Fold the short sides toward the middle.

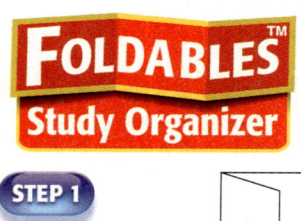

STEP 2
Fold Again
Fold the top to the bottom.

STEP 3
Cut
Open. Cut along the second fold to make four tabs.

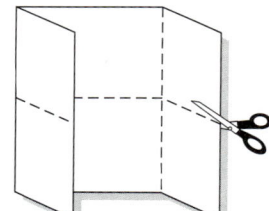

STEP 4
Label
Label each of the tabs as shown.

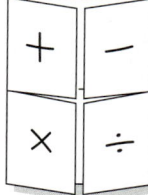

LOOK FOR...
FOLDABLES™
on these pages:
5, 53, 105, 149, 195, 239, 287, 333, 369, 411, 469, and 511.

Need to Cover Your Book?
Inside the back cover are directions for a Foldable that you can use to cover your math book quickly and easily!

DOING YOUR HOMEWORK

Regardless of how well you paid attention in class, by the time you arrive at home, your notes may no longer make any sense and your homework seems impossible. It's during these times that your book can be most useful.

- Each lesson has example problems, solved step-by-step, so you can review the day's lesson material.

- A Web site has extra examples to coach you through solving those difficult problems.

- Each exercise set has Homework Help boxes that show you which examples may help with your homework problems.

- Answers to the odd-numbered problems are in the back of the book. Use them to see if you are solving the problems correctly. If you have difficulty on an even problem, do the odd problem next to it. That should give you a hint about how to proceed with the even problem.

Look For...
The Web site with extra examples on these pages in Chapter 1: 7, 11, 15, 19, 25, 31, 35, 39, and 44.
Homework Help boxes on these pages in Chapter 1: 9, 12, 16, 20, 26, 32, 36, 40, and 45.
Selected Answers starting on page 645.

HOW TO...
Use Your Math Book

BEFORE A TEST

Admit it! You think there is no way to study for a math test! However, there *are* ways to review before a test. Your book offers help with this also.

- Review all of the new vocabulary words and be sure you understand their definitions. These can be found on the first page of each lesson or highlighted in yellow in the text.

- Review the notes you've taken on your Foldable and write down any questions that you still need answered.

- Practice all of the concepts presented in the chapter by using the chapter Study Guide and Review. It has additional problems for you to try as well as more examples to help you understand. You can also take the Chapter Practice Test.

- Take the self-check quizzes from the Web site.

LOOK FOR...

The Web site with self-check quizzes on these pages in Chapter 1: 9, 13, 16, 21, 27, 33, 36, 38, and 45.

The Study Guide and Review for Chapter 1 on page 46.

SCAVENGER HUNT

LET'S GET STARTED

To help you find the information you need quickly, use the Scavenger Hunt below to learn where things are located in each chapter.

1. What is the title of Chapter 1?

2. How can you tell what you'll learn in Lesson 1-1?

3. Sometimes you may ask, "When am I ever going to use this?" Name a situation that uses the concepts from Lesson 1-2.

4. What is the key concept presented in Lesson 1-3?

5. How many Examples are presented in Lesson 1-3?

6. What is the web address where you could find extra examples?

7. In Lesson 1-4, there is a paragraph that mentions it is common to use the first letter of a word as a variable. What is the main heading above that paragraph?

8. There is a Real-Life Career mentioned in Lesson 1-4. What is it?

9. List the new vocabulary words that are presented in Lesson 1-5.

10. List the math symbols that are presented in Lesson 1-5.

11. Suppose you're doing your homework on page 32 and you get stuck on Exercise 12. Where could you find help?

12. What is the web address that would allow you to take a self-check quiz to be sure you understand the lesson?

13. On what pages will you find the Study Guide and Review?

14. Suppose you can't figure out how to do Exercise 25 in the Study Guide on page 47. Where could you find help?

15. You complete the Practice Test on page 49 to study for your chapter test. Where could you find another test for more practice?

UNIT 1
Decimals, Algebra and Statistics

Chapter 1
Decimal Patterns and Algebra

Chapter 2
Statistics: Analyzing Data

Your study of math requires analysis of data, much of which will be in decimal form. In this unit, you will learn how to evaluate algebraic expressions involving decimals, as well as analyze and graph data.

INTERDISCIPLINARY PROJECT

Lions, Tigers, and Bears

Math and Science Are you ready to join a team of animal experts? As part of your application to be a Zoo's new coordinator, you must complete several challenging tasks. You'll make decisions about what animals to purchase for the zoo based on financial information provided to you. You'll gather specific data about the animals you choose, including their weight and expected lifespan. Finally, you'll present your findings to the hiring committee. So pack up your gear and don't forget your algebra tool kit. This adventure is going to be wild!

WebQuest Log on to **msmath2.net/webquest** to begin your WebQuest.

CHAPTER 1
Decimal Patterns and Algebra

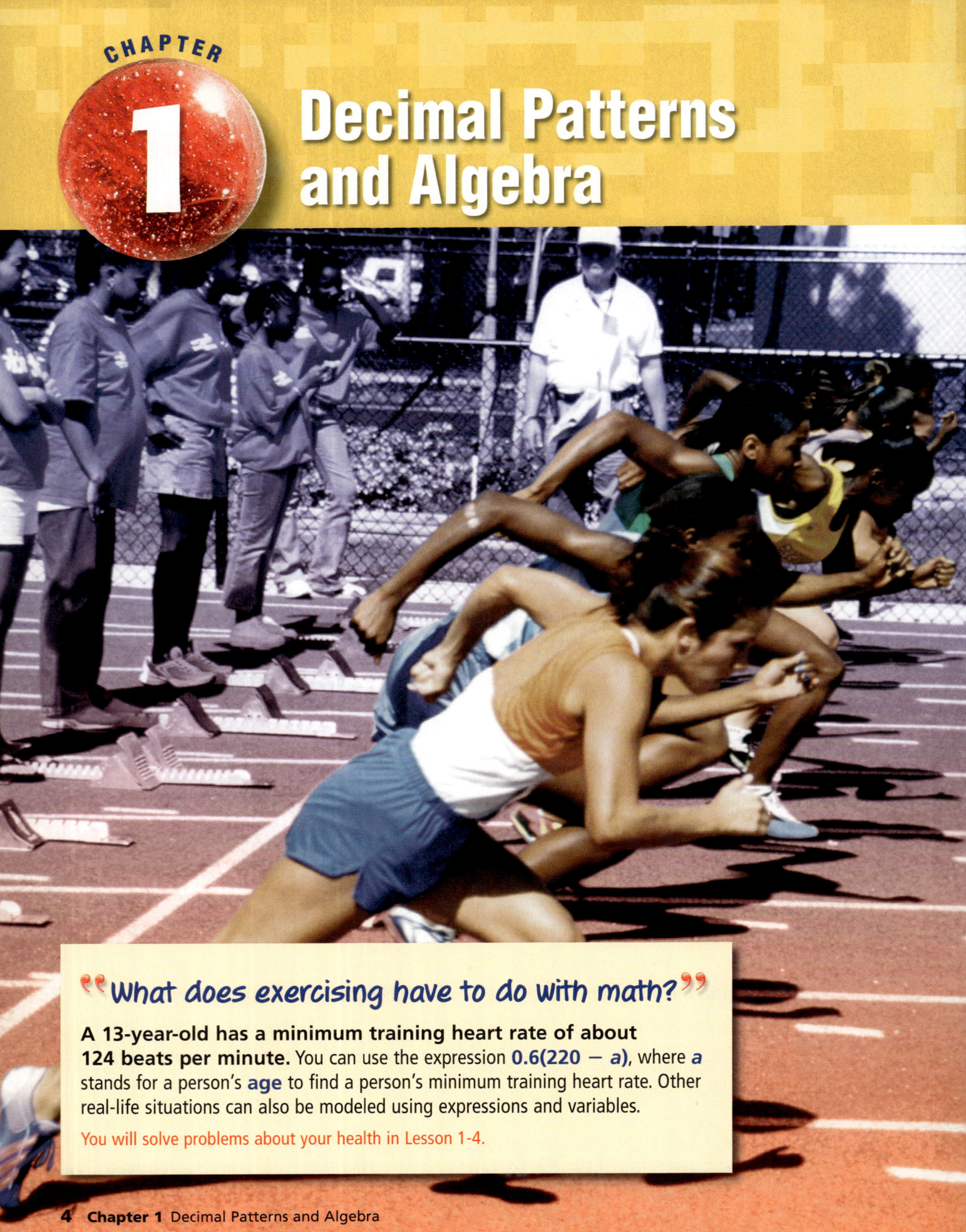

"What does exercising have to do with math?"

A 13-year-old has a minimum training heart rate of about 124 beats per minute. You can use the expression $0.6(220 - a)$, where *a* stands for a person's **age** to find a person's minimum training heart rate. Other real-life situations can also be modeled using expressions and variables.

You will solve problems about your health in Lesson 1-4.

GETTING STARTED

▶ Diagnose Readiness

Take this quiz to see if you are ready to begin Chapter 1. Refer to the page number in parentheses for review.

Vocabulary Review

State whether each sentence is *true* or *false*. If *false*, replace the underlined word to make a true sentence.

1. A decimal point separates the ones place and the <u>tenths</u> place. (Page 555)
2. When <u>multiplying</u> or subtracting decimals, you must first line up the decimal points. (Page 555)

Prerequisite Skills

Add. (Page 559)

3. $89.3 + 16.5$
4. $7.9 + 32.45$
5. $54.25 + 6.39$
6. $10.8 + 2.6$

Subtract. (Page 559)

7. $24.6 - 13.3$
8. $9.1 - 6.6$
9. $30.55 - 2.86$
10. $17.4 - 11.2$

Multiply. (Page 560)

11. 4×7.7
12. 9.8×3
13. 2.7×6.3
14. 8.5×1.2

Divide. (Page 562)

15. $32.6 \div 4$
16. $10.6 \div 2$
17. $5.5 \div 5$
18. $17.84 \div 4$

Multiply or divide. (Page 562)

19. $2.45 \times 1,000$
20. $87.3 \div 100$
21. 0.61×100
22. $10 \div 1,000$

FOLDABLES Study Organizer

Decimal Patterns and Algebra Make this Foldable to help you organize your notes. Begin with ten sheets of notebook paper.

STEP 1 Staple
Staple the ten sheets together to form a booklet.

STEP 2 Cut Tabs
On the second page, make the top tab the width of the white space. On the third page, make the tab 2 lines longer, and so on.

STEP 3 Label
Write the chapter title on the cover and label each tab with the lesson number.

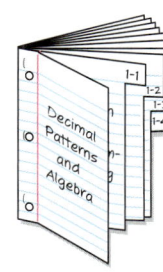

Noteables™ **Chapter Notes** Each time you find this logo throughout the chapter, use your *Noteables™: Interactive Study Notebook with Foldables™* or your own notebook to take notes. Begin your chapter notes with this Foldable activity.

Readiness To prepare yourself for this chapter with another quiz, visit **msmath2.net/chapter_readiness**

1-1 A Plan for Problem Solving

What You'll LEARN

Solve problems using the four-step plan.

WHEN am I ever going to use this?

ANIMALS The graphic shows the top five turkey-producing states in a recent year. What is the total number of turkeys produced in the top five states?

1. Do you have all of the information necessary to solve this problem?

2. Explain how you would solve this problem. Then solve it.

3. Does your answer make sense? Explain.

4. What can you do if your first attempt at solving the problem does not work?

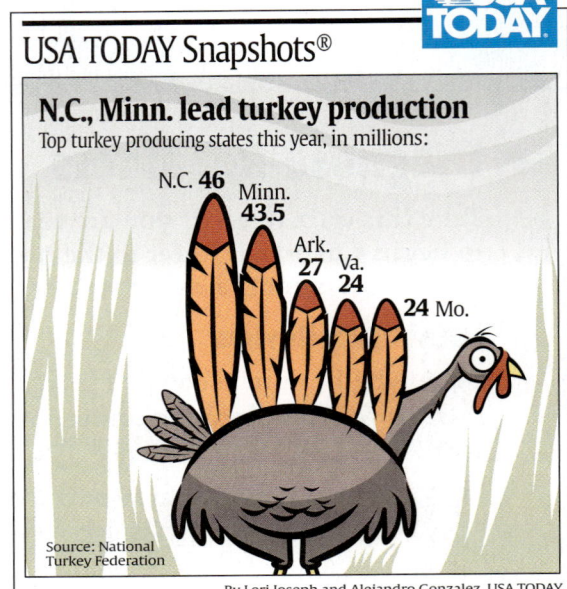

USA TODAY Snapshots®
N.C., Minn. lead turkey production
Top turkey producing states this year, in millions:
N.C. 46, Minn. 43.5, Ark. 27, Va. 24, Mo. 24
Source: National Turkey Federation
By Lori Joseph and Alejandro Gonzalez, USA TODAY

In mathematics, there is a *four-step plan* you can use to help you solve math problems.

1. Explore
- Determine what information is given and what you need to find.
- Do you have all the information to solve the problem?
- Is there any information given that is not needed?

2. Plan
- Select a strategy for solving the problem. There may be several that you can use.
- It is usually helpful to estimate the answer.

3. Solve
- Carry out your plan.
- If your plan doesn't work, keep trying until you find one that does work.
- Make sure your solution contains appropriate units or labels.

4. Examine
- Does your answer make sense given the facts in the problem?
- Is your answer reasonable compared to your estimate?
- If your answer is not correct, make a new plan and start again.

STUDY TIP

Four-Step Plan No matter what strategy you use, you can always use the four-step plan to solve a problem.

6 Chapter 1 Decimal Patterns and Algebra

EXAMPLE   Use the Four-Step Plan

1 TELEVISION Color TV sets were first mass-produced in 1954. In that year, 5,000 color TVs were produced. How many years passed before at least 100 times as many were produced?

Year	Color TVs Produced	Year	Color TVs Produced
1954	5,000	1960	120,000
1955	20,000	1961	140,000
1956	90,000	1962	450,000
1957	85,000	1963	750,000
1958	80,000	1964	1,500,000
1959	90,000		

Source: *Sams Photofact Book on Color TV*

Explore   *What are you trying to find?*
The number of years since 1954 that 100 times as many TVs were produced.

What information do you need to solve the problem?
From the table, you know the number of TVs produced each year from 1954 to 1964.

Plan   You can estimate the number of years that passed before at least 100 times 5,000 TVs, or 500,000 TVs were produced. This happened between 1962 and 1963. So, approximately 10 years had passed.

Solve   100 × 5,000 TVs = 500,000 TVs

In 1963, 750,000 TVs were produced, which is more than 500,000 TVs. Since 1963 − 1954 = 9, nine years passed before production was at least 100 times greater than the number of sets produced in 1954.

Examine   *Is your answer reasonable?*
Look at the data in the table and count the number of years from 1954 to 1963. Also, compare your answer to the estimate.

Your Turn Solve using the four-step plan.

a. How many times more TVs were produced in 1960 than in 1954?
b. Between which two years did TV production increase the most?

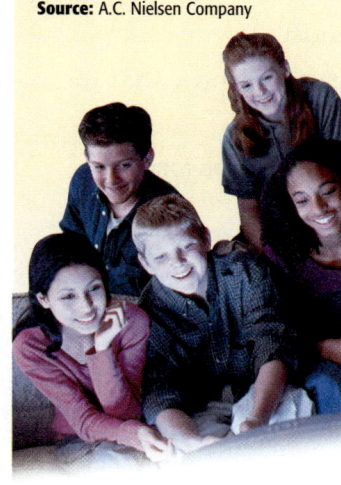

REAL-LIFE MATH

TELEVISION The average American watches more than 4 hours of TV per day.
Source: A.C. Nielsen Company

Problems can be solved using different operations or strategies.

Noteables™   **Key Concept: Problem-Solving Strategies**

guess and check	use a graph
look for a pattern	work backward
make an organized list	eliminate possibilities
draw a diagram	estimate reasonable answers
act it out	use logical reasoning
solve a simpler problem	make a model

msmath2.net/extra_examples   **Lesson 1-1** A Plan for Problem Solving   **7**

EXAMPLE   Use a Strategy in the Four-Step Plan

2 **SPORTS** In a basketball tournament, 128 teams play in the first round. Then there are 64 teams left to play in the second round, 32 teams left to play in the third round, and so on. How many rounds does it take to determine a winner?

Explore You know the number of teams at the start of the tournament and how many teams play in each of the first three rounds.

Plan You can look for a pattern by organizing the information in a table. Then continue the pattern until one team is left.

Solve

Round	1	2	3	4	5	6	7
Number of Teams	128	64	32	16	8	4	2

÷2 ÷2

So, a winner is determined in 7 rounds.

Examine You could complete the diagram below to show that the answer is reasonable.

Round 1 → 128 teams: 64 teams win, 64 teams lose
Round 2 → 64 teams: 32 teams win, 32 teams lose
Round 3 → 32 teams: 16 teams win, 16 teams lose

Your Turn

c. How many rounds would it take to determine a winner if 512 teams participated in the tournament?

Skill and Concept Check

1. **Explain** why it is important to plan before solving a problem.
2. **Writing Math** Describe what to do if your solution to a problem seems unrealistic.
3. **OPEN ENDED** Write a real-life problem that can be solved by adding 79 and 42.

GUIDED PRACTICE

Use the four-step plan to solve each problem.

4. **MONEY** To attend the class picnic, each student will have to pay $5.00 for transportation and $6.50 for food. If there are 300 students in the class, how much money will be collected for the picnic?

5. **PATTERNS** If Isaac receives an E-mail every 20 minutes during the workday, how many E-mails would he expect to receive between 8:00 A.M. and noon?

Practice and Applications

Use the four-step plan to solve each problem.

6. **VIDEO RENTALS** A video store took in $5,400 in video rentals during July. January sales are expected to be double that amount. If videos rent for $4, how many video rentals are expected in January?

7. **RACING** The Indianapolis Motor Speedway has 3,200,000 bricks that lie beneath the 2.5 miles of track. Find how many bricks lie under one mile of track at the race track.

8. **COOKING** Mr. Sanchez is serving Cajun fried turkey at 3:00 P.M. The 15-pound turkey has to remain in the fryer 3 minutes for every pound and must cool for at least 45 minutes before it is carved. What is the latest time he can start frying?

9. **GEOMETRY** Draw the next two figures in the pattern.

10. **MULTI STEP** Kishi wants to buy a DVD player that costs $250 with tax. So far, she has saved $145. If she saves $5 every week, in how many weeks will she be able to purchase the DVD player?

11. **COLLECTIONS** In 2001, Dustin started collecting 6 die-cast cars every year. In 2003, his brother Logan started collecting 9 cars per year. In what year did Dustin and Logan have the same number of cars?

12. **CRITICAL THINKING** Use the digits 1, 2, 3, 4, and 5 to form a two-digit and a three-digit number so that their product is the least product possible. Use each digit only once.

HOMEWORK HELP

For Exercises	See Examples
6–9	1
10–11	2

Extra Practice
See pages 564, 596.

Spiral Review with Standardized Test Practice

13. **SHORT RESPONSE** Jeannie has $86 to spend on birthday gifts for her mother. Which three items from the table could Jeannie purchase for her mother, not including tax?

14. **MULTIPLE CHOICE** In a field near Duns, Scotland, a reproduction of the painting *Sunflowers* was created with 250,000 plants and flowers. The "painting" covered a 46,000-square foot area. What is a reasonable number of plants and flowers per square foot?

    Ⓐ 5   Ⓑ 50   Ⓒ 100   Ⓓ 500

Item	Cost
sweater	$29.50
gloves	$22.75
purse	$32.00
chocolates	$15.00
movie passes	$27.75

GETTING READY FOR THE NEXT LESSON

BASIC SKILL Multiply.

15. 10×10 16. $3 \times 3 \times 3$ 17. $5 \times 5 \times 5 \times 5$ 18. $2 \times 2 \times 2 \times 2 \times 2$

msmath2.net/self_check_quiz

1-2 Powers and Exponents

WHEN am I ever going to use this?

TECHNOLOGY Computer engineer Gordon E. Moore made this observation in 1964. Moore's Law says the amount of available storage space on a computer chip doubles every year.

1. How is doubling shown in the table?
2. If the pattern continued, how much storage space would be available by year 6?
3. What is the relationship between the number of 2s and the year?

Year	Amount of Storage Space (units)	
1	2	= 2
2	2 × 2	= 4
3	2 × 2 × 2	= 8
4	2 × 2 × 2 × 2	= 16

What You'll LEARN
Use powers and exponents.

NEW Vocabulary
- factors
- exponent
- base
- powers
- squared
- cubed
- evaluate
- standard form
- exponential form

Two or more numbers that are multiplied together to form a product are called **factors**. The amount of storage space on a computer chip in year 4 can be written using only factors of 2. When the same factor is used, you may use an exponent to simplify the notation.

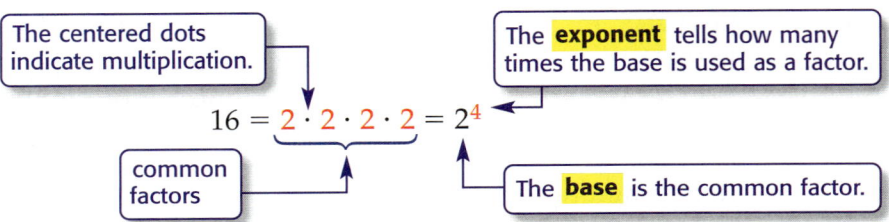

- The centered dots indicate multiplication.
- common factors
- The **exponent** tells how many times the base is used as a factor.
- The **base** is the common factor.

$$16 = 2 \cdot 2 \cdot 2 \cdot 2 = 2^4$$

Numbers expressed using exponents are called **powers**. You can say that 2^4 is a power of 2. This and the powers 5^2 and 4^3 are read as follows.

Symbols	Words
5^2	five to the second power or five **squared**
4^3	four to the third power or four **cubed**
2^4	two to the fourth power

EXAMPLES Write Powers as Products

Write each power as a product of the same factor.

 7^5

The base is 7. The exponent 5 means that 7 is used as a factor five times.

$7^5 = 7 \cdot 7 \cdot 7 \cdot 7 \cdot 7$

 3^2

The base is 3. The exponent 2 means the 3 is used as a factor twice.

$3^2 = 3 \cdot 3$

READING Math
First Power When a number does not have an exponent, it is understood to be 1. For example, $8 = 8^1$.

You can **evaluate**, or find the value of, powers by multiplying the factors. Numbers written without exponents are in **standard form**.

EXAMPLES Write Powers in Standard Form

Evaluate each expression.

3 2^5

$2^5 = 2 \cdot 2 \cdot 2 \cdot 2 \cdot 2$
$ = 32$

4 4^3

$4^3 = 4 \cdot 4 \cdot 4$
$ = 64$

Your Turn Evaluate each expression.

a. 10^2 b. 7^3 c. 5^4

Numbers written with exponents are in **exponential form**.

EXAMPLE Write Numbers in Exponential Form

5 Write $3 \cdot 3 \cdot 3 \cdot 3$ in exponential form.

3 is the base. It is used as a factor 4 times. So, the exponent is 4.

$3 \cdot 3 \cdot 3 \cdot 3 = 3^4$

Your Turn Write each product in exponential form.

d. $5 \cdot 5 \cdot 5$ e. $12 \cdot 12 \cdot 12 \cdot 12 \cdot 12 \cdot 12$

Skill and Concept Check

1. **Writing Math** Explain what *five to the fourth power* means.
2. Write 7^5 in words.
3. **OPEN ENDED** Write one number in exponential form and another number in standard form.
4. **Which One Doesn't Belong?** Identify the number that cannot be written as a power with an exponent greater than 1. Explain your reasoning.

   | 4 | 9 | 16 | 50 |

GUIDED PRACTICE

Write each power as a product of the same factor.

5. 6^2 6. 4^4 7. 8^5

Evaluate each expression.

8. 3^4 9. 5^5 10. 10^3

Write each product in exponential form.

11. $5 \cdot 5 \cdot 5 \cdot 5 \cdot 5 \cdot 5$ 12. $1 \cdot 1 \cdot 1 \cdot 1$

13. Evaluate *eleven to the third power*.

msmath2.net/extra_examples   Lesson 1-2 Powers and Exponents **11**

Practice and Applications

Write each power as a product of the same factor.

14. 1^5
15. 4^2
16. 3^8
17. 9^3
18. 10^4
19. 11^3

Evaluate each expression.

20. 2^3
21. 2^6
22. 4^3
23. 5^4
24. 1^{10}
25. 10^1
26. 3^8
27. 4^6
28. 10^5
29. 12^4
30. 20^4
31. 50^3

HOMEWORK HELP

For Exercises	See Examples
14–19	1, 2
20–31, 38, 42–47, 54	3, 4
32–37, 39, 57, 59, 61	5

Extra Practice
See pages 564, 596.

Write each product in exponential form.

32. $3 \cdot 3$
33. $7 \cdot 7 \cdot 7 \cdot 7$
34. $1 \cdot 1 \cdot 1$
35. $6 \cdot 6 \cdot 6 \cdot 6 \cdot 6$
36. $4 \cdot 4 \cdot 4 \cdot 4 \cdot 4 \cdot 4 \cdot 4$
37. $13 \cdot 13 \cdot 13 \cdot 13 \cdot 13$

38. Find the value of *6 cubed*.
39. Express 256 in exponential form as a power of 2.

GEOMETRY For Exercises 40 and 41, use the figures below.

40. Find the number of unit cubes that make up each large cube. Write your answers using exponents.
41. Why do you think the expression 2^3 is sometimes read as *2 cubed*?

Use a calculator to determine whether each sentence is *true* or *false*.

42. $3^7 > 7^3$
43. $14^4 = 19^2$
44. $927 > 9^3$
45. $3^2 = 2^3$
46. $6^3 < 4^4$
47. $3^4 < 9^2$

Order the following numbers from least to greatest.

48. $6^5, 1^{14}, 4^{10}, 17^3$
49. $2^8, 15^2, 6^3, 3^5$
50. $5^3, 4^6, 2^{11}, 7^2$
51. $6^4, 3^6, 5^5, 8^3$
52. $3^7, 10^3, 8^2, 2^{10}$
53. $6^5, 9^4, 11^3, 5^5$

54. **POPULATION** There are approximately 7^{10} people living in the United States. About how many people is this?
55. Explain why a *quadrillion*, 10^{15}, is usually written in exponential form instead of standard form.
56. Tell whether the following statement is *true* or *false*. Explain your reasoning. *The number 64 can be written in exponential form in only one way.*
57. Use exponents to write $5 \cdot 5 \cdot 5 \cdot 5 \cdot 4 \cdot 4 \cdot 4$ in the shortest form possible.

12 Chapter 1 Decimal Patterns and Algebra

58. **MULTI STEP** Use a calculator to express 1,679,616 in exponential form as a power of 6.

59. **POPULATION** The population of Fort Worth, Texas, is approximately 534,000. Find a number less than 534,000 and a number greater than 534,000 that can be expressed in exponential form.

60. **RESEARCH** Our numbering system is based on the powers of 10. The ancient Mayan system is based on another power. Use the Internet or another source to write a paragraph about the Mayan numbering system.

61. **PHYSICAL SCIENCE** The speed of sound varies as it travels through different substances. At ordinary temperatures, sound travels through water at a rate of 5,000 feet per second. Write 5,000 using exponents and the factors 5 and 10.

62. **CRITICAL THINKING** Based on the number pattern shown at the right, write a convincing argument that any number, except 0, raised to the 0 power equals 1.

$2^4 = 16$
$2^3 = 8$
$2^2 = 4$
$2^1 = 2$
$2^0 = ?$

Spiral Review with Standardized Test Practice

63. **MULTIPLE CHOICE** Which is *two to the sixth power* written in standard form?

    A. 26
    B. $6 \cdot 6$
    C. 64
    D. $2 \cdot 2 \cdot 2 \cdot 2 \cdot 2 \cdot 2$

64. **GRID IN** How many zeros does the value of 10^8 have?

65. **RACING** The graph shows which numbered-cars have the most wins at the Indianapolis 500. How many more wins did the number 3 car have than the number 14 car? (Lesson 1-1)

66. **PRODUCTION** A machine on a production line fills 8 soft drink cans per minute. How many cans does it fill in 8 hours? (Lesson 1-1)

67. *True* or *False*? In the four-step plan, the *Solve* step comes last. (Lesson 1-1)

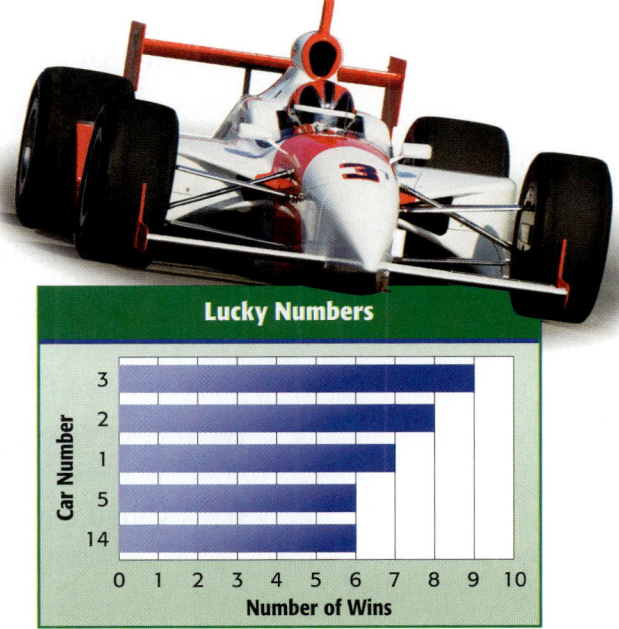

Source: Indianapolis 500 Record Book

GETTING READY FOR THE NEXT LESSON

PREREQUISITE SKILL Add, subtract, multiply, or divide. (Pages 559, 560, 562)

68. $13.4 + 8.7$
69. $13.6 - 10.2$
70. 1.5×6.8
71. $8.6 - 5.53$
72. $36 \div 3$
73. $1.7 + 32.28$
74. 3.2×6
75. $24.5 \div 7$

1-3 Order of Operations

What You'll LEARN
Evaluate expressions using the order of operations.

NEW Vocabulary
numerical expression
order of operations

WHEN am I ever going to use this?

GAMES Kaitlyn and Percy are each writing as many different expressions as possible using the numbers rolled on a number cube. In the first round, 3, 5, 4, and 6 were rolled in order.

Kaitlyn
$3 \cdot 5 - (4 + 6) = 3 \cdot 5 - 10$
$= 15 - 10 \text{ or } 5$

Percy
$3 \cdot (5 - 4) + 6 = 3 \cdot 1 + 6$
$= 3 + 6 \text{ or } 9$

1. List the similarities and differences between the two expressions.
2. What was Kaitlyn's first step in simplifying her expression? Percy's first step?
3. **Make a conjecture** about what should be the first step in simplifying $(3 \cdot 5) - 4 + 6$.

The expressions $3 \cdot 5 - (4 + 6)$ and $3 \cdot (5 - 4) + 6$ are **numerical expressions**. Mathematicians have agreed upon steps called the **order of operations** to find the values of such expressions. These rules ensure that numerical expressions have only one value.

Noteables™ Key Concept: Order of Operations
1. Do all operations within grouping symbols first.
2. Evaluate all powers before other operations.
3. Multiply and divide in order from left to right.
4. Add and subtract in order from left to right.

You can use the order of operations to evaluate numerical expressions.

EXAMPLES Evaluate Expressions

1 Evaluate $5 + (12 - 3)$.

$5 + (12 - 3) = 5 + 9$ Subtract first since $12 - 3$ is in parentheses.
$ = 14$ Add 5 and 9.

2 Evaluate $8 - 3 \cdot 2 + 7$.

$8 - 3 \cdot 2 + 7 = 8 - 6 + 7$ Multiply 3 and 2.
$ = 2 + 7$ Subtract 6 from 8.
$ = 9$ Add 2 and 7.

Your Turn Evaluate each expression.

a. $39 \div (9 + 4)$ b. $10 + 8 \div 2 - 6$

READING in the Content Area
For strategies in reading this lesson, visit **msmath2.net/reading**.

STUDY TIP

Powers If an exponent lies outside a grouping symbol, complete the operations within the grouping symbol before applying the power.

EXAMPLES Evaluate Expressions with Powers

3 Evaluate 15×10^3.

$15 \times 10^3 = 15 \times 1{,}000$ Find the value of 10^3.
$ = 15{,}000$ Multiply.

4 Evaluate $36 \div (1 + 2)^2$.

$36 \div (1 + 2)^2 = 36 \div 3^2$ Add 1 and 2 inside the parentheses.
$ = 36 \div 9$ Find the value of 3^2.
$ = 4$ Divide.

Your Turn Evaluate each expression.

c. 3×10^4 d. $(5 - 1)^3 \div 4$

In addition to using the symbols $\times$ and $\cdot$, multiplication can be indicated by using parentheses.

$2(3 + 5)$ means $2 \times (3 + 5)$ or $2 \cdot (3 + 5)$.
$(4 - 2)3$ means $(4 - 2) \times 3$ or $(4 - 2) \cdot 3$.

EXAMPLE Evaluate an Expression

5 Evaluate $14 + 3(7 - 2) - 2 \cdot 5$.

$14 + 3(7 - 2) - 2 \cdot 5$
$= 14 + 3(5) - 2 \cdot 5$ Subtract 2 from 7.
$= 14 + 15 - 2 \cdot 5$ Multiply from left to right, $3 \cdot 5 = 15$.
$= 14 + 15 - 10$ Multiply from left to right, $2 \cdot 5 = 10$.
$= 29 - 10$ Add from left to right, $14 + 15 = 29$.
$= 19$ Subtract 10 from 29.

REAL-LIFE MATH

VIDEO GAMES A team of 500 engineers and technicians built the first video-game console in 1966.

Source: www.geekcomix.com

EXAMPLE Use an Expression to Solve a Problem

6 **VIDEO GAMES** Evita is buying a video game station, three extra controllers, and four new video games. What is the total cost?

Item	Quantity	Unit Cost
game station	1	$180.00
controller	3	$24.95
game	4	$35.99

$\underbrace{\text{cost of game station}}_{\$180} + \underbrace{\text{number of controllers}}_{3} \times \underbrace{\text{cost of controller}}_{\$24.95} + \underbrace{\text{number of games}}_{4} \times \underbrace{\text{cost of game}}_{\$35.99}$

$= 180 + 74.85 + 143.96$ Multiply from left to right.
$= 398.81$ Add.

So, the total cost is $398.81.

Check Check the reasonableness of the answer by estimating. The cost is about $180 + (25 \times 3) + (40 \times 4) = 180 + 75 + 160$, or $415. So, the solution of $398.81 is reasonable.

Skill and Concept Check

1. **Identify** the operation that should be done first in each expression.
   a. $9 \div 3 + (14 - 7)$
   b. $3 + 24 \div 3 \cdot 4$

2. **OPEN ENDED** Write an expression containing five numbers that is evaluated by first multiplying.

3. **FIND THE ERROR** Yutaka and Cynthia are evaluating $16 - 24 \div 6 \cdot 2$. Who is correct? Explain your reasoning.

   Yutaka
   $16 - 24 \div 6 \cdot 2$
   $= 16 - 24 \div 12$
   $= 16 - 2 \text{ or } 14$

   Cynthia
   $16 - 24 \div 6 \cdot 2$
   $= 16 - 4 \cdot 2$
   $= 16 - 8 \text{ or } 8$

GUIDED PRACTICE

Evaluate each expression.

4. $11 - (3 \cdot 2)$
5. $25 \div (9 - 4)$
6. $8 - 4 + 3.7$
7. $2 + 5 \cdot 5$
8. $14 \div 2 \cdot 6$
9. $8 \cdot 5 - 4 \cdot 3$
10. 4×10^2
11. $3.5 \times 5 + 6^2$
12. $7 + 4(5.6 - 2) - 9$

13. Evaluate $(16 \div 4)^3 - 6$.
14. Find the value of $(6 + 8) \div (10 - 8)$.

Practice and Applications

Evaluate each expression.

15. $(1 + 8) \times 3$
16. $10 - (3 + 4)$
17. $(25 \div 5) + 8$
18. $(11 - 2) \div 9$
19. $3 \cdot 2 + 7$
20. $15 \div 3 + 4$
21. $12 + 6.6 \div 3$
22. $18 - 3 \cdot 6$
23. $8 - 7.2 + 5$
24. $28 \div 7(5)$
25. $(17 + 3) \div (4 + 1)$
26. $(6 + 5) \cdot (8 - 6)$
27. $21 \div 3 \times 2 - 4$
28. $35 \div 5 + 56 \div 7$
29. $2 \times 9 - 4^2$
30. $24 \div 3 + 5^3$
31. $7 + (8 - 7 + 2)^4$
32. $(2 + 10)^2 \div 4$
33. 6×10^2
34. 18×10^3
35. 1.95×10^2
36. 3.7×10^4
37. $6 + 2(9.4 - 1)$
38. $3(4.5 + 7.2) - 5 \cdot 4$
39. $72 \div 3 - 5(8.8 - 6) + 9$
40. $9 \div 3 \cdot 14(10 - 8) - 60$

HOMEWORK HELP

For Exercises	See Examples
15–28	1, 2
29–36, 43–45	3, 4
37–40, 46–48	5
41–42, 50	6

Extra Practice
See pages 564, 596.

41. **GEOMETRY** The distance around a geometric figure is called its *perimeter*. Write a numerical expression to find the perimeter of the figure at the right. Then evaluate the expression.

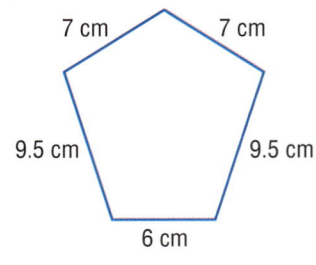

42. **MARATHONS** On Mondays, Wednesdays, and Thursdays, Jacob trains for a marathon for 3.5 hours. On Tuesdays and Fridays, he trains for 2 hours, and on Saturdays, he trains for 4.5 hours. How many hours does Jacob train per week?

Find the value of each expression.

43. 4.5×10^2 **44.** 3.08×10^4 **45.** 2.965×10^5

46. $3 \cdot 4(5 - 3.8) + 2.7$ **47.** $7.1 \times 9 - (4 + 3) + 1$ **48.** $(6 + 1)^2 - 2.5(3)$

49. Insert parentheses to make $72 \div 9 + 27 - 2 = 0$ true.

50. MONEY MATTERS Luke is ordering nine reams of paper, four boxes of pens, and two rolls of tape. Use the table to find the cost of his order.

Item	Cost
ream of paper	$3.95
box of pens	$7.49
roll of tape	$1.29

CRITICAL THINKING Some calculators are programmed to follow the order of operations.

51. Ruby evaluates $12 \div 2 + 2 \cdot 5$ and gets 15. When she checks using her calculator, she gets 16. Which is correct? Explain.

52. Explain how you could tell whether your calculator follows the order of operations.

Spiral Review with Standardized Test Practice

53. MULTIPLE CHOICE Which is the first step in evaluating $10 - 3 \cdot 2 + 4(15 - 5)$?

A $10 - 3$ **B** $3 \cdot 2$ **C** $2 + 4$ **D** $15 - 5$

54. MULTIPLE CHOICE Anthony has 2 boxes of cookies containing 24 cookies each and 3 packages of brownies containing 15 brownies each. Which expression *cannot* be used to find the total number of items he has?

F $3 \times 15 + 2 \times 24$ **G** $5 \times (24 + 15)$
H $2(24) + 3(15)$ **I** $15 + 15 + 15 + 24 + 24$

55. TESTING Use the graphic to estimate how much longer educators spend preparing students in grades 6–8 for tests than students in grades 9–12. (Lesson 1-1)

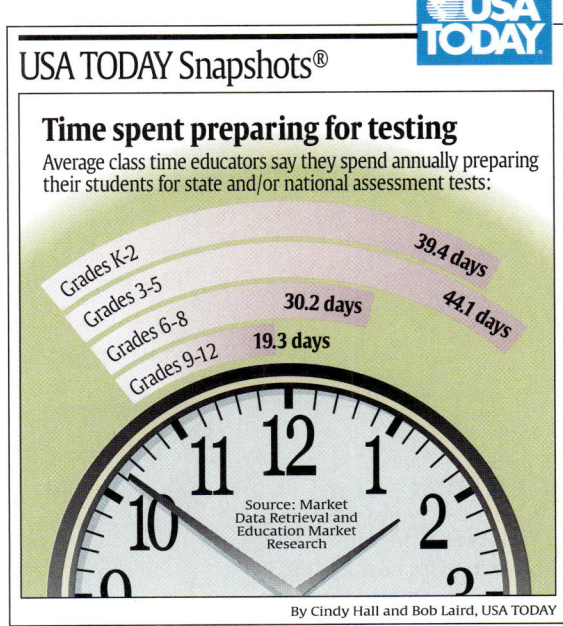

USA TODAY Snapshots®
Time spent preparing for testing
Average class time educators say they spend annually preparing their students for state and/or national assessment tests:
Grades K-2: 39.4 days
Grades 3-5: 44.1 days
Grades 6-8: 30.2 days
Grades 9-12: 19.3 days
Source: Market Data Retrieval and Education Market Research
By Cindy Hall and Bob Laird, USA TODAY

Evaluate each expression. (Lesson 1-2)

56. 2^4 **57.** 11^2
58. 3^6 **59.** 4^5

Write each power as a product of the same factor. (Lesson 1-2)

60. 3^4 **61.** 8^5
62. 6^2 **63.** 4^6
64. 1^4 **65.** 9^1

GETTING READY FOR THE NEXT LESSON

BASIC SKILL Divide.

66. $45 \div 15$ **67.** $90 \div 5$ **68.** $110 \div 10$ **69.** $52 \div 4$

1-4 Algebra: Variables and Expressions

What You'll LEARN
Evaluate simple algebraic expressions.

NEW Vocabulary
variable
algebraic expression
algebra
term
coefficient
constant

REVIEW Vocabulary
evaluate: find the value (Lesson 1-2)

HANDS-ON Mini Lab

Materials
- isometric dot paper

Work with a partner.

The pattern below is made up of triangles, each with side lengths of 1.

1. Draw the next three figures in the pattern.
2. Find the perimeter of each figure and record your data in a table like the one shown below. The first three are completed for you.

Number of Triangles	1	2	3	4	5	6
Perimeter	3	4	5	?	?	?

3. Without drawing the figure, determine the perimeter of a figure made up of 10 triangles. Check by making a drawing.
4. Find a relationship between the number of triangles and the perimeter at each stage of the pattern.

In the Mini Lab, you found that the perimeter of the figure is two more than the number of triangles. You can use a placeholder, or **variable**, to represent the number of triangles.

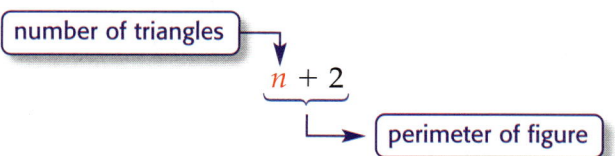

The expression $n + 2$ is called an **algebraic expression** because it contains variables, numbers, and at least one operation. The branch of mathematics that involves expressions with variables is called **algebra**.

EXAMPLE Evaluate an Expression

1 Evaluate $n + 3$ if $n = 4$.

$n + 3 = 4 + 3$ Replace n with 4.
$ = 7$

Your Turn Evaluate each expression if $c = 8$.

a. $c + 2$ b. $c - 3$ c. $15 - c$

READING Math
Variables The letter x is often used as a variable. It is also common to use the first letter of a word as the variable. For example, use t for time or n for number.

18 Chapter 1 Decimal Patterns and Algebra

The following symbols are used for multiplication and division with variables. Note that the multiplication sign is usually omitted.

$6d$ means $6 \times d$ or $6 \cdot d$

ef means $e \times f$

$9st$ means $9 \times s \times t$

$\dfrac{c}{d}$ means $c \div d$

When plus or minus signs separate an algebraic expression into parts, each part is a **term**. The numerical factor of a term that contains a variable is called a **coefficient**. A term that does not contain a variable is called a **constant**.

8 is the coefficient of 8x. → $8x + y + 3$ ← constant

1 is the coefficient of y because $1y = y$.

EXAMPLES Evaluate Expressions

2 Evaluate $8w - 2v$ if $w = 5$ and $v = 3$.

$8w - 2v = 8(5) - 2(3)$ Replace w with 5 and v with 3.
$\qquad\quad\; = 40 - 6$ Use the order of operations.
$\qquad\quad\; = 34$ Subtract 6 from 40.

3 Evaluate $\dfrac{mn}{6}$ if $m = 8$ and $n = 12$.

$\dfrac{mn}{6} = \dfrac{(8)(12)}{6}$ Replace m with 8 and n with 12.

$\qquad = \dfrac{96}{6}$ The fraction bar is like a grouping symbol.

$\qquad = 16$ Divide.

4 Evaluate $y^2 + 2$ if $y = 3$.

$y^2 + 2 = 3^2 + 2$ Replace y with 3.
$\qquad\;\; = 9 + 2$ or 11 Use the order of operations.

Your Turn Evaluate each expression if $a = 4$ and $b = 3$.
d. $9a - 6b$ e. $\dfrac{ab}{2}$ f. $a^2 + 5$

REAL-LIFE CAREERS

How Does an Athletic Trainer Use Math?
Athletic trainers use the formula $(220 - a) \times 0.6$, where a is a person's age, to find their minimum training heart rate.

Math Online Research
For information about a career as an athletic trainer, visit: msmath2.net/careers

EXAMPLE Use an Expression to Solve a Problem

5 **HEALTH** Use the formula at the left to find Kaylee's minimum training heart rate if she is 17 years old.

$(220 - a) \times 0.6 = (220 - 17) \times 0.6$ Replace a with 17.
$\qquad\qquad\qquad\;\; = 203 \times 0.6$ or 121.8 Use the order of operations.

Kaylee's minimum training heart rate of 121.8 beats per minute is rounded up to 122 because her heart cannot beat 0.8 times.

msmath2.net/extra_examples   Lesson 1-4 Algebra: Variables and Expressions

Skill and Concept Check

1. **Identify** the coefficient and constant in the expression $2x - 6$.
2. **OPEN ENDED** Write an algebraic expression that involves multiplication and addition.
3. **Which One Doesn't Belong?** Identify the expression that does not have the same characteristic as the other three. Explain your reasoning.

| $4x$ | ab | $r \cdot 8$ | $\dfrac{x}{y}$ |

GUIDED PRACTICE

Evaluate each expression if $a = 3$, $b = 5$, and $c = 4$.

4. $a + 7$
5. $4b$
6. $b + c$
7. $6c - a$
8. $\dfrac{c}{4}$
9. $\dfrac{5c + b}{5}$
10. $4b^2$
11. $\dfrac{3c^2}{a}$

12. What operations are used in the algebraic expression $\dfrac{xyz}{2}$?
13. Evaluate $n^2 + 1$ if $n = 2$.

Practice and Applications

Evaluate each expression if $d = 8$, $e = 3$, $f = 4$, and $g = 1$.

14. $d + 9$
15. $10 - e$
16. $4f$
17. $8g$
18. $f - e$
19. $d + f$
20. $10g - 6$
21. $8 + 5d$
22. $7e + 2d$
23. $\dfrac{d}{5}$
24. $\dfrac{16}{f}$
25. $\dfrac{5d - 25}{5}$
26. e^2
27. $\dfrac{d^2}{2}$
28. $6f^2$
29. $11.9d - 10e$
30. $\dfrac{2f}{g^2}$
31. $\dfrac{2.5f + 5}{3}$
32. $7e^2 + 5$
33. $\dfrac{(5 + g)^2}{2}$

HOMEWORK HELP

For Exercises	See Examples
14–19	1
20–22	2
23–25	3
26–35	4
38–40, 43–45	5

Extra Practice
See pages 565, 596.

34. Evaluate $9h + h^2$ if $h = 3.5$.
35. For what value of x is the expression x^2 equal to 9?

Determine whether each statement is *sometimes*, *always*, or *never* true. Explain.

36. The expressions $x - 3$ and $y - 3$ represent the same value.
37. The expression ab means the same as the expression $a \times b$.

BASEBALL For Exercises 38–40, use the following information.
Charlie earns $16 per game for umpiring Little League baseball games.

38. Make a table that shows how much money Charlie will earn for umpiring 1, 2, 3, 4, and 5 baseball games.
39. Write an expression to find how much money Charlie will earn for any number of games. Let n be the number of games he umpires.
40. Charlie umpires 18 games during the summer. How much will he earn?

Complete each table.

41.
Quarts (q)	Gallons $\left(\frac{q}{4}\right)$
4	1
6	1.5
8	?
10	?
12	?
14	?

42.
Dollars (d)	Quarters (4d)
1	4
2	8
3	?
4	?
5	?
6	?

43. **HEALTH** The expression $110 + \frac{A}{2}$, where A stands for a person's age, is used to estimate a person's normal systolic blood pressure. Estimate the normal systolic blood pressure for a 16-year old.

BOATING For Exercises 44 and 45, use the information below and at the right.
The Lerman family is vacationing at the lake and wants to rent a boat.

44. Write an expression showing the cost of renting a boat for h hours.

45. How much would it cost if they rented the boat for 2, 3, or 4 hours?

46. **WRITE A PROBLEM** Write an algebraic expression with the variable x that has a value of 3 when evaluated.

47. **CRITICAL THINKING** Find values of x and y so that the value of $7x + 2$ is greater than the value of $3y + 23$.

Boat Rentals
$45 per hour
with a
$65 nonrefundable deposit.

Spiral Review with Standardized Test Practice

48. **MULTIPLE CHOICE** Which expression could be used to find the cost of buying b baseball bats at $75 each and g baseball gloves at $98 each?

    Ⓐ $75b + 98g$ Ⓑ $b + g$ Ⓒ $75 + 98$ Ⓓ $75b \div 98g$

49. **SHORT RESPONSE** Juan opened a bank account one month with $40 and deposited $15 each month after that. He was able to save $100 by December. In which month did he open the bank account?

Evaluate each expression. (Lesson 1-3)

50. $6(5) - 2$ 51. $9 + 9 \div 3$ 52. $4 \cdot 2(8 - 1)$ 53. $(17 + 3) \div 5$

54. Write 2^5 as a product of the same factor. (Lesson 1-2)

GETTING READY FOR THE NEXT LESSON

PREREQUISITE SKILL Determine whether each sentence is *true* or *false*.
(Lesson 1-3)

55. $15 - 2(3) = 9$ 56. $20 \div 5 \times 4 = 1$ 57. $4^2 + 6 \cdot 7 = 154$ 58. $24 \div (7 - 3) = 6$

1-5a Problem-Solving Strategy
A Preview of Lesson 1-5

Guess and Check

What You'll LEARN

Solve problems using the guess and check strategy.

The total ticket sales for the school play were $263.50. Amber, how many adult and student tickets were sold?

Adult tickets are $7.50 and student tickets are $4.00. Also, twice as many students bought tickets as adults. Let's **guess and check** to find out.

Explore	We know adult tickets are $7.50, student tickets are $4.00, and twice as many students bought tickets than did adults.
Plan	Let's make a guess and check if it is correct. Then adjust the guess until we get the correct answer.
Solve	Make a guess. 10 adults, 20 students 7.5(10) + 4(20) = $155 *too low* Adjust the guess upward. 20 adults, 40 students 7.5(20) + 4(40) = $310 *too high* Adjust the guess downward slightly. 18 adults, 36 students 7.5(18) + 4(36) = $279 *still too high* Adjust the guess downward again. 17 adults, 34 students 7.5(17) + 4(34) = $263.50 *correct* There were 17 adult and 34 student tickets sold.
Examine	Seventeen adult tickets cost $127.50, and 34 student tickets cost $136. Since $127.50 + $136 = $263.50 and 34 tickets are twice as many as 17 tickets, the guess is correct.

Analyze the Strategy

1. **Explain** why you must keep a careful record of each of your guesses and their results in the *solve* step of the problem-solving plan.
2. **Write** a problem that could be solved by guess and check. Then write the steps you would take to find the solution to your problem.

22 Chapter 1 Decimal Patterns and Algebra

Apply the Strategy

Solve. Use the guess and check strategy.

3. **CAR WASH** The band sponsored a car wash to help pay for uniforms. They charged $4 for a car and $6 for an SUV. During the first hour, they washed 16 vehicles and earned $78. How many of each type of vehicle did they wash?

4. **SOUVENIRS** The Pike's Peak souvenir shop sells standard-sized postcards in packages of 5 and large-sized postcards in packages of 3. If Juan bought 16 postcards, how many packages of each did he buy?

Mixed Problem Solving

Solve. Use any strategy.

5. **MUSIC** Flor is burning a CD for her friend Derek. The CD will hold 30 minutes of music. Which songs should she select from the list at the right to record the maximum time on the CD without going over?

SONG	TIME
A	3 min 30 s
B	6 min 0 s
C	4 min 45 s
D	5 min 30 s
E	5 min 6 s
F	4 min 15 s
G	4 min 30 s
H	3 min 48 s

6. **NUMBER THEORY** A number is squared, and the result is 256. Find the number.

7. **PATTERNS** Copy the table below. Extend the pattern to complete the table.

Input	1	2	3	4	5	6	7
Output	1	8	27	?	?	?	?

8. **MULTI STEP** The Wicked Twister roller coaster can accommodate about 1,000 passengers per hour. The coaster has 8 vehicles that each carry 4 passengers. Approximately how many runs are made in 1 hour?

9. **PHYSICAL SCIENCE** Telephone calls travel through optical fibers at the speed of light, which is 186,000 miles per second. A millisecond is 0.001 of a second. How far can your voice travel over an optical line in 1 millisecond?

10. **MONEY MATTERS** Syreeta needs to buy four markers to make posters for a science project. She has $4. Does she have enough money if the cost of each marker, including tax, is 89¢? Explain.

SNOWFALL For Exercises 11 and 12, use the graph below.

Source: www.usatoday.com

11. What is the total average snowfall for October through April?

12. Between which two consecutive months is the difference in snowfall the least?

13. **STANDARDIZED TEST PRACTICE**
Kelly has $2.80 in change in her purse. If the change is made up of an equal number of nickels, dimes, and quarters, how many of each does she have?

Ⓐ 4   Ⓑ 5   Ⓒ 6   Ⓓ 7

1-5 Algebra: Equations

What You'll LEARN

Solve equations using mental math.

NEW Vocabulary

equation
solution
solving an equation
defining the variable

MATH Symbols

= is equal to
≠ is not equal to

WHEN am I ever going to use this?

BASKETBALL The table shows the number of wins for six WNBA teams after playing 34 games each.

1. How many losses did each team have?

2. Write a rule to describe how you found the number of losses.

3. Let w represent the number of wins and ℓ represent the number of losses. Rewrite your rule using numbers, variables, and an equals sign.

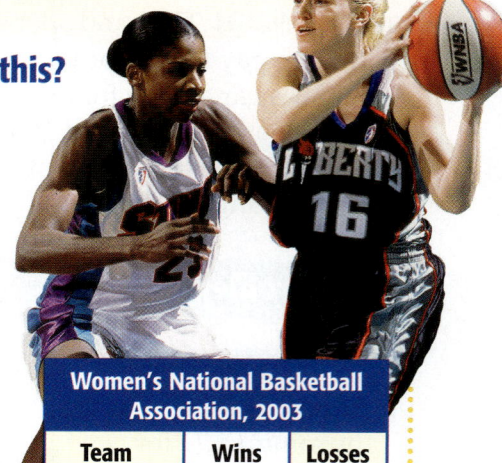

Women's National Basketball Association, 2003

Team	Wins	Losses
Detroit	25	?
Los Angeles	24	?
Houston	20	?
Charlotte	18	?
New York	16	?
Phoenix	8	?

Source: wnba.com

An **equation** is a sentence in mathematics that contains an equals sign. The equals sign tells you that the expression on the left is equivalent to the expression on the right.

$$4 + 3 = 8 - 1 \qquad\qquad 3(4) = 24 \div 2 \qquad\qquad 17 = 13 + 2 + 2$$

An equation that contains a variable is neither true nor false until the variable is replaced with a number. The **solution** of an equation is a number that makes the sentence true.

The process of finding a solution is called **solving an equation**. Some equations are easy to solve using mental mathematics.

EXAMPLE Solve an Equation Mentally

1 Solve $18 = 14 + t$ mentally.

$18 = 14 + t$ Write the equation.
$18 = 14 + 4$ You know that $14 + 4$ is 18.
$18 = 18$ Simplify.

The solution is 4.

 Your Turn Solve each equation mentally.

a. $p - 5 = 20$ b. $8 = y \div 3$ c. $7h = 56$

24 Chapter 1 Decimal Patterns and Algebra

EXAMPLE — Graph the Solution of an Equation

② Graph the solution of the equation in Example 1.

Locate the point named by the solution on a number line. Then draw a dot at the solution, 4.

When you write an equation that represents a real-life problem, you are *modeling* the problem. First, choose a variable to represent one of the unknowns. This is called **defining the variable**.

EXAMPLE — Write an Equation to Solve a Problem

Study Tip

Defining the Variable Use the first letter of the word you are defining as a variable. For example: h stands for the cost of a hamburger.

③ **FOOD** The total cost of a hamburger, fries, and soft drink is $5.50. If the fries and drink cost $2.50 together, what is the cost of the hamburger?

Words	The cost of a hamburger, fries, and soft drink is $5.50.
Variable	Let h represent the cost of the hamburger.
Equation	$h + 2.50 = 5.50$

$h + 2.50 = 5.50$ Write the equation.
$3 + 2.50 = 5.50$ Replace h with 3 to make the equation true.
$5.50 = 5.50$ Simplify.

The number 3 is the solution of the equation. So, the cost of the hamburger is $3.00.

EXAMPLE — Find a Solution of an Equation

Standardized Test Practice

 MULTIPLE-CHOICE TEST ITEM

What value of x is a solution of $x + 14.6 = 30.2$?

A 14.9   **B** 15.2   **C** 15.6   **D** 16.1

Test-Taking Tip

Solutions Since an equation with one variable has only one solution, the value of 16.1, or choice D, does not need to be checked.

Read the Test Item
Substitute each value for x to determine which makes the left side of the equation equivalent to the right side.

Solve the Test Item

Replace x with 14.9.   Replace x with 15.2.   Replace x with 15.6.
$x + 14.6 = 30.2$ $x + 14.6 = 30.2$ $x + 14.6 = 30.2$
$14.9 + 14.6 \stackrel{?}{=} 30.2$ $15.2 + 14.6 \stackrel{?}{=} 30.2$ $15.6 + 14.6 \stackrel{?}{=} 30.2$
$29.5 \neq 30.2$ false $29.8 \neq 30.2$ false $30.2 = 30.2$ true ✓

The value 15.6 makes the equation true. So, the answer is C.

msmath2.net/extra_examples

Lesson 1-5 Algebra: Equations **25**

Skill and Concept Check

1. **Writing Math** Explain what it means to solve an equation.

2. **OPEN ENDED** Write an example of an equation that is easily solved using mental mathematics and an equation that is not as easily solved using mental mathematics.

3. **FIND THE ERROR** Joshua and Ivan had a contest to see who could solve $w - 25 = 50$ the fastest. Their solutions are shown below. Who is correct? Explain.

   Joshua
   w = 25

   Ivan
   w = 75

GUIDED PRACTICE

Solve each equation mentally.

4. $y - 18 = 20$
5. $5 + a = 22$
6. $w \div 4 = 7$
7. $\frac{r}{9} = 6$
8. $75 = w + 72$
9. $69 = 3f$

10. Graph the solution of $x + 5 = 11$.

Name the number that is the solution of the given equation.

11. $x + 15 = 19$; 4, 5, 6
12. $13k = 80.6$; 5.5, 5.8, 6.2

13. **BASKETBALL** Jason scored 14 more points than David in the last basketball game. If David scored 9 points, how many did Jason score?

Practice and Applications

Solve each equation mentally.

14. $b + 7 = 18$
15. $8 + x = 15$
16. $y - 15 = 71$
17. $a - 18 = 20$
18. $25 - 19 = n$
19. $24 + 39 = m$
20. $12 \cdot 5 = s$
21. $n = \frac{30}{6}$
22. $b - 42 = 84$
23. $7t = 77$
24. $3d = 99$
25. $22 + h = 42$
26. $20 = y \div 5$
27. $16 = \frac{u}{4}$
28. $\frac{z}{7} = 12$
29. $18j = 360$
30. $m - 36 = 123$
31. $108 = 9b$

HOMEWORK HELP

For Exercises	See Examples
14–31	1
32–35	2
40–46	3
36–39	4

Extra Practice
See pages 565, 596.

Graph the solution of each equation on a number line.

32. $w + 3 = 10$
33. $x - 7 = 11$
34. $12y = 48$
35. $11 = 132 \div z$

Name the number that is the solution of the given equation.

36. $13.4 \cdot 9 = h$; 117.8, 118.7, 120.6
37. $n \div 10 = 4$; 20, 30, 40
38. $9.9 + r = 24.2$; 12.7, 14.3, 16.3
39. $c - 8 = 17$; 15, 25, 35

40. **ORNAMENTS** Last year, Desiree had 25 glass ornaments. This year, she has 38 ornaments. Solve $25 + n = 38$ to find how many new ornaments Desiree collected.

41. **MONEY** Maria was paid $9 per hour and earned $67.50. How many hours did she work? Use the equation $67.50 = 9h$, where h is hours worked.

WHALES For Exercises 42 and 43, use the information below.
Each winter, Humpback whales migrate about 1,500 miles north to the Indian Ocean. However, scientists tracked one whale that migrated 5,000 miles in one season.

42. Write an equation that can be used to find how many more miles the whale migrated compared to the usual migration.

43. How much farther did the whale travel than the usual migration?

44. **MULTI STEP** A long-running TV show aired 179 episodes from October 1951 to September 1961. About how many episodes were filmed each year?

45. **PENGUINS** On average, how much taller are Emperor penguins than Adelie penguins? Write an equation and solve.

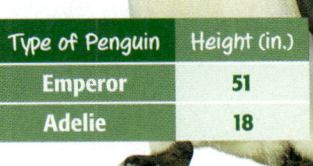

Type of Penguin	Height (in.)
Emperor	51
Adelie	18

 Data Update How do the heights of other types of penguins compare to Adelie penguins? Visit **msmath2.net/data_update** to learn more.

46. **BASKETBALL** During one game of his rookie year, LeBron James scored 41 of the Cleveland Cavalier's 107 points. How many points did the rest of the team score?

47. **CRITICAL THINKING** Consider the equation $0 \times a = b$. Find the values of a and b.

Spiral Review with Standardized Test Practice

48. **MULTIPLE CHOICE** Choose the solution of the equation $\frac{48}{c} = 4 + 2$.

    Ⓐ 2   Ⓑ 4   Ⓒ 6   Ⓓ 8

49. **GRID IN** Use the table to determine how much farther in air miles Phoenix is from New Orleans than Memphis.

From	To	Distance (air mi)
Memphis, TN	New Orleans, LA	358
Phoenix, AZ	New Orleans, LA	1,316

50. **ALGEBRA** Evaluate $3a + b^2$ if $a = 2$ and $b = 3$. (Lesson 1-4)

Evaluate each expression. (Lesson 1-3)

51. $11 \cdot 6 \div 3 + 9$ 52. $5 \cdot 13 - 6^2$ 53. $1 + 2(8 - 5)^2$

54. **DINING** Four hundred sixty people are scheduled to attend a banquet. If each table seats 8 people, how many tables are needed? (Lesson 1-1)

GETTING READY FOR THE NEXT LESSON

BASIC SKILL Multiply.

55. $2 \cdot (4 + 10)$ 56. $(9 \cdot 1) \cdot 8$ 57. $(5 \cdot 3)(5 \cdot 2)$ 58. $(6 + 8) \cdot 12$

 msmath2.net/self_check_quiz

Chapter 1 Mid-Chapter Practice Test

Vocabulary and Concepts

1. **List** the steps in the four-step problem-solving plan. (Lesson 1-1)

2. **OPEN ENDED** Write an algebraic expression and identify any coefficients, variables, and constants. (Lesson 1-4)

Skills and Applications

3. **SPORTS** A baseball stadium holds 20,000 people. If 3,650 people can be seated in the bleachers, how many seats are available in the rest of the stadium? (Lesson 1-1)

4. Write $2 \cdot 2 \cdot 2 \cdot 2 \cdot 2 \cdot 2 \cdot 2$ in exponential form. (Lesson 1-2)

Evaluate each expression. (Lessons 1-2 and 1-3)

5. 4^5
6. 9^6
7. 3^{10}
8. $25 - (3 + 2 \cdot 5)$
9. $(2^3 + 4) + 3^2$
10. $\dfrac{2(7-3)}{2^2}$

Evaluate each expression if $x = 12$, $y = 4$, and $z = 8$. (Lesson 1-4)

11. $x - 5$
12. $3y + 10z$
13. $\dfrac{yz}{2}$
14. $\dfrac{(y+8)^2}{x}$

Solve each equation mentally. (Lesson 1-5)

15. $x + 16 = 23$
16. $36 = 3y$
17. $\dfrac{65}{z} = 5$

18. **GEOMETRY** The perimeter of a rectangle is given by the expression $2(\ell + w)$ where ℓ = length and w = width. Find the perimeter of the rectangle. (Lesson 1-4)

5.4 ft

2.3 ft

Standardized Test Practice

19. **MULTIPLE CHOICE** A cycling club is planning a 1,800-mile trip. They can average 15 miles per hour for 6 hours each day. How many days will it take them to complete the trip? (Lesson 1-1)

    Ⓐ 20 days   Ⓑ 21 days
    Ⓒ 22 days   Ⓓ 23 days

20. **GRID IN** The expression $\dfrac{w}{30}$, where w is a person's weight in pounds, is used to find the approximate number of quarts of blood in the person's body. How much blood does a 120-pound person have? (Lesson 1-4)

The Game Zone

A Place To Practice Your Math Skills

Math Skill: Evaluating Expressions

Express Yourself!

GET READY!
Players: two, three, or four
Materials: scissors, 18 index cards cut in half, one number cube

GET SET!
- On each card, write a different expression containing only one variable. The coefficients and constants are to be values less than or equal to ten and any exponents should be less than four. Four examples are shown at the right.

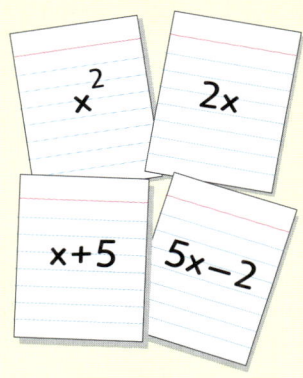

GO!
- Deal all cards to the players. The dealer then rolls the number cube. The number rolled is the value of the variable for the first round.
- The player to the left of the dealer puts a card of his or her choice faceup on the table, evaluates the expression on the card, and announces its value. Play continues until all players have placed one card on the table. This is the end of the first round. The person whose card has the greatest value wins all of the cards for that round.
- The player to the left of the dealer rolls the number cube. This is the value of the variable for the next round. Play continues until all the cards are played.
- **Who wins?** The person who has the most cards at the end of the game is the winner.

1-6 Algebra: Properties

What You'll LEARN

Use addition and multiplication properties to solve problems.

NEW Vocabulary

equivalent expressions
properties

Link to READING

Everyday Meaning of Distribute: to divide among several, as in distribute a deck of cards

WHEN am I ever going to use this?

RESTAURANTS Land-Ho! Fish Market is having a Friday night special.

FISH-BAKE DINNER
- 2 FISH FILLETS
- GARLIC MASHED POTATOES
- SIDE OF MIXED GREENS
- BEVERAGE $8.95
ADD A SLICE OF CHEESECAKE FOR ONLY $2.15

1. Find the total cost for a 5-member family, without tax and tip, if each one orders a fish-bake dinner and cheesecake.
2. Describe the method used to find the total cost.
3. Is there more than one way to find the total cost?

Here are two ways to find the total cost of the dinner:

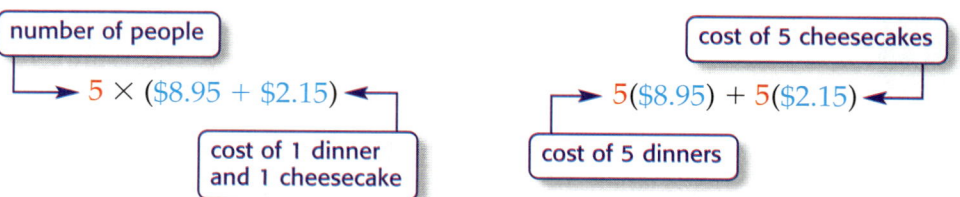

The expressions 5($8.95 + $2.15) and 5($8.95) + 5($2.15) are **equivalent expressions** because they have the same value, $55.50. This shows how the **Distributive Property** combines addition and multiplication.

Noteables™   Key Concept: Distributive Property

Words   To multiply a sum by a number, multiply each addend of the sum by the number outside the parentheses.

Symbols   Arithmetic   Algebra

$3(4 + 6) = 3(4) + 3(6)$ $a(b + c) = a(b) + a(c)$

EXAMPLES   Use the Distributive Property

Use the Distributive Property to write each expression as an equivalent expression. Then evaluate the expression.

1 $5(3 + 2)$

$5(3 + 2) = 5 \cdot 3 + 5 \cdot 2$
$\qquad = 15 + 10$ Multiply.
$\qquad = 25$ Add.

2 $(7 + 4)3$

$(7 + 4)3 = 7 \cdot 3 + 4 \cdot 3$
$\qquad = 21 + 12$ Multiply.
$\qquad = 33$ Add.

30 Chapter 1 Decimal Patterns and Algebra

EXAMPLE Use the Distributive Property

3 HISTORY The Pony Express riders carried mail from St. Joseph, Missouri, to Sacramento, California, in eight days. On average, the riders covered 250 miles each day. About how far did the Pony Express riders travel over the eight-day period?

You can find how many miles the riders traveled over the eight-day period by finding 8×250. You can use the Distributive Property to multiply mentally.

$$
\begin{aligned}
8(250) &= 8(200 + 50) && \text{Rewrite 250 as 200 + 50.} \\
&= 8(200) + 8(50) && \text{Distributive Property} \\
&= 1{,}600 + 400 && \text{Multiply.} \\
&= 2{,}000 && \text{Add.}
\end{aligned}
$$

So, the riders on the Pony Express traveled about 2,000 miles over an eight-day period.

Properties are statements that are true for any number or variable.

STUDY TIP

Properties The *Commutative Property* shows how the terms move and change places.

The *Associative Property* shows how the parentheses move and change places.

Noteables™   Key Concept: Properties

Commutative Property	The order in which two numbers are added or multiplied does not change their sum or product.
Associative Property	The way in which three numbers are grouped when they are added or multiplied does not change their sum or product.
Identity Property	The sum of an addend and 0 is the addend. The product of a factor and 1 is the factor.

EXAMPLES Identify Properties

Name the property shown by each statement.

4 $6 + (2 + 7) = (6 + 2) + 7$ Associative Property of Addition

5 $15 \times 10 = 10 \times 15$ Commutative Property of Multiplication

6 $4 \times 1 = 4$ Identity Property of Multiplication

7 $4(6 + 8) = 4(6) + 4(8)$ Distributive Property

Your Turn Name the property shown by each statement.

a. $1 \times (3 \times 4) = (1 \times 3) \times 4$ b. $a + 0 = a$

Skill and Concept Check

1. **Rewrite** $p(q) + p(r)$ using the Distributive Property.

2. **OPEN ENDED** Write an equation that illustrates the Associative Property of Addition.

3. **NUMBER SENSE** *True* or *False*? Explain your answer.
   $(11 + 18) \times 5 = 11 + 18 \times 5$

GUIDED PRACTICE

Use the Distributive Property to write each expression as an equivalent expression. Then evaluate the expression.

4. $7(4 + 3)$
5. $(10 + 8)2$
6. $5(6 - 2)$

Name the property shown by each statement.

7. $5 + 4 = 4 + 5$
8. $(2 \times 3) \times 7 = 2 \times (3 \times 7)$
9. $6(9 + 3) = 6(9) + 6(3)$
10. $5 \times 1 = 5$

Practice and Applications

Use the Distributive Property to write each expression as an equivalent expression. Then evaluate the expression.

11. $2(6 + 7)$
12. $(3 + 8)4$
13. $(11 + 3)8$
14. $2(5 - 4)$
15. $7(8 - 6)$
16. $6(12 + 5)$

17. Which expression is easier to evaluate mentally: $3(452)$ or $3(400 + 50 + 2)$? Explain.

18. Is $5 + 15 = 5(1 + 3)$ a true equation? Explain your reasoning.

HOMEWORK HELP

For Exercises	See Examples
11–16	1, 2
39, 40	3
19–30	4, 5, 6, 7

Extra Practice
See pages 565, 596.

Name the property shown by each statement.

19. $1 \times (5 \times 9) = (1 \times 5) \times 9$
20. $4 + b = b + 4$
21. $m \times n = n \times m$
22. $8 + 0 = 8$
23. $(1 + r) + s = 1 + (r + s)$
24. $(2.5 + 7)e = 2.5e + 7e$
25. $x(y + z) = xy + xz$
26. $19 \times 1 = 19$

TRAVEL For Exercises 27 and 28, use the table at the right.

27. Write a sentence comparing the mileage from Jacksonville to Charleston to Norfolk and the mileage from Norfolk to Charleston to Jacksonville.

28. Name the property that is shown in the sentence.

COIN COLLECTING Mrs. Jackson has collected 152 rare coins.

29. If she does not add any coins to her collection, write a sentence that represents this situation.

30. Name the property that is illustrated.

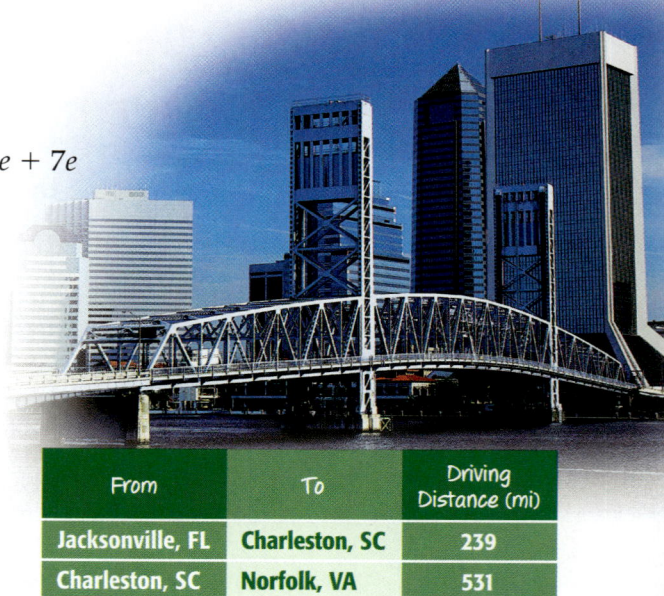

From	To	Driving Distance (mi)
Jacksonville, FL	Charleston, SC	239
Charleston, SC	Norfolk, VA	531

Use properties to rewrite each expression as an equivalent expression without parentheses.

31. $(y + 1) + 4$ **32.** $2 + (x + 4)$ **33.** $4(8b)$ **34.** $(3a)2$

35. $2(x + 3)$ **36.** $4(2 + b)$ **37.** $6(c + 1) - c$ **38.** $3(f + 4) + 2f$

39. MENTAL MATH Use one of the properties to find 6(48) mentally. Which property did you use?

40. BASKETBALL Use mental math to find the total number of NBA draftees that have come from the schools in the graph.

41. WRITE A PROBLEM Write about a real-life situation that can be solved using the Distributive, Commutative, Associative, or Identity Properties. Then use the property to solve the problem.

42. CRITICAL THINKING A *counterexample* is an example showing that a statement is not true. Provide a counterexample to the following statement. *Division of whole numbers is associative.*

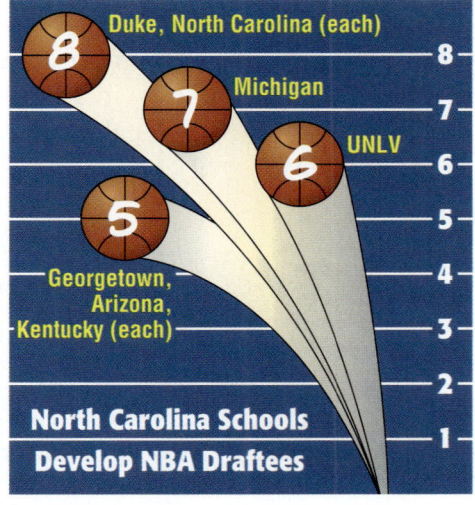

Source: nbcsports.com

Spiral Review with Standardized Test Practice

43. MULTIPLE CHOICE Rewrite $p \times (q \times r)$ using the Associative Property.

A $(p \times q) \times r$ **B** $p \times q \times r$

C $(p + q) + r$ **D** $p \times q \times r \times 1 = p \times q \times r$

44. SHORT RESPONSE Rewrite $6(9 + 8)$ using the Distributive Property.

Name the number that is the solution of the given equation. (Lesson 1-5)

45. $x + 12 = 20$; 8, 9, 10   **46.** $7.3 = t - 4$; 10.3, 11.3, 12.3

47. $35.5 = 5n$; 5.1, 7.1, 9.1   **48.** $z \div 6 = 18$; 88, 98, 108

LIFE SCIENCE For Exercises 49 and 50, use the information below.
It is believed that a dog ages 7 human years for every calendar year. (Lesson 1-4)

49. Write an expression for determining a dog's age in human years. Let y represent the number of calendar years the dog has lived.

50. Find the human age of a dog that has lived for 12 calendar years.

51. Evaluate $(14 - 9)^4$. (Lesson 1-3)

GETTING READY FOR THE NEXT LESSON

BASIC SKILL Find the next number in each pattern.

52. 2, 4, 6, 8, ?   **53.** 10, 21, 32, 43, ?   **54.** 1.4, 2.2, 3.0, 3.8, ?

55. 4, 8, 12, 16, ?   **56.** 64, 32, 16, 8, ?   **57.** 0.4, 1.2, 3.6, 10.8, ?

msmath2.net/self_check_quiz

1-7 Sequences

What You'll LEARN
Recognize and extend patterns for sequences.

NEW Vocabulary
sequence
term
arithmetic sequence
geometric sequence

Link to READING
Everyday Meaning of Sequence: a connected series, as in a sequence of events

HANDS-ON Mini Lab

Materials
- straightedge

Work with a partner.

A *diagonal* connects two nonconsecutive vertices in a figure, as shown at the right.

3 sides
0 diagonals

4 sides
2 diagonals

5 sides
5 diagonals

1. Draw a six-sided figure and all of its diagonals.
2. How many diagonals does the figure have?
3. Describe the pattern formed by the number of diagonals: 0, 2, 5, 9, … .
4. **Make a prediction** about how many diagonals a 7-sided, an 8-sided, and a 9-sided figure would have.

A **sequence** is an ordered list of numbers. Each number in a sequence is called a **term**.

8, 11, 14, 17, 20, …
+3 +3 +3 +3

In an **arithmetic sequence**, each term is found by *adding* the same number to the previous term.

3, 6, 12, 24, 48, …
×2 ×2 ×2 ×2

In a **geometric sequence**, each term is found by *multiplying* the previous term by the same number.

EXAMPLES Describe Patterns in Sequences

Describe the pattern in each sequence and identify the sequence as *arithmetic*, *geometric*, or *neither*.

1 8, 13, 18, 23, …

8, 13, 18, 23, …
+5 +5 +5

Each term is found by adding 5 to the previous term. This sequence is arithmetic.

2 1, 4, 16, 64, …

1, 4, 16, 64, …
×4 ×4 ×4

Each term is found by multiplying the previous term by 4. This sequence is geometric.

Your Turn Describe the pattern in each sequence and identify the sequence as *arithmetic*, *geometric*, or *neither*.

a. 4, 12, 36, 108, …
b. 1, 2, 4, 7, 11, …

34 Chapter 1 Decimal Patterns and Algebra

If you know the pattern of a sequence, you can use it to determine the terms.

EXAMPLES **Determine Terms in Sequences**

Write the next three terms of each sequence.

3) 0, 13, 26, 39, …

0, 13, 26, 39, …   Each term is 13 greater
 +13 +13 +13   than the previous term.

Continue the pattern to find the next three terms.

39 + 13 = 52   52 + 13 = 65   65 + 13 = 78

The next three terms are 52, 65, and 78.

4) 0.16, 0.8, 4, 20, …

0.16, 0.8, 4, 20, …   Each term is 5 times the previous term.
 ×5 ×5 ×5

Continue the pattern to find the next three terms.

20 × 5 = 100   100 × 5 = 500   500 × 5 = 2,500

The next three terms are 100, 500, and 2,500.

Your Turn Write the next three terms of each sequence.

c. 0.5, 1.5, 2.5, 3.5, …   d. 0.5, 1, 2, 4, …

Skill and Concept Check

1. **Writing Math** Compare and contrast arithmetic and geometric sequences.

2. **Explain** how you would find the next term in the sequence 14, 22, 30, 38, 46, … .

3. **OPEN ENDED** Write five terms of an arithmetic sequence and describe the rule for finding the terms.

GUIDED PRACTICE

Describe the pattern in each sequence and identify the sequence as *arithmetic*, *geometric*, or *neither*.

4. 0, 9, 18, 27, …   5. 200, 202, 206, 212, …   6. 36, 144, 576, 2,304, …

Write the next three terms of each sequence.

7. 4, 12, 36, 108, …   8. 22, 33, 44, 55, …   9. 5, 5.4, 5.8, 6.2, …

10. Find the missing terms in the sequence 26, 33, 40, __?__, 54, 61, __?__, … .

11. Create a sequence of four terms in which the first term is 70 and the next terms are found by multiplying 0.1.

msmath2.net/extra_examples   Lesson 1-7 Sequences

Practice and Applications

Describe the pattern in each sequence and identify the sequence as *arithmetic*, *geometric*, or *neither*.

12. 1, 6, 36, 216, …
13. 19, 31, 43, 55, …
14. 2, 14, 98, 686, …
15. 3.5, 10.5, 31.5, 94.5, …
16. 2.0, 3.1, 4.2, 5.3, …
17. 1, 2, 6, 24, 120, …

HOMEWORK HELP

For Exercises	See Examples
12–17, 26	1, 2
18–25	3, 4

Extra Practice
See pages 566, 596.

Write the next three terms of each sequence.

18. 33, 38, 43, 48, …
19. 4.6, 8.6, 12.6, 16.6, …
20. 0.05, 0.4, 3.2, 25.6, …
21. 0.1, 0.4, 0.7, 1.0, …
22. 3, 12, 48, 192, …
23. 125, 25, 5, 1, …

24. **INSECTS** Every 17 years, billions of insects called *cicada* emerge from hibernation. The last time they emerged was in 1996. Find the next three years in which they will come out of hibernation.

GEOMETRY For Exercises 25 and 26, use the following information.
Numbers that can be represented by a triangular arrangement of dots are called *triangular numbers*. The first five triangular numbers are shown below.

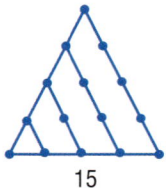

1   3   6   10   15

25. Write a sequence formed by the first eight triangular numbers.
26. Write a rule for generating the sequence.

27. **CRITICAL THINKING** You can find any term in an arithmetic sequence by using the expression $a + (n - 1)d$, where a is the first term, d is the difference between each pair of consecutive terms, and n is the position in the sequence. Find the eleventh term in the sequence 6, 13, 20, 27, 34, … .

Spiral Review with Standardized Test Practice

28. **SHORT RESPONSE** The table shows the height of a bamboo plant after each number of hours. How tall will the plant be after 6 hours?

29. **GRID IN** Find the next term in the sequence 44, 58, 72, 86, … .

Time (h)	Height (in.)
0	5
1	27
2	49
3	71

Name the property shown by each statement. (Lesson 1-6)

30. $8 + 4.1 = 4.1 + 8$
31. $3(n + 6) = 3n + 3 \cdot 6$

32. Solve $30 + y = 50$ mentally. (Lesson 1-5)

GETTING READY FOR THE NEXT LESSON

PREREQUISITE SKILL Multiply or divide. (Pages 560, 562)

33. 52×10
34. $400 \div 1{,}000$
35. $32 \div 10$
36. 0.31×100

msmath2.net/self_check_quiz

1-7b Hands-On Lab

A Follow-Up of Lesson 1-7

Exploring Sequences

What You'll LEARN

Explore patterns in sequences using paper folding.

Materials
- calculator
- piece of paper
- recording table
- colored pencils

INVESTIGATE *Work with a partner.*

STEP 1 Fold a piece of paper in half. Make a table like the one at the right and record the number of layers of paper.

Number of Folds	Layers of Paper	Fraction of Paper *Not* Shaded
1	2	$\frac{1}{2}$
2	4	$\frac{1}{4}$
3	8	
4		
⋮		

STEP 2 Shade one side of the folded paper.

STEP 3 Open the piece of paper. Record the fractional part of the paper that is *not* shaded. Refold the paper.

STEP 4 Fold your paper in half again so that the unshaded side is on the outside. Record the number of layers of paper.

STEP 5 Shade one side of the folded paper.

STEP 6 Open the piece of paper. Record the fractional part of the paper that is *not* shaded. Completely refold the paper.

STEP 7 Continue this folding, shading, and recording process for five folds.

Writing Math

Work with a partner.

1. **Examine** the sequence of numbers in the "Layers" column of your table. Is the sequence arithmetic or geometric? Then write a rule to find the next three terms.

2. **Examine** the sequence of numbers in the "Fraction" column of your table. Is the sequence arithmetic or geometric? Then write a rule to find the next three terms.

3. Assume you could continue the paper-folding process indefinitely. Suppose your unfolded piece of paper is 0.002 inch thick. Add a column to your table and find the thickness of the paper for the first five folds.

4. How many folds would it take until the paper is as tall as you?

5. **Explain** the relationship between the number of layers and the fraction of the shaded region.

1-8 Measurement: The Metric System

What You'll LEARN
Change metric units of length, capacity, and mass.

NEW Vocabulary
meter
metric system
gram
kilogram
liter

HANDS-ON Mini Lab

Materials
- centimeter ruler
- computer diskette
- CD case
- meterstick

Work with a partner.

The widths of two objects are shown below.

Object	Width (millimeters)	Width (centimeters)
computer diskette	?	8.9
CD case	?	14.4

1. Find three other objects. Use a ruler to find and record the width of all five objects to the nearest millimeter and tenth of a centimeter. Add your measurements to the table.

2. Compare the measurements of each object, and write a rule that describes how to convert from millimeters to centimeters.

3. Use a meterstick to measure the length of your classroom in meters. **Make a conjecture** about which operation you would use to convert this measure to centimeters. Explain.

The **meter** (m) is the base unit of length in the **metric system**. A meter is about the distance from the floor to a doorknob, or a little more than a yard. One kilometer is equivalent to 1,000 meters, and 1 meter is equivalent to 100 centimeters and also to 1,000 millimeters.

Prefix	Meaning in Words	Meaning in Numbers
kilo-	thousands	1,000
centi-	hundredths	0.01
milli-	thousandths	0.001

Metric prefixes indicate the decimal place-value position of the measurement.

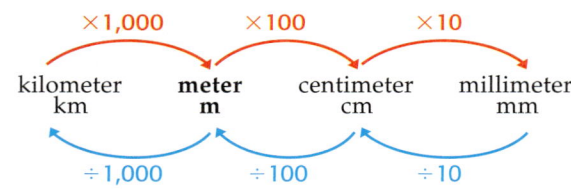

MULTIPLY to convert from larger units to smaller units.

DIVIDE to convert from smaller units to larger units.

STUDY TIP

Mental Math To multiply or divide by a power of ten, you can move the decimal point right or left.

EXAMPLES Convert Units of Length

Complete.

1 4.5 m = __?__ cm
To convert from meters to centimeters, multiply by 100.
4.5 × 100 = 450
4.5 m = 450 cm

2 2,600 m = __?__ km
To convert from meters to kilometers, divide by 1,000.
2,600 ÷ 1,000 = 2.6
2,600 m = 2.6 km

READING Math

Prefixes The kilogram is the only base unit in the metric system that has a prefix as part of its name.

The **gram** (g) measures *mass*, the amount of matter in an object. A paper clip has a mass of about one gram. The base unit of mass in the metric system is the **kilogram**, which is equivalent to 1,000 grams. One gram is equivalent to 1,000 milligrams.

EXAMPLES Convert Units of Mass

Complete.

3 8.43 kg = _?_ g
To convert from kilograms to grams, multiply by 1,000.
$8.43 \times 1,000 = 8,430$
8.43 kg = 8,430 g

4 500 mg = _?_ g
To convert from milligrams to grams, divide by 1,000.
$500 \div 1,000 = 0.5$
500 mg = 0.5 g

Your Turn Complete.

a. 23.5 g = _?_ mg
b. 9,014 g = _?_ kg

The **liter** (L) is widely used and accepted as the metric measure for capacity. *Capacity* is the amount of dry or liquid material an object can hold. Some soft drinks are sold in one-liter bottles. Kiloliter, liter, and milliliter are related in a manner similar to kilometer, meter, and millimeter.

EXAMPLES Convert Units of Capacity

Complete.

5 3.2 L = _?_ mL
To convert from liters to milliliters, multiply by 1,000.
$3.2 \times 1,000 = 3,200$
3.2 L = 3,200 mL

6 1,640 L = _?_ kL
To convert from liters to kiloliters, divide by 1,000.
$1,640 \div 1,000 = 1.64$
1,640 L = 1.64 kL

Your Turn Complete.

c. 0.75 kL = _?_ L
d. 8,800 mL = _?_ L

The following table is a summary of how to convert measures in the metric system.

Concept Summary — Converting Units

	Larger Units → Smaller Units (multiply by a power of 10)	Smaller Units → Larger Units (divide by a power of 10)
Units of Length	1 km = 1,000 m 1 m = 100 cm 1 cm = 10 mm	1 mm = 0.1 cm 1 cm = 0.01 m 1 m = 0.001 km
Units of Mass	1 kg = 1,000 g 1 g = 1,000 mg	1 mg = 0.001 g 1 g = 0.001 kg
Units of Capacity	1 kL = 1,000 L 1 L = 1,000 mL	1 mL = 0.001 L 1 L = 0.001 kL

Skill and Concept Check

1. **Writing Math** Explain why it makes sense to multiply when changing from a larger unit to a smaller unit.

2. **OPEN ENDED** List a set of measurement references, made up of everyday objects, that will help you remember the basic units in the metric system.

3. **FIND THE ERROR** Hunter and Arturo are converting 45.7 milligrams to grams. Who is correct? Explain.

   Hunter
   45.7 mg = 0.0457 g

   Arturo
   45.7 mg = 45,700 g

GUIDED PRACTICE

Complete.

4. 550 m = __?__ km
5. 45 g = __?__ mg
6. 1,460 mL = __?__ L
7. 3.7 m = __?__ cm
8. 3,800 mg = __?__ g
9. 2.34 kL = __?__ L

10. How many milliliters are in 0.04 liter?

Practice and Applications

Complete.

11. 720 cm = __?__ m
12. 983 mm = __?__ m
13. 64 kg = __?__ g
14. 7.5 g = __?__ mg
15. 925 mg = __?__ g
16. 345 mL = __?__ L
17. 0.85 cm = __?__ mm
18. 3.2 m = __?__ cm
19. 9.1 L = __?__ mL
20. 73.2 g = __?__ mg
21. 43 L = __?__ kL
22. 130.5 kL = __?__ L
23. 997 g = __?__ kg
24. 0.046 kL = __?__ L
25. 82.1 g = __?__ kg
26. 0.03 m = __?__ mm
27. 57,000 mL = __?__ L
28. 0.22 mm = __?__ cm

HOMEWORK HELP

For Exercises	See Examples
11–41	1–6

Extra Practice
See pages 566, 596.

MONUMENTS For Exercises 29 and 30, use the information about the Washington Monument at the right.

29. What is the height of the monument in kilometers?
30. What is the height in centimeters? in millimeters?

31. Order 0.02 km, 50 m, and 3,000 cm from least to greatest length.
32. Order 0.05 kg, 32 g, and 430,000 mg from greatest to least mass.
33. Order 660 mL, 0.06 L, and 6.6 kL from least to greatest capacity.

169 m

Complete.

34. 0.05 km = __?__ mm
35. 0.93 km = __?__ cm
36. 23,000 mg = __?__ kg
37. 40,000 mL = __?__ kL

40 Chapter 1 Decimal Patterns and Algebra

38. **TRACK AND FIELD** Refer to the table. Convert the distance of each event to kilometers.

Event	Number of Kilometers
10,000-meter run	?
110-meter hurdles	?
1,600-meter relay	?
3,000-meter steeplechase	?

39. **MULTI STEP** Heather needs a 2.5-meter pole for a birdfeeder that she is building. How many centimeters will she need to cut off a 3-meter pole in order to use it for the birdfeeder?

40. **MULTI STEP** Margarita is running the school concession stand. The fruit punch machine holds 15 liters. Each drink cup holds 500 milliliters of punch. How many cups of punch can Margarita sell before the fountain is empty?

41. **BAKING** Evan is making peach cobblers for the fair. The recipe requires 900 grams of peaches per cobbler. If Evan has 4.5 kilograms of peaches, how many cobblers can he make?

42. **RESEARCH** Use the Internet or another source to investigate relationships between metric units of mass and capacity. Explain the relationships and give examples of how they are used in real life.

CRITICAL THINKING The metric prefix *giga* refers to something that is one billion times larger than the base unit. The metric prefix *nano* refers to something that is one billion times smaller than the base unit.

43. How many meters are in 3.9 gigameters?

44. How many meters are in 3.9 nanometers?

Spiral Review with Standardized Test Practice

45. **MULTIPLE CHOICE** Choose the best estimate for the mass of a golf ball.
    A 4.5 mg   **B** 4.5 g   **C** 45 g   **D** 45 kg

46. **GRID IN** What is the mass in grams of a 200-milligram dosage of medication?

Describe the pattern in each sequence and identify the sequence as *arithmetic*, *geometric*, **or** *neither*. (Lesson 1-7)

47. 27, 36, 45, 54, …   48. 6, 36, 216, 1,296, …   49. 1.1, 2.2, 3.3, …   50. 1, 4, 9, 16, …

Use the Distributive Property to write each expression as an equivalent expression. Then evaluate the expression. (Lesson 1-6)

51. $5(9 + 7)$ 52. $(12 + 4)4$ 53. $8(7 - 2)$ 54. $(6 - 5)10$

GETTING READY FOR THE NEXT LESSON

PREREQUISITE SKILL Multiply. (Lesson 1-3)

55. $8 \cdot 10^2$ 56. $5 \cdot 10^3$ 57. $4 \cdot 10^5$

58. $2.7 \cdot 10^3$ 59. $65 \cdot 10^2$ 60. $1.35 \cdot 10^4$

Study Skill

Use Power Notes

Taking Good Notes

Do you ever have trouble organizing your notes? Try using power notes. They can help you organize the main ideas and details in a lesson.

Power notes are similar to lesson outlines, but they are simpler to organize. Power notes use the numbers 1, 2, 3, and so on.

Power 1: This is the main idea.
  Power 2: This provides details about the main idea.
    Power 3: This provides details about Power 2.
      and so on….

Here's a sample of power notes from Lesson 1-8. Notice that you can even add drawings or examples to your power notes.

SKILL PRACTICE

1. **Explain** why the four-step plan in Lesson 1-1 is a form of power notes.

2. Complete the power notes above for units of mass and capacity.

Use power notes to make an outline for each concept.

3. order of operations (*Hint:* See Lesson 1-3.)

4. addition and multiplication properties (*Hint:* See Lesson 1-6.)

5. algebraic and geometric sequences (*Hint:* See Lesson 1-7.)

1-9 Scientific Notation

What You'll LEARN
Write numbers greater than 100 in scientific notation and in standard form.

NEW Vocabulary
scientific notation

REVIEW Vocabulary
standard form: numbers written without exponents (Lesson 1-2)

WHEN am I ever going to use this?

ENTERTAINMENT The Mall of America in Bloomington, Minnesota, is the largest shopping and entertainment center in the United States. The table shows two features of the mall.

	Standard Form	Abbreviated Form
Number of Parking Spaces	?	13 thousand
Area of Mall (square feet)	4,200,000	?

1. Complete the table.
2. Write both numbers as the product of a number and a power of ten.
3. Rewrite both powers of ten using exponents.

When you work with very large numbers, it is difficult to keep track of the place value. You can write numbers like 4.2 million in **scientific notation** by using a power of ten.

$$4.2 \text{ million} = 4.2 \times 1,000,000$$
$$= 4.2 \times 10^6 \qquad 10^6 = 1,000,000$$

Noteables™   Key Concept: Scientific Notation

Words   A number is expressed in scientific notation when it is written as the product of a number and a power of ten. The number must be greater than or equal to 1 and less than 10.

Examples $60,000 = 6.0 \times 10^4$ $3.5 \text{ billion} = 3.5 \times 10^9$

In Lesson 1-3, you used the order of operations to evaluate expressions like 5.2×10^5. You can also use the order of operations to write a number in standard form.

EXAMPLE   Write a Number in Standard Form

1 Write 5.2×10^5 in standard form.

$5.2 \times 10^5 = 5.2 \times 100,000 \qquad 10^5 = 100,000$
$ = 520000 \qquad$ Move the decimal point 5 places to the right.
$ = 520,000$

Your Turn Write each number in standard form.

   a. 7.2×10^4 b. 3.8×10^9 c. 1.0×10^8

To write a number in scientific notation, move the decimal to the right of the first nonzero number. Then find the power of ten.

$13{,}000 = 1.3 \times 10^4$

Move the decimal point to the right of the first nonzero digit.

The decimal point was moved 4 places.

EXAMPLE Write a Number in Scientific Notation

2 Write 23,500,000 in scientific notation.

$23{,}500{,}000 = 2.35 \times 10{,}000{,}000$ Move the decimal point 7 places to find a number between 1 and 10.

$= 2.35 \times 10^7$

Your Turn Write each number in scientific notation.

d. 38,000,000   e. 19,000   f. 86.1 billion

When you use a calculator to compute large numbers, the numbers are often displayed in scientific notation.

STUDY TIP

Technology To enter 3.42×10^{10} on a calculator, press 3.42 [EE] 10 [ENTER]. The screen would look like this: **3.42** $\times 10^{10}$

EXAMPLE Compute with Large Numbers

3 **RESERVOIRS** Lake Francis Case in South Dakota contains 5.7 billion cubic meters of water. Lake Mead in Nevada has about 6 times as much water. Find the approximate amount of water in Lake Mead.

To find the amount of water in Lake Mead, multiply 5.7 billion by 6.

5.7 billion = 5,700,000,000

Enter 5700000000 [×] 6 [ENTER].

The number **3.42** $\times 10^{10}$ on the calculator represents 3.42×10^{10}.

So, Lake Mead has about 3.42×10^{10} cubic meters of water.

Skill and Concept Check

1. **Writing Math** Explain why scientific notation is used with large numbers.

2. **OPEN ENDED** Write a number in standard form and then in scientific notation.

3. **NUMBER SENSE** Which number describes the population of Chicago, 3.0×10^6 or 3.0×10^3? Explain your reasoning.

GUIDED PRACTICE

Write each number in standard form.

4. 5.1×10^4 5. 1.0×10^5 6. 8.05×10^6

Write each number in scientific notation.

7. 70,000   8. 510,000   9. 17,500,000

Practice and Applications

Write each number in standard form.

10. 4.2×10^3
11. 7.3×10^4
12. 6.9×10^6
13. 3.38×10^5
14. 1.1×10^8
15. 8.98×10^6
16. 5.16×10^4
17. 7.98×10^5
18. 1.05×10^7
19. 9.90×10^8
20. 5.33×10^{10}
21. 8.06×10^7

HOMEWORK HELP

For Exercises	See Examples
10–21	1
24–35, 42	2
22, 43	3

Extra Practice
See pages 566, 596.

22. Triple the number 435,000 and write it in scientific notation.
23. Order the numbers 53,400, 4.98×10^6, 5.03×10^5, and 4,980,100 from least to greatest.

Write each number in scientific notation.

24. 46,000
25. 19,800
26. 169,000
27. 9,970,000
28. 3,720,000
29. 505,000
30. 607,000
31. 23,600,000
32. 7 million
33. 1.1 million
34. 40 billion
35. 60.8 billion

NUMBER NAMES For Exercises 36 and 37, use the table.

36. Write 13 quintillion in scientific notation.
37. Write the name of the number 5.43×10^{16}.

Power of 10	Name
10^6	million
10^9	billion
10^{12}	trillion
10^{15}	quadrillion
10^{18}	quintillion

Replace each ● with <, >, or = to make a true sentence.

38. 5,000 ● 5.0×10^3
39. 600 ● 6.0×10^1
40. 81,900 ● 8.19×10^5
41. 1 million ● 1.1×10^6

42. **GEOGRAPHY** The Mariana Trench is the deepest part of the Pacific Ocean with a depth of about 36,000 feet. Write this depth in scientific notation.

43. **EARTH SCIENCE** A light-year, the distance light travels in one year, is about 5,880,000,000,000 miles. How many miles is 50 light years?

44. **CRITICAL THINKING** The distance from Earth to the Sun is 1.55×10^8 kilometers. If the speed of light is 3×10^5 kilometers per second, about how long does it take for light to travel from the Sun to Earth? (*Hint*: distance = rate × time)

Spiral Review with Standardized Test Practice

45. **MULTIPLE CHOICE** Asia has an area of 44,579,000 square kilometers. Write 44,579,000 in scientific notation.
    Ⓐ 4.4579×10^6 Ⓑ 4.4579×10^7 Ⓒ 4.4579×10^8 Ⓓ 4.4579×10^9

46. **SHORT RESPONSE** Write the standard form of 5.56×10^{12}.

Complete. (Lesson 1-8)

47. 1.01 kg = __?__ g
48. 36 mL = __?__ L
49. 2.33 km = __?__ m

50. **SOCCER** The World Cup is played every four years. If the World Cup was played in 2002, when are the next three years that it will be played? (Lesson 1-7)

CHAPTER 1 Study Guide and Review

Vocabulary and Concept Check

algebra (p. 18)
algebraic expression (p. 18)
arithmetic sequence (p. 34)
base (p. 10)
coefficient (p. 19)
constant (p. 19)
cubed (p. 10)
defining the variable (p. 25)
equation (p. 24)
equivalent expressions (p. 30)
evaluate (p. 11)

exponent (p. 10)
exponential form (p. 11)
factors (p. 10)
geometric sequence (p. 34)
gram (p. 39)
kilogram (p. 39)
liter (p. 39)
meter (p. 38)
metric system (p. 38)
numerical expression (p. 14)
order of operations (p. 14)

powers (p. 10)
properties (p. 31)
scientific notation (p. 43)
sequence (p. 34)
solution (p. 24)
solving an equation (p. 24)
squared (p. 10)
standard form (p. 11)
term (pp. 19, 34)
variable (p. 18)

Choose the letter of the term that best matches each phrase.

1. a number that makes an equation true
2. numbers expressed using exponents
3. two or more numbers that are multiplied together to form a product
4. rules used to ensure that numerical expressions have only one value
5. tells how many times the base is used as a factor

a. algebra
b. solution
c. order of operations
d. factors
e. powers
f. exponent

Lesson-by-Lesson Exercises and Examples

1-1 A Plan for Problem Solving (pp. 6–9)

Use the four-step plan to solve each problem.

6. **CELLULAR PHONES** Sherita's service charges a monthly fee of $20.00 plus $0.15 per minute. One monthly bill is $31.25. How many minutes did Sherita use during the month?

7. **CAR RENTAL** ABC Car Rental charges $25 per day to rent a mid-sized car plus $0.20 per mile driven. Mr. Ruiz rents a mid-sized car for 3 days and drives a total of 72 miles. Find the amount of Mr. Ruiz's bill.

Example 1 Use the four-step plan to solve the problem.

GARDENING A bag of mulch covers 25 square feet of garden space. Taylor uses 7 bags of mulch to cover her garden. How large is Taylor's garden?

Explore Taylor used 7 bags of mulch, each covering 25 square feet.
Plan Multiply 25 by 7.
Solve 25 · 7 = 175
Taylor's garden is 175 square feet.
Examine Add 25 seven times.

46 Chapter 1 Decimal Patterns and Algebra

msmath2.net/vocabulary_review

1-2 Powers and Exponents (pp. 10–13)

Evaluate each expression.

8. 3^5
9. 7^9
10. 2^8
11. 18^2
12. 9^4
13. 10^4
14. 20^3
15. 100^1

16. Write $15 \cdot 15 \cdot 15$ in exponential form.

Example 2 Evaluate 4^5.
The base is 4. The exponent 5 means that 4 is used as a factor 5 times.
$4^5 = 4 \cdot 4 \cdot 4 \cdot 4 \cdot 4$
$ = 1{,}024$

1-3 Order of Operations (pp. 14–17)

Evaluate each expression.

17. $24 - 8 + 3^2$
18. $48 \div 6 + 2 \cdot 5$
19. $5 \cdot 9 - (2 + 3)^2$
20. $(14 - 4) \div 2$
21. $9 + 3(7 - 5)^3$
22. 12×10^4
23. $42 \div 6 - 63 \div 9$
24. $15 + 9 \div 3 - 7$

Example 3 Evaluate $24 - (8 \div 4)^4$.
$24 - (8 \div 4)^4 = 24 - 2^4$ Divide first since $8 \div 4$ is in parentheses.
$ = 24 - 16$ Find the value of 2^4.
$ = 8$ Subtract.

1-4 Algebra: Variables and Expressions (pp. 18–21)

Evaluate each expression if $a = 10$, $b = 4$, and $c = 8$.

25. $b + 7$
26. $7b$
27. $(a - b)^2$
28. $ab \div c$
29. $3b^2 + c$
30. $\dfrac{(b + c)^2}{3}$

31. **PRODUCTION** The cost of producing T-shirts is given by the expression $350 + 0.82x$, where x is the number of T-shirts produced. Find the cost of producing 750 T-shirts.

Example 4 Evaluate $2m^2 - 5n$ if $m = 4$ and $n = 3$.
$2m^2 - 5n = 2(4)^2 - 5(3)$ Replace m with 4 and n with 3.
$ = 2(16) - 5(3)$ Find the value of 4^2.
$ = 32 - 15$ Multiply.
$ = 17$ Subtract.

1-5 Algebra: Equations (pp. 24–27)

Solve each equation mentally.

32. $h + 9 = 17$
33. $5 = c - 3$
34. $32 - 19 = w$
35. $8a = 56$
36. $31 - y = 8$
37. $\dfrac{t}{9} = 12$
38. $100 = 20g$
39. $p - 49 = 18$

Example 5 Solve $14 = 5 + x$ mentally.
$14 = 5 + x$ Write the equation.
$14 = 5 + 9$ You know that $5 + 9 = 14$.
$14 = 14$ Simplify.
The solution is 9.

Study Guide and Review continued

Mixed Problem Solving
For mixed problem-solving practice, see page 596.

1-6 Algebra: Properties (pp. 30–33)

Name the property shown by each statement.

40. $3 + 8 = 8 + 3$
41. $3 \cdot (2 \cdot 7) = (3 \cdot 2) \cdot 7$
42. $6(1 + 4) = 6(1) + 6(4)$
43. $29 \cdot 1 = 29$

Example 6 Name the property shown by the statement $12 + (8 + 2) = (12 + 8) + 2$.

This statement shows that the way in which three numbers are grouped when they are added does not change their sum. This is an example of the Associative Property of Addition.

1-7 Sequences (pp. 34–36)

Describe the pattern in each sequence and identify the sequence as *arithmetic*, *geometric*, or *neither*. Then find the next three terms.

44. 5, 15, 45, 135, …
45. 2.0, 3.2, 4.4, 5.6, …
46. 1, 3, 7, 13, …

47. **CONTESTS** The prizes for first, second, and third places in a spelling bee are $1,000, $500, and $250, respectively. If this pattern continues, what would the fourth, fifth, and sixth prizes be?

Example 7 Describe the pattern in the sequence 2, 4, 8, 16, 32, …, and identify the sequence as *arithmetic*, *geometric*, or *neither*. Then find the next three terms.

This is a geometric sequence created by multiplying the previous term by 2.

The next three terms are 64, 128, and 256.

1-8 Measurement: The Metric System (pp. 38–41)

Complete.

48. 3.6 km = __?__ m
49. 29 L = __?__ mL
50. 237 mg = __?__ g
51. 7 mL = __?__ L
52. 3,200 cm = __?__ m

Example 8 Complete.
6,750 m = __?__ km

Divide 6,750 by 1,000.
$6{,}750 \div 1{,}000 = 6.75$
So, 6,750 meters is equivalent to 6.75 kilometers.

1-9 Scientific Notation (pp. 43–45)

Write each number in scientific notation.

53. 59,000
54. 84,600,000
55. 324,000
56. 10,000,000,000
57. 1,030,000
58. 333,000
59. 9.1 million
60. 6 billion

Example 9 Write 23,100,000 in scientific notation.

Move the decimal point 7 places to find a number between 1 and 10.

$23{,}100{,}000 = 2.31 \times 10^7$

48 Chapter 1 Decimal Patterns and Algebra

CHAPTER 1 Practice Test

Vocabulary and Concepts

1. **List** the order of operations used to find the value of an expression.
2. **Explain** the difference between an arithmetic sequence and a geometric sequence.

Skills and Applications

Evaluate each expression.

3. 3^5
4. 15^4
5. $18 - 3 + 5$
6. $8 + (12 \div 3)^3$

Evaluate each expression if $x = 12$, $y = 5$, and $z = 3$.

7. $7 - x \div z$
8. $yz + 23$
9. $(y - z)^5$
10. $\dfrac{xz}{y + 13}$

Solve each equation mentally.

11. $9 + m = 16$
12. $d - 14 = 37$
13. $23 = \dfrac{92}{t}$
14. $6x = 126$

Name the property shown by each statement.

15. $2 + (3 + 7) = (2 + 3) + 7$
16. $5(w + 2) = 5w + 10$

Complete.

17. $5.45 \text{ m} = \underline{\ ?\ } \text{ cm}$
18. $27 \text{ mL} = \underline{\ ?\ } \text{ L}$
19. $8{,}200 \text{ g} = \underline{\ ?\ } \text{ kg}$

20. **CAPACITY** The table lists the capacity of three different containers. Determine which container has the largest capacity.

Container	Capacity
#1	5400 mL
#2	0.008 kL
#3	4.78 L

21. **ANIMALS** Irene left both her dog and cat at a kennel for 3 nights. The kennel charges $8 per night for the dog and $5 per night for the cat. Find Irene's bill.

Write each number in scientific notation.

22. 42,300
23. 3,780,000
24. 502,000,000

Standardized Test Practice

25. Ms. Carter runs a pizza parlor. Her daily costs are $125 for rent, employee wages, and utilities, plus $2 for every pizza. If n represents the number of pizzas the pizza parlor makes one day, which expression represents Ms. Carter's total cost for that day?

   Ⓐ $125 + 2n$
   Ⓑ $125n + \$2$
   Ⓒ $\$125n + 2n$
   Ⓓ $\$127$

msmath2.net/chapter_test

CHAPTER 1 Standardized Test Practice

PART 1 Multiple Choice

Record your answers on the answer sheet provided by your teacher or on a sheet of paper.

1. The strategy of guess and check would be *most* useful in solving which problem? (Lesson 1-1)

   (A) About how long does it take to travel 62 miles at 58 miles per hour?

   (B) What is the area of a rectangle if it has a length of 4.5 centimeters and a width of 9 centimeters?

   (C) It costs $3 to cross the town bridge to enter town, but it is free of charge to cross it to leave town. How much does a person who makes 150 trips across the bridge into town and 50 trips across the bridge to leave town spend in bridge tolls?

   (D) The product of two consecutive even numbers is 728. Find the two numbers.

2. Ciro earns $8 per hour plus tips cleaning rooms at Sandman's Inn. Which expression represents Ciro's earnings for a week in which he works 36 hours and collects $25 in tips? (Lesson 1-3)

   (F) $(36 \times 8) + 25$

   (G) $(36 + 25) \times 8$

   (H) $(8 \times 25) + 36$

   (I) $(36 \div 8) + 25$

3. To estimate the number of miles a thunderstorm is from you, count the number of seconds between the thunder and the lightning. Then divide the number of seconds by 5. Which expression can be used to estimate the distance? (Lesson 1-4)

   (A) $5 \div s$ (B) $s - 5$

   (C) $5s$ (D) $s \div 5$

4. Which value of p makes $64 \div p = 8$ a true sentence? (Lesson 1-5)

   (F) 2   (G) 2^2

   (H) 2^3 (I) 3^2

5. Which is the graph of the solution of $x - 15 = 33$? (Lesson 1-5)

   (A)

   (B)

   (C)

   (D)

6. Identify the property shown by the sentence $16 + (4 + 7) = (16 + 4) + 7$. (Lesson 1-6)

   (F) Commutative Property of Addition

   (G) Associative Property of Addition

   (H) Associative Property of Multiplication

   (I) Identity Property of Multiplication

7. Which is the next term of the sequence 6, 12, 18, 24, … ? (Lesson 1-7)

   (A) 28   (B) 30

   (C) 36   (D) 48

8. Every morning, Morgan swims 0.5 kilometer. How many meters does she swim? (Lesson 1-8)

   (F) 5 m   (G) 50 m

   (H) 100 m   (I) 500 m

9. Which is 5,300,000 written in scientific notation? (Lesson 1-9)

   (A) 0.53×10^6 (B) 53×10^5

   (C) 5.3×10^6 (D) 5.3×10^7

Preparing for Standardized Tests

For test-taking strategies and more practice, see pages 608–625.

PART 2 Short Response/Grid In

Record your answers on the answer sheet provided by your teacher or on a sheet of paper.

10. What value is represented by $(1 \times 10^4) + (7 \times 10^3) + (3 \times 10^2) + (0 \times 10^1)$? (Prerequisite Skill, p. 561)

11. The graph shows the number of World Series Championships for selected teams from 1903–2001. How many total championships have been won by these teams? (Lesson 1-1)

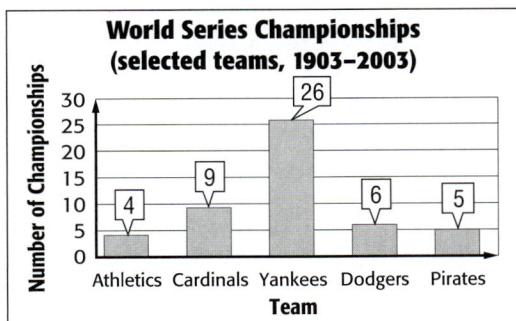

Source: The World Almanac

12. Write x^6 as a product of the same factor. (Lessons 1-2 and 1-4)

13. What is the value of $2^2 + 4^3$? (Lesson 1-3)

14. Tyler works at the library 3 times as many hours as Lugo and Lydia combined. Write an algebraic expression that represents the number of hours Tyler works per week if h represents the number of hours Lugo and Lydia work combined. (Lesson 1-4)

15. The town of Holden's water tank can hold 10 kiloliters. The town of Jasper's water tank can hold half the amount of Holden's tank. How many liters of water can Jasper's water tank hold? (Lesson 1-8)

TEST-TAKING TIP

Question 13 Fill in the answer bubbles below each numeral completely on the answer grid-in sheet. That way your answer will be clear.

16. Four countries and their approximate populations in 2001 are shown in the table. Which country has the least population? (Lesson 1-9)

Country	Population, 2003
Honduras	6.7×10^6
Australia	2.0×10^7
Croatia	4.3×10^6
Denmark	5.4×10^6

Source: www.factmonster.com

PART 3 Extended Response

Record your answers on a sheet of paper. Show your work.

17. Figures 1, 2, and 3 form a pattern made of equal-sized squares. Each side of the square in Figure 1 is 2 units in length. (Lesson 1-7)

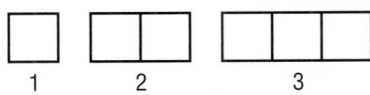

a. Extend the pattern to find the perimeters of Figures 3–8.

b. Describe the sequence of perimeters and identify the sequence as *arithmetic*, *geometric*, or *neither*.

Figure	Perimeter
1	8
2	12
3	?
4	?
5	?
6	?
7	?
8	?

c. The perimeter of a term above is given by the expression $4f + 4$, where f is the figure number in the pattern. What is the perimeter of figure 26?

18. Marlene made her special punch by mixing 4 liters of ginger ale and 2,650 milliliters of cranberry juice. (Lesson 1-8)

a. How many liters are in the punch bowl when the punch is complete?

b. About how many 200-mL servings of punch can be served?

CHAPTER 2
Statistics: Analyzing Data

"What do bridges have to do with math?"

Twenty-two of the longest suspension bridges in North America are 1,500 feet to 4,260 feet long. You can use a box-and-whisker plot to summarize the data about bridges. In mathematics, you will use tables, graphs, and diagrams to organize lists of numbers that describe real-life events.

You will solve problems by organizing data in box-and-whisker plots in Lesson 2-6.

GETTING STARTED

▶ Diagnose Readiness

Take this quiz to see if you are ready to begin Chapter 2. Refer to the page number in parentheses for review.

Vocabulary Review

Choose the correct term to complete each sentence.

1. 173.02 and 37.125 have the same digit in the (hundreds, hundredths, tenths) place. (Page 555)
2. The number 0.02 is (less than, equal to, greater than) 0.002. (Page 556)
3. The numbers 10.88, 10.089, and 1.08 are in order from (least to greatest, greatest to least). (Page 556)

Prerequisite Skills

Name the place value of each underlined digit. (Page 555)

4. 7.9̲5
5. 2̲,064
6. 48.0̲5
7. 163̲.3

Order from least to greatest. (Page 556)

8. 96.2, 96.02, 95.89
9. 5.61, 5.062, 5.16
10. 22.02, 22, 22.012
11. 3.98, 39.8, 0.398

Order from greatest to least. (Page 556)

12. 74.65, 74.67, 74.7
13. 1.26, 1.026, 10.26
14. 3.304, 3.04, 3.340
15. 68.9, 69.79, 68.99

Add or divide. (Pages 559 and 562)

16. 88.21 + 4.75 + 62.4 + 39.60
17. 215 ÷ 2.5
18. 5.9 + 28.1 + 9.2
19. 333 ÷ 0.6
20. 51 ÷ 1.7

Statistics Make this Foldable to help you organize information about analyzing data. Begin with eight sheets of notebook paper.

STEP 1 Fold
Fold 8 sheets of paper in half along the width.

STEP 2 Cut
Cut a 1" tab along the left edge through one thickness.

STEP 3 Glue and Label
Glue the 1" tab down. Write the lesson number and title on the front tab.

STEP 4 Repeat and Staple
Repeat Steps 2 and 3 for the remaining sheets. Staple them together on the glued tabs to form a booklet.

 Chapter Notes Each time you find this logo throughout the chapter, use your *Noteables™: Interactive Study Notebook with Foldables™* or your own notebook to take notes. Begin your chapter notes with this Foldable activity.

Readiness To prepare yourself for this chapter with another quiz, visit msmath2.net/chapter_readiness

2-1 Frequency Tables

What You'll LEARN

Organize and interpret data in a frequency table.

NEW Vocabulary

statistics
data
frequency table
scale
interval

WHEN am I ever going to use this?

SCOOTERS Riding a scooter is a fun way to get exercise. The table shows prices for 35 types of nonmotorized scooters found at an online store.

1. What is the cost of the least expensive scooter? the most expensive?
2. How many scooters cost $51 to $75?
3. How could you reorganize the prices so that they are easier to find and read?

Prices of Scooters ($)				
60	25	29	25	40
35	50	80	30	70
80	90	80	55	70
40	100	52	45	60
99	60	100	89	11
99	100	49	35	90
92	20	49	80	50

Statistics deals with collecting, organizing, and interpreting data. **Data** are pieces of information, which are often numerical. Large amounts of data can be organized in a frequency table. A **frequency table** shows the number of pieces of data that fall within given intervals.

The **scale** allows you to record all of the data. It includes the least value, 11, and the greatest value, 100. The scale is 1 to 100.

Price ($)	Tally	Frequency
1–25	IIII	4
26–50	|||| ||||		11
51–75	|||| ||	7
76–100	|||| |||| |||	13

The **interval** separates the scale into equal parts. The interval is 25.

Tally marks are counters used to record items in a group.

From the frequency table, you can see that just over half the scooters cost between $51 and $100.

EXAMPLE Make a Frequency Table

1 RACING Cyclo-cross is cycle racing that takes place on short off-road courses. The table shows the total points earned by the top 20 racers in the cyclo-cross World Cup events. Make a frequency table of the data.

Cyclo-Cross Points			
310	179	116	91
225	173	107	76
223	150	105	55
218	149	100	52
190	123	100	50

Source: www.cyclo-cross.com

Step 1 Choose an appropriate interval and scale for the data. The scale should include the least value, 50, and the greatest value, 310.

interval: 50
scale: 50 to 349

The scale includes all of the data, and the interval separates it into equal parts.

54 Chapter 2 Statistics: Analyzing Data

Step 2 Draw a table with three columns and label the columns *Points*, *Tally*, and *Frequency*.

Step 3 Complete the table.

Points	Tally	Frequency
50–99	|||||	5
100–149	||||| ||	7
150–199	||||	4
200–249	|||	3
250–299		0
300–349			1

Frequency tables can also have categories without scales and intervals.

EXAMPLE Make and Use a Frequency Table

2 **OLYMPICS** Enrique asked his classmates to name their favorite winter Olympic event. The results are shown at the right. Make a frequency table of the data. Then determine the overall favorite event.

Draw a table with three columns. In the first column, list the events. Then complete the rest of the table.

Favorite Winter Olympic Events

S	F	F	S	B	F
F	S	H	J	B	S
J	F	J	F	H	F
F	B	S	F	S	H
S	B	H	F	J	J

B = bobsleigh, F = figure skating,
H = hockey, J = ski jumping,
S = snowboarding

Event	Tally	Frequency
bobsleigh	||||	4
figure skating	||||| |||||	10
hockey	||||	4
ski jumping	|||||	5
snowboarding	||||| ||	7

Figure skating, with 10 tallies, was the favorite event. Snowboarding was the second favorite, with 7 tallies.

REAL-LIFE MATH

OLYMPICS In 2002, Vonetta Flowers became the first African-American to win a gold medal in the Winter Olympics. She and Jill Bakken won the first ever 2-woman bobsleigh event with a two-mile time of 1 minute 37.76 seconds.

Source: www.olympic-usa.org

Frequency tables are useful for interpreting data.

EXAMPLE Interpret Data

3 **HISTORY** The frequency table shows the reigns of the last forty rulers of England. How many of them ruled England for more than 30 years?

There are three categories with more than 30 years.

31–45 years = 6 rulers

46–60 years = 4 rulers

61–75 years = 1 ruler

So, 6 + 4 + 1 or 11 of the last forty rulers each ruled England for more than 30 years.

Rulers of England

Reign (years)	Tally	Frequency
1–15	||||| ||||| ||||| |||	18
16–30	||||| |||||		11
31–45	|||||		6
46–60	||||	4
61–75			1

Source: www.scotlandroyalty.org/kings.html

Skill and Concept Check

1. **Writing Math** Describe an advantage and a disadvantage of using a frequency table to display data.

2. **OPEN ENDED** Make a frequency table that displays thirty data items and has an interval of 5.

3. **DATA SENSE** Explain why the frequency table in Example 1 does not have 300–310 as its last category when the greatest data value is 310.

GUIDED PRACTICE

For Exercises 4 and 5, use the table below.

Number of Movie Rentals During the Month								
10	9	2	17	12	9	11	10
4	14	3	15	0	6	5	9
20	10	11	8	3	1	13	4

4. Choose an appropriate interval and scale to make a frequency table.

5. Make a frequency table of the data.

For Exercises 6 and 7, use the table below.

Favorite Breakfast Foods					
e	c	e	p	c	c
c	p	p	e	c	e
c	e	c	p	p	e

c = cereal, e = eggs, p = pancakes

6. Make a frequency table of the data.

7. What is the favorite breakfast food?

Practice and Applications

Make a frequency table of each set of data.

8.

Ages of Students in Ms. Warren's Class				
12	13	13	14	14
14	14	13	13	14
13	14	13	13	12
14	13	14	12	13

9.

Years of Major Hurricanes				
1900	1975	1985	1995	1996
1938	1944	1957	1980	1992
1961	1935	1989	1972	1928
1955	1969	1979	1915	1964
1970	1960	1954	1926	1967
1999	1998			

Source: The World Almanac

HOMEWORK HELP

For Exercises	See Examples
8–12	1
13–18	2, 3

Extra Practice
See pages 567, 597.

10.

How Do You Listen to the Most Music?				
I	C	M	I	R
R	M	I	M	C
R	C	C	R	I
M	C	C	R	C
R	I	R	R	C

C = CDs, R = radio, I = Internet, M = music videos

11.

Which Chore Do Students Like Least?					
D	L	Y	L	V	R	L
D	L	R	D	R	D	Y
V	D	V	R	R	Y	R
R	R	Y	L	D	V	R
D	R	V	R	R	D	L

D = wash dishes, V = vacuum, R = clean room, L = do laundry, Y = do yard work

12. **Determine** whether the categories 0–100, 100–200, and 200–300 could be used in a frequency table displaying data. Explain.

56 Chapter 2 Statistics: Analyzing Data

MOVIES For Exercises 13–15, use the table at the right. It shows when the 20 top-grossing movies of all time were released as of 2002.

Years of Top-Grossing Movies			
1997	1993	1983	1990
1977	1994	1996	2001
1982	2001	2002	1975
1999	1994	1999	2000
2002	2001	1980	2001

Source: www.movieweb.com

13. Make a frequency table of the data. Use the intervals 1970–1979, 1980–1989, 1990–1999, and 2000–2009.

14. In which years were the greatest number of top-grossing movies released?

15. **RESEARCH** Use the Internet or another source to find the current 20 top-grossing movies. Make a frequency table of the data and compare it to the table you made in Exercise 13.

CEREAL For Exercises 16–18, use the table at the right. It shows the number of Calories contained in one serving of students' favorite cereals.

Number of Calories per Serving of Dry Cereal				
70	120	100	100	140
70	120	100	100	100
110	110	110	120	100
110	100	100	110	150
90	110	120	110	100
140	130	100	80	100
100	110	80	90	110

16. Make a frequency table of the data.

17. How many Calories are in one serving of most of the cereals?

18. If a student eats two bowls of his favorite cereal, what is the greatest and least number of Calories possible that he consumes?

19. **CRITICAL THINKING** Make a frequency table of the scooter data at the beginning of this lesson using a different scale and interval. Analyze the data in the new table and describe an advantage and a disadvantage of using this scale and interval.

Spiral Review with Standardized Test Practice

20. **MULTIPLE CHOICE** The following are scores of the Duke University women's basketball team in recent conference games: 107, 76, 90, 91, 81, 73, 86, 102, 102, 86, 66, 83, 77, 81, 88, 90. Choose the best scale for making a frequency table of the scores.

    A 66–105   B 68–102   C 66–110   D 70–108

21. **MULTIPLE CHOICE** If there are 24 students in math class, how many received a score on the exam in the interval 81–90?

    F 7   G 6   H 5   I 4

Score	Frequency
71–80	8
81–90	?
91–100	9

Write each number in standard form. (Lesson 1-9)

22. 1.8×10^3 23. 5.33×10^5 24. 2.4×10^4 25. 6.0×10^7

26. **MEASUREMENT** How many centimeters are in 8.05 meters? (Lesson 1-8)

GETTING READY FOR THE NEXT LESSON

PREREQUISITE SKILL

27. At 5:00 P.M. the outside temperature was 81°F; at 6:00 P.M. it was 80°F; at 7:00 P.M. it was 79°F. If the pattern continues, predict the outside temperature at 8:00 P.M. (Lesson 1-7)

msmath2.net/self_check_quiz

2-2a Problem-Solving Strategy
A Preview of Lesson 2-2

Use a Graph

What You'll LEARN

Solve problems by using graphs.

How many home runs do you think the Pittsburgh Pirates will get in 2004?

By just looking at the table, I can't really tell. Let's **use a graph** to get a better picture.

Pittsburgh Pirates Home Runs													
Year	'91	'92	'93	'94	'95	'96	'97	'98	'99	'00	'01	'02	'03
Home Runs	126	106	110	80	125	138	129	107	171	168	161	142	163

Source: www.baseball-almanac.com

Explore	We know the number of home runs from 1991 to 2002. We need to predict how many there will be in 2004.
Plan	If we organize the information in a graph, we may be able to see if there are any trends in the number of home runs.
Solve	We can see from the graph that generally, as the years progress, the number of home runs increases. We can predict that the number of home runs in 2004 will be about 190.
Examine	Draw a line through the points that is close to as many of the points as possible, as shown above. The estimate is close to the line, so the prediction is reasonable.

Analyze the Strategy

1. **Explain** why analyzing a graph is a useful way to quickly make conclusions about a set of data.

2. The last step of the four-step plan asks you to *examine* the solution. **Write** a problem in which using a graph would be a useful way to examine the solution.

58 Chapter 2 Statistics: Analyzing Data

Apply the Strategy

Solve. Use a graph.

EARTH SCIENCE For Exercises 3 and 4, refer to the graph at the right. It shows the average high and low temperatures each month from January to June in Memphis, Tennessee. Suppose the trends continue through July.

3. Predict the average high temperature for the month of July.

4. Predict the average low temperature for the month of July.

Source: www.weather.com

Mixed Problem Solving

Solve. Use any strategy.

5. **READING** Maya read 10 pages of a 150-page book on Monday and plans to read twice as many pages each day than she did the previous day. On what day will she finish the book?

6. **SCHOOL** The graph shows the number of energy bars sold in a school's cafeteria during one week.

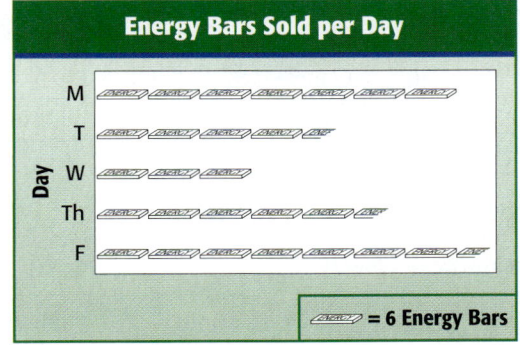

According to the graph, did the cafeteria sell more energy bars on Monday or Tuesday?

7. **SAFETY** An elevator sign reads *Do not exceed 2,500 pounds.* How many people, each weighing about 150 pounds, can be in the elevator at the same time?

8. **NUMBER THEORY** Find two numbers whose sum is 56 and whose product is 783.

9. **BOWLING** Tariq and three of his friends are going bowling, and they have a total of $70 to spend. Suppose they buy a large pizza, four beverages, and each rent bowling shoes. How many games can they bowl if they all bowl the same number of games?

Bowling Costs	
Item	Price
large pizza	$13.75
beverage	$1.50
shoe rental	$3.50
game	$4.00

10. **STANDARDIZED TEST PRACTICE**

The following sequence forms a pattern.
4, 8, 12, 16, …

If this pattern continues, which expression can be used to find the *sixth* term?

- Ⓐ 6 + 4
- Ⓑ 6 − 4
- Ⓒ 6 ÷ 4
- Ⓓ 6 × 4

Lesson 2-2a Problem-Solving Strategy: Use a Graph **59**

2-2 Making Predictions

What You'll LEARN
Make predictions from graphs.

NEW Vocabulary
line graph
scatter plot

HANDS-ON Mini Lab

Materials
- measuring cup
- drinking glass
- metric ruler
- marbles
- grid paper

Work with a partner.

- Pour 1 cup of water into the drinking glass.
- Measure the height of the water and record it in a table like the one below.
- Place five marbles in the glass. Measure the height of the water and record it in your table.
- Continue adding marbles, five at a time, until there are twenty marbles in the glass. After each time, measure and record the height of the water.

Number of Marbles	Height of Water (cm)
0	
5	
10	
15	
20	

1. By how much did the height of the water change after each addition of five marbles?
2. Predict the height of the water when thirty marbles are in the drinking glass. Explain how you made your prediction.
3. Test your prediction by placing ten more marbles in the glass.
4. Draw a graph of the data that you recorded in the table.

You created a line graph in the Mini Lab. **Line graphs** can be useful in predicting future events when they show trends over time.

EXAMPLE   Use a Line Graph to Predict

 STATUE OF LIBERTY The table shows the time it takes Kyle to climb the Statue of Liberty. Make a line graph and predict the total time it will take him to climb 354 steps to the top.

Time (min)	0	2	4	6	8	10
Number of Steps	0	51	114	163	210	275

The data values go from 0 to 275. You want to predict the time when 354 steps are climbed. So, make the scale go from 0 to 400 with an interval of 100.

Climbing the Statue of Liberty

Graph the data and connect the points.

Continue the graph with a dotted line in the same direction until you reach a horizontal position of 354 steps.

It will take Kyle about 13 minutes to climb 354 steps.

60 Chapter 2 Statistics: Analyzing Data

READING Math

Scatter Plots Another name for scatter plot is *scattergram*.

A <mark>scatter plot</mark> displays two sets of data on the same graph. Like line graphs, scatter plots are useful for making predictions because they show trends in data. If the points on a scatter plot come close to lying on a straight line, the two sets of data are related.

EXAMPLE Use a Scatter Plot to Predict

2 SWIMMING The scatter plot shows the winning times for the women's 100-meter butterfly event at the Summer Olympics from 1968 to 2000. Use it to predict a winning time for this event at the 2012 Olympics.

Source: *The World Almanac*

By looking at the pattern in the graph, we can predict that the winning time at the 2012 Olympics will be about 54 seconds.

Skill and Concept Check

1. **Writing Math** Explain how a graph can be used to make predictions.

2. **OPEN ENDED** Name two sets of data that can be graphed on a scatter plot.

3. **Which One Doesn't Belong?** Identify the term that does not have the same characteristic as the other three. Explain your reasoning.

| line graph | tally mark | frequency table | scatter plot |

GUIDED PRACTICE

4. **SNACKS** The graph shows the sales, in billions of dollars, of salty snacks like potato chips and pretzels for several years. If the trend continues, about what will the sales be in 2006?

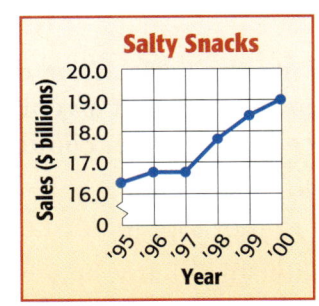

Source: *American Demographics*

Lesson 2-2 Making Predictions **61**

Practice and Applications

5. **RUNNING** Mahkah has been training for several weeks for cross-country tryouts. His progress is shown in the graph at the right. To make the team, he must be able to run 1 mile in less than 8 minutes. If the trend continues, will Mahkah make the team? Explain.

HOMEWORK HELP

For Exercises	See Examples
5–9	1
10–12	2

Extra Practice
See pages 567, 597.

6. **SCHOOLS** The line graph shows public school teachers' salaries for the past few years. If the trend continues, about what will the average annual salary be in 2005?

7. **RESEARCH** Use the Internet or another source to find a real-life example of a line graph. Write a description of what the graph displays and extend the graph to show where the data will be in the future.

OLYMPICS For Exercises 8 and 9, use the table below. It shows the number of nations that participated in the Summer Olympics from 1948 to 2000.

Year	Number of Nations	Year	Number of Nations
1948	59	1976	92
1952	69	1980	80
1956	72	1984	140
1960	83	1988	159
1964	93	1992	169
1968	112	1996	197
1972	121	2000	199

Source: www.factmonster.com

8. Make a line graph of the data.
9. Predict the number of nations that will participate in the 2008 Summer Olympics.

For Exercises 10–12, use the table at the right. It shows the relationship between the speed of a vehicle and the distance required to stop in certain weather conditions.

Speed (mph)	Stopping Distance (ft)
55	273
65	355
75	447

10. Make a scatter plot of the data. Use the speed on the horizontal axis and the stopping distance on the vertical axis.
11. Describe the relationship, if any, between the two sets of data.
12. Predict the stopping distance for a car traveling 45 miles per hour.

13. **CRITICAL THINKING** What can you conclude about the relationship between students' birth month and height shown in the scatter plot at the right?

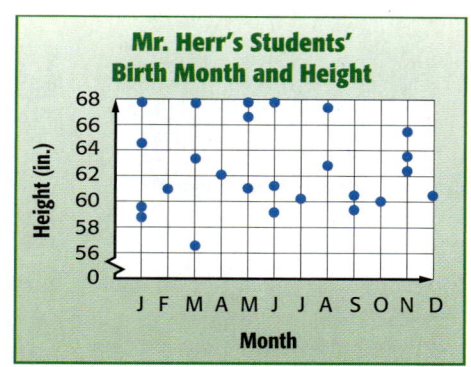

14. **EXTENDING THE LESSON** The graph at the right is called a *multiple-line graph* because it contains more than one line graph.

    The multiple-line graph shows the population of Miami-Dade and Broward Counties in Florida from 1950 to 2000. Do you think that the population of Broward County will catch up to the population of Miami-Dade County in the next census of 2010? Explain.

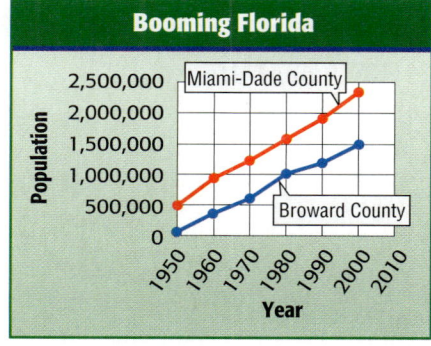

Source: U.S. Census Bureau

Spiral Review with Standardized Test Practice

For Exercises 15 and 16, refer to the graph at the right that shows the number of customers at Clothes Palace at different times during the day.

15. **MULTIPLE CHOICE** Extra workers are needed when the number of customers in the store exceeds 50. Between which hours is extra help needed?

    Ⓐ 12:00 P.M.–3:00 P.M. Ⓑ 11:00 A.M.–4:00 P.M.

    Ⓒ 1:00 P.M.–4:00 P.M.   Ⓓ 1:00 P.M.–3:00 P.M.

16. **SHORT RESPONSE** Predict about how many customers are in the store at 8:00 P.M.

17. **STATISTICS** Choose an appropriate interval and scale for the following set of data: 35, 42, 18, 25, 32, 47, 34. (Lesson 2-1)

18. Write 72,000 in scientific notation. (Lesson 1-9)

Write each product in exponential form. (Lesson 1-2)

19. $9 \cdot 9$ 20. $14 \cdot 14 \cdot 14 \cdot 14$ 21. $7 \cdot 7 \cdot 7$ 22. $5 \cdot 5 \cdot 5 \cdot 5 \cdot 5$

GETTING READY FOR THE NEXT LESSON

PREREQUISITE SKILL The following data are the heights of the players, in feet, on the Miami Heat basketball team: 6.8, 7.0, 6.2, 6.7, 6.8, 6.4, 6.7, 6.1, 6.5, 6.2, 6.5, 6.8, 6.8, 7.0, 6.2. (Page 556)

23. Order the heights from least to greatest.

2-3 Line Plots

What You'll LEARN

Construct and interpret line plots.

NEW Vocabulary

line plot
cluster
outlier
range

Link to READING

Everyday Meaning of Range: in music, all the notes between the high and low notes, as in a singer with a wide range

WHEN am I ever going to use this?

BUILDINGS The table shows the number of stories in the 32 tallest buildings in Los Angeles.

1. Do any of the values seem much greater or much less than the other data values?
2. Do some of the buildings have the same number of stories? Is this easy to see? Explain.
3. What better ways can you suggest for organizing these data?

Number of Stories

28	52	44	39
17	44	36	73
32	44	42	53
40	33	33	37
42	62	36	52
34	55	34	28
31	48	41	52
52	54	52	34

Source: The World Almanac

One way to show how data are spread out is to use a line plot. A **line plot** is a diagram that shows the frequency of data on a number line. An "×" is placed above a number on a number line each time that data value occurs.

EXAMPLE Make a Line Plot

1 BUILDINGS Make a line plot of the data shown above.

Step 1 Draw a number line. Because the shortest building in the table has 17 stories, and the tallest has 73, you can use a scale of 15 to 75 and an interval of 5. Other scales and intervals could also be used.

Step 2 Put an × above the number that represents the number of stories in each building.

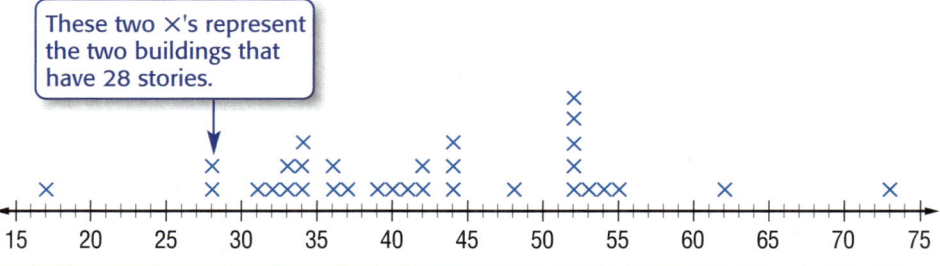

These two ×'s represent the two buildings that have 28 stories.

READING in the Content Area

For strategies in reading this lesson, visit **msmath2.net/reading**.

You can see on the line plot how many buildings have the same number of stories. This information is not easily found using the table above.

64 Chapter 2 Statistics: Analyzing Data

You can make some observations about the *distribution* of data, or how data are grouped together or spread out, using the line plot in Example 1.

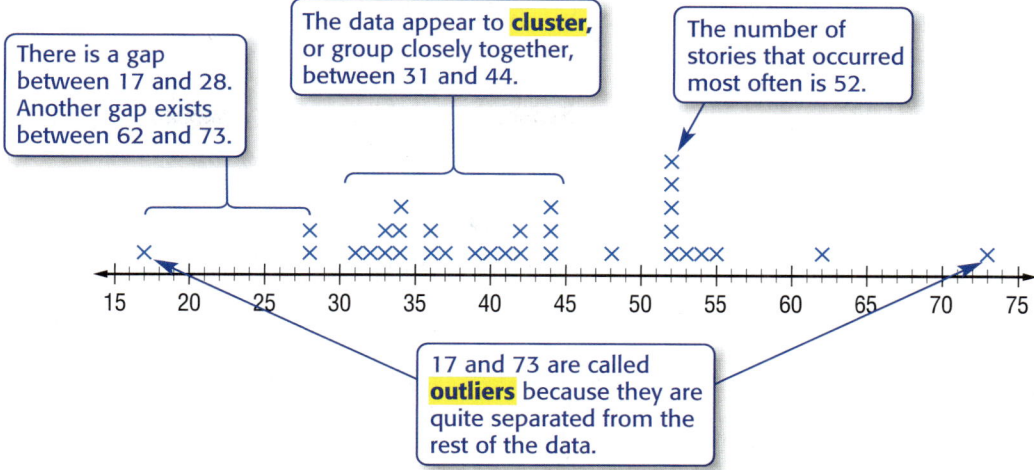

There is a gap between 17 and 28. Another gap exists between 62 and 73.

The data appear to **cluster**, or group closely together, between 31 and 44.

The number of stories that occurred most often is 52.

17 and 73 are called **outliers** because they are quite separated from the rest of the data.

In a line plot, you can also easily find the **range** of the data, or the difference between the greatest and least numbers in the data set. This value is helpful in seeing how spread out the data are.

EXAMPLES   Use a Line Plot to Analyze Data

2 **SHOES** The line plot below shows prices for different basketball shoes. What is the range of the data?

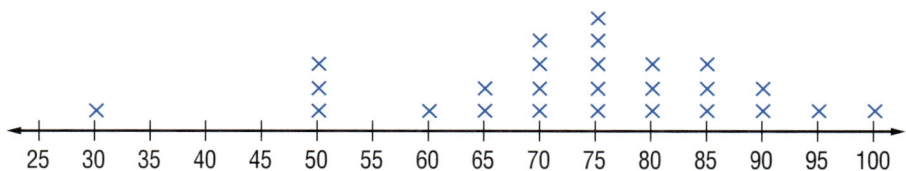

The greatest price is $100, and the lowest price is $30.

range: 100 − 30 or 70

The range of the prices is $70.

STUDY TIP

Clusters You can describe a cluster by stating a range of values or by giving a single value around which the data appear to be grouped.

3 Identify any clusters, gaps, and outliers in the data in Example 2 and explain what they mean.

Many of the data cluster around $75. You could say that most of the shoes cost from $70 to $85.

There is a gap from $30 to $50, so there were no shoes in this price range.

The number 30 appears removed from the rest of the data, so it could be considered an outlier. This means that $30 is an extremely low price and not representative of the whole data set.

Your Turn

a. Identify the range of the data in Example 1.

msmath2.net/extra_examples   Lesson 2-3 Line Plots **65**

Skill and Concept Check

1. **List** the information you can identify from a line plot that is not obvious in an unorganized list of data.

2. **OPEN ENDED** Draw a line plot that displays fifteen pieces of data. Arrange the data so that there is a cluster and one outlier.

3. **FIND THE ERROR** Ryan and Darnell are analyzing the data shown in the line plot at the right. Who is correct? Explain.

   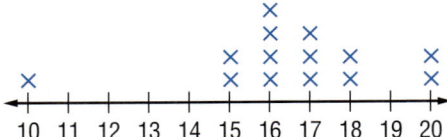

   Ryan
   greatest data value: 16
   least data value: 10

   Darnell
   greatest data value: 20
   least data value: 10

4. **DATA SENSE** Refer to the line plot in Exercise 3. Find the range and identify any clusters, gaps, or outliers.

GUIDED PRACTICE

For Exercises 5–7, use the table at the right. It shows the number of music CDs owned by students in a classroom.

Number of Music CDs				
12	9	11	8	12
0	7	10	20	11
6	4	10	10	8
10	9	8	11	6

5. Make a line plot of the data.
6. Identify any clusters, gaps, or outliers.
7. Summarize the data in a sentence or two.

WEATHER For Exercises 8–11, use the line plot below. It shows the record high temperatures recorded by weather stations in each of the fifty states.

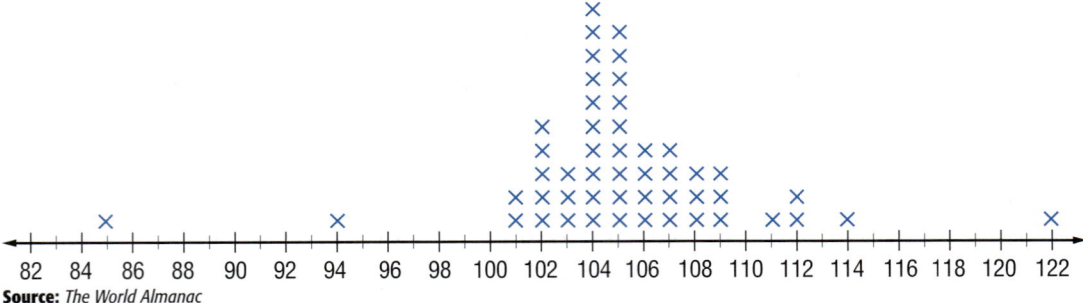

Source: The World Almanac

8. What is the range of the data?
9. What temperature occurred most often?
10. Identify any clusters, gaps, or outliers and analyze the data by describing what these values represent.
11. Describe how the range of the data would change if 122 were not part of the data set.

Data Update What are the record high temperatures for some of the cities in your state? Visit msmath2.net/data_update to learn more.

66 Chapter 2 Statistics: Analyzing Data

Practice and Applications

Make a line plot for each set of data. Identify any clusters, gaps, or outliers.

12.
Heights of Desert Cacti (ft)				
30	10	1	15	10
10	10	10	2	10
20	3	2	15	5

Source: ag.arizona.edu/pubs/garden/mg/cacti

13.
Test Scores					
98	90	97	85	86	92
92	93	95	79	91	92
91	94	88	90	93	92

HOMEWORK HELP

For Exercises	See Examples
12–13, 19–22	1–3
14–18	2, 3

Extra Practice
See pages 567, 597.

Determine whether each statement is *sometimes*, *always*, or *never* true. Explain.

14. You can determine the number of items in a data set from a line plot.

15. If there is a cluster, it will appear in the center of the line plot.

SURVEY For Exercises 16–18, use the line plot at the right and the information below.

Jamie asked her classmates how many cans of soda they drink on a typical day. The results are shown on the line plot.

16. What was the most frequent response?

17. What is the range?

18. Summarize the data in a sentence or two.

ANIMALS For Exercises 19–22, use the table below.

Average Life Spans					
Animal	Span (yr)	Animal	Span (yr)	Animal	Span (yr)
black bear	18	giraffe	10	lion	15
camel	12	gorilla	20	pig	10
cat	12	grizzly bear	25	polar bear	20
chipmunk	6	horse	20	rhesus monkey	15
dog	12	kangaroo	7	squirrel	10
elephant	40	leopard	12	white rhinoceros	20

Source: *The World Almanac for Kids*

19. Make a line plot of the data.

20. Find the range and determine any clusters or outliers.

21. Summarize the data on the line plot in a sentence or two.

22. The following data are the *maximum* life spans, in order, of the animals in the table: 36, 50, 28, 10, 20, 77, 36, 54, 50, 50, 24, 23, 30, 27, 45, 37, 24, 50. Make a line plot of the data and compare it to the line plot you made in Exercise 19. Include a discussion about clusters, outliers, range, and gaps in data.

msmath2.net/self_check_quiz

Lesson 2-3 Line Plots **67**

23. **DATA SENSE** The last fourteen bowling scores of two people are shown on the line plots below. Describe which person is more consistent and explain how you know.

Bowler A

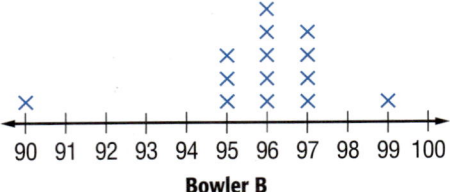
Bowler B

24. **CRITICAL THINKING** Compare and contrast line plots and frequency tables. Include a discussion about what they show and when it is better to use each one.

Spiral Review with Standardized Test Practice

For Exercises 25 and 26, refer to the line plot shown below.

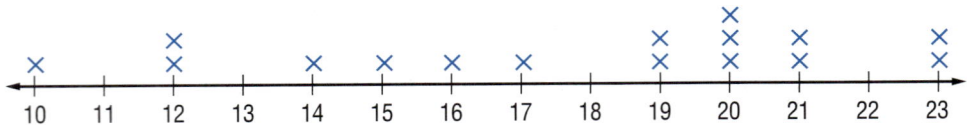

25. **MULTIPLE CHOICE** Which of the following describes a cluster of data?

    Ⓐ 20–23   Ⓑ 14–17   Ⓒ 12–14   Ⓓ 19–21

26. **MULTIPLE CHOICE** Determine which sentence is *not* true.

    Ⓕ The number that occurs most often is 20.
    Ⓖ There are 23 numbers in the data set.
    Ⓗ The majority of the data is greater than 18.
    Ⓘ The range is 13.

27. **RAIN** The scatter plot shows the height of grass on several lawns and the amount of rain that has fallen on each lawn. Write a conclusion based on the graph. (Lesson 2-2)

28. *True* or *False*? A frequency table is a good way to organize large amounts of data. (Lesson 2-1)

Solve each equation mentally. (Lesson 1-5)

29. $\frac{t}{5} = 15$

30. $m + 18 = 33$

31. $25 = a - 4$

32. $12x = 120$

GETTING READY FOR THE NEXT LESSON

BASIC SKILL Add or divide. Round to the nearest tenth if necessary.

33. $16 + 14 + 17$
34. $32 + 34 + 36$
35. $4.6 + 2.5 + 9$
36. $77 + 71 + 74$
37. $\frac{202}{16}$
38. $\frac{56}{3}$
39. $\frac{255}{7}$
40. $\frac{155.3}{30}$

68 Chapter 2 Statistics: Analyzing Data

2-4 Mean, Median, and Mode

What You'll LEARN

Find the mean, median, and mode of a set of data.

NEW Vocabulary

measures of central tendency
mean
median
mode

Link to READING

Everyday Meaning of Median: the middle paved or planted section of a highway, as in median strip

HANDS-ON Mini Lab

Materials
- centimeter cubes

Work with a partner.

In five days, it snowed 4 inches, 3 inches, 5 inches, 1 inch, and 2 inches.

- Make a stack of cubes to represent the snowfall for each day, as shown at the right.
- Move the cubes from stack to stack until each stack has the same number of cubes.

4 in. 3 in. 5 in. 1 in. 2 in.

1. How many cubes are in each stack?
2. In the past five days, it snowed an average of ? inches per day.
3. Suppose on the sixth day it snowed 9 inches. If you moved the cubes again, how many cubes would be in each stack?

If you had to choose one number to represent the data set 1, 2, 3, 4, 5, a good choice would be the middle number, 3. When analyzing data, it is often helpful to use a single number to describe the whole set.

Numbers called **measures of central tendency** can be used to describe the *center* of data. The most common of these measures are mean, median, and mode.

Noteables™   Key Concept: Mean

Words   The **mean** of a set of data is the sum of the data divided by the number of items in the data set.

Example   data set: 1, 1, 5, 2, 2, 4, 2, 5

mean: $\dfrac{1+1+5+2+2+4+2+5}{8}$ or 2.75

EXAMPLE   Find the Mean

1 NUTRITION The table shows the grams of sugar in twenty different breakfast bars. Find the mean.

Grams of Sugar in Breakfast Bars				
14	11	7	14	15
22	18	18	21	10
10	10	11	9	11
11	9	12	8	12

Source: www.cspinet.org

mean = $\dfrac{14 + 11 + \ldots + 12}{20}$ ← sum of data
← number of data items

= $\dfrac{253}{20}$ or 12.65

The mean amount of sugar for the breakfast bars is 12.65 grams.

msmath2.net/extra_examples

Noteables™

Key Concept: Median

Words   The **median** of a set of data is the middle number of the ordered data, or the mean of the middle two numbers.

Example   data set: 7, 11, 15, 17, 20, 20

median: $\frac{15 + 17}{2}$ or 16   The median divides the data in half.

Key Concept: Mode

Words   The **mode** or modes of a set of data is the number or numbers that occur most often.

Example   data set: 50, 45, 45, 52, 49, 56, 56

modes: 45 and 56

REAL-LIFE MATH

SPACE The International Space Station measures 356 feet by 290 feet, and contains almost an acre of solar panels.

Source: *The World Almanac*

EXAMPLE Find the Mean, Median, and Mode

② SPACE Twenty-seven countries have sent people into space. The table shows the number of individuals from each country. Find the mean, median, and mode of the data.

People in Space								
267	1	9	8	1	1	1	1	1
97	1	1	1	3	1	1	2	1
11	2	1	1	5	1	1	1	1

Source: *The World Almanac*

mean: sum of data divided by 27, or 15.6
median: 14th number of the ordered data, or 1
mode: number appearing most often, or 1

The mean of the data in Example 2 does not accurately describe the data set because it is greater than most of the data items. This happens when one or more of the data is an outlier.

EXAMPLE Analyze Data

③ TESTS The line plot shows the test scores of the students in Mrs. Hiroshi's math class. Would the mean, median, or mode best represent the test scores?

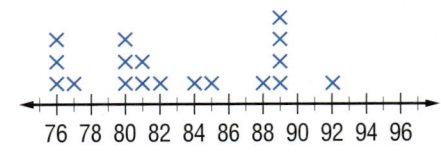

mean: $\frac{76 + 76 + 76 + \ldots + 92}{18}$ or 83

median: $\frac{9\text{th term} + 10\text{th term}}{2} = \frac{81 + 82}{2}$ or 81.5

mode: 89

The mode of 89 misrepresents the scores as being higher than they really were. Either the mean of 83 or the median of 81.5 could be used to represent the scores.

STUDY TIP

Median When there is an odd number of data, the median is the middle number of the ordered set. When there is an even number of data, the median is the mean of the two middle numbers.

---

70 Chapter 2 Statistics: Analyzing Data

You can find measures of central tendency from charts, tables, graphs, and line plots. Below are some general guidelines for using these measures.

Concept Summary — Mean, Median, and Mode

Measure	Most Useful When . . .
Mean	• data set has no outliers
Median	• data set has outliers • there are no big gaps in the middle of the data
Mode	• data set has many identical numbers

Skill and Concept Check

1. **Writing Math** Determine whether the median is *sometimes*, *always*, or *never* part of the data set. Explain.

2. **OPEN ENDED** Write a problem in which the mean is not the best representation of the data set. Explain why not.

3. **FIND THE ERROR** Jared and Marissa are analyzing the data in the frequency table at the right. Who is correct? Explain.

   Jared
   median: 2 min
   mode: 3 min

   Marissa
   median: 2.5 min
   mode: 9 min

Time (min)	Tally	Frequency
1							5
2								6
3											9
4			1

GUIDED PRACTICE

Find the mean, median, and mode for each set of data. Round to the nearest tenth if necessary.

4. 14, 5, 16, 7, 4, 15, 9, 7, 2, 14, 6

5. 29, 14, 80, 59, 78, 30, 59, 69, 55, 50

6.

7.

Number of Wins	Tally	Frequency
7														12
8						4
9				2
10									7

SCHOOL For Exercises 8 and 9, use the line plot. It shows the number of times per day that students drink from the water fountain at school.

8. Find the mean, median, and mode.

9. Which measure best describes the data? Explain.

Lesson 2-4 Mean, Median, and Mode

Practice and Applications

Find the mean, median, and mode for each set of data. Round to the nearest tenth if necessary.

10. 65, 56, 57, 75, 76, 66, 64
11. 9.3, 8.4, 8.0, 9.1, 9.4, 9.0, 7.8, 9.3, 8.0
12. 95, 90, 92, 94, 91, 90, 98, 88, 89, 100
13. $10, $18, $11, $6, $6, $5, $10, $11, $46, $7, $6, $8

14.

15.
Price	Tally	Frequency
$7.50	||||| |||	8
$8.00	|||	3
$8.50	|||||		6

HOMEWORK HELP
For Exercises	See Examples
10–16	1, 2
17, 18	3

Extra Practice
See pages 568, 597.

MUSIC For Exercises 16 and 17, use the data at the right. It shows the number of weeks that songs have been on the *Top 20 Hot Country Singles & Tracks* chart.

16. Find the mean, median, and mode of the data.
17. Determine which measure best represents the data. Explain.

Number of Weeks in Top 20
8	15	14	24	21
28	12	20	19	18
12	7	11	30	25
15	21	22	18	31

Source: www.billboard.com

Data Update What is the mean, median, and mode of the number of weeks of the current songs that are on the *Top 20 Hot Country Singles & Tracks* chart? Visit msmath2.net/data_update to learn more.

18. **DATA SENSE** According to the U.S. Census Bureau, the average number of family members per household in Texas is 3.28. State whether this average is a mean or mode. Explain how you know.

19. **CRITICAL THINKING** Without calculating, would the mean, median, or mode be most affected by eliminating 1,000 from the list below? Which would be the least affected? Explain.
 50, 100, 75, 60, 75, 1,000, 90, 100

Spiral Review with Standardized Test Practice

20. **MULTIPLE CHOICE** Which measure describes the center of a set of data?
 Ⓐ mean Ⓑ median Ⓒ mode Ⓓ all of the above

21. **GRID IN** The Cardinals baseball team played 8 games and scored a total of 96 runs. What was the mean number of runs scored per game?

22. **NUTRITION** The grams of fiber in one serving of fifteen different cereals are 5, 5, 4, 3, 3, 3, 1, 1, 1, 2, 1, 1, 1, 1, and 0. Make a line plot of the data. *(Lesson 2-3)*

23. What type of graph often shows trends over time? *(Lesson 2-2)*

GETTING READY FOR THE NEXT LESSON

BASIC SKILL Name the place value of the highlighted digit.

24. **5**81   25. 6,29**5**   26. **4**,369   27. 2**8**4

 msmath2.net/self_check_quiz

2-4b Hands-On Lab

A Follow-Up of Lesson 2-4

Are You Average?

What You'll LEARN

Use mean, median, mode, and range to describe a set of data.

Materials
- markers
- ruler

INVESTIGATE *Work in groups of four.*

What do you think of when you hear the word "average"? How would you describe an "average" student in your school? In this lab, you will investigate what an average student in your math class is like.

STEP 1 Your group will create a survey of ten appropriate questions to ask your classmates that will help you describe the "average" student in your class. For example:

- What is your height in inches?
- How many hours per week do you spend doing homework?
- What is your favorite subject in school?
- What is your favorite extracurricular activity?

STEP 2 After your teacher approves your list of questions, have your classmates complete your survey.

STEP 3 Compile the data in a frequency table. Find the mean, median, mode, and range of the data for each question if appropriate. Which measure best describes each set of data? Justify your answer.

STEP 4 Create appropriate graphs to display your data from all ten questions. Each group will present their graphs with a description of what the "average" student in your classroom is like.

Writing Math

1. **Explain** why results may vary if you survey another class in your school.
2. **Explain** why your group selected the graph that you did to display your data. Could you have used another type of graph? Explain.
3. The word *bias* means to influence. **Describe** any factors that could have unfairly influenced the responses given by your classmates. Is there a way to limit bias in a survey?

Chapter 2 Mid-Chapter Practice Test

Vocabulary and Concepts

1. **Explain** what a frequency table shows. (Lesson 2-1)
2. **Describe** a situation in which a line graph would be useful. (Lesson 2-2)
3. **List** the information that you can identify from a line plot that is not obvious in an unorganized list of data. (Lesson 2-3)
4. **Define** *mean*. (Lesson 2-4)

Skills and Applications

For Exercises 5–10, use the data below that shows the ages of twenty people when they got their driver's licenses. (Lessons 2-1, 2-3, and 2-4)

16 17 16 16 18 21 16 16 18 18 17 25 16 17 17 17 17 16 20 16

5. Make a frequency table of the data.
6. Make a line plot of the data.
7. Identify any clusters or gaps.
8. Identify any outliers.
9. Find the mean, median, and mode.
10. What is the range of the data?

11. **HEATING** Use the scatter plot to make a conclusion about the average monthly outside temperature and heating costs of a home. (Lesson 2-2)

Find the mean, median, and mode for each set of data. (Lesson 2-4)

12. 27, 32, 18, 24, 32, 39, 41
13. 7.6, 6.8, 6.5, 7, 7.9, 6.8, 7, 7, 6.8, 8.1

Standardized Test Practice

14. **GRID IN** Below are the quiz scores of a science class.

    89 92 67 75 95 89 82 92 89
    89 92 96 85 94 90 79 94 87

    What is the median quiz score? (Lesson 2-4)

15. **MULTIPLE CHOICE** Which may *not* show individual pieces of data? (Lessons 2-1, 2-2, and 2-3)

    Ⓐ scatter plot
    Ⓑ frequency table
    Ⓒ line plot
    Ⓓ an ordered list

The Game Zone

A Place To Practice Your Math Skills

Math Skill: Finding The Mean

Can You Guess?

- **GET READY!**
  Players: the entire class divided into three-person teams
  Materials: cup, coins, water dropper, self-adhesive notes

- **GET SET!**
  - Draw a number line on the chalkboard with a scale from 10 to 25.
  - Each team should fill a cup with water.

- **GO!**
  - Use the water dropper and find how many drops of water your team can get to stay on the head of a dime. Write the results on a sticky note. Post it on the chalkboard above the corresponding number on the number line. Find the mean of the numbers.

  - Next, use the mean to predict how many drops of water will fit on the head of a penny. Record your team's prediction on a piece of paper.
  - Find how many drops of water will stay on the head of a penny. Record the result on a sticky note and post it on the chalkboard. Find the mean of all of the results for the pennies. If your prediction is within three drops of the mean, your team gets to stay in the game.
  - Continue predicting and testing using a nickel and a quarter.
  - **Who Wins?** The teams left at the end of the game are the winners.

2-5 Stem-and-Leaf Plots

What You'll LEARN

Construct and interpret stem-and-leaf plots.

NEW Vocabulary

stem-and-leaf plot
leaf
stem

WHEN am I ever going to use this?

HISTORY Benjamin Franklin was the oldest signer of the Declaration of Independence. The table shows the ages in years of all the people who signed the Declaration of Independence.

1. What was the age of the youngest signer?
2. What methods do you already have for showing the distribution of ages?

Ages of Signers of Declaration of Independence

40	70	69	34	46	53	31
38	50	30	45	46	39	38
35	50	37	48	41	41	52
39	50	65	46	29	34	45
33	41	44	63	60	26	42
42	52	37	35	45	35	42
47	26	55	57	45	33	60
62	35	33	53	49	50	46

Source: *The World Almanac*

A useful way to organize data as you collect it is a stem-and-leaf plot. In a **stem-and-leaf plot**, the data are organized from least to greatest. The digits of the least place value usually form the **leaves**, and the next place value digits form the **stems**.

EXAMPLE Construct a Stem-and-Leaf Plot

1 **HISTORY** Make a stem-and-leaf plot of the data above.

Step 1 The digits in the least place value will form the leaves and the remaining digits will form the stems. In this data, 26 is the least value, and 70 is the greatest. So, the ones digits will form the leaves and the tens digits will form the stems.

Step 2 List the stems 2 to 7 in order from least to greatest in the *Stem* column. Write the leaves, the ones digits of the ages, to the right of the corresponding stems. The ages in the first column of the chart are shown at the right.

Stem	Leaf
2	
3	8 5 9 3
4	0 2 7
5	
6	2
7	

Step 3 Order the leaves and write a *key* that explains how to read the stems and leaves.

Stem	Leaf
2	6 6 9
3	0 1 3 3 3 4 4 5 5 5 5 7 7 8 8 9 9
4	0 1 1 1 2 2 2 2 4 5 5 5 5 6 6 6 6 7 8 9
5	0 0 0 0 2 2 3 3 5 7
6	0 0 2 3 5 9
7	0

3 | 1 = 31 yr

The tens digits of the data form the stems.

A key shows how the digits are related.

The ones digits of the data form the leaves.

76 Chapter 2 Statistics: Analyzing Data

Stem-and-leaf plots are useful in analyzing data because you can see all the data values, including the greatest, least, mode, and median value.

EXAMPLE Analyze Data

2 **PETS** The stem-and-leaf plot shows the average chick weight in grams of sixteen different species of pet birds. Find the range, median, and mode of the data.

Stem	Leaf
0	5 6 7
1	0 0 1 1 2 2 2 3 8 9
2	0 1 1 5 1|2 = 12 g

Source: www.upatsix.com

range: greatest weight − least weight = 25 − 5 or 20 grams

median: middle value, or 12 grams

mode: most frequent value, or 12 grams

Stem-and-leaf plots are also useful for making conclusions about data. Most of the data in the plot above are clustered in the 10–19 interval. So, you could say that most of those birds weigh 10 to 19 grams as chicks.

EXAMPLE Make Conclusions About Data

3 **NUTRITION** The table shows the average amounts of pasta that people in different countries consume each year. Make a stem-and-leaf plot of the data. Then use it to describe how the data are spread out.

Country	Pasta per Person (lb)	Country	Pasta per Person (lb)
Argentina	15	Netherlands	9
Australia	5	Portugal	15
Canada	14	Russia	15
Egypt	15	Spain	9
France	15	Sweden	8
Germany	10	Switzerland	20
Greece	19	Turkey	11
Italy	59	United Kingdom	4
Japan	3	United States	20
Mexico	8	Venezuela	28

Source: National Pasta & National Restaurant Associations

The least value is 3, and the greatest value is 59. So, the tens digits form the stems, and the ones digits form the leaves.

Stem	Leaf
0	3 4 5 8 8 9 9
1	0 1 4 5 5 5 5 5 9
2	0 0 8
3	
4	
5	9 2|8 = 28 lb

The data that do not have a digit in the tens place have 0 for a stem.

It is easy to see in the stem-and-leaf plot that people in most of the countries consume less than 20 pounds of pasta per year. You can also see that the mode is 15 and that 59 is an outlier, with a significant gap between that value and the next smaller value.

REAL-LIFE CAREERS

How Does a Nutritionist Use Math?
Nutritionists use statistical graphs to summarize and compare the nutritional values of various foods.

Research
For information about a career as a nutritionist, visit:
msmath2.net/careers

Skill and Concept Check

1. **Writing Math** Compare and contrast a stem-and-leaf plot and a frequency table.

2. **OPEN ENDED** Make a stem-and-leaf plot in which the median of the data is 25.

3. **Write** the stems that would be used in a stem-and-leaf plot for the following set of data: 21, 8, 12, 7, 18, 9, 22, 31, 1, 14, 19, 11, 2.

4. **FIND THE ERROR** Selena and Lauren are analyzing the stem-and-leaf plot at the right. Who is correct? Explain.

Stem	Leaf
2	6 6 9
3	
4	6
5	3 6 2|6 = 26 in.

   Selena
   None of the data have 3 in the tens place.

   Lauren
   Thirty is one of the data values.

GUIDED PRACTICE

Make a stem-and-leaf plot for each set of data.

5. 42, 47, 19, 16, 21, 23, 25, 25, 29, 31, 33, 34, 35, 39, 48

6. 225, 227, 230, 229, 246, 243, 269, 269, 267, 278, 278, 278

CHESS For Exercises 7–10, use the stem-and-leaf plot at the right. It shows the number of matches won by each member of the Avery Middle School Chess Team.

Stem	Leaf
0	8 8 9
1	9
2	0 0 2 4 4 8 9
3	1 1 2 4 5 5 6 6 7 7 8
4	0 0 0 3 8 9
5	2 4
6	1 3|2 = 32 wins

7. How many members are on the chess team?

8. What is the range of the number of matches won?

9. Find the mean, median, and mode for these data.

10. Describe how the data are spread out.

Practice and Applications

Make a stem-and-leaf plot for each set of data.

11. 15, 13, 28, 32, 38, 30, 31, 13, 36, 35, 38, 32, 38, 24

12. 56, 70, 96, 62, 80, 80, 69, 93, 66, 55, 95, 63, 90, 93, 60, 91, 67, 60

13. 451, 458, 439, 403, 411, 434, 433, 454, 457, 442, 449, 459, 454

14. 8.2, 9.3, 10, 9.9, 8, 9.2, 8.7, 8, 8.2, 9, 9.9, 8.7, 8.5, 8.1, 8.8, 9.3

HOMEWORK HELP

For Exercises	See Examples
11–14, 19	1
15–17, 20	2
18, 21	3

Extra Practice
See pages 568, 597.

SHOPPING For Exercises 15–18, use the stem-and-leaf plot at the right. It shows the prices of portable CD players found on an Internet shopping site.

Stem	Leaf
7	0
8	2 5 5
9	9 9
10	0 0 2 5 6 8
11	0 0 5 5 5 9 9
12	5 7 7
13	
14	
15	0 0 0 5 11|5 = $115

15. How many CD players are represented on the stem-and-leaf plot?

16. What is the range of the prices?

17. Find the median and mode for these data.

18. Write a sentence or two to describe how the prices of the portable CD players are spread out.

78 Chapter 2 Statistics: Analyzing Data

BASKETBALL For Exercises 19–21, use the table at the right. It shows the points scored in each post-season tournament game by the Indiana Hoosiers men's basketball teams from 1990 to 2003.

Points Scored in Post-Season Games

63	79	82	65	94	52
106	78	97	73	82	67
67	68	60	51	62	52
85	84	68	61	89	
108	77	94	57	73	
75	76	74	81	73	

Source: www.iuhoosiers.com

19. Make a stem-and leaf-plot of the data.
20. Find the mean, median, mode, and range of the data.
21. Write a sentence or two to describe how the points shown in the stem-and-leaf plot are spread out.

22. **RESEARCH** Use the Internet or another source to find an interesting set of data that can be displayed using a stem-and-leaf plot. Then make a stem-and-leaf plot of the data.

23. **CRITICAL THINKING** Make a frequency table, a line plot, and a stem-and-leaf plot of the data at the right. Describe the similarities and differences among the representations and state which of the three you prefer to make and use. Explain your reasoning.

Price of Jeans ($)

40	45	38	30
35	32	33	24
26	36	56	36
26	38	49	34
28	40	40	35

Spiral Review with Standardized Test Practice

24. **MULTIPLE CHOICE** The table shows how many minutes the students in Mr. Davis' class spent doing their homework one evening. Choose the correct stems for a stem-and-leaf plot of the data.

Ⓐ 1, 2, 3, 4, 5, 6, 7
Ⓑ 0, 1, 2, 3, 4, 5, 6, 7
Ⓒ 0, 2, 3, 4, 5, 7
Ⓓ 2, 3, 4, 5, 6, 7

Homework Time (min)

42	5	75	30	55
45	47	0	24	75
45	51	56	23	45
39	30	49	58	35

25. **GRID IN** Determine the mode of the data shown in the stem-and-leaf plot at the right.

Stem	Leaf
7	2 3 5 5
8	0 0 0 1 2
9	6 7 7 7 7 7|2 = 72

Find the mean, median, and mode for each set of data. (Lesson 2-4)

26. 80, 23, 55, 58, 45, 32, 40, 55, 50
27. 3.6, 2.4, 3.0, 7.9, 7.8, 2.4, 3.6, 3.9

28. Make a line plot of the following test scores. (Lesson 2-3)
83, 94, 78, 78, 85, 86, 88, 83, 82, 92, 90, 77, 83,
81, 89, 90, 88, 87, 88, 85, 84, 81, 83, 85, 91

Evaluate each expression. (Lesson 1-3)

29. $18 + 9 \div 3$
30. $24 \div (6 + 2)$
31. $3(7 - 3) + 1$
32. $(3 + 5) \div 4 - 2$

GETTING READY FOR THE NEXT LESSON

PREREQUISITE SKILL Order each set of decimals from least to greatest. (Page 556)

33. 6, 8.4, 8, 7.9
34. 4.0, 2.3, 4.5, 3.0, 3.8
35. 0.14, 0.33, 0.7, 0.09
36. 5.01, 5.8, 4.99, 5.64, 4.25

msmath2.net/self_check_quiz   Lesson 2-5 Stem-and-Leaf Plots **79**

2-6 Box-and-Whisker Plots

What You'll LEARN

Construct and interpret box-and-whisker plots.

NEW Vocabulary

lower quartile
upper quartile
box-and-whisker plot
lower extreme
upper extreme
interquartile range

REVIEW Vocabulary

outlier: a number separated from the rest of the data (Lesson 2-3)

WHEN am I ever going to use this?

NUTRITION The Calories per serving of items from the meat, poultry, and fish food group are shown in the table.

1. What is the median of the data?
2. Into how many parts does the median divide the data?

Nutrition Facts

Item	Calories	Item	Calories
bacon	110	ham	205
beef steak	240	pork chop	275
bologna	180	roast beef	135
crabmeat	135	salmon	120
fish sticks	70	sardines	175
fried shrimp	200	trout	175
ground beef	245	tuna	165

Source: *The World Almanac*

A median divides a data set into two parts. To divide the data into four parts, find the median of the lower half, called the **lower quartile (LQ)**, and the median of the upper half, called the **upper quartile (UQ)**. A **box-and-whisker plot** is a diagram that summarizes data by dividing it into four parts.

EXAMPLE Construct a Box-and-Whisker Plot

1 NUTRITION Make a box-and-whisker plot of the data in the table above.

Step 1 Order the data from least to greatest.

Step 2 Find the median and the quartiles.

Step 3 Draw a number line. The scale should include the median, the quartiles, and the least and greatest values, which are called the **lower extreme** and the **upper extreme**, respectively. Graph the values as points above the line.

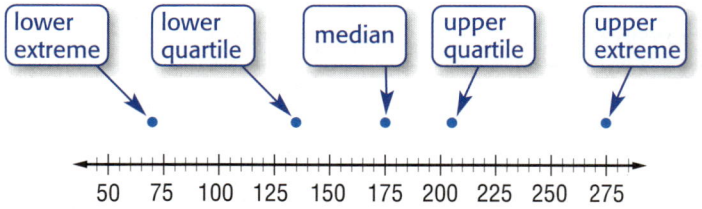

80 Chapter 2 Statistics: Analyzing Data

Step 4 Draw the box and whiskers.

The box contains the LQ and UQ. Draw a line through the median.

The whiskers extend from the quartiles to the extreme values.

Each of the four parts of a box-and-whisker plot contain one fourth of the data.

STUDY TIP

Length of Parts The parts of a box-and-whisker plot are different lengths because of how the data are clustered or spread out.
- short part: data are clustered together
- long part: data are spread out

EXAMPLE   Analyze Data

2 ARCHITECTURE The table shows the heights of the ten tallest buildings in St. Louis, Missouri. Make a box-and-whisker plot of the data. Then use it to describe how the data are spread out.

Find the median, the quartiles, and the extremes. Then construct the plot.

median = $\frac{485 + 420}{2}$ or 452.5

LQ = 398   UQ = 557

lower extreme = 375

upper extreme = 593

Building	Height (ft)
Metropolitan Square Tower	593
One Bell Center	588
Thomas F. Eagleton Federal Courthouse	557
Mercantile Center Tower	540
Firstar Center	485
Boatmen's Plaza	420
Laclede Gas Building	400
SW Bell Telephone Building	398
Civil Courts Building	390
One City Center	375

Source: *The World Almanac*

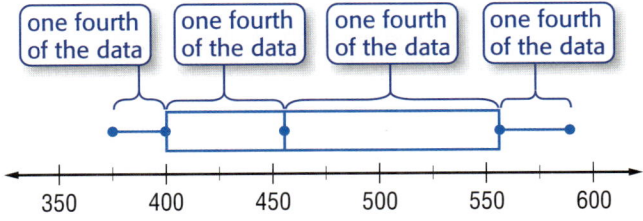

The graph shows that half of the tallest buildings are between 398 and 557 feet high. The largest range of the four quartiles is from 452.5 to 557 feet. One-fourth of the tallest buildings are within these heights.

READING Math

Quartiles A *quartile* can also refer to any of the four parts of a box-and-whisker plot. So, each quartile contains one-fourth, or 25%, of the data.

Another way to describe how data are spread out is to use the **interquartile range**, or the difference between the upper quartile and the lower quartile.

In Lesson 2-3, you learned that outliers are values separated from the rest of the data. In a box-and-whisker plot, outliers are data that are more than 1.5 times the interquartile range from the quartiles. They are not included in the whiskers of a box-and-whisker plot.

Lesson 2-6 Box-and-Whisker Plots **81**

EXAMPLE Identify and Plot Outliers

3 **PRESIDENTS** The table shows the ages of the last 21 presidents when they first took office. Make a box-and-whisker plot of the data.

Ages of Presidents

47	55	54	42	51	56	55
51	54	51	60	62	43	55
56	61	52	69	64	46	54

Find the median and the quartiles. Plot the points and draw a box.

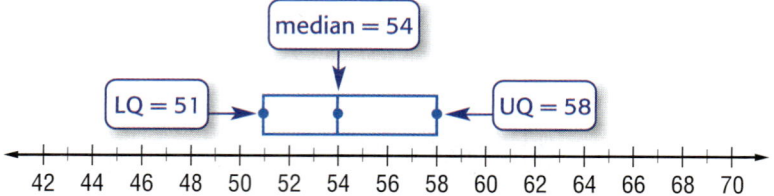

Next, determine whether there are any outliers.

interquartile range: UQ − LQ = 58 − 51 or 7

So, outliers are data more than 1.5(7) or 10.5 from the quartile.

lower limit: LQ − 10.5 = 51 − 10.5 or 40.5
upper limit: UQ + 10.5 = 58 + 10.5 or 68.5

40.5 and 68.5 are limits for the outliers.

Any data point that is less than 40.5 or greater than 68.5 is an outlier. So, 69 is an outlier. Plot the outlier using a dot. Then draw the lower whisker to the lower extreme, 42, and the upper whisker to the greatest value that is not an outlier, 64.

REAL-LIFE MATH

PRESIDENTS The state where the greatest number of presidents was born is Virginia, with 8 presidents.
Source: www.learningnetwork.com

George Washington   Woodrow Wilson

Skill and Concept Check

1. **Writing Math** Describe advantages of using a stem-and-leaf plot and advantages of using a box-and-whisker plot to display data.

2. **OPEN ENDED** Draw a box-and-whisker plot in which the data between the lower quartile and the median are clustered and the data between the median and the upper quartile are spread out.

GUIDED PRACTICE

For Exercises 3–6, refer to the data at the right. It gives the number of Calories in common snacks you can buy at the movies.

Movie Snack Calories

380	286	588
350	1,221	901
492	360	500
363	270	

3. Find the lower extreme, LQ, median, UQ, and upper extreme.
4. Are there any outliers? If so, identify them.
5. Draw a box-and-whisker plot of the data.
6. What is the largest range of the four quartiles? Explain what this means.

82 Chapter 2 Statistics: Analyzing Data

Practice and Applications

SPORTS For Exercises 7–9, refer to the table at the right. It shows the regular season games won by each men's professional basketball team in a recent season.

Number of Wins					
25	36	46	15	30	53
40	32	17	45	41	31
56	50	52	47	26	48
43	56	51	50	55	58
44	47	53	23	19	

HOMEWORK HELP

For Exercises	See Examples
7, 8, 12	1
9, 13	2
10, 11	3

Extra Practice
See pages 568, 597.

7. Find the lower extreme, LQ, median, UQ, and upper extreme.

8. Draw a box-and-whisker plot of the data.

9. What fraction of the data are between 45 and 51.5?

BRIDGES For Exercises 10–13, refer to table at the right. It shows the lengths of 22 of the longest suspension bridges in North America.

Length of Bridges (ft)			
4,260	2,310	1,800	1,595
4,200	2,300	1,750	1,550
3,800	2,190	1,632	1,500
3,500	2,150	1,600	1,500
2,800	2,000	1,600	
2,800	1,850	1,600	

10. Determine the interquartile range.

11. What are the limits on outliers? Are there any outliers?

12. Draw a box-and-whisker plot of the data.

13. What fraction of the bridges are between 1,600 and 2,800 feet long?

14. Compare and contrast the data represented in the box-and-whisker plots at the right.

15. **CRITICAL THINKING** Describe a set of data in which there is only one whisker in its box-and-whisker plot.

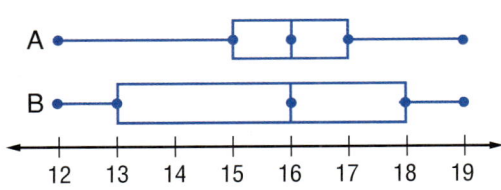

Spiral Review with Standardized Test Practice

16. **MULTIPLE CHOICE** What is the lower quartile of the data in the box-and-whisker plot at the right?

   Ⓐ 60   Ⓑ 50   Ⓒ 30   Ⓓ 20

17. **GRID IN** Find the interquartile range of the following set of data.
25, 36, 46, 15, 30, 53, 40, 32, 17, 45, 41, 31, 56, 50, 52, 47, 26

RAINFALL For Exercises 18 and 19, refer to the table at the right.

18. Make a stem-and-leaf plot of these data. (Lesson 2-5)

19. Find the mean, median, and mode. Round to the nearest tenth if necessary.
(Lesson 2-4)

Days of Rain per Year		
17	26	29
30	32	37
42	36	35

GETTING READY FOR THE NEXT LESSON

PREREQUISITE SKILL Choose an appropriate interval and scale to make a frequency table for each set of data. (Lesson 2-1)

20. 9, 0, 18, 19, 2, 9, 8, 13, 4

21. 30, 20, 60, 80, 90, 110, 40

msmath2.net/self_check_quiz

Graphing Calculator Investigation
A Follow-Up of Lesson 2-6

What You'll LEARN

Use a graphing calculator to make box-and-whisker plots.

Box-and-Whisker Plots

You can create a box-and-whisker plot using a TI-83/84 Plus graphing calculator.

Make a box-and-whisker plot of the data at the right. It shows the grades on Miss Romero's last Math 7 test.

Miss Romero's Math 7 Test							
78	94	85	92	72	56	89	92
90	84	98	82	75	100	94	87
92	85	94	70	78	95	70	80

STEP 1 **Enter the data.**
Begin by clearing any existing data in the first list, **L1**.
Press [STAT] [ENTER] [▲] [CLEAR] [ENTER].

Then enter the data into **L1**. Input each number and press [ENTER].

STEP 2 **Choose the type of graph.**
Press [2nd] [STAT PLOT] [ENTER] to choose the first plot. Highlight On, the modified box-and-whisker plot for the type, **L1** for the **Xlist**, and 1 as the frequency.

STEP 3 **Choose the display window.**
Press [WINDOW] and choose appropriate range settings for the *x* values. The window 50 to 110 with a scale of 5 includes all of this data.

STEP 4 **Display the graph.**
Press [GRAPH]. Press [TRACE] and the arrow keys to determine the five key data points of your graph.

EXERCISES

1. What are the values of the five key data points of the graph? What do they represent?

2. What percent of the test scores are below 78?

3. What percent of the test scores are above the median? What percent of the test scores are below the median?

4. What percent of the scores are between 56 and 86?

5. Suppose you earned a grade of 80. Describe what percent of students scored higher and what percent scored lower than you.

Bar Graphs and Histograms

What You'll LEARN

Construct and interpret bar graphs and histograms.

NEW Vocabulary

bar graph
histogram

REVIEW Vocabulary

frequency table: a table that shows how often an item occurs (Lesson 2-1)

WHEN am I ever going to use this?

ANIMALS The cheetah is the fastest known land animal. The table shows its fastest speed and the top speeds of four other land animals.

1. What are the fastest and slowest speeds recorded in the table?
2. How can you create a visual representation to summarize the data?
3. Do any of these representations show both the animal name and its speed?

Animal	Speed (mph)
cheetah	70
wildebeest	50
lion	50
elk	45
zebra	40

Source: The World Almanac

A **bar graph** is one method of comparing data by using solid bars to represent quantities.

EXAMPLE Construct a Bar Graph

1 Make a bar graph to display the data in the table above.

Step 1 Draw a horizontal axis and a vertical axis. Label the axes as shown. In this case, the scale on the vertical axis is chosen so that it includes all the speeds. Add a title.

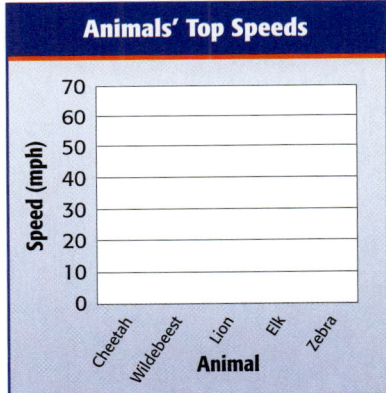

Step 2 Draw a bar to represent each category. In this case, a bar is used to represent the speed of each animal.

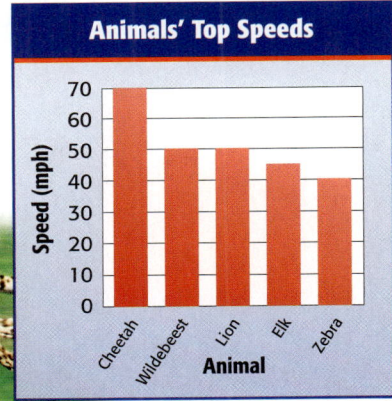

Lesson 2-7 Bar Graphs and Histograms **85**

A special kind of bar graph, called a **histogram**, uses bars to represent the frequency of numerical data that have been organized in intervals.

EXAMPLE — Construct a Histogram

2 ANIMALS The top speeds of thirty different land animals have been organized into a frequency table. Make a histogram of the data.

Maximum Speed (mph)	Frequency
11–20	5
21–30	5
31–40	10
41–50	8
51–60	0
61–70	2

Source: *The World Almanac*

Step 1 Draw and label horizontal and vertical axes. Add a title.

Step 2 Draw a bar to represent the frequency of each interval.

> **STUDY TIP**
>
> **Histograms** Because the intervals are equal, all of the bars have the same width, with no space between them. The space at 51–60 indicates that there are no data values on that interval.

The two highest bars represent a majority of the data. From the graph, you can easily see that most of the animals have a top speed of 31–50 miles per hour.

EXAMPLE — Compare Bar Graphs and Histograms

3 TEXTBOOKS Refer to the graphs below.

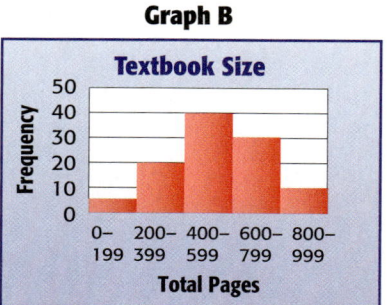

86 Chapter 2 Statistics: Analyzing Data

Which graph would you use to determine how many textbooks have less than 600 pages?

Graph B

Which graph would you use to compare the size of a science book and a history book?

Graph A

EXAMPLE   Use Graphs to Solve a Problem

Standardized Test Practice

Test-Taking Tip

Read a Graph
When analyzing data in a histogram, ask yourself questions such as, "What does each bar represent?"

4 **MULTIPLE-CHOICE TEST ITEM** Which conclusion *cannot* be made about the data in Graph B in Example 3?

- **A** There are 105 books in the data set.
- **B** Twenty books have between 200 and 399 pages.
- **C** Health books have the fewest pages.
- **D** The interval with the fewest books is 0–199.

Read the Test Item
Determine which of the four statements is *not* correct.

Solve the Test Item
A is correct; 5 books have 0–199 pages, 20 books have 200–399 pages, and so on, for a total of 105 books.

B is correct; the bar representing 200–399 has a value of 20.

C is incorrect; the graph does not describe individual books.

D is correct; only 5 books have 0 to 199 pages, which is less than the other intervals. The answer is C.

Skill and Concept Check

1. **Describe** how to determine the number of values in a data set using a histogram.

2. **Writing Math** Can any data set be displayed using a histogram? If yes, explain why. If no, give a counterexample and explain why not.

GUIDED PRACTICE

For Exercises 3–6, use the table at the right.

3. Describe the data in the table.
4. Make a bar graph of the data.
5. What do the horizontal and vertical axes represent?
6. Which country won the most titles? the least titles?

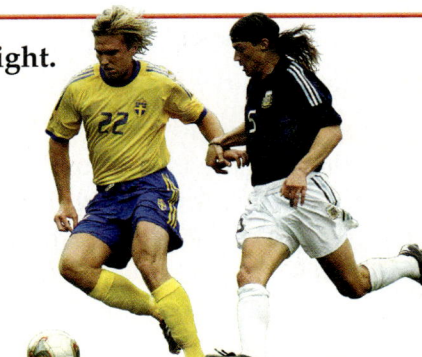

Men's World Cup Soccer Titles, 1930–2002	
Country	**Titles Won**
Uruguay	2
Italy	3
Germany	3
Brazil	5
England	1
Argentina	2
France	1

Source: fifaworldcup.yahoo.com/en

Practice and Applications

For Exercises 7–10, use the table below.

7. Describe the data in the table.
8. What does each interval represent?
9. Make a histogram of the data.
10. Describe general patterns in the histogram.

Candies in Snack-Size Packages	
Number of Candies	Frequency
9–11	8
12–14	15
15–17	24
18–20	3

HOMEWORK HELP

For Exercises	See Examples
7–9	1, 2
10–16	3

Extra Practice
See pages 569, 597.

TECHNOLOGY For Exercises 11–13, use the bar graph below.

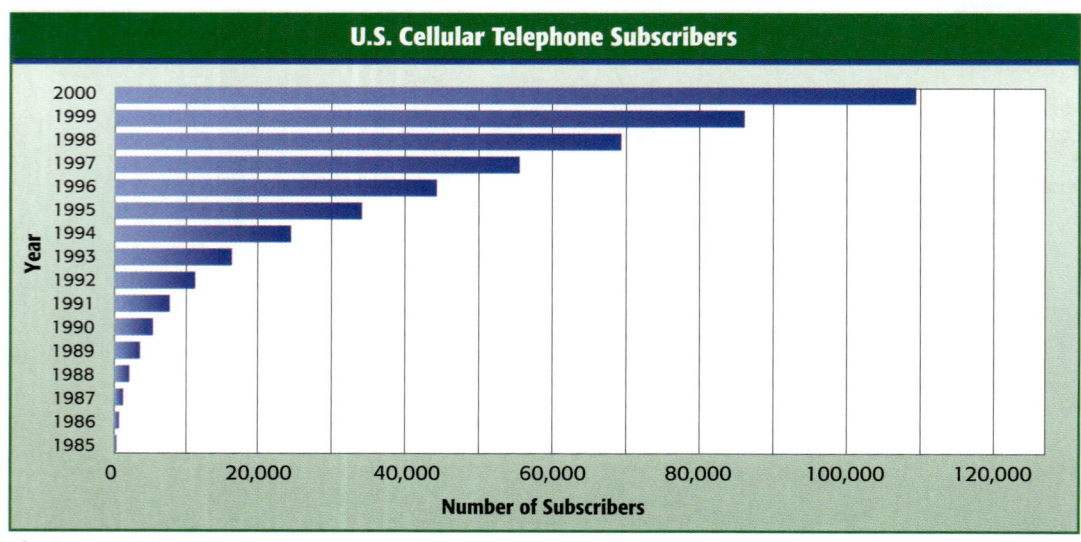

Source: *The World Almanac*

11. How is this bar graph different from the other bar graphs in the lesson?
12. Describe how the number of cell phone subscribers grew over the 15-year period.
13. Use the graph to predict the number of cell phone subscribers in 2006.

CIVICS For Exercises 14–16, use the histogram at the right and the information below.

Congress is made up of senators and representatives. Each state has 2 senators, and the number of representatives per state is determined by the state's population.

14. What does each interval represent?
15. Describe general patterns in the histogram.
16. There are 435 members in the House of Representatives. Determine the mean number of representatives per state. Compare this value to the value of the median number, according to the histogram. Discuss the significance of 38 states having 10 or fewer representatives.

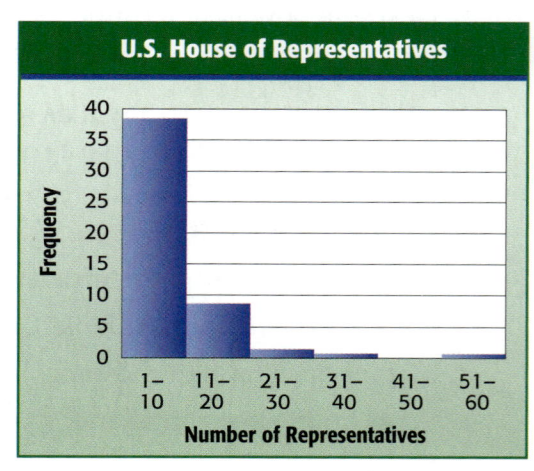

88 Chapter 2 Statistics: Analyzing Data

17. **CRITICAL THINKING** The histograms show players' salaries to the nearest million dollars for two major league baseball teams. Compare the salary distributions of the two teams.

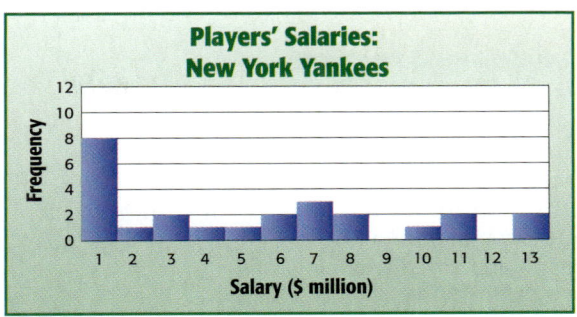

EXTENDING THE LESSON

A *multiple-bar graph* shows two or more sets of data on the same graph. The multiple-bar graph compares the annual costs of using old model appliances and using equivalent new model appliances.

18. For which appliance is the difference in electricity costs between the old and new model the greatest? Explain.

19. Describe an advantage of using a multiple-bar graph rather than two separate graphs to compare data.

Source: Association of Home Appliance Manufacturers

Spiral Review with Standardized Test Practice

20. **MULTIPLE CHOICE** The histogram at the right shows the number of bordering states for each of the fifty states. Which is *not* a true statement?

    Ⓐ The mode of the data is 4.

    Ⓑ The greatest number of bordering states is 8.

    Ⓒ Ten states have five bordering states.

    Ⓓ No states have only 1 bordering state.

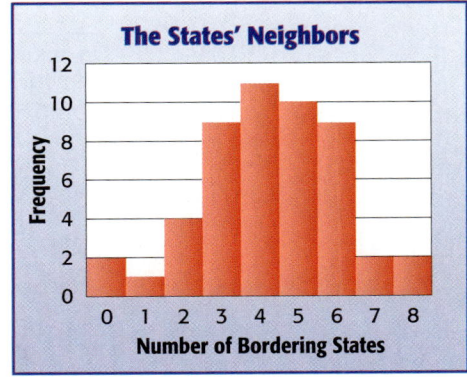

21. **SHORT RESPONSE** Which type of graph can be drawn using the data in a frequency table?

For Exercises 22 and 23, use the data 12, 17, 23, 5, 9, 25, 13, 16, 2, 25, 31, and 10.

22. Make a box-and-whisker plot of the data. (Lesson 2-6)

23. Make a stem-and-leaf plot of the data. (Lesson 2-5)

GETTING READY FOR THE NEXT LESSON

PREREQUISITE SKILL Determine whether each statement is *true* or *false*. (Lesson 2-2)

24. The vertical scale on a line graph must have equal intervals.

25. You do not need to label the axes of a line graph.

msmath2.net/self_check_quiz

2-7b Spreadsheet Investigation
A Follow-Up of Lesson 2-2

Multiple-Line and Multiple-Bar Graphs

What You'll LEARN

Use a spreadsheet to make a multiple-line graph and a multiple-bar graph.

In Lessons 2-2 and 2-7, you interpreted data in a multiple-line graph and in a multiple-bar graph, respectively. You can use a Microsoft® Excel® spreadsheet to make these two types of graphs.

ACTIVITY 1

The stopping distances on dry pavement and on wet pavement are shown in the table at the right.

Speed (mph)	Stopping Distance (ft)	
	Dry Pavement	Wet Pavement
50	200	250
60	271	333
70	342	430
80	422	532

Source: Continental Teves

Set up a spreadsheet like the one shown below.

In column A, enter the driving speeds.

In columns B and C enter the stopping distances on dry and wet pavement, respectively.

The next step is to "tell" the spreadsheet to make a double-line graph for the data.

1. Highlight the data in columns B and C, from B2 through C6.
2. Click on the Chart Wizard icon.
3. Choose the line graph and click Next.

   This tells the spreadsheet to read the data in columns B and C.

4. To set the *x*-axis, choose the Series tab and press the icon next to the Category (X) axis labels.
5. On the spreadsheet, highlight the data in column A, from A3 through A6.
6. Press the icon on the bottom of the Chart Wizard box to automatically paste the information.
7. Click Next and enter the chart title and labels for the *x*- and *y*-axes.
8. Click Next and then Finish.

90 Chapter 2 Statistics: Analyzing Data

ACTIVITY 2

Use the same data to make a multiple-bar graph. Set up the spreadsheet shown in Activity 1. The next step is to "tell" the spreadsheet to make a bar graph for the data.

- Highlight the data in columns B and C, from B2 through C6.
- Click on the Chart Wizard icon.
- Click on Column and Next to choose the vertical bar graph.
- Complete steps 4–8 from Activity 1.

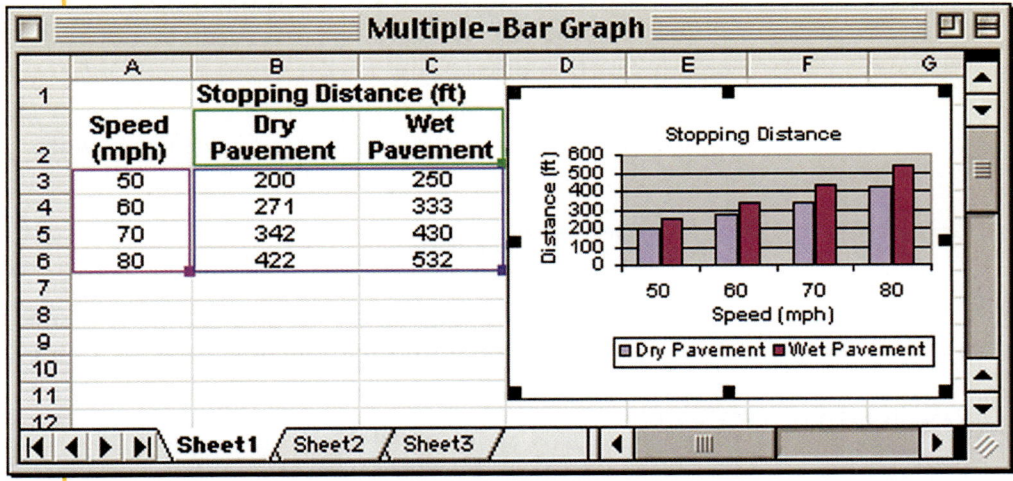

EXERCISES

1. **Explain** the steps you would take to make a multiple-line graph of the stopping distances that include the speeds 55, 65, and 75.

2. **OPEN ENDED** Collect some data that can be displayed in a multiple-line graph or a multiple-bar graph. Record the data in a spreadsheet. Then make a graph to display the data.

3. **Discuss** the advantages and disadvantages of using a spreadsheet to make a multiple-bar graph.

Lesson 2-7b Spreadsheet Investigation: Multiple-Line and Multiple-Bar Graphs **91**

2-8 Misleading Statistics

What You'll LEARN

Recognize when statistics and graphs are misleading.

WHEN am I ever going to use this?

TRANSPORTATION A graph like the one at the right appeared in a brochure describing various modes of transportation.

1. Approximately how many more passengers per lane can a 40-foot bus transport in an hour than a car can transport?
2. Is the bus on the graph about twice as large as the car? Explain.
3. Do you think the graph appeared in a brochure for a train/bus transit company or for a car company? What makes you think so?

The graph above shows that a car can transport about 5,000 people per lane, per hour. A train can transport about 4 times that many. However, the train on the graph is much more than 4 times larger than the car. For this reason, the graph could be considered misleading.

Other ways that graphs could be misleading are if there is no title, there are no labels on the axes or scales, or the intervals on a scale are not equal.

EXAMPLE Misleading Graphs

1 BUSINESS The line graphs below show monthly profits of a company from October to March. Which graph is misleading? Explain.

Graph B is misleading because the vertical scale has intervals of both $500 and $100. The increase in profits are exaggerated.

92 Chapter 2 Statistics: Analyzing Data

Statistical averages used to describe data may also be misleading. For example, whenever there are outliers, the mean may not be a good way to describe a data set because it is affected the most by outliers.

EXAMPLE Misleading Statistics

2. SALARY ABC Corporation claims the average salary for its employees is more than $60,000, while the average salary at XYZ Incorporated is only $25,000. Use the table to explain their reasoning and determine where you would prefer to work.

Position	Salary ($)	
	ABC Corp.	XYZ Inc.
President	500,000	120,000
1st Vice President	400,000	85,000
2nd Vice President	240,000	75,000
Sales Staff (5)	20,000	40,000
Supporting Staff (8)	15,000	25,000
Catalog Staff (7)	9,000	22,500

	ABC Corp.	XYZ Inc.
mean salary	$61,870	$36,413
median salary	$15,000	$25,000

ABC Corporation used the mean to represent its average salary. They used the median to represent the average salary at XYZ Inc.

Unless you could be a president or vice president, it would be better to work for XYZ Inc. because its 20 lowest-paid employees are all better paid than any of the 20 lowest-paid employees at ABC Corporation.

Skill and Concept Check

1. **Writing Math** Describe two ways in which the presentation of data can be misleading.
2. **OPEN ENDED** Find an example of a graph in a newspaper or magazine that could be considered misleading.

GUIDED PRACTICE

For Exercises 3–6, use the graph at the right. It shows the prices for natural gas charged by an Illinois natural gas supplier.

3. What trend seems to be revealed in the line graph?
4. What problems might there be in using this graph to predict future prices of natural gas?
5. Which measure of central tendency might the gas supplier use to advertise its prices? Explain.
6. Which measure of central tendency could competitors use to discourage customers from choosing that natural gas supplier? Explain.

Practice and Applications

7. **ENTERTAINMENT** Both bar graphs show the percent of viewers that watch network television. Which graph makes it appear that the number of viewers is decreasing rapidly? Explain.

HOMEWORK HELP

For Exercises	See Examples
7–8	1
9–13	2

Extra Practice
See pages 569, 597.

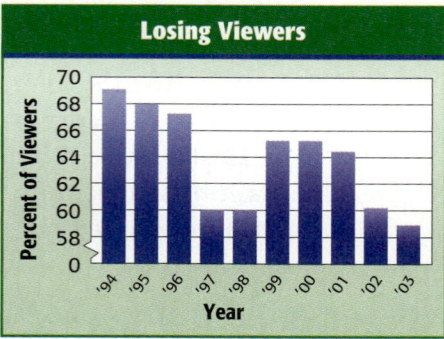

8. **UTILITIES** The line graph shows the monthly electric bill for the condominium that Toshiko is interested in renting. Why is the graph misleading?

GEOGRAPHY For Exercises 9–11, use the table at the right.

9. Find the mean, mode, and median of the data.
10. Which measure of central tendency would be misleading in describing the size of the islands? Explain.
11. Which measure would most accurately describe the data?

Caribbean Islands			
Island	Area (sq mi)	Island	Area (sq mi)
Antigua	108	Martinique	425
Aruba	75	Puerto Rico	3,339
Barbados	166	Tobago	116
Curacao	171	Virgin Islands, UL	59
Dominica	290	Virgin Islands U.S.	134

Source: U.S. Census Bureau

TESTS For Exercises 12 and 13, use Miguel's test scores shown in the line plot.

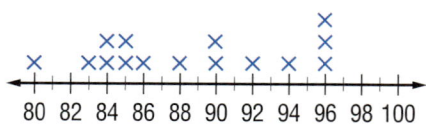

12. Which measure of central tendency might he use to emphasize the high scores?
13. Would this number be misleading? Explain.

14. **CRITICAL THINKING** Do values that are much greater or much less than the other values affect the median of a set of data? Give an example to support your answer.

Spiral Review with Standardized Test Practice

15. **MULTIPLE CHOICE** The table below shows the population of ten counties in Arizona from the 2000 Census. Which measure of central tendency, if any, is misleading in representing the data?

County	Population	County	Population
Apache	69,423	La Paz	19,715
Cochise	117,755	Maricopa	3,072,149
Coconino	116,320	Mohave	155,032
Gila	51,335	Navajo	97,470
Graham	33,489	Santa Cruz	38,381

    Source: U.S. Census Bureau

    Ⓐ mode   Ⓑ mean   Ⓒ median   Ⓓ none of them

16. **SHORT RESPONSE** Describe one way that a graph could be misleading.

17. **SKATING** Describe what the height of the fourth bar in the graph represents.
    (Lesson 2-7)

18. Make a box-and-whisker plot of the following data. 12, 15, 20, 12, 13, 14, 15, 14, 17, 16, 16, 18, 15 (Lesson 2-6)

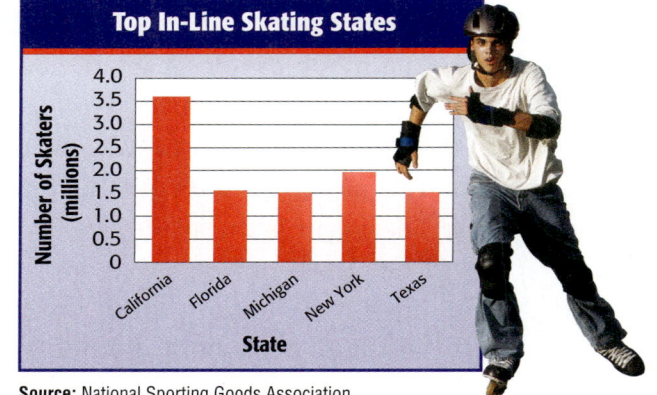

MOVIES For Exercises 19 and 20, use the movie ticket prices below.

$7.50, $6.50, $7.50, $8.00, $8.50, $8.00, $7.00

19. Find the mean, median, and mode. Round to the nearest tenth if necessary. (Lesson 2-4)

20. Find the range of the data. (Lesson 2-3)

Complete. (Lesson 1-8)

21. 8.5 g = __?__ mg
22. 19 cm = __?__ mm
23. 643 mg = __?__ g
24. 5,375 mL = __?__ L

INTERDISCIPLINARY PROJECT

Lions, Tigers, and Bears

Math and Science It's time to complete your project. Use the information and data you have gathered about the animals you selected to prepare a video or poster. Be sure to include a spreadsheet and all the necessary graphs with your project.

WebQuest   msmath2.net/webquest

CHAPTER 2 Study Guide and Review

Vocabulary and Concept Check

bar graph (p. 85)
box-and-whisker plot (p. 80)
cluster (p. 65)
data (p. 54)
frequency table (p. 54)
histogram (p. 86)
interquartile range (p. 81)
interval (p. 54)
leaf (p. 76)
line graph (p. 60)
line plot (p. 64)
lower extreme (p. 80)
lower quartile (p. 80)
mean (p. 69)
measures of central tendency (p. 69)
median (p. 70)
mode (p. 70)
outlier (p. 65)
range (p. 65)
scale (p. 54)
scatter plot (p. 61)
statistics (p. 54)
stem (p. 76)
stem-and-leaf plot (p. 76)
upper extreme (p. 80)
upper quartile (p. 80)

State whether each sentence is *true* or *false*. If *false*, replace the underlined word to make a true sentence.

1. The <u>range</u> is the difference between the greatest and the least values in a set of data.
2. The <u>mode</u> divides a set of data in half.
3. A graph that uses bars to make comparisons is a <u>bar graph</u>.
4. A <u>scatter plot</u> shows two sets of related data on the same graph.
5. The <u>median</u> is a data value that is quite separated from the rest of the data.
6. The <u>mean</u> is the arithmetic average of a set of data.
7. The number or item that appears most often in a set of data is the <u>mode</u>.
8. The <u>lower quartile</u> is the middle number in a set of data when the data are arranged in numerical order.

Lesson-by-Lesson Exercises and Examples

2-1 Frequency Tables (pp. 54–57)

Make a frequency table of each set of data.

9. **Hours of Sleep Per Night**

9	6	7	8	9	5	8	7
6	7	6	7	7	8	9	8

10. **Favorite Sport**

B	F	H	S	K	F	S	B	B	F
B	F	K	S	F	F	B	K	F	S

B = baseball, F = football, H = hockey,
S = soccer, K = basketball

Example 1 Make a frequency table of the data at the right.

Age of Pet (yr)

4	2	2	3	8	9
5	9	8	7	10	4

interval: 4
scale: 0–11

Age (yr)	Tally	Frequency
0–3	III	3
4–7	IIII	4
8–11	HHT	5

96 Chapter 2 Statistics: Analyzing Data

msmath2.net/vocabulary_review

2-2 Making Predictions (pp. 60–63)

11. **ENTERTAINMENT** The graph shows the attendance at a theme park. If the trend continues, predict the attendence in 2005.

Example 2

SPORTS The graph shows the 400-meter winning times in the Summer Olympics. Predict the winning time in 2004.

By extending the graph, it appears that the winning time in 2004 will be about 42.1 seconds.

2-3 Line Plots (pp. 64–68)

Make a line plot for each set of data. Identify any clusters, gaps, or outliers.

12. 10, 12, 10, 8, 13, 10, 8, 12
13. 7, 8, 8, 9, 14, 9, 8, 7
14. 43, 41, 42, 45, 43, 42, 43, 46, 44, 44
15. 21, 23, 19, 21, 22, 28, 20, 22, 23, 19

Example 3 Make a line plot for the data set 72, 75, 73, 72, 74, 68, 73, 74, 74, 75, and 73. Identify any clusters, gaps, or outliers.

```
                    ×  ×
          ×      ×  ×  ×
          ×  ×  ×  ×  ×
 ×        ×  ×  ×  ×  ×
─┼──┼──┼──┼──┼──┼──┼──┼→
 68 69 70 71 72 73 74 75
```

cluster: 72–75, gap: 68–72, outlier: 68

2-4 Mean, Median, and Mode (pp. 69–72)

Find the mean, median, and mode for each set of data.

16. 2, 3, 4, 3, 4, 3, 8, 7, 2
17. 31, 24, 26, 18, 23, 31, 18
18. 89, 76, 93, 100, 72, 86, 74
19. 125, 108, 172, 136, 121, 112, 148, 136

Example 4 Find the mean, median, and mode for the data set 23, 22, 19, 19, and 20.

mean: $\frac{23 + 22 + 19 + 19 + 20}{5}$ or 20.6

median: 20   middle value of the ordered set

mode: 19   value that occurs most often

2-5 Stem-and-Leaf Plots (pp. 76–79)

Make a stem-and-leaf plot for each set of data.

20. 29, 54, 31, 26, 38, 46, 23, 21, 32, 37
21. 75, 83, 78, 85, 87, 92, 78, 53, 87, 89, 91
22. 9, 5, 12, 21, 18, 7, 16, 24, 11, 10, 3, 14
23. 234, 218, 229, 204, 221, 219, 201, 225

Example 5 Make a stem-and-leaf plot for the data set 12, 15, 17, 20, 22, 22, 23, 25, 27, and 45.

The tens digits form the stems, and the ones digits form the leaves.

Stem	Leaf
1	2 5 7
2	0 2 2 3 5 7
3	
4	5

$2 | 3 = 23$

Chapter 2 Study Guide and Review **97**

Study Guide and Review continued

Mixed Problem Solving
For mixed problem-solving practice, see page 597.

2-6 Box-and-Whisker Plots (pp. 80–83)

GRADES For Exercises 24–27, use the following math test scores.

83, 92, 75, 59, 94, 82, 88, 79, 92, 90, 85, 96

24. Find the lower extreme, LQ, median, UQ, and upper extreme.
25. Determine the interquartile range.
26. Identify any outliers.
27. Draw a box-and-whisker plot of the data.

Example 6 Make a box-and-whisker plot of the data set 12, 12, 13, 14, 14, 15, 15, 15, 16, 16, 17, 18, and 20.

lower extreme = 12   upper extreme = 20
median = 15   LQ = 13.5   UQ = 16.5

2-7 Bar Graphs and Histograms (pp. 85–89)

For Exercises 28 and 29, refer to the graph at the right.

28. How many students are 50 to 59 inches tall?
29. How many students are at least 60 inches tall?

Example 7 The graph shows the heights to the nearest inch of students in a classroom. Is this a bar graph or a histogram? Explain.

This is a histogram because it shows the number of students having the heights described by each interval.

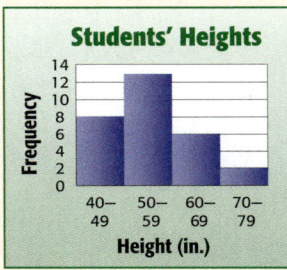

2-8 Misleading Statistics (pp. 92–95)

30. **SALES** The graph below shows the monthly CD sales at the Music Barn. Why is the graph misleading?

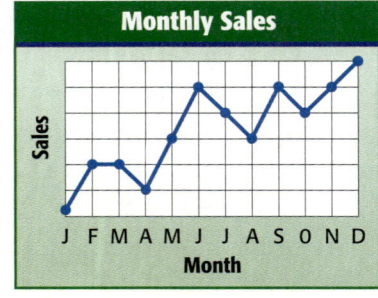

Example 8 The graph shows the pounds of aluminum cans that students recycled. Why is this graph misleading?

The vertical scale does not have equal intervals. It has intervals of 200 and 50.

98 Chapter 2 Statistics: Analyzing Data

Chapter 2 Practice Test

Vocabulary and Concepts

1. **Explain** how the range is different from the mean, median, and mode.
2. **Describe** how quartiles separate a set of data.

Skills and Applications

TESTS For Exercises 3–5, use the French test grades shown below.

95, 76, 82, 90, 83, 76, 79, 82, 95, 85, 93, 81, 63

3. Make a frequency table of the data.
4. Make a line plot of the data.
5. Identify any clusters, gaps, or outliers.

6. **EMPLOYMENT** The line graph shows the percent of women who had jobs outside the home from 1975 to 2000. Use the graph to predict the percent of women who will have jobs outside the home in 2005.

Find the mean, median, and mode for each set of data. Round to the nearest tenth if necessary.

7. 4, 6, 11, 7, 4, 11, 4
8. 12.4, 17.9, 16.5, 10.2

Make a stem-and-leaf plot for each set of data.

9. 37, 59, 26, 42, 57, 53, 31, 58
10. 461, 422, 430, 425, 425, 467, 429

ALLOWANCES For Exercises 11 and 12, use the monthly allowances below.

$25, $32, $40, $10, $18, $15, $30, $40, $25, $36, $20

11. Find the lower extreme, LQ, median, UQ, and upper extreme.
12. Make a box-and-whisker plot of the data.
13. **SALES** The line graph shows ship sales at Marvin's Marina in thousands of dollars. Why is the graph misleading?
14. Make a histogram for the French test grades given for Exercises 3–5.

Standardized Test Practice

15. **MULTIPLE CHOICE** Choose the statement that is *not* correct.
    - Ⓐ The bars on a histogram represent the frequency of data in intervals.
    - Ⓑ The bars on a bar graph represent the number of data items in each interval.
    - Ⓒ The bars on a bar graph have space between them.
    - Ⓓ There is no space between the bars on a histogram.

CHAPTER 2 Standardized Test Practice

PART 1 Multiple Choice

Record your answers on the answer sheet provided by your teacher or on a sheet of paper.

1. Which is greater than 6.8? (Prerequisite Skill, p. 556)
   - A. 6.08
   - B. 6.79
   - C. 6.80
   - D. 6.9

2. What is the value of the expression $5 + 2 \times 4^2 - 9$? (Lesson 1-3)
   - F. 4
   - G. 19
   - H. 28
   - I. 103

3. Which is 430,000 written in scientific notation? (Lesson 1-9)
   - A. 4.3×10^5
   - B. 4.3×10^6
   - C. 43×10^5
   - D. 43×10^6

4. The line graph shows a baby's length during the first 8 months. If the pattern of growth continues, predict the baby's length at 9 months old. (Lesson 2-2)

   - F. 26 in.
   - G. 27 in.
   - H. 28 in.
   - I. 29 in.

5. The line plot shows the number of minutes ten different students spent on homework last night. What is the range of the data? (Lesson 2-3)

   - A. 20
   - B. 25
   - C. 30
   - D. 50

6. The table shows the length of the phone calls made in one household on a given day. What is the median? (Lesson 2-4)
   - F. 2 min
   - G. 6 min
   - H. 7 min
   - I. 8 min

Length of Phone Calls (min)					
2	3	10	2	8	12
6	7	25	7	6	7

7. The stem-and-leaf plot shows the average monthly temperature in one city for a year. What is the mode? (Lesson 2-5)

Stem	Leaf
4	2 2 2 5
5	5 6
6	4
7	0 8 8
8	2 5

   $5 | 6 = 56°F$

   - A. 42
   - B. 55
   - C. 65
   - D. 78

8. The box-and-whisker plot below summarizes the ages of Leila's cousins. Which statement is *not* true? (Lesson 2-6)

   - F. One fourth of Leila's cousins are between 10 and 16 years old.
   - G. Half of Leila's cousins are between 16 and 21 years old.
   - H. The range of ages is 18 years.
   - I. Half of Leila's cousins are older than 21.

100 Chapter 2 Statistics: Analyzing Data

Preparing for Standardized Tests

For test-taking strategies and more practice, see pages 608–625.

PART 2 Short Response/Grid In

Record your answers on the answer sheet provided by your teacher or on a sheet of paper.

9. What is the value of $14 + x^2$ if $x = 3$? (Lesson 1-4)

10. Find the missing number in the sequence below. (Lesson 1-7)

    24.7, 25.0, __?__, 25.6, 25.9, …

11. The frequency table summarizes the number of miles Dino ran daily to train for a race. What interval shows the number of miles that he ran most frequently? (Lesson 2-1)

Number of Miles	Tally	Frequency
0–4	|||| ||||	9
5–9	|||| |||| |||	13
10–14	||||	4
15–19			1

12. If the number 18 was taken out of the number data set 7, 18, 46, 24, 2, 44, 31, 11, would the median or the mean change the most? (Lesson 2-4)

13. What is the upper extreme of the box-and-whisker plot below? (Lesson 2-6)

14. The graph below shows the sales of compact discs during a five-day sale. How many more compact discs were sold on Saturday than on Wednesday? (Lesson 2-7)

TEST-TAKING TIP

Question 15 Make sure you have answered every part of an open-ended question. If you find that you cannot answer the question entirely, do as much as you can, because you may earn partial credit.

PART 3 Extended Response

Record your answers on a sheet of paper. Show your work.

15. The table shows how the value of a painting increased over ten years. (Lesson 2-2)

Year	Value	Year	Value
1997	$350	2002	$1,851
1998	$650	2003	$2,151
1999	$950	2004	$2,451
2000	$1,250	2005	$2,752
2001	$1,551	2006	$3,052

    a. Make a line graph of the data.

    b. Use the graph to predict what the value of the painting will be in 2010.

16. Dominique had swimming practice before school one week and after school the following week. Her timed results are in the table below. (Lesson 2-7)

Day	Time Before School (s)	Time After School (s)
Monday	45.2	41.5
Tuesday	41.3	38.7
Wednesday	40.6	37.4
Thursday	42.9	35.2
Friday	40.7	38.9

    a. Make a double-line graph of the data. Let the x-axis represent the days and let the y-axis represent the times. Title the graph and label each part.

    b. Write a sentence comparing the data.

UNIT 2
Integers and Algebra

Chapter 3
Algebra: Integers

Chapter 4
Algebra: Linear Equations and Functions

Your study of math includes more than just whole numbers and decimals. In this unit, you will use negative numbers to describe many real-life situations and you will solve and graph equations that represent them.

INTERDISCIPLINARY PROJECT

The Wide World of Soccer

Math and Geography Soccer fans, get up on your feet! You've been selected by an elite committee to join us on a world-wide soccer adventure. Along the way, you'll be gathering data about the geography of countries where soccer is the favorite sport. You'll also make some predictions about the future of soccer in the United States. We will be leaving on our adventure very shortly, so pack your math tools and your thinking cap. This is one adventure you don't want to miss.

 Log on to **msmath2.net/webquest** to begin your WebQuest.

Unit 2 Integers and Algebra

CHAPTER 3
Algebra: Integers

"What does lightning have to do with math?"

Have you heard of the phrase "opposites attract"? During thunderstorms, negatively-charged electrons in the clouds are attracted to positively-charged protons on the ground. This opposite attraction causes lightning. In mathematics, you can use opposites to help you add and subtract positive and negative integers.

You will solve problems about thunderstorms in Lesson 3-4.

GETTING STARTED

▶ Diagnose Readiness

Take this quiz to see if you are ready to begin Chapter 3. Refer to the lesson or page number in parentheses for review.

Vocabulary Review

State whether each sentence is *true* or *false*. If *false*, replace the underlined word to make a true sentence.

1. The <u>mean</u> of 1, 3, and 6 is 3.
   (Lesson 2-4)

2. The difference between the greatest number and the least number in a set of data is called the <u>range</u>. (Lesson 2-3)

Prerequisite Skills

Replace each ● with < or > to make a true sentence. (Page 556)

3. 1,458 ● 1,548
4. 36 ● 34
5. 1.02 ● 1.20
6. 76.7 ● 77.6

Add.

7. 84 + 39
8. 198 + 289
9. 826 + 904
10. 3,068 + 5,294

Multiply.

11. 2 · 5 · 3
12. 18 · 9
13. 15 · 6
14. 10 · 4 · 7

Divide.

15. 63 ÷ 9
16. 96 ÷ 12
17. 125 ÷ 5
18. 187 ÷ 17

Find the mean and range for each set of data. (Lessons 2-3 and 2-4)

19. 12, 8, 25, 16, 9
20. 34, 57, 60, 45

Integers Make this Foldable to help you organize information about integers. Begin with two sheets of $8\frac{1}{2}'' \times 11''$ paper.

STEP 1 Fold and Cut One Sheet
Fold in half from top to bottom. Cut along fold from edges to margin.

STEP 2 Fold and Cut the Other Sheet
Fold in half from top to bottom. Cut along fold between margins.

STEP 3 Fold
Insert first sheet through second sheet and align folds.

STEP 4 Label
Label each page with a lesson number and title.

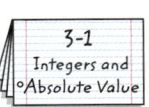

Chapter Notes Each time you find this logo throughout the chapter, use your *Noteables™: Interactive Study Notebook with Foldables™* or your own notebook to take notes. Begin your chapter notes with this Foldable activity.

Readiness To prepare yourself for this chapter with another quiz, visit **msmath2.net/chapter_readiness**

3-1 Integers and Absolute Value

What You'll LEARN

Read and write integers, and find the absolute value of an integer.

NEW Vocabulary

integer
graph
positive integer
negative integer
absolute value

MATH Symbols

+3   positive three
−3   negative three
|3| absolute value of three

WHEN am I ever going to use this?

FOOTBALL The graph shows the number of yards the Bears gained or lost on the first four downs. A value of −3 represents a 3-yard loss.

1. What does a value of −2 represent?
2. On which down did they lose the most yards?
3. How can you represent a gain of 9 yards?

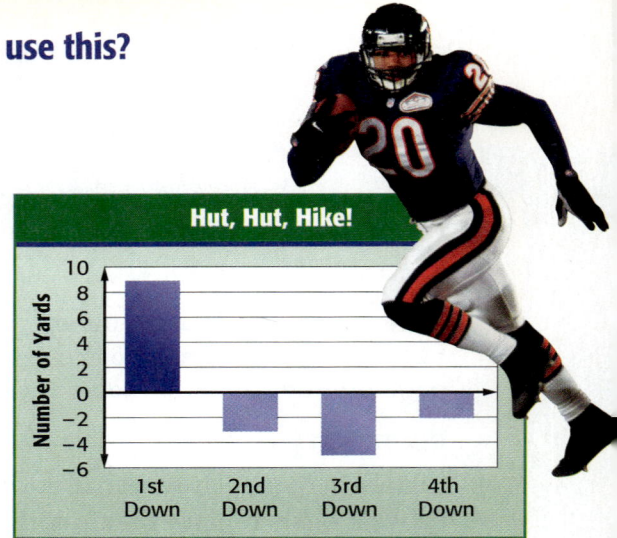

Numbers like 9 and −2 are called integers. An **integer** is any number from the set {…, −4, −3, −2, −1, 0, 1, 2, 3, 4, …}. Integers can be graphed on a number line. To **graph** a point on the number line, draw a point on the line at its location.

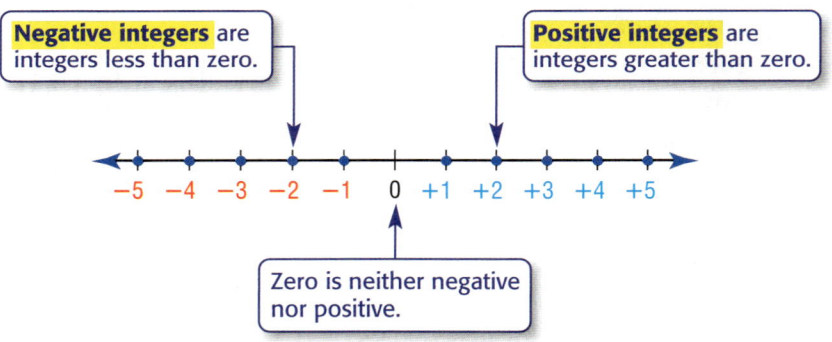

EXAMPLES Write Integers for Real-Life Situations

WEATHER Write an integer for each situation.

1 The average temperature in Tennessee for May was 5 degrees below normal.

Because it represents *below* normal, the integer is −5.

2 The average rainfall in Virginia for November was 5 inches above normal.

Because it represents *above* normal, the integer is +5 or 5.

Your Turn Write an integer for each situation.

a. 6 degrees above normal   b. 2 inches below normal

106 Chapter 3 Algebra: Integers

READING Math

Set Theory The number 5 is an *element*, or member, of the set of integers. The set {−5, 5} is a *subset* of the set of integers.

You can represent the integers from Examples 1 and 2 on a number line.

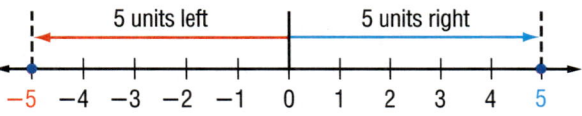

The numbers −5 and 5 are the same distance from 0, but on opposite sides of 0. So, −5 and 5 have the same **absolute value**.

Noteables™ — Key Concept: Absolute Value

Words The absolute value of an integer is the distance the number is from zero on a number line.

Examples $|6| = 6$ $|-6| = 6$

EXAMPLES Evaluate Expressions

Evaluate each expression.

3 $|-4|$

On the number line, the graph of −4 is 4 units from 0.
So, $|-4| = 4$.

4 $|-5| - |2|$

$|-5| - |2| = 5 - 2$ $|-5| = 5, |2| = 2$
$ = 3$ Subtract.

Skill and Concept Check

1. **Writing Math** Describe a situation in everyday life where negative numbers are used.

2. **OPEN ENDED** On a number line, graph two different points that have the same absolute value.

3. **Which One Doesn't Belong?** Identify the expression that does not have the same value as the other three. Explain your reasoning.

   $|-3|$ −3 $|3|$ 3

GUIDED PRACTICE

Write an integer for each situation.

4. 6°F below 0
5. a loss of 11 yards
6. a deposit of $16

Evaluate each expression.

7. $|7|$
8. $|-4|$
9. $|-7| - |1|$

10. **STOCK MARKET** The price of a company's stock fell 21 points in two days. Write an integer to represent the amount the stock price fell.

Practice and Applications

Write an integer for each situation.

11. a profit of $9
12. 53°C below 0
13. no gain on first down
14. an elevator goes up 12 floors
15. 2008 A.D.
16. 160 feet above sea level
17. a bank withdrawal of $50
18. 1000 B.C.

HOMEWORK HELP

For Exercises	See Examples
11–18, 29	1, 2
19–22, 27, 28	3
23–26	4

Extra Practice
See pages 569, 598.

Evaluate each expression.

19. $|6|$
20. $|-12|$
21. $|-9|$
22. $|21|$
23. $|12| - |-8|$
24. $|-10| - 5$
25. $|-9| + |5|$
26. $|26| + |-4|$

27. What is the absolute value of 0?
28. Find $|x|$ if $x = -6$.

29. **STATIC ELECTRICITY** Electrical charges are made up of positively-charged protons and negatively-charged electrons. Suppose you rub a balloon through your hair to make the balloon stick to a wall. There are 2 protons on the wall and 5 electrons on the balloon. Write an integer for each charge.

Graph each set of integers on a number line.

30. {0, 1, −3}
31. {−4, 5, 4}
32. {−5, −1, 10, −9}
33. {−2, −4, −6, −8}

34. **WEATHER** A meteorologist reports a 20° change in the temperature from yesterday to today. Describe what this could mean.

CRITICAL THINKING Determine whether each statement is *true* or *false*. If *false*, give a counterexample.

35. Every integer has an absolute value.
36. The absolute value of every integer is positive.

Spiral Review with Standardized Test Practice

37. **MULTIPLE CHOICE** Identify the point that represents −5.
    Ⓐ A   Ⓑ B   Ⓒ C   Ⓓ D

38. **SHORT RESPONSE** Write an integer for 23°F below 0.

STATISTICS For Exercises 39 and 40, use the following information.
The mean income for a group of accountants is $36,266.67. Their incomes are $27,500, $36,100, $29,800, $33,400, $31,300, and $59,500.

39. In what way is the mean misleading? (Lesson 2-8)
40. Draw a bar graph of the data. (Lesson 2-7)

GETTING READY FOR THE NEXT LESSON

PREREQUISITE SKILL Replace each with < or > to make a true sentence. (Page 556)

41. 16 ● 6
42. 2.3 ● 3.2
43. 101 ● 111
44. 87.3 ● 83.7
45. 1,051 ● 1,015

108 Chapter 3 Algebra: Integers

3-2 Comparing and Ordering Integers

What You'll LEARN
Compare and order integers.

REVIEW Vocabulary
median: the middle number in an ordered data set (Lesson 2-4)

MATH Symbols
< is less than
> is greater than

WHEN am I ever going to use this?

WEATHER The Wind Chill Temperature Index table shows how cold air feels on human skin.

1. What is the wind chill if there is a wind at 20 miles per hour and the temperature is 5°?
2. Which is colder, a temperature of 15° with a 20 mile-per-hour wind or a temperature of 10° with a 10 mile-per-hour wind?
3. Graph both wind chills found in Exercise 2 on a number line.

WIND CHILL

Wind (mph)	Temperature (°F)				
	15	10	5	0	−5
5	7	1	−5	−11	−16
10	3	−4	−10	−16	−22
15	0	−7	−13	−19	−26
20	−2	−9	−15	−22	−29

When two numbers are graphed on a number line, the number to the left is always less than the number to the right. The number to the right is always greater than the number to the left.

Noteables™ **Key Concept: Compare Integers**

Model
Words −4 is less than −2.   −2 is greater than −4.
Symbols −4 < −2   −2 > −4

The symbol points to the lesser number.

EXAMPLE Compare Integers

1 Replace the ● with < or > to make −5 ● −3 a true sentence.

Graph each integer on a number line.

Since −5 is to the left of −3, −5 < −3.

Your Turn Replace each ● with < or > to make a true sentence.

a. −8 ● −4   b. 5 ● −1   c. −10 ● −13

READING in the Content Area
For strategies in reading this lesson, visit **msmath2.net/reading**.

Integers are used to compare numbers in many real-life situations.

Standardized Test Practice

EXAMPLE Order Integers

2 **MULTIPLE-CHOICE TEST ITEM** The lowest temperatures in Alaska, Florida, Hawaii, and Montana are listed in the table. Order the temperatures from least to greatest.

- Ⓐ −80, −70, 12, −2
- Ⓑ −80, −70, −2, 12
- Ⓒ 12, −2, −70, −80
- Ⓓ −2, 12, −70, −80

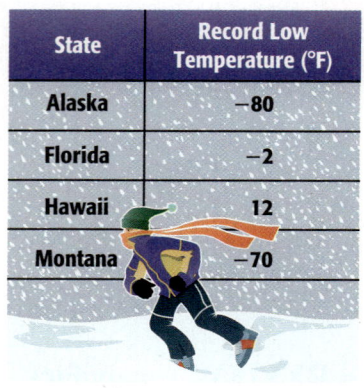

State	Record Low Temperature (°F)
Alaska	−80
Florida	−2
Hawaii	12
Montana	−70

Source: *The World Almanac and Book of Facts*

Test-Taking Tip

Eliminating Answer Choices
If you are unsure of the correct answer, eliminate the choices you know are incorrect. Then consider the remaining choices. You can eliminate choice C since the list begins with a positive number.

Read the Test Item To order the integers, graph them on a number line.

Solve the Test Item

```
     A  M              F  H
  +--+--+--+--+--+--+--+--+--+--+--+
 −90 −80 −70 −60 −50 −40 −30 −20 −10  0  10  20
```

Order the integers from least to greatest by reading from left to right: −80, −70, −2, 12. So, the answer is B.

Skill and Concept Check

1. **Draw** a number line to show that −5 is less than −1.

2. **OPEN ENDED** Write an integer that is less than −9. Explain.

3. **NUMBER SENSE** Complete the sentence: −7 is greater than −12 because −7 lies to the __?__ of −12 on a number line.

4. **Name** the greatest negative integer.

GUIDED PRACTICE

Replace each ● with < or > to make a true sentence.

5. −5 ● −6
6. −2 ● 8
7. 0 ● −10

8. Order 51, −63, 49, −24, −38, and 38 from least to greatest.

MONEY For Exercises 9 and 10, use the information below and at the right.
Marva is saving money for a new bike and has already saved $21. She begins a log to keep track of her money.

9. Write each entry as an integer.

10. Order the integers from least to greatest.

Money Log
—earned $15 raking leaves
—loaned $13 to a friend
—received allowance of $10
—spent $4 on snacks
—friend paid back $13 loan
—bought $5 lunch

110 Chapter 3 Algebra: Integers

Practice and Applications

Replace each ● with < or > to make a true sentence.

11. $-17 \bullet -20$
12. $-21 \bullet -12$
13. $3 \bullet -10$
14. $-5 \bullet 17$
15. $4 \bullet -4$
16. $-25 \bullet -20$
17. $-52 \bullet -72$
18. $100 \bullet -10$
19. $|-8| \bullet 0$
20. $-13 \bullet |-14|$
21. $|36| \bullet -37$
22. $|-29| \bullet |92|$

HOMEWORK HELP

For Exercises	See Examples
11–28	1
29, 31–32	2

Extra Practice
See pages 570, 598.

Determine whether each sentence is *true* or *false*. If *false*, change one number to make the sentence true.

23. $-8 > 5$
24. $-7 < 0$
25. $|-9| = 9$
26. $|5| < -6$
27. $10 > |-8|$
28. $7 > |-7|$

WEATHER For Exercises 29 and 30, use the information in the table. It shows the record low temperatures in Indianapolis, Indiana, for March 1–7 of a recent year.

29. Arrange the dates from the coldest temperature to the warmest.
30. Find the median temperature.
31. Order $-7, 5, -6, -4, 1,$ and 3 from least to greatest.
32. Order $|51|, -53, |-52|, 55, -56,$ and -57 from greatest to least.
33. **CRITICAL THINKING** If 0 is the greatest integer in a set of five integers, what can you conclude about the other four integers?

Day	Temperature (°F)
Mar. 1	−4
Mar. 2	−7
Mar. 3	1
Mar. 4	3
Mar. 5	−1
Mar. 6	−6
Mar. 7	6

Source: www.weather.com

Spiral Review with Standardized Test Practice

34. **MULTIPLE CHOICE** The table shows the inventions of several toys. Order the inventions from earliest to most recent.

    Ⓐ chess, yo-yo, teddy bear, checkers
    Ⓑ checkers, yo-yo, chess, teddy bear
    Ⓒ yo-yo, teddy bear, checkers, chess
    Ⓓ chess, teddy bear, yo-yo, checkers

Toy	Year
Yo-Yo	1000 B.C.
Teddy Bear	1902 A.D.
Chess	600 A.D.
Checkers	2000 B.C.

35. **GRID IN** Which is greater, -12 or 7?

Write an integer for each situation. (Lesson 3-1)

36. 9°C below 0
37. a gain of 20 feet

38. **PROFITS** The daily profits of T-shirts sold last week were $55, $35, $25, $30, and $55. Which average might be misleading: the mode, the median, or the mean? Explain. (Lesson 2-8)

GETTING READY FOR THE NEXT LESSON

PREREQUISITE SKILL Graph the solution of each equation on a number line. (Lesson 1-5)

39. $x + 3 = 5$
40. $x - 4 = 8$
41. $3x = 9$
42. $5x = 30$

msmath2.net/self_check_quiz

3-3 Geometry: The Coordinate Plane

What You'll LEARN

Graph points on a coordinate plane.

NEW Vocabulary

coordinate plane
coordinate grid
x-axis
y-axis
origin
ordered pair
x-coordinate
y-coordinate
quadrant

WHEN am I ever going to use this?

MAPS A map of Terrell's neighborhood is shown.

1. Suppose Terrell starts at the corner of Russel and Main and walks 1 block north and 2 blocks east. Name the intersection of his location.

2. Using the words *north*, *south*, *west*, and *east*, write directions to go from the corner of School and Highland to the corner of Main and Oak.

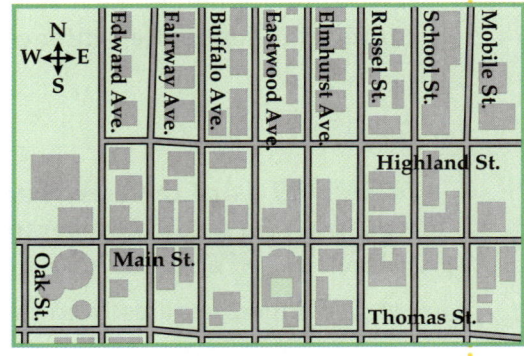

A **coordinate plane** is used to locate points. It is a plane in which a horizontal number line and a vertical number line intersect at their zero points. A coordinate plane is also called a **coordinate grid**.

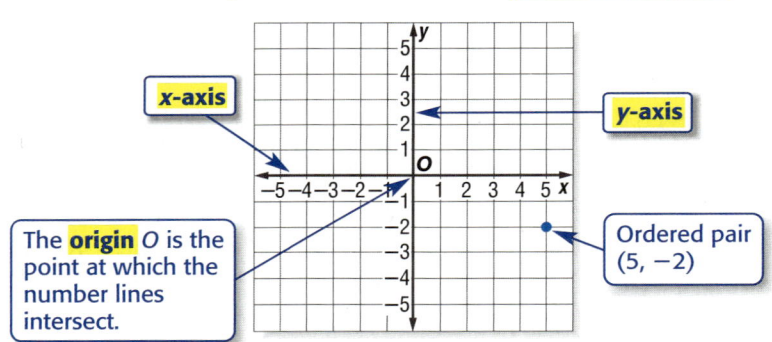

An **ordered pair** is a pair of numbers such as (5, −2) used to locate a point in the coordinate plane.

| The **x-coordinate** corresponds to a number on the *x*-axis. | (5, −2) | The **y-coordinate** corresponds to a number on the *y*-axis. |

EXAMPLE Name an Ordered Pair

1 Name the ordered pair for point *P*.

- Start at the origin.
- Move left to find the *x*-coordinate of point *P*, which is −4.
- Move up to find the *y*-coordinate, which is 2.

So, the ordered pair for point *P* is (−4, 2).

Graphing Moving right or up on a coordinate plane is in the *positive direction*. Moving left or down is in the *negative direction*.

112 Chapter 3 Algebra: Integers

READING Math

Scale When no numbers are shown on the *x*- or *y*-axis, you can assume that each grid is 1 unit long on each side.

EXAMPLE Graph an Ordered Pair

② Graph and label the point $Q(2, -5)$.
- Draw a coordinate plane.
- Move 2 units to the right. Then move 5 units down.
- Draw a dot and label it $Q(2, -5)$.

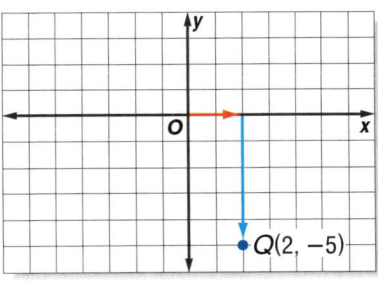

Your Turn Graph each point.

a. $A(6, 0)$ b. $B(-5, -3)$

The coordinate plane is separated into four sections called **quadrants**.

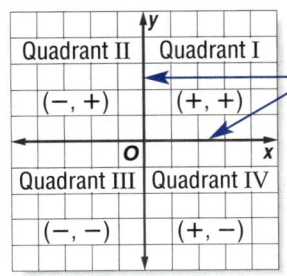

Axes is the plural form of axis. The axes are not located in any of the quadrants.

REAL-LIFE CAREERS

How Does a Cartographer Use Math?
Cartographers use coordinates to prepare geographic, political, and cultural maps of large areas.

Math Online
Research
For more information about a career as a cartographer, visit:
msmath2.net/careers

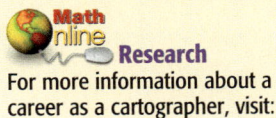

EXAMPLES Identify Quadrants

③ **GEOGRAPHY** The world map can be divided into a coordinate grid where (x, y) represents (degrees longitude, degrees latitude). In which quadrant is the United States located?

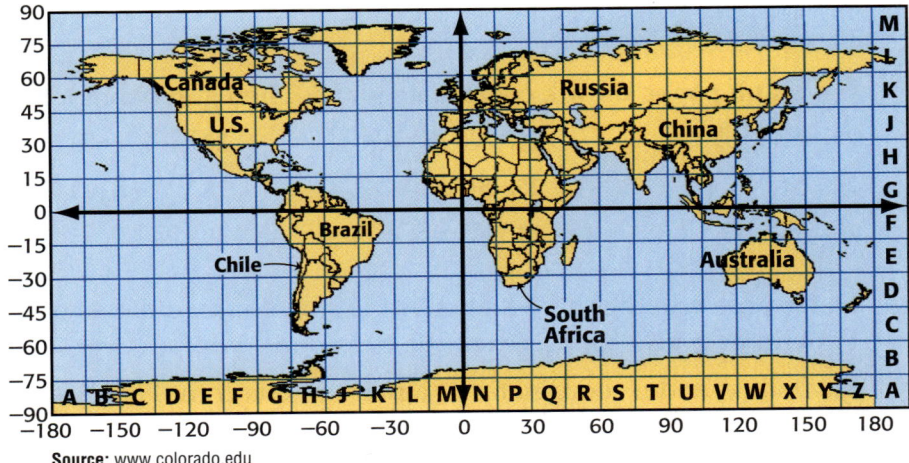

Source: www.colorado.edu

The United States is located in the upper left quadrant, quadrant II.

④ Name a country from the map that is located in quadrant III.

Quadrant III is the bottom-left quadrant. So, Chile is in quadrant III.

msmath2.net/extra_examples   Lesson 3-3 Geometry: The Coordinate Plane **113**

Skill and Concept Check

1. Explain why point $A(1, -2)$ is different from point $B(-2, 1)$.

2. **OPEN ENDED** Name and graph a point in quadrant IV.

GUIDED PRACTICE

Name the ordered pair for each point graphed at the right. Then identify the quadrant in which each point lies.

3. P
4. Q
5. R

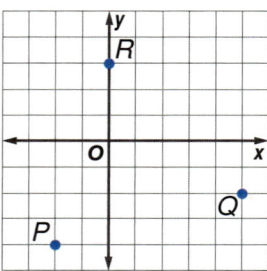

On graph paper, draw a coordinate plane. Then graph and label each point.

6. $S(2, 3)$
7. $T(-4, 6)$
8. $U(-5, 0)$

Practice and Applications

Name the ordered pair for each point graphed at the right. Then identify the quadrant in which each point lies.

9. A 10. B
11. C 12. D
13. E 14. F
15. G 16. H
17. I 18. J

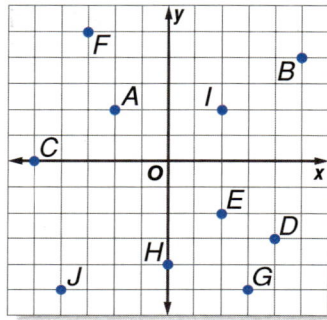

HOMEWORK HELP

For Examples	See Examples
9–18	1, 3
19–31	2
35–42	3, 4

Extra Practice
See pages 570, 598.

19. Write the ordered pair for the point that lies on the y-axis and is 32 units down from the origin.

On graph paper, draw a coordinate plane. Then graph and label each point.

20. $M(5, 6)$
21. $N(-2, 10)$
22. $P(7, -8)$
23. $Q(3, 0)$
24. $R(-1, -7)$
25. $S(0, 6)$
26. $T(-3, 7)$
27. $U(5, -2)$
28. $V(8, 1)$
29. $W(-5, -7)$
30. $X(1.5, -3)$
31. $Y(-6.5, 6.5)$

Determine whether each statement is *sometimes*, *always*, or *never* true. Explain or give a counterexample to support your answer.

32. Both x- and y-coordinates of a point in quadrant I are negative.
33. The x-coordinate of a point that lies on the x-axis is negative.
34. The y-coordinate of a point in quadrant IV is negative.

GEOGRAPHY For Exercises 35 and 36, use the map in Example 3.

35. In what country is the point (105° longitude, 30° latitude) located?
36. Find an ordered pair that can represent the location of California.

114 Chapter 3 Algebra: Integers

For Exercises 37–41, use the map of the Brookfield Zoo.

Brookfield Zoo, Brookfield, IL

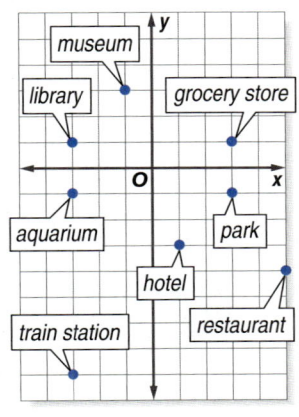

Source: www.brookfieldzoo.org

37. What exhibit is located at (4, −2)?

38. In which quadrant is the Dragonfly Marsh exhibit located?

39. Find the ordered pair that represents the location of Baboon Island.

40. What is located at the origin?

41. Describe how you would walk from the entrance of the Pachyderm House at (−2, 2) to the entrance of The Swamp at (−1, −2).

42. **GEOMETRY** Graph the points $A(-3, 2)$, $B(2, 2)$, $C(2, -4)$, and $D(-3, -4)$ on the same coordinate plane. Connect the points from A to B, B to C, C to D, and D to A. Name the figure.

43. **CRITICAL THINKING** Find the possible locations for any ordered pair whose x- and y-coordinates are always the same integer. Explain.

Spiral Review with Standardized Test Practice

For Exercises 44–46, use the coordinate plane at the right.

44. **MULTIPLE CHOICE** Which building has the coordinates of $(-3, -1)$?
    A museum **B** park **C** library **D** aquarium

45. **MULTIPLE CHOICE** What are the coordinates of the point that shows the location of the hotel?
    F $(-1, -3)$ **G** $(1, -3)$ **H** $(-1, 3)$ **I** $(1, 3)$

46. **SHORT RESPONSE** In which quadrant is the grocery store located?

Replace each ● with < or > to make a true sentence. (Lesson 3-2)

47. $14 \; ● \; |-15|$
48. $-8 \; ● \; -3$
49. $26 \; ● \; -30$
50. $-40 \; ● \; |40|$

51. Find the absolute value of -101. (Lesson 3-1)

52. **SPORTS** A triathlon competition consists of swimming 3 miles, running 10 miles, and bicycling 35 miles. How many miles does an athlete travel during the competition? (Lesson 1-1)

GETTING READY FOR THE NEXT LESSON

BASIC SKILL Add.

53. $138 + 246$
54. $814 + 512$
55. $2,653 + 4,817$
56. $6,003 + 5,734$

Chapter 3 Mid-Chapter Practice Test

Vocabulary and Concepts

1. **Define** *absolute value*. (Lesson 3-1)

2. **Write** the ordered pair which identifies a point 4 units to the left of the y-axis and three units above the x-axis. (Lesson 3-3)

3. **Draw** a coordinate plane, and label the quadrants. (Lesson 3-3)

Skills and Applications

Write an integer for each situation. (Lesson 3-1)

4. 45 feet below sea level
5. a deposit of $100
6. a gain of 8 yards
7. lost a $5 bill

Replace each ● with < or > to make a true sentence. (Lesson 3-2)

8. -12 ● -9
9. -4 ● 4
10. $|-14|$ ● $|3|$

11. **FOOTBALL** The Tigers have recorded the following yardage on the past six plays: 9, −2, 5, 0, 12, and −7. Order the integers from least to greatest. (Lesson 3-2)

On graph paper, draw a coordinate plane. Then graph and label each point. (Lesson 3-3)

12. $D(4, -3)$
13. $E(1, 3)$
14. $F(0, -5)$

Standardized Test Practice

15. **MULTIPLE CHOICE** Which of the following points represents a number and its absolute value? (Lesson 3-1)

  A. B and E
  B. C and F
  C. B and D
  D. A and E

16. **SHORT RESPONSE** The table shows the number of inches of monthly precipitation above or below normal for a midwestern city in a recent year. Find the median monthly precipitation above or below normal. (Lesson 3-2)

Monthly Precipitation Above or Below Normal					
J	F	M	A	M	J
4	−1	6	−2	−3	1
J	A	S	O	N	D
−2	−1	2	−3	1	−3

The Game Zone

A Place To Practice Your Math Skills

Math Skill: Graphing Points on a Coordinate Plane

Tic-Tac-Toe

- **GET READY!**
  Players: two
  Materials: grid paper

- **GET SET!**
  - Draw a coordinate plane on grid paper.
  - This game is similar to tic-tac-toe, except players must get four Xs or four Os in a row.

- **GO!**
  - Player 1 chooses two numbers: the first number is the x-coordinate of an ordered pair, and the second number is the y-coordinate. Each number must be between −5 and 5. Then Player 1 announces the ordered pair and plots the X or O on the coordinate plane.
  - Player 2 then chooses his or her numbers, announces them, and plots the points.
  - An ordered pair cannot be changed after it has been announced.
  - If a player announces an ordered pair that has already been used or graphs an ordered pair incorrectly, the player loses a turn.
  - **Who Wins?** The first player to get four Xs or Os in a row is the winner.

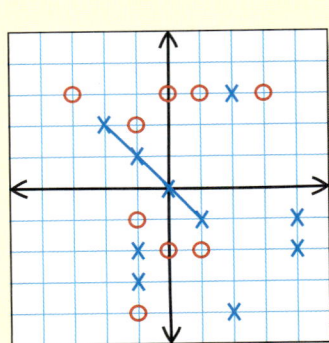

The Game Zone: Graphing Points on a Coordinate Plane 117

3-4a Hands-On Lab

A Preview of Lesson 3-4

Adding Integers

What You'll LEARN

Use counters to model the addition of integers.

Materials
- counters
- integer mat

You can use positive and negative counters to model the addition of integers. The counter ⊕ represents 1, and the counter ⊖ represents −1. Remember that addition means *combining* two sets.

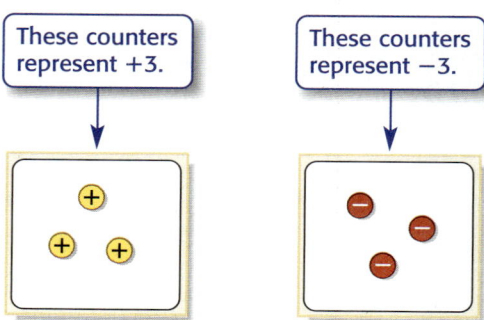

These counters represent +3.

These counters represent −3.

ACTIVITY   Work with a partner.

1 Use counters to find $-2 + (-4)$.

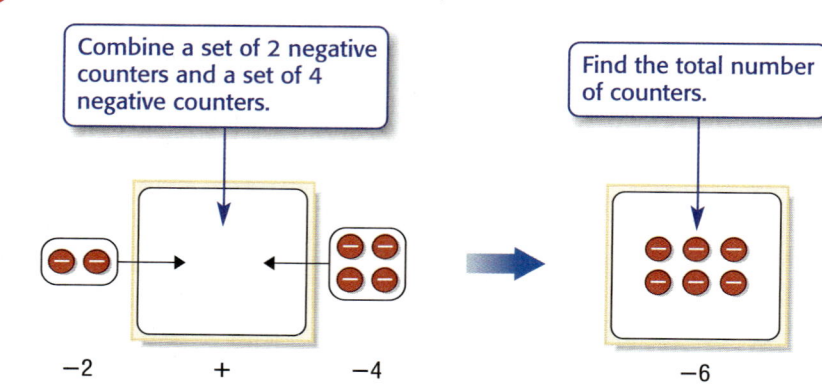

Combine a set of 2 negative counters and a set of 4 negative counters.

Find the total number of counters.

−2   +   −4   −6

So, $-2 + (-4) = -6$.

READING Math

Addends and Sums The numbers you add are called *addends*. The result is called the *sum*.

Your Turn Use counters to find each sum.

a. $5 + 6$
b. $-3 + (-5)$
c. $-5 + (-4)$
d. $7 + 3$
e. $-2 + (-5)$
f. $-8 + (-6)$

The following two properties are important when modeling operations with integers.

- When one positive counter is paired with one negative counter, the result is called a **zero pair**. The value of a zero pair is 0.

- You can add or remove zero pairs from a mat because adding or removing zero does not change the value of the counters on the mat.

You will use zero pairs in Activity 2 and Activity 3.

ACTIVITIES Work with a partner.

Use counters to find each sum.

2 −5 + 3

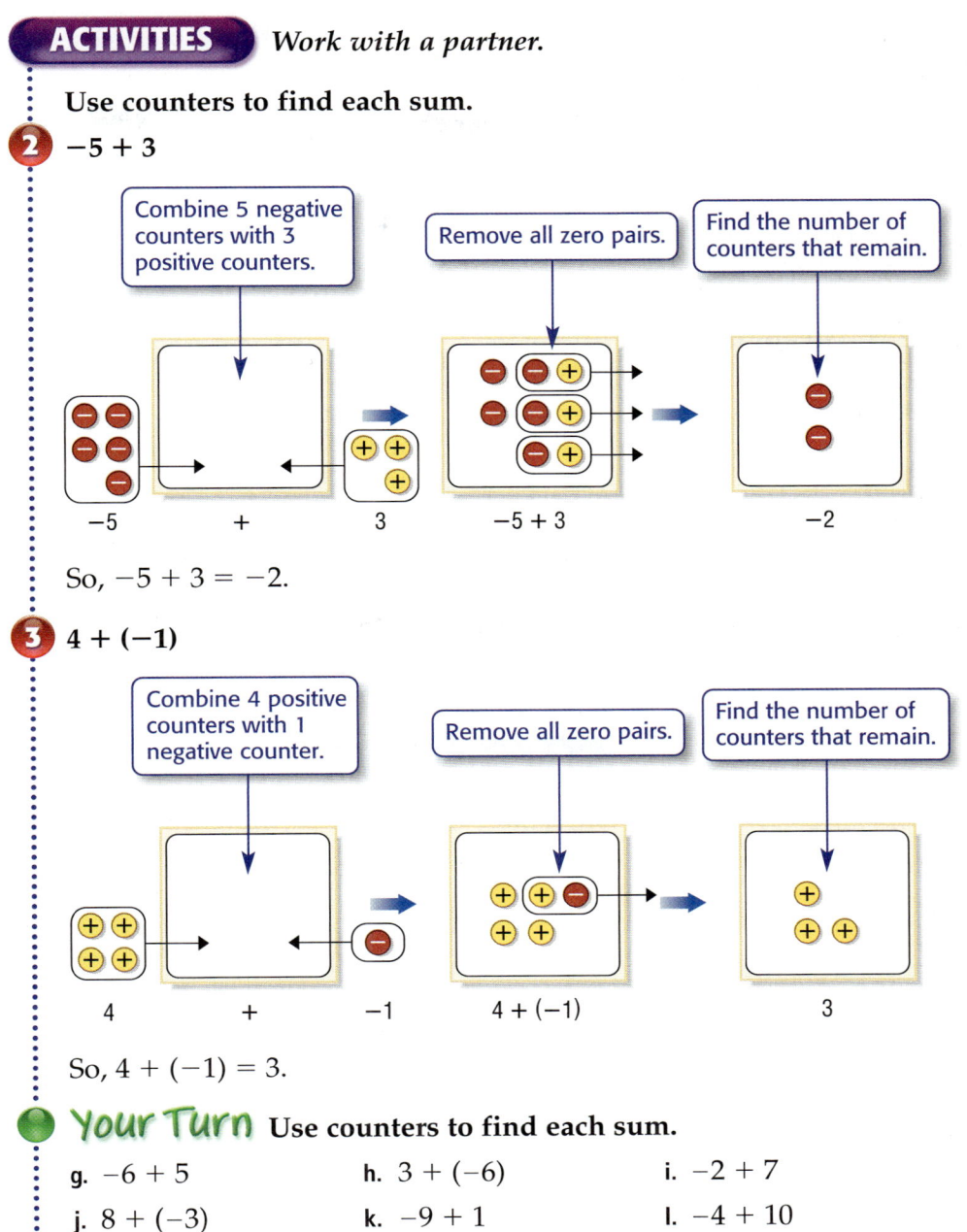

So, −5 + 3 = −2.

3 4 + (−1)

So, 4 + (−1) = 3.

Your Turn Use counters to find each sum.

g. −6 + 5
h. 3 + (−6)
i. −2 + 7
j. 8 + (−3)
k. −9 + 1
l. −4 + 10

Writing Math

1. **Write** two addition sentences where the sum is positive. In each sentence, one addend should be positive and the other negative.
2. **Write** two addition sentences where the sum is negative. In each sentence, one addend should be positive and the other negative.
3. **MAKE A CONJECTURE** Write a rule that will help you determine the sign when finding the sum of integers.

3-4 Adding Integers

What You'll LEARN
Add integers.

NEW Vocabulary
opposites
additive inverse

Link to READING
Everyday Meaning of Opposite: something that is across from or is facing the other way, as in running the opposite way

WHEN am I ever going to use this?

EARTH SCIENCE Thunderstorms are made of both positive and negative electrical charges. The negative charges (electrons) are at the bottom of a thundercloud, and positive charges (protons) are at the top.

1. What is the charge at the top of a cloud where there are more protons than electrons?

2. What is the charge at the bottom of a cloud where there are more electrons than protons?

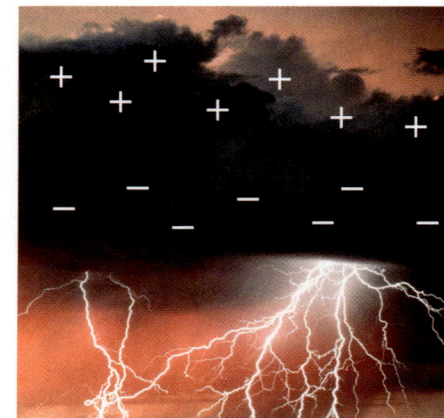

Combining positive and negative electrical charges in a thunderstorm is similar to adding integers.

EXAMPLE Add Integers with the Same Sign

1 Find $-3 + (-2)$.

Use a number line.
- Start at 0.
- Move 3 units left to show -3.
- From there, move 2 units left to show -2.

So, $-3 + (-2) = -5$.

Noteables™ Key Concept: Add Integers with the Same Sign

Words The sum of two positive integers is positive.
The sum of two negative integers is negative.

Examples $7 + 4 = 11$ $-7 + (-4) = -11$

STUDY TIP

Technology To enter a negative integer on a calculator, use the (−) key. For example, to enter -5, press (−) 5.

EXAMPLE Add Integers with the Same Sign

2 Find $-26 + (-17)$.

$-26 + (-17) = -43$ The sum of two negative integers is negative.

Your Turn Add.

a. $-14 + (-16)$
b. $23 + 38$
c. $-35 + (-49)$

120 Chapter 3 Algebra: Integers

The integers 43 and −43 are called <mark>opposites</mark> of each other because they are the same distance from 0, but on opposite sides of 0. Two integers that are opposites are also called <mark>additive inverses</mark>.

> **Noteables™** — **Key Concept: Additive Inverse Property**
>
> **Words** The sum of any number and its additive inverse is 0.
>
> **Example** $5 + (−5) = 0$

EXAMPLES — Add Integers with Different Signs

3 Find $5 + (−3)$.

Use counters.

Remove all zero pairs.

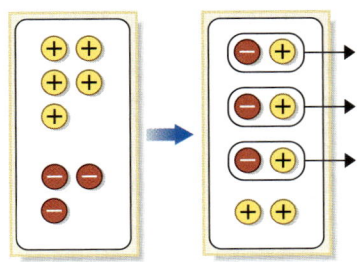

So, $5 + (−3) = 2$.

4 Find $−3 + 2$.

Use a number line.
- Start at 0.
- Move 3 units left.
- Then move 2 units right.

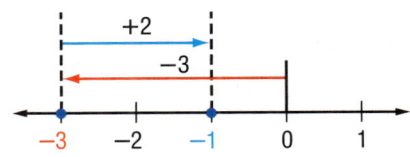

So, $−3 + 2 = −1$.

STUDY TIP

Look Back You can review absolute value in Lesson 3-1.

> **Noteables™** — **Key Concept: Add Integers with Different Signs**
>
> **Words** To add integers with different signs, subtract their absolute values. The sum is:
> - positive if the positive integer has the greater absolute value.
> - negative if the negative integer has the greater absolute value.
>
> **Examples** $9 + (−4) = 5$ $−9 + 4 = −5$

EXAMPLES — Add Integers with Different Signs

5 Find $7 + (−1)$.

$7 + (−1) = 6$ Subtract. Keep the sign of the integer with the greater absolute value.

6 Find $−8 + 4$.

$−8 + 4 = −4$ Subtract. Keep the sign of the integer with the greater absolute value.

Your Turn Add.

d. $10 + (−12)$ e. $−16 + 9$ f. $−13 + 18$

msmath2.net/extra_examples   Lesson 3-4 Adding Integers **121**

EXAMPLE **Simplify an Expression with Integers**

7 **ALGEBRA** Simplify $12 + x + (-20)$.

$12 + x + (-20) = 12 + (-20) + x$ Commutative Property of Addition
$ = -8 + x$ Add.

Your Turn Simplify each expression.

g. $1 + y + (-5)$ h. $z + (-2) + 9$

REAL-LIFE MATH

OCEANOGRAPHY The Great Barrier Reef off the coast of Australia is home to 1,500 species of fish and 215 species of birds.

Source: www.ozramp.net.au

EXAMPLE **Use Integers to Solve a Problem**

8 **OCEANOGRAPHY** Anna was scuba diving near the Great Barrier Reef 16 meters below the surface of the water. She saw a dolphin swim by 7 meters above her. What was the depth of the dolphin?

Anna is 16 meters underwater, and the dolphin is 7 meters above her. So, the depth of the dolphin can be represented by the expression $-16 + 7$, or -9.

The dolphin is 9 meters below the surface of the water.

Skill and Concept Check

1. **Draw** a model to show $2 + (-7)$.

2. **OPEN ENDED** Give an example of integers that are additive inverses.

3. **FIND THE ERROR** Brooke and Javier are finding $-12 + 13$. Who is correct? Explain.

   Brooke
   $-12 + 13 = 1$

   Javier
   $-12 + 13 = -1$

4. **NUMBER SENSE** Tell whether each sum is *positive, negative,* or *zero* without adding.

   a. $-6 + (-7)$ b. $-8 + 10$ c. $-14 + 14$

GUIDED PRACTICE

Add.

5. $-6 + (-8)$ 6. $-3 + 10$ 7. $7 + (-11)$ 8. $9 + (-9)$

9. **MONEY** You pay your brother $42 that you owe him. The same week you earn $35 dog-sitting for the neighbors. Do you have more or less money than at the beginning of the week?

10. **ALGEBRA** Simplify $12 + y + (-8)$.

Practice and Applications

HOMEWORK HELP

For Exercises	See Examples
11–42	1–6
46–57	7
43–45, 58	8

Extra Practice See pages 570, 598.

Add.

11. $-8 + 8$
12. $-9 + 11$
13. $13 + (-19)$
14. $6 + 10$
15. $-10 + (-15)$
16. $-12 + 10$
17. $-30 + 16$
18. $18 + (-5)$
19. $21 + (-21)$
20. $18 + (-20)$
21. $-22 + (-16)$
22. $-24 + 19$
23. $-11 + 13 + 6$
24. $-16 + (-21) + 15$
25. $12 + (-17) + (-25)$
26. $20 + (-30) + (-40)$

Write an addition expression to describe each situation. Then find each sum.

27. **WEATHER** The temperature outside is $-3°$F. The temperature drops $6°$.

28. **SUBMARINE** A submarine dives 106 feet below the water. Then, it rises 63 feet.

29. **SKATEBOARDING** Hakeem starts at the bottom of a half pipe 6 feet below street level. He rises 14 feet at the top of his kickturn.

30. **MONEY MATTERS** Stephanie has $43 in the bank. She withdraws $35.

ALGEBRA Evaluate each expression if $x = -10$, $y = 7$, and $z = -8$.

31. $x + 14$
32. $6 + y$
33. $z + (-5)$
34. $-17 + y$
35. $20 + z$
36. $-10 + x$
37. $z + 8$
38. $15 + x$
39. $x + y$
40. $y + z$
41. $x + z$
42. $x + y + z$

GOLF For Exercises 43–45, use the information below.

Scores over par in a golf tournament are recorded as positive integers. Scores under par are recorded as negative integers. Even par is recorded as 0. The person with the lowest total score wins. The table shows the top two finishers in the 2004 LPGA Championship.

43. Find Annika Sorenstam's final score.
44. Find Shi Hyun Ahn's final score.
45. Who had the better score? Explain.

	Round 1	Round 2	Round 3	Round 4
Annika Sorenstam	−3	−4	−7	+1
Shi Hyun Ahn	−2	−1	−2	−5

Source: www.lpga.com

 Data Update What were the four-round scores of the latest winners of the LPGA Championship? Visit **msmath2.net/data_update** to learn more.

ALGEBRA Simplify.

46. $x + (-5) + 1$
47. $4 + y + (-2)$
48. $-9 + m + (-6)$
49. $8 + (-8) + n$
50. $-1 + a + 7$
51. $f + (-19) + 11$

Lesson 3-4 Adding Integers

Explain how the Commutative and Associative Properties of Addition can help you find each sum mentally. Then find each sum.

52. $7 + (-2) + (-7)$
53. $-6 + 9 + (-4)$
54. $-5 + (-6) + (-3)$
55. $8 + 10 + (-8)$
56. $-5 + (-7) + (-10)$
57. $8 + (-9) + 9$

58. **STOCK MARKET** The members of the Investment Club purchased a stock for $50. The next day the price of the stock dropped $18. On the second and third days, the price dropped another $16 and then rose $21. How much was the stock worth at the end of the third day?

59. **WRITE A PROBLEM** Write about a real-life problem using the addition sentence $-8 + 11 = t$. Then solve the equation and explain what the solution represents.

CRITICAL THINKING For Exercises 60 and 61, use a number line to find each sum. Does the order of the addends make a difference? Explain.

60. $3 + (-8)$ and $-8 + 3$
61. $[7 + (-3)] + (-6)$ and $7 + [-3 + (-6)]$

Spiral Review with Standardized Test Practice

62. **MULTIPLE CHOICE** In a game with a standard deck of cards and the scoring system at the right, three cards are dealt and added together to get a final score. Dylan is dealt the 4 of hearts, the king of spades, and the 3 of diamonds. What is his final score?

    Ⓐ -11 Ⓑ -3 Ⓒ 3 Ⓓ 9

Card	Points
Face cards	10
Aces	1
2–10	equal their value
Spades and clubs (black)	positive value
Hearts and diamonds (red)	negative value

63. **SHORT RESPONSE** Jeremy owes his sister $5. Then he borrows $6 more from her. Write the total amount he owes as an integer.

Name the ordered pair for each point graphed at the right. Then identify the quadrant in which each point lies. (Lesson 3-3)

64. J
65. K
66. L
67. M

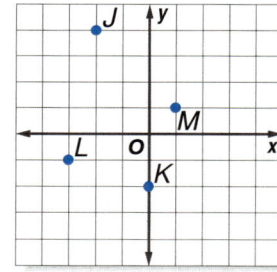

68. Order $6, -3, 0, 4, -8, 1,$ and -4 from least to greatest. (Lesson 3-2)

69. **STATISTICS** Construct a line plot for the following test scores: 81, 83, 75, 81, 82, 81, 75, 82, 82, 86, 83, 81, and 79. (Lesson 2-3)

GETTING READY FOR THE NEXT LESSON

PREREQUISITE SKILL Find the range for each set of data. (Lesson 2-3)

70. 13, 7, 6, 22, 21
71. 54, 32, 43, 49, 30
72. 62, 59, 85, 74, 82

Study Skill

Use a Flowchart

Taking Good Notes

Have you ever tried to solve a math problem and then realized you left out an important step? Try using a flowchart when you take notes to map out the steps you should follow.

A *flowchart* is like a map that tells you how to get from the beginning of a problem to the end.

Flowchart Symbols	
◇	A diamond contains a question. You need to stop and make a decision.
▭	A rectangle tells you what to do.
⬭	An oval indicates the beginning or end.

Here's a flowchart for adding two integers. Just follow the arrows.

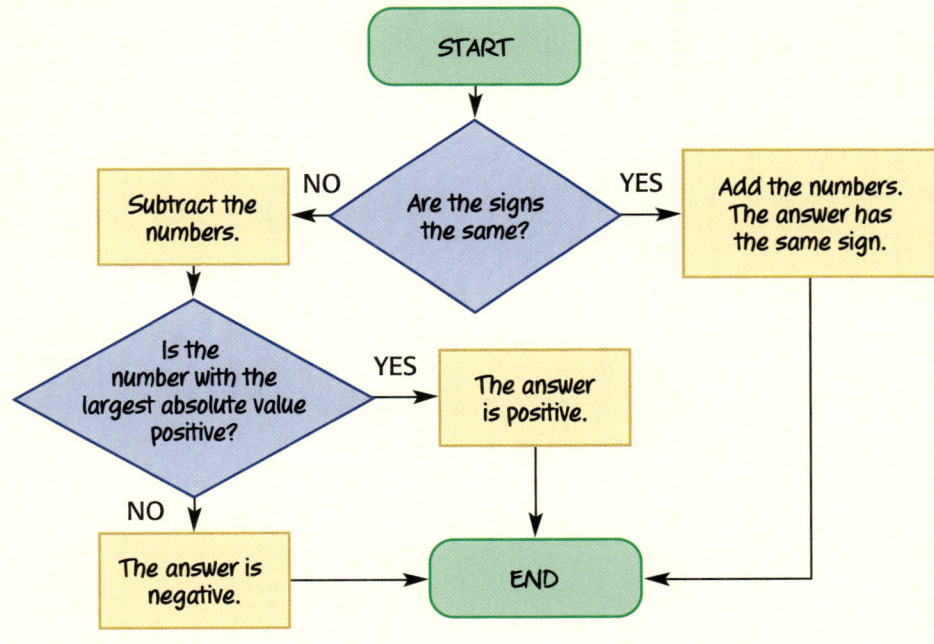

SKILL PRACTICE
Make a flowchart for each kind of problem.

1. rounding a decimal to a given place (See page 557.)
2. evaluating an expression using order of operations (See Lesson 1-3.)

3-5a HANDS-ON LAB
A Preview of Lesson 3-5

Subtracting Integers

You can also use counters to model subtraction of integers. Remember one meaning of subtraction is to *take away*.

What You'll LEARN

Use counters to model the subtraction of integers.

Materials
- counters
- integer mat

ACTIVITIES *Work with a partner.*

Use counters to find each difference.

1) 6 − 4

Place 6 positive counters on the mat. Remove 4 positive counters.

So, 6 − 4 = 2.

2) 3 − (−2)

Place 3 positive counters on the mat. Remove 2 negative counters. However, there are 0 negative counters.

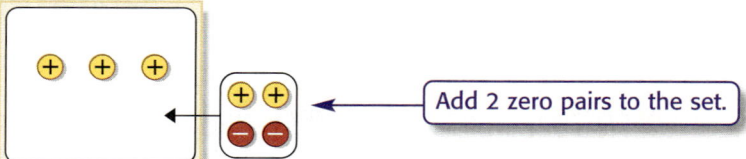

Add 2 zero pairs to the set.

Now you can remove 2 negative counters. Find the remaining number of counters.

So, 3 − (−2) = 5.

Your Turn Use counters to find each difference.

    a. 7 − 6   **b.** 5 − (−3)   **c.** 6 − (−3)   **d.** 5 − 8

ACTIVITIES Work with a partner.

Use counters to find each difference.

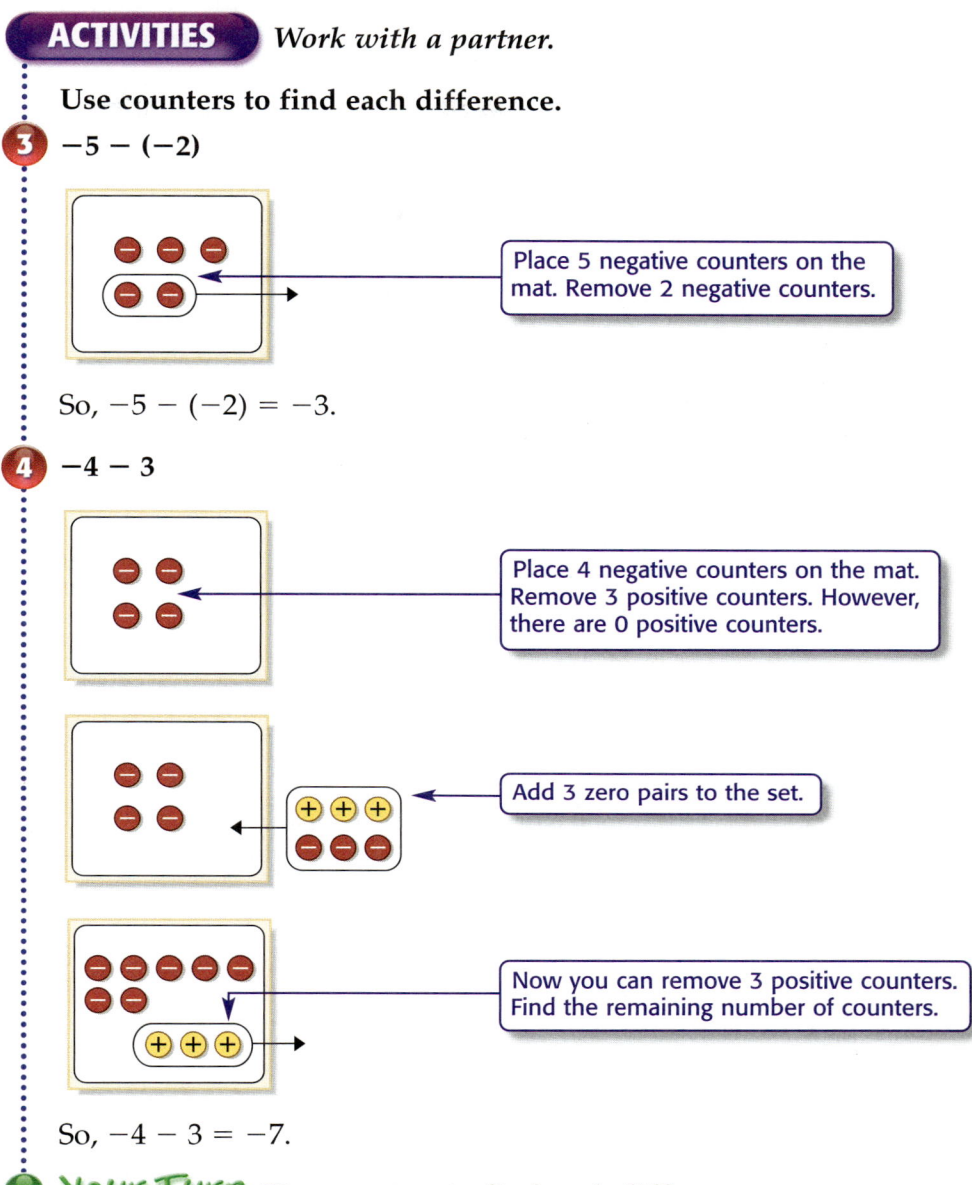

3 $-5 - (-2)$

Place 5 negative counters on the mat. Remove 2 negative counters.

So, $-5 - (-2) = -3$.

4 $-4 - 3$

Place 4 negative counters on the mat. Remove 3 positive counters. However, there are 0 positive counters.

Add 3 zero pairs to the set.

Now you can remove 3 positive counters. Find the remaining number of counters.

So, $-4 - 3 = -7$.

READING Math

Minuends, Subtrahends, and Differences In the subtraction sentence $-4 - 3 = -7$, -4 is the *minuend*, 3 is the *subtrahend*, and -7 is the *difference*.

Your Turn Use counters to find each difference.

e. $-6 - (-3)$ f. $-7 - 3$ g. $-5 - (-7)$

Writing Math

Work with a partner.

1. **Write** two subtraction sentences where the difference is positive. Make sure you use a combination of positive and negative integers.

2. **Write** two subtraction sentences where the difference is negative. Make sure you use a combination of positive and negative integers.

3. **MAKE A CONJECTURE** Write a rule that will help you determine the sign of the difference of two integers.

3-5 Subtracting Integers

What You'll LEARN

Subtract integers.

REVIEW Vocabulary

range: the difference between the greatest number and the least number in a set of data (Lesson 2-3)

HANDS-ON Mini Lab

Materials
- graph paper

Work with a partner.

The subtraction problems below are modeled on number lines.

$3 - 5 = -2$

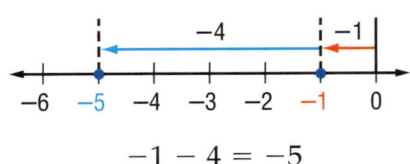

$-1 - 4 = -5$

1. Write a related addition sentence for each subtraction sentence.

Use a number line to find each difference. Write an equivalent addition sentence for each.

2. $1 - 5$ 3. $-2 - 1$ 4. $-3 - 4$ 5. $0 - 5$

6. **Compare and contrast** subtraction sentences with their related addition sentences.

When you subtract 5, as shown in the Mini Lab, the result is the same as adding -5. When you subtract 4, the result is the same as adding -4.

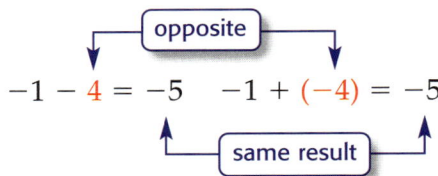

Noteables™ — Key Concept: Subtract Integers

Words To subtract an integer, add its opposite.

Example $4 - 9 = 4 + (-9) = -5$

EXAMPLES Subtract Positive Integers

Subtract.

1) $8 - 13$

$8 - 13 = 8 + (-13)$ To subtract 13, add -13.

$= -5$ Simplify.

2) $-10 - 7$

$-10 - 7 = -10 + (-7)$ To subtract 7, add -7.

$= -17$ Simplify.

128 Chapter 3 Algebra: Integers

EXAMPLES Subtract Negative Integers

3 Find $1 - (-2)$.

$1 - (-2) = 1 + 2$ To subtract -2, add 2.
$ = 3$ Simplify.

4 Find $-10 - (-7)$.

$-10 - (-7) = -10 + 7$ To subtract -7, add 7.
$ = -3$ Simplify.

STUDY TIP

Technology Use the $\boxed{-}$ key to subtract a number. Use the $\boxed{(-)}$ key to enter a negative number.

Your Turn Subtract.

a. $4 - (-12)$ b. $-15 - (-5)$ c. $18 - (-6)$

EXAMPLE Evaluate an Expression

5 **ALGEBRA** Evaluate $x - y$ if $x = -6$ and $y = 5$.

$x - y = -6 - 5$ Replace x with -6 and y with 5.
$ = -6 + (-5)$ To subtract 5, add -5.
$ = -11$ Simplify.

Your Turn Evaluate each expression if $a = 5$, $b = -8$, and $c = -9$.

d. $b - 10$ e. $a - b$ f. $c - a$

EXAMPLE Use Integers to Solve a Problem

6 **EARTH SCIENCE** The legend on the sea-surface temperature map shows the minimum temperature at $-2°C$ and the maximum temperature at $31°C$. What is the range of temperatures on the map?

To find the range, or difference in temperatures, subtract the lowest temperature from the highest temperature.

$31 - (-2) = 31 + 2$ To subtract -2, add 2.
$ = 33$ Simplify.

So, the range of temperatures is $33°C$.

msmath2.net/extra_examples   **Lesson 3-5** Subtracting Integers **129**

Skill and Concept Check

1. **Writing Math** Explain how additive inverses are used in subtraction.

2. **OPEN ENDED** Write a subtraction sentence using integers. Then, write the equivalent addition sentence and find the sum.

3. **FIND THE ERROR** Bradley and Mitsu are finding $-16 - (-19)$. Who is correct? Explain.

   Bradley
   $-16 - (-19) = -16 + (-19)$
   $= -35$

   Mitsu
   $-16 - (-19) = -16 + (19)$
   $= 3$

GUIDED PRACTICE

Subtract.

4. $-4 - 8$
5. $14 - 17$
6. $14 - (-10)$
7. $-3 - (-1)$

ALGEBRA Evaluate each expression if $p = 8$, $q = -14$, and $r = -6$.

8. $p - q$
9. $q - r$
10. $r - p$

11. **METEOROLOGY** The highest temperature ever recorded on Earth was 136°F in Libya. The lowest temperature was -129°F in Antarctica. What is the range of the highest and lowest temperatures on Earth?

Practice and Applications

Subtract.

12. $-9 - 5$
13. $0 - 10$
14. $-8 - 9$
15. $17 - 13$
16. $27 - (-8)$
17. $-25 - (-5)$
18. $12 - 26$
19. $4 - (-19)$
20. $-11 - 42$
21. $15 - (-14)$
22. $-27 - (-19)$
23. $-18 - (-20)$
24. $31 - 48$
25. $-33 - (-27)$
26. $52 - (-52)$
27. $-44 - (-41)$
28. $-2 - 9 + 7$
29. $6 + (-1) - 4$

30. What is -3 minus 4?
31. Find $-23 - (-19)$.

HOMEWORK HELP

For Exercises	See Examples
12–31	1–4
32–44	5
45–48	6

Extra Practice
See pages 571, 598.

ALGEBRA Evaluate each expression if $f = -6$, $g = 7$, and $h = 9$.

32. $5 - f$
33. $h - (-9)$
34. $f - g$
35. $g - 7$
36. $h - f$
37. $f - 6$
38. $g - h$
39. $4 - (-g)$
40. $-h - 10$
41. $-f - h$
42. $f - g - h$
43. $h - g - f$

44. **ALGEBRA** Find $|a - b|$ when $a = -7$ and $b = 11$.

45. **GEOGRAPHY** The Dead Sea's deepest part is 799 meters below sea level. A plateau to the east of the Dead Sea rises to about 1,340 meters above sea level. What is the difference between the deepest part of the Dead Sea and the top of the plateau?

130 Chapter 3 Algebra: Integers

HISTORY For Exercises 46–48, use the timeline that shows the lives of three rulers of Rome.

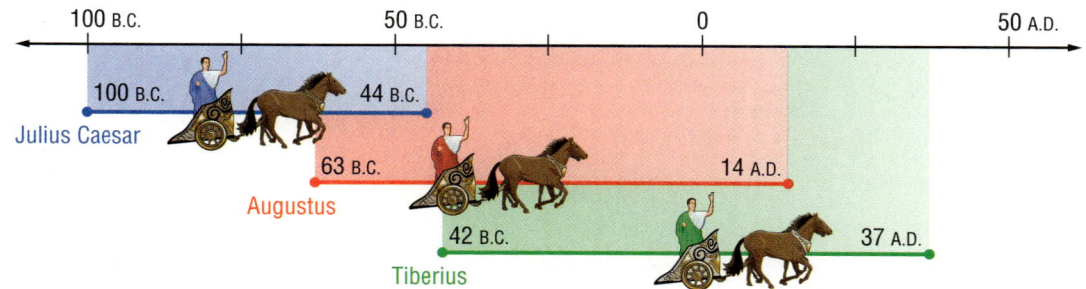

46. How old was Augustus when he died?
47. Who lived the longest? How old was he when he died?
48. How many years were there between when Julius Caesar was born and when Tiberius died?

Determine whether each statement is *sometimes*, *always*, or *never* true. Give an example or counterexample for each answer.

49. negative − positive = negative
50. negative − negative = positive
51. positive − positive = positive
52. positive − negative = negative

53. **CRITICAL THINKING**
    True or *False*? When n is a negative integer, $n - n = 0$.

Spiral Review with Standardized Test Practice

54. **MULTIPLE CHOICE** Find the correct subtraction sentence shown in the model.

    A $2 - (-6) = -4$ **B** $-6 - 2 = -4$
    C $-2 - 4 = -6$ **D** $2 - 6 = -4$

55. **SHORT RESPONSE** The temperatures on the moon vary from −173°C to 127°C. Find the range of temperatures.

Add. (Lesson 3-4)

56. $10 + (-3)$
57. $-2 + (-9)$
58. $-7 + (-6)$
59. $-18 + 4$

60. In which quadrant do ordered pairs with a positive *x*-coordinate and a negative *y*-coordinate lie? (Lesson 3-3)

GETTING READY FOR THE NEXT LESSON

BASIC SKILL Multiply.

61. $14 \cdot 5$
62. $9 \cdot 16$
63. $6 \cdot 8 \cdot 4$
64. $11 \cdot 7 \cdot 7$

Problem-Solving Strategy
A Preview of Lesson 3-6

Look for a Pattern

What You'll LEARN

Solve problems using the look for a pattern strategy.

We've already saved $155 in four months. If we keep saving our money at the same rate, how long do you think it will take to save enough money to buy a DVD player that costs $330?

I found the table where we listed our savings each month. Let's **look for a pattern** to figure it out.

Explore	We began with $50 and added more money to our savings every month. We need to find the number of months when we will have $330 to buy the DVD player.
Plan	Let's look for a pattern or rule that increases the balance each month. Then use the rule to extend the pattern and find the solution.
Solve	After the initial $50, we saved $35 per month. To extend the pattern, add $35 to each monthly balance until the balance equals $330. We will have enough money saved after 9 months.
Examine	We saved about 2 × $155, or $310 in 8 months. So, 9 months is a reasonable answer.

Month	Balance
1	$50
2	$85
3	$120
4	$155

Month	Balance
1	$50
2	$85
3	$120
4	$155
5	$190
6	$225
7	$260
8	$295
9	$330

Analyze the Strategy

1. **Explain** when you would use the look for a pattern strategy to solve a problem.
2. **Describe** how to solve a problem using the look for a pattern method as a problem-solving strategy.
3. **Write** a problem that could be solved by looking for a pattern. Explain your answer.

Apply the Strategy

Solve. Use the look for a pattern strategy.

4. **LIFE SCIENCE** The table shows about how many times a firefly flashes at different temperatures. Estimate how many times a firefly will flash when the temperature is 36°C.

Outside Temperature (°C)	Flashes per Minute
16	8
20	9
24	11
28	14

5. **CEREAL** A display of cereal boxes is stacked in the shape of a pyramid. There are 4 boxes in the top row, 6 boxes in the second row, 8 boxes in the next row, and so on. The display contains 7 rows of boxes. How many boxes are in the display?

Mixed Problem Solving

Solve. Use any strategy.

LIFE SCIENCE For Exercises 6–8, use the information and the graph.

6. What does 0 on this graph represent?

7. Write an integer to represent the rainfall for each month.

Summer Rainfall (compared to normal)

8. Write a sentence that summarizes the message this graph conveys about this summer's rainfall.

9. **EARTH SCIENCE** Hydrothermal vents are similar to geysers, but are found on the ocean floor. A hydrothermal vent chimney can grow at an average rate of 9 meters in 18 months. What is the average rate of growth per month?

10. **MULTI STEP** Francisco is on vacation and is planning to send postcards and letters to his friends. He has $3.04 to spend on postage. A stamp for a letter costs 37¢, and a stamp for a postcard costs 23¢. If he is going to spend the entire $3.04 on postage, how many postcards and letters can he send?

11. **BASKETBALL** Laura makes 3 free throws out of every 5 she attempts. Find the number of free throws she will make after 15, 20, and 30 attempts.

12. **COINS** Olivia has seven coins that total $1.32. What are the coins?

13. **FOOD** The school cafeteria added a breakfast special to their menu. The table shows the foods that are part of the special and the number of Calories. Estimate how many Calories there are in the special.

Food	Calories
whole-wheat bagel	156
skim milk	90
nonfat strawberry yogurt	183
fresh fruit salad	68

14. **STANDARDIZED TEST PRACTICE**
    The total land area of Illinois is about 55,593 square miles. According to the 2000 U.S. Census Bureau, about 223.4 persons per square mile were living in Illinois. What was the approximate population of Illinois in 2000?
    - **A** 124,000
    - **B** 1,240,000
    - **C** 12,400,000
    - **D** 124,000,000

3-6 Multiplying Integers

What You'll LEARN

Multiply integers.

Hands-On Mini Lab

Materials
- counters
- integer mat

Work with a partner.

Counters can be used to multiply positive and negative integers.

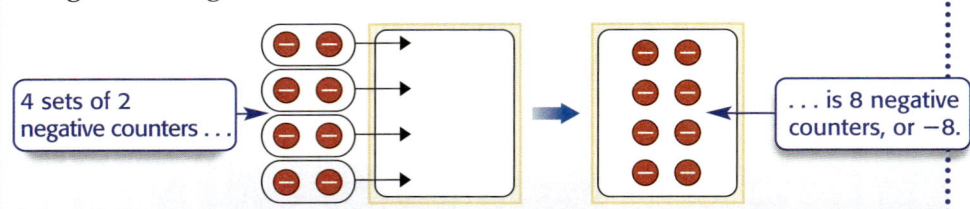

4 sets of 2 negative counters . . . . . . is 8 negative counters, or −8.

1. Write a multiplication sentence that describes the model above.

Find each product using counters.

2. 3(−2)   3. 4(−3)   4. 1(−7)   5. 5(−2)

6. Write a rule for finding the sign of the product of a positive and negative integer.

Remember that multiplication is the same as repeated addition. The multiplication expression 4(−2) in the Mini Lab means that −2 is used as an addend four times.

$4(-2) = (-2) + (-2) + (-2) + (-2)$
$\quad\quad\quad = -8$

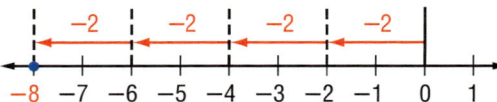

By the Commutative Property of Multiplication, 4(−2) = −2(4). When two integers have different signs, the following rule applies.

Noteables™ Key Concept: Multiply Integers with Different Signs

Words   The product of two integers with different signs is negative.
Examples   6(−4) = −24   −5(7) = −35

EXAMPLES Multiply Integers with Different Signs

Multiply.

1 3(−5)

3(−5) = −15   The integers have different signs. The product is negative.

2 −6(8)

−6(8) = −48   The integers have different signs. The product is negative.

134 Chapter 3 Algebra: Integers

STUDY TIP

Zero The Multiplicative Property of Zero states that when any number is multiplied by zero, the product is zero.

The product of two positive integers is positive. You can use a pattern to find the sign of the product of two negative integers.

When two integers have the same sign, the following rule applies.

Noteables™ Key Concept: Multiply Integers with Same Sign

Words   The product of two integers with the same sign is positive.
Examples $2(6) = 12$ $-10(-6) = 60$

EXAMPLES Multiply Integers with the Same Sign

Multiply.

3 $-11(-9)$

$-11(-9) = 99$ The integers have the same sign. The product is positive.

4 $(-4)^2$

$(-4)^2 = (-4)(-4)$ There are two factors of -4.
$ = 16$ The product is positive.

STUDY TIP

Look Back You can review **exponents** in Lesson 1-2.

Your Turn Multiply.

a. $-12(-4)$ b. $(-5)^2$ c. $(-2)^3$

EXAMPLES Simplify and Evaluate Expressions

5 **ALGEBRA** Simplify $-2(3x)$.

$-2(3x) = (-2 \cdot 3)x$ Associative Property of Multiplication
$ = -6x$ Simplify.

6 **ALGEBRA** Evaluate pqr if $p = -3$, $q = 4$, and $r = -1$.

$pqr = (-3)(4)(-1)$ Replace p with -3, q with 4, and r with -1.
$ = (-12)(-1)$ Multiply -3 and 4.
$ = 12$ Multiply -12 and -1.

Your Turn

d. Simplify $-5(2y)$.

e. Evaluate xyz if $x = -7$, $y = -4$, and $z = 2$.

Skill and Concept Check

1. **Model** the product of 2 and -3 using counters. Then write the multiplication sentence.

2. **OPEN ENDED** Name two integers whose product is negative.

3. **NUMBER SENSE** What is the sign of the product of three negative integers? Give an example.

GUIDED PRACTICE

Multiply.

4. $6(-10)$
5. $-15(-3)$
6. $(-2)^2$

ALGEBRA Simplify each expression.

7. $-5(2a)$
8. $3(-6b)$
9. $-5(-9c)$

ALGEBRA Evaluate each expression if $f = -1$, $g = 7$, and $h = -10$.

10. $5f$
11. fgh
12. $-h^2$

13. **SUBMARINES** A submarine is diving from the surface of the water at a rate of 125 feet per minute. What is the depth of the submarine after 7 minutes?

Practice and Applications

Multiply.

14. $8(-13)$
15. $-16(-5)$
16. $(-9)^2$
17. $-10(-17)$
18. $-7(16)$
19. $(-6)^2$
20. $-20(-8)$
21. $-15(30)$
22. $-31(-5)$
23. $11(-20)$
24. $-(7^2)$
25. $(-4)^3$

26. Find the product of -13 and 13.

27. Find -7 squared.

HOMEWORK HELP

For Exercises	See Examples
14–27, 44	1–4
28–35	5
36–43	6

Extra Practice
See pages 571, 598.

ALGEBRA Simplify each expression.

28. $-3(6c)$
29. $-7(10d)$
30. $5(-4e)$
31. $9(-8f)$
32. $-2(-3g)$
33. $-6(-4h)$
34. $(2x)(-3y)$
35. $(-5r)(2s)$

ALGEBRA Evaluate each expression if $w = 7$, $x = -8$, $y = 5$, and $z = 10$.

36. $-4w$
37. xy
38. $-2xz$
39. xyz
40. $-7wy$
41. $-3z^2$
42. $12x^2$
43. $-wz^2$

44. **VOLUNTEERING** The Volunteer Club raked leaves at several senior citizens' homes in the neighborhood. If each group of three students could remove 8 cubic meters of leaves in one hour, find an integer to represent the number of cubic meters of leaves 12 students could remove in five hours.

45. **PATTERNS** Find the next two numbers in the pattern $1, -2, 4, -8, 16, \ldots$. Then describe the pattern.

GEOMETRY For Exercises 46–48, use the graph at the right.

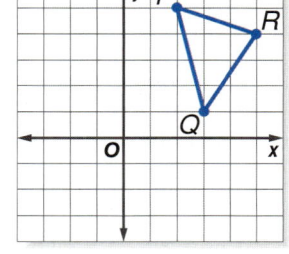

46. Name the ordered pairs for *P*, *Q*, and *R*. Multiply each *x*- and *y*-coordinate by −1 to get three new ordered pairs.

47. Graph the ordered pairs and connect them to form a new triangle. Describe its position with respect to the original triangle.

48. In which quadrant does a triangle lie if only the *y*-coordinates of the original triangle are multiplied by −2?

49. **CRITICAL THINKING** For what values of *n* is $(-2)^n$ positive?

Spiral Review with Standardized Test Practice

50. **MULTIPLE CHOICE** An oil rig is drilling into the ground at a rate of 7 feet per minute. What integer represents the position of the oil rig after 42 minutes?

    (A) −294 ft   (B) −35 ft   (C) 35 ft   (D) 294 ft

51. **MULTIPLE CHOICE** Monifa has 100 shares of stock each worth $15. If the price drops to $8, what integer represents the change in Monifa's current investment?

    (F) −$700   (G) −$7   (H) $7   (I) $700

52. **GRID IN** Evaluate $-6[-2(3) + 0(-5)] + (-4)^2$.

53. Find $-25 - (-33)$. (Lesson 3-5)

ALGEBRA Evaluate each expression if $x = -4$, $y = 6$, and $z = 1$. (Lesson 3-4)

54. $x + (-2)$ 55. $-1 + z$ 56. $-15 + y$ 57. $x + y$

58. **EARTH SCIENCE** The low temperatures in degrees Fahrenheit for ten cities on January 23 were −3, 27, 13, −6, −14, 36, 47, 52, −2, and 0. Order these temperatures from greatest to least. (Lesson 3-2)

MILITARY For Exercises 59 and 60, use the double-bar graph at the right. (Lesson 2-7)

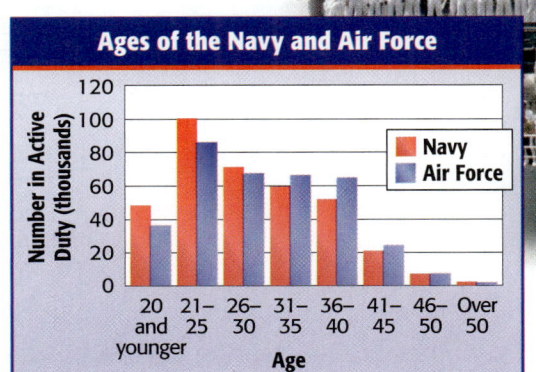

59. In which age groups are there more members in the Navy than members in the Air Force?

60. About how many 36–40 year olds are in the Air Force?

GETTING READY FOR THE NEXT LESSON

BASIC SKILL Divide.

61. $72 \div 9$ 62. $120 \div 6$ 63. $84 \div 21$ 64. $215 \div 43$

3-7 Dividing Integers

HANDS-ON Mini Lab

Materials
- counters
- integer mat

What You'll LEARN

Divide integers.

Work with a partner.

You can use counters to model division of integers. Follow these steps to find $-10 \div 5$.

Place 10 negative counters on the mat.

Separate the counters into 5 equal groups.

There are 2 negative counters in each group. So, $-10 \div 5 = -2$.

Find each quotient using counters.

1. $-6 \div 2$
2. $-12 \div 3$

Division of integers is related to multiplication. When finding the quotient of two integers, you can use a related multiplication sentence.

The factor in the multiplication sentence . . .

$2(-6) = -12 \rightarrow -12 \div 2 = -6$
$4(-5) = -20 \rightarrow -20 \div 4 = -5$

. . . is the quotient in the division sentence.

Since multiplication and division sentences are related, you can use them to find the quotient of integers with different signs.

different signs

$8(-9) = -72 \rightarrow -72 \div 8 = -9$
$-8(-9) = 72 \rightarrow 72 \div (-8) = -9$

negative quotient

different signs

These related sentences lead to the following rule.

Noteables™ Key Concept: Divide Integers with Different Signs

Words The quotient of two integers with different signs is negative.

Examples $33 \div (-11) = -3$ $-64 \div 8 = -8$

138 Chapter 3 Algebra: Integers

READING Math

Division Sentence
In the division sentence $80 \div (-10) = -8$, 80 is called the *dividend*, -10 is called the *divisor*, and -8 is called the *quotient*.

EXAMPLES — Dividing Integers with Different Signs

Divide.

1 $80 \div (-10)$ The integers have different signs.
$80 \div (-10) = -8$ The quotient is negative.

2 $\dfrac{-55}{11}$ The integers have different signs.
$\dfrac{-55}{11} = -5$ The quotient is negative.

Your Turn Divide.

a. $20 \div (-4)$ b. $\dfrac{-81}{9}$ c. $-45 \div 9$

The quotient of two integers with the same sign is positive.

Noteables™   **Key Concept: Divide Integers with the Same Sign**

Words   The quotient of two integers with the same sign is positive.
Examples $15 \div 5 = 3$ $-21 \div (-3) = 7$

EXAMPLE — Dividing Integers with the Same Sign

Divide.

3 $-14 \div (-7)$ The integers have the same sign.
$-14 \div (-7) = 2$ The quotient is positive.

Your Turn Divide.

d. $-24 \div (-4)$ e. $-9 \div (-3)$ f. $\dfrac{-28}{-7}$

EXAMPLE — Evaluate an Expression

4 **ASTRONOMY** The average surface temperature on Mars is $-81°$F. Use the expression $\dfrac{5(F - 32)}{9}$, where F represents the number of degrees Fahrenheit, to find the temperature on Mars in degrees Celsius.

$\dfrac{5(F - 32)}{9} = \dfrac{5(-81 - 32)}{9}$ Replace F with -81.

$= \dfrac{5(-113)}{9}$ Subtract 32 from -81.

$= \dfrac{-565}{9}$ Multiply 5 and -113.

≈ -62.8 Divide.

The average temperature on the surface of Mars is about $-63°$C.

REAL-LIFE MATH

ASTRONOMY In January 2004, Mars Exploration Rover successfully sent signals to Earth after landing on Mars.
Source: nasa.gov

msmath2.net/extra_examples   **Lesson 3-7** Dividing Integers

Skill and Concept Check

1. **Write** two division sentences related to the multiplication sentence $-6 \cdot 7 = -42$.

2. **OPEN ENDED** Write a division sentence. Then draw a model to show how the quotient can be found.

3. **Which One Doesn't Belong?** Identify the division expression whose quotient does not have the same sign as the other three. Explain your reasoning.

| $-24 \div 6$ | $-18 \div (-9)$ | $28 \div (-7)$ | $-22 \div 11$ |

GUIDED PRACTICE

Divide.

4. $32 \div (-8)$ 5. $-16 \div 2$ 6. $-60 \div (-5)$ 7. $\dfrac{-6}{6}$

ALGEBRA Evaluate each expression if $d = -9$, $e = 36$, and $f = -6$.

8. $-108 \div f$ 9. $e \div d$ 10. $\dfrac{e - f}{f}$

Practice and Applications

Divide.

11. $-18 \div 9$
12. $50 \div (-5)$
13. $-15 \div (-3)$
14. $\dfrac{21}{-7}$
15. $56 \div (-8)$
16. $\dfrac{0}{-5}$
17. $-52 \div (-13)$
18. $-34 \div 2$
19. $\dfrac{90}{6}$
20. $-300 \div 25$
21. $99 \div (-99)$
22. $-184 \div (-23)$

HOMEWORK HELP

For Exercises	See Examples
11–24	1–3
25–37	4

Extra Practice
See pages 571, 598.

23. Find the quotient of -65 and 13.

24. Divide 200 by -100.

ALGEBRA Evaluate each expression if $r = 12$, $s = -4$, and $t = -6$.

25. $-12 \div r$
26. $72 \div t$
27. $r \div s$
28. $rs \div 16$
29. $\dfrac{-r}{t}$
30. $\dfrac{16 - (-r)}{-s}$
31. $t^2 \div r$
32. $\dfrac{r^2}{s^2}$

33. **FOOTBALL** During the fourth quarter, the Colts were penalized 3 times for the same amount for a total of 45 yards. Write a division sentence to represent this situation. Then find the number of yards for each penalty.

34. **EARTH SCIENCE** Use the expression $\dfrac{5(F - 32)}{9}$, where F represents the number of degrees Fahrenheit, to convert 5°F to degrees Celsius.

35. **PATTERNS** Find the next two numbers in the pattern 729, -243, 81, -27, 9, ... Explain your reasoning.

140 Chapter 3 Algebra: Integers

36. **SALES** The graph shows five magazines that had losses in a recent year. The numbers represent the profit the magazines made in 2006 compared to 2005. What is the mean of the losses for these five magazines?

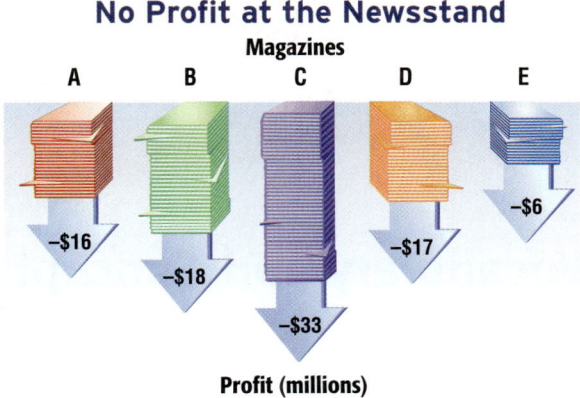

37. **MULTI STEP** The sea otter population is increasing. There were 2,377 sea otters in 1995. The population rose to 2,505 in 2003. Find the average rate of change for the sea otter population between 1995 and 2003.

38. **WRITE A PROBLEM** Write about a situation in your life in which you used positive and negative integers. Create a problem and solve it using any of the four operations.

39. **CRITICAL THINKING** List all of the numbers by which -20 is divisible.

Spiral Review with Standardized Test Practice

40. **MULTIPLE CHOICE** Shenequa is driving cross-country. If she expects to drive between 350 and 450 miles per day, which number of days is reasonable for her to drive 3,800 miles?

    A fewer than 6 days
    B between 6 and 8 days
    C between 9 and 11 days
    D more than 11 days

41. **MULTIPLE CHOICE** The width of a beach at 8:00 P.M. is 107 feet. At 3:00 A.M., the width of the beach narrows to 23 feet due to the high tide. What is the average rate of change per hour of the beach width?

    F -15 ft/h   G -13 ft/h   H -12 ft/h   I -10 ft/h

42. **SHORT RESPONSE** Find the mean of the following low temperatures for a 5-day period: $-2°F$, $-3°F$, $1°F$, $4°F$, and $-5°F$.

Multiply. (Lesson 3–6)

43. $14(-2)$ 44. $(-9)^2$ 45. $-20(-3)$ 46. $-5(7)$

47. Find $6 - (-12)$. (Lesson 3-5)

48. **DIVING** Mark jumped into 12 feet of water and touched the bottom of the pool before he surfaced. Which integer describes where Mark was in relation to the surface of the water? (Lesson 3-1)

49. **PHYSICAL SCIENCE** A chemistry experiment requires 3 milligrams of potassium chloride. How many grams of potassium chloride are needed? (Lesson 1-8)

CHAPTER 3 Study Guide and Review

Vocabulary and Concept Check

absolute value (p. 107)
additive inverse (p. 121)
coordinate grid (p. 112)
coordinate plane (p. 112)
graph (p. 106)
integer (p. 106)
negative integer (p. 106)
opposite (p. 121)
ordered pair (p. 112)
origin (p. 112)
positive integer (p. 106)
quadrant (p. 113)
x-axis (p. 112)
x-coordinate (p. 112)
y-axis (p. 112)
y-coordinate (p. 112)

Choose the correct term or number to complete each sentence.

1. Integers less than zero are (positive, negative) integers.
2. Two numbers represented by points that are the same distance from 0 are (opposites, integers).
3. The absolute value of 7 is (7, −7).
4. The opposite of (−12, 12) is −12.
5. The (coordinate plane, origin) is the point where the horizontal and vertical number lines intersect.
6. The x-axis and the y-axis separate the plane into four (quadrants, coordinates).
7. The first number in an ordered pair is the (x-coordinate, y-coordinate).
8. The sum of two (positive, negative) integers is negative.
9. The product of a positive and a negative integer is (positive, negative).
10. The quotient of a negative integer and a (positive, negative) integer is negative.

Lesson-by-Lesson Exercises and Examples

3-1 Integers and Absolute Value (pp. 106–108)

Write an integer for each situation.

11. a loss of $150
12. 350 feet above sea level
13. a gain of 8 yards
14. 12°F below 0

Evaluate each expression.

15. $|-11|$
16. $|100|$
17. $|5|$
18. $|-32|$
19. $|-16| + |9|$

Example 1 Write an integer for 8 feet below sea level.

Since this situation represents an elevation *below* sea level, −8 represents the situation.

Example 2 Evaluate $|-10|$.

On the number line, the graph of −10 is 10 units from 0.

So, $|-10| = 10$.

142 Chapter 3 Algebra: Integers

 msmath2.net/vocabulary_review

3-2 Comparing and Ordering Integers (pp. 109–111)

Replace each ● with < or > to make a true sentence.

20. −3 ● −9
21. 8 ● −12
22. −3 ● 3
23. |−10| ● |−13|
24. 25 ● |8|
25. 0 ● |−4|

Order each set of integers from least to greatest.

26. −3, 8, −10, 0, 5, −12, 9
27. −21, 19, −23, 14, −32, 25
28. |−17|, −18, 18, |15|, −16, |16|

29. **EARTH SCIENCE** The predicted low temperatures for Monday through Friday are 3°, −1°, −2°, 0°, and 1°. Order the temperatures from greatest to least.

Example 3 Replace ● with < or > to make −4 ● −7 a true sentence.

Graph each integer on a number line.

Since −4 is to the right of −7, −4 > −7.

Example 4 Order the integers −4, −3, 5, 3, 0, −2 from least to greatest.

Graph the integers on a number line.

Order the integers by reading from left to right: −4, −3, −2, 0, 3, 5.

3-3 Geometry: The Coordinate Plane (pp. 112–115)

Name the ordered pair for each point graphed at the right. Then identify the quadrant in which each point lies.

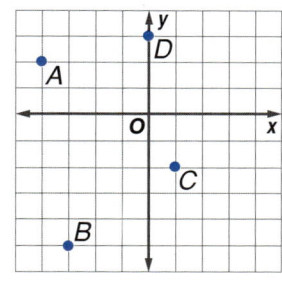

30. A
31. B
32. C
33. D

On graph paper, draw a coordinate plane. Then graph and label each point.

34. E(1, −4)
35. F(−5, 2)
36. G(−2, −3)
37. H(4, 0)

Example 5 Name the ordered pair for point W graphed at the right. Then identify the quadrant in which point W lies.

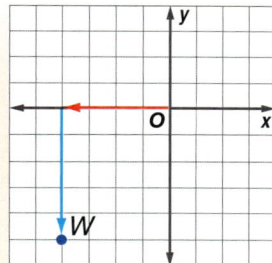

The ordered pair is (−4, −5). Point W is in quadrant III.

Example 6 Graph and label the point S(3, −1).

Draw a coordinate plane. Move 3 units to the right. Then move 1 unit down. Draw a dot and label it S(3, −1).

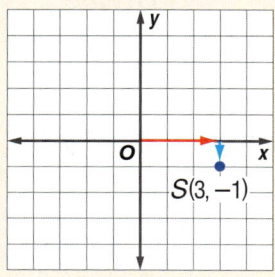

Study Guide and Review continued

Mixed Problem Solving
For mixed problem-solving practice, see page 598.

3-4 Adding Integers (pp. 120–124)

Add.
38. $-6 + 8$
39. $-4 + (-9)$
40. $7 + (-12)$
41. $-18 + 18$

42. **FOOTBALL** On the first play of the game, the Bulldogs lost 8 yards. On the second and third plays, they gained 5 yards and then lost 2 yards. Find the result of the first three plays.

Example 7 Find $-4 + 3$.

So, $-4 + 3 = -1$.

3-5 Subtracting Integers (pp. 128–131)

Subtract.
43. $-5 - 8$
44. $3 - 6$
45. $5 - (-2)$
46. $-4 - (-8)$

Example 8 Find $-3 - 9$.
$-3 - 9 = -3 + (-9)$ To subtract 9, add -9.
$ = -12$ Simplify.

3-6 Multiplying Integers (pp. 134–137)

Multiply.
47. $-4(3)$
48. $8(-6)$
49. $-5(-7)$
50. $-2(40)$

ALGEBRA Evaluate each expression if $a = -4$, $b = -7$, and $c = 5$.
51. ab
52. $-3c$
53. bc
54. abc

Example 9 Find $-4(3)$.
$-4(3) = -12$ The integers have different signs. The product is negative.

Example 10 Evaluate xyz if $x = -6$, $y = 11$, and $z = -10$.
$xyz = (-6)(11)(-10)$ $x = -6, y = 11, z = -10$
$ = (-66)(-10)$ Multiply -6 and 11.
$ = 660$ Multiply -66 and -10.

3-7 Dividing Integers (pp. 138–141)

Divide.
55. $-45 \div (-9)$
56. $36 \div (-12)$
57. $-12 \div 6$
58. $-81 \div (-9)$

59. **HIKING** Marta started a hike at sea level and ended the hike 6 hours later at 300 feet below sea level. If Marta hiked at the same pace during the trip, how far did she travel each hour?

Example 11 Find $-72 \div (-9)$.
$-72 \div (-9) = 8$ The integers have the same sign. The quotient is positive.

CHAPTER 3 Practice Test

Vocabulary and Concepts

1. **Explain** what it means for two numbers to be opposites.
2. **Name** the rule for dividing integers with different signs.

Skills and Applications

Write an integer for each situation.

3. a stock increased by $5
4. 1000 B.C.
5. an elevator goes down 11 floors

Replace each ● with < or > to make a true sentence.

6. -3 ● -9
7. $|9|$ ● $|-12|$
8. $|-7|$ ● 9

9. **WEATHER** The local weather service records the following changes in temperature during the last week: 4, −7, −3, 2, 9, −8, 1. Order these temperature changes from greatest to least.

Name the ordered pair for each point graphed at the right. Then identify the quadrant in which each point lies.

10. P
11. Q
12. R

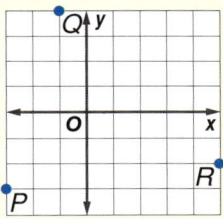

Add, subtract, multiply, or divide.

13. $-3 + 8$
14. $12 + (-19)$
15. $-3 - 8$
16. $-7 - (-20)$
17. $-7(-3)$
18. $5(-11)$
19. $-24 \div 8$
20. $-36 \div (-9)$

ALGEBRA Evaluate each expression if $a = -5$, $b = 4$, and $c = -12$.

21. $c - a$
22. ab
23. $ac \div b$

24. **STOCK MARKET** The value of a stock went down $3 each week for a period of seven weeks. Describe the change in the value of the stock at the end of the seven week period.

Standardized Test Practice

25. **MULTIPLE CHOICE** Choose the graph that shows the ordered pair (2, −1).

CHAPTER 3 Standardized Test Practice

PART 1 Multiple Choice

Record your answers on the answer sheet provided by your teacher or on a sheet of paper.

1. The table shows the five most common languages spoken in the United States other than English. How many more people speak Chinese than French? (Lesson 1-1)

Language	Number of People
Spanish	26,745,067
Chinese	1,976,564
French	1,914,918
German	1,224,213
Tagalog	1,184,939

   Source: U.S. Census Bureau

   - Ⓐ 61,646
   - Ⓑ 1,914,918
   - Ⓒ 1,976,564
   - Ⓓ 3,891,482

2. Which is equivalent to 5^4? (Lesson 1-2)
   - Ⓕ 20
   - Ⓖ 125
   - Ⓗ $5 \cdot 5 \cdot 5 \cdot 5$
   - Ⓘ $4 \cdot 4 \cdot 4 \cdot 4 \cdot 4$

3. How many millimeters are in 13 centimeters? (Lesson 1-8)
   - Ⓐ 0.13
   - Ⓑ 1.3
   - Ⓒ 13
   - Ⓓ 130

4. Salvador recorded the number of minutes it took him to drive to work each day for a week. Find the mean for the following times: 12, 23, 10, 14, and 11. (Lesson 2-4)
   - Ⓕ 12 min
   - Ⓖ 13 min
   - Ⓗ 14 min
   - Ⓘ 15 min

5. Find the interquartile range of the data in the box-and-whisker plot. (Lesson 2-6)

   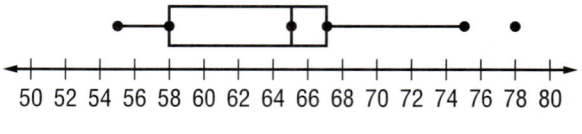

   - Ⓐ 9
   - Ⓑ 23
   - Ⓒ 58
   - Ⓓ 65

6. What is the value of $|-2|$? (Lesson 3-1)
   - Ⓕ -2
   - Ⓖ -1
   - Ⓗ 0
   - Ⓘ 2

7. Which of these is the correct order of the integers from greatest to least? (Lesson 3-2)
   - Ⓐ $0, 1, -2, -5$
   - Ⓑ $1, 0, -2, -5$
   - Ⓒ $-5, -2, 0, 1$
   - Ⓓ $1, -2, -5, 0$

8. Which of these are the coordinates of point Z? (Lesson 3-3)
   - Ⓕ (2, 2)
   - Ⓖ (3, 3)
   - Ⓗ (−2, −2)
   - Ⓘ (−3, −3)

   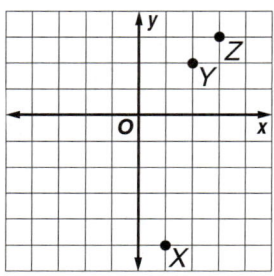

9. If you graph and connect the following points on a coordinate plane, what shape would you make? (Lesson 3-3)

   $(0, -3), (0, 4), (4, 4), (4, -3)$

   - Ⓐ rectangle
   - Ⓑ triangle
   - Ⓒ circle
   - Ⓓ parallelogram

10. Which of the following points on the number line represents opposites? (Lesson 3-4)

    - Ⓕ C and D
    - Ⓖ B and D
    - Ⓗ A and D
    - Ⓘ A and C

TEST-TAKING TIP

Question 10 Always be sure to check every answer choice of a multiple-choice question. Start with answer choice F. Each time you find an incorrect answer choice, cross it off so you remember that you've eliminated it.

PART 2 Short Response/Grid In

Record your answers on the answer sheet provided by your teacher or on a sheet of paper.

11. Wallpaper costs $16 per roll, and border costs $9 per roll. If 12 rolls of wallpaper and 6 rolls of border are needed for one room, find the total cost of the wallpaper and border. (Lesson 1-3)

12. Each triangle in the figure below is made from three toothpicks. Extend the pattern. Find the number of toothpicks in the fifth figure. (Lesson 1-7)

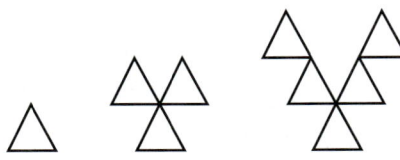

13. Write 63.5 million in scientific notation. (Lesson 1-9)

For Questions 14 and 15, use the information below.
The graph shows the number of orders taken each hour one day at a fast food restaurant. (Lesson 2-2)

14. About how many orders were taken at 11:00 A.M.?

15. What time of the day appears to be the least busy at this restaurant?

For Questions 16 and 17, use the information below.
The temperature of the liquid in Connor's beaker changed drastically. The temperatures he recorded were $-28°F$, $59°F$, $1°F$, $-16°F$, $24°F$, and $8°F$. (Lesson 3-2)

16. Order the temperatures from least to greatest.

17. Find the mean temperature.

18. When Elena went hiking at 8 A.M., she started at an elevation of 16 meters below sea level. But at 2 P.M., she was 12 meters higher than when she started. What was Elena's elevation at 2 P.M.? (Lesson 3-5)

19. Find $\dfrac{-12 + 8(-3)}{-2 - 7}$. (Lesson 3-7)

PART 3 Extended Response

Record your answers on a sheet of paper. Show your work.

20. Use the table of ordered pairs at the right to answer the questions below. (Lesson 3-3)

x	y
-2	1
3	2
-1	-4

    a. Plot the ordered pairs in the table on a coordinate plane.

    b. In which quadrant is a point on the graph not represented?

    c. Connect the points on the graph. What shape do they form?

    d. How would you double the size of the figure you drew above?

    e. Add four more columns to the table. Label the columns $x + y$, $x - y$, xy, and $x \div y$. Complete the table.

Chapter 4
Algebra: Linear Equations and Functions

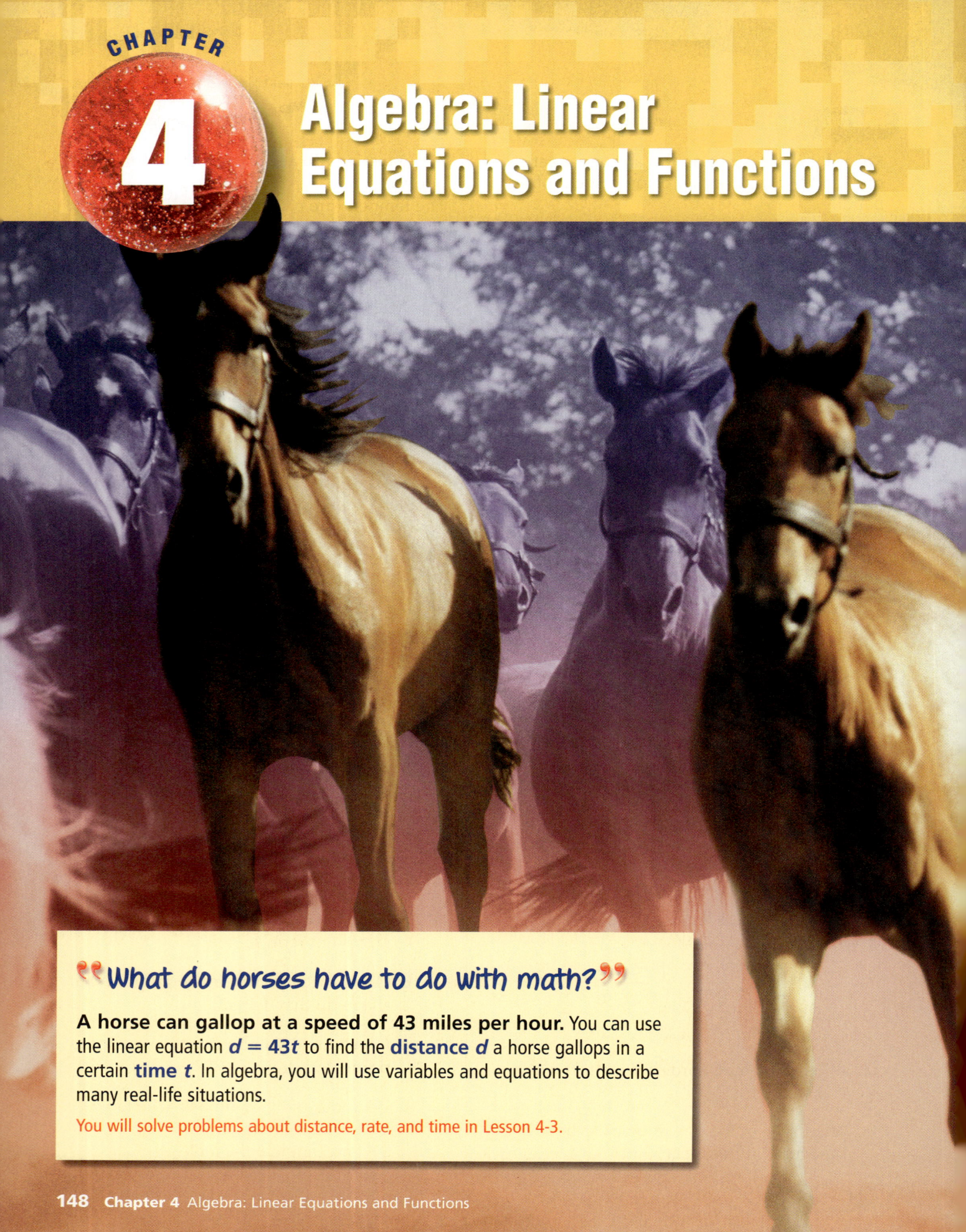

"What do horses have to do with math?"

A horse can gallop at a speed of 43 miles per hour. You can use the linear equation $d = 43t$ to find the **distance d** a horse gallops in a certain **time t**. In algebra, you will use variables and equations to describe many real-life situations.

You will solve problems about distance, rate, and time in Lesson 4-3.

GETTING STARTED

▶ Diagnose Readiness

Take this quiz to see if you are ready to begin Chapter 4. Refer to the lesson number in parentheses for review.

Vocabulary Review

Complete each sentence.

1. When you replace the variable with a number that makes an equation true, you have __?__ the equation. (Lesson 1-5)
2. The first number in an ordered pair is the __?__, and the second number is the __?__. (Lesson 3-3)

Prerequisite Skills

Name the number that is the solution of the given equation. (Lesson 1-5)

3. $a + 15 = 19$; 4, 5, 6
4. $11a = 77$; 6, 7, 8
5. $x + 9 = -2$; 7, -11, 11

Graph each point on a coordinate plane. (Lesson 3-3)

6. $(-4, 3)$
7. $(-2, -1)$

Add. (Lesson 3-4)

8. $-3 + (-5)$
9. $-8 + 3$
10. $9 + (-5)$
11. $-10 + 15$

Subtract. (Lesson 3-5)

12. $-5 - 6$
13. $8 - 10$
14. $8 - (-6)$
15. $-3 - (-1)$

Divide. (Lesson 3-7)

16. $-6 \div (-3)$
17. $-12 \div 3$
18. $10 \div (-5)$
19. $-24 \div (-4)$

Solving Equations Make this Foldable to help you organize your notes. Begin with a sheet of $8\frac{1}{2}" \times 11"$ paper.

 Fold
Fold the short sides toward the middle.

 Fold Again
Fold the top to the bottom.

 Cut
Open. Cut along the second fold to make four tabs.

 Label
Label each of the tabs as shown.

 Chapter Notes Each time you find this logo throughout the chapter, use your *Noteables™: Interactive Study Notebook with Foldables™* or your own notebook to take notes. Begin your chapter notes with this Foldable activity.

 Readiness To prepare yourself for this chapter with another quiz, visit **msmath2.net/chapter_readiness**

4-1 Writing Expressions and Equations

What You'll LEARN

Write verbal phrases and sentences as simple algebraic expressions and equations.

REVIEW Vocabulary

expression: a combination of variables, numbers, and at least one operation (Lesson 1-4)

WHEN am I ever going to use this?

COMICS Even these characters from *Peanuts* are using algebra.

1. Suppose the daughter is 12 years old. How old is the son?
2. What operation did you use to find the son's age? Explain.
3. Suppose the comic said that *the son is twice as old as the daughter.* If the daughter is 12 years old, how old is the son?
4. What operation did you use to find the son's age? Explain.

Words and phrases in problems often suggest addition, subtraction, multiplication, and division. Here are some examples.

Addition and Subtraction		Multiplication and Division	
plus	minus	times	divided
sum	difference	product	quotient
more than	less than	multiplied	per
increased by	less	twice	separate
in all	decreased by	each	

EXAMPLE   Write a Phrase as an Expression

1 Write the phrase *five dollars less than Jennifer earned* as an algebraic expression.

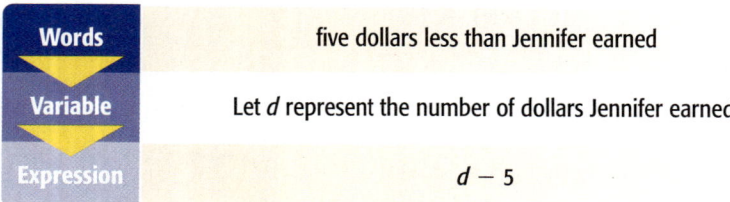

Your Turn Write each phrase as an algebraic expression.

a. twice as many tomatoes as last year
b. 3 more runs than the Pirates scored

150 Chapter 4 Algebra: Linear Equations and Functions

Remember, an equation is a sentence in mathematics that contains an equals sign. When you write a verbal sentence as an equation, you can use the equals sign (=) for the words *equals* or *is*.

EXAMPLES Write Sentences as Equations

Write each sentence as an algebraic equation.

Sentence	Equation
2 Five more than a number is 20.	$n + 5 = 20$
3 Three times Jack's age equals 12.	$3a = 12$

4 **FOOD** It is estimated that 12.4 million pounds of potato chips were consumed during a recent Super Bowl. This was 3.1 million pounds more than the number of pounds of tortilla chips consumed. Write an equation that models this situation.

Words	Potato chips were 3.1 million more than tortilla chips.
Variable	Let t = number of millions of pounds of tortilla chips.
Equation	$12.4 = 3.1 + t$

The equation is $12.4 = 3.1 + t$.

Skill and Concept Check

1. **OPEN ENDED** Write a verbal sentence that translates into the equation $n + 5 = 8$.

2. **FIND THE ERROR** Antonio and Julia are writing an algebraic expression for the phrase *2 less than a number*. Who is correct? Explain.

Antonio
2 − n

Julia
n − 2

GUIDED PRACTICE

Write each phrase as an algebraic expression.

3. eight more than x
4. nine less than t
5. ten times as many hours
6. -15 divided by some number

Write each sentence as an algebraic equation.

7. The sum of a number and four is -9.
8. Twice a number equals -10.

9. **POPULATION** The median age of people living in the United States was 35.3 years in 2000. This is 2.4 years older than the median age in 1990. Write an equation that models this situation.

Practice and Applications

Write each phrase as an algebraic expression.

10. fifteen increased by t
11. five years older than Luis
12. nine dollars less than j
13. a number less six
14. the product of r and 8
15. twice as many oranges
16. Emily's age divided by 3
17. a number divided by -12

HOMEWORK HELP

For Exercises	See Examples
10–17, 26–27	1
18–25, 28	2–4

Extra Practice
See pages 572, 599.

Write each sentence as an algebraic equation.

18. The sum of a number and four is -8.
19. Two more than the number of cookies is 4.
20. The product of a number and five is -20.
21. Ten times the number of students is 280.
22. Ten inches less than her height equals 26.
23. Five less than a number equals 31.
24. Seven more than twice his age is 51.
25. Three more than twice a number is 15.

MONEY For Exercises 26 and 27, use the table.
The table shows the average lifespan of several kinds of paper currency in the United States. Let y represent the average lifespan of a $5 bill.

26. Which lifespan can be represented by $2y$?
27. Write an expression to represent the lifespan of a $50 bill.

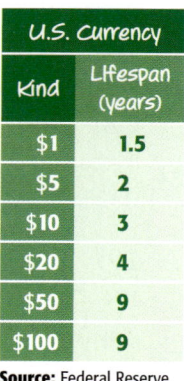

U.S. Currency	
Kind	Lifespan (years)
$1	1.5
$5	2
$10	3
$20	4
$50	9
$100	9

Source: Federal Reserve System

28. **TOURISM** The Washington Monument is 555 feet tall. It is 75 feet shorter than the Gateway to the West Arch. Write an equation that models this situation.

29. **CRITICAL THINKING** If x is an odd number, how would you represent the odd number immediately following it? preceding it?

Spiral Review with Standardized Test Practice

30. **MULTIPLE CHOICE** A mechanic charges a $35 initial fee and $32.50 for each hour he works. Which equation could be used to find the cost c of a repair job that lasts h hours?

   Ⓐ $c = 32.5 + 35h$
   Ⓑ $c = 35 + 32.5h$
   Ⓒ $c = 32.5 - 35h$
   Ⓓ $c = 32.5(35 - h)$

31. **MULTIPLE CHOICE** Translate *12 more than d* into an algebraic expression.

   Ⓕ $12d$
   Ⓖ $d - 12$
   Ⓗ $d + 12$
   Ⓘ $12 - d$

Divide. (Lesson 3-7)

32. $-18 \div 3$
33. $25 \div (-5)$
34. $-14 \div (-7)$
35. $72 \div (-9)$

GETTING READY FOR THE NEXT LESSON

PREREQUISITE SKILL Add. (Lesson 3-4)

36. $-8 + (-3)$
37. $-10 + 9$
38. $12 + (-20)$
39. $-15 + 15$

Study Skill

Simplify the Problem

Reading Math Problems

Have you ever tried to solve a long word problem and didn't know where to start? Try to rewrite the problem using fewer and fewer words. Then translate the problem into an equation.

▶ **READ** the problem.

> Shopping networks on television are a popular way to shop. In addition to the cost of the items, you usually pay a shipping fee. Kylie wants to order several pairs of running shorts that cost $12 each. The total shipping fee is $7. How many shorts can she order with $55?

▶ **REWRITE** the problem to make it simpler. Keep all of the important information but use fewer words.

Kylie has $55 to spend on some shorts that cost $12 each plus a shipping fee of $7. How many can she buy?

▶ **REWRITE** the problem using even fewer words. Write a variable for the unknown.

The total cost of x shorts at $12 each plus $7 is $55.

▶ **TRANSLATE** the words into an equation.

$12x + 7 = 55$

SKILL PRACTICE
Use the method above to write an equation for each word problem.

1. **FLYING** Orville and Wilbur Wright flew their airplane called Flyer I in Kitty Hawk, North Carolina, on December 17, 1903. Wilbur's flight was 364 feet. This was 120 feet longer than Orville's flight. How far was Orville's flight?

2. **ANIMALS** The cougars that are found in the colder parts of North and South America are about 75 inches long. They are about 1.5 times longer than the cougars that are found in the tropical jungles of Central America. Find the length of the tropical cougar.

3. **MONEY** Akira is saving money to buy a scooter that costs $125. He has already saved $80 and plans to save an additional $5 each week. In how many weeks will he have enough money for the scooter?

4. **TRAVEL** A taxi company charges $1.50 per mile plus a $10 fee. Suppose Olivia can afford to spend $19 for a taxi ride from her apartment to the mall. How far can she travel by taxi with $19?

4-2a Hands-On Lab

A Preview of Lesson 4-2

Solving Equations Using Models

INVESTIGATE The scale at the right is balanced, and the bag contains a certain number of blocks.

1. Suppose you cannot look in the bag. How can you find the number of blocks in the bag?
2. In what way is a balanced scale like an equation?
3. What does it mean to *solve an equation*?

What You'll LEARN
Solve equations using models.

REVIEW Vocabulary
equation: a mathematical sentence that contains an equals sign (Lesson 1-5)

To solve an equation using models, you can use these steps.

- You can add or subtract the same number of counters from each side of the mat.
- You can add or subtract zero from each side of the mat.

Materials
- cups and counters
- equation mat

ACTIVITY *Work with a partner.*

1 Solve $x + 2 = 5$ using models.

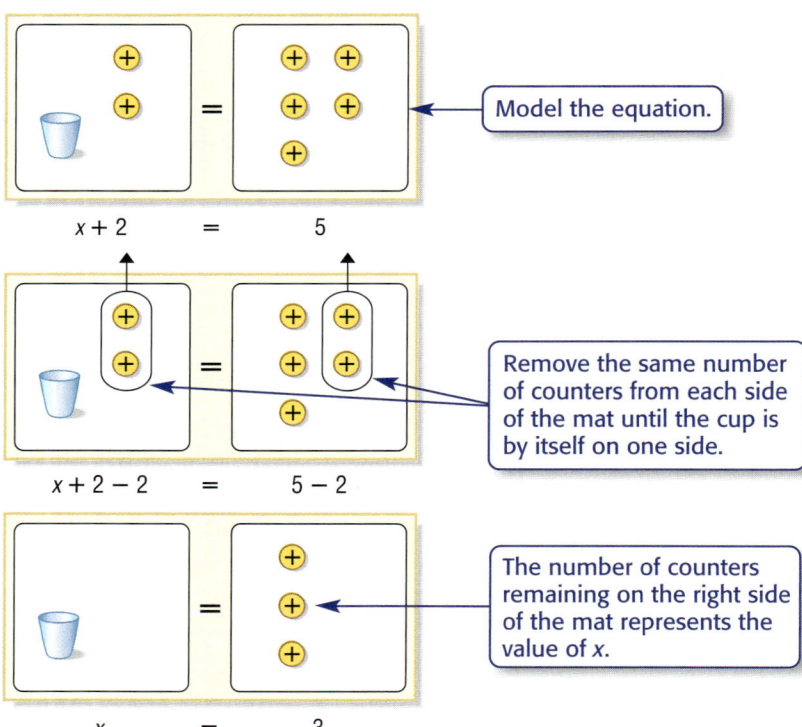

Therefore, $x = 3$. Since $3 + 2 = 5$, the solution is correct.

Your Turn Solve each equation using models.

a. $x + 1 = 3$ b. $x + 3 = 7$ c. $x + 4 = 4$

154 Chapter 4 Algebra: Linear Equations and Functions

STUDY TIP

Look Back You can review **zero pairs** in Lesson 3-4.

Some equations are solved by using **zero pairs**. You can add or subtract a zero pair from either side of an equation without changing its value, because the value of a zero pair is zero.

ACTIVITY Work with a partner.

2 Solve $x + 2 = -1$ using models.

Therefore, $x = -3$. Since $-3 + 2 = -1$, the solution is correct.

Your Turn Solve each equation using models.

d. $x + 3 = -2$ e. $x + 4 = 1$ f. $-2 = x + 1$
g. $x - 3 = -2$ h. $x - 1 = -3$ i. $4 = x - 2$

Writing Math

1. How is solving an equation similar to keeping a scale in balance?
2. For any equation, how can you determine how many counters to add or subtract from each side?
3. **Identify** the property of numbers that is illustrated by a zero pair.
4. **Identify** the property of numbers that allows you to add or subtract zero without changing the value of a number.
5. **MAKE A CONJECTURE** Write a rule that you can use to solve an equation like $x + 3 = 2$ without using models.

Lesson 4-2a Hands-On Lab: Solving Equations Using Models

4-2 Solving Addition and Subtraction Equations

What You'll LEARN
Solve addition and subtraction equations.

NEW Vocabulary
inverse operations

REVIEW Vocabulary
solve: finding a number that makes the equation true (Lesson 1-5)

WHEN am I ever going to use this?

SCIENCE Bottle-nosed dolphins and killer whales are the best-known species of the dolphin family. A killer whale, which can grow to a length of 9 meters, is 4 meters longer than a bottle-nosed dolphin.

1. What does x represent in the figure?
2. Write an expression to represent *4 meters longer than a dolphin*.
3. Write an addition equation you could use to find the length of a dolphin.

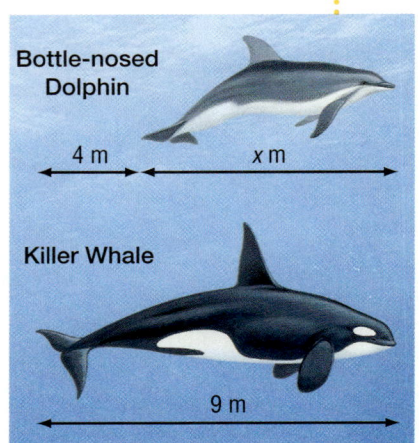

You can use inverse operations to solve the equation $x + 4 = 9$. **Inverse operations** "undo" each other. To undo the addition of 4 in the equation $x + 4 = 9$, you would subtract 4 from each side of the equation.

EXAMPLE Solve an Addition Equation

 Solve $x + 4 = 9$.

Method 1 Use symbols.

$x + 4 = 9$ Write the equation.

$\begin{aligned} x + 4 &= 9 \\ -4 &= -4 \end{aligned}$ Subtract 4 from each side.

$\begin{aligned} x + 4 &= 9 \\ -4 &= -4 \\ \hline x &= 5 \end{aligned}$ Simplify.

The solution is 5.

Method 2 Use models.

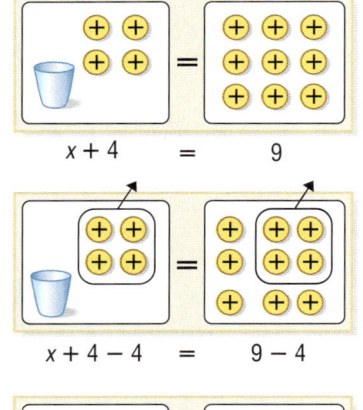

READING in the Content Area
For strategies in reading this lesson, visit **msmath2.net/reading**.

Your Turn Solve each equation.

a. $y + 6 = 9$ b. $x + 3 = 1$ c. $-3 = a + 4$

156 Chapter 4 Algebra: Linear Equations and Functions

You can use the **Subtraction Property of Equality** and the **Addition Property of Equality** to solve equations like $x + 4 = 9$ and $x - 2 = 1$.

Noteables™ Key Concept: Subtraction Property of Equality

Words If you subtract the same number from each side of an equation, the two sides remain equal.

Symbols

Arithmetic	Algebra
$5 = 5$	$x + 2 = 3$
$-3 = -3$	$-2 = -2$
$2 = 2$	$x = 1$

Noteables™ Key Concept: Addition Property of Equality

Words If you add the same number to each side of an equation, the two sides remain equal.

Symbols

Arithmetic	Algebra
$5 = 5$	$x - 2 = 4$
$+3 = +3$	$+2 = +2$
$8 = 8$	$x = 6$

EXAMPLE Solve a Subtraction Equation

2 Solve $1 = x - 2$. Check your solution.

Method 1 Use symbols.

$1 = x - 2$ Write the equation.

$1 = x - 2$
$+2 = +2$ Add 2 to each side.

$1 = x - 2$
$+2 = +2$
$3 = x$ Simplify.

Check $1 = x - 2$ Write the original equation.
$1 \stackrel{?}{=} 3 - 2$ Replace x with 3.
$1 = 1$ ✓ This sentence is true.

The solution is 3.

Method 2 Use models.

$1 = x - 2$

$1 + 2 = x - 2 + 2$

$3 = x$

Study Tip

Checking Solutions It is always wise to check your solution. You can often use arithmetic facts to check the solutions of simple equations.

Your Turn Solve each equation.

d. $y - 3 = 4$ **e.** $r - 4 = -2$ **f.** $-9 = q - 8$

EXAMPLE   Use an Equation to Solve a Problem

3 SPORTS Tiger Woods won the 2002 Masters Championship with a final score of 12 under par, or −12. His scores for the first three rounds were −2, −3, and −6. What was his score for the fourth round?

STUDY TIP

Look Back To review adding integers, see Lesson 3-4.

Words	The sum of the scores for all of the rounds was −12.
Variable	Let s represent the score for the fourth round.
Equation	scores for the first three rounds score for the fourth round final score $-2 + (-3) + (-6)\quad + \quad s \quad = \quad -12$

$-2 + (-3) + (-6) + s = -12$ Write the equation.

$ -11 + s = -12 \quad -2 + (-3) + (-6) = -11$

$ \underline{+11 = +11} \quad$ Add 11 to each side.

$ s = -1 \quad$ Simplify.

Check You can check the solution by adding.
$-2 + (-3) + (-6) + (-1) = -12$ ✓

Skill and Concept Check

1. **Tell** what property you would use to solve $x - 4 = -2$.

2. **OPEN ENDED** Write two different equations that have −2 as a solution.

3. **Which One Doesn't Belong?** Identify the equation that does not have the same solution as the other three. Explain your reasoning.

| $x + 3 = 4$ | $b + 5 = 4$ | $7 + y = 8$ | $8 + a = 9$ |

GUIDED PRACTICE

Solve each equation. Check your solution.

4. $n + 6 = 9$
5. $m + 5 = 3$
6. $-2 = a + 6$
7. $x - 5 = 6$
8. $-1 = c - 6$
9. $s - 4 = -7$

10. **GEOMETRY** The sum of the measures of the angles of a triangle is 180°. Find the missing measure.

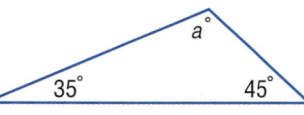

158 Chapter 4 Algebra: Linear Equations and Functions

Practice and Applications

Solve each equation. Check your solution.

11. $a + 3 = 10$
12. $y + 5 = 11$
13. $x + 8 = 5$
14. $y + 15 = 11$
15. $r + 6 = -3$
16. $k + 3 = -9$
17. $s - 8 = 9$
18. $w - 7 = 11$
19. $q - 8 = -1$
20. $p - 13 = -2$
21. $x - 5 = -9$
22. $w - 9 = -12$
23. $34 + r = 95$
24. $64 + y = 84$
25. $-23 = x - 18$
26. $-59 = m - 11$
27. $-18 + c = -30$
28. $-34 = t + 9$

29. The sum of a number and 3 is -2. Find the number.
30. If you decrease a number by 4, the result is -5. Find the number.

Solve each equation. Check your solution.

31. $a - 3.5 = 14.9$
32. $x - 2.8 = 9.5$
33. $r - 8.5 = -2.1$
34. $z - 9.4 = -3.6$
35. $n + 1.4 = 0.72$
36. $b + 2.25 = 1$

HOMEWORK HELP

For Exercises	See Examples
11–16	1
17–22	2
29–30, 37–38	3

Extra Practice
See pages 572, 599.

ROLLER COASTERS For Exercises 37 and 38, use the table.

37. *Superman The Escape* is 105 feet taller than *Millennium Force*. Write and solve an addition equation that you could use to find the height of *Millennium Force*.

38. The difference in the speeds of *Superman The Escape* and *Millennium Force* is 7 miles per hour. If *Superman The Escape* has the greater speed, write and solve a subtraction equation to find its speed.

Tallest Steel Roller Coasters in North America	Height (feet)	Speed (mph)
Superman The Escape	415	?
Millenium Force	?	93
Titan	245	85
Goliath	235	85
Nitro	230	80

Source: www.rcdb.com

39. **CRITICAL THINKING** Suppose $x + y = 10$ and the value of x increases by 3. What must happen to the value of y so that $x + y = 10$ is still a true sentence?

Spiral Review with Standardized Test Practice

40. **MULTIPLE CHOICE** The Sears Tower in Chicago is 1,454 feet tall. It is 204 feet taller than the Empire State Building in New York City. Use the equation $e + 204 = 1,454$ to find the height of the Empire State Building.

   Ⓐ 1,250 ft Ⓑ 1,350 ft Ⓒ 1,450 ft Ⓓ 1,650 ft

41. **GRID IN** If $x + 3 = 2$, what is the value of $2x + 5$?

42. **ALGEBRA** Write an expression for 6 inches less than w. (Lesson 4-1)

Divide. (Lesson 3-7)

43. $-15 \div 3$
44. $36 \div (-9)$
45. $-63 \div (-7)$
46. $-27 \div 9$

GETTING READY FOR THE NEXT LESSON

PREREQUISITE SKILL Divide. (Page 562)

47. $15.6 \div 13$
48. $8.84 \div 3.4$
49. $75.25 \div 0.25$
50. $0.76 \div 0.5$

4-3 Solving Multiplication Equations

What You'll LEARN

Solve multiplication equations.

REVIEW Vocabulary

coefficient: the numerical factor of a multiplication expression (Lesson 1-4)

HANDS-ON Mini Lab

Materials
- cups and counters
- equation mat

Work with a partner.

Equations like $2x = -6$ are called multiplication equations because the expression $2x$ means *2 times the value of x.* Follow these steps to solve $2x = -6$.

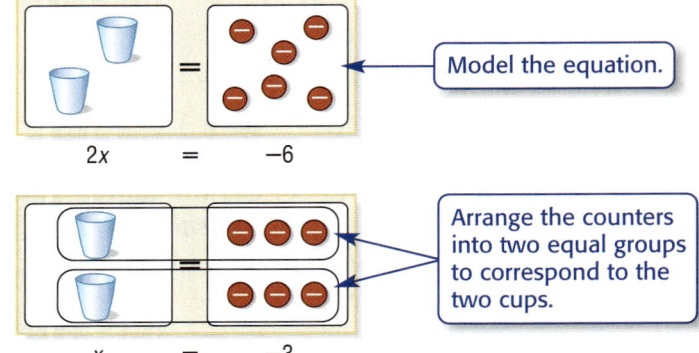

Model the equation.

$2x = -6$

Arrange the counters into two equal groups to correspond to the two cups.

$x = -3$

Each cup is matched with 3 negative counters. So, $x = -3$.

Solve each equation using models.

1. $4x = 8$
2. $-8 = 2x$
3. $3x = 3$
4. What operation did you use to find each solution?
5. How can you use the coefficient of x to solve $4x = 12$?
6. How can you use the coefficient of x to solve $-5x = 10$ without using cups and counters?

In the Mini Lab above, you matched each cup with an equal number of counters. This suggests the operation of division. You can use the **Division Property of Equality** to solve equations like $2x = -6$.

Noteables™ Key Concept: Division Property of Equality

Words If you divide each side of an equation by the same nonzero number, the two sides remain equal.

Symbols

Arithmetic
$$8 = 8$$
$$\frac{8}{2} = \frac{8}{2}$$
$$4 = 4$$

Algebra
$$2x = -6$$
$$\frac{2x}{2} = \frac{-6}{2}$$
$$x = -3$$

Notice that the division expression $8 \div 2$ can be written as the fraction $\frac{8}{2}$.

160 Chapter 4 Algebra: Linear Equations and Functions

EXAMPLES — Solve Multiplication Equations

1 Solve $20 = 4x$. Check your solution.

$20 = 4x$ Write the equation.

$\dfrac{20}{4} = \dfrac{4x}{4}$ Divide each side of the equation by 4.

$5 = x$ $20 \div 4 = 5$

Check $20 = 4x$ Write the original equation.

$20 \stackrel{?}{=} 4(5)$ Replace x with 5. Is this sentence true?

$20 = 20$ ✓

The solution is 5.

2 Solve $-8y = 24$. Check your solution.

$-8y = 24$ Write the equation.

$\dfrac{-8y}{-8} = \dfrac{24}{-8}$ Divide each side by -8.

$y = -3$ $24 \div (-8) = -3$

Check $-8y = 24$ Write the original equation.

$-8(-3) \stackrel{?}{=} 24$ Replace y with -3. Is this sentence true?

$24 = 24$ ✓

The solution is -3.

Your Turn Solve each equation.

a. $30 = 6x$ b. $-6a = 36$ c. $-9d = -72$

The equation $d = rt$ shows the relationship between the variables d (distance), r (rate or speed), and t (time).

EXAMPLE — Use an Equation to Solve a Problem

REAL-LIFE MATH

ANIMALS The tortoise is one of the slowest land animals, reaching a top speed of about 0.25 mile per hour. Dragonflies, the fastest insect, can fly at a speed of about 18 miles per hour.

Source: *Incredible Comparisons*

3 **ANIMALS** Refer to the information at the left. At this speed, how long will it take a tortoise to travel 1.5 miles?

Words	Distance is equal to the rate times the time.
Variables	$d = r \cdot t$
Equation	$1.5 = 0.25t$

$1.5 = 0.25t$ Write the equation.

$\dfrac{1.5}{0.25} = \dfrac{0.25t}{0.25}$ Divide each side by 0.25.

$6 = t$ $1.5 \div 0.25 = 6$

At this speed, it would take a tortoise 6 hours to travel 1.5 miles.   **Check this solution.**

Skill and Concept Check

1. **Tell** whether -4 is a solution of $-3x = -12$. Explain.

2. **OPEN ENDED** Write two different multiplication equations that have a negative integer as a solution.

3. **FIND THE ERROR** Jesse and Haley are solving $-5x = 30$. Who is correct? Explain.

Jesse
$-5x = 30$
$\dfrac{-5x}{-5} = \dfrac{30}{-5}$
$x = -6$

Haley
$-5x = 30$
$\dfrac{-5x}{5} = \dfrac{30}{5}$
$x = -6$

GUIDED PRACTICE

Solve each equation. Check your solution.

4. $6c = 18$
5. $10y = 20$
6. $-6s = 24$
7. $-9r = 36$
8. $-8z = -40$
9. $-11r = -77$
10. $15 = 5z$
11. $72 = 12r$

12. The product of a number and -4 is 64. Find the number.

13. If you multiply a number by 3, the result is -21. What is the number?

Practice and Applications

Solve each equation. Check your solution.

14. $7a = 49$
15. $9e = 27$
16. $35 = 5v$
17. $112 = 8p$
18. $4j = -36$
19. $12y = -60$
20. $48 = -6r$
21. $266 = -2t$
22. $-3w = -36$
23. $-10g = -100$
24. $-28 = -7f$
25. $-275 = -5s$

HOMEWORK HELP

For Exercises	See Examples
14–19	1
20–25	2
26–31, 38–39	3

Extra Practice
See pages 572, 599.

26. When a number is multiplied by -12, the result is -168. Find the number.

27. The product of a number and 25 is 1,000. What is the number?

28. **BABY-SITTING** Gracia earns \$5 per hour when she baby-sits. How many hours does she need to work to earn \$75?

29. **TRAVEL** A Boeing 747 aircraft has a cruising speed of about 600 miles per hour. At that speed, how long will it take to travel 1,500 miles? Use the formula $d = rt$.

KITES For Exercises 30 and 31, use the following information.
In a simple kite, the length of the longer stick should be 1.5 times the length of the shorter stick.

30. Suppose the length of the longer stick is 36 inches. Write a multiplication equation to find the length of the shorter stick.

31. Solve the equation.

162 Chapter 4 Algebra: Linear Equations and Functions

Solve each equation. Check your solution.

32. $4x = 9.2$
33. $9y = 13.5$
34. $-5.4 = 0.3p$
35. $-9.72 = 1.8a$
36. $3.9y = 18.33$
37. $2.6b = 2.08$

EARTHQUAKES For Exercises 38 and 39, use the following information.
Scientists determine the epicenter of an earthquake by measuring the time it takes for surface waves to travel between two places. Surface waves travel about 6 kilometers per second through Earth's crust.

38. The distance from Los Angeles, California, to Phoenix, Arizona, is 600 kilometers. Write a multiplication equation to find how long it would take surface waves to travel from Los Angeles to Phoenix.

39. Solve the equation.

ROLLER COASTERS For Exercises 40 and 41 use the table.

40. Without calculating, explain whether the *Blue Streak* or *Magnum* has the greater speed.

41. Find the speed of each roller coaster in feet per second.

42. **CRITICAL THINKING** Solve $3|x| = 12$.

Name	Track Length (ft)	Time of Ride
Blue Streak	2,558	1 min 45 s
Corkscrew	2,050	2 min
Magnum	5,106	2 min
Mean Streak	5,427	2 min 45 s

Source: www.rcdb.com

Spiral Review with Standardized Test Practice

43. **MULTIPLE CHOICE** A car is traveling at a speed of 75 feet per second. How many yards will the car travel in 90 seconds if it maintains the same speed?

    Ⓐ 75 yd   Ⓑ 270 yd   Ⓒ 2,250 yd   Ⓓ 6,750 yd

44. **MULTIPLE CHOICE** Use the formula $A = bh$ to find the height of a parallelogram with a base of 34 millimeters and an area of 612 square millimeters.

    Ⓕ 20,800 mm   Ⓖ 646 mm
    Ⓗ 578 mm   Ⓘ 18 mm

ALGEBRA Solve each equation. (Lesson 4-2)

45. $y + 8 = -2$
46. $x - 7 = -2$
47. $20 = z + 23$

ALGEBRA Write an algebraic expression for each phrase. (Lesson 4-1)

48. the product of -3 and y
49. 5 years older than Rafael
50. 10 fewer students than last year
51. twice as many runs as the Marlins scored

GETTING READY FOR THE NEXT LESSON

PREREQUISITE SKILL Subtract. (Lesson 3-5)

52. $8 - (-2)$
53. $-7 - 7$
54. $-3 - (-9)$
55. $-3 - 18$

Problem-Solving Strategy
A Preview of Lesson 4-4

Work Backward

What You'll LEARN

Solve problems using the work backward strategy.

"Alicia, how much money did we take to the movies today?"

"Well, we spent $9 on movie tickets. Then we spent $5 on candy, and one half of what was left on popcorn. We have $3 left. Let's **work backward** to figure it out."

Explore	We know we have $3 left and the amounts spent. We need to find the initial amount we had.	
Plan	Let's start with the end result and work backward.	
Solve	We have $3 left. **Undo** the half spent on popcorn.	3 × 2 ――― 6
	Undo the $5 spent on candy.	+ 5 ――― 11
	Undo the $9 spent on movie tickets. So, we initially had $20.	+ 9 ――― 20
Examine	Assume that we started with $20. After buying movie tickets, we had $20 − $9 or $11. We spent $5 on candy, so we had $11 − $5 or $6. Then we spent one half of the remaining money on popcorn, so we had $6 ÷ 2 or $3. So, our answer of $20 is correct.	

Analyze the Strategy

1. **Explain** when you would use the work backward strategy to solve a problem.
2. **Describe** how to solve a problem by working backward.
3. **Write** a problem that can be solved by working backward. Then write the steps you would take to find the solution to your problem.

164 Chapter 4 Algebra: Linear Equations and Functions

Apply the Strategy

Solve. Use the work backward strategy.

4. **MONEY** Mia spent $4.50 at the bakery and then spent four times that amount at the grocery. She had $12.80 left. How much money did she have initially?

5. **NUMBER THEORY** A number is multiplied by 4. Then 6 is subtracted from the product. After adding 5, the result is 15. What is the number?

Mixed Problem Solving

Solve. Use any strategy.

6. **PATTERNS** Draw the sixth figure in the pattern shown.

7. **FOOD** Mateo goes to the grocery store and buys a ham for $24.98 and a vegetable tray for $17.49. There is no tax. He gives the cashier one bill and receives less than $10 in change. What was the denomination of the bill Mateo gave the cashier?

8. **TIME** A shuttle bus schedule is shown. What is the earliest time after noon when the bus departs?

Departs	Arrives
8:55 A.M.	9:20 A.M.
9:43 A.M.	10:08 A.M.
10:31 A.M.	10:56 A.M.
11:19 A.M.	11:44 A.M.

9. **GEOGRAPHY** The land area of Texas is 267,277 square miles. This is about 5 times the land area of Arkansas. Estimate the land area of Arkansas.

10. **NUMBER THEORY** How many different two-digit numbers can you make using the digits 2, 4, and 9 if no digit is repeated within a number?

11. **AGE** Maya is two years older than her sister Jenna. Jenna is 5 years older than their brother Trevor, who is 9 years younger than their brother Trent. Trent is 17 years old. How old is Maya?

FOOD For Exercises 12 and 13, use the graph below.

12. How much more garlic was eaten in 2000 than in 1990?

13. Find the average monthly consumption of garlic in 2000.

14. **VIDEOS** Carlos rented 2 times as many DVDs as Ashley last month. Ashley rented 4 fewer than Greg, but four more than Grace. Greg rented 9 DVDs. How many DVDs did each person rent?

15. **STANDARDIZED TEST PRACTICE** Which of the following is the most reasonable total amount for the items purchased?

    A $17
    B $20
    C $26
    D $30

Pens	$ 2.09
Ruler	$ 0.99
Paper	$ 1.49
Book	$14.99
Candy	$ 0.49
Glue	$ 0.89
Folder	$ 1.19
Erasers	$ 1.99
Pencils	$ 1.87

Lesson 4-4a Problem-Solving Strategy: Work Backward

4-4 Solving Two-Step Equations

What You'll LEARN
Solve two-step equations.

NEW Vocabulary
two-step equation

HANDS-ON Mini Lab

Materials
- cups and counters
- equation mat

Work with a partner.

A **two-step equation** has two different operations. Follow these steps to solve $2x - 3 = 1$.

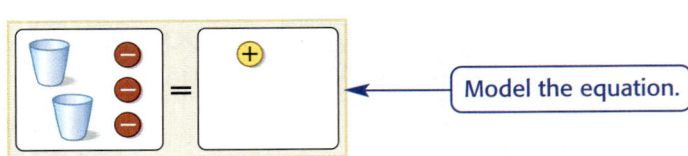

Model the equation.

$2x - 3 = 1$

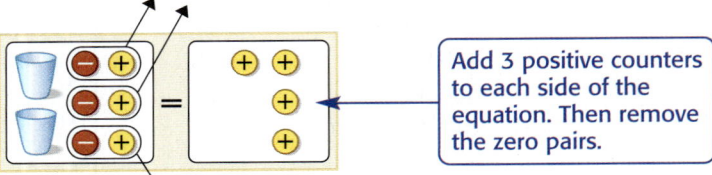

Add 3 positive counters to each side of the equation. Then remove the zero pairs.

$2x - 3 + 3 = 1 + 3$

Arrange the cups and counters into two equal groups.

$2x = 4$

Each cup is matched with 2 positive counters. So, $x = 2$.

Solve each equation using models.

1. $3x + 1 = 7$
2. $2x - 4 = 2$
3. $2x + 3 = -3$

To solve two-step equations, "undo" the operations in reverse order of the order of operations. You are using the *work backward* strategy.

EXAMPLE Solve a Two-Step Equation

1 Solve $3x + 2 = 23$. Check your solution.

$$3x + 2 = 23 \quad \text{Write the equation.}$$
$$\underline{\;-2 = -2} \quad \text{Subtract 2 from each side.}$$
$$3x = 21 \quad \text{Simplify.}$$
$$\frac{3x}{3} = \frac{21}{3} \quad \text{Divide each side by 3.}$$
$$x = 7 \quad \text{Simplify.}$$

The solution is 7. Check the solution.

166 Chapter 4 Algebra: Linear Equations and Functions

Study Tip

Solving Equations When you solve a simple equation like $-2y = 10$, you can mentally divide each side by -2.

EXAMPLES — Solve Two-Step Equations

Solve each equation. Check your solution.

2 $-2y - 7 = 3$

$$-2y - 7 = 3 \quad \text{Write the equation.}$$
$$\underline{+7 = +7} \quad \text{Add 7 to each side.}$$
$$-2y = 10 \quad \text{Simplify.}$$
$$\frac{-2y}{-2} = \frac{10}{-2} \quad \text{Divide each side by } -2.$$
$$y = -5 \quad \text{Simplify.}$$

The solution is -5. Check the solution.

3 $-11 = 4 + 5r$

$$-11 = 4 + 5r \quad \text{Write the equation.}$$
$$\underline{-4 = -4} \quad \text{Subtract 4 from each side.}$$
$$-15 = 5r \quad \text{Simplify.}$$
$$\frac{-15}{5} = \frac{5r}{5} \quad \text{Divide each side by 5.}$$
$$-3 = r \quad \text{Simplify.}$$

The solution is -3. Check the solution.

Your Turn Solve each equation.

a. $4x + 5 = 13$ b. $-3n - 8 = 7$ c. $1 + 2y = -3$

REAL-LIFE MATH

BOWLING *Handicapping* allows bowlers with varying skills to score about the same. Your handicap h is added to your score s, so your final score is $h + s$.
Source: www.bowl.com

Some problems start with a given amount and increase at a certain rate.

EXAMPLE — Use an Equation to Solve a Problem

4 **BOWLING** Suppose you spend $6 to rent bowling shoes at The Bowling Alley. Each game costs $3.50. How many games can you bowl if you have $20 to spend?

Variable Let $x =$ the number of games.

Words Cost of the shoes plus the cost of the games equals $20.

Equation $6 + 3.50x = 20$

$$6 + 3.50x = 20 \quad \text{Write the equation.}$$
$$\underline{-6 = -6} \quad \text{Subtract 6 from each side.}$$
$$3.50x = 14 \quad \text{Simplify.}$$
$$\frac{3.50x}{3.50} = \frac{14}{3.50} \quad \text{Divide each side by 3.50.}$$
$$x = 4 \quad 14 \div 3.50 = 4$$

You can bowl 4 games. Is this answer reasonable?

msmath2.net/extra_examples   Lesson 4-4 Solving Two-Step Equations

Skill and Concept Check

1. **OPEN ENDED** Write a two-step equation that has -2 as the solution.

2. **FIND THE ERROR** Jackson and Michele are solving $4 + 2x = -8$. Who is correct? Explain.

Jackson
$4 + 2x = -8$
$4 + \dfrac{2x}{2} = \dfrac{-8}{2}$
$4 + x = -4$
$x = -8$

Michele
$4 + 2x = -8$
$\underline{-4 = -4}$
$2x = -12$
$x = -6$

GUIDED PRACTICE

Solve each equation. Check your solution.

3. $3x + 1 = 7$
4. $4h - 6 = 22$
5. $-6r + 1 = -17$
6. $13 = 1 + 4s$

7. Five more than three times a number is 23. Find the number.

Practice and Applications

Solve each equation. Check your solution.

8. $3x + 1 = 10$
9. $5x + 4 = 19$
10. $2t + 7 = -1$
11. $6m + 1 = -23$
12. $-4w - 4 = 8$
13. $-7y + 3 = -25$
14. $-8s + 1 = 33$
15. $-2x + 5 = -13$
16. $3 + 8n = -5$
17. $5 + 4d = 37$
18. $14 + 2p = 8$
19. $25 + 2y = 47$
20. $2 = 3t - 13$
21. $57 = -8x - 7$
22. $18 = 9d - 18$
23. $4 = 4 + 7f$
24. $21 + 11x = -1$
25. $15x + 4 = 49$

HOMEWORK HELP

For Exercises	See Examples
8–11	1
12–15	2
16–19	3
26–29, 38–41	4

Extra Practice
See pages 573, 599.

26. Three more than the product of a number and 4 is 15. Find the number.
27. Five less than three times a number is 1. Find the number.
28. The product of 2 and a number is increased by 9. The result is -17. Find the number.
29. If you subtract 3 from twice a number, the result is 25. Find the number.

Solve each equation. Check your solution.

30. $2r - 3.1 = 1.7$
31. $4t + 3.5 = 12.5$
32. $16b - 6.5 = 9.5$
33. $5w + 9.2 = 19.7$
34. $16 = 0.5r - 8$
35. $0.2n + 3 = 8.6$
36. $7.5s - 2 = 28$
37. $1.5v - 16 = 8$

38. **MONEY MATTERS** Joshua has saved $74 toward a new sound system that costs $149. He plans on saving an additional $15 each week. How many weeks will it take Joshua to save enough money to buy the sound system?

39. **WEATHER** The temperature is 20°F. It is expected to rise at a rate of 4° each hour for the next several hours. In how many hours will the temperature be 32°?

168 Chapter 4 Algebra: Linear Equations and Functions

TEMPERATURE For Exercises 40 and 41, use the following information and the graph.

Temperature is usually measured on the Fahrenheit scale (°F) or the Celsius scale (°C). Use the formula $F = 1.8C + 32$ to convert from one scale to the other.

40. The highest temperature ever recorded in Virginia Beach, Virginia, was 104°F. Find this temperature in degrees Celsius.

41. **MULTI STEP** The lowest temperature ever recorded in Virginia Beach was −3°F. Is this temperature greater or less than the lowest temperature ever recorded in Paris, France?

Data Update What are the record high and low temperatures for a large city near you? Visit **msmath2.net/data_update** to learn more.

42. **CRITICAL THINKING** Is there a temperature at which the number of Celsius degrees is the same as the number of Fahrenheit degrees? If so, find it. If not, explain why not.

Spiral Review with Standardized Test Practice

43. **MULTIPLE CHOICE** A taxi driver charges $2.00 plus $0.80 for each mile traveled. Which expression could be used to find the cost of a taxi ride if m miles are traveled?

   A $2.80m$ **B** $2m + 0.80$ **C** $2 + 0.80m$ **D** $2m + 0.80m$

44. **SHORT RESPONSE** In 2004, the first-class postage rates were $0.37 for the first ounce plus an additional cost for each ounce.

Weight (oz)	1	2	3	4	5	6
Postage (dollars)	0.37	0.60	0.83	1.06	1.29	1.52

What is the cost, in dollars, for a 12-ounce letter?

ALGEBRA Solve each equation. Check your solution. (Lessons 4-2 and 4-3)

45. $4f = 28$ 46. $6p = -72$ 47. $p - 14 = 27$ 48. $26 = n + 2$

49. **DIVING** Find the distance between two divers if one diver is 27 feet below sea level and the other diver is 13 feet below sea level. (Lesson 3-5)

GETTING READY FOR THE NEXT LESSON

PREREQUISITE SKILL Replace each ● with < or > to make a true sentence. (Lesson 3-2)

50. $-3 \; ● \; -12$ 51. $1 \; ● \; -1$ 52. $7 \; ● \; -18$ 53. $-120 \; ● \; -30$

Chapter 4 Mid-Chapter Practice Test

Vocabulary and Concepts

1. **State** the property of equality used to solve $a - 7 = -2$. (Lesson 4-2)
2. **Write** a two-step equation. Then solve your equation. (Lesson 4-4)
3. **Define** *inverse operation* and give an example. (Lesson 4-2)

Skills and Applications

Write each sentence as an algebraic equation. (Lesson 4-1)

4. The product of a number and 3 is -16.
5. 10 less than a number is 45.

Solve each equation. Check your solution. (Lessons 4-2, 4-3, and 4-4)

6. $21 + m = 33$
7. $a - 5 = -12$
8. $7y = 63$
9. $5f = -75$
10. $-28 = -2d$
11. $-1.6w = 4.8$
12. $3z - 7 = 17$
13. $2g - 9 = -5$
14. $-4c - 1 = 11$

15. **FLYING** An airplane is flying at an altitude of t feet before it increases its altitude by 1,000 feet. Write an expression for its new altitude. (Lesson 4-1)

16. **GEOMETRY** The sum of the measures of the angles of a triangle is 180°. Find the missing measure. (Lesson 4-2)

LAWN SERVICE For Exercises 17 and 18, use the following information.
Alex earned $326 this summer mowing lawns. The total was 4 times more than what he earned last summer. (Lesson 4-3)

17. Write a multiplication equation to find how much Alex earned last summer.
18. Solve the equation.

Standardized Test Practice

19. **MULTIPLE CHOICE** Kim's time for the 5K race was four minutes less than Tanya's time. If Tanya's time is t, which expression represents Kim's time? (Lesson 4-1)

    Ⓐ $4 - t$
    Ⓑ $t - 4$
    Ⓒ $t + 4$
    Ⓓ $4t$

20. **GRID IN** Last baseball season, Ryan had four less than twice the number of hits that Marcus had. Ryan had 48 hits. How many hits did Marcus have last season? (Lesson 4-4)

The Game Zone

A Place To Practice Your Math Skills

Math Skill: Solving Equations

Math-O

GET READY!

Players: two, three, or four
Materials: scissors, 26 index cards, and 4 different colored markers

GET SET!

- Cut each index card in half, making 52 cards.
- To make a set of four cards, use the markers to put a different-colored stripe at the top of each card.
- Then write a different equation on each card. The solution of each equation should be 1.
- Continue to make sets of four cards having equations with solutions of 2, 3, 4, 5, 6, 0, −1, −2, −3, −4, and −5.
- Mark the remaining set of four cards "Wild."

GO!

- The dealer shuffles the cards and deals five to each person. The remaining cards are placed in a pile facedown in the middle of the table. The dealer turns the top card faceup.
- The player to the left of the dealer plays a card with the same color or solution as the faceup card. Wild cards can be played any time. If the player cannot play a card, he or she takes a card from the pile and plays it, if possible. If it is not possible to play, the player places the card in his or her hand, and it is the next player's turn.
- **Who Wins?** The first person to play all cards in his or her hand is the winner.

4-5 Inequalities

What You'll LEARN
Solve inequalities.

NEW Vocabulary
inequality

MATH Symbols
< is less than
> is greater than
≤ is less than or equal to
≥ is greater than or equal to

WHEN am I ever going to use this?

BREAKFAST The table shows the nutrition requirements for a healthy breakfast cereal with milk.

Healthy Breakfast Cereals (per serving)	
Fat	less than 3 g
Protein	more than 5 g
Fiber	at least 3 g
Sugar	at most 5 g

1. Suppose your favorite cereal has 2 grams of fat, 7 grams of protein, 4 grams of fiber, and 3 grams of sugar. Is it a healthy cereal? Explain.

2. Is a cereal with 3 grams of fiber considered healthy? Explain.

3. Is a cereal with 5 grams of sugar considered healthy? Explain.

An **inequality** is a mathematical sentence that contains the symbols $<, >, \leq,$ or $\geq$.

Inequalities				
Words	is less than	is greater than	is less than or equal to is at most	is greater than or equal to is at least
Symbols	<	>	≤	≥

Any number that makes the inequality true is a solution of the inequality. Inequalities may have many solutions. The solutions are shown by shading a number line.

EXAMPLES Graph Solutions of Inequalities

Graph each inequality on a number line.

① $x < 3$

② $x \geq 3$

The open circle means the number is *not* included in the solution.

The closed circle means the number *is* included in the solution.

③ $x > 5$

④ $x \leq 5$

  Your Turn Graph each inequality on a number line.

a. $x < -1$ b. $x \geq -2$ c. $x > 0$ d. $x \leq 3$

172 Chapter 4 Algebra: Linear Equations and Functions

To solve an inequality, follow the same steps you use to solve an equation.

EXAMPLES — Solve One-Step Inequalities

5 Solve $x + 4 > 9$. Check your solution. Then graph the solution.

$$\begin{aligned} x + 4 &> 9 &&\text{Write the inequality.} \\ -4 &= -4 &&\text{Subtract 4 from each side.} \\ x &> 5 &&\text{Simplify.} \end{aligned}$$

Check Try 6, a number greater than 5.

$$\begin{aligned} x + 4 &> 9 &&\text{Write the inequality.} \\ 6 + 4 &\stackrel{?}{>} 9 &&\text{Replace } x \text{ with 6. Is this sentence true?} \\ 10 &> 9 \checkmark \end{aligned}$$

The solution is all numbers greater than 5.

Study Tip

Checking Solutions
You may also want to check $x = 5$. The solution of the equation is the *boundary* between numbers that are solutions and numbers that are not solutions.

6 Solve $4y \leq 8$. Graph the solution.

$$\begin{aligned} 4y &\leq 8 &&\text{Write the inequality.} \\ \frac{4y}{4} &\leq \frac{8}{4} &&\text{Divide each side by 4.} \\ y &\leq 2 &&\text{Check this solution.} \end{aligned}$$

The solution is all numbers less than or equal to 2.

Your Turn
Solve each inequality. Graph the solution.

e. $x + 6 > 8$ f. $x - 4 \geq -7$ g. $5x < 25$

EXAMPLE — Use an Inequality to Solve a Problem

REAL-LIFE CAREERS

How Does a Nurse Use Math?
Nurses use inequalities when they record test results like a person's blood pressure.

 Research
For information about a career as a nurse, visit:
msmath2.net/careers

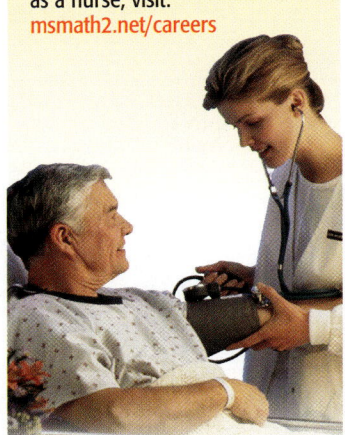

7 HEALTH The formula $110 + 0.5A < P$ is used to estimate whether a person has high blood pressure. In the formula, A is the person's age and P is the blood pressure. For which ages is a blood pressure of 120 considered high?

$$\begin{aligned} 110 + 0.5A &< P &&\text{Write the formula.} \\ 110 + 0.5A &< 120 &&\text{Replace } P \text{ with 120.} \\ -110 &= -110 &&\text{Subtract 110 from each side.} \\ 0.5A &< 10 &&\text{Simplify.} \\ \frac{0.5A}{0.5} &< \frac{10}{0.5} &&\text{Divide each side by 0.5.} \\ A &< 20 &&10 \div 0.5 = 20 \end{aligned}$$

A blood pressure of 120 is considered high for a person who is younger than 20 years old.

msmath2.net/extra_examples

Skill and Concept Check

1. **Draw** a number line that shows all numbers less than or equal to −4.

2. **Writing Math** Explain the difference between $x > 3$ and $x \geq 3$.

3. **OPEN ENDED** Write two different inequalities whose solution is $x < 2$.

4. **FIND THE ERROR** Courtney and Diego are writing an inequality for the expression *at least 2 hours of homework*. Who is correct? Explain.

   Courtney
   $h \leq 2$

   Diego
   $h \geq 2$

GUIDED PRACTICE

Graph each inequality on a number line.

5. $y > -3$
6. $x \geq 0$
7. $d < 7$

Solve each inequality.

8. $x + 3 > -4$
9. $6d \geq 24$
10. $3x + 8 < 15$

Practice and Applications

Graph each inequality on a number line.

11. $s > -4$
12. $y > 3$
13. $t \geq -4$
14. $a \geq 3$
15. $h < 2$
16. $g < -5$
17. $z \leq -1$
18. $p \leq 0$
19. $w \leq -6$

HOMEWORK HELP

For Exercises	See Examples
11–19	1–4
20–25, 32–33	5–6
26–27, 40–41	7

Extra Practice
See pages 573, 599.

Solve each inequality.

20. $y + 5 < 14$
21. $x + 6 < 0$
22. $a - 3 \geq -5$
23. $g - 5 > 2$
24. $3r \leq 18$
25. $6u \geq 36$
26. $2x + 8 < 24$
27. $3y + 1 \leq 5$

Write the inequality for each graph.

28. number line with open circle at 2, arrow right (−1 0 1 2 3)
29. number line with closed circle at −2, arrow left (−4 −3 −2 −1 0)
30. number line with closed circle at 3, arrow right (2 3 4 5 6)
31. number line with open circle at 0, arrow left (−2 −1 0 1 2)

Write an inequality for each sentence. Then solve the inequality.

32. Five times a number is greater than 25.

33. The sum of a number and 1 is at least 5.

34. **WEATHER** A tropical depression has maximum sustained winds of less than 39 miles per hour. Write an inequality showing the wind speeds.

35. **DRIVING** In Ohio, you can get a driver's license if you are at least 16 years old. Write an inequality showing the age of all drivers in Ohio.

Data Update What are the minimum ages for getting a driver's license in other states? Visit **msmath2.net/data_update** to learn more.

SPORTS For Exercises 36–39, use the graphic.

36. In which sport(s) are more than 400,000 children hurt?

37. In which sport(s) are at least 185,000 children hurt?

38. Of the sports listed, which have less than 300,000 injuries?

39. Write an inequality that describes the number of children hurt each year playing sports.

SHOPPING For Exercises 40 and 41, use the following information.
Suppose a pair of jeans costs $29 and a necklace costs $8. You have $70 to spend on both.

40. Write an inequality to find how many pairs of jeans you can buy along with one necklace.

41. Solve the inequality.

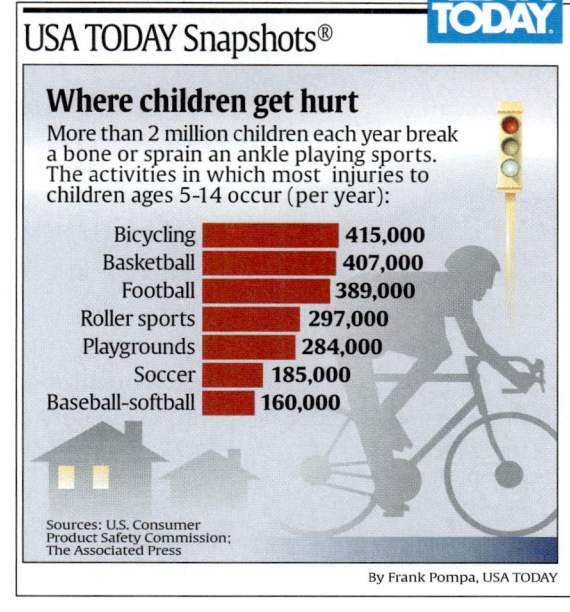

42. **CRITICAL THINKING** A *compound inequality* is formed by two inequalities connected by the words *and* or *or*.

Type	Example	Solution	Graph
Intersection	$x > -3$ and $x < 4$	Both inequalities are true. $\{x \mid -3 < x < 4\}$	−5−4−3−2−1 0 1 2 3 4 5
Union	$x < 2$ or $x > 5$	Either inequality is true. $\{x \mid x < 2 \text{ or } x > 5\}$	−5−4−3−2−1 0 1 2 3 4 5

Identify each compound inequality as an intersection or union. Then find and graph the solution.

a. $x > 2$ and $x < 7$ b. $x < -4$ or $x > 4$ c. $x \geq -3$ or $x < -6$ d. $x \leq 10$ and $x > 0$

Spiral Review with Standardized Test Practice

43. **MULTIPLE CHOICE** Which inequality is graphed on the number line?

    Ⓐ $x < 3,000$ Ⓑ $x \leq 3,000$
    Ⓒ $x > 3,000$ Ⓓ $x \geq 3,000$

44. **MULTIPLE CHOICE** Ayano has $26 to spend. A hat costs $8. Which inequality shows how much he can spend for a T-shirt?

    Ⓕ $s + 8 < 26$ Ⓖ $s + 8 > 26$ Ⓗ $s + 8 \leq 26$ Ⓘ $s + 8 \geq 26$

ALGEBRA Solve each equation. Check your solution. (Lessons 4-2, 4-3, and 4-4)

45. $13 = s + 5$ 46. $18x = -54$ 47. $2q + 6 = -20$ 48. $-7q + 4.6 = -0.3$

GETTING READY FOR THE NEXT LESSON

PREREQUISITE SKILL Graph each point on a coordinate plane. (Lesson 3-3)

49. $(-4, 2)$ 50. $(3, -1)$ 51. $(-3, -4)$ 52. $(2, 0)$

 msmath2.net/self_check_quiz

4-6a HANDS-ON LAB

A Preview of Lesson 4-6

Functions and Graphs

INVESTIGATE *Work as a class.*

Have you ever been at a sporting event when the crowd does the "wave"? In this Lab, you will investigate how long it would take the students at your school to complete the "wave."

What You'll LEARN

Graph a function on a scatter plot.

Materials
- stopwatch
- grid paper
- uncooked spaghetti

STEP 1 Begin with five students sitting in a row.

STEP 2 At the timer's signal, the first student stands up, waves his or her arms overhead, and sits down. Each student repeats the wave in order.

STEP 3 When the last student sits down, the timer records the time in seconds.

STEP 4 Repeat for 6, 7, 8, and so on, up to 25 students.

Writing Math

Work with a partner.

1. **Graph** the ordered pairs (number of students, time) on a coordinate grid like the one at the right.

2. **Describe** how the points appear on your graph.

3. Place one piece of uncooked spaghetti on your graph so that it covers as many of the points as possible. **Predict** how long it would take 30 students to complete the "wave." Make a prediction for 50 students.

4. **Find a pattern** in the data and use the pattern to predict how long it would take the students in your school to complete the "wave." Explain your reasoning.

5. A *function* describes the relationship between two quantities. In a function, one quantity *depends on* the other. **Complete the sentence:** The time it takes to do the "wave" depends on ___?___.

176 Chapter 4 Algebra: Linear Equations and Functions

Functions and Linear Equations

What You'll LEARN

Graph linear equations.

NEW Vocabulary

function
function table
domain
range
linear equation

REVIEW Vocabulary

ordered pair: a pair of numbers used to locate a point in a coordinate system (Lesson 3-3)

WHEN am I ever going to use this?

FAST FOOD Suppose you can buy hamburgers for $2 each.

1. Copy and complete the table to find the cost of 2, 3, and 4 hamburgers.
2. On grid paper, graph the ordered pairs (number, cost). Describe how the points appear on the grid.
3. What happens to the cost as the number of hamburgers increases?

Hamburgers		
Number	Multiply by 2	Cost ($)
1	2 × 1	2
2		
3		
4		

The total cost of the hamburgers depends on the number of hamburgers. A relationship where one thing depends on another is called a **function**. In a function, you start with an *input* number, perform one or more operations on it, and get an *output* number.

You can organize the input numbers, output numbers, and the function rule in a **function table**.

EXAMPLE Make a Function Table

1) MONEY MATTERS Suppose you earn $5 each week. Make a function table that shows your total earnings after 1, 2, 3, and 4 weeks.

Input	Function Rule	Output
Number of Weeks	Multiply by 5	Total Earnings ($)
1	5 × 1	5
2	5 × 2	10
3	5 × 3	15
4	5 × 4	20

The set of input values is called the **domain**, and the set of output values is called the **range**. In Example 1, the domain is {1, 2, 3, 4}, and the range is {5, 10, 15, 20}.

Lesson 4-6 Functions and Linear Equations **177**

Functions are often written as equations with two variables—one to represent the input and one to represent the output. Here's an equation for the situation in Example 1.

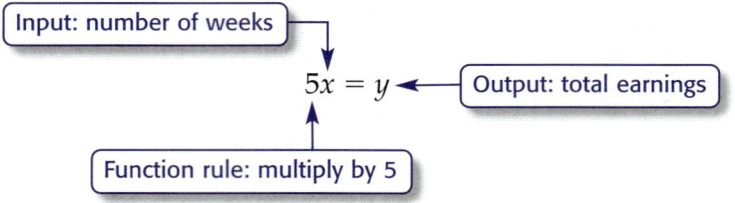

STUDY TIP

Input and Output
When x and y are used in an equation, x usually represents the input, and y usually represents the output.

The solution of an equation with two variables consists of two numbers, one for each variable, that make the equation true. The solution is usually written as an ordered pair (x, y), which can be graphed.

EXAMPLE Graph Solutions of Linear Equations

2 Graph $y = 2x + 1$.

Select any four values for the input x. We chose 2, 1, 0, and -1. Substitute these values for x to find the output y.

x	2x + 1	y	(x, y)
2	2(2) + 1	5	(2, 5)
1	2(1) + 1	3	(1, 3)
0	2(0) + 1	1	(0, 1)
−1	2(−1) + 1	−1	(−1, −1)

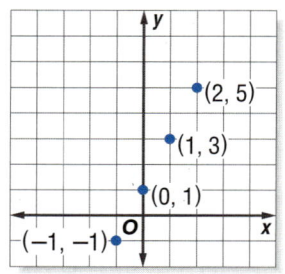

Four solutions are (2, 5), (1, 3), (0, 1), and (−1, −1). The graph is shown above at the right.

Your Turn Graph each equation.

a. $y = x - 3$ b. $y = -3x$ c. $y = -3x + 2$

STUDY TIP

Graphing Equations
Only two points are needed to graph the line. However, graph more points to check accuracy.

Notice that all four points in the graph lie on a line. Draw a line through the points to graph *all* solutions of the equation $y = 2x + 1$. The graph of (3, 7) is also on the line.

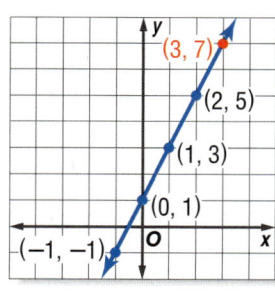

$y = 2x + 1$ Write the equation.

$7 \stackrel{?}{=} 2(3) + 1$ Replace x with 3 and y with 7.

$7 = 7$ ✓   This sentence is true.

So, (3, 7) is also a solution of $y = 2x + 1$.

An equation like $y = 2x + 1$ is called a **linear equation** because its graph is a straight line.

178 Chapter 4 Algebra: Linear Equations and Functions

EXAMPLE Represent Real-World Functions

REAL-LIFE MATH

SKATEBOARDING About one fourth of all skateboarding purchases are made by teenagers from 14 to 17 years old.

Source: *Statistical Abstract of the United States*

3 SKATEBOARDING The top speed reached by a standing skateboarder is 80 feet per second. The equation $d = 80t$ describes the distance d that a skateboarder can travel in time t. Represent the function with a graph.

Step 1 Select any four values for t. Select only positive numbers because t represents time. Make a function table.

t	80t	d	(t, d)
1	80(1)	80	(1, 80)
2	80(2)	160	(2, 160)
3	80(3)	240	(3, 240)
4	80(4)	320	(4, 320)

Step 2 Graph the ordered pairs and draw a line through the points.

Skill and Concept Check

1. **OPEN ENDED** Write an equation that has (1, 2) as a solution.

2. **Writing Math** Explain the relationship among *input*, *output*, and *function rule*.

GUIDED PRACTICE

Copy and complete each function table. Identify the domain and range.

3. $y = x - 2$

x	x − 2	y
1		
2		
3		
4		

4. $y = 4x$

x	4x	y
−1		
0		
1		
2		

Graph each equation.

5. $y = x - 1$

6. $y = -1x$

7. $y = -2x + 3$

Practice and Applications

Copy and complete each function table. Identify the domain and range.

8. $y = x - 4$

x	x − 4	y
1		
2		
3		
4		

9. $y = x + 5$

x	x + 5	y
1		
2		
3		
4		

10. $y = 2x$

x	2x	y
−1		
0		
1		
2		

11. $y = -6x$

x	−6x	y
−1		
0		
1		
2		

12. $y = 2x - 1$

x	2x − 1	y
1		
2		
3		
4		

13. $y = -2x - 2$

x	−2x − 2	y
−1		
0		
1		
2		

HOMEWORK HELP

For Example	See Examples
8–13	1
14–25	2
29–35	3

Extra Practice
See pages 573, 599.

Graph each equation.

14. $y = x + 1$
15. $y = x + 3$
16. $y = x$
17. $y = -2x$
18. $y = 2x + 3$
19. $y = 3x - 1$
20. $y = 4x - 2$
21. $y = 2x + 5$
22. $y = x + 0.5$
23. $y = 0.25x$
24. $y = 0.5x - 1$
25. $y = 2x - 1.5$

Make a function table for each sentence. Then write an equation using x to represent the first number and y to represent the second number.

26. The second number is three more than the first number.
27. The second number is five less than the first number.
28. The second number is ten times the first number.

SPENDING For Exercises 29–31, use the graph.

29. Make a function table that shows the total average defense spending per person for 1, 2, 3, and 4 days.
30. Write an equation in which x represents the days and y represents the total spending.
31. Graph the equation.

INTERNET For Exercises 32–34, use the following information.
An Internet provider charges $20 each month.

32. Make a function table that shows the total charge for 1, 2, 3, and 4 months of service.
33. Write an equation in which x represents months and y represents the total charge.
34. Graph the equation.

USA TODAY Snapshots®

Health research spending small

About 4.6 cents of each U.S. health-care dollar is spent on medical/health research. By comparison, federal spending per person per day on:

Social Security — $3.94
Defense — $2.67
Medicare — $2.05

Source: Research America

By Lori Joseph, Keith Simmons and Marcy E. Mullins, USA TODAY

35. **GEOMETRY** The formula for the area A of a rectangle whose length is 5 units is $A = 5w$, where w is the width. Graph the function.

36. **WRITE A PROBLEM** Write about a real-life situation that can be represented by the equation $y = 3x$.

CRITICAL THINKING For Exercises 37–40, write an equation for the function shown in each function table.

37.
x	y
1	3
2	4
3	5
4	6

38.
x	y
2	6
4	12
6	18
8	24

39.
x	y
−1	4
0	0
1	−4
2	−8

40.
x	y
1	3
2	5
3	7
4	9

EXTENDING THE LESSON Not all equations have graphs that are straight lines. The graphs at the right show two *nonlinear equations*.

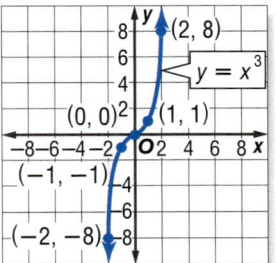

Graph each equation. Determine whether it is *linear* or *nonlinear*.

41. $y = x^2 - 1$
42. $y = x + 1$
43. $y = 2x$
44. $xy = 12$
45. $y = 2x^2$
46. $y = x^3 - 1$

Spiral Review with Standardized Test Practice

47. **MULTIPLE CHOICE** Choose the equation that is graphed at the right.

    A) $y = 2x$ B) $y = x$
    C) $y = x - 1$ D) $y = 2x - 2$

48. **MULTIPLE CHOICE** Which ordered pairs are solutions of the equation $y = 2x - 4$?

    F) $(-2, -3), (0, 2)$ G) $(-2, -1), (2, -3)$
    H) $(1, -2), (3, 2)$ I) $(-3, 2), (0, -4)$

49. **CIVICS** To serve as a U.S. Representative, a person must be at least 25 years old and a citizen of the United States for at least 7 years. Write an inequality showing the age of a person who may be a U.S. Representative. (Lesson 4-5)

ALGEBRA Solve each equation. Check your solution. (Lesson 4-4)

50. $8 = 3h - 1$ 51. $2q + 6 = -20$ 52. $32 = -4 + 9m$

GETTING READY FOR THE NEXT LESSON

PREREQUISITE SKILL Divide. (Lesson 3-7)

53. $-4 \div 2$ 54. $10 \div (-5)$ 55. $-12 \div (-4)$ 56. $-16 \div 16$

4-7 Lines and Slope

What You'll LEARN
Find the slope of a line.

NEW Vocabulary
slope

Link to READING
Everyday Meaning of Slope: ground that forms an incline, as in a ski slope

WHEN am I ever going to use this?

COST OF GASOLINE In recent years, the cost of one gallon of gasoline has varied from a low of about $1 per gallon to a high of about $3 per gallon. The equations $y = 1x$ and $y = 3x$ are graphed.

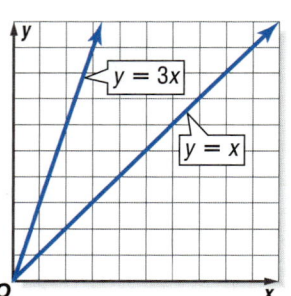

1. Which line is steeper?
2. What causes one line to be steeper?
3. **Make a conjecture** about where the line showing a cost of $2 per gallon would be graphed. Explain.

The function table shows the total cost y of x gallons of gasoline at $2 per gallon. The equation $y = 2x$ is graphed below.

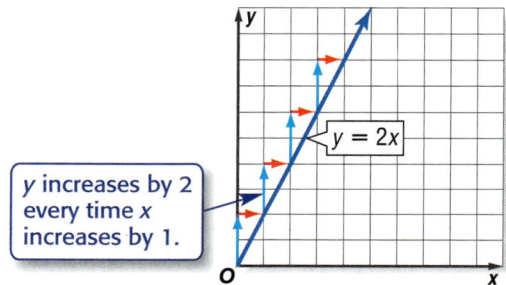

y increases by 2 every time *x* increases by 1.

The change in y with respect to the change in x is called the **slope** of a line. Slope is a number that tells how steep the line is.

$$\text{slope} = \frac{\text{change in } y}{\text{change in } x} \quad \leftarrow \text{vertical change} \\ \leftarrow \text{horizontal change}$$

$$= \frac{2}{1} \text{ or } 2$$

The slope is the same for any two points on a straight line.

EXAMPLE Positive Slope

1 Find the slope of the line.

$$\text{slope} = \frac{\text{change in } y}{\text{change in } x}$$

$$= \frac{4}{1} \text{ or } 4$$

The slope of the line is 4.

When the value of y decreases as the value of x increases, the slope is a negative number. The slope of a line can also be a fraction.

STUDY TIP

Slope A line with a *positive slope* rises to the right. A line with a *negative slope* falls to the right.

EXAMPLES Negative Slope

2 Find the slope of each line.

slope = $\dfrac{\text{change in } y}{\text{change in } x}$

$= \dfrac{-2}{1}$ or -2

The slope of the line is -2.

3

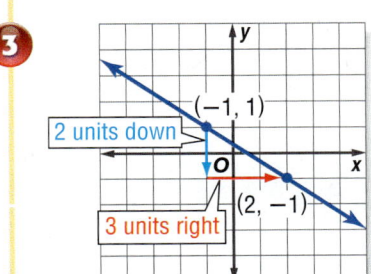

slope = $\dfrac{\text{change in } y}{\text{change in } x}$

$= \dfrac{-2}{3}$ or $-\dfrac{2}{3}$

The slope of the line is $-\dfrac{2}{3}$.

Your Turn Find the slope of the line that passes through each pair of points.

a. $(-2, -4), (1, 5)$ b. $(0, 3), (4, -1)$ c. $(2, 2), (5, 3)$

Standardized Test Practice

EXAMPLE Compare Slopes

4 MULTIPLE-CHOICE TEST ITEM The table shows information about the rise and run of three ski slopes in Pennsylvania. Which has the steepest slope?

Ski Slope	Rise (ft)	Run (ft)
Giant Boulder	750	4,000
Giant Steps	750	3,200
Gunner	750	3,360

Ⓐ Giant Boulder   Ⓑ Giant Steps
Ⓒ Gunner   Ⓓ They have the same slope.

Test-Taking Tip

Make a Drawing Whenever possible, make a drawing of the problem. Then use the drawing to estimate the answer.

Read the Test Item
The *rise* corresponds to the vertical change, or change in y.
The *run* corresponds to the horizontal change, or change in x.

Solve the Test Item
Giant Boulder: $\dfrac{750}{4,000}$ Giant Steps: $\dfrac{750}{3,200}$ Gunner: $\dfrac{750}{3,360}$

All three fractions have the same rise. But Giant Steps has the shortest run. Therefore, its slope is greater. The answer is B.

msmath2.net/extra_examples   Lesson 4-7 Lines and Slope **183**

Skill and Concept Check

1. Define *slope*.
2. **Writing Math** Explain how a line can have a negative slope.
3. **OPEN ENDED** On a coordinate plane, draw a line with a slope of $\frac{1}{2}$.

GUIDED PRACTICE

Find the slope of the line that passes through each pair of points.

4.
5.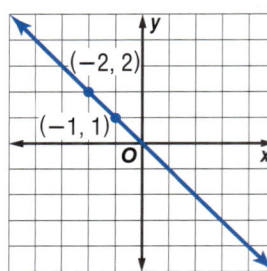
6. $(0, 0), (1, 3)$

Practice and Applications

Find the slope of the line that passes through each pair of points.

7.
8.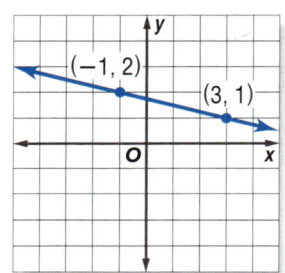

HOMEWORK HELP

For Exercises	See Examples
7–16	1–3
17–20	4

Extra Practice
See pages 574, 599.

9. $(-2, 0), (1, 3)$
10. $(3, 4), (4, 6)$
11. $(1, 0), (2, -3)$
12. $(-2, 2), (-1, -2)$
13. $(1, 1), (4, 2)$
14. $(-2, 1), (3, 3)$
15. $(1, -1), (3, -2)$
16. $(0, 0), (3, -2)$

EARNINGS For Exercises 17–19, use the table at the right.

17. Suppose each of the functions in the table was graphed on a coordinate plane. Which line is steeper? Explain.
18. Find the slope of each line.
19. What does the slope of each line represent?

Hours Worked	Earnings ($)	
	Greg	Monica
1	4	5
2	8	10
3	12	15
4	16	20

20. **SKIING** Aerial skiers launch themselves into the air from a ramp like the one shown at the right. Is the slope of the ramp greater than one or less than one?

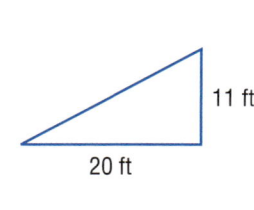

184 Chapter 4 Algebra: Linear Equations and Functions

CAMPING For Exercises 21–23, use the graph.

21. Which section of the graph shows the greatest increase in sales of camping gear? Describe the slope of this part of the graph.

22. What happened to sales between 1995 and 1996? Describe the slope of this part of the graph.

23. What happened to sales between 1997 and 1998? Describe the slope of this part of the graph.

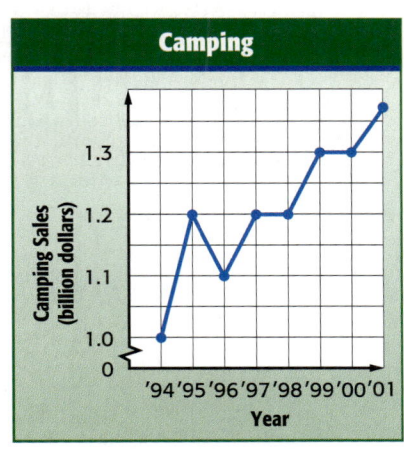

Source: National Sporting Goods Association

CRITICAL THINKING A linear function has a constant slope. Determine whether each function is *linear* or *nonlinear*.

24.
x	y
1	5
2	6
3	7
4	8

25.
x	y
1	3
2	4
3	6
4	9

26.
x	y
−1	1
0	0
1	1
2	4

EXTENDING THE LESSON You can graph a line if you know the slope and the coordinates of a point on the line. The figure at the right shows how to graph a line with slope 2 that passes through (1, −1).

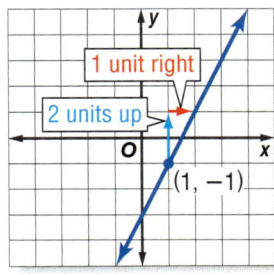

Graph each line with the given slope that passes through the given point.

27. slope = 3; (2, 3)

28. slope = −1; (−3, 2)

29. slope = −2; (−4, −1)

30. slope = 5; (0, −4)

Spiral Review with Standardized Test Practice

31. **MULTIPLE CHOICE** What is the slope of the line in the graph?

    Ⓐ 3 Ⓑ −3 Ⓒ $\frac{1}{3}$ Ⓓ $-\frac{1}{3}$

32. **GRID IN** The graph of a straight line contains the points (0, −3), (1, 2), and (2, y). What is the value of y?

Graph each equation. (Lesson 4-6)

33. $y = 2x - 1$

34. $y = -3x$

35. $y = x + 4$

36. Solve $2x + 6 < 8$. (Lesson 4-5)

INTERDISCIPLINARY PROJECT

The Wide World of Soccer

Math and Geography It is time to complete your project. Use the information and data you have gathered about countries where soccer is a favorite sport to prepare a Web page or poster. Be sure to include a graph with your project.

WebQuest msmath2.net/webquest

Lesson 4-7 Lines and Slope **185**

Chapter 4 Study Guide and Review

Vocabulary and Concept Check

domain (p. 177)
function (p. 177)
function table (p. 177)
inequality (p. 172)
inverse operations (p. 156)
linear equation (p. 178)
range (p. 177)
slope (p. 182)
two-step equation (p. 166)
work backward strategy (p. 164)

State whether each sentence is *true* or *false*. If *false*, replace the underlined word or number to make a true sentence.

1. <u>Inverse operations</u> "undo" each other.
2. The words "more than" sometimes suggest the operation of <u>multiplication</u>.
3. An <u>inequality</u> is a mathematical sentence that contains the symbols $<$, $>$, $\leq$, or $\geq$.
4. When graphing $t < 2$ on a number line, an <u>open circle</u> should be used to show that 2 is not included in the solution.
5. <u>Slope</u> is a number that tells how steep a line is.
6. An equation is called a linear equation if its graph is a <u>point</u>.
7. The solution of $m + 5 = 12$ is <u>17</u>.
8. The solution of $g - 4 = 18$ is <u>22</u>.
9. <u>Six</u> is the solution of $-8w = 48$.
10. The solution of $2y \leq 18$ is <u>$y \leq 16$</u>.

Lesson-by-Lesson Exercises and Examples

Writing Expressions and Equations (pp. 150–152)

Write each phrase as an algebraic expression.

11. the sum of a number and 5
12. six inches less than her height
13. twice as many apples
14. four times the number of dollars

Write each sentence as an algebraic equation.

15. Ten years older than Theresa's age is 23.
16. Four less than a number is 19.
17. The quotient of 56 and a number is 14.
18. Eight times the number of students is 64.

Example 1 Write the phrase as an algebraic expression.

four times the price

Let p represent the price.
The algebraic expression is $4p$.

Example 2 Write the sentence as an algebraic equation.

Six less than the number of cookies is 24.

Let c represent the number of cookies.
The equation is $c - 6 = 24$.

186 Chapter 4 Algebra: Linear Equations and Functions

msmath2.net/vocabulary_review

4-2 Solving Addition and Subtraction Equations (pp. 156–159)

Solve each equation. Check your solution.

19. $x + 5 = 8$
20. $y + 4 = 12$
21. $a + 6 = 5$
22. $r + 8 = 2$
23. $p + 9 = -4$
24. $d + 14 = 23$
25. $s - 8 = 15$
26. $t - 6 = 7$
27. $n - 1 = -3$
28. $w - 9 = 28$

Example 3 Solve $x + 6 = 4$.

$$\begin{aligned} x + 6 &= 4 \\ -6 &= -6 \quad \text{Subtract 6 from each side.} \\ x &= -2 \end{aligned}$$

Example 4 Solve $y - 3 = -2$.

$$\begin{aligned} y - 3 &= -2 \\ +3 &= +3 \quad \text{Add 3 to each side.} \\ y &= 1 \end{aligned}$$

4-3 Solving Multiplication Equations (pp. 160–163)

Solve each equation. Check your solution.

29. $7c = 28$
30. $9y = 45$
31. $-2h = 24$
32. $-8w = 72$
33. $10y = -90$
34. $6q = -18$
35. $-11f = -121$
36. $-12r = -36$

Example 5 Solve $-4b = 32$.

$$\begin{aligned} -4b &= 32 \\ \frac{-4b}{-4} &= \frac{32}{-4} \quad \text{Divide each side by } -4. \\ b &= -8 \end{aligned}$$

4-4 Solving Two-Step Equations (pp. 166–169)

Solve each equation. Check your solution.

37. $3y - 12 = 6$
38. $6x - 4 = 20$
39. $2x + 5 = 3$
40. $5m + 6 = -4$
41. $10c - 8 = 90$
42. $3r - 20 = -5$

43. Ten more than five times a number is 25. Find the number.

Example 6 Solve $3p - 4 = 8$.

$$\begin{aligned} 3p - 4 &= 8 \\ +4 &= +4 \quad \text{Add 4 to each side.} \\ 3p &= 12 \\ \frac{3p}{3} &= \frac{12}{3} \quad \text{Divide each side by 3.} \\ p &= 4 \end{aligned}$$

4-5 Inequalities (pp. 172–175)

Solve each inequality. Graph the solution.

44. $x + 3 < 8$
45. $y + 2 > 5$
46. $a + 4 \geq 10$
47. $d + 1 \leq 6$
48. $h - 5 \geq 7$
49. $s - 2 \leq 9$
50. $y + 2 < -3$
51. $m - 7 > -10$
52. $b + 9 \leq -11$
53. $t - 10 \geq -8$

Example 7 Solve $g + 8 \leq 10$. Graph your solution.

$$\begin{aligned} g + 8 &\leq 10 \\ -8 &= -8 \quad \text{Subtract 8 from each side.} \\ g &\leq 2 \end{aligned}$$

number line from −5 to 5 with solution $g \leq 2$ graphed

Study Guide and Review continued

Mixed Problem Solving
For mixed problem-solving practice, see page 599.

4-6 Functions and Linear Equations (pp. 177–181)

Graph each equation.

54. $y = x + 5$
55. $y = x - 4$
56. $y = 2x$
57. $y = -1x$
58. $y = 3x + 2$
59. $y = -2x + 3$

MONEY MATTERS For Exercises 60–62, use the following infomation.

Angel earns $6 per hour working at the Ice Cream Shop.

60. Make a table that shows her total earnings for working 3, 5, 7, and 9 hours.
61. Write an equation in which x represents the number of hours and y represents Angel's total earnings.
62. Graph the equation.

Example 8 Graph $y = x + 3$.

Select four values for x. Substitute these values for x to find values for y.

x	$x + 3$	y
−1	−1 + 3	2
0	0 + 3	3
1	1 + 3	4
2	2 + 3	5

Four solutions are (−1, 2), (0, 3), (1, 4), and (2, 5). The graph is shown below.

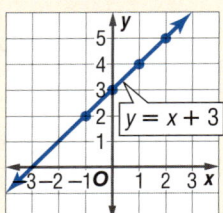

4-7 Lines and Slope (pp. 182–185)

Find the slope of the line that passes through each pair of points.

63.

64.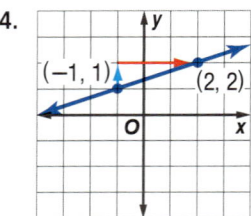

65. (1, 1), (2, 0)
66. (2, 3), (4, −1)

Example 9 Find the slope of the line.

$$\text{slope} = \frac{\text{change in } y}{\text{change in } x}$$

$$= \frac{3}{1} \text{ or } 3$$

The slope of the line is 3.

188 Chapter 4 Algebra: Linear Equations and Functions

CHAPTER 4 Practice Test

Vocabulary and Concepts

1. **Explain** the difference between an equation and an inequality.
2. **Define** *function* and give an example.

Skills and Applications

Write each phrase as an algebraic expression.

3. $5 less than Matt has
4. 4 years older than Hana

Solve each equation. Check your solution.

5. $x + 5 = -8$
6. $y - 11 = 15$
7. $9z = -81$
8. $-6k - 4 = 38$

Solve each inequality. Graph the solution.

9. $p - 4 \geq -3$
10. $j + 5 > 2$

SHOPPING For Exercises 11 and 12, use the following information.
Suppose you want to buy 3 CDs and a new CD case that costs $7. Each CD costs the same amount.

11. If you spend $46, write an equation to find the cost of each CD.
12. Solve the equation.

Graph each equation.

13. $y = 3x - 2$
14. $y = -2x + 4$
15. $y = 0.5x$

MOVIES For Exercises 16–18, use the following information.
A student ticket to the movies costs $3.

16. Make a table that shows the total cost of 2, 4, and 6 tickets.
17. Write an equation in which x represents the number of tickets, and y represents the total cost.
18. Graph the equation.
19. Find the slope of the line that passes through $(-2, 3)$ and $(-1, 2)$.

Standardized Test Practice

20. **MULTIPLE CHOICE** Which line has a slope of 2?

A
B
C
D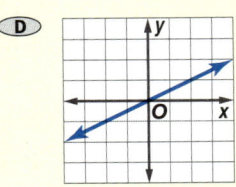

msmath2.net/chapter_test

CHAPTER 4 Standardized Test Practice

PART 1 Multiple Choice

Record your answers on the answer sheet provided by your teacher or on a sheet of paper.

1. Felicia's family wants to buy her a gift for $220. Her parents will pay half. Her older sister will pay $50. If her three other siblings split the remaining cost, how much will each pay? (Lesson 1-1)
   - A $20
   - B $60
   - C $130
   - D $170

2. Which is equivalent to 3^6? (Lesson 1-2)
   - F 18
   - G 36
   - H $6 \cdot 6 \cdot 6$
   - I $3 \cdot 3 \cdot 3 \cdot 3 \cdot 3 \cdot 3$

3. Suppose you need 0.65 liter of water for a science experiment, but the container is measured in milliliters. How many milliliters of water do you need? (Lesson 1-5)
   - A 0.0065 mL
   - B 0.65 mL
   - C 6.5 mL
   - D 650 mL

4. If 18 is added to the data set below, which statement is true? (Lesson 2-4)

   16, 14, 22, 16, 16, 18, 15, 25

   - F The mode increases.
   - G The mean decreases.
   - H The mean increases.
   - I The median increases.

5. The stem-and-leaf plot shows the number of points scored by the Bears in each of their basketball games this season. In how many games did they score at least 30 points? (Lesson 2-5)

Stem	Leaf
1	8 9
2	0 2 3 3 6 8 8 9
3	0 1 4 4 5 6 8 9
4	0 1 2

   1 | 8 = 18 points

   - A 8
   - B 9
   - C 11
   - D 20

6. Suppose points given by (x, y) in the table are graphed. Which statement is true about the graphs? (Lesson 3-3)

x	y
−4	5
−5	10
−2	6

   - F The graphs of the points are located in Quadrant I.
   - G The graphs of the points are located in Quadrant II.
   - H The graphs of the points are located in Quadrant III.
   - I The graphs of the points are located in Quadrant IV.

7. The temperature at 6:00 A.M. was −5°F. What was the temperature at 8:00 A.M. if it had risen 7 degrees? (Lesson 3-4)
   - A −12°F
   - B −2°F
   - C 2°F
   - D 12°F

8. The Tigers scored four more runs than the Giants scored. Which expression represents the number of runs the Giants scored if the Tigers scored n runs? (Lesson 4-1)
   - F $n + 4$
   - G $n - 4$
   - H $4 - n$
   - I $4n$

9. Which is the graph of the equation $y = 3x - 2$? (Lesson 4-6)

   A
   B
   C
   D

PART 2 Short Response/Grid In

Record your answers on the answer sheet provided by your teacher or on a sheet of paper.

10. The charge to enter a park is a flat amount per vehicle plus a fee for each person in the vehicle. The table shows the charge for vehicles holding up to 4 people.

Number of people	Charge (dollars)
1	2.00
2	2.50
3	3.00
4	3.50

    What is the charge, in dollars, for a vehicle holding 8 people? (Lesson 1-1)

11. Evaluate $2(8 + 5^2)$. (Lesson 1-3)

12. The line plot shows how far in kilometers some students live from the school. How many students are represented in the plot? (Lesson 2-3)

13. The ordered pairs (1, 2), (6, 2), and (1, −5) are coordinates of three of the vertices of a rectangle. What is the y-coordinate of the ordered pair that represents the fourth vertex? (Lesson 3-3)

14. What number should replace y in the table? (Lesson 3-6)

x	2x − 5
1	−3
2	y
4	3
8	11
16	27

    TEST-TAKING TIP

    Question 13 This question does not include a drawing. Make one. It can help you quickly see how to solve the problem.

15. What value of x makes $x - 2 = -4$ a true statement? (Lesson 4-2)

16. Twice a number plus 7 is 35. Find the number. (Lesson 4-4)

17. Winston earns $7 an hour landscaping. He wants to buy a DVD player that costs $140. Write an inequality for the least number of hours h he needs to work to reach his goal. (Lesson 4-5)

18. The graph of a line contains (1, 2) and (3, y). What is the value of y if the slope of the line is 2? (Lesson 4-7)

PART 3 Extended Response

Record your answers on a sheet of paper. Show your work.

19. Pete's Paints charges a $20 fee plus $7.50 per hour to rent a paint sprayer.

    a. Write an equation that could be used to determine the cost c of renting a paint sprayer for h hours.

    b. Suppose you want to spend no more than $50 to rent the paint sprayer. Write an inequality for this situation.

    c. Solve the inequality in part b. Explain the meaning of the solution.

20. The distances traveled by a bicycle rider are given in the table. (Lesson 4-6)

    a. Graph the ordered pairs.

    b. Write an equation that relates the time t to the distance d.

    c. Use your equation to predict the distance traveled in 3.5 hours.

Time (hours)	Distance (miles)
1	10
2	20
3	30
4	40
5	50

UNIT 3
Fractions

Chapter 5
Fractions, Decimals, and Percents

Chapter 6
Applying Fractions

Fractions, decimals, and percents are different ways of representing the same values. In this unit, you will describe real-life situations, perform basic operations, and solve equations with fractions.

INTERDISCIPLINARY PROJECT

A Well-Balanced Diet

Math and Health You are what you eat! So are you ice cream or broccoli? You're on a mission to find out! Along the way, you'll collect and analyze data about what you eat over a period of five days. You'll take into account serving sizes, Calories, and fat grams. You'll also take on the roll of a nutritionist, researching the Food Pyramid and creating a healthy meal plan. So bring a hearty appetite and your math tool kit. This adventure will tantalize your taste buds!

WebQuest Log on to **msmath2.net/webquest** to begin your WebQuest.

CHAPTER 5
Fractions, Decimals, and Percents

"What does music have to do with math?"

The *frequencies* of two musical notes can be compared using a fraction. If the fraction can be simplified, the two notes are harmonious. In mathematics, you will use fractions to describe many real-life situations.

You will solve problems about music in Lesson 5-3.

GETTING STARTED

▶ Diagnose Readiness

Take this quiz to see if you are ready to begin Chapter 5. Refer to the lesson or page number in parentheses for review.

Vocabulary Review

Complete each sentence.

1. After comparing the decimals 0.41 and 0.04, you find that 0.41 is __?__ than 0.04. (Page 556)

2. Two or more numbers that are multiplied together to form a product are called __?__. (Lesson 1-2)

Prerequisite Skills

State which decimal is greater. (Page 556)

3. 0.6, 0.61
4. 1.25, 1.52
5. 0.33, 0.13
6. 1.08, 10.8

Use divisibility rules to determine whether each number is divisible by 2, 3, 5, 6, or 10. (Page 554)

7. 125
8. 78
9. 37

Divide. (Page 562)

10. $5\overline{)2.0}$
11. $4\overline{)3.00}$

Write each power as a product of the same factor. (Lesson 1-2)

12. 2^3
13. 5^5
14. 7^2

Write each product in exponential form. (Lesson 1-2)

15. $4 \times 4 \times 4$
16. 8×8
17. $1 \times 1 \times 1 \times 1 \times 1$
18. $7 \times 7 \times 7 \times 7$

Fractions, Decimals, and Percents Begin with a sheet of $8\frac{1}{2}''$ by $11''$ construction paper and two sheets of notebook paper.

STEP 1 **Fold and Label** Fold the construction paper in half lengthwise. Label the chapter title on the outside.

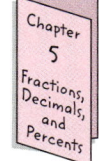

STEP 2 **Fold** Fold the sheets of notebook paper in half lengthwise. Then fold top to bottom twice.

STEP 3 **Cut** Open the notebook paper. Cut along the second folds to make four tabs.

STEP 4 **Glue and Label** Glue the uncut notebook paper side by side onto the construction paper. Label each tab with the lesson number.

 Chapter Notes Each time you find this logo throughout the chapter, use your *Noteables™: Interactive Study Notebook with Foldables™* or your own notebook to take notes. Begin your chapter notes with this Foldable activity.

 Readiness To prepare yourself for this chapter with another quiz, visit **msmath2.net/chapter_readiness**

5-1a HANDS-ON LAB
A Preview of Lesson 5-1

Exploring Factors

INVESTIGATE *Work as a class.*

What You'll LEARN
Discover factors of whole numbers.

Materials
- 15 index cards cut in half
- markers

STEP 1 Number each index card consecutively with the numbers 1 through 30.

STEP 2 In order around the classroom, give each of thirty students one index card. Have each student stand up and write the number 1 on the back of his or her card.

STEP 3 Begin with the student holding the "2" card. Have this student and every second student sit down and write the number 2 on the back of his or her card. The other students remain standing.

STEP 4 Next, start with the student holding the "3" card. Have this student and every third student stand up or sit down (depending on whether the student is already sitting or standing) and write the number 3 on the back of his or her card.

STEP 5 Continue this process for each of the remaining numbers up to 30. The thirtieth student ends the activity by standing or sitting, and writing the number 30 on the back of his or her card.

Writing Math

Work as a class.

1. How many students are standing at the end of the activity? Which cards are they holding?
2. **LOOK FOR A PATTERN** Suppose there were 100 students holding index cards. Extend the pattern in Exercise 1 to predict the numbers that would be held by students standing at the end of the activity.
3. **Explain** the relationship between the numbers on the front and the back of the cards.
4. Separate the cards into two groups: one group with exactly two numbers on the back of the card and one group with more than two numbers. **Describe** any special characteristics of each group.

196 Chapter 5 Fractions, Decimals, and Percents

5-1 Prime Factorization

What You'll LEARN

Find the prime factorization of a composite number.

NEW Vocabulary

prime number
composite number
prime factorization
factor tree

REVIEW Vocabulary

factors: two or more numbers that are multiplied together to form a product (Lesson 1-2)

HANDS-ON Mini Lab

Materials
- grid paper

Work with a partner.

There is only one way that 2 can be expressed as the product of whole numbers. The figure shows that there is only one way that two squares can form a rectangle.

1. Using your grid paper, draw as many different rectangles as possible containing 3, 4, 5, 6, 7, 8, 9, and 10 squares.
2. Which numbers of squares can be drawn in only one rectangle? In more than one rectangle?

The numbers of squares that can only be drawn in exactly one rectangle are prime numbers. The numbers that can be drawn in more than one rectangle are composite numbers.

Whole Numbers	Factors
2	1, 2
3	1, 3
5	1, 5
7	1, 7
4	1, 2, 4
6	1, 2, 3, 6
8	1, 2, 4, 8
9	1, 3, 9
10	1, 2, 5, 10
0	many
1	1

A **prime number** is a whole number greater than 1 that has exactly two factors, 1 and itself.

A **composite number** is a whole number greater than 1 that has more than two factors.

The numbers 0 and 1 are neither prime nor composite.

EXAMPLES Identify Numbers as Prime or Composite

Determine whether each number is *prime* or *composite*.

1 17

The number 17 has only two factors, 1 and 17, so it is prime.

2 12

The number 12 has six factors: 1, 2, 3, 4, 6, and 12. So, it is composite.

Your Turn Determine whether each number is *prime* or *composite*.

a. 11   b. 15   c. 24

STUDY TIP

Commutative Property Multiplication is commutative, so the order of the factors does not matter.

Every composite number can be written as a product of prime numbers in exactly one way. The product is called the **prime factorization** of the number. You can use a **factor tree** to find the prime factorization.

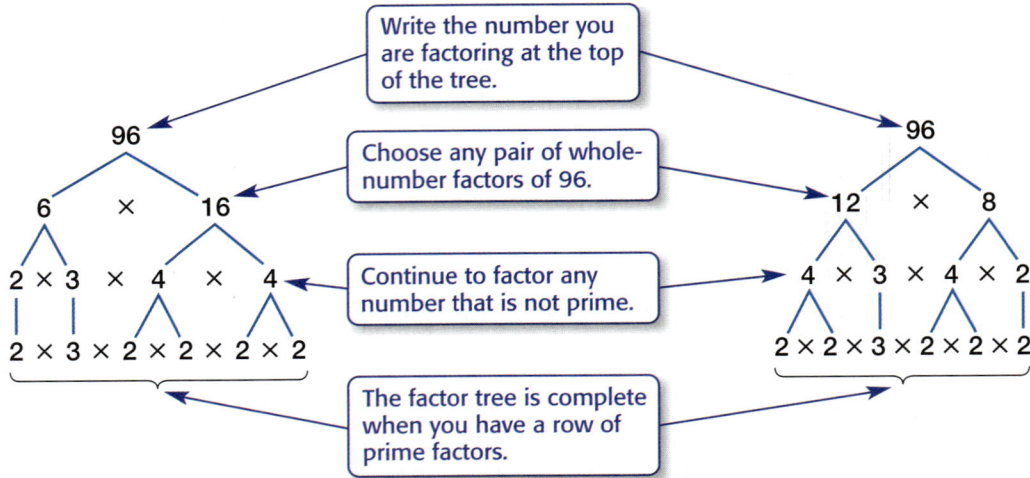

STUDY TIP

Mental Math You can use divisibility rules to find prime factors. See page 554.

Both trees give the same prime factorization of 96, but in different orders. The prime factorization of 96 is $2 \times 2 \times 2 \times 2 \times 2 \times 3$, or $2^5 \times 3$ using exponents.

EXAMPLE Find the Prime Factorization

3 Find the prime factorization of 24.

Method 1 Use a factor tree.

```
        24
       /  \
     12  ×  2
    /  \
   4  × 3 × 2
  / \
 2 × 2 × 3 × 2
```

Method 2 Divide by prime numbers.

$$\begin{array}{r} 2 \\ 2\overline{)4} \\ 3\overline{)12} \\ 2\overline{)24} \end{array}$$ ← Start here.

The divisors are 2, 2, 3, and 2.

The prime factorization of 24 is $2 \times 2 \times 3 \times 2$ or $2^3 \times 3$.

Your Turn Find the prime factorization of each number.

d. 18   e. 28   f. 16

You can factor an expression like $6ab$ as the product of prime numbers and variables.

EXAMPLE Factor an Algebraic Expression

4 ALGEBRA Factor $6ab$.

```
      6ab
     /   \
    6  ×  ab
   / \    / \
  2 × 3 × a × b
```

$6ab = 2 \cdot 3 \cdot a \cdot b$

Recall that a raised dot means multiplication.

198 Chapter 5 Fractions, Decimals, and Percents

Skill and Concept Check

1. **Writing Math** Compare and contrast prime and composite numbers.
2. **OPEN ENDED** Draw two different factor trees to find the prime factorization of 36.
3. **Explain** why $2^3 \times 9$ is *not* the prime factorization of 72.
4. **Which One Doesn't Belong?** Identify the number that is not prime.

   | 11 | 37 | 17 | 51 |

GUIDED PRACTICE

Determine whether each number is *prime* or *composite*.

5. 7
6. 50
7. 67

Find the prime factorization of each number.

8. 34
9. 30
10. 12

Factor each expression.

11. $10ac$
12. $16x^2$

Practice and Applications

Determine whether each number is *prime* or *composite*.

13. 22
14. 44
15. 13
16. 39
17. 81
18. 31
19. 97
20. 43

Find the prime factorization of each number.

21. 36
22. 42
23. 99
24. 64
25. 210
26. 180
27. 126
28. 375

29. **ALGEBRA** Is the value of $3a + 6b$ *prime* or *composite* if $a = 1$ and $b = 5$?

Factor each expression.

30. $15mn$
31. $20pq$
32. $34jkl$
33. $49y^2$
34. $52gh^2$
35. $48a^2b^2$

HOMEWORK HELP

For Exercises	See Examples
13–20, 29	1, 2
21–28, 36–38, 45–47	3
30–35	4

Extra Practice
See pages 574, 600.

PACKAGING For Exercises 36–38, use the following information.
A juice company is designing a package that holds 36 juice cans. For example, a package could be 1 can high, 6 cans wide, and 6 cans long. Such a package would contain $1 \times 6 \times 6$, or 36 cans.

36. List two other ways to arrange 36 cans.
37. How many different ways can you arrange 36 cans?
38. In which arrangement should the 36 cans be packaged? Explain your reasoning.

Lesson 5-1 Prime Factorization

Replace each ■ with prime factors to make a true sentence.

39. $2^3 \times ■ \times 11 = 616$
40. $2 \times ■ \times 5^2 = 450$
41. $3 \times 2^4 \times ■ = 1{,}200$

42. **NUMBER THEORY** *True* or *False*? All prime numbers are odd. If false, give a counterexample.

43. **NUMBER THEORY** Primes that differ by two are called *twin primes*. For example, 59 and 61 are twin primes. Give three examples of twin primes that are less than 50.

LANDSCAPING For Exercises 44–47, use the following information.

Mrs. Franks is building a patio that covers an area measuring 8 feet by 15 feet. She will use rectangular concrete tiles for the patio.

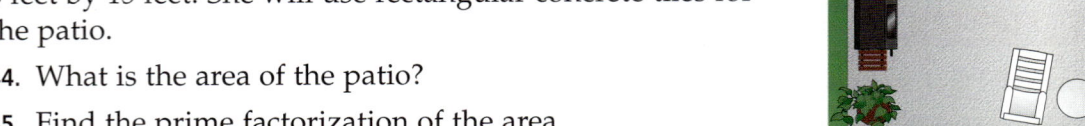

44. What is the area of the patio?
45. Find the prime factorization of the area.
46. If she can only buy tiles with dimensions that are prime numbers, name the dimensions of the concrete tiles available for her patio.
47. How many tiles will she need to cover the patio?

48. **NUMBER THEORY** Suppose n represents a whole number. Is $2n$ prime or composite? Explain.

49. **CRITICAL THINKING** Find the mystery number from the following clues.
    - This whole number is between 30 and 40.
    - It has only two prime factors.
    - The sum of its prime factors is 5.

Spiral Review with Standardized Test Practice

50. **MULTIPLE CHOICE** Which expression represents the prime factorization of 126?

    Ⓐ $2^2 \times 3 \times 7$ Ⓑ $2 \times 3^2 \times 7$ Ⓒ $2 \times 3 \times 7^2$ Ⓓ $2^2 \times 3^2 \times 7^2$

51. **SHORT RESPONSE** Determine whether the value of the expression $x^2 - 3x + 11$ is prime or composite when $x = 4$.

52. **ALGEBRA** What is the slope of the line that passes through points at $(-2, 5)$ and $(3, -1)$? (Lesson 4-7)

53. **ALGEBRA** Graph $y = 3x + 1$. (Lesson 4-6)

Tell whether each sum is *positive*, *negative*, or *zero*. (Lesson 3-4)

54. $6 + (-4)$
55. $-13 + 9$
56. $25 + (-26)$
57. $-5 + 5$

GETTING READY FOR THE NEXT LESSON

PREREQUISITE SKILL State whether each number is divisible by 2, 3, 5, 6, 9, or 10. (Page 554)

58. 24
59. 70
60. 120
61. 99
62. 125

5-2a Problem-Solving Strategy
A Preview of Lesson 5-2

Make an Organized List

What You'll LEARN

Solve problems by making an organized list.

I wonder how many different ways a woodwind trio can be made if either a bass clarinet or a bassoon fills the first position and either a clarinet, oboe, or flute fills the other two positions?

Let's **make an organized list** of all of the possibilities.

Explore We know that either a bass clarinet or a bassoon must fill the first position. The other two positions are to be filled with two out of the three remaining instruments. We have to find all possible ways the trio can be made.

Plan Let's make an organized list.

Solve

	Trio 1	Trio 2	Trio 3	Trio 4	Trio 5	Trio 6
Position 1	bass clarinet	bass clarinet	bass clarinet	bassoon	bassoon	bassoon
Position 2	clarinet	oboe	flute	clarinet	oboe	flute
Position 3	oboe	flute	clarinet	oboe	flute	clarinet

There are six possibilities.

Examine We can draw a tree diagram to check our result.

Analyze the Strategy

1. **Explain** why making an organized list was a useful strategy in solving this problem.
2. **Explain** why a tree diagram is another method for making an organized list.
3. **Write** a problem that could be solved by making an organized list. Explain.

Lesson 5-2a Problem-Solving Strategy: Make an Organized List **201**

Apply the Strategy

Solve. Use the make an organized list strategy.

4. **FOOD** Daniel is making a peanut butter and jelly sandwich. His choices are creamy or crunchy peanut butter, white or wheat bread, and grape, apple, or strawberry jelly. How many different types of sandwiches can Daniel make?

5. **GAMES** On the game board, you plan to move two spaces away from square A. You can either move horizontally, vertically, or diagonally. How many different moves can you make from square A? List them.

Mixed Problem Solving

Solve. Use any strategy.

6. **SAVINGS** The graph below shows deposits and withdrawals. During which week was the difference between deposits and withdrawals the greatest?

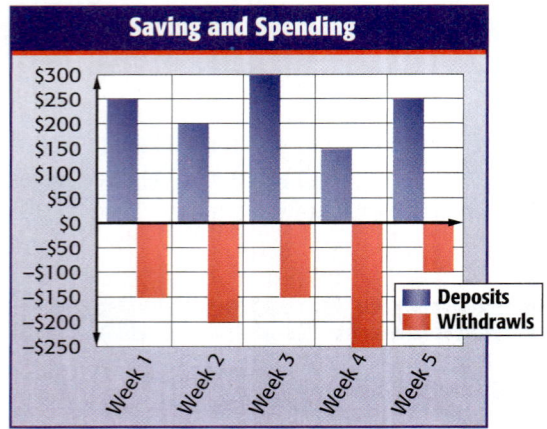

7. **SCIENCE FAIR** Mrs. Cassidy is taking a picture with the first, second, and third place winners of the science fair. If Mrs. Cassidy always stands on the far left, how many different ways can the students arrange themselves for the picture?

8. **SCHOOL** Members of Student Council are raising money to attend a conference. The total costs of the conference are shown in the table. If 24 students are attending the conference, how much does each student have to raise?

Transportation	$1,200
Registration	$288
Luncheon	$360

9. **NUMBER THEORY** Find two consecutive odd numbers that when added equal 56, and when multiplied equal 783.

10. **MULTI STEP** Terrez took a bag of cookies to play rehearsal. Half were given to the musicians and five to the director of the play. Terrez was left with 15 cookies. How many cookies did he take to rehearsal?

11. **EARTH SCIENCE** Giant kelp seaweed is found in the Pacific Ocean. One plant grows 3 feet the first two days. If it continues to grow at the same rate, what would be the length of the seaweed at the end of 80 days?

12. **STANDARDIZED TEST PRACTICE** Mary has to make deliveries to three neighbors. She lives at house b on the map. Which is the shortest route to make the deliveries and return home?

Ⓐ b-a-d-c-b
Ⓑ b-a-c-d-b
Ⓒ b-c-a-d-b
Ⓓ b-d-a-c-b

5-2 Greatest Common Factor

What You'll LEARN

Find the greatest common factor of two or more numbers.

NEW Vocabulary

Venn diagram
greatest common factor (GCF)

WHEN am I ever going to use this?

INTERNET A group of friends spent time in two Internet chat rooms. The diagram shows the chat rooms Angel, Sydney, Ian, Candace, and Christine visited. The friends were able to stay in one chat room or go to the other one.

1. Who visited the Fashion Chat Room?
2. Who visited the Music Chat Room?
3. Who visited both chat rooms?

The diagram above is called a **Venn diagram**. It uses circles to show how elements among sets of numbers or objects are related. The region where circles overlap represents items that are common to two or more sets. It shows that Ian visited both chat rooms.

Venn diagrams can also show factors that are common to two or more numbers. The greatest of these common factors is called the **greatest common factor (GCF)**. The GCF of prime numbers is 1.

EXAMPLE Find the GCF by Listing Factors

1 Find the GCF of 20 and 24.

First, list the factors of 20 and 24.
factors of 20: 1, 2, 4, 5, 10, 20
factors of 24: 1, 2, 3, 4, 6, 8, 12, 24   } common factors: 1, 2, 4

Notice that 1, 2, and 4 are common factors of 20 and 24. So, the GCF is 4.

Check You can draw a Venn diagram to check your answer.

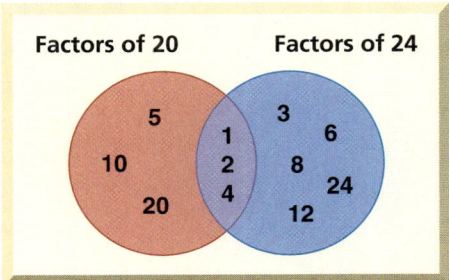

READING in the Content Area

For strategies in reading this lesson, visit **msmath2.net/reading**.

Your Turn Find the GCF of each pair of numbers.

a. 8 and 10   b. 6 and 12   c. 10, 17

msmath2.net/extra_examples

Lesson 5-2 Greatest Common Factor **203**

EXAMPLES — Find the GCF Using Prime Factors

Find the GCF of each set of numbers.

2) 18, 48

Method 1 Write the prime factorization.

$18 = \boxed{2} \times 3 \times \boxed{3}$
$48 = 2 \times 2 \times \boxed{2} \times 2 \times \boxed{3}$

Method 2 Divide by prime numbers.

Divide both 18 and 48 by 2. Then divide the quotients by 3.

$$\begin{array}{r} 3 \quad\; 8 \\ 3\overline{)9 \quad 24} \\ 2\overline{)18 \quad 48} \end{array}$$ ← Start here.

The common prime factors are 2 and 3.
So, the GCF of 18 and 48 is 2×3, or 6.

3) 12, 24, 60

$12 = \boxed{2} \times \boxed{2} \times \boxed{3}$
$24 = 2 \times \boxed{2} \times \boxed{2} \times \boxed{3}$ Circle the common factors.
$60 = \boxed{2} \times \boxed{2} \times \boxed{3} \times 5$

The common prime factors are 2, 2, and 3.
So, the GCF is $2 \times 2 \times 3$, or 12.

STUDY TIP

Choosing a Method
To find the GCF of two numbers, it is easier to:
- list factors for small numbers, and
- find the prime factorization or use division for large numbers.

Your Turn Find the GCF of each set of numbers.

d. 36, 48   e. 10, 35   f. 30, 45, 75

EXAMPLE — Find the GCF of an Algebraic Expression

4) ALGEBRA Find the GCF of $10a$ and $15a^2$.

Factor each expression.

$10a = 2 \cdot \boxed{5} \cdot \boxed{a}$
$15a^2 = 3 \cdot \boxed{5} \cdot \boxed{a} \cdot a$ Circle the common factors.

The GCF is $5 \cdot a$, or $5a$.

EXAMPLE — Use the GCF to Solve a Problem

5) PARADES In a parade, 64 eighth-graders are to march in front of 88 seventh-graders. Both groups should have the same number of students in each row. Find the greatest number of students in each row.

The greatest number of students in each row is the GCF of the number of students in each group.

$64 = \boxed{2} \times \boxed{2} \times \boxed{2} \times 2 \times 2 \times 2 = 2^6$
$88 = \boxed{2} \times \boxed{2} \times \boxed{2} \times 11 = 2^3 \times 11$

The GCF of 64 and 88 is 2^3, or 8.

So, there should be 8 students in each row.

Chapter 5 Fractions, Decimals, and Percents

Skill and Concept Check

1. **OPEN ENDED** Find two numbers whose GCF is 16.

2. **Draw** a Venn diagram showing the factors of 30 and 42. Name the GCF.

3. **FIND THE ERROR** Charles and Tiffany both found the GCF of 12, 18, and 27. Who is correct? Explain.

Charles
12 = 2 · 2 · 3
18 = 2 · 3 · 3
27 = 3 · 3 · 3
GCF: 2 · 3 · 3 = 18

Tiffany
12 = 2 · 2 · 3
18 = 2 · 3 · 3
27 = 3 · 3 · 3
GCF: 3

GUIDED PRACTICE

Find the GCF of each set of numbers.

4. 18, 30
5. 45, 60
6. 6, 8, 12
7. 18, 42, 60

8. Find the GCF of $14xy$ and $7x^2$.

Practice and Applications

Find the GCF of each set of numbers.

9. 12, 78
10. 40, 50
11. 20, 45
12. 32, 48
13. 24, 48
14. 45, 75
15. 56, 96
16. 40, 125
17. 18, 24, 30
18. 36, 60, 84
19. 35, 49, 84
20. 36, 50, 130

21. What is the GCF of $2^4 \times 5$ and $5^2 \cdot 7$?

22. Find the GCF of $2^3 \times 3^2 \times 7$ and $2^2 \times 3 \times 11$.

HOMEWORK HELP

For Exercises	See Examples
9–16, 21–22	1, 2
17–20	3
26–29	4
30–33	5

Extra Practice
See pages 574, 600.

Determine whether each statement is *sometimes*, *always*, or *never* true.

23. The GCF of two numbers is greater than both numbers.

24. If two numbers have no common prime factors, the GCF is 1.

25. The GCF of two numbers is one of the numbers.

Find the GCF of each set of algebraic expressions.

26. $24a, 6a$
27. $27b^2, 36b$
28. $16r^3, 28r$
29. $20x^2, 50xy^2$

INDUSTRIAL TECHNOLOGY For Exercises 30 and 31, use the following information.
Kibbe is building scale models of high-rise buildings for his class project. He designs the models using one-inch cubes. He is planning to build three buildings, the first with 108 blue cubes, the second with 270 red cubes, and the third with 225 yellow cubes. All of the buildings must be the same height, but not necessarily the same length and width.

30. What is the maximum height of each high-rise Kibbe can build?

31. What are the dimensions of all three buildings?

FOOD For Exercises 32 and 33, use the following information.
The track-and-field coaches threw a party at the end of the season and bought a 32-inch, a 20-inch, and a 12-inch submarine sandwich. Suppose the sandwiches are cut into equal-sized pieces.

32. What is the length of the longest piece that can be cut?

33. If there are 18 people on the team, will everyone receive a sandwich? Explain.

34. **RESEARCH** Use the Internet or another source to find information about the Sieve of Eratosthenes. Explain how you can find the GCF of 18 and 54 using the Sieve.

CRITICAL THINKING For Exercises 35–37, use the following information.
Numbers that have a GCF of 1 are *relatively prime*. Use this definition to determine whether each statement is true or false. Explain your answer.

35. Any two prime numbers are relatively prime.

36. If two numbers are relatively prime, one of them must be prime.

37. Two consecutive numbers are always relatively prime.

Spiral Review with Standardized Test Practice

38. **MULTIPLE CHOICE** Felisa is making shelves to store her books. How many shelves measuring 8 inches by 12 inches can be cut from a 36-inch-by-56-inch piece of plywood so that there is no waste?

    A 17   B 18   C 20   D 21

39. **MULTIPLE CHOICE** Which set of numbers does *not* have a GCF of 6?

    F 42, 150   G 54, 18   H 24, 6   I 18, 30

40. **SHORT RESPONSE** Find the GCF of $20st^2$ and $50s^2$.

Determine whether each number is *prime* or *composite*. (Lesson 5-1)

41. 21   42. 31   43. 65   44. 129

45. **ALGEBRA** Find the slope of the line that passes through the pair of points at the right. (Lesson 4-7)

46. **GEOMETRY** Write the ordered pair that is 16 units up from and 11 units to the left of the origin. (Lesson 3-3)

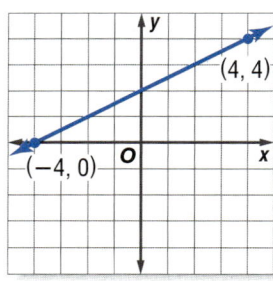

47. **MEASUREMENT** Twenty milliliters is equal to how many liters? (Lesson 1-8)

GETTING READY FOR THE NEXT LESSON

BASIC SKILL Divide.

48. $27 \div 3$ 49. $48 \div 2$ 50. $160 \div 40$ 51. $96 \div 6$

5-3 Simplifying Fractions

HANDS-ON Mini Lab

Materials
- grid paper
- colored pencils

What You'll LEARN
Write fractions in simplest form.

NEW Vocabulary
equivalent fractions
simplest form

Work with a partner.

On grid paper, draw the two figures at the right. Both figures should be the same size. Shade 8 out of the 20 squares in one figure. Shade 2 out of the 5 rectangles in the other.

1. Write a fraction to describe each figure: $\frac{\text{number of shaded parts}}{\text{total number of parts}}$.
2. Which figure has a greater portion of its parts shaded?
3. What can you conclude about the fractions you wrote above?

The fractions $\frac{8}{20}$ and $\frac{2}{5}$ are **equivalent fractions** because they have the same value. The fraction $\frac{2}{5}$ is in **simplest form** because the GCF of the numerator and denominator is 1.

EXAMPLES Write Fractions in Simplest Form

Write each fraction in simplest form.

1 $\frac{8}{20}$

First, find the GCF of the numerator and denominator.

factors of 8: 1, 2, 4, 8
factors of 20: 1, 2, 4, 5, 10, 20   } The GCF of 8 and 20 is 4.

Then, divide the numerator and the denominator by the GCF.

$\frac{8}{20} = \frac{8 \div 4}{20 \div 4} = \frac{2}{5}$ So, $\frac{8}{20}$ written in simplest form is $\frac{2}{5}$.

Check Multiply the numerator and denominator of the answer by the GCF. The result should be the original fraction.

STUDY TIP

Simplest Form
A fraction is in simplest form when the GCF of the numerator and denominator is 1.

2 $\frac{18}{30}$

$18 = \boxed{2} \cdot \boxed{3} \cdot 3$
$30 = \boxed{2} \cdot \boxed{3} \cdot 5$

GCF: $2 \cdot 3 = 6$

$\frac{18}{30} = \frac{18 \div 6}{30 \div 6} = \frac{3}{5}$

So, $\frac{18}{30}$ written in simplest form is $\frac{3}{5}$.

 =

$\frac{18}{30}$ = $\frac{3}{5}$

msmath2.net/extra_examples

EXAMPLE — Use Fractions to Solve a Problem

3 **MUSIC** If the fraction of the frequencies of two notes can be simplified, the two notes are harmonious. Use the graphic below to find the simplified fraction of the frequency of notes C and E.

Frequency Chart (Hertz)

$$\frac{\text{frequency of note C}}{\text{frequency of note E}} = \frac{264}{330}$$

$$= \frac{\overset{1}{2} \times 2 \times 2 \times \overset{1}{3} \times \overset{1}{11}}{\underset{1}{2} \times \underset{1}{3} \times 5 \times \underset{1}{11}}$$

$$= \frac{4}{5}$$

The slashes mean that part of the numerator and part of the denominator are both divided by the same number. For example, $3 \div 3 = 1$.

The fraction of the frequency of notes C and E is $\frac{4}{5}$.

Your Turn Find the simplified fraction of the frequency of each pair of notes.

a. D and G
b. E and F

REAL-LIFE CAREERS

How Does a Music Composer Use Math?
Music composers write notes that can vary in length. There are whole notes, half notes, quarter notes, eighth notes, and sixteenth notes.

Math Online Research
For information about a career as a music composer, visit: msmath2.net/careers

---

Skill and Concept Check

1. **Writing Math** Explain how to determine if a fraction is in simplest form.

2. **OPEN ENDED** Write a fraction that is *not* in simplest form. Then, simplify it.

3. **FIND THE ERROR** Seki and Luther both wrote $\frac{16}{36}$ in simplest form. Who is correct? Explain.

   Seki
   $\frac{16}{36} = \frac{16 \div 4}{36 \div 4} = \frac{4}{9}$

   Luther
   $\frac{16}{36} = \frac{16 \div 2}{36 \div 2} = \frac{8}{18}$

GUIDED PRACTICE

Write each fraction in simplest form.

4. $\frac{3}{9}$
5. $\frac{18}{30}$
6. $\frac{81}{90}$

7. **AGRICULTURE** George Washington Carver produced about 100 products from sweet potatoes and about 325 from peanuts. Write the number of sweet potato products as a fraction of the number of peanut products in simplest form.

Practice and Applications

Write each fraction in simplest form.

8. $\dfrac{9}{12}$
9. $\dfrac{25}{35}$
10. $\dfrac{16}{32}$
11. $\dfrac{14}{20}$
12. $\dfrac{10}{20}$
13. $\dfrac{12}{21}$
14. $\dfrac{15}{25}$
15. $\dfrac{24}{28}$
16. $\dfrac{48}{64}$
17. $\dfrac{32}{32}$
18. $\dfrac{50}{300}$
19. $\dfrac{80}{96}$

HOMEWORK HELP

For Exercises	See Examples
8–23	1, 2
24–28	3

Extra Practice
See pages 575, 600.

Write two fractions that are equivalent to each fraction.

20. $\dfrac{1}{2}$
21. $\dfrac{3}{5}$
22. $\dfrac{6}{7}$
23. $\dfrac{5}{8}$

24. **PARKING** New York holds the number one spot when it comes to the cost for parking your car. Use the graphic to express the cost of parking a car in Chicago as a fraction of the cost of parking a car in Boston in simplest form.

25. **MUSIC** Find the simplified fraction of the frequency of notes C and A in Example 3 on page 208.

Where parking costs most
Five most expensive markets for unreserved monthly parking:

New York midtown	$520
New York downtown	$400
Boston	$375
San Francisco	$325
Chicago	$300

Note: Reserved parking cost 33% more on average.

Source: Colliers International
By Shannon Reilly and Bob Laird, USA TODAY

Write each fraction in simplest form.

26. Fifteen minutes is what part of one hour?
27. Nine inches is what part of one foot?
28. Four days is what part of the month of April?

29. **CRITICAL THINKING** Both the numerator and denominator of a fraction are even. Is the fraction in simplest form? Explain.

Spiral Review with Standardized Test Practice

30. **MULTIPLE CHOICE** The drive to the football game took 56 minutes. Twenty-four minutes of the time was spent stopped in traffic. What fraction of the drive was spent stopped in traffic?

    Ⓐ $\dfrac{3}{7}$ Ⓑ $\dfrac{6}{14}$ Ⓒ $\dfrac{15}{35}$ Ⓓ $\dfrac{1}{3}$

31. **SHORT RESPONSE** Draw a model that shows $\dfrac{15}{25} = \dfrac{3}{5}$.

32. Find the GCF of 345, 253, and 115. (Lesson 5-2)

33. Find the prime factorization of 630. (Lesson 5-1)

GETTING READY FOR THE NEXT LESSON

PREREQUISITE SKILL Divide. (Page 562)

34. $2\overline{)1.0}$
35. $4\overline{)1.00}$
36. $10\overline{)7.0}$
37. $8\overline{)3.000}$

5-4 Fractions and Decimals

WHEN am I ever going to use this?

SOFTBALL The United States Women's softball team won the gold medal in the 2000 Summer Olympic Games. The table shows the results of the games they played in Sydney, Australia. The number of runs scored by the United States is listed first.

Opponent	Score
Japan	2–1
Australia	1–0
China	3–0
Italy	6–0
New Zealand	2–0
Australia	1–2
China	0–2
Japan	1–2
Cuba	3–0
Canada	6–0

Source: www.usasoftball.com

1. How many games did the USA softball team win? How many did they play?
2. Write a fraction comparing the number of times the team won to the total number of games played.

What You'll LEARN
Write fractions as terminating or repeating decimals and write decimals as fractions.

NEW Vocabulary
terminating decimals
repeating decimals
bar notation

Link to READING
Everyday Meaning of terminate: coming to an end, as in terminate a game

The softball team won $\frac{7}{10}$ of the games played. When the denominator is a power of 10, you can use place value to write the fraction as a decimal.

Words	Model	Fraction	Decimal
seven tenths		$\frac{7}{10}$	0.7

If the denominator is *not* a power of 10, you can write the fraction as a decimal using division.

EXAMPLES Write Fractions as Decimals

Write each fraction or mixed number as a decimal.

1 $\frac{1}{4}$

The fraction $\frac{1}{4}$ indicates $1 \div 4$.

Method 1 Use paper and pencil.

$$\begin{array}{r} 0.25 \\ 4\overline{)1.00} \\ \underline{8} \\ 20 \\ \underline{20} \\ 0 \end{array}$$

Division ends when the remainder is 0.

Method 2 Use a calculator.

1 4 ENTER 0.25

So, $\frac{1}{4} = 0.25$.

2 $5\frac{3}{4}$

The mixed number $5\frac{3}{4}$ is $5 + \frac{3}{4}$.

Method 1 Use paper and pencil.

$5\frac{3}{4} = 5 + \frac{3}{4}$ Write as a sum.
$\phantom{5\frac{3}{4}} = 5 + 0.75$ Write $\frac{3}{4}$ as 0.75.
$\phantom{5\frac{3}{4}} = 5.75$ Add.

Method 2 Use a calculator.

3 4 + 5 ENTER 5.75

So, $5\frac{3}{4} = 5.75$.

210 Chapter 5 Fractions, Decimals, and Percents

In Example 1, the division ends, or terminates, when the remainder is zero. So, 0.25 is called a **terminating decimal**.

Repeating decimals have a pattern in the digits that repeats forever.

Consider $\frac{1}{3}$.

You can use **bar notation** to indicate that a number repeats forever. A bar is written over the digits that repeat.

$0.33333\ldots = 0.\overline{3}$ $0.121212\ldots = 0.\overline{12}$ $11.3858585\ldots = 11.3\overline{85}$

EXAMPLES — Write Fractions as Repeating Decimals

Write each fraction or mixed number as a decimal.

3) $\frac{7}{9}$

Method 1 Use paper and pencil.

Method 2 Use a calculator.

7 ÷ 9 ENTER 0.7777…

So, $\frac{7}{9} = 0.\overline{7}$.

4) $8\frac{1}{3}$

Method 1 Use paper and pencil.

$8\frac{1}{3} = 8 + \frac{1}{3}$ Write as a sum.

$\phantom{8\frac{1}{3}} = 8 + 0.\overline{3}$ Write $\frac{1}{3}$ as $0.\overline{3}$.

$\phantom{8\frac{1}{3}} = 8.\overline{3}$ Add.

Method 2 Use a calculator.

1 ÷ 3 + 8 ENTER 8.3333…

So, $8\frac{1}{3} = 8.\overline{3}$.

Your Turn Write each fraction or mixed number as a decimal. Use bar notation if the decimal is a repeating decimal.

a. $\frac{5}{8}$ b. $\frac{2}{3}$ c. $1\frac{3}{11}$ d. $6\frac{2}{5}$ e. $4\frac{5}{6}$

STUDY TIP

Mental Math It will be helpful to memorize the following fraction-decimal equivalencies.

$\frac{1}{2} = 0.5$

$\frac{1}{3} = 0.\overline{3}$ $\frac{2}{3} = 0.\overline{6}$

$\frac{1}{4} = 0.25$ $\frac{3}{4} = 0.75$

$\frac{1}{5} = 0.2$ $\frac{1}{10} = 0.1$

$\frac{1}{8} = 0.125$

You can use a power of 10 to write a decimal as a fraction. Use the place value of the final digit as the denominator.

EXAMPLE — Write Decimals as Fractions

5) Write 0.48 as a fraction in simplest form.

$0.48 = \frac{48}{100}$ The 8 is in the hundredths place.

$ = \frac{12}{25}$ Simplify.

Your Turn Write each number as a fraction in simplest form.

f. 0.56   g. 0.3   h. 2.8

msmath2.net/extra_examples   Lesson 5-4 Fractions and Decimals **211**

Skill and Concept Check

1. **OPEN ENDED** Write a fraction that is equivalent to a terminating decimal and one that is equivalent to a repeating decimal.

2. **Writing Math** Compare 1 out of 3 and 3 out of 10. Are these values equal? Explain your reasoning.

GUIDED PRACTICE

Write each repeating decimal using bar notation.

3. 0.6333…
4. 5.313131…
5. 12.470470470…

Write each fraction or mixed number as a decimal. Use bar notation if the decimal is a repeating decimal.

6. $\frac{2}{5}$
7. $3\frac{5}{8}$
8. $\frac{5}{9}$
9. $1\frac{5}{6}$

Write each decimal as a fraction in simplest form.

10. 0.22
11. 0.1
12. 4.6

13. **SOCCER** Cirilo surveyed his physical education class and found that 12 out of 23 students chose soccer as their favorite sport. Write the fraction of students who chose soccer as a decimal to the nearest thousandth.

Practice and Applications

Write each repeating decimal using bar notation.

14. 0.999…
15. 5.92111…
16. 2.010101…
17. 13.1464646…
18. 26.993993…
19. 30.6841841841…

HOMEWORK HELP

For Exercises	See Examples
20–21, 24–27, 32, 43	1, 2
14–19, 22–23, 28–31, 44	3, 4
33–42	5

Extra Practice
See pages 575, 600.

Write each fraction or mixed number as a decimal. Use bar notation if the decimal is a repeating decimal.

20. $\frac{3}{8}$
21. $\frac{4}{5}$
22. $\frac{4}{9}$
23. $\frac{8}{11}$
24. $\frac{5}{32}$
25. $\frac{3}{20}$
26. $1\frac{7}{20}$
27. $6\frac{17}{40}$
28. $\frac{13}{24}$
29. $\frac{5}{7}$
30. $8\frac{1}{33}$
31. $4\frac{203}{999}$

32. **DOGS** There are an estimated 53 billion dogs in the U.S. The table shows the results of a survey taken recently of dog owners. Write the fraction of people who pet and hug their dogs daily as a decimal.

Activity	Fraction of People
Play with your dog daily	$\frac{23}{25}$
Take your dog on vacation	$\frac{9}{20}$
Pet and hug your dog daily	$\frac{19}{20}$
Celebrate your dog's birthday	$\frac{43}{100}$

Source: Ralston Purina

Write each decimal as a fraction in simplest form.

33. 0.75
34. 0.34
35. 0.2
36. 0.9
37. 5.96
38. 2.66

BASKETBALL Use the graph to write the number of rebounds and blocks per game for each player as a mixed number in simplest form.

39. Wallace   40. Divac

41. Gasol   42. O'Neal

43. **MULTI STEP** Kelsey practiced playing the cello for 2 hours and 18 minutes. Write the time Kelsey spent practicing in hours as a decimal.

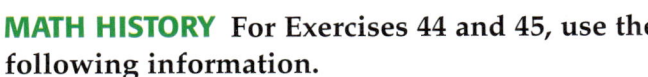

MATH HISTORY For Exercises 44 and 45, use the following information.
The value of pi (π) is 3.1415927… . Pi is a nonrepeating, nonterminating decimal. Mathematicians have used many methods to find the value of π.

44. Archimedes believed that π was between $3\frac{1}{7}$ and $3\frac{10}{71}$. Write each fraction as a decimal rounded to the nearest hundred-thousandth. Was Archimedes correct?

45. The Rhind Papyrus states that the Egyptians used $\frac{256}{81}$ for π. Write the fraction as a decimal rounded to the nearest hundred-thousandth. Which is closer to the actual value of π, Archimedes' value or the Egyptians' value?

46. **CRITICAL THINKING** Fractions with denominators of 2, 4, 8, 16, and 32 produce terminating decimals. Fractions with denominators of 6, 12, 18, and 24 produce repeating decimals. What do you think causes the difference? Explain.

Spiral Review with Standardized Test Practice

47. **MULTIPLE CHOICE** Express $\frac{9}{25}$ as a decimal.
    A 0.036   **B** 0.36   **C** $0.\overline{36}$ **D** 3.6

48. **SHORT RESPONSE** In 2003, Ichiro Suzuki had 212 hits in 679 at-bats. The batting average of a baseball player is the number of hits divided by the number of at-bats. Find Suzuki's batting average to the nearest thousandth.

Write each fraction in simplest form. (Lesson 5-3)

49. $\frac{10}{24}$ 50. $\frac{39}{81}$ 51. $\frac{28}{98}$ 52. $\frac{51}{68}$

53. Find the GCF of 36 and 48. (Lesson 5-2)

GETTING READY FOR THE NEXT LESSON

BASIC SKILL Write a fraction for the number of squares shaded to the total number of squares.

54.   55.   56.   57.

Chapter 5 Mid-Chapter Practice Test

Vocabulary and Concepts

1. **Define** *prime number* and give an example. (Lesson 5-1)

2. **Describe** how you know when a fraction is in simplest form. (Lesson 5-3)

Skills and Applications

3. Determine whether 24 is *prime* or *composite*. (Lesson 5-1)

Factor each expression. (Lesson 5-1)

4. 30
5. 120
6. $14x^2y$
7. $50mn$

Find the GCF of each set of numbers or algebraic expressions. (Lesson 5-2)

8. 16, 40
9. 65, 100
10. $12x, 20x^3$
11. $45a^2b^2, 81ab^3$

12. **CYCLING** The *gear ratio* of a bicycle is the comparison of the number of teeth on a chainwheel to the number on a freewheel. If the gear ratio for Alexis' 10-speed bike is $\frac{52}{16}$, write this fraction in simplest form. (Lesson 5-3)

Write each fraction or mixed number as a decimal. Use bar notation if the decimal is a repeating decimal. (Lesson 5-4)

13. $\frac{7}{8}$
14. $\frac{2}{9}$
15. $3\frac{13}{20}$

Write each decimal as a fraction in simplest form. (Lesson 5-4)

16. 0.6
17. 0.48
18. 7.02

Standardized Test Practice

19. **MULTIPLE CHOICE** What is the greatest number of crayons in each row of an 8-, a 64-, and a 96-crayon box if all rows have the same number? (Lesson 5-2)

   Ⓐ 2 Ⓑ 4 Ⓒ 8 Ⓓ 12

20. **GRID IN** Amy spent 35 minutes in the afternoon practicing the piano. What part of one hour did she spend practicing? Write as a fraction in simplest form. (Lesson 5-3)

The Game Zone

A Place To Practice Your Math Skills

Math Skill: Finding Factors

The Factor Fair

GET READY!
Players: two teams
Materials: 36 index cards, tape

GET SET!
- Number the index cards 1 through 36.
- Tape the index cards in order on the chalkboard.
- Divide the class into two teams.

GO!
- Team A chooses one of the index cards, such as the 8 card, and takes it off the chalkboard. Team A gets 8 points. Team B gets all the cards that are factors of 8 that have not yet been taken. The sum of the factors are the points they receive. For example, Team B would receive 1 + 2 + 4, or 7 points.

Team A	Team B
8	1
	2
	+ 4
	7

- Team B then chooses a card and gets those points.
- A team loses a turn if it selects an illegal number. A number is considered illegal if it does not have at least one factor available.
- Teams continue to take turns until there are no legal plays remaining.
- **Who Wins?** The team with the most points is the winner.

5-5 Fractions and Percents

What You'll LEARN
Write fractions as percents and percents as fractions.

NEW Vocabulary
ratio
percent

REVIEW Vocabulary
simplest form: a fraction whose numerator and denominator have a GCF of 1 (Lesson 5-3)

WHEN am I ever going to use this?

EDUCATION The table shows the results of a survey in which students were asked to choose methods that make learning new subjects more interesting.

1. Shade a 10 × 10 grid that represents the number of students that chose each method.
2. What fraction of the students chose the Internet as the method that makes learning more interesting?

Method	Number of Students
Internet	34 out of 100
Teacher	29 out of 100
TV Program	24 out of 100
Textbook	12 out of 100

Source: Opinion Research Corporation

A **ratio** is a comparison of two numbers by division. Ratios, like the ones above, can be written in several different ways.

$$34 \text{ out of } 100 \qquad 34:100 \qquad \frac{34}{100}$$

When a ratio compares a number to 100, it can be written as a **percent**.

Noteables™ — Key Concept: Percent

Words   A percent is a ratio that compares a number to 100.

Symbols $\frac{n}{100} = n\%$

Ratio   34 to 100

Model

EXAMPLES Write Ratios as Percents

Write each ratio as a percent.

1 Annie answered 90 out of 100 questions correctly.

You can represent 90 out of 100 with a model.

$\frac{90}{100} = 90\%$

2 On average, 50.5 out of 100 students own a pet.

$\frac{50.5}{100} = 50.5\%$

Your Turn Write each ratio as a percent.

a. $\frac{45}{100}$ b. 19.2 out of 100 c. $3.30:$100

READING Math

Percent *Percent* means per *hundred* or *hundredths*. The symbol % means percent.

Fractions and percents are ratios that can represent the same number. You can write a fraction as a percent by finding an equivalent fraction with a denominator of 100.

EXAMPLE Write a Fraction as a Percent

3. Write $\frac{3}{20}$ as a percent.

Since $100 \div 20 = 5$, ... $\frac{3}{20} = \frac{15}{100}$ ... multiply the numerator and denominator by 5.

$\frac{15}{100} = 15\%$

So, $\frac{3}{20} = 15\%$.

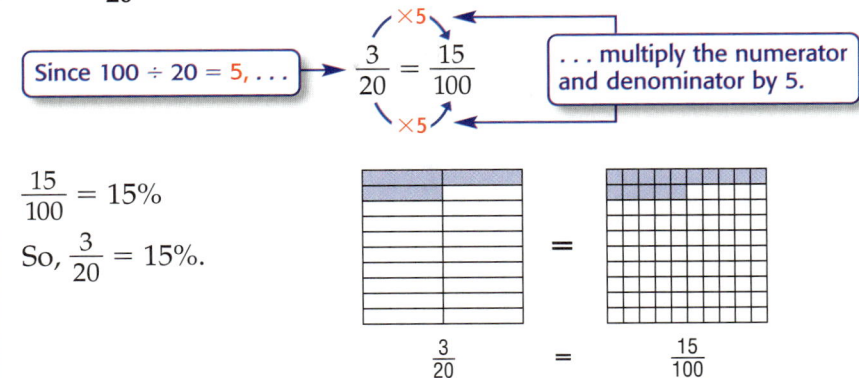

$\frac{3}{20} = \frac{15}{100}$

Your Turn Write each fraction as a percent.

d. $\frac{17}{20}$ e. $\frac{3}{5}$ f. $\frac{2}{25}$

You can also use the meaning of percent to write percents as fractions.

EXAMPLE Write a Percent as a Fraction

4. Write 48% as a fraction in simplest form.

$48\% = \frac{48}{100}$ Definition of percent

$= \frac{12}{25}$ Simplify.

Your Turn Write each percent as a fraction.

g. 40%   h. 6%   i. 24%

REAL-LIFE MATH

COINS Each new quarter has a diameter of 24.26 millimeters, a thickness of 1.75 millimeters, and weighs 5.670 grams.

Source: www.usmint.gov

EXAMPLE Use Percent to Solve a Problem

5. **COINS** The 50 State Quarters® Program allows the United States Mint to release a new quarter every ten weeks from 1999 through 2008 commemorating the 50 states. By the end of 2004, 30 state coins were released. What percent of the coins is this?

By the end of 2004, 30 out of 50 state coins were released.

$\frac{30}{50} = \frac{60}{100}$ Write an equivalent fraction with a denominator of 100.

$= 60\%$ $\frac{60}{100} = 60\%$

So, 60% of the state coins were released by the end of 2004.

msmath2.net/extra_examples

Lesson 5-5 Fractions and Percents

Skill and Concept Check

1. **Write** a fraction in simplest form and a percent to represent the shaded area at the right.

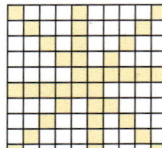

2. **OPEN ENDED** Draw a model that shows 25%.

3. **Which One Doesn't Belong?** Identify the number that does not have the same value as the other three. Explain your reasoning.

| $\frac{10}{100}$ | 1% | $\frac{1}{10}$ | 100 out of 1,000 |

GUIDED PRACTICE

Write each ratio as a percent.

4. $\frac{57}{100}$ 5. $29.20 per $100   6. 11 teams out of 100

Write each fraction as a percent.

7. $\frac{1}{4}$ 8. $\frac{6}{10}$ 9. $\frac{4}{5}$ 10. $\frac{17}{20}$

Write each percent as a fraction in simplest form.

11. 90%   12. 80%   13. 75%   14. 22%

15. **ELECTRICITY** Electrical power is measured in a unit called the watt. Write a percent that compares the amount of power given off by a 75-watt bulb to the power given off by a 100-watt bulb.

Practice and Applications

Write each ratio as a percent.

16. $\frac{87}{100}$ 17. $\frac{12.2}{100}$ 18. 42 per 100

19. $11\frac{2}{7}$ out of 100   20. $66\frac{2}{3}$:100   21. 99.9:100

22. Fifty-one out of every 100 households have at least one computer.

23. Twenty out of every 100 adults use the Internet to check on news, weather, and sports.

Write each fraction as a percent.

24. $\frac{7}{10}$ 25. $\frac{16}{20}$ 26. $\frac{15}{25}$ 27. $\frac{37}{50}$

28. $\frac{73}{100}$ 29. $\frac{13}{50}$ 30. $\frac{1}{5}$ 31. $\frac{19}{20}$

32. $\frac{3}{5}$ 33. $\frac{3}{25}$ 34. $\frac{10}{10}$ 35. $\frac{19}{25}$

36. **SPECIAL OLYMPICS** Mila and Randi are volunteers for their school district's Special Olympics competition. They are expecting 100 participants, but only 67 have registered so far. What percent of the Special Olympics participants still need to register before the competition begins?

HOMEWORK HELP

For Exercises	See Examples
16–23	1, 2
24–35	3
39–50	4
36–38	5

Extra Practice
See pages 575, 600.

218 Chapter 5 Fractions, Decimals, and Percents

CHORES For Exercises 37 and 38, use the graphic. It shows the percent of chores that parents have to remind their children to do the most.

37. What fraction of parents remind their children to clean their rooms?

38. If 1,500 parents were surveyed, how many have to remind their children to put their dirty clothes in the hamper?

Source: Opinion Research Corp. for The Soap and Detergent Association

Write each percent as a fraction in simplest form.

39. 45%   40. 30%   41. 50%
42. 20%   43. 62%   44. 88%
45. 68%   46. 13%   47. 44%
48. 52%   49. 2%   50. 14%

51. **GEOMETRY** What percent of the larger rectangle at the right is shaded?

52. What percent of 100 is 1?

53. 17% of 100 is what number?

54. **CRITICAL THINKING** Kenneth and Rico both finished a 1-minute typing test on the computer. Kenneth misspelled one word and received a score of 96% correct. Rico also misspelled one word but received a score of 90% correct. Explain why both students received different scores with only one misspelled word.

Spiral Review with Standardized Test Practice

55. **MULTIPLE CHOICE** Find the ratio that represents 20%.
    - Ⓐ $\frac{20}{100}$
    - Ⓑ 2:10
    - Ⓒ 200 per 1,000
    - Ⓓ all of the above

56. **SHORT RESPONSE** Write a percent for a $7.50 donation from a $100 check.

Write each repeating decimal using bar notation. (Lesson 5-4)

57. 0.6555…   58. 4.232323…   59. 0.414141…

60. **MUSIC** In a survey, 12 of the 78 people preferred classical music. Write this ratio as a fraction in simplest form. (Lesson 5-3)

GETTING READY FOR THE NEXT LESSON

BASIC SKILL Multiply or divide.

61. 16.2 × 10   62. 0.71 × 100   63. 14.4 ÷ 100   64. 791 ÷ 1,000

5-6 Percents and Decimals

What You'll LEARN

Write percents as decimals and decimals as percents.

WHEN am I ever going to use this?

READING The graphic shows the reasons that students in 6th through 12th grades read.

1. Write the percent of students who read for fun as a fraction.
2. Write the fraction as a decimal.

Source: National Education Association, *American Demographics*

In Lesson 5-4, you learned that any fraction can be written as a decimal. You can use this fact to write percents as decimals.

EXAMPLES Write Percents as Decimals

1 **GEOGRAPHY** Alaska is the largest state, making up more than 16% of the land area of the United States. Write 16% as a decimal.

$16\% = \frac{16}{100}$ Write the percent as a fraction.

$= 0.16$ Write the fraction as a decimal.

So, $16\% = 0.16$.

2 Write 85.3% as a decimal.

$85.3\% = \frac{85.3}{100}$ Write the percent as a fraction.

$= \frac{85.3 \times 10}{100 \times 10}$ Multiply by 10 to remove the decimal in the numerator.

$= \frac{853}{1,000}$ Simplify.

$= 0.853$ Write the fraction as a decimal.

So, $85.3\% = 0.853$.

Study the pattern in the percents and the equivalent decimals in Examples 1 and 2. Notice that you can write the percent as a decimal by dividing the number by 100 and removing the percent symbol.

Noteables™ Key Concept: Write Percents as Decimals

Words   To write a percent as a decimal, divide the percent by 100 and remove the percent symbol.

Symbols $25\% = 25 = 0.25$

220 Chapter 5 Fractions, Decimals, and Percents

EXAMPLES   Write Percents as Decimals

3 Write 23% as a decimal.

$23\% = 23$ Divide by 100.
$= 0.23$ Remove the %.
So, $23\% = 0.23$.

4 Write $3\frac{1}{4}\%$ as a decimal.

$3\frac{1}{4}\% = 3.25\%$ Write $\frac{1}{4}$ as 0.25.
$= 03.25$ Divide by 100.
$= 0.0325$ Remove the %.
So, $3\frac{1}{4}\% = 0.0325$.

Your Turn Write each percent as a decimal.

a. 76%   b. 8.5%   c. $92\frac{1}{2}\%$

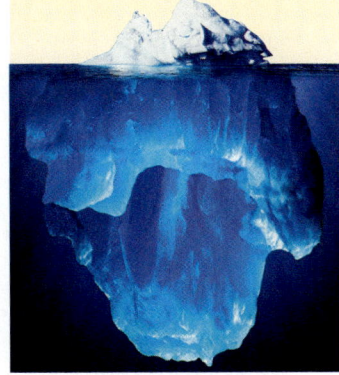

REAL-LIFE MATH

GEOGRAPHY A large Antarctic iceberg can weigh 400 million tons, rise 10 stories above the surface of the water, and supply water for 3 million people for a year.

Source: www.chennaionline.com

You can also use fractions to help you write decimals as percents.

EXAMPLES   Write Decimals as Percents

5 **GEOGRAPHY** Nearly 0.02 of the world's fresh water comes from the Antarctic Icecap. Write 0.02 as a percent.

$0.02 = \frac{2}{100}$ Definition of decimal
$= 2\%$ Definition of percent

So, $0.02 = 2\%$.

6 Write 0.347 as a percent.

$0.347 = \frac{347}{1{,}000}$ Definition of decimal

$= \frac{34.7}{100}$ Divide both numerator and denominator by 10.

$= 34.7\%$ Definition of percent

So, $0.347 = 34.7\%$.

Study the pattern in the decimals and the equivalent percents in Examples 5 and 6. Notice that you can write the decimal as a percent by multiplying the number by 100 and adding the percent symbol.

Noteables™   Key Concept: Write Decimals as Percents

Words   To write a decimal as a percent, multiply the percent by 100 and add a percent symbol.

Symbols $0.58 = 0.58 = 58\%$

STUDY TIP

Mental Math
• To write a percent as a decimal, move the decimal point two places to the left.
• To write a decimal as a percent, move the decimal point two places to the right.

EXAMPLES   Write Decimals as Percents

7 Write 0.64 as a percent.

$0.64 = 0.64$ Multiply by 100.
$= 64\%$ Add the %.
So, $0.64 = 64\%$.

8 Write 0.875 as a percent.

$0.875 = 0.875$ Multiply by 100.
$= 87.5\%$ Add the %.
So, $0.875 = 87.5\%$.

msmath2.net/extra_examples   Lesson 5-6 Percents and Decimals

Skill and Concept Check

1. **OPEN ENDED** Choose a decimal. Then write it as a fraction in simplest form and as a percent.

2. **FIND THE ERROR** Jessica and Gregorio both wrote 0.881 as a percent. Who is correct? Explain.

   Jessica
   0.881 = 0.881
   = 88.1%

   Gregorio
   0.881 = 000.881
   = 0.00881%

GUIDED PRACTICE

Write each percent as a decimal.

3. 68%
4. 3%
5. 27.6%
6. $45\frac{1}{2}\%$

Write each decimal as a percent.

7. 0.56
8. 0.12
9. 0.08
10. 0.399

11. **GEOGRAPHY** Pennsylvania makes up 1.2% of the landmass of the United States. Write this percent as a decimal.

Practice and Applications

Write each percent as a decimal.

12. 27%
13. 70%
14. 6%
15. 1%
16. 18.5%
17. 2.2%
18. $15\frac{1}{2}\%$
19. $30\frac{1}{4}\%$

HOMEWORK HELP

For Exercises	See Examples
12–19, 34–37	1, 2, 3, 4
20–33	5, 6, 7, 8

Extra Practice
See pages 576, 600.

Write each decimal as a percent.

20. 0.95
21. 0.08
22. 0.17
23. 0.6
24. 0.675
25. 0.145
26. 0.012
27. 0.7025

TENNIS For Exercises 28–32, use the graphic at the right. Write each decimal as a percent for the following players.

28. Monica Seles
29. Martina Hingis
30. Venus Williams
31. Jennifer Capriati
32. Based on her winning percentage, which player wins 413 out of every 500 games played?

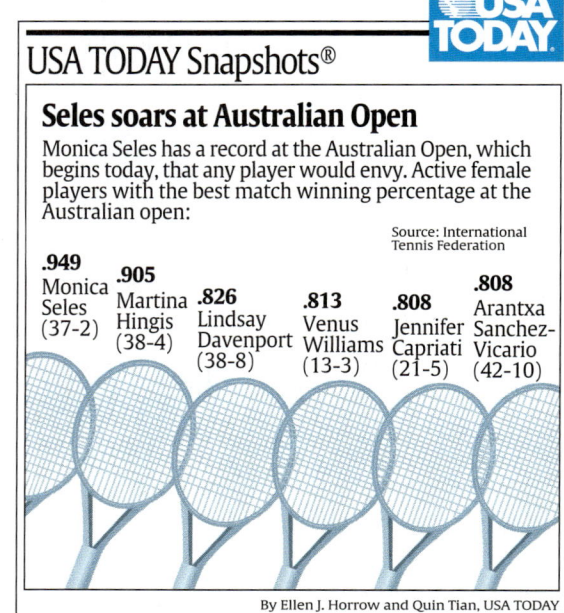

USA TODAY Snapshots®

Seles soars at Australian Open

Monica Seles has a record at the Australian Open, which begins today, that any player would envy. Active female players with the best match winning percentage at the Australian open:

Source: International Tennis Federation

.949 Monica Seles (37-2)
.905 Martina Hingis (38-4)
.826 Lindsay Davenport (38-8)
.813 Venus Williams (13-3)
.808 Jennifer Capriati (21-5)
.808 Arantxa Sanchez-Vicario (42-10)

By Ellen J. Horrow and Quin Tian, USA TODAY

33. **BIRDS** Birds shown in a field guide are 0.10 to 0.67 their actual size. What percent of their actual size are they?

MODEL TRAINS For Exercises 34–37, use the table at the right.

34. The five scales at the right can be used to build a model train based on the life-size original. For example, a G-scale train is about 4% of the size of the original. So, the model of a 75-foot-long locomotive measures 40 inches. Find the decimal equivalents for each scale.

35. Which scale is the smallest?

36. Find the approximate length of a 75-foot train in the N scale.

37. Name the scale used when a model of a 75-foot train measures approximately 18.75 inches.

Scale	Approximate Percent Equivalent
G	4.4%
O	2.083%
HO	1.15%
N	0.625%
Z	0.45%

Source: www.internettrains.com

CRITICAL THINKING For Exercises 38–43, choose the greater number in each pair.

38. $1\frac{1}{4}\%$, 0.125
39. 0.76, 76.5%
40. 42%, 4.2
41. 99%, 0.985
42. $\frac{13}{40}$, 0.30
43. 0.56, $\frac{45}{80}$

Spiral Review with Standardized Test Practice

44. **MULTIPLE CHOICE** Which is 0.08 written as a percent?
    Ⓐ 0.08%   Ⓑ 0.8%   Ⓒ 8%   Ⓓ 80%

45. **MULTIPLE CHOICE** Which is 96.3% written as a decimal?
    Ⓕ 0.0963   Ⓖ 0.963   Ⓗ 9.63   Ⓘ 963

46. **GRID IN** Write $78\frac{1}{2}\%$ as a decimal.

Write each ratio as a percent. (Lesson 5-5)

47. $\frac{72}{100}$
48. 90.9:100
49. 18 per 100
50. 23.1 out of 100

51. Write $9\frac{3}{8}$ as a decimal. (Lesson 5-4)

52. **BALLOONS** Write an integer to represent a hot-air balloon descending 83 feet. (Lesson 3-1)

TESTS For Exercises 53 and 54, use the stem-and-leaf plot of Chapter 5 test scores at the right.

53. How many students took the test? (Lesson 2-5)
54. What was the median score? (Lesson 2-4)

Stem	Leaf
10	0
9	2 3 3 4 5 9
8	0 1 1 2 3 5 9 9
7	0 4 6 8 8 9
6	4 8|1 = 81 points

GETTING READY FOR THE NEXT LESSON

PREREQUISITE SKILL Use exponents to write the prime factorization of each number. (Lesson 5-1)

55. 50
56. 32
57. 76
58. 105

5-7 Least Common Multiple

What You'll LEARN

Find the least common multiple of two or more numbers.

NEW Vocabulary

multiple
least common multiple (LCM)

HANDS-ON Mini Lab

Materials
- centimeter cubes

Work with a partner.

Use 9 centimeter cubes to model the first floor of Building 1 and 12 centimeter cubes to model the first floor of Building 2, as shown.

Building 1

Building 2

1. Add a second floor to each building. Record the total number of cubes used in a table like the one shown below.

Number of Floors	1	2	3	4	5
Number of Cubes in Building 1	9	?	?	?	?
Number of Cubes in Building 2	12	?	?	?	?

2. Continue adding floors until each building has five floors. Record your results.
3. Describe two buildings that have the same number of cubes.
4. If you keep adding floors, will the two buildings have the same number of cubes again? Explain.

In the Mini Lab, you listed multiples of 9 and 12. A **multiple** is the product of a number and any whole number. The **least common multiple**, or **(LCM)**, of two or more numbers is the least of their common multiples, excluding zero. So, the LCM of 9 and 12 is 36.

You can use several strategies to find the LCM of two or more numbers.

EXAMPLE Find the LCM by Listing Multiples

1 Find the LCM of 6 and 10.

First, list the multiples of 6 and 10.

multiples of 6:   6, 12, 18, 24, 30, 36, 42, 48, 54, 60, …
multiples of 10: 10, 20, 30, 40, 50, 60, 70, 80, …

Notice that 30, 60, …, are common multiples. So, the LCM of 6 and 10 is 30.

Your Turn Find the LCM of each pair of numbers.

a. 3, 12   b. 10, 12   c. 4, 5

You can also use prime factorization to find the LCM. The LCM is the smallest product that contains the prime factors of each number.

EXAMPLES — Find the LCM Using Prime Factors

2 Find the LCM of 24 and 36.

Method 1 Write the prime factorization.

$24 = 2 \times 2 \times 2 \times 3 = 2^3 \times 3$
$36 = 2 \times 2 \times 3 \times 3 = 2^2 \times 3^2$

The prime factors of 24 and 36 are 2 and 3. Multiply the greatest power of both 2 and 3.

The LCM of 24 and 36 is $2^3 \times 3^2$, or 72.

Method 2 Divide by prime numbers.

```
      2     3
   3)6     9
   2)12   18
   2)24   36   ← Start here.
```

Start dividing by prime factors until both numbers cannot be divided by the same divisor. Then multiply the divisors and quotients to find the LCM.

STUDY TIP

Choosing a Method
To find the LCM of two or more numbers, it is easier to:
- list multiples for small numbers, and
- use prime factorization or division for large numbers.

3 Find the LCM of 4, 10, and 15.

$4 = 2^2$
$10 = 2 \times 5$
$15 = 3 \times 5$
LCM: $2^2 \times 3 \times 5 = 60$

So, the LCM of 4, 10, and 15 is 60.

Your Turn Find the LCM of each set of numbers.

d. 9, 15   e. 10, 20, 30   f. 6, 17, 34

Skill and Concept Check

1. **Writing Math** Describe the relationship between 12, 4, and 3 using the words *multiple* and *factor*.

2. **OPEN ENDED** Find three numbers that each have a multiple of 30.

3. **NUMBER SENSE** Determine whether the LCM for all pairs of odd numbers is *sometimes*, *always*, or *never* their product. Explain.

GUIDED PRACTICE

Find the LCM of each set of numbers.

4. 4, 10   5. 6, 7   6. 3, 5, 12

7. **PICNIC** Flavio is having a cookout for his class and serving hot dogs. Hot dogs come in packages of 10. Hot dog buns come in packages of 8. If Flavio wants to have the same number of hot dogs and buns, what is the least number of each that he will have to buy?

Practice and Applications

Find the LCM of each set of numbers.

8. 6, 8
9. 12, 16
10. 30, 45
11. 8, 18
12. 11, 12
13. 45, 63
14. 2, 3, 5
15. 6, 8, 9
16. 8, 12, 16
17. 12, 15, 28
18. 22, 33, 44
19. 12, 16, 36

HOMEWORK HELP

For Exercises	See Examples
8–13, 20, 24	1, 2
14–19, 22–23	3

Extra Practice
See pages 576, 600.

20. Find the LCM of 2×3^2 and $2^3 \times 3$.

21. **PATTERNS** Write a rule that describes the common multiples of 9 and 15 that are greater than 180.

PLANETS For Exercises 22 and 23, use the following information and the table at the right.

The nine planets in our solar system revolve around the Sun at different orbital speeds.

22. When will the number of miles revolved by Earth, Mercury, and Pluto be the same?

23. In how many seconds does Earth revolve the number of miles in Exercise 22? Mercury? Pluto?

Planet	Orbital Speed (mi/s)
Earth	18
Mercury	30
Pluto	3

Source: infoplease.com

24. **NUMBER THEORY** The LCM of two consecutive numbers is greater than 200 and is a multiple of 7. Name the numbers.

25. **CRITICAL THINKING** Two numbers have an LCM of $2^2 \times 3 \times 5^2$. Their GCF is 2×5. If one of the numbers is $2 \times 3 \times 5$, what is the other number?

Spiral Review with Standardized Test Practice

26. **MULTIPLE CHOICE** Find the LCM of 27 and 30.
    - Ⓐ 810
    - Ⓑ 300
    - Ⓒ 270
    - Ⓓ 3

27. **SHORT RESPONSE** Presidents are elected every four years. Senators are elected every six years. If a senator was elected in the presidential election year 2000, in what year will she campaign again during a presidential election year?

28. Write 55.7% as a decimal to the nearest hundredth. *(Lesson 5-6)*

DIAMONDS The graph shows the percent of diamond engagement rings that are different shapes. Write each percent as a fraction in simplest form. *(Lesson 5-5)*

29. round
30. marquise
31. others

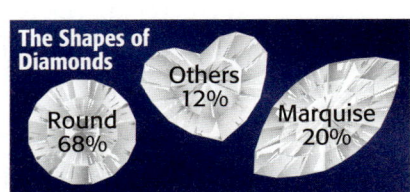

Source: Diamond Information Center

GETTING READY FOR THE NEXT LESSON

PREREQUISITE SKILL Replace each ● with < or > to make a true sentence. *(Page 556)*

32. 6.85 ● 5.68
33. 2.34 ● 2.43
34. 6.9 ● 5.99
35. 1.001 ● 1.1

226 Chapter 5 Fractions, Decimals, and Percents

5-8 Comparing and Ordering Rational Numbers

What You'll LEARN

Compare and order fractions, decimals, and percents.

NEW Vocabulary

common denominator
least common denominator (LCD)
rational numbers

WHEN am I ever going to use this?

SOFTBALL The table shows how Latanya and Patrick did during the first month of the softball season.

Player	At-bats	Hits
Latanya	12	7
Patrick	18	8

1. A batting average is the ratio of hits to at-bats. Write each player's batting average as a fraction.
2. Estimate which fraction is greater than $\frac{1}{2}$. Which is less than $\frac{1}{2}$?
3. Which player has the better batting average?

To compare and order fractions like the ones above, you can rewrite each fraction with a common denominator and then compare the numerators.

A **common denominator** is a common multiple of the denominators of two or more fractions. The **least common denominator (LCD)** is the LCM of the denominators.

EXAMPLE   Compare Fractions

1 Who has the better batting average, Latanya or Patrick?

Method 1 Rename using the LCD. Then compare numerators.

Latanya: $\frac{7}{12} = \frac{7 \times 3}{12 \times 3} = \frac{21}{36}$

Patrick: $\frac{8}{18} = \frac{8 \times 2}{18 \times 2} = \frac{16}{36}$

The LCD is 36.

Since $\frac{21}{36} > \frac{16}{36}$, then $\frac{7}{12} > \frac{8}{18}$. Latanya has the better average.

Method 2 Write each fraction as a decimal. Then compare decimals.

Latanya: $\frac{7}{12} = 0.58\overline{3}$

Patrick: $\frac{8}{18} = 0.\overline{4}$

Since $0.58\overline{3} > 0.\overline{4}$, then $\frac{7}{12} > \frac{8}{18}$. Latanya has the better average.

Your Turn Replace each ● with <, >, or = to make a true sentence.

a. $\frac{5}{6}$ ● $\frac{7}{9}$ b. $\frac{1}{5}$ ● $\frac{7}{50}$ c. $\frac{9}{16}$ ● $\frac{7}{10}$

Lesson 5-8 Comparing and Ordering Rational Numbers **227**

REAL-LIFE MATH

MOUNTAIN BIKING More than 7.5 million people participate in off-road mountain biking each year.

Source: National Sporting Goods Association

EXAMPLE Compare Ratios

2 **MOUNTAIN BIKING** In Coach Ito's first period class, 19 out of 32 students owned a mountain bike. In his seventh period class, 16 out of 28 students owned a mountain bike. Did a greater fraction of students own a mountain bike in the first period class or the seventh period class?

Estimate Both fractions are slightly greater than $\frac{1}{2}$.

Since the denominators are large, write $\frac{19}{32}$ and $\frac{16}{28}$ as decimals and then compare.

$19 \div 32 \approx 0.5938$ $16 \div 28 \approx 0.5714$ Use a calculator.

Since $0.5938 > 0.5714$, then $\frac{19}{32} > \frac{16}{28}$. So, a greater fraction of students in the first period class owned a mountain bike.

Some fractions, decimals, and percents are used more frequently than others. So, it is a good idea to be familiar with these common fraction-decimal-percent equivalents.

Concept Summary — Fraction-Decimal-Percent Equivalents

$\frac{1}{5} = 0.2 = 20\%$	$\frac{1}{10} = 0.1 = 10\%$	$\frac{1}{4} = 0.25 = 25\%$
$\frac{2}{5} = 0.4 = 40\%$	$\frac{3}{10} = 0.3 = 30\%$	$\frac{1}{2} = 0.5 = 50\%$
$\frac{3}{5} = 0.6 = 60\%$	$\frac{7}{10} = 0.7 = 70\%$	$\frac{3}{4} = 0.75 = 75\%$
$\frac{4}{5} = 0.8 = 80\%$	$\frac{9}{10} = 0.9 = 90\%$	

EXAMPLE Order Ratios

3 Order 0.6, 48%, and $\frac{1}{2}$ from least to greatest.

Use estimation to find that 48% is the least. Then, compare $\frac{1}{2}$ and 0.6 using decimals.

$\frac{1}{2} = 0.5$ and $0.5 < 0.6$ So, $48\% < \frac{1}{2} < 0.6$.

Check You can change 48% and 0.6 to fractions, then compare all three fractions using the LCD.

$0.6 = \frac{60}{100}$ $48\% = \frac{48}{100}$ $\frac{1}{2} = \frac{50}{100}$

Since $\frac{48}{100} < \frac{50}{100} < \frac{60}{100}$, $48\% < \frac{1}{2} < 0.6$.

Your Turn Order each set of ratios from least to greatest.

d. 22%, 0.3, $\frac{2}{10}$ e. $\frac{1}{5}$, 2%, 0.18   f. 0.74, $\frac{3}{4}$, 70%

228 Chapter 5 Fractions, Decimals, and Percents

Fractions, terminating and repeating decimals, and integers are called **rational numbers** because they can be written as fractions.

Noteables™ — Key Concept: Rational Numbers

Words Rational numbers are numbers that can be written as fractions.

Symbols $\frac{a}{b}$, where a and b are integers and $b \neq 0$

Model

EXAMPLE — Identify Rational Numbers

Standardized Test Practice

Test-Taking Tip

Eliminate Possibilities Eliminate the possibilities that you know are incorrect. Then consider the choices that are left.

4 MULTIPLE-CHOICE TEST ITEM

Find the number that is *not* rational.

   Ⓐ -8 Ⓑ $0.0\overline{9}$ Ⓒ $\frac{3}{2}$ Ⓓ 2.010010001…

Read the Test Item To find the number that is not rational, identify three rational numbers.

Solve the Test Item -8, $0.0\overline{9}$, and $\frac{3}{2}$ can be expressed as fractions. So, 2.010010001… is not rational, and the answer is D.

Skill and Concept Check

1. **OPEN ENDED** Write two fractions whose LCD is 30.

2. **Which One Doesn't Belong?** Identify the ratio that does not have the same value as the other three. Explain your reasoning.

| 9 out of 15 | 0.06 | 60% | $\frac{3}{5}$ |

GUIDED PRACTICE

Find the LCD for each pair of fractions.

3. $\frac{4}{5}, \frac{3}{9}$ 4. $\frac{3}{8}, \frac{11}{32}$

Replace each ● with <, >, or = to make a true sentence.

5. $\frac{6}{15}$ ● $\frac{3}{8}$ 6. $\frac{1}{6}$ ● $\frac{3}{18}$ 7. $\frac{5}{4}$ ● $\frac{8}{7}$ 8. $\frac{1}{8}$ ● 0.2

Order each set of ratios from least to greatest.

9. 60%, $\frac{31}{50}$, 0.59   10. $\frac{3}{20}$, 0.02, 16%

Determine whether each number is rational. Write *yes* or *no*.

11. $\frac{1}{11}$ 12. 1.121121112…

Practice and Applications

Find the LCD for each pair of fractions.

13. $\dfrac{1}{4}, \dfrac{3}{10}$
14. $\dfrac{5}{6}, \dfrac{7}{15}$
15. $\dfrac{9}{14}, \dfrac{2}{5}$
16. $\dfrac{1}{6}, \dfrac{3}{8}$
17. $\dfrac{7}{12}, \dfrac{13}{36}$
18. $\dfrac{13}{17}, \dfrac{3}{4}$

Replace each ● with <, >, or = to make a true sentence.

19. $\dfrac{7}{10}$ ● $\dfrac{2}{3}$
20. $\dfrac{5}{14}$ ● $\dfrac{3}{7}$
21. $\dfrac{4}{7}$ ● $\dfrac{5}{8}$
22. $\dfrac{2}{3}$ ● $\dfrac{10}{15}$
23. $\dfrac{16}{20}$ ● $\dfrac{40}{50}$
24. $\dfrac{11}{15}$ ● $\dfrac{5}{9}$
25. $\dfrac{9}{13}$ ● $\dfrac{14}{20}$
26. $\dfrac{17}{24}$ ● $\dfrac{11}{12}$
27. $\dfrac{7}{5}$ ● $\dfrac{14}{11}$
28. $2\dfrac{3}{4}$ ● $2\dfrac{2}{3}$
29. $-\dfrac{11}{16}$ ● $-\dfrac{7}{8}$
30. $-\dfrac{3}{2}$ ● $-1\dfrac{1}{2}$
31. 0.82 ● $\dfrac{5}{6}$
32. $\dfrac{9}{20}$ ● 0.45
33. $\dfrac{3}{5}$ ● 59%
34. 40% ● $\dfrac{11}{25}$

35. Which average is better for a soccer goalie: 3 saves out of 4 or 7 saves out of 11? Explain.

36. Order the fractions from greatest to least: $\dfrac{7}{8}, \dfrac{4}{5}, \dfrac{22}{25}$.

Order each set of ratios from least to greatest.

37. $0.23, 19\%, \dfrac{1}{5}$
38. $\dfrac{8}{10}, 81\%, 0.805$
39. $0.615, \dfrac{5}{8}, 62\%$
40. $0.4, \dfrac{1}{25}, 0.25$
41. $-0.49, -\dfrac{49}{50}, -0.5$
42. $-\dfrac{4}{7}, -44\%, -0.47$

43. **DARTS** Bianca and Christopher were playing darts. Bianca hit the bull's-eye 5 out of 18 times. Christopher missed the bull's-eye 4 out of 15 times. Who hit the bull's-eye a greater fraction of the time?

Determine whether each number is rational. Write yes or no.

44. $\dfrac{1}{9}$
45. $1.141141114\ldots$
46. $1.2345\ldots$
47. π
48. $-3\dfrac{4}{5}$
49. $5.\overline{23}$

50. **WEATHER** Refer to the table at the right that shows about how much rain falls in Albuquerque and Denver. Which city has the greater fraction of inches of rain per day? Explain.

City	Amount of Rain (in.)	Number of Days
Albuquerque, NM	9	60
Denver, CO	15	90

Source: www.weather.com

51. **BAKING** Ofelia needs $2\dfrac{3}{4}$ cups of water for a cake recipe. If her measuring cup only shows decimals in increments of tenths, how will Ofelia measure out the water for the recipe?

52. **MULTI STEP** Find one terminating decimal and one repeating decimal between $\dfrac{2}{3}$ and $\dfrac{3}{4}$.

HOMEWORK HELP

For Exercises	See Examples
13–30, 36, 51–53	1
31–35, 43, 50, 54–55	2
37–42	3
44–49	4

Extra Practice
See pages 576, 600.

53. NUMBER SENSE Is $1\frac{15}{16}$, $\frac{17}{8}$, or $\frac{63}{32}$ nearest to 2? Explain.

BASEBALL For Exercises 54–56, use the following information and the table below.

Mark McGwire of the St. Louis Cardinals and Sammy Sosa of the Chicago Cubs competed in a home run race during the 1998 baseball season.

54. Write the ratio of home runs to games as a decimal for both players. Who had a better chance of hitting a home run during a game?

55. Write the ratio of home runs to at-bats as a decimal for both players. Who had a better chance of hitting a home run during an at-bat?

56. WRITE A PROBLEM In 2001, Barry Bonds of the San Francisco Giants set the single-season home run record. He had 476 at-bats and hit 73 home runs in 153 games. Use this information and the information in Exercises 54 and 55 to write and then solve a problem involving ratios.

Player	Games	At-bats	Home Runs
McGwire	155	509	70
Sosa	159	643	66

 Data Update Has anyone surpassed Bonds' single-season mark of 73 home runs? Visit msmath2.net/data_update to learn more.

57. CRITICAL THINKING How do you order fractions that have the same numerator? Explain.

Spiral Review with Standardized Test Practice

58. MULTIPLE CHOICE Find the least fraction.

Ⓐ $\frac{3}{4}$ Ⓑ $\frac{7}{9}$ Ⓒ $\frac{13}{18}$ Ⓓ $\frac{11}{15}$

59. SHORT RESPONSE Order the ratios from least to greatest: $47.\overline{4}$, $\frac{4}{7}$, $47\frac{2}{5}$, 47.41, 47%.

60. Find the LCM of 14 and 21. (Lesson 5-7)

FOOD The graph shows what percent of Americans chose the given pizza toppings as their favorite. Write the percent of people who chose each topping as a decimal. (Lesson 5-6)

61. onions

62. vegetables

63. mushrooms

64. ALGEBRA Translate the following sentence into an algebraic equation. *Four less than two times a number is 30.* (Lesson 4-1)

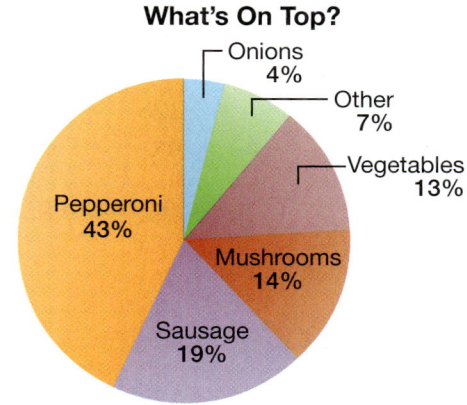

What's On Top?

Source: Market Facts for Bolla

CHAPTER 5 Study Guide and Review

Vocabulary and Concept Check

bar notation (p. 211)
common denominator (p. 227)
composite number (p. 197)
equivalent fractions (p. 207)
factor tree (p. 198)
greatest common factor (GCF) (p. 203)
least common denominator (LCD) (p. 227)
least common multiple (LCM) (p. 224)
multiple (p. 224)
percent (p. 216)
prime factorization (p. 198)
prime number (p. 197)
ratio (p. 216)
rational numbers (p. 229)
repeating decimals (p. 211)
simplest form (p. 207)
terminating decimals (p. 211)
Venn diagram (p. 203)

Choose the letter of the term or number that best matches each phrase.

1. a common multiple of the denominators of two or more fractions
2. a fraction in which the GCF of the numerator and denominator is 1
3. a ratio that compares a number to 100
4. a whole number greater than 1 that has exactly two factors, 1 and itself
5. a comparison of two numbers by division

a. ratio
b. simplest form
c. common denominator
d. percent
e. prime number
f. multiple
g. bar notation

Lesson-by-Lesson Exercises and Examples

5-1 Prime Factorization (pp. 197–200)

Determine whether each number is *prime* or *composite*.

6. 45
7. 37
8. 91
9. 118

Find the prime factorization of each number.

10. 54
11. 128
12. 68
13. 95

14. **ALGEBRA** Factor the expression $36x^2yz^3$.

Example 1 Determine whether 21 is *prime* or *composite*.
The number 21 has factors 1, 3, 7, and 21. So, 21 is a composite number.

Example 2 Find the prime factorization of 18.
Use a factor tree.

```
      18
     /  \
    6  × 3
   / \
  2 × 3 × 3
```

The prime factorization of 18 is 2×3^2.

msmath2.net/vocabulary_review

232 Chapter 5 Fractions, Decimals, and Percents

5-2 Greatest Common Factor (pp. 203–206)

Find the GCF of each set of numbers.
15. 18, 27
16. 30, 72
17. 28, 70, 98
18. 42, 63, 105

Find the GCF of each set of algebraic expressions.
19. $18w, 54w^2y$
20. $21de^2, 35d^3e^3$

Example 3 Find the GCF of 24 and 56.
First, make a list of all of the factors of 24 and 56.

factors of 24: 1, 2, 3, 4, 6, 8, 12, 24
factors of 56: 1, 2, 4, 7, 8, 14, 28, 56
common factors: 1, 2, 4, 8

The GCF of 24 and 56 is 8.

5-3 Simplifying Fractions (pp. 207–209)

Write each fraction in simplest form.
21. $\frac{12}{15}$
22. $\frac{35}{60}$
23. $\frac{11}{121}$
24. $\frac{14}{63}$
25. $\frac{37}{45}$
26. $\frac{55}{110}$

27. **LIFE SCIENCE** The human body has 60,000 miles of blood vessels. The distance across the United States from east to west is 3,000 miles. Write a fraction in simplest form that compares the distance across the United States to the miles of blood vessels.

Example 4 Write $\frac{24}{32}$ in simplest form.
Find the GCF of the numerator and denominator.

$24 = 2 \times 2 \times 2 \times 3$
$32 = 2 \times 2 \times 2 \times 2 \times 2$
GCF: $2 \times 2 \times 2 = 8$

$\frac{24}{32} = \frac{24 \div 8}{32 \div 8} = \frac{3}{4}$ Divide the numerator and denominator by the GCF.

So, $\frac{24}{32}$ in simplest form is $\frac{3}{4}$.

5-4 Fractions and Decimals (pp. 210–213)

Write each fraction or mixed number as a decimal. Use bar notation if the decimal is a repeating decimal.
28. $\frac{3}{4}$
29. $\frac{7}{8}$
30. $\frac{5}{9}$
31. $4\frac{1}{3}$
32. $6\frac{2}{5}$
33. $1\frac{6}{7}$

Write each decimal as a fraction in simplest form.
34. 0.7
35. 0.44
36. 0.05
37. 0.18
38. 0.54
39. 0.08

Example 5 Write $\frac{3}{8}$ as a decimal.

```
  0.375
8)3.000
  24
  ‾‾
   60
   56
   ‾‾
    40
    40
    ‾‾
     0
```

So, $\frac{3}{8} = 0.375$.

Example 6 Write 0.64 as a fraction in simplest form.

$0.64 = \frac{64}{100}$ Write as a fraction with a denominator of 100.

$= \frac{16}{25}$ Simplify.

Study Guide and Review continued

Mixed Problem Solving
For mixed problem-solving practice, see page 600.

5-5 Fractions and Percents (pp. 216–219)

Write each fraction as a percent.

40. $\frac{32}{100}$ 41. $\frac{11}{25}$ 42. $\frac{47}{50}$ 43. $\frac{8}{20}$

Write each percent as a fraction in simplest form.

44. 68% 45. 95% 46. 42% 47. 16%

48. **FOOD** A recent survey showed that 58% of school children chose peanut butter and jelly sandwiches as their favorite lunch food. Write this percent as a fraction in simplest form.

Example 7 Write $\frac{27}{50}$ as a percent.

$\frac{27}{50} = \frac{54}{100}$ Write an equivalent fraction with a denominator of 100.

$= 54\%$ Definition of percent

Example 8 Write 96% as a fraction in simplest form.

$96\% = \frac{96}{100}$ Definition of percent

$= \frac{24}{25}$ Simplify.

5-6 Percents and Decimals (pp. 220–223)

Write each percent as a decimal.

49. 48% 50. 7% 51. 12.5% 52. $75\frac{1}{4}\%$

Write each decimal as a percent.

53. 0.61 54. 0.055 55. 0.19 56. 0.999

57. **WEATHER** Twenty-three percent of the days last month were rainy. Write this percent as a decimal.

Example 9 Write 35% as a decimal.

$35\% = \frac{35}{100}$ Write the percent as a fraction.

$= 0.35$ Write the fraction as a decimal.

Example 10 Write 0.625 as a percent.

$0.625 = 0.625$ Multiply by 100.

$= 62.5\%$ Add the %.

5-7 Least Common Multiple (pp. 224–226)

Find the LCM of each set of numbers.

58. 9, 15 59. 4, 8
60. 16, 24 61. 3, 8, 12
62. 4, 9, 12 63. 15, 24, 30

Example 11 Find the LCM of 8 and 36.
Write each prime factorization.
$8 = 2^3 \qquad 36 = 2^2 \times 3^2$
LCM: $2^3 \times 3^2 = 72$
The LCM of 8 and 36 is 72.

5-8 Comparing and Ordering Rational Numbers (pp. 227–231)

Replace each ● with <, >, or = to make a true sentence.

64. $\frac{3}{8}$ ● $\frac{2}{3}$ 65. 0.45 ● $\frac{9}{20}$

66. $\frac{8}{9}$ ● 85%   67. $3\frac{3}{4}$ ● $3\frac{5}{8}$

68. 72% ● $\frac{8}{11}$ 69. $\frac{5}{7}$ ● $\frac{60}{84}$

Example 12 Replace ● with <, >, or = to make $\frac{3}{5}$ ● $\frac{5}{8}$ a true sentence.

Find equivalent fractions. The LCD is 40.

$\frac{3}{5} = \frac{3 \times 8}{5 \times 8} = \frac{24}{40} \qquad \frac{5}{8} = \frac{5 \times 5}{8 \times 5} = \frac{25}{40}$

Since $\frac{24}{40} < \frac{25}{40}$, then $\frac{3}{5} < \frac{5}{8}$.

Chapter 5 Practice Test

Vocabulary and Concepts

1. Which of the models shown at the right represents a greater fraction?

2. **Explain** how to write a percent as a decimal.

Skills and Applications

3. Find the prime factorization of 72.

4. Find the GCF of 24 and 40.

Write each fraction in simplest form.

5. $\dfrac{24}{60}$

6. $\dfrac{64}{72}$

Write each fraction, mixed number, or percent as a decimal. Use bar notation if the decimal is a repeating decimal.

7. $\dfrac{7}{9}$

8. $4\dfrac{5}{8}$

9. 91%

Write each decimal or percent as a fraction in simplest form.

10. 0.84

11. 0.006

12. 34%

Write each fraction or decimal as a percent.

13. $\dfrac{15}{25}$

14. 0.26

15. 0.135

16. **PAINTING** Suppose you are painting a wall that measures 100 square feet. You have already painted 46 square feet. Write a percent for the portion of the wall that remains unpainted.

Replace each ● with < , > , or = to make a true sentence.

17. $\dfrac{3}{5}$ ● $\dfrac{5}{9}$

18. $\dfrac{7}{12}$ ● $\dfrac{6}{8}$

19. $\dfrac{13}{20}$ ● 65%

Standardized Test Practice

20. **MULTIPLE CHOICE** One type of cicada emerges from hibernation every 17 years. Another type emerges every 13 years. If both types come out of hibernation one year, in how many years would this happen again?

Ⓐ 30 yr
Ⓑ 68 yr
Ⓒ 120 yr
Ⓓ 221 yr

msmath2.net/chapter_test

Chapter 5 Standardized Test Practice

PART 1 Multiple Choice

Record your answers on the answer sheet provided by your teacher or on a sheet of paper.

1. Which number is divisible by 2, 3, 5, 6, 9, and 10? (Prerequisite Skill, p. 554)
   - A) 90
   - B) 120
   - C) 150
   - D) 200

2. Students at Lincoln High School have collected money this year for charity. According to the graph, how much did all four classes collect this year? (Lesson 1-1)

   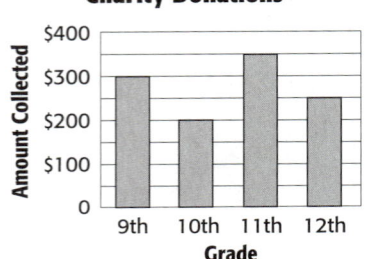

   Charity Donations

   - F) $950
   - G) $1,000
   - H) $1,050
   - I) $1,100

3. Which expression is the same as k^6? (Lesson 1-2)
   - A) $k + k + k + k + k + k$
   - B) $6k$
   - C) $k^3 + k^3$
   - D) $k \times k \times k \times k \times k \times k$

TEST-TAKING TIP

Question 2 When an item includes a graph, scan the graph to see what kind of information it includes and how the information is organized. Read each answer choice and compare it with the graph to see if the information in the answer choice is correct. Eliminate any wrong answer choices.

4. Which sentence shows the Commutative Property of Multiplication? (Lesson 1-6)
   - F) $f \cdot (g \cdot h) = (f \cdot g) \cdot h$
   - G) $q \cdot (r + s) = q \cdot r + q \cdot s$
   - H) $x \cdot y = y \cdot x$
   - I) $a + b = b + a$

5. In which quadrant is $(-5, -3)$ located? (Lesson 3-3)
   - A) quadrant I
   - B) quadrant II
   - C) quadrant III
   - D) quadrant IV

6. Write an addition sentence modeled by the number line. (Lesson 3-4)

   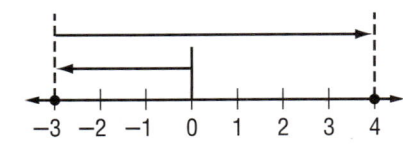

   - F) $-3 + 7 = 4$
   - G) $3 + 4 = 7$
   - H) $-7 + 3 = -4$
   - I) $3 + (-7) = -4$

7. Which of the ordered pairs is *not* a solution of $y = -3x + 7$? (Lesson 4-6)
   - A) $(-1, 10)$
   - B) $(1, 4)$
   - C) $(2, 1)$
   - D) $(3, 2)$

8. Find the percent that represents the shaded area. (Lesson 5-5)
   - F) 75%
   - G) 60%
   - H) 50%
   - I) 40%

9. Mrs. Benitez hired a contractor for a kitchen renovation project. The contractor renovated 60% of the kitchen the first week. What part of the project is left to do? (Lesson 5-6)
   - A) 0.04
   - B) 0.4
   - C) 0.44
   - D) 4.00

236 Chapter 5 Fractions, Decimals, and Percents

PART 2 Short Response/Grid In

Record your answers on the answer sheet provided by your teacher or on a sheet of paper.

10. Find the eighth term in the following sequence. 1, 3, 7, 13, 21, … . (Lesson 1-7)

For Questions 11 and 12, use the information below.
The table gives monthly water-level readings of Snail Creek Lake over an 8-month period. The readings indicate the number of inches above or below flood level.

Month	J	F	M	A	M	J	J	A
Water Level	−8	−8	−6	−5	−2	−5	0	3

11. Find the median of the readings. (Lesson 3-2)

12. Find the range of readings. (Lesson 3-5)

13. The equation $4.5h + 75 = 165$, where h is the number of hours, describes how long Hannah must work to earn $165. How many hours must she work to earn $165? (Lesson 4-4)

14. Write an equation for the function shown in the table. (Lesson 4-6)

x	−1	0	1	2
y	3	1	−1	−3

15. Aaron invited 30 friends to a party, but only 22 of them were able to attend. What fraction of the friends could not attend? (Lesson 5-5)

16. A string of red, green, and blue lights blink at different rates. The red lights blink once every 4 seconds, the green lights blink once every 7 seconds, and the blue lights blink once every 9 seconds. How many seconds will go by between times when all three colors will blink at the same time? (Lesson 5-7)

17. Tino painted a mural on the side of his shed. He used $\frac{1}{7}$ can of yellow paint, $\frac{1}{4}$ can of red paint, and $\frac{2}{9}$ can of white paint. Of which color did he use the least? (Lesson 5-8)

PART 3 Extended Response

Record your answers on a sheet of paper. Show your work.

18. A soft drink company sells its beverages in packages of 12 and 24 plastic bottles. It added a package with 30 bottles. (Lessons 5-2 and 5-7)

   a. Suppose the company designed boxes so that all three packages have the same number of bottles in each row. What is the greatest number of bottles in each row?

   b. The time it takes to seal a package is equal to 1 second per bottle. If three assembly lines containing the 12, 24, and 30 packages begin at the same time, when is the first time that all three packages are sealed at the same time?

19. The table shows the number of students who volunteered at the food bank. (Lesson 5-8)

Classroom	Number of Students Volunteering	Total Number of Students
A	7	24
B	3	8
C	6	18

   a. Model each ratio of the number of students volunteering to the total number of students on grid paper.

   b. Which classroom had the greatest number of students volunteering?

   c. Which classroom had the greatest fraction of students volunteering?

   d. Write each ratio as a fraction in simplest form, as a decimal, and as a percent.

CHAPTER 6
Applying Fractions

"What do Ferris wheels have to do with math?"

A Ferris wheel follows a circular path. To find the distance that you travel when you go one time around a Ferris wheel, you can use the formula $C = \pi d$, where π is approximately $\frac{22}{7}$, or 3.14. In geometry, you will use formulas to solve many real-life problems.

You will solve problems about Ferris wheels in Lesson 6-9.

GETTING STARTED

▶ Diagnose Readiness

Take this quiz to see if you are ready to begin Chapter 6. Refer to the lesson or page number in parentheses for review.

Vocabulary Review

Choose the correct term to complete each sentence.

1. The Division (Identity, Property) of Equality states that if you divide each side of an equation by the same nonzero number, the two sides remain equal. (Lesson 4-3)

2. The (GCF, LCD) of 12 and 16 is 4. (Lessons 5-2 and 5-7)

Prerequisite Skills

Find the LCD of each pair of fractions. (Lesson 5-8)

3. $\frac{5}{7}, \frac{3}{5}$
4. $\frac{1}{2}, \frac{4}{9}$
5. $\frac{8}{15}, \frac{1}{6}$
6. $\frac{3}{4}, \frac{7}{10}$

Multiply or divide. (Pages 560, 562)

7. 1.8×12
8. $99 \div 12$
9. $83 \div 100$
10. 4.6×0.3

Complete to show equivalent mixed numbers. (Page 563)

11. $3\frac{1}{5} = 2\frac{\blacksquare}{5}$
12. $9\frac{2}{3} = \blacksquare\frac{5}{3}$
13. $6\frac{1}{4} = 5\frac{\blacksquare}{4}$
14. $8\frac{6}{7} = 7\frac{\blacksquare}{7}$

Write each mixed number as an improper fraction. (Page 563)

15. $10\frac{3}{4}$
16. $1\frac{7}{8}$
17. $4\frac{2}{5}$
18. $7\frac{2}{9}$

Fractions Make this Foldable to help you organize your notes. Begin with a sheet of $8\frac{1}{2}''$ by $11''$ paper, four index cards, and glue.

STEP 1 Fold
Fold the paper in half widthwise.

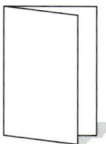

STEP 2 Open and Fold Again
Open and fold along the length about $2\frac{1}{2}''$ from the bottom.

STEP 3 Glue
Glue the edges on either side to form two pockets.

STEP 4 Label
Label the pockets *Fractions* and *Mixed Numbers*, respectively. Place two index cards in each pocket.

Chapter Notes Each time you find this logo throughout the chapter, use your *Noteables™: Interactive Study Notebook with Foldables™* or your own notebook to take notes. Begin your chapter notes with this Foldable activity.

Readiness To prepare yourself for this chapter with another quiz, visit **msmath2.net/chapter_readiness**

6-1 Estimating with Fractions

WHEN am I ever going to use this?

KITES For a kite to have balance while flying, the left and right sides of the horizontal support must each be $\frac{2}{3}$ as long as the bottom of the vertical support. Also, the top must be $\frac{1}{3}$ as long as the bottom portion.

1. Suppose the bottom portion of the vertical support is $2\frac{3}{4}$ feet. Round this length to the nearest foot.
2. About how long should the upper portion of the vertical support be?
3. About how long should the left and right sides of the horizontal support be?

What You'll LEARN

Estimate sums, differences, products, and quotients of fractions and mixed numbers.

NEW Vocabulary

compatible numbers

MATH Symbols

≈ is approximately equal to

To estimate the sum, difference, product, or quotient of mixed numbers, round the mixed numbers to the nearest whole number.

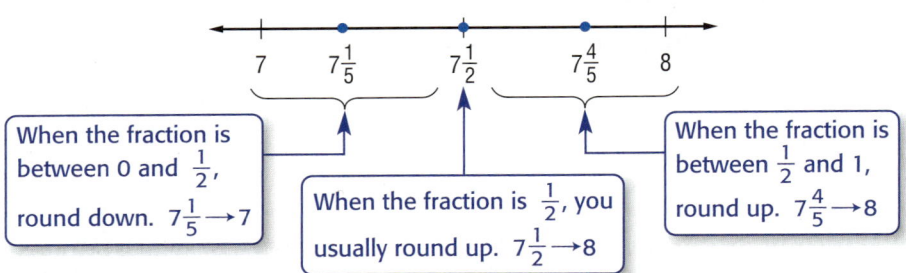

When the fraction is between 0 and $\frac{1}{2}$, round down. $7\frac{1}{5} \rightarrow 7$

When the fraction is $\frac{1}{2}$, you usually round up. $7\frac{1}{2} \rightarrow 8$

When the fraction is between $\frac{1}{2}$ and 1, round up. $7\frac{4}{5} \rightarrow 8$

STUDY TIP

Front-End Estimation Add the integers, estimate the sum of the fractions, and find the total sum.

$3\frac{2}{3} + 5\frac{1}{6}$
$\approx 8 + 1$
≈ 9

EXAMPLES Estimate with Mixed Numbers

Estimate.

1 $3\frac{2}{3} + 5\frac{1}{6}$

$3\frac{2}{3} + 5\frac{1}{6} \rightarrow 4 + 5 = 9$

The sum is *about* 9.

2 $6\frac{2}{5} \times 1\frac{7}{8}$

$6\frac{2}{5} \times 1\frac{7}{8} \rightarrow 6 \times 2 = 12$

The product is *about* 12.

Your Turn Estimate.

a. $2\frac{1}{5} + 3\frac{1}{2}$ b. $4\frac{3}{8} \times 5\frac{1}{4}$ c. $8\frac{7}{9} \div 2\frac{3}{4}$

240 Chapter 6 Applying Fractions

To estimate the sum, difference, product, or quotient of fractions, round each fraction to 0, $\frac{1}{2}$, or 1, whichever is closest. Number lines and fraction models can help you decide how to round.

Fractions Close to 0	Fractions Close to $\frac{1}{2}$	Fractions Close to 1
number line with $\frac{1}{6}$ marked near 0	number line with $\frac{5}{8}$ marked near $\frac{1}{2}$	number line with $\frac{9}{10}$ marked near 1
fraction bar $\frac{1}{7}$	fraction bar $\frac{4}{9}$	fraction bar $\frac{5}{6}$

EXAMPLES Estimate with Fractions

Estimate.

3 $\frac{1}{8} + \frac{2}{3}$

$\frac{1}{8}$ is about 0.

$\frac{2}{3}$ is about $\frac{1}{2}$.

$\frac{1}{8} + \frac{2}{3} \rightarrow 0 + \frac{1}{2} = \frac{1}{2}$ The sum is *about* $\frac{1}{2}$.

4 $\frac{6}{7} - \frac{7}{10}$

$\frac{6}{7}$ is about 1.

$\frac{7}{10}$ is about $\frac{1}{2}$.

$\frac{6}{7} - \frac{7}{10} \rightarrow 1 - \frac{1}{2} = \frac{1}{2}$ The difference is *about* $\frac{1}{2}$.

STUDY TIP

Fractions When the numerator and the denominator of a fraction are very close in value, such as $\frac{6}{7}$ or $\frac{8}{9}$, the fraction is close to 1.

5 $\frac{8}{9} \div \frac{5}{6}$

$\frac{8}{9} \div \frac{5}{6} \rightarrow 1 \div 1 = 1$ $\frac{8}{9}$ is about 1, and $\frac{5}{6}$ is about 1.

The quotient is *about* 1.

Your Turn Estimate.

d. $\frac{4}{5} + \frac{2}{7}$ e. $\frac{5}{8} - \frac{3}{7}$ f. $\frac{3}{5} \times \frac{11}{12}$

Sometimes it makes sense to round fractions to the nearest $\frac{1}{2}$, or mixed numbers to the nearest integer. Other times, it is useful to use **compatible numbers**, or numbers that are easy to compute mentally.

EXAMPLES Use Compatible Numbers

Estimate.

6 $\frac{1}{3} \times 14$ **THINK** What is $\frac{1}{3}$ of 14?

$\frac{1}{3} \times 14 \rightarrow \frac{1}{3} \times 15 = 5$ Round 14 to 15, since 15 is divisible by 3.

The product is *about* 5.

7 $9\frac{7}{8} \div 4\frac{1}{5}$

$9\frac{7}{8} \div 4\frac{1}{5} \rightarrow 10 \div 4\frac{1}{5}$ Round $9\frac{7}{8}$ to 10.

$\rightarrow 10 \div 5 = 2$ Round $4\frac{1}{5}$ to 5, since 10 is divisible by 5.

The quotient is *about* 2.

STUDY TIP

Compatible Numbers When dividing mixed numbers, round so that the dividend is a multiple of the divisor.

Your Turn Estimate.

g. $\frac{1}{4} \cdot 21$ h. $\frac{1}{2} \times 17$ i. $12 \div 6\frac{2}{3}$

Skill and Concept Check

1. **Writing Math** Explain how models are useful when estimating with fractions.

2. **OPEN ENDED** Describe when estimation is a better method for solving a problem rather than using pencil and paper, a calculator, or a computer. Then give a real-life example.

3. **NUMBER SENSE** Determine which of the following has a sum that is greater than 1. Write *yes* or *no* and explain.

   a. $\frac{1}{2} + \frac{4}{7}$ b. $\frac{3}{4} + \frac{5}{8}$ c. $\frac{2}{5} + \frac{1}{6}$

GUIDED PRACTICE

Estimate.

4. $8\frac{3}{8} + 1\frac{4}{5}$ 5. $5\frac{5}{7} \times 2\frac{7}{8}$ 6. $\frac{1}{6} + \frac{2}{5}$

7. $\frac{6}{7} - \frac{1}{5}$ 8. $\frac{1}{4} \cdot 15$ 9. $21\frac{5}{6} \div 9\frac{3}{4}$

10. **CONSTRUCTION** About how many bookcase shelves shown at the right can a carpenter cut from a board that is 1 foot wide and 12 feet long?

    1 ft
    $3\frac{1}{2}$ ft

242 Chapter 6 Applying Fractions

Practice and Applications

Estimate.

11. $3\frac{3}{4} + 4\frac{5}{6}$
12. $1\frac{1}{8} + 5\frac{11}{12}$
13. $5\frac{1}{3} - 3\frac{1}{6}$
14. $4\frac{2}{5} - 1\frac{1}{2}$
15. $2\frac{2}{3} \cdot 6\frac{1}{3}$
16. $6\frac{1}{8} \div 1\frac{2}{3}$
17. $\frac{3}{4} + \frac{3}{8}$
18. $\frac{5}{8} + \frac{3}{7}$
19. $\frac{5}{9} - \frac{1}{6}$
20. $\frac{3}{4} - \frac{3}{5}$
21. $\frac{1}{8} \times \frac{3}{4}$
22. $\frac{4}{9} \cdot \frac{11}{12}$
23. $\frac{4}{5} \div \frac{7}{8}$
24. $\frac{1}{10} \div \frac{5}{6}$
25. $\frac{1}{2} \times 13$
26. $\frac{1}{4} \times 39$
27. $25\frac{3}{10} \div 5\frac{2}{3}$
28. $27\frac{5}{8} \div 6\frac{1}{5}$
29. $12\frac{2}{7} \div 4\frac{1}{3}$
30. $5\frac{9}{10} \cdot 4\frac{1}{6}$

HOMEWORK HELP

For Exercises	See Examples
11–16, 37–38	1, 2
17–24	3–5
25–28	6–7

Extra Practice
See pages 577, 601.

31. Estimate $23\frac{2}{9}$ divided by 3.

32. Estimate the sum of $4\frac{1}{8}$, $1\frac{5}{6}$, and $\frac{7}{9}$.

Estimate.

33. $-3\frac{2}{7} \times 8\frac{3}{4}$
34. $\frac{4}{5}\left(-\frac{5}{8}\right)$
35. $-\frac{1}{6} \times (-65)$
36. $12\frac{1}{4} \cdot 2\frac{7}{9}$

37. **BAKING** Kayla wants to make the bread recipe shown at the right, but she has only $1\frac{1}{3}$ cups of flour. About how much more flour does she need?

38. **SEWING** A skirt that is $15\frac{7}{8}$ inches long has a hem of $1\frac{1}{8}$ inches. Approximately how long will the skirt be if the hem is let down?

39. **CRITICAL THINKING** If a number being divided is rounded up and the divisor is rounded down, what is the effect on the quotient?

Spiral Review with Standardized Test Practice

40. **MULTIPLE CHOICE** Choose the best estimate for $2\frac{1}{5} + 3\frac{3}{4}$.
    - A) 6
    - B) 5
    - C) 4
    - D) 2

41. **MULTIPLE CHOICE** If Lucas's car gets $23\frac{1}{3}$ miles per gallon, about how many miles can he drive on $1\frac{3}{4}$ gallons?
    - F) 0.46 mi
    - G) 4.6 mi
    - H) 46 mi
    - I) 460 mi

Replace each with <, >, or = to make a true sentence. (Lesson 5-8)

42. $\frac{7}{8}$ ● 0.75
43. $\frac{4}{5}$ ● $\frac{5}{7}$
44. $2\frac{1}{3}$ ● $\frac{7}{3}$
45. $\frac{6}{11}$ ● $\frac{9}{14}$

46. Find the LCM of 9 and 12. (Lesson 5-7)

GETTING READY FOR THE NEXT LESSON

PREREQUISITE SKILL Find the LCD of each pair of fractions. (Lesson 5-8)

47. $\frac{3}{4}, \frac{5}{12}$
48. $\frac{1}{2}, \frac{7}{10}$
49. $\frac{1}{6}, \frac{1}{8}$
50. $\frac{4}{5}, \frac{2}{3}$

6-2 Adding and Subtracting Fractions

What You'll LEARN

Add and subtract fractions.

REVIEW Vocabulary

LCD: the least common multiple of the denominators of two or more fractions (Lesson 5-8)

HANDS-ON Mini Lab

Materials
- ruler

Work with a partner.

1. Find $\frac{3}{8}$ inch on a ruler. From that point, add $\frac{4}{8}$ inch. What is the result?

2. Use a ruler to add $\frac{1}{4}$ inch and $\frac{2}{4}$ inch.

3. **Make a conjecture** about how to find each sum. Check using a ruler.

   a. $\frac{5}{8} + \frac{7}{8}$ b. $\frac{3}{16} + \frac{1}{16}$ c. $\frac{1}{2} + \frac{3}{4}$

Fractions that have the same denominators are called *like fractions*.

Noteables™ Key Concept: Add and Subtract Like Fractions

Words To add or subtract like fractions, add or subtract the numerators and write the result over the denominator. Simplify if necessary.

Symbols

Arithmetic

$\frac{1}{5} + \frac{3}{5} = \frac{4}{5}$

$\frac{11}{12} - \frac{7}{12} = \frac{4}{12}$ or $\frac{1}{3}$

Algebra

$\frac{a}{c} + \frac{b}{c} = \frac{a+b}{c}$, where $c \neq 0$

$\frac{a}{c} - \frac{b}{c} = \frac{a-b}{c}$, where $c \neq 0$

EXAMPLES Add and Subtract Like Fractions

Add or subtract. Write in simplest form.

1 $\frac{5}{9} + \frac{2}{9}$

$\frac{5}{9} + \frac{2}{9} = \frac{5+2}{9}$ Add the numerators.

$= \frac{7}{9}$ Write the sum over the denominator.

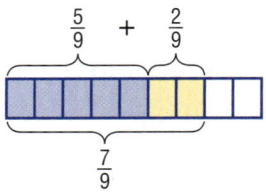

2 $\frac{9}{10} - \frac{1}{10}$

$\frac{9}{10} - \frac{1}{10} = \frac{9-1}{10}$ Subtract the numerators.

$= \frac{8}{10}$ Write the difference over the denominator.

$= \frac{4}{5}$ Simplify.

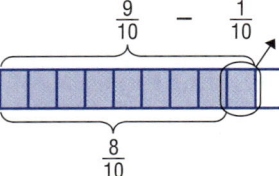

READING in the Content Area

For strategies in reading this lesson, visit **msmath2.net/reading**.

To add or subtract *unlike fractions*, or fractions with different denominators, rename the fractions using the LCD. Then add or subtract as with like fractions.

EXAMPLES Add and Subtract Unlike Fractions

Add or subtract. Write in simplest form.

3 $\frac{1}{2} + \frac{1}{6}$ **Estimate** $\frac{1}{2} + 0 = \frac{1}{2}$

The least common denominator of 2 and 6 is 6.

$\frac{1}{2} = \frac{1 \times 3}{2 \times 3} = \frac{3}{6}$ Rename $\frac{1}{2}$ using the LCD.

$$\begin{array}{r}\frac{1}{2} \rightarrow \frac{3}{6} \\ +\frac{1}{6} \rightarrow +\frac{1}{6} \\ \hline \frac{4}{6} \text{ or } \frac{2}{3}\end{array}$$

$\frac{2}{3}$ is close to the estimate. So, $\frac{1}{2} + \frac{1}{6} = \frac{2}{3}$.

4 $-\frac{5}{6} + \frac{7}{8}$ **Estimate** $-1 + 1 = 0$

The LCD of 6 and 8 is 24.
$$\begin{array}{r}-\frac{5}{6} \rightarrow -\frac{5 \times 4}{6 \times 4} \rightarrow -\frac{20}{24} \\ +\frac{7}{8} \rightarrow \frac{7 \times 3}{8 \times 3} \rightarrow +\frac{21}{24} \\ \hline \frac{1}{24}\end{array}$$

So, $-\frac{5}{6} + \frac{7}{8} = \frac{1}{24}$. Compare to the estimate.

5 **WEATHER** The average precipitation for November and December in Grand Junction, Colorado, is $\frac{7}{10}$ inch and $\frac{3}{5}$ inch, respectively. What is the difference of the average precipitation for these two months?

Estimate $\frac{1}{2} - \frac{1}{2} = 0$

$\frac{7}{10} - \frac{3}{5} = \frac{7}{10} - \frac{3 \times 2}{5 \times 2}$ The LCD of 10 and 5 is 10.

$\phantom{\frac{7}{10} - \frac{3}{5}} = \frac{7}{10} - \frac{6}{10}$ Rename the fractions with the LCD.

$\phantom{\frac{7}{10} - \frac{3}{5}} = \frac{1}{10}$ Subtract the numerators.

The difference in the precipitation is $\frac{1}{10}$ inch.

Your Turn Add or subtract. Write in simplest form.

a. $\frac{8}{9} - \frac{2}{9}$ b. $-\frac{3}{8} + \frac{5}{6}$ c. $\frac{7}{8} + \frac{3}{4}$ d. $\frac{2}{3} - \frac{1}{6}$

REAL-LIFE MATH

WEATHER The average precipitation for November and December in Seattle, Washington, is $5\frac{7}{10}$ inches and 6 inches, respectively.
Source: *The World Almanac*

Skill and Concept Check

1. **Draw** a model to show $\frac{3}{8} + \frac{1}{4}$.

2. **OPEN ENDED** Write a subtraction problem with fractions in which the difference is $\frac{2}{7}$.

3. **FIND THE ERROR** Marissa and Jacinta are finding $\frac{5}{9} + \frac{1}{6}$. Who is correct? Explain.

> Marissa
> $\frac{5}{9} + \frac{1}{6} = \frac{5+9}{9+6}$

> Jacinta
> $\frac{5}{9} + \frac{1}{6} = \frac{5 \times 2}{9 \times 2} + \frac{1 \times 3}{6 \times 3}$

GUIDED PRACTICE

Add or subtract. Write in simplest form.

4. $\frac{4}{5} - \frac{2}{5}$

5. $\frac{6}{7} + \frac{3}{7}$

6. $\frac{1}{6} + \frac{3}{8}$

7. $\frac{5}{6} - \frac{7}{12}$

8. $\frac{4}{9} + \frac{2}{9}$

9. $\frac{3}{8} - \frac{1}{8}$

10. $\frac{2}{3} + \frac{5}{6}$

11. $\frac{5}{6} + \frac{4}{9}$

12. $-\frac{1}{3} + \frac{3}{4}$

13. $-\frac{1}{6} + \frac{7}{10}$

14. **PRESIDENTS** Of the United States presidents, $\frac{5}{14}$ were born in either Virginia or Ohio, and $\frac{1}{6}$ were born in either Massachusetts or New York. What fraction were born in one of these four states?

Practice and Applications

Add or subtract. Write in simplest form.

15. $\frac{3}{7} + \frac{1}{7}$

16. $\frac{5}{8} + \frac{7}{8}$

17. $\frac{5}{6} - \frac{1}{6}$

18. $\frac{7}{10} - \frac{3}{10}$

19. $-\frac{1}{15} + \frac{3}{5}$

20. $-\frac{7}{9} - \frac{1}{3}$

21. $\frac{4}{5} - \frac{1}{6}$

22. $\frac{7}{12} + \frac{7}{10}$

23. $\frac{4}{5} - \frac{3}{5}$

24. $\frac{8}{9} - \frac{5}{9}$

25. $\frac{3}{8} + \frac{7}{8}$

26. $\frac{5}{6} + \frac{5}{6}$

27. $\frac{5}{8} - \frac{7}{12}$

28. $\frac{2}{15} + \frac{4}{9}$

29. $-\frac{5}{8} + \frac{11}{12}$

30. $-\frac{3}{8} - \frac{1}{12}$

31. $\frac{3}{10} - \frac{1}{4}$

32. $\frac{9}{10} + \frac{4}{15}$

33. $-\frac{9}{10} - \frac{1}{6}$

34. $-\frac{7}{9} + \left(-\frac{5}{6}\right)$

HOMEWORK HELP

For Exercises	See Examples
15–18, 23–26	1, 2
19–22, 27–41	3–5

Extra Practice
See pages 577, 601.

ALGEBRA Evaluate each expression if $a = \frac{3}{4}$ and $b = \frac{5}{6}$.

35. $\frac{1}{12} + a$

36. $b - \frac{7}{10}$

37. $b - a$

38. $-a + b$

39. **MULTI STEP** After 1 hour, Jon had finished $\frac{5}{6}$ of a long-distance race, and Ling had finished $\frac{7}{9}$ of it. At that time, who had finished a greater fraction of the race, and by how much?

MONEY For Exercises 40 and 41, use the following information and the table at the right.

Sierra and Jacob each receive an equal allowance. The table shows the fraction of their allowance that they each deposit into their savings account and the fraction they each spend at the mall.

Where Money Goes	Fraction of Allowance	
	Sierra	Jacob
savings account	$\frac{1}{2}$	$\frac{1}{3}$
spend at mall	$\frac{1}{4}$	$\frac{3}{5}$
left over	?	?

40. What fraction of Jacob's allowance goes into his savings account or is spent at the mall?

41. Who has more money left over? Explain.

42. **CRITICAL THINKING** Does $\frac{1}{3} + \frac{5}{9} - \frac{5}{12} = \frac{5}{9} + \frac{5}{12} - \frac{1}{3}$? Explain.

Spiral Review with Standardized Test Practice

43. **MULTIPLE CHOICE** Makayla uses $\frac{1}{5}$ pound of ham and $\frac{1}{8}$ pound of turkey for her sandwich. How much meat does she use in all?

    A. $\frac{1}{13}$ lb   B. $\frac{2}{13}$ lb   C. $\frac{13}{40}$ lb   D. $\frac{7}{20}$ lb

44. **MULTIPLE CHOICE** Jamal used a bucket that was $\frac{7}{9}$ full with soapy water to wash his mother's car. After washing the car, the bucket was only $\frac{1}{6}$ full. What part of the bucket of soapy water did Jamal use?

    F. $\frac{1}{9}$ G. $\frac{8}{15}$ H. $\frac{11}{18}$ I. $\frac{17}{18}$

Estimate. (Lesson 6-1)

45. $\frac{6}{7} - \frac{5}{12}$ 46. $4\frac{1}{9} + 3\frac{3}{4}$ 47. $16\frac{2}{3} \div 8\frac{1}{5}$ 48. $5\frac{4}{5} \cdot 3\frac{1}{3}$

PETS For Exercises 49 and 50, refer to the table at the right. It shows where pet owners get their pets. (Lesson 5-8)

49. Where do the greatest number of people get their pets?

50. Of the sources listed, where do the fewest people get their pets?

Pet Source	Portion of Pet Owners
animal shelter	$\frac{3}{20}$
friend/family	$\frac{21}{50}$
pet store	0.07
find as stray	0.14

Source: Yankelovich Partners

GETTING READY FOR THE NEXT LESSON

BASIC SKILL Complete.

Example: $8\frac{1}{2} = 7\frac{3}{2}$

51. $5\frac{2}{3} = 4\frac{\blacksquare}{3}$ 52. $7\frac{8}{9} = 6\frac{\blacksquare}{9}$ 53. $12\frac{1}{5} = \blacksquare\frac{6}{5}$ 54. $4\frac{3}{8} = \blacksquare\frac{11}{8}$

6-3 Adding and Subtracting Mixed Numbers

What You'll LEARN
Add and subtract mixed numbers.

WHEN am I ever going to use this?

ASTRONOMY Astronomers use *astronomical units* (AU) to represent large distances in space. One AU is the average distance from Earth to the Sun. Mercury is about $\frac{2}{5}$ AU from the Sun.

Diagram is not drawn to scale.

1. Jupiter is $5\frac{2}{5}$ AU from the Sun and Saturn's distance is $9\frac{1}{2}$ AU. Write an expression to find how much closer to the Sun Jupiter is than Saturn.

2. Find the difference of the fractional parts of the mixed numbers.

3. Find the difference of the whole numbers.

4. **Make a conjecture** about how to find $9\frac{1}{2} - 5\frac{2}{5}$. Then use your conjecture to find the difference.

To add or subtract mixed numbers, first add or subtract the fractions. If necessary, rename them using the LCD. Then add or subtract the whole numbers and simplify if necessary.

EXAMPLES Add and Subtract Mixed Numbers

Add or subtract. Write in simplest form.

1 $7\frac{4}{9} + 10\frac{2}{9}$

Estimate $7 + 10 = 17$

$$\begin{array}{r} 7\frac{4}{9} \\ + 10\frac{2}{9} \\ \hline 17\frac{6}{9} \text{ or } 17\frac{2}{3} \end{array}$$

Add the whole numbers and fractions separately. Simplify.

Compare each sum to its estimate.

2 $8\frac{5}{6} - 2\frac{1}{3}$

Estimate $9 - 2 = 7$

$$\begin{array}{r} 8\frac{5}{6} \rightarrow 8\frac{5}{6} \\ -2\frac{1}{3} \rightarrow -2\frac{2}{6} \\ \hline 6\frac{3}{6} \text{ or } 6\frac{1}{2} \end{array}$$

Rename the fraction using the LCD. Simplify.

Your Turn Add or subtract. Write in simplest form.

a. $6\frac{1}{8} + 2\frac{5}{8}$ b. $13\frac{7}{8} - 9\frac{3}{4}$ c. $1\frac{5}{9} + 4\frac{1}{6}$

REAL-LIFE CAREERS

How Does a Plumber Use Math?
Plumbers add and subtract mixed numbers when calculating the dimensions for installing house fixtures such as shower stalls and sinks.

Research For information about a career as a plumber, visit: msmath2.net/careers

EXAMPLE — Use Mixed Numbers to Solve a Problem

3 BUILDING There should be $2\frac{2}{3}$ feet of clearance space in front of a bathtub. What is the total width of the bathtub and clearance space shown at the right?

$2\frac{1}{2} + 2\frac{2}{3} = 2\frac{3}{6} + 2\frac{4}{6}$ Rename the fractions.

$\quad = 4 + \frac{7}{6}$ Add the whole numbers and add the fractions.

$\quad = 4 + 1\frac{1}{6}$ Rename $\frac{7}{6}$ as $1\frac{1}{6}$.

$\quad = 5\frac{1}{6}$ Simplify.

The total width is $5\frac{1}{6}$ feet.

Sometimes when you subtract mixed numbers, the fraction in the first mixed number is less than the fraction in the second.

EXAMPLES — Rename Numbers to Subtract

4 Find $2\frac{1}{3} - 1\frac{2}{3}$.

Rename $2\frac{1}{3}$ before subtracting.

Change 1 to $\frac{3}{3}$.

$2\frac{1}{3} = 1\frac{3}{3} + \frac{1}{3}$ or $1\frac{4}{3}$

$2\frac{1}{3} \rightarrow 1\frac{4}{3}$ Rename $2\frac{1}{3}$ as $1\frac{4}{3}$.
$-1\frac{2}{3} \rightarrow -1\frac{2}{3}$ First subtract the whole numbers and then the fractions.
$\quad\quad\quad\quad \frac{2}{3}$ So, $2\frac{1}{3} - 1\frac{2}{3} = \frac{2}{3}$.

5 Find $8 - 3\frac{3}{4}$.   **Estimate** $8 - 4 = 4$

$8 \rightarrow 7\frac{4}{4}$ $7 + \frac{4}{4} = 8$, so rename 8 as $7\frac{4}{4}$.
$-3\frac{3}{4} \rightarrow -3\frac{3}{4}$ Subtract.
$\quad\quad\quad 4\frac{1}{4}$ The difference is close to the estimate.

Your Turn Subtract. Write in simplest form.

d. $11\frac{2}{5} - 2\frac{3}{5}$ e. $5\frac{3}{8} - 4\frac{11}{12}$ f. $7 - 1\frac{1}{2}$

msmath2.net/extra_examples   Lesson 6-3 Adding and Subtracting Mixed Numbers

Skill and Concept Check

1. **OPEN ENDED** Write a problem that can be solved by finding $8\frac{1}{2} + 2\frac{1}{3}$. Explain what the numbers represent and find the sum.

2. **FIND THE ERROR** Evan and José are finding $3\frac{3}{4} - 1\frac{7}{8}$. Who is correct? Explain.

   Evan
   $3\frac{3}{4} - 1\frac{7}{8} = 3\frac{6}{8} - 1\frac{7}{8}$
   $= 2\frac{1}{8}$

   José
   $3\frac{3}{4} - 1\frac{7}{8} = 3\frac{6}{8} - 1\frac{7}{8}$
   $= 2\frac{14}{8} - 1\frac{7}{8}$ or $1\frac{7}{8}$

3. **NUMBER SENSE** Use estimation to determine whether $6\frac{3}{4} + \frac{4}{5}$ is *greater than, less than,* or *equal to* $2\frac{1}{9} + 6\frac{7}{8}$. Explain.

GUIDED PRACTICE

Add or subtract. Write in simplest form.

4. $1\frac{5}{7} + 8\frac{1}{7}$
5. $7\frac{5}{6} - 3\frac{1}{6}$
6. $8\frac{1}{2} + 3\frac{4}{5}$
7. $6\frac{3}{4} + 2\frac{9}{10}$

8. $9\frac{4}{5} - 2\frac{3}{5}$
9. $2\frac{3}{8} + 5\frac{7}{8}$
10. $4\frac{3}{10} + 4\frac{9}{10}$
11. $7\frac{5}{6} + 9\frac{3}{8}$
12. $3\frac{1}{4} - 1\frac{3}{4}$
13. $11 - 6\frac{3}{8}$

14. **AQUARIUMS** A fish tank that holds $18\frac{2}{3}$ gallons of water has $10\frac{1}{2}$ gallons so far. How much more water can be added?

Practice and Applications

Add or subtract. Write in simplest form.

15. $2\frac{1}{9} + 7\frac{4}{9}$
16. $10\frac{4}{5} - 2\frac{1}{5}$
17. $3\frac{1}{6} + 5\frac{1}{6}$
18. $6\frac{5}{7} + 8\frac{6}{7}$

19. $11\frac{3}{4} - 4\frac{1}{3}$
20. $4\frac{3}{8} + 10\frac{5}{12}$
21. $9\frac{4}{5} - 2\frac{3}{10}$
22. $6\frac{5}{6} - 2\frac{1}{3}$

23. $2\frac{3}{8} + 5\frac{7}{8}$
24. $8\frac{3}{7} + 1\frac{4}{7}$
25. $6\frac{2}{3} - 1\frac{3}{5}$
26. $4\frac{3}{10} - 1\frac{3}{4}$
27. $14\frac{1}{6} - 7\frac{1}{3}$
28. $3\frac{7}{9} + 3\frac{5}{9}$
29. $8\frac{5}{12} + 11\frac{1}{4}$
30. $7\frac{7}{8} + 10\frac{5}{6}$
31. $9\frac{1}{5} - 2\frac{3}{5}$
32. $12\frac{1}{2} - 6\frac{5}{8}$
33. $8 - 3\frac{2}{3}$
34. $13 - 5\frac{5}{6}$

HOMEWORK HELP

For Exercises	See Examples
15–25, 28–30, 35–38	1, 2
26–27, 31–34	3
	4, 5

Extra Practice
See pages 577, 601.

STOCK MARKET For Exercises 35–37, use the following information.
Until several years ago, stock prices were listed as mixed numbers. Find the difference between the high and low price of each restaurant chain stock shown in the table.

Stock Prices		
Company	High Price	Low Price
Restaurant A	$52\frac{5}{16}$	$21\frac{1}{8}$
Restaurant B	$42\frac{1}{4}$	$30\frac{3}{4}$
Restaurant C	$68\frac{3}{8}$	$29\frac{3}{4}$

35. Restaurant A
36. Restaurant B
37. Restaurant C

38. **MONUMENTS** The Washington Monument is 555 feet $5\frac{1}{8}$ inches tall. The San Jacinto Monument near Houston, Texas, is 14 feet $6\frac{7}{8}$ inches taller. How tall is the San Jacinto Monument?

Add or subtract. Write in simplest form.

39. $10 - 3\frac{5}{11}$
40. $24 - 8\frac{3}{4}$
41. $6\frac{1}{6} + 1\frac{2}{3} + 5\frac{5}{9}$
42. $3\frac{1}{4} + 2\frac{5}{6} - 4\frac{1}{3}$

43. **CRITICAL THINKING** A string is cut in half, and one of the halves is used to bundle newspapers. Then one fifth of the remaining string is cut off. The piece left is 8 feet long. How long was the string originally?

Spiral Review with Standardized Test Practice

44. **MULTIPLE CHOICE** What is the sum of $7\frac{1}{2}$ and $2\frac{1}{6}$?
    Ⓐ $5\frac{1}{3}$ Ⓑ $9\frac{1}{2}$ Ⓒ $9\frac{2}{3}$ Ⓓ $10\frac{2}{3}$

45. **MULTIPLE CHOICE** Melanie had $4\frac{2}{3}$ pounds of chopped walnuts. She used $1\frac{1}{4}$ pounds in a recipe. How many pounds of chopped walnuts did she have left?
    Ⓕ $2\frac{1}{3}$ lb   Ⓖ $2\frac{5}{12}$ lb   Ⓗ $3\frac{5}{12}$ lb   Ⓘ $3\frac{1}{2}$ lb

46. Find $\frac{7}{10} - \frac{1}{3}$. Write in simplest form. (Lesson 6-2)

Estimate. (Lesson 6-1)
47. $\frac{8}{9} \div \frac{9}{10}$
48. $3\frac{1}{2} + 6\frac{2}{3}$
49. $8\frac{4}{5} \times 7\frac{1}{9}$
50. $4\frac{2}{9} - 1\frac{1}{4}$

Replace each ● with <, >, or = to make a true sentence. (Lesson 5-8)
51. $\frac{4}{5}$ ● $\frac{7}{9}$
52. $\frac{2}{3}$ ● $\frac{5}{6}$
53. $\frac{1}{8}$ ● 0.15
54. $\frac{3}{7}$ ● 0.4

GETTING READY FOR THE NEXT LESSON

PREREQUISITE SKILL Write each mixed number as an improper fraction. (Page 563)

55. $2\frac{3}{8}$
56. $1\frac{2}{7}$
57. $5\frac{1}{10}$
58. $6\frac{4}{5}$

6-3b Problem-Solving Strategy
A Follow-Up of Lesson 6-3

Eliminate Possibilities

What You'll LEARN

Solve problems by eliminating possibilities.

I recorded $3\frac{1}{4}$ hours of a miniseries on a videotape that can record 6 hours of programming. What is the most that I can record on the rest of the same tape—2 hours, $2\frac{1}{2}$ hours, or 3 hours?

Well, we can **eliminate** some **possibilities** by estimating.

Explore	We know the combined hours of programming must be less than or equal to 6 hours.
Plan	Let's eliminate answers that are not reasonable.
Solve	You couldn't record 3 more hours on the tape because $3\frac{1}{4} + 3 = 6\frac{1}{4}$. So, we can eliminate that choice. Now let's check the choice of $2\frac{1}{2}$ hours. $3\frac{1}{4} + 2\frac{1}{2} = 5\frac{3}{4}$ Since this is less than 6 hours, this choice is correct. You could record $2\frac{1}{2}$ more hours on the tape.
Examine	Recording 2 more hours would give $3\frac{1}{4} + 2$ or $5\frac{1}{4}$ hours. This is less than the 6-hour maximum, but not the most that you could record.

Analyze the Strategy

1. **Describe** different ways that you can eliminate possibilities when solving problems.
2. **Explain** how the strategy of eliminating possibilities is useful for taking multiple choice tests.
3. **Write** a problem that could be solved by eliminating the possibilities. Explain your answer.

252 Chapter 6 Applying Fractions

Apply the Strategy

Solve. Use the eliminate possibilities strategy.

4. **JUICE** Lauren has a 3-gallon cooler with $1\frac{3}{4}$ gallons of juice in it. If she wants the cooler full for her soccer game, how much juice does she need to add?
   - A) 4 gal
   - B) $3\frac{1}{4}$ gal
   - C) $1\frac{1}{4}$ gal
   - D) $\frac{1}{4}$ gal

5. **ELEPHANTS** An elephant in a zoo eats 58 cabbages in a week. About how many cabbages does an elephant eat in one year?
   - F) 7
   - G) 700
   - H) 1,500
   - I) 3,000

Mixed Problem Solving

Solve. Use any strategy.

6. **RAIN FOREST** In some areas of the rain forest, 325 inches of rain may fall in a year. Which is the *best* estimate for the average rainfall per day in such an area?
   - A) $\frac{1}{3}$ in.
   - B) 1 in.
   - C) 5 in.
   - D) 33 in.

7. **GRADES** Explain why the graph showing a student's science grades is misleading.

8. **ELEVATORS** An elevator can hold a maximum weight of 3,500 pounds. Which is the *best* estimate for the number of adults that the elevator can hold?
   - F) 10 adults
   - G) 20 adults
   - H) 35 adults
   - I) 80 adults

9. **SUPPLIES** Vanessa has $55 to buy school supplies. She bought a backpack that costs $23.50, a combination lock that costs $6.25, and 4 binders that are $3.99 each. If mechanical pencils are $2.50 per pack, how many packs can she buy?

10. **SHOPPING** Abby bought the items at the right for a party. Which is the *best* estimate of the total cost of the items, not including tax?

Item	Price
balloons	$2.95
cups and plates	$9.30
streamers	$4.50

    - A) less than $15
    - B) between $15 and $25
    - C) between $25 and $35
    - D) more than $35

11. **TRAVEL** Mr. Rollins drove 780 miles on a five-day trip. He rented a car for $23 per day plus $0.15 per mile after 500 free miles. About how much did the rental car cost?
    - F) $100
    - G) $130
    - H) $160
    - I) $180

12. **STANDARDIZED TEST PRACTICE** If the trend in the graph continued, which is the best estimate for the average movie ticket price in the United States in 2006?
    - A) $5.50
    - B) $6.00
    - C) $6.25
    - D) $6.75

6-4 Multiplying Fractions and Mixed Numbers

What You'll LEARN

Multiply fractions and mixed numbers.

REVIEW Vocabulary

GCF: the greatest of the common factors of two or more numbers (Lesson 5-2)

WHEN am I ever going to use this?

EARTH SCIENCE About $\frac{1}{3}$ of the land in the United States is forests. About $\frac{2}{5}$ of U.S. forests are publicly owned.

1. What part of the rectangle represents $\frac{1}{3}$?
2. What part of the rectangle represents $\frac{2}{5}$ of $\frac{1}{3}$?
3. **Make a conjecture** about what fraction of U.S. land is publicly owned forests.

You can multiply fractions by using the rule below.

Noteables™

Key Concept: Multiply Fractions

Words To multiply fractions, multiply the numerators and multiply the denominators.

Symbols

Arithmetic

$$\frac{1}{3} \times \frac{2}{5} = \frac{1 \times 2}{3 \times 5} = \frac{2}{15}$$

Algebra

$$\frac{a}{b} \times \frac{c}{d} = \frac{a \times c}{b \times d} = \frac{ac}{bd}$$

EXAMPLES Multiply Fractions

Multiply. Write in simplest form.

1 $\frac{1}{2} \times \frac{1}{3}$

$\frac{1}{2} \times \frac{1}{3} = \frac{1 \times 1}{2 \times 3}$ ← Multiply the numerators.
← Multiply the denominators.

$= \frac{1}{6}$ Simplify.

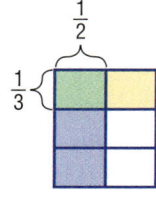

2 $2 \times \frac{3}{4}$

$2 \times \frac{3}{4} = \frac{2}{1} \times \frac{3}{4}$ Write 2 as $\frac{2}{1}$.

$= \frac{2 \times 3}{1 \times 4}$ Multiply the numerators and multiply the denominators.

$= \frac{6}{4}$ or $1\frac{1}{2}$ Simplify.

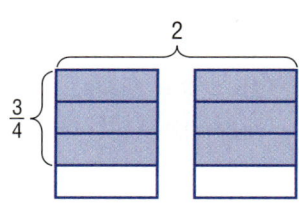

Your Turn Multiply. Write in simplest form.

a. $\frac{3}{5} \times \frac{1}{2}$ b. $\frac{1}{3} \times \frac{3}{4}$ c. $\frac{2}{3} \times 4$

254 Chapter 6 Applying Fractions

If the numerator and denominator of either fraction have common factors, you can simplify before multiplying.

STUDY TIP

Simplifying You can simplify by dividing a numerator and a denominator by the same factor. Doing this step before you multiply is easier than doing it after you multiply.

EXAMPLE **Simplify Before Multiplying**

3 Find $-\frac{2}{7} \times \frac{3}{8}$. Write in simplest form.

$$-\frac{2}{7} \times \frac{3}{8} = -\frac{\overset{1}{\cancel{2}}}{7} \times \frac{3}{\underset{4}{\cancel{8}}} \quad \text{Divide 2 and 8 by their GCF, 2.}$$

$$= -\frac{1 \times 3}{7 \times 4} \quad \text{Multiply the numerators and multiply the denominators.}$$

$$= -\frac{3}{28} \quad \text{Simplify.}$$

Your Turn Multiply. Write in simplest form.

d. $\frac{1}{3} \times \frac{3}{7}$ e. $-\frac{4}{9} \times \frac{1}{8}$ f. $\frac{5}{6} \times \frac{3}{5}$

EXAMPLE **Multiply Mixed Numbers**

4 Find $\frac{1}{2} \times 4\frac{2}{5}$. Write in simplest form. **Estimate** $\frac{1}{2} \times 4 = 2$

Method 1 Rename the mixed number.

$$\frac{1}{2} \times 4\frac{2}{5} = \frac{1}{\underset{1}{\cancel{2}}} \times \frac{\overset{11}{\cancel{22}}}{5} \quad \text{Rename } 4\frac{2}{5} \text{ as an improper fraction, } \frac{22}{5}.$$

$$= \frac{1 \times 11}{1 \times 5} \quad \text{Multiply.}$$

$$= \frac{11}{5} \text{ or } 2\frac{1}{5} \quad \text{Simplify.}$$

The product is close to the estimate.

Method 2 Use mental math.

STUDY TIP

Mental Math When you see a problem like $\frac{1}{2} \times 4\frac{2}{5}$, you can use the Distributive Property. Think, "What is $\frac{1}{2}$ of 4 and what is $\frac{1}{2}$ of $\frac{2}{5}$?" This is equal to $\frac{1}{2}(4 + \frac{2}{5})$.

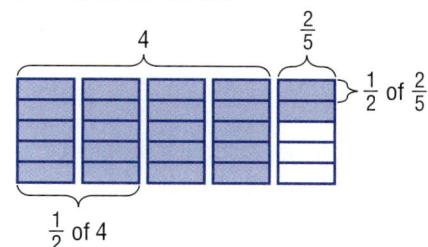

$$\frac{1}{2} \times 4\frac{2}{5} = \frac{1}{2} \times \left(4 + \frac{2}{5}\right) \quad \text{Write } 4\frac{2}{5} \text{ as a sum of its parts.}$$

$$= \left(\frac{1}{2} \times 4\right) + \left(\frac{1}{2} \times \frac{2}{5}\right) \quad \text{Distributive Property}$$

$$= 2 + \frac{1}{5} \text{ or } 2\frac{1}{5} \quad \text{Multiply.}$$

Compare this product to the model shown above.

Your Turn Multiply. Write in simplest form.

g. $\frac{1}{4} \times 8\frac{4}{9}$ h. $3 \times 5\frac{1}{3}$ i. $-1\frac{7}{8} \times \left(-2\frac{2}{5}\right)$

Lesson 6-4 Multiplying Fractions and Mixed Numbers 255

Skill and Concept Check

1. **OPEN ENDED** Write a pair of fractions whose product is $\frac{8}{15}$.

2. **Which One Doesn't Belong?** Identify the expression that does not have the same value as the other three. Explain your reasoning.

   | $\frac{1}{5}$ of 25 | $25 \times \frac{1}{5}$ | $5 \div 25$ | $\frac{1}{5} \cdot \frac{25}{1}$ |

3. **NUMBER SENSE** Is $18 \times \frac{4}{5}$ *greater than*, *less than*, or *equal to* 18? Explain.

GUIDED PRACTICE

Multiply. Write in simplest form.

4. $\frac{2}{3} \times \frac{1}{3}$
5. $\frac{1}{2} \times \frac{2}{5}$
6. $-\frac{2}{3} \times \frac{3}{8}$
7. $-\frac{1}{6} \times 4$
8. $2\frac{1}{4} \times \frac{2}{3}$
9. $1\frac{5}{6} \times 3\frac{3}{5}$

10. **FOOD** An average slice of American cheese is about $\frac{1}{8}$ inch thick. What is the height of a package containing 20 slices?

Practice and Applications

Multiply. Write in simplest form.

11. $\frac{3}{4} \times \frac{1}{8}$
12. $\frac{2}{5} \times \frac{2}{3}$
13. $\frac{1}{5} \times \frac{5}{6}$
14. $\frac{4}{9} \times \frac{1}{4}$
15. $-\frac{2}{3} \times \frac{1}{4}$
16. $-\frac{1}{12} \times \frac{3}{5}$
17. $\frac{4}{7} \times \frac{7}{8}$
18. $\frac{2}{5} \times \frac{15}{16}$
19. $\frac{3}{8} \times \frac{10}{27}$
20. $\frac{9}{10} \times \frac{5}{6}$
21. $-9 \times \left(-\frac{1}{2}\right)$
22. $-\frac{4}{5} \times (-6)$

HOMEWORK HELP

For Exercises	See Examples
11–12	1
21–24, 34	2
13–20, 33	3
25–32, 35–36	4

Extra Practice
See pages 578, 601.

23. **ELECTIONS** In an election in which 4,500 votes were cast, one candidate received $\frac{3}{5}$ of the votes. How many votes did the candidate receive?

24. **PACKAGING** The plastic cases used to store compact disks and DVDs are about $\frac{1}{5}$-inch thick. A company wants to sell 10 of these cases in plastic wrapping. What is the height of 10 cases?

$\frac{1}{5}$ in.

Multiply. Write in simplest form.

25. $4\frac{2}{3} \times \frac{4}{7}$
26. $\frac{5}{8} \times 2\frac{1}{2}$
27. $14 \times 1\frac{1}{7}$
28. $3\frac{3}{4} \times 8$
29. $-9 \times 4\frac{2}{3}$
30. $-4 \times 7\frac{5}{6}$
31. $3\frac{1}{4} \times 2\frac{2}{3}$
32. $5\frac{1}{3} \times 3\frac{3}{4}$

33. **TELEVISION** A media research survey showed that one evening, $\frac{2}{3}$ of all U.S. households had their TVs on, and $\frac{3}{8}$ of them were watching a World Series baseball game. What fraction of U.S. households was watching the game?

256 Chapter 6 Applying Fractions

34. **ANIMALS** Komodo dragons are the largest lizards in the world. A 250-pound komodo dragon can eat enough at one time to increase its weight by $\frac{3}{4}$. Find $\frac{3}{4} \times 250$ to determine how much weight a 250-pound komodo dragon could gain after eating.

35. **TURTLES** A giant tortoise can travel about one tenth of a kilometer in an hour. At this rate, how far can it travel in $1\frac{3}{4}$ hours?

36. **FLAGS** By law, the length of an official United States flag must be $1\frac{9}{10}$ times its width. What is the length of the flag shown at the right?

$3\frac{1}{2}$ ft

37. **CRITICAL THINKING** Two positive fractions which are *not* improper are multiplied. Is the product *sometimes*, *always*, or *never* less than 1? Explain.

Spiral Review with Standardized Test Practice

38. **MULTIPLE CHOICE** A box of books weighs $8\frac{2}{3}$ pounds. How much do $4\frac{1}{2}$ boxes weigh?

    Ⓐ 19 lb Ⓑ $27\frac{2}{3}$ lb Ⓒ $32\frac{1}{2}$ lb Ⓓ 39 lb

39. **GRID IN** Jeanette and Vanesa are each taking half of the leftover pizza shown at the right. What fraction of the whole pizza does each person take?

40. **LIFE SCIENCE** A female anglerfish is approximately $2\frac{1}{2}$ inches long, and a male is about $\frac{2}{5}$ inch long. How much longer is the female than the male? (Lesson 6-3)

Add or subtract. Write in simplest form. (Lesson 6-2)

41. $\frac{2}{7} + \frac{4}{7}$ 42. $\frac{1}{2} - \frac{1}{8}$ 43. $\frac{5}{9} + \frac{5}{6}$ 44. $\frac{3}{4} - \frac{1}{6}$

Find the greatest common factor of each set of numbers. (Lesson 5-2)

45. 44, 60   46. 25, 75   47. 6, 15, 27   48. 12, 30, 48

49. **ALGEBRA** On a January day in Buffalo, New York, the temperature dropped to $-20.8°C$. Find this temperature in degrees Fahrenheit by using the formula $C = 0.6F - 17.8$. (Lesson 4-4)

GETTING READY FOR THE NEXT LESSON

PREREQUISITE SKILL Multiply. (Page 560)

50. $2.8 \cdot 5$ 51. $1.9 \cdot 33$ 52. $7 \cdot 12.5$ 53. $3.6 \cdot 0.8$

6-5 Algebra: Solving Equations

What You'll LEARN
Solve equations with rational number solutions.

NEW Vocabulary
multiplicative inverse
reciprocal

Link to READING
Everyday Meaning of Inverse: opposite in order, as in an inverse statement in logic

HANDS-ON Mini Lab

Materials
- grid paper
- colored pencils

Work with a partner.

The model below shows $\frac{1}{2} \cdot 2$.

1. What is the product?
2. Use grid paper to model $\frac{1}{3} \cdot 3$. What is the product?
3. Copy and complete the table below.

$\frac{1}{2} \cdot 2 = \underline{?}$	$\frac{2}{3} \cdot \frac{3}{2} = \underline{?}$	$\frac{5}{6} \cdot \underline{?} = 1$
$\frac{1}{3} \cdot 3 = \underline{?}$	$\frac{3}{5} \cdot \frac{5}{3} = \underline{?}$	$\frac{9}{20} \cdot \underline{?} = 1$
$\frac{1}{4} \cdot \frac{4}{1} = \underline{?}$	$\frac{11}{12} \cdot \frac{12}{11} = \underline{?}$	$\underline{?} \cdot \frac{17}{8} = 1$

Two numbers whose product is 1 are called **multiplicative inverses**, or **reciprocals**.

Noteables™ Key Concept: Multiplicative Inverse Property

Words The product of a number and its multiplicative inverse is 1.

Symbols

Arithmetic
$$\frac{3}{4} \times \frac{4}{3} = 1$$

Algebra
$$\frac{a}{b} \times \frac{b}{a} = 1, \text{ for } a, b \neq 0$$

EXAMPLES Find Multiplicative Inverses

Find the multiplicative inverse of each number.

1 $\frac{2}{5}$

$\frac{2}{5} \cdot \frac{5}{2} = 1$ Multiply $\frac{2}{5}$ by $\frac{5}{2}$ to get the product 1.

The multiplicative inverse of $\frac{2}{5}$ is $\frac{5}{2}$, or $2\frac{1}{2}$.

2 $2\frac{1}{3}$

$2\frac{1}{3} = \frac{7}{3}$ Rename the mixed number as an improper fraction.

$\frac{7}{3} \cdot \frac{3}{7} = 1$ Multiply $\frac{7}{3}$ by $\frac{3}{7}$ to get the product 1.

The multiplicative inverse of $2\frac{1}{3}$ is $\frac{3}{7}$.

Your Turn Find the multiplicative inverse of each number.

a. $\frac{5}{6}$ b. $1\frac{1}{2}$ c. 8   d. $-\frac{4}{3}$

258 Chapter 6 Applying Fractions

Study Tip

Look Back You can review the other properties of equality in Lessons 4-2 and 4-3.

In Chapter 4, you learned to solve equations using the Addition, Subtraction, and Division Properties of Equality. You can also solve equations by multiplying each side by the same number. This is called the **Multiplication Property of Equality**.

Noteables **Key Concept: Multiplication Property of Equality**

Words If you multiply each side of an equation by the same nonzero number, the two sides remain equal.

Symbols

Arithmetic

$5 = 5$
$5 \cdot 2 = 5 \cdot 2$
$10 = 10$

Algebra

$\frac{x}{2} = -3$ $\quad\quad\quad\quad \frac{2}{3}x = 4$
$\frac{x}{2}(2) = -3(2)$ $\quad\quad \frac{3}{2} \cdot \frac{2}{3}x = \frac{3}{2} \cdot 4$
$x = -6$ $\quad\quad\quad\quad x = 6$

EXAMPLE Solve a Division Equation

 Solve $7 = \frac{n}{4}$. Check your solution.

$7 = \frac{n}{4}$ Write the equation.

$7 \cdot 4 = \frac{n}{4} \cdot 4$ Multiply each side of the equation by 4.

$28 = n$ Simplify.

The solution is 28.

When the coefficient of x is a fraction, multiply each side of the equation by its reciprocal.

EXAMPLE Use a Reciprocal to Solve an Equation

Standardized Test Practice

4 MULTIPLE-CHOICE TEST ITEM

What value of x makes $\frac{2}{3}x = -9$ a true sentence?

Ⓐ $\frac{3}{2}$ Ⓑ $-\frac{18}{27}$ Ⓒ $-\frac{9}{1}$ Ⓓ $-13\frac{1}{2}$

Read the Test Item To find the value of x, solve the equation.

Test-Taking Tip

Backsolving Sometimes it may be easier to substitute each of the choices into the given equation until you get a true sentence.

Solve the Test Item

$\frac{2}{3}x = -9$ Write the equation.

$\left(\frac{3}{2}\right)\frac{2}{3}x = \left(\frac{3}{2}\right)(-9)$ Multiply each side by the reciprocal of $\frac{2}{3}$, $\frac{3}{2}$.

$x = -\frac{27}{2}$ or $-13\frac{1}{2}$ Simplify.

The answer is D.

Your Turn

e. What value of b makes $24 = \frac{3}{4}b$ a true sentence?

Ⓐ 6   Ⓑ 18   Ⓒ 32   Ⓓ 72

 msmath2.net/extra_examples

Lesson 6-5 Algebra: Solving Equations **259**

Skill and Concept Check

1. **Writing Math** Tell whether 8 is a solution of $\frac{n}{3} = 24$. Explain.

2. **OPEN ENDED** Write a division equation that can be solved by multiplying each side by $\frac{9}{4}$.

3. **Which One Doesn't Belong?** Identify the pair of numbers that does not have the same relationship as the other three. Explain your reasoning.

| $\frac{7}{8}, \frac{8}{7}$ | $5, \frac{1}{5}$ | $\frac{2}{3}, 3$ | $\frac{10}{3}, \frac{3}{10}$ |

GUIDED PRACTICE

Find the multiplicative inverse of each number.

4. $\frac{8}{5}$
5. $\frac{2}{9}$
6. -9
7. $5\frac{4}{5}$

Solve each equation. Check your solution.

8. $\frac{k}{16} = 2$
9. $-4 = \frac{y}{3}$
10. $6 = \frac{4}{7}u$
11. $\frac{1}{4}t = \frac{3}{8}$
12. $\frac{5}{7}y = -1.5$
13. $\frac{b}{8.2} = 2.5$

14. **MEASUREMENT** The weight in pounds p of an object with a mass m of 25 kilograms is given by the equation $\frac{p}{m} = 2.2$. How many pounds does the object weigh?

Practice and Applications

Find the multiplicative inverse of each number.

15. $\frac{11}{2}$
16. $-\frac{9}{5}$
17. $-\frac{3}{8}$
18. $\frac{1}{6}$
19. 3
20. -14
21. $4\frac{2}{5}$
22. $6\frac{2}{3}$

HOMEWORK HELP

For Exercises	See Examples
15–22	1, 2
23–36	3, 4

Extra Practice
See pages 578, 601.

Solve each equation. Check your solution.

23. $\frac{x}{12} = 3$
24. $\frac{d}{4} = 28$
25. $-\frac{2}{5}t = -12$
26. $-24 = \frac{3}{4}a$
27. $\frac{7}{8}k = -21$
28. $14 = \frac{8}{3}b$
29. $\frac{1}{2}z = -\frac{2}{5}$
30. $\frac{3}{5} = \frac{3}{7}r$
31. $35.1 = \frac{5}{6}m$
32. $-\frac{a}{3.2} = 5$
33. $0.8 = \frac{h}{3.6}$
34. $\frac{m}{4.6} = 2.8$

35. **VACATION** The distance Katie travels in her car while driving 55 miles per hour for 2.5 hours is given by the equation $\frac{d}{2.5} = 55$. How far did she travel?

36. **MONEY** Based on recent exchange rates, the equation $d = \frac{31}{50}c$ shows the value in U.S. dollars d for an amount of Canadian dollars c. To the nearest cent, find the value in Canadian currency for $250 in U.S. dollars.

 Data Update What is the value in Canadian currency for $250 in U.S. dollars today? Visit msmath2.net/data_update to learn more.

37. **CRITICAL THINKING** In Lesson 11-5, you will learn that the area of a triangle A is given by the equation $A = \frac{1}{2}bh$, where b is the base of the triangle and h is the height. Explain how you can use the properties of equality to find the value of b in terms of A and h. Then solve for b.

Spiral Review with Standardized Test Practice

38. **MULTIPLE CHOICE** What is the reciprocal of $2\frac{1}{5}$?

    A. $2\frac{5}{1}$
    B. $\frac{11}{5}$
    C. $\frac{5}{11}$
    D. $\frac{1}{10}$

39. **GRID IN** Aaron is serving a 12-pound turkey at a dinner party. As a rule, you should allow about $\frac{3}{4}$ of a pound of meat per person. Use $\frac{3}{4}p = 12$ to find the number of people p that can be served at the dinner party.

Multiply. Write in simplest form. (Lesson 6-4)

40. $\frac{3}{8} \times \frac{4}{9}$
41. $1\frac{1}{2} \times 6$
42. $2\frac{2}{5} \times \frac{1}{6}$
43. $1\frac{1}{2} \times 1\frac{7}{9}$

44. Find $7\frac{1}{3} - 3\frac{5}{9}$. (Lesson 6-3)

For Exercises 45–47, use the graphic at the right. Write a fraction that compares the number of women champions to the total number of champions for each college. Write in simplest form. (Lesson 5-3)

45. Villanova
46. Texas–El Paso
47. Texas

GETTING READY FOR THE NEXT LESSON

PREREQUISITE SKILL Estimate. (Lesson 6-1)

48. $18\frac{1}{6} \div 3$
49. $24\frac{3}{8} \div 11\frac{7}{9}$
50. $\frac{2}{11} \div \frac{11}{12}$
51. $\frac{9}{10} \div \frac{6}{7}$

Chapter 6 Mid-Chapter Practice Test

Vocabulary and Concepts

1. **Write** an addition expression involving fractions shown by the model at the right. Then find the sum. Write in simplest form. (Lesson 6-2)

2. **Define** reciprocals. (Lesson 6-5)

Skills and Applications

Estimate. (Lesson 6-1)

3. $7\frac{1}{9} + 1\frac{1}{6}$
4. $13\frac{1}{2} \div 7\frac{2}{9}$
5. $\frac{11}{20} - \frac{5}{8}$

6. **SAVINGS** Jessica saves $\frac{1}{3}$ of the money she earns baby-sitting. If she earns $25 one evening, estimate the amount she saves. (Lesson 6-1)

Add, subtract, or multiply. Write in simplest form. (Lessons 6-2, 6-3, and 6-4)

7. $\frac{11}{15} - \frac{1}{15}$
8. $\frac{4}{7} + \left(-\frac{3}{14}\right)$
9. $\frac{5}{8} + \frac{3}{4}$
10. $5\frac{1}{6} - 1\frac{1}{3}$
11. $\frac{7}{12} \times \frac{4}{9}$
12. $2\frac{3}{5} + 6\frac{13}{15}$
13. $2\frac{3}{4} \times 12$
14. $4\frac{2}{7} \times 5\frac{5}{6}$

15. **AIRPLANES** The aircraft *Voyager* weighed 2,000 pounds. In 1986, it carried about $3\frac{1}{2}$ times its weight in fuel to fly nonstop around the world. How many pounds of fuel did *Voyager* carry? (Lesson 6-4)

Solve each equation. Check your solution. (Lesson 6-5)

16. $\frac{t}{5} = -11$
17. $2 = \frac{3}{8}y$
18. $16.2 = \frac{3}{4}k$

Standardized Test Practice

19. **MULTIPLE CHOICE** One batch of cookies uses $2\frac{1}{2}$ cups of flour and $1\frac{2}{3}$ cups of sugar. Which is the best estimate of the total amount of flour and sugar used in eight batches of cookies? (Lesson 6-1)

    Ⓐ less than 30 c
    Ⓑ between 30 c and 45 c
    Ⓒ between 45 c and 55 c
    Ⓓ more than 55 c

20. **MULTIPLE CHOICE** How much does a $12\frac{3}{4}$-pound package weigh after a $3\frac{5}{8}$-pound book is taken out of it? (Lesson 6-3)

    Ⓕ $8\frac{1}{8}$ lb
    Ⓖ 9 lb
    Ⓗ $9\frac{1}{8}$ lb
    Ⓘ 15 lb

262 Chapter 6 Applying Fractions

The Game Zone
A Place To Practice Your Math Skills

Math Skill: Multiplying Fractions

Totally Mental

- **GET READY!**
  Players: two
  Materials: 2 index cards, spinner with the digits 1 through 9

- **GET SET!**
  - Each player should make a game sheet on an index card like the one shown at the right.

- **GO!**
  - One player spins the spinner. The number that is spun should be written in one of the four boxes on his or her game sheet.

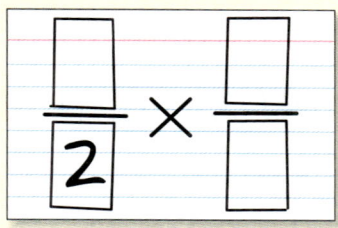

  - The second player spins the spinner and writes the number from the spinner in one of the boxes on his or her game sheet.
  - Continue until each person has spun the spinner four times.
  - Find the product of the fractions on your game sheet.
  - **Who Wins?** After four spins, the player with the greatest product is the winner.

The Game Zone: Multiplying Fractions 263

msmath2.net/extra_examples   Lesson 6-6 Dividing Fractions and Mixed Numbers 265

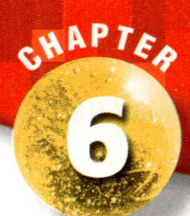

6-6 Dividing Fractions

CHAPTER 6 Study Guide and Review

Vocabulary and Concept Check

area (p. 271)	diameter (p. 275)	pint (p. 267)
center (p. 275)	formula (p. 271)	pound (p. 267)
circle (p. 275)	gallon (p. 267)	quart (p. 267)
circumference (p. 275)	multiplicative inverse (p. 258)	radius (p. 275)
compatible numbers (p. 242)	ounce (p. 267)	reciprocal (p. 258)
cup (p. 267)	perimeter (p. 270)	ton (p. 267)

Choose the correct term or number to complete each sentence.
1. The (radius, diameter) is the distance across a circle through its center.
2. To add like fractions, add the (numerators, denominators).
3. (Perimeter, Area) is the measure of the surface enclosed by a figure.
4. Fractions with different denominators are called (like, unlike) fractions.
5. When dividing by a fraction, multiply by its (value, reciprocal).
6. One cup is equivalent to (8, 12) fluid ounces.

Lesson-by-Lesson Exercises and Examples

6-1 Estimating with Fractions (pp. 240–243)

Estimate.

7. $2\frac{9}{10} \div 1\frac{1}{8}$
8. $6\frac{2}{9} - 5\frac{1}{7}$
9. $\frac{13}{15} \times \frac{1}{5}$
10. $\frac{1}{2} + \frac{3}{8}$
11. $\frac{1}{2} \cdot 25$
12. $15\frac{6}{7} \div 7\frac{1}{3}$

Example 1 Estimate $5\frac{1}{12} + 2\frac{5}{6}$.

$5\frac{1}{12} + 2\frac{5}{6} \rightarrow 5 + 3 = 8$

Example 2 Estimate $\frac{7}{8} - \frac{4}{7}$.

$\frac{7}{8} - \frac{4}{7} \rightarrow 1 - \frac{1}{2} = \frac{1}{2}$

6-2 Adding and Subtracting Fractions (pp. 244–247)

Add or subtract. Write in simplest form.

13. $\frac{2}{6} - \frac{1}{6}$
14. $\frac{3}{7} + \frac{9}{14}$
15. $\frac{5}{6} - \frac{3}{4}$
16. $\frac{1}{9} + \frac{5}{9}$
17. $\frac{4}{5} + \frac{4}{5}$
18. $\frac{9}{10} - \frac{3}{10}$
19. $\frac{5}{8} - \frac{5}{12}$
20. $\frac{11}{12} - \frac{1}{6}$
21. $-\frac{3}{4} + \frac{7}{20}$

Example 3 Find $\frac{1}{8} + \frac{3}{8}$.

$\frac{1}{8} + \frac{3}{8} = \frac{1+3}{8}$ Add the numerators.

$= \frac{4}{8}$ or $\frac{1}{2}$ Simplify.

Example 4 Find $\frac{3}{10} - \frac{1}{4}$.

$\frac{3}{10} - \frac{1}{4} = \frac{6}{20} - \frac{5}{20}$ or $\frac{1}{20}$

msmath2.net/vocabulary_review

6-3 Adding and Subtracting Mixed Numbers (pp. 248–251)

Add or subtract. Write in simplest form.

22. $3\frac{2}{15} + 6\frac{9}{15}$

23. $9\frac{4}{5} - 2\frac{1}{5}$

24. $4\frac{1}{3} - 2\frac{2}{3}$

25. $8\frac{2}{7} + 1\frac{6}{7}$

26. $7\frac{11}{12} - 4\frac{3}{12}$

27. $7\frac{3}{5} - 5\frac{1}{3}$

28. $5\frac{3}{4} + 1\frac{1}{6}$

29. $3\frac{5}{8} + 11\frac{1}{2}$

30. $4\frac{3}{10} - 2\frac{4}{5}$

Example 5 Find $5\frac{2}{3} + 3\frac{1}{2}$.

$5\frac{2}{3} + 3\frac{1}{2} = 5\frac{4}{6} + 3\frac{3}{6}$ Rename the fractions.

$= 8\frac{7}{6}$ or $9\frac{1}{6}$ Add the whole numbers and add the fractions.

Example 6 Find $4\frac{1}{5} - 2\frac{3}{5}$.

$4\frac{1}{5} - 2\frac{3}{5} = 3\frac{6}{5} - 2\frac{3}{5}$ Rename $4\frac{1}{5}$ as $3\frac{6}{5}$.

$= 1\frac{3}{5}$ Subtract the whole numbers and subtract the fractions.

6-4 Multiplying Fractions and Mixed Numbers (pp. 254–257)

Multiply. Write in simplest form.

31. $\frac{3}{5} \times \frac{2}{7}$

32. $\frac{5}{12} \times \frac{4}{9}$

33. $\frac{3}{5} \times \frac{10}{21}$

34. $4 \times \frac{13}{20}$

35. $-2\frac{1}{3} \times \frac{3}{4}$

36. $4\frac{1}{2} \times 2\frac{1}{12}$

37. **TRACK AND FIELD** One lap around the high school track is $\frac{3}{8}$ of a mile. If Matthew runs $4\frac{1}{2}$ laps, how far does he run?

Example 7 Find $\frac{5}{9} \times \frac{2}{3}$.

$\frac{5}{9} \times \frac{2}{3} = \frac{5 \times 2}{9 \times 3}$ Multiply the numerators and multiply the denominators.

$= \frac{10}{27}$ Simplify.

Example 8 Find $3\frac{1}{2} \times 2\frac{3}{4}$.

$3\frac{1}{2} \times 2\frac{3}{4} = \frac{7}{2} \times \frac{11}{4}$ Rename $3\frac{1}{2}$ and $2\frac{3}{4}$.

$= \frac{7 \times 11}{2 \times 4}$ Multiply the numerators and multiply the denominators.

$= \frac{77}{8}$ or $9\frac{5}{8}$ Simplify.

6-5 Algebra: Solving Equations (pp. 258–261)

Find the multiplicative inverse of each number.

38. $\frac{7}{12}$

39. 5

40. $3\frac{1}{3}$

Solve each equation. Check your solution.

41. $8 = \frac{w}{2}$

42. $\frac{4}{5}b = 12$

43. $-7.6 = \frac{n}{3}$

44. **EARTH SCIENCE** In 1996, a new planet was discovered. Earth's diameter, 7,970 miles, is only $\frac{5}{86}$ the size of this planet's diameter. Solve $\frac{5}{86}d = 7,970$ to find d, the diameter of this planet in miles.

Example 9 Find the multiplicative inverse of $\frac{9}{5}$.

$\frac{9}{5} \cdot \frac{5}{9} = 1$ The product of $\frac{9}{5}$ and $\frac{5}{9}$ is 1.

The multiplicative inverse of $\frac{9}{5}$ is $\frac{5}{9}$.

Example 10 Solve $\frac{3}{4}g = 2$.

$\frac{3}{4}g = 2$ Write the equation.

$\frac{4}{3} \cdot \frac{3}{4}g = \frac{4}{3} \cdot 2$ Multiply each side by the reciprocal of $\frac{3}{4}$.

$g = \frac{8}{3}$ or $2\frac{2}{3}$ Simplify.

Chapter 6 Study Guide and Review

Study Guide and Review continued

Mixed Problem Solving For mixed problem-solving practice, see page 601.

6-6 Dividing Fractions and Mixed Numbers (pp. 264–266)

Divide. Write in simplest form.

45. $\frac{3}{5} \div \frac{6}{7}$
46. $\frac{1}{2} \div \frac{1}{3}$
47. $5 \div \frac{10}{13}$
48. $4 \div \frac{2}{3}$
49. $2\frac{3}{4} \div \frac{5}{6}$
50. $-\frac{2}{5} \div 3$
51. $\frac{6}{11} \div 4$
52. $4\frac{3}{10} \div 2\frac{1}{5}$
53. $-\frac{2}{7} \div \frac{8}{21}$

Example 11 Find $2\frac{4}{5} \div \frac{7}{10}$.

$2\frac{4}{5} \div \frac{7}{10} = \frac{14}{5} \div \frac{7}{10}$ Rename $2\frac{4}{5}$.

$= \frac{\overset{2}{\cancel{14}}}{\underset{1}{\cancel{5}}} \cdot \frac{\overset{2}{\cancel{10}}}{\underset{1}{\cancel{7}}}$ Multiply by the reciprocal of $\frac{7}{10}$.

$= \frac{4}{1}$ or 4   Simplify.

6-7 Measurement: Changing Customary Units (pp. 267–269)

Complete.

54. 4 qt = __?__ pt
55. 6 gal = __?__ qt
56. 48 oz = __?__ lb
57. 8,000 lb = __?__ T
58. 9 c = __?__ pt
59. 36 in. = __?__ ft

Example 12 Complete: 32 qt = __?__ gal

Since 4 quarts are in 1 gallon, divide by 4.

$32 \div 4 = 8$

32 quarts = 8 gallons

6-8 Geometry: Area and Perimeter (pp. 270–273)

Find the perimeter and area of each rectangle.

60. 13 cm, 6 cm

61. 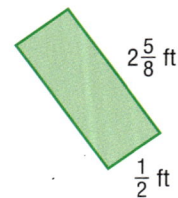 $2\frac{5}{8}$ ft, $\frac{1}{2}$ ft

62. $\ell = 9$ cm, $w = 4$ cm
63. $\ell = 5$ in., $w = \frac{1}{2}$ in.
64. $\ell = 3.2$ m, $w = 6$ m
65. $\ell = 4\frac{1}{2}$ ft, $w = 2\frac{1}{3}$ ft

Example 13 Find the perimeter and area of the rectangle.

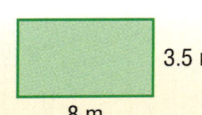 3.5 m, 8 m

$P = 2\ell + 2w$ Perimeter of a rectangle
$P = 2(8) + 2(3.5)$ Substitution
$P = 23$ Simplify.

The perimeter is 23 meters.

$A = \ell \cdot w$ Area of a rectangle
$A = 8 \cdot 3.5$ Replace ℓ with 8 and w with 3.5.
$A = 28$ Multiply.

The area is 28 square meters.

6-9 Geometry: Circles and Circumference (pp. 275–277)

Find the circumference of each circle. Use 3.14 or $\frac{22}{7}$ for π. Round to the nearest tenth if necessary.

66. $r = 4.2$ cm
67. $d = 8$ yd
68. $r = \frac{7}{11}$ ft
69. $d = 8\frac{2}{5}$ ft

Example 14 Find the circumference of a circle with a diameter of 12.2 meters. Round to the nearest tenth.

$C = \pi d$ Circumference of a circle
$C \approx 3.14(12.2)$ $\pi \approx 3.14$ and $d = 12.2$
$C \approx 38.308$ Multiply.

The circumference is about 38.3 meters.

Chapter 6 Practice Test

Vocabulary and Concepts

1. **Explain** how to add unlike fractions.
2. **Define** perimeter.

Skills and Applications

Estimate.

3. $5\frac{7}{9} - 1\frac{2}{13}$
4. $3\frac{1}{12} + 6\frac{5}{7}$
5. $\frac{3}{7} \times \frac{13}{15}$

Add, subtract, multiply, or divide. Write in simplest form.

6. $\frac{4}{15} + \frac{8}{15}$
7. $\frac{7}{10} - \frac{1}{6}$
8. $\frac{5}{8} \times \frac{2}{5}$
9. $6 \times \frac{8}{21}$
10. $4\frac{5}{12} - 2\frac{1}{12}$
11. $6\frac{7}{9} + 3\frac{5}{12}$
12. $8\frac{2}{7} - 1\frac{5}{14}$
13. $-\frac{5}{6} \div \frac{2}{3}$
14. $\frac{8}{9} \div 5\frac{1}{3}$

15. **COOKING** Taylor wants to make $2\frac{1}{2}$ times the quantity given in a recipe. The recipe calls for $1\frac{3}{4}$ cups of flour. How much flour will Taylor need?

16. **FLAG DAY** A giant cake decorated as an American flag measured 60 feet by 90 feet. What was the perimeter of the cake?

Solve each equation. Check your solution.

17. $\frac{y}{3} = 8$
18. $-6 = \frac{2}{5}m$
19. $\frac{3}{4} = \frac{5}{8}x$

Complete.

20. 42 ft = __?__ yd
21. 9 qt = __?__ pt
22. 7,600 lb = __?__ T

23. Find the perimeter and area of the rectangle.

24. Find the circumference of a circle with a radius of 5 meters. Round to the nearest tenth.

2 in.

$4\frac{1}{2}$ in.

Standardized Test Practice

25. **MULTIPLE CHOICE** In the 1999–2000 school year, the average backpack weighed $7\frac{1}{2}$ pounds. In the 2001–2002 school year, the average backpack weighed $7\frac{1}{5}$ pounds. By how much did the average backpack weight decrease?

    Ⓐ $\frac{1}{5}$ lb
    Ⓑ $\frac{3}{10}$ lb
    Ⓒ $\frac{1}{2}$ lb
    Ⓓ $\frac{7}{10}$ lb

CHAPTER 6 Standardized Test Practice

PART 1 Multiple Choice

Record your answers on the answer sheet provided by your teacher or on a sheet of paper.

1. The table shows four major rivers that run through Texas. Which is the most appropriate way to display this information?
   (Lessons 2-2 and 2-7)

River	Length (mi)
Brazos	950
Sabine	380
Trinity	360
Washita	500

   Source: *The World Almanac*

   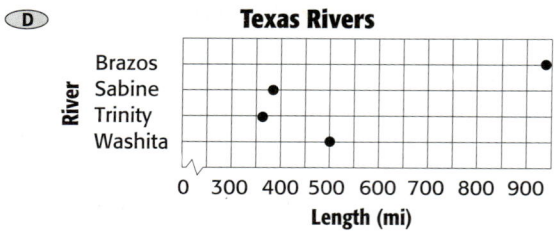

2. Which is the solution of $2x - 5 > 13$? (Lesson 4-5)

3. A jellyfish's body is made up of 95% water. What is 95% written as a decimal? (Lesson 5-6)
   - A 0.095
   - B 0.95
   - C 9.5
   - D 95

4. Dimitri has $6\frac{1}{8}$ ounces of orange juice and $10\frac{2}{3}$ ounces of cranberry juice. What is the best estimate of the total ounces of juice that he has? (Lesson 6-1)
   - F 15 oz
   - G 16 oz
   - H 17 oz
   - I 18 oz

5. Cole had $\frac{7}{8}$ of a tank of gas in the lawn mower. After mowing the grass, he had $\frac{1}{4}$ of a tank. What fraction of a tank did Cole use mowing the lawn? (Lesson 6-2)
   - A $\frac{1}{8}$
   - B $\frac{3}{8}$
   - C $\frac{5}{8}$
   - D $\frac{3}{4}$

6. What is the solution of $27 = \frac{3}{4}t$? (Lesson 6-5)
   - F 108
   - G 36
   - H 20.25
   - I 9

7. Kaylee bought 16 gallons of fruit juices and soda for the school's graduation picnic. How many quarts of drinks did Kaylee buy? (Lesson 6-7)
   - A 4 qt
   - B 16 qt
   - C 18 qt
   - D 64 qt

282 Chapter 6 Applying Fractions

PART 2 Short Response/Grid In

Record your answers on the answer sheet provided by your teacher or on a sheet of paper.

8. If 4 computers are needed for every 7 students in a grade, how many computers are needed for 280 students? (Lesson 1-1)

9. What are the coordinates of point *K*? (Lesson 3-3)

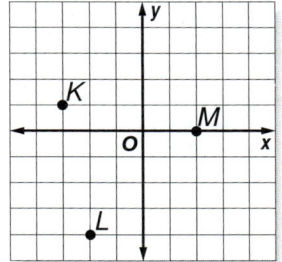

10. The graph shows the average number of greeting cards purchased yearly by the average person in the United States.

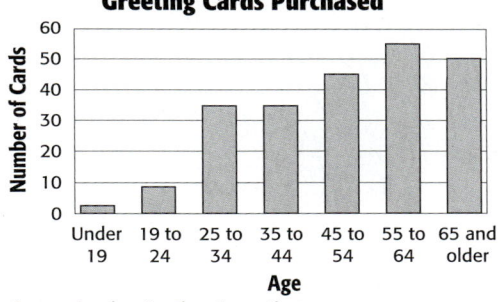

Source: American Greetings Corporation

Let *c* represent the number of cards purchased by Americans ages 35 to 44. Write an expression using *c* to represent the number of cards purchased by Americans ages 45 to 54. (Lesson 4-1)

11. Write a percent to represent the shaded area. (Lesson 5-5)

12. Find $7\frac{2}{7} + 3\frac{1}{4}$. (Lesson 6-3)

13. To make one batch of cookies, you need $\frac{3}{4}$ cup of butter. How much butter would you need to make $2\frac{1}{2}$ batches of cookies? (Lesson 6-4)

14. A box of laundry detergent contains 35 cups. If you use $1\frac{1}{4}$ cups per load of laundry, how many loads can you wash with 1 box? (Lesson 6-6)

15. What is the area of the rectangle? (Lesson 6-8)

5 in.

8 in.

PART 3 Extended Response

Record your answers on a sheet of paper. Show your work.

16. Suppose you want to build a 4-foot wide deck around a circular swimming pool that has a radius of 66 inches. You also want to put a fence around the deck. (Lesson 6-9)

   a. Make a drawing of the problem. Include labels.

   b. About how much fencing will you need to the nearest foot?

   c. The fence costs $10 per foot. How much would you save if you put the fence just around the pool instead of the deck? Explain.

TEST-TAKING TIP

Question 16 Many standardized tests include any necessary formulas in the test booklet. It helps to be familiar with formulas such as the area of a rectangle and the circumference of a circle, but use any formulas that are given to you with the test.

UNIT 4
Proportional Reasoning

Chapter 7
Ratios and Proportions

Chapter 8
Applying Percent

Chapter 9
Probability

In Unit 3, you learned about percents. In this unit, you will solve real-life problems involving proportions and percents. You will also discover how percents are used to describe probabilities.

INTERDISCIPLINARY PROJECT

Step Right Up and Win A Prize

Math and Recreation What are my chances of winning? You probably ask yourself that question any time you play a game that offers a prize for winning. You're about to embark on a gaming adventure. You'll investigate the mathematical probabilities of winning various carnival games. You'll also research and design a game of your own. So, come on and take a chance! Sharpen up that hand-eye coordination and grab your probability tool kit. This adventure is a win-win situation!

WebQuest Log on to **msmath2.net/webquest** to begin your WebQuest.

Unit 4 Proportional Reasoning

CHAPTER 7

Ratios and Proportions

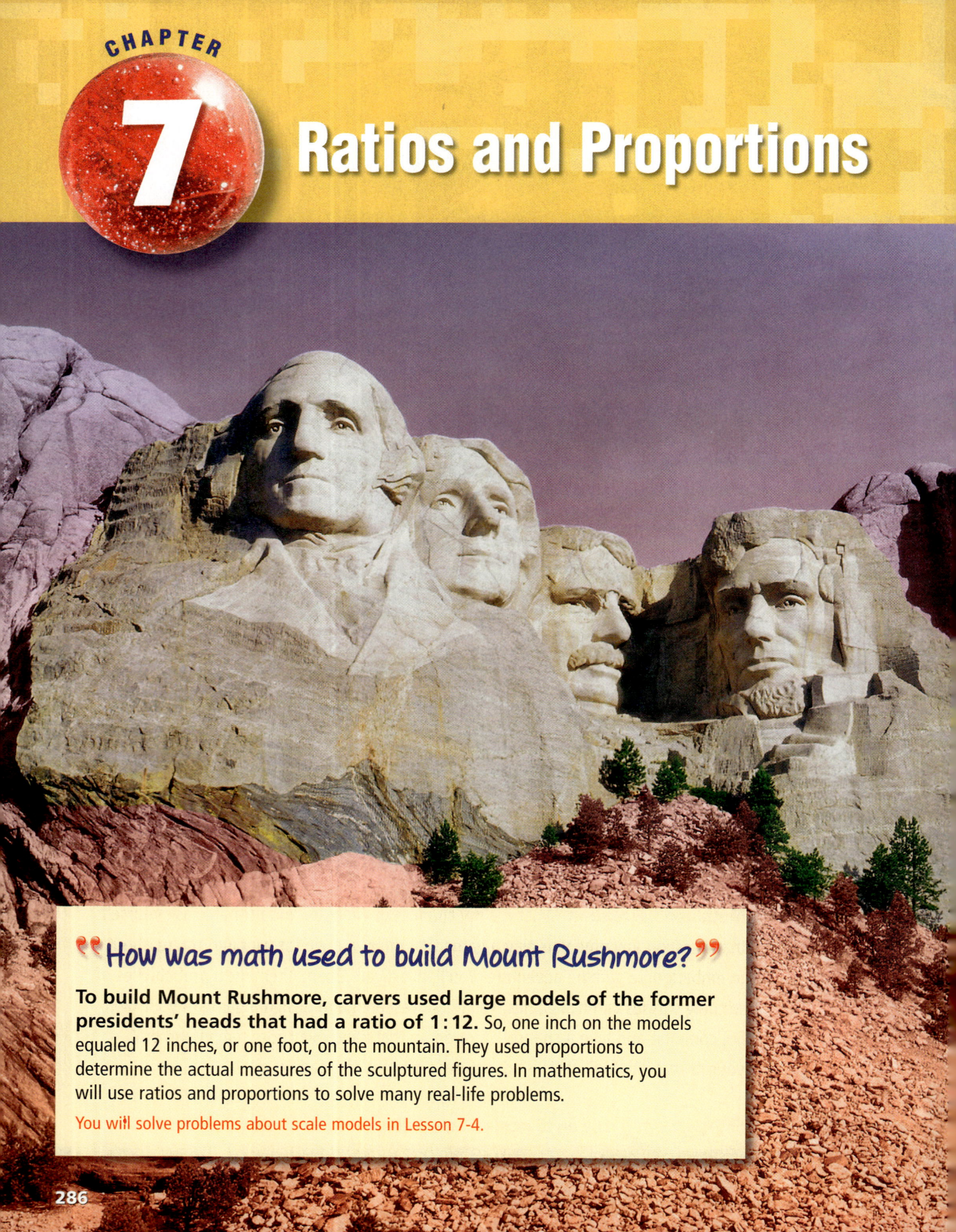

"How was math used to build Mount Rushmore?"

To build Mount Rushmore, carvers used large models of the former presidents' heads that had a ratio of 1:12. So, one inch on the models equaled 12 inches, or one foot, on the mountain. They used proportions to determine the actual measures of the sculptured figures. In mathematics, you will use ratios and proportions to solve many real-life problems.

You will solve problems about scale models in Lesson 7-4.

GETTING STARTED

▶ Diagnose Readiness

Take this quiz to see if you are ready to begin Chapter 7. Refer to the lesson number in parentheses for review.

Vocabulary Review
Complete each sentence.

1. A comparison of two numbers by division is a ____. (Lesson 5-5)

2. A ratio that compares a number to 100 is called a ____. (Lesson 5-5)

Prerequisite Skills
Evaluate each expression. Round to the nearest tenth if necessary. (Lesson 1-3)

3. $100 \times 25 \div 52$
4. $10 \div 4 \times 31$
5. $\frac{63 \times 4}{34}$
6. $\frac{2 \times 100}{68}$

Write each fraction in simplest form. (Lesson 5-3)

7. $\frac{9}{45}$
8. $\frac{16}{24}$
9. $\frac{38}{46}$

Write each decimal as a fraction in simplest form. (Lesson 5-4)

10. 0.78
11. 0.320
12. 0.06

Solve each equation. (Lesson 6-5)

13. $\frac{1}{4}x = 6$
14. $\frac{a}{5} = 8$
15. $4r = 22$
16. $2 = \frac{3}{5}z$

Complete. (Lesson 6-7)

17. 2 yd = __?__ ft
18. 48 oz = __?__ lb
19. $4\frac{1}{2}$ ft = __?__ yd
20. $3\frac{1}{4}$ h = __?__ min

Ratios and Proportions Make this Foldable to help you organize your notes. Begin with a sheet of notebook paper.

 Fold
Fold lengthwise to the holes.

 Cut
Cut along the top line and then make equal cuts to form 10 tabs.

 Label
Label the major topics as shown.

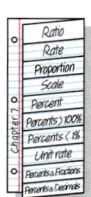

Chapter Notes Each time you find this logo throughout the chapter, use your *Noteables™: Interactive Study Notebook with Foldables™* or your own notebook to take notes. Begin your chapter notes with this Foldable activity.

 Readiness To prepare yourself for this chapter with another quiz, visit **msmath2.net/chapter_readiness**

7-1 Ratios

What You'll LEARN
Write ratios as fractions and determine whether two ratios are equivalent.

NEW Vocabulary
equivalent ratios

REVIEW Vocabulary
ratio: a comparison of two numbers by division (Lesson 5-5)

WHEN am I ever going to use this?

GEARS Gears are used in objects that have spinning parts, such as bicycles and pendulum clocks. Suppose you have a larger gear that has 80 teeth being turned by a smaller gear that has 20 teeth. The comparison of the size of the larger gear to the size of the smaller gear is called a *gear ratio*.

1. Express the gear ratio as a fraction. Then write it as a fraction with a denominator of 1.
2. How many times does the smaller gear turn for every turn of the larger gear?
3. Describe some of the possible sizes of two gears that have a gear ratio of 3 to 1.

In Lesson 5-5, you learned that ratios can be written in several different ways. Ratios can also be simplified as shown below.

$$80 \text{ teeth} : 20 \text{ teeth} = \frac{\overset{4}{\cancel{80 \text{ teeth}}}}{\underset{1}{\cancel{20 \text{ teeth}}}} = \frac{4}{1}$$

EXAMPLES Write Ratios in Simplest Form

Write each ratio as a fraction in simplest form.

① 25 to 10

$25 \text{ to } 10 = \frac{25}{10}$ Write the ratio as a fraction.

$\phantom{25 \text{ to } 10} = \frac{5}{2}$ Simplify.

Written as a fraction in simplest form, the ratio 25 to 10 is $\frac{5}{2}$. For ratios, leave the fraction as an improper fraction.

② 6 : 18

$6 : 18 = \frac{6}{18}$ Write the ratio as a fraction.

$ = \frac{1}{3}$ Simplify.

The ratio 6 : 18 is $\frac{1}{3}$ in simplest form.

READING in the Content Area
For strategies in reading this lesson, visit msmath2.net/reading.

Your Turn Write each ratio as a fraction in simplest form.
a. 9 to 12   b. 27 : 15   c. 8 to 56

288 Chapter 7 Ratios and Proportions

When writing ratios comparing units of length, units of time, units of weight, and so on, both measures should have the same unit.

STUDY TIP

Canceling Units
You can cancel units that are the same in the numerator and in the denominator.

EXAMPLE — Write a Ratio by Converting Units

3 Write the ratio 2 feet to 10 inches as a fraction in simplest form.

$$\frac{2 \text{ feet}}{10 \text{ inches}} = \frac{24 \text{ inches}}{10 \text{ inches}} \quad \text{Convert 2 feet to inches.}$$

$$= \frac{\cancel{24 \text{ inches}}^{12}}{\cancel{10 \text{ inches}}_{5}} \quad \text{Divide by the GCF, 2 inches.}$$

$$= \frac{12}{5} \quad \text{Simplify.}$$

So, the ratio 2 feet to 10 inches can be written as $\frac{12}{5}$.

Your Turn Write each ratio as a fraction in simplest form.

d. 4 feet : 4 yards
e. 15 ounces to 3 pounds

Two ratios that have the same value are **equivalent ratios**.

EXAMPLES — Compare Ratios

4 Determine whether 6 : 8 and 36 : 48 are equivalent.

Write each ratio as a fraction in simplest form.

$6 : 8 = \frac{6 \div 2}{8 \div 2}$ or $\frac{3}{4}$ The GCF of 6 and 8 is 2. $36 : 48 = \frac{36 \div 12}{48 \div 12}$ or $\frac{3}{4}$ The GCF of 36 and 48 is 12.

The ratios in simplest form both equal $\frac{3}{4}$. So, 6 : 8 and 36 : 48 are equivalent ratios.

5 **PONDS** It is recommended that for every 9 square feet of surface, a pond should have 2 fish. A pond that has a surface of 45 square feet contains 6 fish. Is this the correct number of fish based on the above recommendation?

9 to 2 or $\frac{9}{2}$ Recommended ratio

45 to 6 or $\frac{45}{6} = \frac{15}{2}$ Actual ratio

Since $\frac{9}{2} \neq \frac{15}{2}$, the ratios are not equivalent.

So, the number of fish is not correct for the pond.

REAL-LIFE MATH

PONDS The recommended number of water lilies for a 9-square-foot pond is 1.

Your Turn Determine whether the ratios are equivalent.

f. $\frac{3}{8}$ and $\frac{6}{12}$
g. 35 students : 5 adults and 14 students : 2 adults

msmath2.net/extra_examples   Lesson 7-1 Ratios **289**

Skill and Concept Check

1. **Write** the ratio 8 feet out of 15 feet in three different ways.

2. **OPEN ENDED** Write two different ratios that are equivalent to 3:2.

3. **Which One Doesn't Belong?** Identify the item that does not have the same ratio as the other three. Explain your reasoning.

   | $\frac{30}{45}$ | 4 to 6 | 26:39 | 3 to 2 |

4. **FIND THE ERROR** Marcus and Nicole are writing the ratio 2 days to 18 hours as a fraction in simplest form. Who is correct? Explain.

   Marcus
   $2:18 = \frac{2}{18}$
   $= \frac{1}{9}$

   Nicole
   $48:18 = \frac{48}{18}$
   $= \frac{8}{3}$

GUIDED PRACTICE

Write each ratio as a fraction in simplest form.

5. 8:30
6. 27 to 36
7. 11 to 4
8. 5 meters:1 meter
9. 28 feet:7 yards
10. 9 hours to 3 days

Determine whether the ratios are equivalent. Explain.

11. $\frac{3}{5}$ and $\frac{9}{15}$
12. 2:3 and 3:9

13. **HEIGHTS** Megan is 5 feet 3 inches tall, and her brother Troy is 5 feet 9 inches tall. Write a ratio in simplest form that compares Megan's height to Troy's height in inches.

Practice and Applications

Write each ratio as a fraction in simplest form.

14. 24 to 9
15. 14 to 70
16. 18:19
17. 45:21
18. 120:30
19. 33 to 90
20. 66 inches:72 inches
21. 12 pounds to 64 pounds
22. 4 pounds to 12 ounces
23. $8\frac{1}{3}$ yards:5 feet
24. 18 weeks:1 year
25. 45 minutes to 2 hours

HOMEWORK HELP

For Exercises	See Examples
14–19	1, 2
20–25	3
26–31, 33–37	4, 5

Extra Practice
See pages 580, 602.

Determine whether the ratios are equivalent. Explain.

26. $\frac{6}{9}$ and $\frac{2}{3}$
27. $\frac{9}{15}$ and $\frac{6}{10}$
28. 4:7 and 16:49
29. 8:21 and 16:42
30. 14 in.:9 ft and 16 ft:12 yd
31. 6 lb:72 oz and 2 lb:24 oz

290 Chapter 7 Ratios and Proportions

32. **WRITE A PROBLEM** Write a problem that could be represented by using a ratio.

33. **MOVIES** One week in 2002, a movie about the mathematician John Nash earned 6 million dollars at the box office. Total revenues for all movies that week were 78 million dollars. Write a ratio to compare the money earned by that movie to the money earned by all the movies.

PHOTOGRAPHS The *aspect ratio* of a photograph is a ratio comparing the length and width. A 35 mm negative has an aspect ratio of 1.5 : 1. Photo sizes with the same aspect ratio can be printed full frame without cropping. Determine which size photos can be printed full frame from a 35 mm negative.

34. 10″ × 15″   35. 8″ × 10″   36. 5″ × 7.5″

37. **MONEY** In simplest form, write the ratio of the cost in cents to time in years for each type of currency shown in the table. Are they equivalent? Explain.

38. **FIND A PATTERN** In decimal form, write the ratios of the first eight consecutive square numbers $1^2 : 2^2$, $2^2 : 3^2$, $3^2 : 4^2$, … . What do you notice about the ratios?

Currency	Cost to Make	Time in Use
Sacagawea dollar coin	12¢	30 years
Dollar bill	3.5¢	18 months

Source: www.sciencenews.org

39. **CRITICAL THINKING** Find the next number in the pattern 480, 240, 80, 20, __?__. Explain your reasoning. (*Hint:* Look at the ratios of successive numbers.)

Spiral Review with Standardized Test Practice

40. **MULTIPLE CHOICE** Express the ratio of shaded squares to total squares as a fraction in simplest form.

    Ⓐ $\frac{3}{4}$ Ⓑ $\frac{15}{20}$ Ⓒ $\frac{7}{10}$ Ⓓ $\frac{3}{5}$

41. **MULTIPLE CHOICE** In the seventh grade, 16 of the 120 students are left-handed. Write the ratio of left-handed students to right-handed students as a fraction in simplest form.

    Ⓕ $\frac{2}{15}$ Ⓖ $\frac{2}{13}$ Ⓗ $\frac{13}{2}$ Ⓘ $\frac{15}{2}$

42. **TREES** A giant sequoia tree has a diameter of about 30 feet. What is the circumference? Round to the nearest foot. (Lesson 6-9)

Find the perimeter and area of each rectangle. (Lesson 6-8)

43. $\ell = 9$ ft, $w = 4$ ft
44. $\ell = 12$ cm, $w = 10.6$ cm
45. $\ell = 7.5$ cm, $w = 2$ cm
46. $\ell = 1\frac{2}{5}$ in., $w = \frac{1}{2}$ in.

GETTING READY FOR THE NEXT LESSON

PREREQUISITE SKILL Divide. Round to the nearest hundredth if necessary. (Page 562)

47. $9.8 \div 2$ 48. $\$4.30 \div 5$ 49. $\$12.49 \div 40$ 50. $27.36 \div 3.2$

msmath2.net/self_check_quiz

7-2 Rates

HANDS-ON Mini Lab

Materials
- textbook
- computer
- stopwatch

What You'll LEARN

Determine unit rates.

NEW Vocabulary

rate
unit rate

Work with a partner.

Choose a page in a textbook and take turns typing as much as possible in 2 minutes.

1. Count the number of words that each of you typed.
2. Write the ratio *number of words to number of minutes* as a fraction.
3. Simplify the fractions by dividing the numerator and the denominator by 2.

A ratio that compares two quantities with different kinds of units is called a **rate**.

$\frac{100 \text{ words}}{2 \text{ minutes}}$ — The units *words* and *minutes* are different.

When a rate is simplified so that it has a denominator of 1 unit, it is called a **unit rate**.

$\frac{50 \text{ words}}{1 \text{ minute}}$ — The denominator is 1 unit.

The unit rate $\frac{50 \text{ words}}{1 \text{ minute}}$ can be read as *50 words per minute*. The table below shows some common unit rates used in everyday life.

Ratio	Unit Rate	Abbreviation	Name
$\frac{\text{number of miles}}{1 \text{ hour}}$	miles per hour	mi/h or mph	speed
$\frac{\text{number of miles}}{1 \text{ gallon}}$	miles per gallon	mi/gal or mpg	gas mileage
$\frac{\text{number of dollars}}{1 \text{ pound}}$	price per pound	dollars/lb	unit price
$\frac{\text{number of dollars}}{1 \text{ hour}}$	dollars per hour	dollars/h	hourly wage

EXAMPLE   Find a Unit Rate

1 **RUNNING** Alethia ran 24 miles in 3 hours. What is her average speed in miles per hour?

Write the rate as a fraction. Then find an equivalent rate with a denominator of 1.

24 miles in 3 hours $= \frac{24 \text{ mi}}{3 \text{ h}}$ Write the rate as a fraction.

$= \frac{24 \text{ mi} \div 3}{3 \text{ h} \div 3}$ Divide the numerator and the denominator by 3.

$= \frac{8 \text{ mi}}{1 \text{ h}}$ Simplify.

The average speed, or unit rate, is 8 miles per hour.

292 Chapter 7 Ratios and Proportions

Standardized Test Practice

Test-Taking Tip

Grid-In Answers
To fill in the grid, first write the answers in the space above. Then shade the appropriate bubbles. You can align at the right as shown, or start in the left column.

EXAMPLES — Find Unit Rates

2 **GRID-IN TEST ITEM** Write *228 feet in 24 seconds* as a unit rate in feet per second.

Read the Test Item
Write the ratio as a fraction. Then divide to get a denominator of 1.

Solve the Test Item

228 feet in 24 seconds

$= \dfrac{228 \text{ ft}}{24 \text{ s}}$ Write the rate as a fraction.

$= \dfrac{228 \text{ ft} \div 24}{24 \text{ s} \div 24}$ Divide the numerator and the denominator by 24.

$= \dfrac{9.5 \text{ ft}}{1 \text{ s}}$ Simplify.

The unit rate is 9.5 feet per second.

Fill in the Grid

9 . 5

3 Find the unit price per orange if it costs $2 for six oranges. Round to the nearest hundredth if necessary.

$2 for six oranges $= \dfrac{\$2}{6 \text{ oranges}}$ Write the rate as a fraction.

$= \dfrac{\$2 \div 6}{6 \text{ oranges} \div 6}$ Divide the numerator and the denominator by 6.

$\approx \dfrac{\$0.33}{1 \text{ orange}}$ Simplify.

The unit price is about $0.33 per orange.

Your Turn Find each unit rate. Round to the nearest hundredth if necessary.

a. 300 tickets in 6 days   **b.** 220 miles in 8 gallons

In Example 3, you found a special kind of unit rate, called the *unit price*. This is the price per unit and is useful when you want to compare the cost of an item that comes in different sizes.

EXAMPLE — Choose the Best Buy

STUDY TIP

Estimation You can use compatible numbers to estimate the unit prices. For example, $2.49 ÷ 12 oz ≈ $2.40 ÷ 12 oz, or $0.20/oz.

4 The costs of different sizes of peanut butter are shown at the right. Which jar costs the least per ounce?

Find the unit price, or the cost per ounce, of each jar. Divide the price by the number of ounces.

Size	Price
12 oz	$2.49
40 oz	$5.30
80 oz	$10.89

12-ounce jar $2.49 ÷ 12 ounces ≈ $0.21 per ounce

40-ounce jar $5.30 ÷ 40 ounces ≈ $0.13 per ounce

80-ounce jar $10.89 ÷ 80 ounces ≈ $0.14 per ounce

The 40-ounce jar costs the least per ounce.

msmath2.net/extra_examples

Skill and Concept Check

1. **OPEN ENDED** Write a rate and then convert it to a unit rate.

2. **FIND THE ERROR** Mikasi and Julie are determining which size sports drink is the better buy per ounce: a 16-ounce bottle for $1.95 or a 36-ounce bottle for $3.05. Who is correct? Explain.

   Mikasi
   16-oz bottle: 12.2¢ per ounce
   36-oz bottle: 8.5¢ per ounce

   Julie
   16-oz bottle: 8.2¢ per ounce
   36-oz bottle: 11.8¢ per ounce

3. **NUMBER SENSE** In which situation will the rate $\frac{x \text{ ft}}{y \text{ min}}$ increase? Give an example to explain your reasoning.

   a. x increases, y is unchanged
   b. x is unchanged, y increases

GUIDED PRACTICE

Find each unit rate. Round to the nearest hundredth if necessary.

4. 410 miles in 16 gallons
5. 1,500 words in 25 minutes
6. 5 pounds for $2.49
7. 3 cans of juice for $2.95

8. Which has the better unit price: a 6-pack of soda for $2.99 or a 12-pack for $4.50?

9. **MUSIC** Ethan can buy 4 CDs for $71.96 at CD Express or 9 CDs for $134.55 at Music Rox. Which has the better unit cost? Explain.

Practice and Applications

Find each unit rate. Round to the nearest hundredth if necessary.

10. 360 miles in 6 hours
11. 6,840 customers in 45 days
12. 150 people for 5 classes
13. $7.40 for 5 pounds
14. $1.12 for 8.2 ounces
15. 810 Calories in 3 servings
16. 40 meters in 13 seconds
17. 144 miles in 4.5 gallons

HOMEWORK HELP

For Exercises	See Examples
10–17, 21–23	1–3
18–20	4

Extra Practice
See pages 580, 602.

Choose the best unit price.

18. $3.99 for a 16-ounce bag of candy or $2.99 for a 12-ounce bag
19. aspirin sold in bottles of 50 for $5.49, 100 for $8.29, or 150 for $11.99

20. **NUTRITION** Which soft drink described in the table has a lower amount of sodium per ounce? Explain.

21. **FOUNTAINS** The Prometheus Fountain in New York City pumps 60,000 gallons of water every 15 minutes. What is the unit rate?

Soft Drink	Serving Size (oz)	Sodium (mg)
A	12	70
B	8	35

Chapter 7 Ratios and Proportions

22. **WHALES** A humpback whale can migrate 3,000 miles in 30 days. Find a unit rate to describe the average speed per day during migration.

23. **POPULATION** The population of Arkansas is approximately 2.7 million people, and its land area is approximately 52,100 square miles. Find the *population density*, or the population per square mile.

24. **RESEARCH** Use the Internet or another source to find the population and land area of your state. Determine the population density.

Determine whether the following statements are *sometimes*, *always*, or *never* true. Explain by giving an example or a counterexample.

25. A ratio is a rate.

26. A rate is a ratio.

LIFE SCIENCE For Exercises 27 and 28, use the graph. It shows the average number of heartbeats for an adult elephant and an adult human.

27. Whose heart rate is greater? Explain.

28. Whose heart beats more times in one hour?

29. **CRITICAL THINKING** If 8 tickets are sold in 15 minutes, find the rate in tickets per hour.

Spiral Review with Standardized Test Practice

30. **MULTIPLE CHOICE** Which of the following has the lowest unit price?
    - Ⓐ 18 oz: $5.40
    - Ⓑ 16 oz: $4.64
    - Ⓒ 12 oz: $3.72
    - Ⓓ 10 oz: $3.30

31. **MULTIPLE CHOICE** Which rate is the fastest?
    - Ⓕ 589 miles in 11 hours
    - Ⓖ 360 miles in 7 hours
    - Ⓗ 111 miles in 2 hours
    - Ⓘ 283 miles in 5 hours

Write each ratio as a fraction in simplest form. (Lesson 7-1)

32. 11 : 12
33. 24 : 4
34. 15 to 25
35. 20 to 14

36. A serving plate has a radius of 6 inches. Find the circumference to the nearest tenth. (Lesson 6-9)

GETTING READY FOR THE NEXT LESSON

PREREQUISITE SKILL Solve each equation. (Lesson 6-5)

37. $\frac{n}{8} = 7$
38. $\frac{2}{3}x = 5$
39. $9t = 12$
40. $\frac{r}{5.2} = 4$

 msmath2.net/self_check_quiz

7-2b Hands-On Lab

A Follow-Up of Lesson 7-2

Rate of Change

What You'll LEARN
Investigate rate of change.

Materials
- square tiles
- grid paper

INVESTIGATE *Work in groups of three.*

A *rate of change* is a ratio that shows a change in one quantity with respect to a change in another quantity. In this lab, you will use tables and graphs to represent *constant* rates of change.

STEP 1 Use tiles to build the models shown below. Then continue the pattern to build the fourth and fifth models.

1   2   3

STEP 2 For each model, record the number of tiles and the perimeter in a table like the one shown at the right. The first two are done for you.

Model	Number of Tiles (x)	Perimeter (y)
1	1	4
2	3	8
3		
4		
5		

STEP 3 Draw a coordinate plane on grid paper and graph the ordered pairs (x, y).

Writing Math

1. What do you notice about the points?

2. Find the ratio $\frac{\text{change in perimeter}}{\text{change in tiles}}$ between the second and third points, the third and fourth points, and the fourth and fifth points. Each ratio is a rate of change. **Describe** what you observe.

3. **Complete:** As the number of tiles increases by 2 units, the perimeter of the models increases by __?__ units.

4. Refer to the table at the right that appeared in Lesson 4-7. Find the ratio $\frac{\text{change in earnings}}{\text{change in hours worked}}$ for Greg and Monica. How do these values compare to the slopes that you found in Lesson 4-7?

Hours Worked	Earnings ($)	
	Greg	Monica
1	4	5
2	8	10
3	12	15
4	16	20

5. **Make a conjecture** about the relationship between rate of change and a graph of the two quantities.

STUDY TIP
Look Back You can review **slope** in Lesson 4–7.

296 Chapter 7 Ratios and Proportions

7-3 Algebra: Solving Proportions

What You'll LEARN
Solve proportions.

NEW Vocabulary
proportion
cross product

WHEN am I ever going to use this?

NUTRITION As part of a healthy diet, teens need 60 milligrams of vitamin C each day.

1. Write the ratio $\frac{\text{vitamin C}}{\text{amount of cereal}}$ for a half-cup serving of cereal.
2. Rewrite the ratio in Exercise 1 to find the unit rate in milligrams per cup.
3. Simplify $\frac{60 \text{ mg}}{2 \text{ c}}$ to find the unit rate.
4. **Make a conjecture** about the rates $\frac{15 \text{ mg}}{0.5 \text{ c}}$ and $\frac{60 \text{ mg}}{2 \text{ c}}$.

Vitamin C (mg)	Amount of Cereal (c)
15	0.5
60	2

When two ratios are equivalent, they form a **proportion**. Since rates are types of ratios, they can also form proportions.

Noteables™ — Key Concept: Proportions

Words A proportion is an equation stating that two ratios are equivalent.

Symbols

Arithmetic: $\frac{1}{2} = \frac{3}{6}$, $\frac{8 \text{ ft}}{10 \text{ s}} = \frac{4 \text{ ft}}{5 \text{ s}}$

Algebra: $\frac{a}{b} = \frac{c}{d}$, where $b, d \neq 0$

In a proportion, a **cross product** is the product of the numerator of one ratio and the denominator of the other ratio.

1×6 is a cross product. 2×3 is a cross product.

$$\frac{1}{2} = \frac{3}{6}$$
$$1 \times 6 = 2 \times 3$$
$$6 = 6$$

The cross products of a proportion are equal.

EXAMPLE   Identify a Proportion

1 Determine whether $\frac{6}{10}$ and $\frac{3}{5}$ form a proportion.

$\frac{6}{10} \stackrel{?}{=} \frac{3}{5}$ Write a proportion.

$6 \times 5 \stackrel{?}{=} 10 \times 3$ Find the cross products.

$30 = 30$ ✓   Multiply.

The cross products are equal, so the ratios form a proportion.

EXAMPLE — Identify a Proportion

2 Determine whether $\dfrac{\$4}{9\ oz}$ and $\dfrac{\$21}{45\ oz}$ form a proportion.

$\dfrac{4}{9} \stackrel{?}{=} \dfrac{21}{45}$ Write a proportion.

$4 \times 45 \stackrel{?}{=} 9 \times 21$ Find the cross products.

$180 \neq 189$ Multiply.

The cross products are not equal, so the ratios do not form a proportion.

Your Turn Determine whether each pair of ratios forms a proportion.

a. $\dfrac{4}{9}$ and $\dfrac{2}{3}$ b. $\dfrac{15}{9}$ and $\dfrac{10}{6}$ c. $\dfrac{8\ in.}{21\ s}$ and $\dfrac{12\ in.}{31.5\ s}$

You can use cross products to find a missing term in a proportion. This is known as *solving the proportion*. Solving a proportion is similar to solving an equation.

EXAMPLES — Solve Proportions

3 Solve $\dfrac{2.6}{13} = \dfrac{8}{n}$.

$\dfrac{2.6}{13} = \dfrac{8}{n}$ Write the proportion.

$2.6 \cdot n = 13 \cdot 8$ Find the cross products.

$2.6n = 104$ Multiply.

$\dfrac{2.6n}{2.6} = \dfrac{104}{2.6}$ Divide each side by 2.6.

$n = 40$ Simplify.

The solution is 40.

4 **LIFE SCIENCE** The skull length of a crocodile is proportional to its body length. In one species, a 6-foot crocodile has a 2-foot skull. Estimate the length of a crocodile of that species that has a 3.5-foot skull.

body length → $\dfrac{6\ ft}{2\ ft} = \dfrac{c\ ft}{3.5\ ft}$ Write a proportion.
skull length →

$6 \cdot 3.5 = 2 \cdot c$ Find the cross products.

$21 = 2c$ Multiply.

$\dfrac{21}{2} = \dfrac{2c}{2}$ Divide each side by 2.

$10.5 = c$ Simplify.

So, a crocodile with a 3.5-foot skull is about 10.5 feet long.

Your Turn Solve each proportion.

d. $\dfrac{1}{5} = \dfrac{x}{60}$ e. $\dfrac{10}{k} = \dfrac{2.5}{4}$ f. $\dfrac{2\ ft}{6\ min} = \dfrac{5\ ft}{t\ min}$

REAL-LIFE CAREERS

How Does a Paleontologist Use Math?
In 2001, paleontologists discovered a 6-foot skull of a prehistoric crocodile called *SuperCroc*. By using proportions, they estimated its total length as 40 feet.

Source: *National Geographic*

Math Online Research
For information about a career as a paleontologist, visit: msmath2.net/careers

Skill and Concept Check

1. **Writing Math** Explain how to determine if two ratios are equivalent.

2. **OPEN ENDED** Write a proportion that has $\frac{12}{15}$ as one of the ratios.

3. **Which One Doesn't Belong?** Identify the ratio that is not equivalent to the other three. Explain your reasoning.

$$\frac{2}{3} \qquad \frac{6}{9} \qquad \frac{4}{6} \qquad \frac{2}{9}$$

GUIDED PRACTICE

Determine whether each pair of ratios forms a proportion.

4. $\frac{2}{3}$ and $\frac{6}{9}$
5. $\frac{16}{30}$ and $\frac{8}{10.5}$
6. $\frac{5 \text{ ft}}{7 \text{ s}}$ and $\frac{25 \text{ ft}}{49 \text{ s}}$

Solve each proportion.

7. $\frac{24}{6} = \frac{8}{x}$
8. $\frac{n}{5} = \frac{20}{25}$
9. $\frac{7}{t} = \frac{3}{9}$
10. $\frac{8}{12} = \frac{n}{3}$
11. $\frac{a}{6} = \frac{12}{8}$
12. $\frac{13}{25} = \frac{39}{r}$
13. $\frac{21}{5} = \frac{c}{7}$
14. $\frac{7}{x} = \frac{3.5}{9}$

15. **NUTRITION** An 8-ounce glass of orange juice contains 72 milligrams of vitamin C. How much juice contains 60 milligrams of vitamin C?

Practice and Applications

Determine whether each pair of ratios forms a proportion.

16. $\frac{1}{6}$ and $\frac{4}{24}$
17. $\frac{3}{9}$ and $\frac{2}{15}$
18. $\frac{27}{6}$ and $\frac{9}{2}$
19. $\frac{1}{4}$ and $\frac{4}{1}$
20. $\frac{\$8}{20 \text{ lb}}$ and $\frac{\$24}{60 \text{ lb}}$
21. $\frac{21 \text{ mi}}{6 \text{ h}}$ and $\frac{14.3 \text{ mi}}{3 \text{ h}}$

22. If there are 12 wheels, how many bicycles are there?
23. If there are 32 paws, how many kittens are there?

HOMEWORK HELP

For Exercises	See Examples
16–21	1, 2
22–46	3, 4

Extra Practice
See pages 580, 602.

Solve each proportion.

24. $\frac{5}{9} = \frac{10}{x}$
25. $\frac{2}{12} = \frac{a}{36}$
26. $\frac{16}{t} = \frac{2}{3}$
27. $\frac{n}{8} = \frac{3}{4}$
28. $\frac{c}{7} = \frac{18}{42}$
29. $\frac{30}{w} = \frac{8}{20}$
30. $\frac{9}{15} = \frac{b}{10}$
31. $\frac{3}{n} = \frac{27}{18}$
32. $\frac{x}{12} = \frac{12}{4}$
33. $\frac{45}{5} = \frac{t}{7}$
34. $\frac{6}{3} = \frac{5}{w}$
35. $\frac{y}{36} = \frac{15}{24}$
36. $\frac{350}{a} = \frac{2}{10}$
37. $\frac{380}{520} = \frac{760}{n}$
38. $\frac{3}{5} = \frac{0.2}{d}$
39. $\frac{2.5}{4.5} = \frac{7.5}{x}$

40. **MODELS** The toy car shown at the right is modeled after a real car. If the real car is 4.79 meters long, how wide is it? Write the width to the nearest hundredth of a meter.

41. **TESTS** Suppose you received a score of 13 out of 15 on a quiz. Write a proportion to find the percent. Then solve. Round to the nearest percent.

6.6 cm
15.4 cm

COOKING For Exercises 42–44, refer to the table.

Vegetable Equivalents
$\frac{3}{4}$ lb broccoli = 3 c florets
1 lb carrots = 3 c slices
1 stalk celery = $\frac{1}{2}$ c slices
2 peppers = $2\frac{1}{2}$ c rings
1 lb potatoes = $2\frac{3}{4}$ c cubes

42. How many stalks of celery would you need to get $3\frac{1}{2}$ cups of celery slices?

43. How many cups of broccoli florets would 1 pound of broccoli produce?

44. **WRITE A PROBLEM** Write a real-life problem that can be solved using a proportion and one of the vegetable equivalents. Then write a proportion and solve.

45. **MULTI STEP** Amanda can word process 7 words in 6 seconds. At this rate, how many words can she word process in 3 minutes?

46. **MONEY** In 2002, France adopted the *Euro* as its currency, replacing the French franc. At that time, one Euro was worth $0.90 in American currency. In Paris, France, a quart of milk cost 3 Euros. In Los Angeles, a quart of milk cost $1.65. In which city was the quart of milk more expensive?

Data Update What is the current U.S. value of one Euro? Visit msmath2.net/data_update to learn more.

47. **CRITICAL THINKING** In the proportion $\frac{3}{6} = \frac{6}{12}$, the number 6 appears in two of the diagonal positions. This repeated number is called the *geometric mean* of the other two numbers. Find a pair of numbers other than 3 and 12 for which 6 is the geometric mean.

Spiral Review with Standardized Test Practice

48. **MULTIPLE CHOICE** Ming is paid an hourly rate. One week he earned $157.50 by working 30 hours. If he works 35 hours the next week, how much will he earn?

   Ⓐ $135.00   Ⓑ $160.50   Ⓒ $183.75   Ⓓ $210.25

49. **GRID IN** The ratio of boys to girls in the seventh grade is 7 to 8. If there are 120 students in the seventh grade, how many of them are girls?

50. Choose the better unit price: 5 pounds of onions for $2.99 or 3 pounds of onions for $1.29. (Lesson 7-2)

Write each ratio as a fraction in simplest form. (Lesson 7-1)

51. 45 : 81   52. 49 to 14   53. 14 ounces to 5 pounds

MEASUREMENT Complete. (Lesson 6-7)

54. 5 qt = __?__ pt   55. $3\frac{1}{2}$ lb = __?__ oz   56. 28 c = __?__ qt

GETTING READY FOR THE NEXT LESSON

PREREQUISITE SKILL Multiply. Write in simplest form. (Lesson 6-4)

57. $\frac{1}{5} \cdot 8$ 58. $1\frac{3}{4} \cdot 5$ 59. $16 \cdot 3\frac{1}{4}$ 60. $24 \cdot 1\frac{3}{8}$

7-3b HANDS-ON LAB
A Follow-Up of Lesson 7-3

Wildlife Sampling

INVESTIGATE *Work in groups of three.*

Naturalists can estimate the population in a wildlife preserve by using the *capture-recapture* technique. In this lab, you will model this technique using dried beans in a bowl to represent deer in a forest.

What You'll LEARN

Use proportions to estimate.

Materials
- small bowl
- dried beans
- paper cup
- markers

STEP 1 Fill a small bowl with dried beans.

STEP 2 Use the paper cup to scoop out some of the beans. These represent the original *captured* deer. Record the number in a table like the one shown at the right. Mark each bean with an × on both sides. Then return these beans to the bowl and mix well.

Original Number Captured _____

Trial	Sample	Recaptured	P
A			
B			
C			
⋮			
J			
Total			

STEP 3 Scoop another cup of beans from the bowl and count them. This is the *sample* for Trial A. Count the beans with the ×'s. These are the *recaptured* deer. Record both numbers.

STEP 4 Use the proportion below to estimate the total number of beans in the bowl. This represents the total population P. Record the value of P in the table.

$$\frac{\text{captured}}{\text{total population }(P)} = \frac{\text{recaptured}}{\text{sample}}$$

STEP 5 Return all of the beans to the bowl.

STEP 6 Repeat Steps 3–5 nine times.

Writing Math

Work in groups of three.

1. Find the average of the estimates in column P. Is this a good estimate of the number of beans in the bowl? **Explain** your reasoning.

2. **Count** the actual number of beans in the bowl. How does this number compare to your estimate?

7-4a Problem-Solving Strategy
A Preview of Lesson 7-4

Draw a Diagram

What You'll LEARN

Solve problems by drawing a diagram.

Mrs. Dixon said that we've gone about 90 miles, or $\frac{2}{3}$ of the way to the campsite.

So, how much farther do we have to go? I bet if we **draw a diagram**, we could figure it out.

Explore	We know that 90 miles is about $\frac{2}{3}$ of the total distance.
Plan	Let's draw a diagram showing the distance that we've already gone and the fractional part that it represents.
Solve	 If $\frac{2}{3}$ of the distance is 90 miles, then $\frac{1}{3}$ of the distance would be 45 miles. So, the missing third must be another 45 miles. The total distance is 90 + 45 or 135 miles.
Examine	Since $\frac{2}{3}$ of the total distance equals 90 miles, the equation $\frac{2}{3}x = 90$ represents this problem. Solving, we get $x = 135$ miles. So, the solution checks.

Analyze the Strategy

1. **Explain** why drawing a diagram can be a useful problem-solving strategy.
2. **Determine** how far the trip would have been if 90 miles was only $\frac{1}{3}$ of the total distance. Draw a new diagram for this situation.
3. **Write** a problem that could be solved by drawing a diagram. Exchange your problem with a classmate and solve.

Apply the Strategy

Solve. Use the draw a diagram strategy.

4. **PHYSICAL SCIENCE** A ball is dropped from 10 feet above the ground. It hits the ground and bounces up half as high as it fell. This is true for each successive bounce. What height does the ball reach on the fourth bounce?

5. **FAMILY** At Nate's family reunion, 80% of the people are 18 years of age or older. Half of the remaining people are under 12 years old. If 20 children are under 12 years old, how many people are at the reunion?

Mixed Problem Solving

Solve. Use any strategy.

6. **TESTS** The scores on a social studies test are found by adding or subtracting points as shown at the right.

Answer	Points
Correct	+8
Incorrect	−4
No answer	−2

If Mario's score on a 15-question test was 86 points, how many of his answers were correct, incorrect, and blank?

7. **GAMES** Six members of a video game club are having a tournament. In the first round, every player will play a video game against every other player. How many games will be in the first round of the tournament?

8. **BUSINESS** The graph shows the annual spending by five industries.

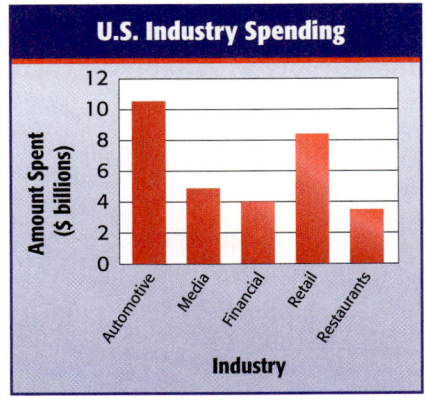

Source: *The Top Ten of Everything*

Estimate how much more the automotive industry spends per year than the retail industry. Write in scientific notation.

9. **ENVIRONMENT** What fraction of landfills is paper and plastic packaging? Write in simplest form.

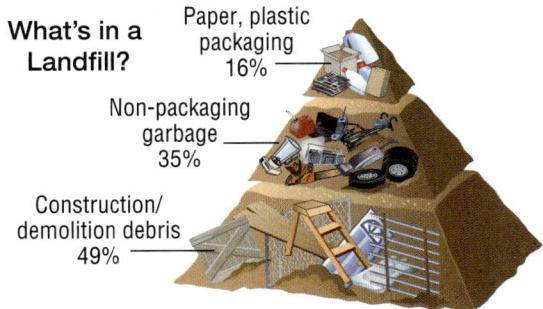

Source: *ULS Report*

10. **MONEY** Mr. Li has $240 in his checking account after writing checks for $15.70, $43.20, and $18. What was his balance before he wrote the three checks?

11. **STANDARDIZED TEST PRACTICE**
Danielle is adding 3 feet to the length and width of her rectangular garden, as shown at the right. Which expression represents the area of land that will be added to the garden?

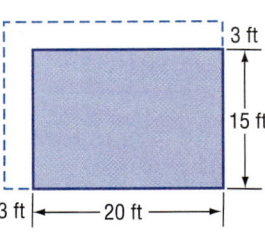

Ⓐ (15 + 3)(20 + 3)

Ⓑ (15 + 3)(20 + 3) − (15)(20)

Ⓒ (15 − 3)(20 − 3)

Ⓓ (15 − 3)(20 − 3) − (15)(20)

7-4 Geometry: Scale Drawings

HANDS-ON Mini Lab

Work with a partner.

- Measure the length of each wall, door, window, and chalkboard in your classroom.
- Record each length to the nearest $\frac{1}{2}$ foot.

Materials
- measuring tape
- straightedge
- $\frac{1}{4}$-inch grid paper

1. Let 1 unit on the grid paper represent 1 foot. So, 6 units = 6 feet. Convert all of your measurements to units.
2. On grid paper, make a drawing of your classroom like the one shown at the right.

What You'll LEARN

Solve problems involving scale drawings.

NEW Vocabulary

scale drawing
scale
scale factor
scale model

A **scale drawing** represents something that is too large or too small to be drawn at actual size.

The map is a scale drawing. Its **scale** gives the relationship between the distance on the map and the actual distance.

The scale is 1 centimeter = 4.5 kilometers.

EXAMPLE Use a Scale Drawing

1 MAPS On the map above, the distance between Coral Springs and Fort Lauderdale is about 4.1 centimeters. What is the actual distance? Round to the nearest tenth.

Let d = the actual distance between the cities. Write and solve a proportion. Use the scale written as a fraction.

$$\begin{array}{r} \text{map} \rightarrow \\ \text{actual} \rightarrow \end{array} \underset{\text{Scale}}{\frac{1 \text{ centimeter}}{4.5 \text{ kilometers}}} = \underset{\substack{\text{Coral Springs to} \\ \text{Ft. Lauderdale}}}{\frac{4.1 \text{ centimeters}}{d \text{ kilometers}}} \begin{array}{l} \leftarrow \text{map} \\ \leftarrow \text{actual} \end{array}$$

$1 \cdot d = 4.5 \cdot 4.1$ Cross products

$d = 18.45$ Simplify.

The distance between Coral Springs and Fort Lauderdale is about 18.5 kilometers.

A blueprint is another example of a scale drawing.

EXAMPLE Read a Scale Drawing

2 **POOLS** On the blueprint of the pool, each square has a side length of $\frac{1}{4}$ inch. What is the actual width of the pool?

The pool on the blueprint is $1\frac{3}{4}$ inches wide. Let w = the actual width of the pool. Write and solve a proportion using the scale.

Scale $\frac{1}{4}$ in. = 2 ft

$$\begin{array}{c}\text{Scale} \quad\quad \text{Width of Pool}\end{array}$$

$$\begin{array}{l}\text{blueprint} \rightarrow \\ \text{actual} \rightarrow \end{array} \dfrac{\frac{1}{4} \text{ inch}}{2 \text{ feet}} = \dfrac{1\frac{3}{4} \text{ inches}}{w \text{ feet}} \begin{array}{l}\leftarrow \text{blueprint}\\ \leftarrow \text{actual}\end{array}$$

$\frac{1}{4} \cdot w = 2 \cdot 1\frac{3}{4}$ Cross products

$\frac{1}{4}w = \frac{14}{4}$ Multiply.

$w = 14$ Simplify. Multiply each side by 4.

The width of the pool is 14 feet.

In Lesson 7-1, you simplified ratios by converting units. You can use the same method to simplify the scale in Example 2. A scale written as a ratio in simplest form is called the **scale factor**.

scale ➤ $\dfrac{\frac{1}{4} \text{ inch}}{2 \text{ feet}} = \dfrac{\frac{1}{4} \text{ inch}}{24 \text{ inches}}$ Convert 2 feet to inches.

$= \dfrac{4}{4} \cdot \dfrac{\frac{1}{4} \text{ inch}}{24 \text{ inches}}$ Multiply by $\frac{4}{4}$ to eliminate the fraction in the numerator.

$= \dfrac{1}{96}$ ← scale factor

EXAMPLE Find the Scale Factor

3 **Find the scale factor of a blueprint if the scale is 1 inch = 8 feet.**

Write the ratio of 1 inch to 8 feet in simplest form.

$\dfrac{1 \text{ inch}}{8 \text{ feet}} = \dfrac{1 \text{ inch}}{96 \text{ inches}}$ Convert 8 feet to inches.

$= \dfrac{1}{96}$ Cancel the units.

The scale factor is $\frac{1}{96}$. That is, each measure on the blueprint is $\frac{1}{96}$ the actual measure.

STUDY TIP

Equivalent Scales
The scales below are equivalent because their scale factors are equal, $\frac{1}{96}$.
- 1 inch = 8 feet
- $\frac{1}{4}$ inch = 2 feet

Your Turn

a. On a scale drawing of a new classroom, the scale is 1 centimeter = 2.5 meters. What is the scale factor?

msmath2.net/extra_examples   Lesson 7-4 Geometry: Scale Drawings **305**

If you know the actual length of an object and the scale, you can build a scale model. A **scale model** can be used to represent something that is too large or too small for an actual-size model.

EXAMPLE — Make a Scale Model

4 COMPUTERS Designers are creating a larger model of the computer memory board to use in design work. If they use a scale of 20 inches = 1 inch, what is the length of the model?

Write a proportion using the scale.

model → $\dfrac{20 \text{ inches}}{1 \text{ inch}} = \dfrac{m \text{ inches}}{5\frac{1}{4} \text{ inches}}$ ← model
actual →   ← actual

$20 \cdot 5\frac{1}{4} = 1 \cdot m$ Find the cross products.

$105 = m$ Multiply.

The scale model is 105 inches long.

STUDY TIP

Scale The scale is the ratio of the drawing/model measure to the actual measure. It is not always the ratio of a smaller measure to a larger measure.

Skill and Concept Check

1. **OPEN ENDED** On grid paper, make a scale drawing of a room in your home. Include the scale that you used.

2. **NUMBER SENSE** Roberto built a model racecar that is 3 inches long and has a scale of 0.5 inch = 2 feet. He determined that the actual racecar must be 3 · 2 or 6 feet long. Is this correct? Explain.

GUIDED PRACTICE

Find the actual distance between each pair of cities. Round to the nearest tenth if necessary.

	Cities	Map Distance	Scale
3.	Kokomo, Indiana, and Chicago, Illinois	8 cm	1 cm = 25 km
4.	Raleigh and Charlotte, North Carolina	$3\frac{1}{4}$ in.	1 in. = 40 mi

BLUEPRINTS For Exercises 5–7, use the blueprint at the right. Each square has a side length of 0.5 centimeter.

5. What is the length ℓ of the kitchen on the blueprint?

6. Find the actual length of the kitchen.

7. Find the scale factor. (*Hint:* 1 ft ≈ 30 cm)

8. **BRIDGES** A bridge is 28 meters long. Find the length of a scale model if the scale is 1 centimeter = 5.5 meters. Round to the nearest tenth.

306 Chapter 7 Ratios and Proportions

Practice and Applications

Find the actual distance between each pair of cities. Round to the nearest tenth if necessary.

	Cities	Map Distance	Scale
9.	Virginia Beach, Virginia, and Washington, D.C.	8.5 cm	1 cm = 30 km
10.	Mobile and Huntsville, Alabama	$14\frac{1}{2}$ in.	1 in. = 20 mi
11.	Knoxville, Tennessee, and Choctaw, Oklahoma	19.28 cm	2 cm = 125 km
12.	Quebec, Canada, and Paris, France	$4\frac{5}{8}$ in.	$\frac{1}{2}$ in. = 355 mi

HOMEWORK HELP

For Exercises	See Examples
9–12, 23	1
17–20	2
13–16, 21, 25	3
22, 24, 26	4

Extra Practice
See pages 581, 602.

Suppose you are making a scale drawing. Find the length of each object on the scale drawing with the given scale. Then find the scale factor.

13. a back yard 120 feet deep; 1 inch : 20 feet
14. an airplane 87 feet long; 2 inches = 15 feet
15. an amusement park ride 36 meters high; 0.5 centimeter = 1.5 meters
16. a surgical instrument $5\frac{7}{8}$ inches long; 1 inch : $\frac{1}{2}$ inch

For Exercises 17–21, refer to the blueprint of the garden shown below. Each square has a side length of $\frac{1}{4}$ inch.

17. What is the length of the garden on the blueprint?
18. Find the actual length of the garden.
19. What is the diameter of the butterfly bush on the blueprint?
20. What is the actual circumference of the butterfly bush? Use 3.14 for π and round to the nearest tenth.
21. Find the scale factor.
22. **BRIDGES** The Natchez Trace Bridge in Franklin, Tennessee, is 1,500 feet long. Suppose you build a model of the bridge using a scale of 1 inch : 25 feet. How long is the model?

23. **GEOGRAPHY** A map of Brasilia has a scale of 1 inch to 5 miles. If the city is $2\frac{7}{16}$ inches across on the map, what is the distance across the actual city? Use estimation to check your answer.

ART For Exercises 24 and 25, refer to the figure at the right and the information below.

Mount Rushmore is a sculpture that was carved using a model with a scale of 1 inch : 1 foot, as shown at the right.

24. How high is George Washington's face on the actual sculpture of Mount Rushmore?

25. What is the scale factor?

26. **MODEL** The diameter of an igloo is 10 feet. If you were to make a model of the igloo with sugar cubes, which might be an appropriate scale: 1 inch = 1 foot or 1 inch = 50 feet? Explain.

27. **CRITICAL THINKING** The distance between Atlanta and Savannah, Georgia, is 245 miles. The distance on a map is $3\frac{1}{2}$ inches. What is the scale of the map?

Spiral Review with Standardized Test Practice

28. **SHORT RESPONSE** A scale model of a new office building is shown at the right. If the scale is 5 inches = 18 feet, how tall is the actual building?

29. **MULTIPLE CHOICE** Which of the following scales does *not* have a scale factor of $\frac{1}{60}$?

   Ⓐ 1 in. = 5 ft   Ⓑ 1 cm = 60 cm
   Ⓒ 1 ft = 20 yd   Ⓓ 1 in. = 60 ft

30. **BIRDS** A binocular has a *magnification factor* of 35 to 1; that is, objects that are actually 35 feet away appear to be only 1 foot away using the binoculars. If a goldfinch is 368 feet away, what will the distance appear to be in the binoculars? Round to the nearest tenth. (Lesson 7-3)

Find each unit rate. (Lesson 7-2)

31. 200 miles in 5 hours   32. 99¢ for 30 ounces
33. $6.20 for 4 pounds   34. 150 meters in 12 seconds

GETTING READY FOR THE NEXT LESSON

PREREQUISITE SKILL Divide. Write in simplest form. (Lesson 6-6)

35. $2\frac{3}{4} \div 10$ 36. $4\frac{1}{3} \div 10$ 37. $30\frac{2}{3} \div 100$ 38. $87\frac{1}{2} \div 100$

7-4b Spreadsheet Investigation
A Follow-Up of Lesson 7-4

What You'll LEARN

Use a spreadsheet to calculate measurements for scale drawings.

Scale Drawings

A computer spreadsheet is a useful tool for calculating measures for scale drawings. You can change the scale factors and the dimensions, and the spreadsheet will automatically calculate the new values.

ACTIVITY

Suppose you want to make a scale drawing of your school. Set up a spreadsheet like the one shown below. In this spreadsheet, the actual measures are in feet, and the scale drawing measures are in inches.

EXERCISES

1. The length of one side of the school building is 100 feet. If you use a scale factor of 1:250, what is the length on your scale drawing?

2. The length of a classroom is 30 feet. What is the scale factor if the length of the classroom on a scale drawing is 3.6 inches?

3. Calculate the length of a 30-foot classroom on a scale drawing if the scale factor is 1:10.

4. Suppose the actual measures of your school are given in meters. Describe how you could use a spreadsheet to calculate the scale drawing measures in centimeters using a scale factor of 1:50.

5. Choose three rooms in your home and use a spreadsheet to make scale drawings. First, choose an appropriate scale and calculate the scale factor. Include a sketch of the furniture in each room.

Lesson 7-4b Spreadsheet Investigation: Scale Drawings

Chapter 7 Mid-Chapter Practice Test

Vocabulary and Concepts

1. **Explain** the difference between a rate and a unit rate and give an example of each. (Lesson 7-2)
2. **Define** cross product. (Lesson 7-3)

Skills and Applications

Write each ratio as a fraction in simplest form. (Lesson 7-1)

3. 18 to 12
4. 7 to 49
5. 20 meters : 44 meters

Determine whether the ratios are equivalent. Explain. (Lesson 7-1)

6. $\frac{6}{9}$ and $\frac{2}{3}$
7. 150 : 15 and 3 : 1
8. 24 : 4 and 72 : 8

Find each unit rate. Round to the nearest hundredth if necessary. (Lesson 7-2)

9. 200 miles in 5 hours
10. $6.20 for 5 pounds

11. **SHOPPING** Which box of cereal shown in the table costs the least per ounce? (Lesson 7-2)

Weight (oz)	Cost ($)
12	2.50
18	3.69
24	4.95

12. Determine whether the ratios $\frac{5}{8}$ and $\frac{25}{40}$ form a proportion. (Lesson 7-3)

Solve each proportion. (Lesson 7-3)

13. $\frac{3}{d} = \frac{12}{20}$
14. $\frac{7}{8} = \frac{m}{48}$
15. $\frac{w}{8} = \frac{1}{3}$

16. **TRUCKS** A truck is 26 feet long. Find the length on a scale model if 1 inch represents $5\frac{1}{2}$ feet. Round to the nearest tenth if necessary. (Lesson 7-4)

Suppose you are making a scale drawing. Find the length of each object on the scale drawing with the given scale. Then find the scale factor. (Lesson 7-4)

17. a bedroom 12 feet long; 1 in. = 3 ft
18. a park area 38 meters wide; 0.5 cm = 1 m

Standardized Test Practice

19. **GRID IN** A train travels 146 miles in 2 hours. At this rate, how many miles will it travel in 3.5 hours? (Lesson 7-3)

20. **SHORT RESPONSE** On a scale drawing, a dog pen is $4\frac{1}{2}$ inches long. If 1 inch = 3 feet, what is the actual length of the dog pen? (Lesson 7-4)

The Game Zone
A Place To Practice Your Math Skills

Racing with Proportions

- **GET READY!**

    Players: two
    Materials: centimeter grid paper, spinner, centimeter cubes

- **GET SET!**
    - Copy the game board at the right onto grid paper.

    - Label equal sections of a spinner with 10, 12, 15, 20, 24, 30, 40, and 60.

- **GO!**
    - Each player places a cube on the *Start* square. (Note that both cubes will not fit on the same square. Place them side by side.)
    - Each player spins the spinner and substitutes the number on the spinner for x in the proportion $\frac{y}{15} = \frac{8}{x}$. The player solves the proportion and moves his or her centimeter cube y spaces.
    - Then each player spins the spinner and substitutes the number on the spinner for y in the given proportion. He or she solves the proportion and moves his or her centimeter cube x spaces.
    - Continue to spin the spinner and solve the proportion for y and x.
    - **Who Wins?** The first person to reach the *End* square wins the round. It is not necessary to land on *End* with an exact roll.

Math Skill
Solving Proportions

The Game Zone: Solving Proportions **311**

7-5 Fractions, Decimals, and Percents

What You'll LEARN

Write percents as fractions, and vice versa.

REVIEW Vocabulary

percent: a ratio that compares a number to 100 (Lesson 5-5)

WHEN am I ever going to use this?

SURVEYS The graph shows the results of a survey in which teens were asked to name the most important invention of the 20th century.

1. What percent of the teens said that the personal computer was the most important invention?
2. How is this percent written as a ratio?
3. Simplify the ratio.

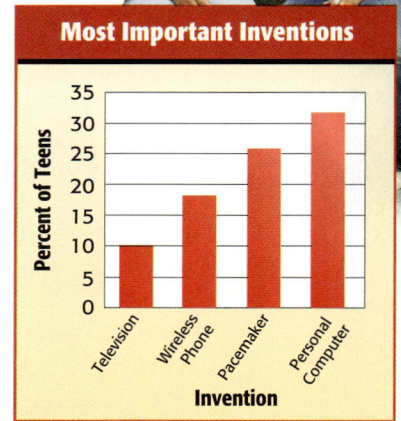

Source: Lemelson-MIT Program

In Lesson 5-5, you wrote percents like 32% as fractions by writing fractions with denominators of 100 and then simplifying. You can use the same method to write percents like 16.8% and $8\frac{1}{3}$% as fractions.

EXAMPLES Percents as Fractions

1 **HOCKEY** In men's college hockey, 16.8% of the players are from Ontario, Canada. What fraction is this? Write in simplest form.

$16.8\% = \frac{16.8}{100}$ Write a fraction with a denominator of 100.

$= \frac{16.8}{100} \cdot \frac{10}{10}$ Multiply by $\frac{10}{10}$ to eliminate the decimal in the numerator.

$= \frac{168}{1,000}$ or $\frac{21}{125}$ Simplify.

So, $\frac{21}{125}$ of the players are from Ontario, Canada.

2 Write $8\frac{1}{3}$% as a fraction in simplest form.

$8\frac{1}{3}\% = \frac{8\frac{1}{3}}{100}$ Write a fraction.

$= 8\frac{1}{3} \div 100$ Divide.

$= \frac{25}{3} \div 100$ Write $8\frac{1}{3}$ as an improper fraction.

$= \frac{25}{3} \times \frac{1}{100}$ Multiply by the reciprocal of 100, which is $\frac{1}{100}$.

$= \frac{25}{300}$ or $\frac{1}{12}$ Simplify.

312 Chapter 7 Ratios and Proportions

To write a fraction like $\frac{8}{25}$ as a percent, multiply the numerator and the denominator by a number so that the denominator is 100. If the denominator is not a factor of 100, you can write fractions as percents by using a proportion.

EXAMPLES Fractions as Percents

3 **TESTS** On a math test, Mary got 7 questions correct out of 8. Find her grade as a percent.

To find her grade, write $\frac{7}{8}$ as a percent.   **Estimate** $\frac{7}{8}$ is greater than $\frac{6}{8} = \frac{3}{4}$, or greater than 75%.

$\frac{7}{8} = \frac{n}{100}$ Write a proportion using $\frac{n}{100}$.

$700 = 8n$ Find the cross products.

$\frac{700}{8} = \frac{8n}{8}$ Divide each side by 8.

$87\frac{1}{2} = n$ Simplify.

So, $\frac{7}{8} = 87\frac{1}{2}\%$ or 87.5%.

This is greater than 75%, which was the estimate.

4 Write $\frac{4}{15}$ as a percent. Round to the nearest hundredth.

Estimate $\frac{4}{15}$ is about $\frac{4}{16}$, which equals $\frac{1}{4}$ or 25%.

$\frac{4}{15} = \frac{n}{100}$ Write a proportion using $\frac{n}{100}$.

$400 = 15n$ Find the cross products.

400 [÷] 15 [ENTER] 26.66666667   Use a calculator to simplify.

So, $\frac{4}{15}$ is about 26.67%. This result is close to the estimate.

Your Turn Write each fraction as a percent. Round to the nearest hundredth if necessary.

a. $\frac{2}{15}$ b. $\frac{7}{16}$ c. $\frac{17}{24}$

STUDY TIP

Choose the Method
To write a fraction as a percent,
- use *multiplication* when a fraction has a denominator that is a factor of 100,
- use a *proportion* for any type of fraction.

In this lesson, you have written percents as fractions and fractions as percents. In Chapter 5, you wrote percents and fractions as decimals. You can also write a fraction as a percent by first writing the fraction as a decimal and then writing the decimal as a percent.

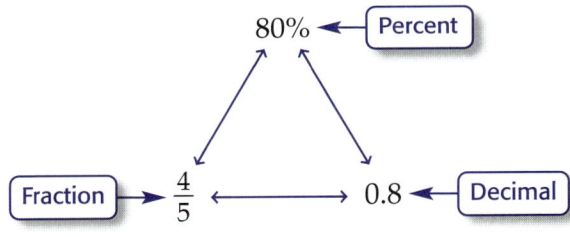

Percents, fractions, and decimals are all different names that represent the same number.

Lesson 7-5 Fractions, Decimals, and Percents

Study Tip

Look Back You can review writing fractions as decimals in Lesson 5-4.

EXAMPLES — Fractions as Percents

5 Write $\frac{1}{8}$ as a percent.

$\frac{1}{8} = 0.125$ Write $\frac{1}{8}$ as a decimal.

$= 12.5\%$ Multiply by 100 and add the %.

6 Write $\frac{5}{6}$ as a percent. Round to the nearest hundredth.

$\frac{5}{6} = 0.833333333\ldots$ Write $\frac{5}{6}$ as a decimal.

$\approx 83.33\%$ Multiply by 100 and add the %.

Your Turn Write each fraction as a percent. Round to the nearest hundredth if necessary.

d. $\frac{5}{16}$ e. $\frac{7}{12}$ f. $\frac{2}{9}$

Some fractions whose denominators are not factors of 100 are used often in everyday situations. It is helpful to be familiar with these fractions and their equivalent decimals and percents.

Noteables — Key Concept: Common Equivalents

Fraction	Decimal	Percent	Fraction	Decimal	Percent
$\frac{1}{3}$	$0.\overline{3}$	$33\frac{1}{3}\%$	$\frac{3}{8}$	0.375	$37\frac{1}{2}\%$
$\frac{2}{3}$	$0.\overline{6}$	$66\frac{2}{3}\%$	$\frac{5}{8}$	0.625	$62\frac{1}{2}\%$
$\frac{1}{8}$	0.125	$12\frac{1}{2}\%$	$\frac{7}{8}$	0.875	$87\frac{1}{2}\%$

Skill and Concept Check

1. **Writing Math** Describe two methods for writing a fraction as a percent.

2. **OPEN ENDED** Write a percent that is between 15.2% and 15.8%. Then write it as a fraction in simplest form.

3. **Explain** why percents are used in statistical graphs rather than fractions.

GUIDED PRACTICE

Write each percent as a fraction in simplest form.

4. 13.5% 5. 18.75% 6. $66\frac{2}{3}\%$ 7. $57\frac{1}{2}\%$

Write each fraction as a percent. Round to the nearest hundredth if necessary.

8. $\frac{4}{25}$ 9. $\frac{5}{8}$ 10. $\frac{9}{40}$ 11. $\frac{4}{11}$

12. **LANGUAGES** In Virginia, 1 person out of every 40 speaks an Asian language at home. What percent is this?

Practice and Applications

Write each percent as a fraction in simplest form.

13. 17.5% 14. 62.5% 15. 6.2% 16. 34.5%

17. 28.75% 18. 56.25% 19. $33\frac{1}{3}$% 20. $16\frac{2}{3}$%

21. $81\frac{1}{4}$% 22. $93\frac{3}{4}$% 23. $78\frac{3}{4}$% 24. $23\frac{1}{3}$%

HOMEWORK HELP

For Exercises	See Examples
13–24, 38	1, 2
25–37, 39	3–6

Extra Practice
See pages 581, 602.

Write each fraction as a percent. Round to the nearest hundredth if necessary.

25. $\frac{11}{20}$ 26. $\frac{18}{25}$ 27. $\frac{3}{8}$ 28. $\frac{21}{40}$ 29. $\frac{29}{30}$ 30. $\frac{8}{9}$

31. $\frac{5}{7}$ 32. $\frac{1}{16}$ 33. $\frac{1}{80}$ 34. $\frac{57}{200}$ 35. $\frac{5}{12}$ 36. $\frac{7}{15}$

37. What percent is equivalent to $\frac{60}{125}$?

CARS For Exercises 38 and 39, use the table at the right. It shows the percent of people who keep the listed items in their car.

38. What fraction of people keep a first-aid kit in their car?

39. Approximately 26 out of 125 people surveyed keep a hairbrush in their car. Is this greater or less than the percent who keep sports equipment? Explain.

40. **CRITICAL THINKING** For what value of x is $\frac{1}{x} = x$% a true sentence?

Items in Car	Percent of People
pen/pencil	73.0%
cassette tapes/CDs	66.1%
first-aid kit	38.2%
sports equipment	28.9%

Source: Penzoil "How's your Automotive Attitude?" survey

Spiral Review with Standardized Test Practice

41. **MULTIPLE CHOICE** Two-thirds of Kodiak Island's 2 million acres has been set aside as a wildlife preserve for the Kodiak brown bear. What percent of the island is a preserve?

    Ⓐ $2\frac{2}{3}$% Ⓑ 23% Ⓒ $33\frac{1}{3}$% Ⓓ $66\frac{2}{3}$%

42. **SHORT RESPONSE** What is $\frac{7}{16}$ written as a percent?

43. **MODELS** On a scale model of a building, 2 inches represents 15 feet. If the model is 9 inches tall, how tall is the actual building? (Lesson 7-4)

Solve each proportion. (Lesson 7-3)

44. $\frac{6}{9} = \frac{4}{x}$ 45. $\frac{3}{t} = \frac{9}{5}$ 46. $\frac{n}{5} = \frac{2.8}{7}$ 47. $\frac{7}{12} = \frac{r}{30}$

GETTING READY FOR THE NEXT LESSON

PREREQUISITE SKILL Write each decimal as a fraction with a denominator of 100. (Lesson 5-4)

48. 0.23 49. 0.09 50. 0.368 51. 0.425

7-6 Percents Greater Than 100% and Percents Less Than 1%

What You'll LEARN

Write percents greater than 100% and percents less than 1% as fractions and as decimals, and vice versa.

HANDS-ON Mini Lab

Materials
- grid paper
- colored pencils

Work with a partner.

- Draw three 10 × 10 squares on a piece of grid paper. Each large square represents 100%, and each small square represents 1%.
- For the first model, use two grids to shade 120 small squares.
- For the second model, shade half of one small square on a grid. Use a different color than on the first model.

1. Which model represents a percent greater than 100%? What is the percent?
2. Which model represents a percent less than 1%? What is the percent?
3. Shade grids to represent each percent.
   a. 150%   b. 215%   c. $\frac{1}{4}$%

Percents greater than 100% or less than 1% can be written as decimals using the methods you learned in Chapter 5. They can also be written as mixed numbers or fractions.

$$250\% = \frac{250}{100}$$
$$= 2.5 \text{ or } 2\frac{1}{2}$$

$$0.5\% = \frac{0.5}{100}$$
$$= 0.005 \text{ or } \frac{1}{200}$$

A percent greater than 100% equals a number greater than 1.

A percent less than 1% equals a number less than 0.01 or $\frac{1}{100}$.

STUDY TIP

Alternate Method
To divide by 100, move the decimal point two places to the left.

410% = 410 or 4.1

0.2% = 00.2 or 0.002

EXAMPLES Percents as Decimals or Fractions

Write each percent as a decimal and as a mixed number or fraction in simplest form.

1) 410%

$410\% = \frac{410}{100}$ Definition of percent

$= 4.1 \text{ or } 4\frac{1}{10}$

2) 0.2%

$0.2\% = \frac{0.2}{100}$ Definition of percent

$= 0.002 \text{ or } \frac{1}{500}$

316 Chapter 7 Ratios and Proportions

msmath2.net/extra_examples

EXAMPLE — Real-Life Percents as Decimals

3 **BEARS** Before bears hibernate, their need for Calories increases 300%. Write 300% as a decimal.

300% = 300   Divide by 100.
     = 3.0 or 3

READING Math

Percents In the example at the right, the solution 3 means *3 times as much*. So, before hibernation bears need 3 times as many Calories as usual.

You can write decimals less than 0.01 and decimals greater than 1 as percents by multiplying by 100 and adding a %.

EXAMPLES — Decimals as Percents

Write each decimal as a percent.

4 1.68

1.68 = 1.68   Multiply by 100.
     = 168%

5 0.0075

0.0075 = 0.0075   Multiply by 100.
       = 0.75%

Your Turn Write each decimal as a percent.

a. 4.5   b. 0.002   c. 0.0016

Skill and Concept Check

1. **Write** the fraction of squares that are shaded as a percent and as a decimal.

2. **OPEN ENDED** Write a number between 6 and 7 as a percent.

3. **Which One Doesn't Belong?** Identify the number that does not have the same value as the other three. Explain your reasoning.

| 63.5% | 0.635 | $63\frac{1}{2}$ | $63\frac{1}{2}\%$ |

GUIDED PRACTICE

Write each percent as a decimal and as a mixed number or fraction in simplest form.

4. 325%   5. 200%   6. 0.7%   7. 0.15%

Write each decimal as a percent.

8. 1.8   9. 0.0015   10. 2.75   11. 0.0005

12. **PHOTOGRAPHY** A photograph developed from 35 mm film is 428% larger than the negative. Express this percent as a decimal.

Lesson 7-6 Percents Greater Than 100% and Percents Less Than 1%

Practice and Applications

Write each percent as a decimal and as a mixed number or fraction in simplest form.

13. 350%
14. 475%
15. 600%
16. 400%
17. 0.6%
18. 0.1%
19. 0.55%
20. 0.48%
21. 260%
22. 195%
23. 0.05%
24. 0.04%

HOMEWORK HELP

For Exercises	See Examples
13–24, 33–37	1, 2
39–40	3
25–32, 38	4, 5

Extra Practice
See pages 581, 602.

Write each decimal as a percent.

25. 8.5
26. 35
27. 0.009
28. 0.003
29. 2.64
30. 1.07
31. 0.0034
32. 0.0077

Write each number as a percent.

33. $3\frac{1}{2}$
34. $9\frac{3}{4}$
35. $\frac{2}{500}$
36. $\frac{1}{400}$

37. Write $1\frac{5}{8}$ as a percent.
38. Write $\frac{1}{8}\%$ as a decimal.

39. **POPULATION** In 2000, about 0.3% of the population in the United States was Japanese. Write this percent as a decimal and as a fraction.

40. **RESEARCH** Use the Internet or another source to find the percent of people in the United States that come from three other countries. Write the percents as decimals.

41. **MULTI STEP** One hour is what percent of one week?

42. **CRITICAL THINKING** Some North African ostriches are 150% of the height of a 6-foot human. What is the height of some ostriches?

Spiral Review with Standardized Test Practice

43. **MULTIPLE CHOICE** A certain stock increased its value by 467% over 10 years. What is this percent written as a decimal?

    Ⓐ 4.67   Ⓑ 46.7   Ⓒ 0.467   Ⓓ 0.00467

44. **MULTIPLE CHOICE** Choose the best estimate for 100.5%.

    Ⓕ 0   Ⓖ 1   Ⓗ 100   Ⓘ 1,000

Write each percent as a fraction in simplest form. (Lesson 7-5)

45. 7.5%
46. 1.2%
47. $6\frac{1}{4}\%$
48. $92\frac{1}{2}\%$

49. **SCALE DRAWING** A garage is 18 feet wide. In a scale drawing, 1 inch = 3 feet. What is the width of the garage on the drawing? (Lesson 7-4)

GETTING READY FOR THE NEXT LESSON

PREREQUISITE SKILL Write each percent as a decimal. (Lesson 5-6)

50. 85%
51. 6.5%
52. 36.9%
53. 12.3%

msmath2.net/self_check_quiz

7-7 Percent of a Number

What You'll LEARN
Find the percent of a number.

WHEN am I ever going to use this?

SURVEYS The graph at the right shows the results of a survey in which 1,016 people were asked to name the things they feared.

1. About how many of them said they were afraid of snakes?
2. How did you estimate?
3. Suppose you pass 50 people at a mall. Based on the results of the survey, do you think more or less than 25 of them are afraid of mice? Explain.

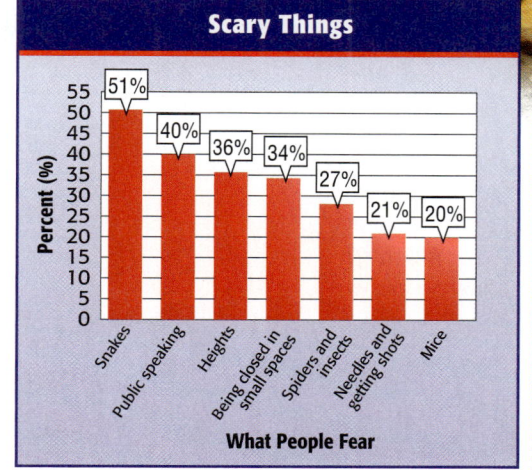

Scary Things

Source: Gallup Poll

One way to find the percent of a number is to use a proportion.

EXAMPLE Use a Proportion to Find a Percent

1 SURVEYS The graphic above shows that 36% of 1,016 adults surveyed were afraid of heights. How many of the people surveyed were afraid of heights?

36% means that 36 out of 100 people were afraid of heights. Find an equivalent ratio x out of 1,016 and write a proportion.

number afraid of heights → $\dfrac{x}{1{,}016} = \dfrac{36}{100}$ } percent of people afraid of heights

Now solve the proportion.

$\dfrac{x}{1{,}016} = \dfrac{36}{100}$ Write the proportion.

$x \cdot 100 = 1{,}016(36)$ Find the cross products.

$100x = 36{,}576$ Multiply.

$\dfrac{100x}{100} = \dfrac{36{,}576}{100}$ Divide each side by 100.

$x = 365.76$ Simplify.

So, about 366 of the 1,016 people surveyed were afraid of heights.

Another method for finding the percent of a number is to use multiplication.

EXAMPLES — Use Multiplication to Find a Percent

2 What number is 140% of 32?

140% of 32 = 140% × 32	Write a multiplication expression.
= 1.4 × 32	Write 140% as a decimal.
= 44.8	Multiply.

So, 140% of 32 is 44.8.

3 Find 75% of $800.

75% of $800 = 75% × $800	Write a multiplication expression.
= 0.75 × 800	Write 75% as a decimal.
= 600	Multiply.

So, 75% of $800 is $600.

> **STUDY TIP**
>
> **Mental Math** You can find 75% of 800 mentally by using the fraction $\frac{3}{4}$.
> $\frac{3}{4}$ of 800 is 600.

Your Turn Find each number.

a. Fifty percent of 140 is what number?

b. Find 37.5% of 64.

Skill and Concept Check

1. **Write** a proportion that can be used to find 0.5% of 22.

2. **NUMBER SENSE** Use mental math to determine whether 30% of $150 is $80. Explain.

3. **OPEN ENDED** Write a problem that can be solved using the graphic at the right. Then solve the problem.

GUIDED PRACTICE

Find each number. Round to the nearest tenth if necessary.

4. Find 95% of $40.

5. Forty-two percent of 263 is what number?

6. What number is 200% of 75?

7. Find 0.25% of 40.

8. **COMMISSION** Ms. Sierra earns a 2% *commission*, or fee paid based on a percent of her sales, on every vacation package that she sells. One day, she sold vacation packages worth $6,500. What was her commission?

USA TODAY Snapshots®

Low-tech tools rule at meetings
1,500 white-collar workers were asked what tools they most commonly used to record information during a meeting.

- Pen/pencil/paper — 93%
- White erasable markerboard — 35%
- PC or laptop — 27%

Source: Steelcase Workplace Index Surveys

By Darryl Haralson and Keith Simmons, USA TODAY

320 Chapter 7 Ratios and Proportions

Practice and Applications

Find each number. Round to the nearest tenth if necessary.

9. What is 65% of 100?
10. Find 45% of $400.
11. Find 25% of $640.
12. What number is 18.5% of 500?
13. 67.5% of 76 is what?
14. 64% of 88 is what number?
15. What is 125% of 10?
16. Find 300% of 14.
17. Find 0.15% of 250
18. What number is 0.5% of 80?
19. Find $20\frac{1}{4}$% of 3.
20. 250.5% of 25 is what number?

SNACKING For Exercises 21 and 22, use the graphic at the right. It shows the results of a poll of 1,746 college students.

21. Write a proportion that can be used to find how many students like to snack at their computer. Then solve.

22. How many students surveyed like to snack at home?

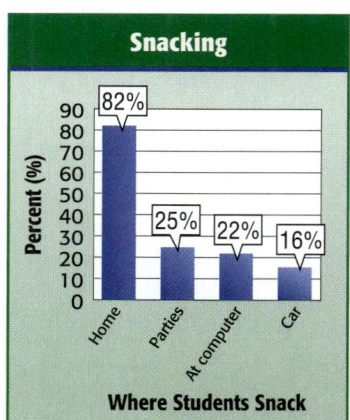

Source: Snack Food Association

23. **DELIVERY** A mountain bike costs $288 plus a 4.5% delivery charge. What is the cost of the bike including the delivery charge?

24. **CRITICAL THINKING** Suppose you add 10% of a number to the number. Then you subtract 10% of the total. Is the result *greater than*, *less than*, or *equal to* the original number? Explain your reasoning.

Spiral Review with Standardized Test Practice

25. **MULTIPLE CHOICE** A football player has made 80% of the field goals he has attempted in his career. If he attempts 5 field goals in a game, how many would he be expected to make?

    Ⓐ 5   Ⓑ 4   Ⓒ 3   Ⓓ 2

26. **SHORT RESPONSE** What number is 12% of 16.5?

Write each decimal as a percent. (Lesson 7-6)

27. 7.5
28. 9
29. 0.0004
30. 0.0018

31. Write $\frac{11}{15}$ as a percent. Round to the nearest hundredth. (Lesson 7-5)

GETTING READY FOR THE NEXT LESSON

PREREQUISITE SKILL Evaluate each expression. Round to the nearest tenth if necessary. (Lesson 1-3)

32. $\frac{36 \cdot 2}{80}$
33. $\frac{14(5)}{3.5}$
34. $\frac{9 \times 100}{56}$
35. $22 \cdot 100 \div 490$

msmath2.net/self_check_quiz

Lesson 7-7 Percent of a Number **321**

7-8a HANDS-ON LAB
A Preview of Lesson 7-8

Using a Percent Model

You can find the percent of a number or the *part* of the whole by using a model.

What You'll LEARN
Use a percent model to find a part.

Materials
- grid paper

ACTIVITY Work with a partner.

Suppose you and a friend eat dinner at the Pizza Palace and your total bill is about $25. If you want to leave a 20% tip, how much should you leave? Use a model to find the amount.

- Draw a 10-unit by 1-unit rectangle on grid paper.
- Label equal units from 0 to 100 to represent percents.
- Label equal units from 0 to 25 to represent parts of the whole quantity, 25.

Draw a vertical line at 20%.

So, you should leave $5 as a tip.

Your Turn Draw a model to find each part.

a. 70% of 50   b. 45% of 20   c. 20% of 75

Writing Math

1. Suppose your whole family had dinner with you and the total bill was $50. How could you change your model to find the tip?

2. **Write** a sentence that describes what is represented by the model at the right.

3. **Write** a percent problem that can be represented and solved using a model.

4. **Make a conjecture** about how you could use a model to estimate the percent represented by the ratio 8 out of 50.

322 Chapter 7 Ratios and Proportions

7-8 The Percent Proportion

What You'll LEARN

Solve problems using the percent proportion.

NEW Vocabulary

percent proportion
part
base

WHEN am I ever going to use this?

SPACE The engine on the space shuttle weighs approximately 19,700 pounds. The entire space shuttle weighs 178,000 pounds.

1. Write the ratio of engine weight to total weight as a fraction.
2. Use a calculator to write the fraction as a decimal to the nearest hundredth.
3. What percent of the space shuttle's weight is the engine?

A **percent proportion** compares **part** of a quantity to the whole quantity, called the **base**, using a percent. You can use a percent proportion to find what percent of a space shuttle's weight is the engine.

part: engine weight → $\dfrac{19{,}700}{178{,}000} = \dfrac{p}{100}$ ← The percent is written as a fraction with a denominator of 100.
base: total weight →

Noteables™   Key Concept: Percent Proportion

Words   The percent proportion is $\dfrac{\text{part}}{\text{base}} = \dfrac{\text{percent}}{100}$.

Symbols $\dfrac{a}{b} = \dfrac{p}{100}$, where a is the part, b is the base, and p is the percent.

To find the percent, write the percent proportion and solve for p.

EXAMPLE   Find the Percent

1 What percent of $15 is $9?

9 is the part, and 15 is the base. You need to find the percent p.

$\dfrac{a}{b} = \dfrac{p}{100}$ Percent proportion

$\dfrac{9}{15} = \dfrac{p}{100}$ Replace a with 9 and b with 15.

$9 \cdot 100 = 15 \cdot p$ Find the cross products.

$900 = 15p$ Simplify.

$\dfrac{900}{15} = \dfrac{15p}{15}$ Divide each side by 15.

$60 = p$ So, 60% of $15 is $9. Compare the answer to the model.

msmath2.net/extra_examples   Lesson 7-8 The Percent Proportion   **323**

Study Tip

Finding the Base In a percent problem, the base follows the word *of*.

You can also use the percent proportion to find a missing part or base.

Concept Summary **Types of Percent Problems**

Type	Example	Proportion
Find the Percent	What percent of 6 is 3?	$\frac{3}{6} = \frac{p}{100}$
Find the Part	What number is 50% of 6?	$\frac{a}{6} = \frac{50}{100}$
Find the Base	3 is 50% of what number?	$\frac{3}{b} = \frac{50}{100}$

To find the part, write the percent proportion and solve for *a*. To find the base, write the percent proportion and solve for *b*.

EXAMPLE Find the Part

2 What number is 40% of 120?

40 is the percent and 120 is the base. You need to find the part.

$\frac{a}{b} = \frac{p}{100}$ Percent proportion

$\frac{a}{120} = \frac{40}{100}$ Replace *b* with 120 and *p* with 40.

$a \cdot 100 = 120 \cdot 40$ Find the cross products.

$100a = 4{,}800$ Simplify.

$\frac{100a}{100} = \frac{4{,}800}{100}$ Divide each side by 100.

$a = 48$ So, 40% of 120 is 48. **Compare to the model.**

EXAMPLE Find the Base

3 18 is 25% of what number?

18 is the part and 25 is the percent. You need to find the base, or the whole quantity.

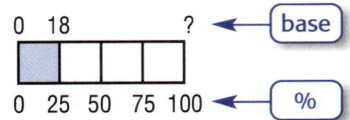

$\frac{a}{b} = \frac{p}{100}$ Percent proportion

$\frac{18}{b} = \frac{25}{100}$ Replace *a* with 18 and *p* with 25.

$18 \cdot 100 = b \cdot 25$ Find the cross products.

$1{,}800 = 25b$ Simplify.

$\frac{1{,}800}{25} = \frac{25b}{25}$ Divide each side by 25.

$72 = b$ So, 18 is 25% of 72.

Check In the model above, the whole quantity is $18 \cdot 4$ or 72.

Your Turn Find each number. Round to the nearest tenth if necessary.

a. What number is 5% of 60?   b. 80 is 75% of what number?

Skill and Concept Check

1. **Choose** the problem that can be solved by using $\frac{8}{b} = \frac{16}{100}$.
   a. What number is 16% of 8?   b. 8 is 16% of what number?

2. **OPEN ENDED** Write a proportion that can be used to find the part if you know the base and the percent.

GUIDED PRACTICE

Find each number. Round to the nearest tenth if necessary.

3. What percent of $50 is $18?
4. What number is 25% of 180?
5. What number is 2% of 35?
6. 62 is 90.5% of what number?

Practice and Applications

Find each number. Round to the nearest tenth if necessary.

7. What percent of 60 is 15?
8. 9 is 45% of what number?
9. What number is 12% of 72?
10. What percent of $12 is $9?
11. 75 is 20% of what number?
12. What number is 40% of 80?
13. $3 is what percent of $40?
14. 12.5% of what number is 24?
15. 8.2% of 50 is what number?
16. What percent of 300 is 0.6?
17. 45 is 150% of what number?
18. What percent of 25 is 30?
19. What number is 0.5% of 8?
20. 6% of what number is $10\frac{1}{2}$?

HOMEWORK HELP
For Exercises	See Examples
7–22	1–3

Extra Practice
See pages 582, 602.

21. **SHOPPING** A paperback book originally priced at $12.50 is on sale for $7.50. What percent of the original cost is the sale price?

22. **MOVIES** Sarah and Monique spent $14, or 35% of their money, on movie tickets. How much money did they have to start with?

23. **CRITICAL THINKING** Without calculating, order the following from greatest to least value. Explain your reasoning.
    20% of 100, 20% of 500, 5% of 100

Spiral Review with Standardized Test Practice

24. **MULTIPLE CHOICE** There are about 53,500 hotels in the United States. Use the table to find how many are located along highways.
    Ⓐ 22,577   Ⓑ 22,470   Ⓒ 17,976   Ⓓ 5,457

U.S. Hotels	
Location	Percent
highways	42.2%
suburbs	33.6%
city	10.2%

25. **SHORT RESPONSE** 95 of 273 students volunteered. About what percent of the students did *not* volunteer?

Find each number. Round to the nearest tenth if necessary. (Lesson 7-7)

26. What is 25% of 120?
27. Find 45% of 70.

28. Write 1.2 as a percent. (Lesson 7-6)

msmath2.net/self_check_quiz

Chapter 7 Study Guide and Review

Vocabulary and Concept Check

base (p. 323)
cross product (p. 297)
equivalent ratios (p. 289)
part (p. 323)
percent proportion (p. 323)
proportion (p. 297)
rate (p. 292)
scale (p. 304)
scale drawing (p. 304)
scale factor (p. 305)
scale model (p. 306)
unit rate (p. 292)

Choose the letter of the term that best matches each phrase.

1. a comparison of two numbers by division
2. two ratios that have the same value
3. a ratio of two measurements with different units
4. an equation that shows that two ratios are equivalent
5. used to represent something that is too large or too small for an actual-size drawing
6. the ratio of the distance on a map to the actual distance
7. the number that is compared to the whole quantity in a percent proportion
8. product of the numerator of one ratio in a proportion and the denominator of the other ratio
9. a rate that is simplified so that it has a denominator of 1
10. the number that represents the whole quantity in a percent proportion

a. rate
b. base
c. scale drawing
d. unit rate
e. ratio
f. scale
g. part
h. proportion
i. equivalent ratios
j. cross products

Lesson-by-Lesson Exercises and Examples

7-1 Ratios (pp. 288–291)

Write each ratio as a fraction in simplest form.

11. 16:12
12. 5 to 25
13. 50 cm to 75 cm
14. 6 ft:12 in.

Determine whether the ratios are equivalent. Explain.

15. $\frac{3}{5}$ and $\frac{21}{35}$
16. $\frac{18}{24}$ and $\frac{5}{20}$
17. 27:15 and 9:5
18. 4:21 and 2:7

Example 1 Write the ratio 32 to 18 as a fraction in simplest form.

32 to 18 = $\frac{32}{18}$ Write the ratio as a fraction.

= $\frac{16}{9}$ Simplify.

Example 2 Determine whether 5:6 and 15:18 are equivalent.

5:6 = $\frac{5}{6}$ 15:18 = $\frac{15}{18}$ or $\frac{5}{6}$

The ratios in simplest form both equal $\frac{5}{6}$. So, 5:6 and 15:18 are equivalent.

326 Chapter 7 Ratios and Proportions

msmath2.net/vocabulary_review

7-2 Rates (pp. 292–295)

Find each unit rate. Round to the nearest hundredth if necessary.

19. $23.75 for 5 pounds
20. 810 miles in 9 days
21. $38 in 4 hours
22. 24 gerbils in 3 cages
23. 14 laps in 4 minutes
24. **SHOPPING** Which bottle of laundry detergent shown at the right costs the least per ounce?

Bottle	Price
12 oz	$2.16
32 oz	$5.60
64 oz	$10.24

Example 3 Find the unit price per ounce of a 16-ounce box of pasta that is on sale for 96¢.

To find the cost per ounce, write the rate as a fraction and then simplify.

$$16\text{-oz box for } 96¢ = \frac{96¢}{16 \text{ ounces}}$$
$$= \frac{96¢ \div 16}{16 \text{ ounces} \div 16}$$
$$= \frac{6¢}{1 \text{ ounce}}$$

The unit price is 6¢ per ounce.

7-3 Algebra: Solving Proportions (pp. 297–300)

Solve each proportion.

25. $\frac{x}{10} = \frac{3}{5}$
26. $\frac{4}{9} = \frac{24}{m}$
27. $\frac{2}{t} = \frac{8}{50}$
28. $\frac{15}{w} = \frac{35}{21}$
29. $\frac{12}{8} = \frac{a}{6}$
30. $\frac{7}{18} = \frac{d}{6}$

31. **MODELS** A Boeing 747 jet is 70.5 meters long and has a wingspan of 60 meters. A model of the 747 has a wingspan of 80 centimeters. What is its length?

Example 4 Solve $\frac{6}{9} = \frac{n}{12}$.

$\frac{6}{9} = \frac{n}{12}$ Write the proportion.

$6 \times 12 = 9 \times n$ Find the cross products.

$72 = 9n$ Multiply.

$\frac{72}{9} = \frac{9n}{9}$ Divide each side by 9.

$8 = n$ Simplify.

The solution is 8.

7-4 Geometry: Scale Drawings (pp. 304–308)

Suppose you are making a scale drawing. Find the length of each object on the drawing with the given scale.

32. a minivan 10 feet long; 2 inches = 1 foot
33. a rose garden 9 meters wide; 3 centimeters = 1 meter
34. a Ferris wheel with a diameter of 42 feet; 2 inches = 3 feet
35. a library 10.4 meters wide; 1 centimeter = 2 meters

Example 5 On a map, the distance between Chicago, Illinois, and Mexico City, Mexico, is 10.9 centimeters. If the scale is 1 centimeter = 250 kilometers, what is the actual distance?

             Chicago to
Scale   Mexico City

$\frac{1 \text{ cm}}{250 \text{ km}} = \frac{10.9 \text{ cm}}{n \text{ km}}$ Write a proportion.

$1 \times n = 250 \times 10.9$ Find the cross products.

$n = 2{,}725$ Simplify.

The actual distance is 2,725 kilometers.

Study Guide and Review continued

Mixed Problem Solving
For mixed problem-solving practice, see page 602.

7-5 Fractions, Decimals, and Percents (pp. 312–315)

Write each percent as a fraction in simplest form.

36. 27.5%   37. 5.4%   38. $45\frac{1}{4}$%

Write each fraction as a percent. Round to the nearest hundredth if necessary.

39. $\frac{1}{8}$ 40. $\frac{5}{6}$ 41. $\frac{7}{40}$

Example 6 Write 82.5% as a fraction in simplest form.

$82.5\% = \frac{82.5}{100}$ Write a fraction with a denominator of 100.

$= \frac{825}{1,000}$ Multiply 82.5 and 100 by 10 to eliminate the decimal.

$= \frac{33}{40}$ Simplify.

7-6 Percents Greater Than 100% and Percents Less Than 1% (pp. 316–318)

Write each percent as a decimal and as a mixed number or fraction in simplest form.

42. 125%   43. 0.75%
44. 563%   45. 0.5%

Write each decimal as a percent.

46. 0.002   47. 4.75
48. 7.5   49. 0.0095

Example 7 Write 235% as a decimal and as a mixed number in simplest form.

$235\% = 2\underset{\smile}{35}$ Definition of percent

$= 2.35$ or $2\frac{7}{20}$

Example 8 Write 0.008 as a percent.

$0.008 = 0.00\underset{\smile}{8}$ Multiply by 100.

$= 0.8\%$

7-7 Percent of a Number (pp. 319–321)

Find each number. Round to the nearest tenth if necessary.

50. Find 78% of 50.
51. 45.5% of 75 is what number?
52. What is 225% of 60?
53. 0.75% of 80 is what number?

Example 9 Find 24% of 200.

24% of 200

$= 24\% \times 200$ Write a multiplication expression.

$= 0.24 \times 200$ Write 24% as a decimal.

$= 48$ Multiply.

So, 24% of 200 is 48.

7-8 The Percent Proportion (pp. 323–325)

Find each number. Round to the nearest tenth if necessary.

54. 6 is what percent of 120?
55. Find 0.8% of 35.
56. What percent of 375 is 40?
57. Find 310% of 42.

Example 10 What percent of 90 is 18?

$\frac{a}{b} = \frac{p}{100}$ Percent proportion

$\frac{18}{90} = \frac{p}{100}$ Replace a with 18 and b with 90.

$18 \cdot 100 = 90 \cdot p$ Find the cross products.

$20 = p$ So, 18 is 20% of 90.

328 Chapter 7 Ratios and Proportions

CHAPTER 7 Practice Test

Vocabulary and Concepts

1. **Define** *equivalent ratios*.
2. **Write** a proportion that can be used to find 76% of 512.

Skills and Applications

Write each ratio as a fraction in simplest form.

3. 45 to 18
4. 24 : 88
5. 15 minutes to 2 hours

Find each unit rate. Round to the nearest hundredth if necessary.

6. 24 cards for $4.80
7. 330 miles on 15 gallons of gasoline

Solve each proportion.

8. $\dfrac{2}{3} = \dfrac{x}{42}$
9. $\dfrac{t}{21} = \dfrac{15}{14}$
10. $\dfrac{9}{m} = \dfrac{12}{36}$

11. **BLUEPRINTS** The dimensions of a rectangular room in a new home are shown in the blueprint at the right. If the scale is 1 centimeter = 2.5 meters, find the length ℓ of the room on the scale drawing. Then find the scale factor.

12. **SWIMMING** Alyssa swims 3 laps in 12 minutes. At this same rate, how many laps will she swim in 30 minutes?

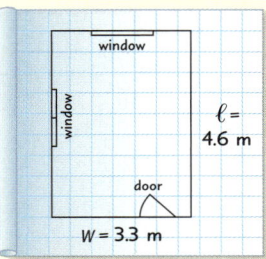

Write each percent as a fraction in simplest form.

13. $56\dfrac{1}{4}\%$
14. $82\dfrac{1}{4}\%$
15. $83\dfrac{1}{3}\%$

Write each fraction as a percent. Round to the nearest hundredth if necessary.

16. $\dfrac{5}{8}$
17. $\dfrac{7}{15}$
18. $\dfrac{33}{40}$

19. Write 0.45% as a decimal and as a fraction in simplest form.
20. Write 8.25 as a percent.

Find each number. Round to the nearest tenth if necessary.

21. What is 18% of 30?
22. What number is 162.2% of 50?
23. What percent of 64 is 12.8?
24. 458 is 105% of what number?

Standardized Test Practice

25. **MULTIPLE CHOICE** The purchase price of a bicycle is $140. The state tax rate is $6\dfrac{1}{2}\%$ of the purchase price. What is the tax rate expressed as a decimal?

    Ⓐ 6.5   Ⓑ 0.65   Ⓒ 0.065   Ⓓ 0.0065

CHAPTER 7 Standardized Test Practice

PART 1 Multiple Choice

Record your answers on the answer sheet provided by your teacher or on a sheet of paper.

1. Andre charges $4 per day for taking care of the neighbor's dog. Which expression represents the amount of money he earns in n days? (Lesson 1-4)

   A) $4n$
   B) $4 + n$
   C) $4 - n$
   D) n^4

2. The box-and-whisker plot shows the prices of single-scoop ice cream cones. What is the median price? (Lesson 2-6)

   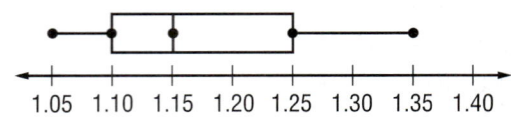

   F) $1.05
   G) $1.15
   H) $1.25
   I) $1.35

3. The graph shows the relationship between which two units of measure?
   (Lessons 4-6 and 6-7)

   A) feet and inches
   B) yards and feet
   C) pounds and ounces
   D) quarts and gallons

4. At Igoe's Campsite, 5 of the 15 sites were rented by campers. Which decimal equals the fraction of sites that were rented? (Lesson 5-4)

   F) $0.\overline{33}$
   G) 0.30
   H) 3.30
   I) 33.30

5. A survey showed that the ratio of tourists at the Grand Canyon who had cameras to those who did not was 8:10. A similar survey was conducted at the Statue of Liberty. If the ratios are equivalent, what could be the ratio at the Statue of Liberty?
   (Lesson 7-1)

   A) 32:30
   B) 30:24
   C) 30:32
   D) 24:30

6. Tre can play 18 holes of golf in 180 minutes. What is his average rate in number of minutes per hole? (Lesson 7-2)

   F) 1.8
   G) 3
   H) 6
   I) 10

7. Which of the following is the value of p in $\frac{p}{7} = \frac{21}{49}$? (Lesson 7-3)

   A) 3
   B) 6
   C) 11
   D) 16

8. A scale model of a computer circuit board is 1 centimeter = 0.2 centimeter. The actual circuit board is 4 centimeters wide. What is the width of the model? (Lesson 7-4)

   F) 0.2 cm
   G) 2 cm
   H) 20 cm
   I) 200 cm

9. The table shows the percent of Internet users in each region of the United States who are 3–17 years old. What fraction of Internet users in the South region are 3–17 years old?
   (Lesson 7-5)

Internet Users 3–17 Years Old	
Region	Percent
Midwest	32
Northeast	36
South	28
West	29

   Source: U.S. Census Bureau

   A) $\frac{13}{50}$
   B) $\frac{7}{25}$
   C) $\frac{8}{25}$
   D) $\frac{36}{50}$

Preparing for Standardized Tests

For test-taking strategies and more practice, see pages 608–625.

PART 2 Short Response/Grid In

Record your answers on the answer sheet provided by your teacher or on a sheet of paper.

10. The graph shows the number of school days per year in four countries. How many more days per year do Japanese students attend school than students in the United States? (Lesson 2-7)

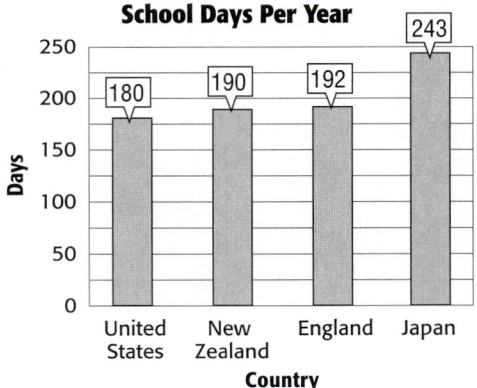

11. A train goes 65 miles per hour and travels 320 miles. How many hours will it take for the train to reach its destination? Round to the nearest tenth. (Lesson 3-7)

12. The table at the right represents a function. What is the missing x-coordinate? (Lesson 4-6)

x	y
3	9
4	12
7	21
?	27

13. Mellia has written $\frac{3}{5}$ of a research paper on fine art and French literature. What percent of the paper has she written? (Lesson 5-5)

14. The area of the rectangle is 18 square inches. What is the width w? (Lesson 6-8)

15. Ron's camera has a shutter speed of $\frac{1}{1,000}$. What is this shutter speed expressed as a percent? (Lesson 7-6)

16. Jean practices her headstand in gymnastics class every day. She has been able to stay in the headstand pose without tipping for 45% of the last 60 classes. How many classes is this? (Lesson 7-7)

17. The table shows the nutrition amounts in one serving of pretzels and the percent of the recommended daily amounts.

Nutrition Facts of Pretzels		
Item	Amount	Percent*
fat	1 g	1.5%
sodium	486 mg	20%
carbohydrate	22 g	7.5%

Source: www.diet-data.com
*based on a 2,000-Calorie diet

How many daily milligrams of sodium are recommended? (Lesson 7-8)

PART 3 Extended Response

Record your answers on a sheet of paper. Show your work.

18. Jessie built a model of her town using items from her family's recycling bin.

   a. One centimeter in Jessie's model equals 10 kilometers in the real town. If Lansdowne Road is 12 centimeters long in the model, how long is the real Lansdowne Road? (Lesson 7-4)

   b. Jessie has cut up milk cartons to construct $\frac{4}{5}$ of her model town. What percent is this? (Lesson 7-5)

   c. Jessie used a total of 70 recycled items to build her model town. Explain how to find the number of items that were milk cartons. Then solve. (Lesson 7-8)

TEST-TAKING TIP

Question 18 When a question involves information from a previous part of a question, make sure to check that information before you move on.

CHAPTER 8 Applying Percent

"What do snowboards have to do with math?"

How much money would you save buying a snowboard that is 35% off its original price of $169? You could use the equation $d = 0.35 \cdot 169$ to find the **discount** d. Equations are useful in finding the total cost of an item including the discount and the sales tax.

You will solve problems involving discount and sales tax in Lesson 8-5.

GETTING STARTED

▶ Diagnose Readiness

Take this quiz to see if you are ready to begin Chapter 8. Refer to the lesson or page number in parentheses for review.

Vocabulary Review

State whether each sentence is *true* or *false*. If *false*, replace the underlined word to make a true sentence.

1. <u>Data</u> refers to pieces of information that are often numerical. (Lesson 2-1)
2. The percent <u>equation</u> is written as $\frac{\text{part}}{\text{base}} = \frac{\text{percent}}{100}$. (Lesson 7-8)

Prerequisite Skills

Multiply. (Page 560)

3. $300 \times 0.02 \times 8$
4. $85 \times 0.25 \times 3$
5. $560 \times 0.6 \times 4.5$
6. $154 \times 0.12 \times 5$

Simplify. Write as a decimal. (Lesson 1-3)

7. $\frac{22 - 8}{8}$
8. $\frac{50 - 33}{50}$
9. $\frac{35 - 7}{35}$

Solve. Round to the nearest tenth if necessary. (Lesson 4-3)

10. $0.4m = 52$
11. $21 = 0.28a$
12. $13 = 0.06s$
13. $0.95z = 37$

Write each percent as a decimal. (Lesson 7-6)

14. 40%
15. 3.25%
16. 7.5%

Use the percent proportion to find each number. Round to the nearest tenth if necessary. (Lesson 7-8)

17. What percent of 86 is 34?
18. 20% of what number is 55?

Percents Make this Foldable to help you organize information about percents. Begin with a piece of 11" by 17" paper.

STEP 1 Fold
Fold a 2" tab along the long side of the paper. Then fold the rest in half.

STEP 2 Open and Fold
Open the paper and fold in half widthwise 3 times to make 8 columns.

STEP 3 Open and Label
Draw lines along the folds and label as shown.

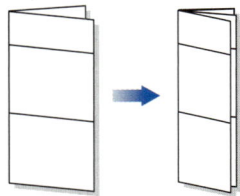

Chapter Notes Each time you find this logo throughout the chapter, use your *Noteables™: Interactive Study Notebook with Foldables™* or your own notebook to take notes. Begin your chapter notes with this Foldable activity.

Readiness To prepare yourself for this chapter with another quiz, visit **msmath2.net/chapter_readiness**

8-1 Percent and Estimation

What You'll LEARN

Estimate percents by using fractions and decimals.

WHEN am I ever going to use this?

GRILLING The graph shows the results of a survey in which people were asked which holiday was their favorite for grilling outdoors. Suppose 80 people were surveyed.

Source: Market Facts for Butterball Turkey

1. What fraction of people surveyed chose Labor Day as their favorite grilling day? How many of the 80 people surveyed is this?

2. About 50% of the people surveyed chose the Fourth of July as their favorite grilling day. Explain how you could use a fraction to estimate the number of people who chose this day. Then estimate.

3. Use a fraction to estimate the number of people surveyed who chose Memorial Day as their favorite grilling day.

Sometimes an exact answer is not needed when using percents. In these cases, you can estimate. One way to estimate the percent of a number is to use a fraction.

EXAMPLES — Use Fractions to Estimate

1 Estimate 48% of 60.

48% is about 50% or $\frac{1}{2}$.

48% of $60 \approx \frac{1}{2} \cdot 60$ Use $\frac{1}{2}$ to estimate.

$\phantom{48\% \text{ of } 60} \approx 30$ Multiply.

So, 48% of 60 is about 30.

2 Estimate 82% of 195.

82% is about 80%, which is $\frac{8}{10}$ or $\frac{4}{5}$.

82% of $195 \approx \frac{4}{5} \cdot 200$ Use $\frac{4}{5}$ to estimate and round 195 to 200.

$\phantom{82\% \text{ of } 195} \approx 160$ Multiply.

So, 82% of 195 is about 160.

Your Turn Estimate by using a fraction.

a. 26% of 80   b. 75% of 23   c. 62% of 507

Another method for estimating the percent of a number is to first find 10% of the number and then multiply. For example, 70% = 7 · 10%. So, 70% of a number equals 7 times 10% of the number.

STUDY TIP

Percents To use decimals in estimating, first round percents to the nearest 10%. To find 10% of a number, move the decimal point one place to the left.

EXAMPLE Estimate by Using 10%

3 Estimate 71% of 300.

Step 1 Find 10% of the number.

10% of 300 = 0.1 · 300   To multiply by 10%, move the decimal point one place to the left.
= 30

Step 2 Multiply.

71% is about 70%.

70% of 300 is 7 times 10% of 300.

7 · 30 = 210

So, 71% of 300 is about 210.

Your Turn Estimate by using 10%.

d. 19% of 40   e. 30% of 217   f. 63% of 91

You can also estimate percents of numbers when the percent is greater than 100 or the percent is less than 1.

EXAMPLES Percents Greater Than 100 or Less Than 1

4 Estimate 122% of 50.

122% is more than 100%, so 122% of 50 is greater than 50.

122% is about 120%.

120% of 50 = (100% of 50) + (20% of 50)   120% = 100% + 20%
= (1 · 50) + ($\frac{1}{5}$ · 50)   100% = 1 and 20% = $\frac{1}{5}$
= 50 + 10 or 60   Simplify.

So, 122% of 50 is about 60.

5 Estimate $\frac{1}{4}$% of 589.

$\frac{1}{4}$% is one fourth of 1%. 589 is about 600.

1% of 600 = 0.01 · 600   To multiply by 1%, move the decimal point two places to the left.
= 6

One fourth of 1% is $\frac{1}{4}$ · 6 or 1.5.

So, $\frac{1}{4}$% of 589 is about 1.5.

Your Turn Estimate.

g. 174% of 200   h. 298% of 45   i. 0.25% of 789

EXAMPLE Estimate Percent to Solve a Problem

6 **POPULATION** In 2000, about 0.5% of the people in Montana were of Asian descent. If Montana had a population of 902,195, estimate the number of people who were of Asian descent.

0.5% is half of 1%.

1% of 900,000 = 0.01 · 900,000   902,195 is about 900,000.

= 9,000

So, 0.5% of 902,195 is about $\frac{1}{2}$ of 9,000 or 4,500.

So, about 4,500 people in Montana were of Asian descent.

Skill and Concept Check

1. **Describe** two different ways to estimate 22% of 136.

2. **OPEN ENDED** Write a problem in which the answer can be found by estimating 12% of 50.

3. **FIND THE ERROR** Ian and Mandy are estimating 1.5% of 420. Who is correct? Explain.

> Ian
> 1.5% of 420
> ≈ 1% of 400 + 0.5% of 400
> = 0.01 · 400 + $\frac{1}{2}$(0.01 · 400)
> = 4 + 2 or 6

> Mandy
> 1.5% of 420
> = 1 · 400 + 0.5 · 400
> = 400 + 200
> = 600

4. **NUMBER SENSE** Explain whether an estimate for the percent of a number is *always*, *sometimes*, or *never* greater than the actual percent of the number. Give an example or a counterexample to support your answer.

GUIDED PRACTICE

Estimate by using fractions.

5. 52% of 160   6. 30% of 79   7. 77% of 22

Estimate by using 10%.

8. 40% of 62   9. 23% of 400   10. 89% of 98

Estimate.

11. 151% of 70   12. 305% of 6   13. $\frac{1}{2}$% of 82

14. **LIFE SCIENCE** The 639 muscles in your body make up about 40% of your total weight. If a person weighs 120 pounds, about how much of the weight is muscle?

Practice and Applications

Estimate by using fractions.

15. 25% of 408
16. 80% of 37
17. 76% of 280
18. 39% of 20
19. 67% of 15.2
20. 10.5% of 238

Estimate by using 10%.

21. 60% of 39
22. 20% of 132
23. 76% of 80
24. 37% of 250
25. 28% of 121
26. 88% of 207

Estimate.

27. 132% of 54
28. 224% of 320
29. 410% of 12
30. 198% of 33
31. 0.4% of 400
32. 0.9% of 74

HOMEWORK HELP

For Exercises	See Examples
15–20	1, 2
21–26	3
27–34	4–6

Extra Practice
See pages 582, 603.

GEOLOGY For Exercises 33 and 34, use the following information.
Granite, a stone found in New Hampshire and Vermont, is 0.8% water.

33. About how many pounds of water are there in 3,000 pounds of granite?

34. About how much water is contained in a 15-pound piece of granite?

35. **CRITICAL THINKING** Explain how you could find $\frac{3}{8}$% of a number.

Spiral Review with Standardized Test Practice

36. **MULTIPLE CHOICE** Estimate 15% of 61.
    - A) 30
    - B) 18
    - C) 15
    - D) 9

37. **MULTIPLE CHOICE** In a survey, 1,031 people were asked to choose the greatest athlete of the 20th century. The top five choices are shown at the right. About how many more people chose Michael Jordan than Muhammad Ali? Choose the best estimate.

    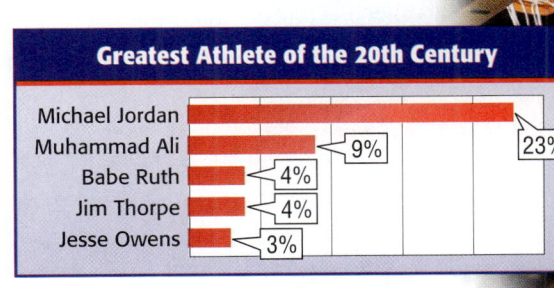
    Source: Gallup Poll

    - F) 90 people
    - G) 100 people
    - H) 150 people
    - I) 200 people

Find each number. Round to the nearest tenth if necessary. (Lesson 7-8)

38. 6 is what percent of 15?
39. Find 72% of 90.
40. What number is 120% of 60?
41. 35% of what number is 55?

42. Find 22% of 85. (Lesson 7-7)

GETTING READY FOR THE NEXT LESSON

PREREQUISITE SKILL Solve. Round to the nearest tenth if necessary. (Lesson 4-3)

43. $40 = 0.8x$
44. $10r = 61$
45. $25 = 0.07t$
46. $56 = 0.32n$

msmath2.net/self_check_quiz

Problem-Solving Strategy
A Follow-Up of Lesson 8-1

Reasonable Answers

What You'll LEARN

Solve problems by determining reasonable answers.

The meals for our group cost a total of $38.95. I think we should leave a 15% tip, which would be about $4.

I think that the tip should be more than that. Let's estimate to find a **reasonable answer**.

Explore	We know that the total bill is $38.95 and we want to leave a 15% tip.
Plan	We can round $38.95 to $40 and then use mental math to find 15% of 40.
Solve	10% of 40 = 0.1 · 40 or 4 10% = 0.1 5% of 40 = $\frac{1}{2}$ · 4 or 2 15% of 40 = (10% of 40) + (5% of 40) = 4 + 2 or 6 So, $6 would be a better amount to leave for a tip.
Examine	Use a calculator to check. .15 [×] 38.95 [ENTER] 5.8425 5.8425 is close to 6, so the answer is reasonable.

Analyze the Strategy

1. The last step of the four-step plan for problem solving asks you to examine your solution. **Explain** how you can use estimation with decimals to help you examine a solution.

2. **Write** a problem that has an unreasonable answer and ask a classmate to explain why they think the answer is unreasonable.

3. **Describe** other problem-solving strategies that you could use to determine whether answers are reasonable.

338 Chapter 8 Applying Percent

Apply the Strategy

Solve. Use the reasonable answer strategy.

4. **COMMUNICATION** Sandra makes a long distance phone call to her grandparents and talks for 45 minutes. The phone company charges a rate of $0.20 per minute. How much does the call cost?

5. **SHOPPING** Suppose you are buying an entertainment system for $1,301.90 and the speakers are 57.6% of the total cost. What is a reasonable estimate for the cost of the speakers? Explain.

Mixed Problem Solving

Solve. Use any strategy.

6. **CHORES** Cameron is using a 2.5-liter container to fill a tank that holds 24 liters of water. How many times will he need to fill the container?

7. **MUSIC** A survey showed that 73% of teens who use computers listen to music at the same time. Suppose there are 410 teens in your school who use computers. Estimate how many of them listen to music while on the computer.

8. **VACATION** The graph below shows how the Mason family spent their time during their summer vacation. What percent of the time shown was spent touring historic sites?

9. **BASKETBALL** Keisha made 18 points in one basketball game. How many possible shot combinations of 2- and 3-pointers could she have made? List the combinations in a table.

10. **POPULATION** About 9.4% of the people in Texas live in Houston. If the population of Texas is about 20,852,000, estimate the population of Houston.

11. **SEAT BELTS** The graph shows the percent of drivers who wore seat belts. Predict the percent of drivers who will wear seat belts in 2006. Explain why your answer is reasonable.

Source: Harris Interactive

12. **GEOMETRY** A rectangle has a length of $4\frac{1}{4}$ inches and a width of $3\frac{2}{5}$ inches. Is between 7 and 8 inches a reasonable estimate for the perimeter? Explain.

13. **STANDARDIZED TEST PRACTICE**
    Mr. Camacho is purchasing carpet for the two rooms described in the table. Which expression shows the number of square yards of carpet that he needs?

Room	Dimensions
living room	15 ft by 18 ft
TV room	18 ft by 20 ft

    Ⓐ $[(15 \times 18) + (18 \times 20)] \div (3 \times 3)$
    Ⓑ $[(15 \times 18(2) \times 20)] \div 3$
    Ⓒ $[(15 \times 18) + (18 \times 20)] \div 3$
    Ⓓ $(15 \times 18 \times 20) \div (3 \times 3)$

8-2 Algebra: The Percent Equation

What You'll LEARN
Solve problems by using the percent equation.

NEW Vocabulary
percent equation

Link to READING
Everyday Meaning of Base: the bottom of something considered as its support, as in the base of a column

WHEN am I ever going to use this?

PHYSICAL SCIENCE The graph shows three main elements that make up most of the human body.

Elements of the Body
- carbon 18%
- other 7%
- oxygen 65%
- hydrogen 10%

Source: *The New York Times Almanac*

1. Suppose a person weighs 120 pounds. Use the percent proportion to find the number of pounds of oxygen, carbon, and hydrogen.
2. Express the percent of each element as a decimal.
3. Multiply each decimal by 120. Record your results.
4. Compare the answers to Exercises 1 and 3.

In Lesson 7-8, you used the percent proportion $\frac{\text{part}}{\text{base}} = \frac{\text{percent}}{100}$ to find the missing part, percent, or base. You can also use an equation.

$\frac{\text{part}}{\text{base}} = \text{percent}$ The percent is written as a decimal.

$\frac{\text{part}}{\text{base}} \cdot \text{base} = \text{percent} \cdot \text{base}$ Multiply each side by the base.

$\text{part} = \text{percent} \cdot \text{base}$ ← This form is called the **percent equation**.

Concept Summary: Types of Percent Problems

Type	Example	Equation
Find the Part	What number is 50% of 6?	$n = 0.5 \cdot 6$
Find the Percent	3 is what percent of 6?	$3 = n \cdot 6$
Find the Base	3 is 50% of what number?	$3 = 0.5 \cdot n$

EXAMPLE Find the Part

1 What number is 12% of 350? **Estimate** $0.1 \cdot 350 = 35$

12% or 0.12 is the percent and 350 is the base. Let n represent the part.

$\underline{\text{part}} = \underline{\text{percent}} \cdot \underline{\text{base}}$

$n = 0.12 \cdot 350$ Write an equation.

$n = 42$ Multiply. The part is 42.

So, 42 is 12% of 350.   This is close to the estimate.

EXAMPLE Find the Percent

2 21 is what percent of 40?   **Estimate** $\frac{21}{40} \approx \frac{1}{2}$ or 50%

Let n represent the percent.

part = percent · base

$21 = n \cdot 40$ Write an equation.

$\frac{21}{40} = \frac{40n}{40}$ Divide each side by 40.

$0.525 = n$ Simplify.

$52.5\% = n$ Write 0.525 as a percent.

So, 21 is 52.5% of 40.   This is close to the estimate.

STUDY TIP

Percent Remember to write the decimal as a percent in your final answer.

EXAMPLE Find the Base

3 13 is 26% of what number?   **Estimate** 13 is 25% or $\frac{1}{4}$ of 52.

Let n represent the base.

part = percent · base

$13 = 0.26 \cdot n$ Write an equation.

$\frac{13}{0.26} = \frac{0.26n}{0.26}$ Divide each side by 0.26.

$50 = n$ The base is 50.

So, 13 is 26% of 50.   Compare to the estimate.

Your Turn Write an equation for each problem. Then solve. Round to the nearest tenth if necessary.

a. What percent of 125 is 75?   b. 39 is 84% of what number?

REAL-LIFE MATH

TECHNOLOGY The following cities in the United States have the highest percent of cell phone users.

Anchorage, AK	56%
Chicago, IL	55%
Honolulu, HI	54%
Atlanta, GA	52%
Detroit, MI	52%

Source: Polk's Research

EXAMPLE Apply the Percent Equation

4 **TECHNOLOGY** Finland has the highest percent of people who have cell phones, 67.8%. If there are about 3,499,000 people with cell phones, what is the population of Finland?

Words	3,499,000 is 67.8% of what number?
Symbols	Let n represent the base.
Equation	$3,499,000 = 0.678 \cdot n$

67.8% is written as the decimal 0.678.

$3,499,000 = 0.678 \cdot n$ Write the equation.

$\frac{3,499,000}{0.678} = \frac{0.678n}{0.678}$ Divide each side by 0.678. Use a calculator.

$5,160,767 \approx n$ Simplify.

The population of Finland is about 5,160,767.

Skill and Concept Check

1. **State** whether the following problem represents a missing part, a missing percent, or a missing base.
   17 is 10% of what number?

2. **OPEN ENDED** Write a real-life problem that can be solved by using the percent equation. State whether you need to find the part, the percent, or the base, and then solve.

GUIDED PRACTICE

Write an equation for each problem. Then solve. Round to the nearest tenth if necessary.

3. What number is 88% of 300?
4. 75 is what percent of 150?
5. 3 is 12% of what number?
6. 84 is 60% of what number?

7. **BASEBALL** In 2003, Derek Jeter had 156 hits in 482 times at bat. What was his *batting average*, or the percent of times at bat that were hits? Round the percent to the nearest tenth. Then write as a decimal.

Data Update How do current batting averages of other baseball players compare to Derek Jeter's average in 2003? Visit msmath2.net/data_update to learn more.

Practice and Applications

Write an equation for each problem. Then solve. Round to the nearest tenth if necessary.

8. What number is 65% of 98?
9. Find 53% of 470.
10. 9 is what percent of 45?
11. 26 is what percent of 96?
12. 84 is 75% of what number?
13. 17 is 40% of what number?
14. Find 24% of 25.
15. What percent of 64 is 30?
16. 98 is what percent of 392?
17. 33% of what number is 1.45?
18. Find 13.5% of 520.
19. What number is 75.2% of 600?
20. What number is 4% of 82.1?
21. 14 is 2.8% of what number?
22. Find 135% of 64.
23. What percent of 200 is 230?

HOMEWORK HELP

For Exercises	See Examples
8–23	1–3
24–28	4

Extra Practice
See pages 583, 603.

24. **SALES** Ms. Allon received a $325 *commission*, which is a fee paid based on a percent of her sales. If her sales totaled $8,125, what is the percent that she earns?

NUMBER THEORY For Exercises 25 and 26, consider the whole numbers 1 through 100.

25. For what percent of these numbers will the digits add to 5?
26. For what percent of these numbers will the digits add to an even number? an odd number?

342 Chapter 8 Applying Percent

CROPS For Exercises 27–29, use the graphic at the right. It shows the acres of cotton planted in the top five states. Round to the nearest percent.

27. About what percent of the cotton planted in the United States is planted in Texas?

28. About what percent of U.S. cotton is planted in North Carolina?

29. What percent of U.S. cotton is planted in states other than those listed on the graphic?

30. **MULTI STEP** A cargo of 18,000 tons of grain is 14% water due to moisture. Another cargo of 26,000 tons of grain is 12.2% water. Which contains more water? How much more?

31. **CRITICAL THINKING** If you need to find the percent of a number, explain how you can predict whether the part will be less than, greater than, or equal to the number.

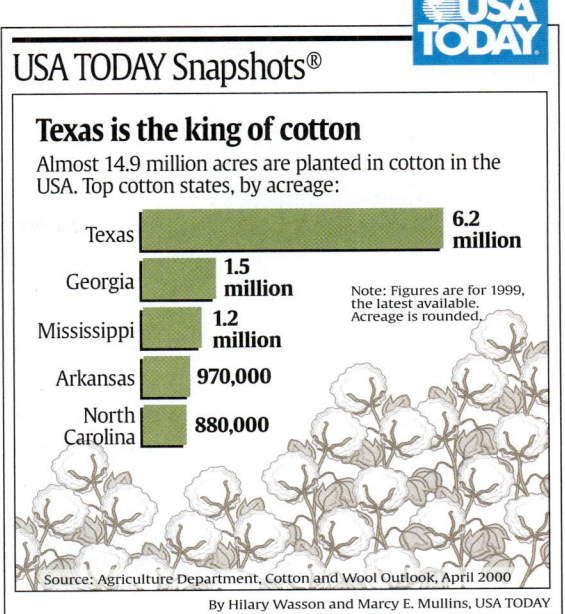

Spiral Review with Standardized Test Practice

32. **MULTIPLE CHOICE** 23 is what percent of 64? Round to the nearest tenth.

    Ⓐ 2.8%   Ⓑ 33.4%   Ⓒ 34.0%   Ⓓ 35.9%

33. **MULTIPLE CHOICE** 56 is 16% of what number? Round to the nearest tenth.

    Ⓕ 300.5   Ⓖ 305.0   Ⓗ 305.5   Ⓘ 350.0

Estimate by using fractions. (Lesson 8-1)

34. 11% of 79   35. 30.5% of 50   36. 48% of 311

37. **NUTRITION** A cup of fruit yogurt has approximately 315 milligrams of calcium. This is about 24% of the recommended daily allowance for people 9 to 18 years old. How many milligrams of calcium is recommended per day for people 9 to 18 years old? (Lesson 7-8)

38. **GEOMETRY** Find the circumference of the circle. Use $\frac{22}{7}$ for π. (Lesson 6-9)

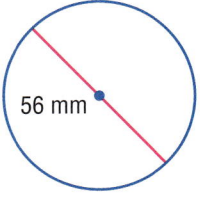

39. Find the greatest common factor of 72 and 270. (Lesson 5-2)

GETTING READY FOR THE NEXT LESSON

PREREQUISITE SKILL Use the percent proportion to find each number. Round to the nearest tenth if necessary. (Lesson 7-9)

40. What number is 45% of 60?
41. What percent of 180 is 30?
42. 20% of what number is 18?
43. 50% of what number is 15.8?

8-3a HANDS-ON LAB

A Preview of Lesson 8-3

Sampling

What You'll LEARN

Use sampling to collect data.

INVESTIGATE *Work as a class.*

To determine the favorite lunches of students at your school, you could ask each student. A more practical method is to survey a representative group of students and use that data to make a conclusion. This method is called **sampling**.

Term	Definition
population	total group of people or items in which the survey is interested
sample	part of the population
sample size	number of people or items in the survey

It is important to obtain a sample that is **unbiased**, that is, not limited or not favoring a particular outcome. Suppose you put the ID numbers of all the students in a box and randomly choose 65 numbers. This sample is unbiased because it is:

- a *random sample*. Each member of the population, or students at your school, has an equal chance of being selected.
- large enough to provide accurate data.

Work in groups of three.

State whether each sample is random. Explain.

1. To determine the favorite spectator sport for women over 25 years old, 350 women over 25 are surveyed at a professional basketball game.

2. To collect data about the study habits of middle school students in the Franklin School District, the name of every middle school student in the district is placed in a bag and 250 names are randomly selected.

For Exercises 3–7, refer to the information below.
Suppose you are to survey students in your school.

3. Formulate a hypothesis about students' activities.

4. Design and conduct a survey. Describe the technique that you used to get a random sample.

5. Organize and display the results of your survey in a table or graph.

6. Analyze ways in which the wording of your questions might have changed the outcome of the survey.

7. Use the results of your survey to evaluate your hypothesis.

Unbiased An unbiased sample is representative of the population.

8-3 Statistics: Using Statistics to Predict

What You'll LEARN

Predict actions of a larger group by using a sample.

NEW Vocabulary

survey
population
random sample

REVIEW Vocabulary

data: pieces of information that are often numerical (Lesson 2-1)

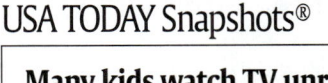 **am I ever going to use this?**

TELEVISION The graphic shows the results of a survey in which children were asked whether they had rules about what they watched on television.

1. What ages of children are represented?
2. Can you tell how many were surveyed? Explain.
3. Describe how you could use the graphic to predict how many students in your school have no restrictions on their television viewing.

Many kids watch TV unrestricted
Nearly half of children 8 to 17 have no parental restrictions on how much TV they watch or what programs or channels they watch.

No restrictions **46%**
Some restrictions **54%**

Source: Statistical Research Inc.

By Hilary Wasson and Marcy E. Mullins, USA TODAY

A **survey** is a question or set of questions designed to collect data about a specific group of people, called the **population**. If a survey uses a **random sample** of a population, or a sample chosen without preference, you can assume that the sample represents the population. Then you can use the results to make predictions about the population.

EXAMPLE Predict Using Percent Proportion

1 TELEVISION Refer to the survey results in the graphic above. Predict how many of the 1,250 Gallatin Middle School students have some restrictions on the television that they watch.

You can use the percent proportion and the survey results to predict what part of the population of students have TV restrictions.

part of the population → $\dfrac{a}{b} = \dfrac{p}{100}$ Percent proportion

entire population → $\dfrac{a}{1{,}250} = \dfrac{54}{100}$ Survey results: $54\% = \dfrac{54}{100}$

$100a = 1{,}250(54)$ Cross products

$a = 675$ Simplify.

So, about 675 of the Gallatin Middle School students have some television restrictions.

READING in the Content Area

For strategies in reading this lesson, visit **msmath2.net/reading**.

msmath2.net/extra_examples

Lesson 8-3 Statistics: Using Statistics to Predict **345**

COMMUNICATION The survey at the right also found that 92% of people 12 to 17 years old use the Internet to send E-mails.

You can also use the percent equation to make predictions using survey results.

EXAMPLE — Predict Using Percent Equation

2 COMMUNICATION A survey showed that 74% of people 12 to 17 years old use the Internet to send instant messages. Predict how many of the 2,450 students at Washington Middle School send instant messages.

You need to predict how many of the 2,450 students send instant messages.

Words	What number is 74% of 2,450?
Variable	Let n represent the number.
Equation	$n = 0.74 \cdot 2,450$

$n = 0.74 \cdot 2,450$ Write the equation.

$n = 1,813$ Multiply.

So, you could predict that about 1,813 students at Washington Middle School use the Internet to send instant messages.

Skill and Concept Check

1. **Writing Math** Explain how to use a sample to predict what a group of people prefer.

2. **OPEN ENDED** Find a newspaper or magazine article that has a table or graph. Identify the population and explain how you think the results were found.

GUIDED PRACTICE

3. **SPENDING** The circle graph shows the results of a poll to which 60,000 teens responded. Predict how many of the approximately 28 million teens in the United States would buy a music CD if they were given $20.

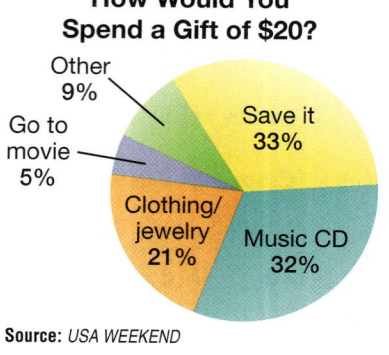

How Would You Spend a Gift of $20?
- Other 9%
- Go to movie 5%
- Save it 33%
- Clothing/jewelry 21%
- Music CD 32%

Source: USA WEEKEND

4. **TRANSPORTATION** The graph shows the results of a survey in which working adults in America were asked how they get to work. Use the data to predict how many working adults in America out of 143 million walk or ride a bicycle to work.

Getting to Work
- Walk or Ride Bike: 4%
- Drive: 87%
- Bus or Subway: 3%
- Other: 6%

Source: Gallup Poll

346 Chapter 8 Applying Percent

Practice and Applications

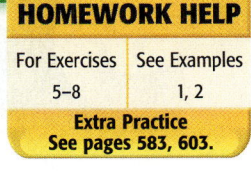

HOMEWORK HELP
For Exercises: 5–8
See Examples: 1, 2
Extra Practice
See pages 583, 603.

5. **COMPUTERS** The table shows the results of a survey in which students were asked how they use a personal computer at home. Use the results to predict how many of the 1,745 Allegheny Valley Middle School students use a PC for homework.

PC Use	Percent
educational programs	93%
homework	80%
games	70%

Source: U.S. Census Bureau

6. **CAMERAS** In a survey, 14% of teens said they own a digital camera. Predict how many of the 420,000 teens in Arizona own digital cameras.

CATS For Exercises 7 and 8, use the graphic at the right. It shows the percent of cat owners who train their cats to prevent each problem.

7. Out of 255 cat owners, how many would you predict have trained their cat not to climb on furniture?

8. Predict how many cat owners in a group of 316 have trained their cat not to claw furniture.

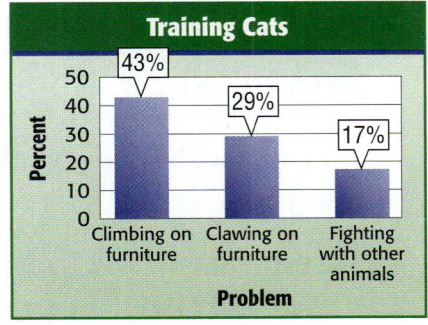

Source: Purina Cat Chow

9. **CRITICAL THINKING** A survey found that 80% of teens enjoy going to the movies in their free time. If 5,200 teens said that they enjoy going to the movies, what was the total number of teens surveyed?

Spiral Review with Standardized Test Practice

10. **MULTIPLE CHOICE** A survey of 80 seventh graders at Lincoln Middle School was taken to find how they get to school each day. The results are shown in the table. Of the 423 seventh graders in the school, predict how many walk to school.

    Ⓐ 23   Ⓑ 64   Ⓒ 123   Ⓓ 394

Getting to School	Percent
take a bus	33%
walk	29%
adult drives	18%
other	20%

11. **SHORT RESPONSE** A survey showed that 16% of the people in Tennessee over the age of 16 belong to a fitness center. Predict how many of the 5 million people in Tennessee over the age of 16 belong to a fitness center.

Write an equation for each problem. Then solve. Round to the nearest tenth if necessary. (Lesson 8-2)

12. What number is 12% of 60?

13. 54 is 72% of what number?

14. Estimate 30% of 149. (Lesson 8-1)

GETTING READY FOR THE NEXT LESSON

PREREQUISITE SKILL Simplify. Write as a decimal. (Lesson 1-3)

15. $\dfrac{10-7}{10}$

16. $\dfrac{50-18}{50}$

17. $\dfrac{22-4}{4}$

18. $\dfrac{39-15}{15}$

msmath2.net/self_check_quiz

Chapter 8 Mid-Chapter Practice Test

Vocabulary and Concepts

1. **Describe** two methods that can be used to estimate the percent of a number. (Lesson 8-1)
2. **Explain** what the *population* is in a survey. (Lesson 8-3)

Skills and Applications

Estimate by using fractions. (Lesson 8-1)

3. 20% of 392
4. 78% of 112
5. 52% of 295

Estimate by using 10%. (Lesson 8-1)

6. 30% of 42
7. 79% of 88
8. 41.5% of 212

Write an equation for each problem. Then solve. Round to the nearest tenth if necessary. (Lesson 8-2)

9. What number is 35% of 72?
10. 16.1 is what percent of 70?
11. 27.2 is 68% of what number?
12. 16% of 32 is what number?

ENTERTAINMENT For Exercises 13 and 14, refer to the graph at the right. It shows the results of a survey of students' favorite TV programs at Morgan Middle School. (Lesson 8-3)

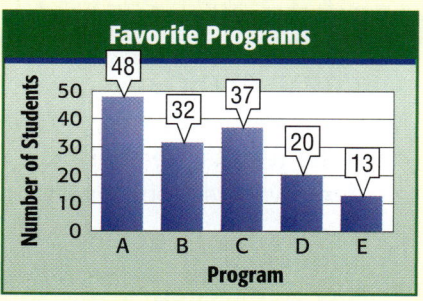

13. What percent of the students surveyed preferred Program A?
14. How many of the 925 students in the school would you expect to choose Program A as their favorite?

Standardized Test Practice

15. **MULTIPLE CHOICE** Miyoki has read 82% of a book that has 214 pages. Which is the best estimate of the number of pages that she has read? (Lesson 8-1)

    Ⓐ 16 pages Ⓑ 50 pages
    Ⓒ 80 pages Ⓓ 160 pages

16. **MULTIPLE CHOICE** A cookie company received 1,600 E-mails in one week. Of those E-mails, 12.5% were people requesting catalogues. How many people requested catalogues that week? (Lesson 8-2)

    Ⓕ 2,000 people Ⓖ 1,280 people
    Ⓗ 200 people Ⓘ 128 people

The Game Zone

A Place To Practice Your Math Skills

Spinning for Percents

GET READY!
Players: two or three
Materials: 6 index cards, 2 spinners

GET SET!
- Copy the numbers 11, 28, 45, 62, 84, and 98 onto index cards, one number per card.
- Shuffle the cards and place them facedown in a pile.
- Label equal sections of two spinners with the digits 0 through 9.
- Decide which spinner will stand for digits in the tens place and which spinner will stand for the ones place. The number formed by spinning both spinners is the percent.

These spinners represent 62%.

GO!
- One player selects the top card from the pile.
- The player spins both spinners and records the percent that is formed. He or she then finds that percent of the number that is on the index card. Round to the nearest whole number if necessary. This is the player's score for this turn.
- Continue in this way, taking turns selecting a card, until no cards remain in the pile.
- **Who Wins?** The player with the greatest total score wins.

8-4 Percent of Change

What You'll LEARN
Find the percent of increase or decrease.

NEW Vocabulary
percent of change
percent of increase
percent of decrease

HANDS-ON Mini Lab

Materials
- paper strips
- scissors
- tape

Work with a partner.

You can use paper strips to model a 50% increase.

- Begin with two paper strips. On each strip, label 0% on the left side and 100% on the right side.

- Fold one of the paper strips in half. Mark 50% in the center.

- Cut the second strip at the 50% mark and tape the piece onto the end of the first strip. The new longer strip represents a 50% increase or 150%.

Model each percent of change.

1. 25% increase
2. 75% increase
3. 30% increase

4. **Describe** a model that represents a 100% increase, a 200% increase, and a 300% increase.

5. **Describe** how this process would change to show percent of decrease.

One way to describe a change in quantities is to use percent of change.

Noteables™ — Key Concept: Percent of Change

Words   A **percent of change** is a ratio that compares the change in quantity to the original amount.

Equation   percent of change = $\dfrac{\text{amount of change}}{\text{original amount}}$

If the original quantity is increased, then it is called a **percent of increase**. If the original quantity is decreased, then it is called a **percent of decrease**.

percent of increase = $\dfrac{\text{amount of increase}}{\text{original amount}}$ ← new amount − original amount

percent of decrease = $\dfrac{\text{amount of decrease}}{\text{original amount}}$ ← original amount − new amount

Chapter 8 Applying Percent

EXAMPLE   Find Percent of Increase

1 TREES Find the percent of change in tree height from year 1 to year 10. Round to the nearest whole percent if necessary.

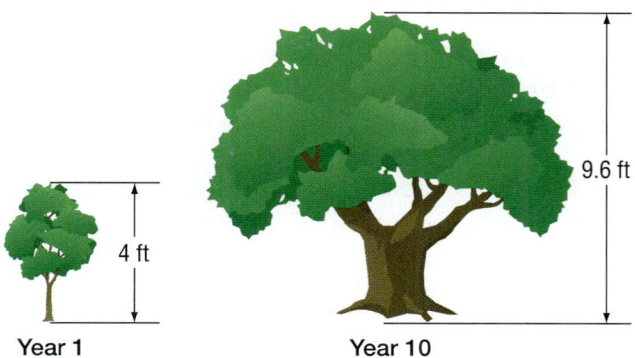

Year 1   Year 10

Since the new height is greater than the original height, this is a percent of increase. The amount of increase is 9.6 − 4 or 5.6 feet.

percent of increase = $\frac{\text{amount of increase}}{\text{original amount}}$

$= \frac{5.6}{4}$ Substitution

$= 1.4$ Simplify.

$= 140\%$ Write 1.4 as a percent.

The percent of increase in the tree height is 140%.

REAL-LIFE CAREERS

How Does a Stockbroker Use Math?
A stockbroker must be able to compare percents of change of different stock prices.

 Research
For information about a career as a stockbroker, visit:
msmath2.net/careers

EXAMPLE   Find Percent of Decrease

2 STOCKS Find the percent of change if the original price of a stock was $75 and the new price is $60. Round to the nearest whole percent if necessary.

Stock	Price
original	$75
new	$60

Since the new price is less than the original price, this is a percent of decrease. The amount of decrease is 75 − 60 or $15.

percent of decrease = $\frac{\text{amount of decrease}}{\text{original amount}}$

$= \frac{15}{75}$ Substitution

$= 0.2$ Simplify.

$= 20\%$ Write 0.2 as a percent.

The percent of decrease of the stock is 20%.

Your Turn Find each percent of change. Round to the nearest whole percent if necessary. State whether the percent of change is an *increase* or *decrease*.

a. original: 10
   new: 13

b. original: 20
   new: 15

msmath2.net/extra_examples   Lesson 8-4 Percent of Change   **351**

Skill and Concept Check

1. **OPEN ENDED** Write a percent of change problem using 14 and 25. State whether there is a percent of increase or decrease. Then solve.

2. **NUMBER SENSE** The costs of two different shelf stereo systems are decreased by $10. The original costs were $90 and $60, respectively. Without calculating, which had greater percent of decrease? Explain.

3. **FIND THE ERROR** Jada and Miranda are finding the percent of change from 46 to 130. Who is correct? Explain.

Jada
$$\frac{130-46}{46} \approx 1.83 \text{ or } 183\%$$

Miranda
$$\frac{130-46}{130} \approx 0.65 \text{ or } 65\%$$

GUIDED PRACTICE

Find each percent of change. Round to the nearest whole percent if necessary. State whether the percent of change is an *increase* or *decrease*.

4. original: 30
   new: 24

5. original: $126
   new: $150

6. original: 624
   new: 702

7. original: $75.80
   new: $94.75

8. original: 1.6
   new: 0.95

9. original: 20.5
   new: 35.5

10. **SPORTS** The table shows the number of people ages 7 to 17 who played soccer. What was the percent of increase in soccer players from 1990 to 2000? Round to the nearest whole percent.

Playing Soccer	
Year	Number (millions)
1990	10.9
2000	12.9

Source: National Sporting Goods Association

Practice and Applications

Find each percent of change. Round to the nearest whole percent if necessary. State whether the percent of change is an *increase* or *decrease*.

11. original: 15
    new: 18

12. original: 100
    new: 140

13. original: $12
    new: $6

14. original: $240
    new: $320

15. original: 48
    new: 14

16. original: 360
    new: 120

17. original: 125
    new: 87.5

18. original: $15.60
    new: $11.70

19. original: $89.50
    new: $105.20

20. original: 132
    new: 125.4

21. original: 782
    new: 789.82

22. original: 12
    new: 60

HOMEWORK HELP

For Exercises	See Examples
11–31	1, 2

Extra Practice
See pages 583, 603.

GEOMETRY For Exercises 23 and 24, refer to the rectangle at the right. Suppose the side lengths are doubled.

23. Find the percent of increase in the perimeter.

24. Find the percent of increase in the area.

8 cm
3 cm

352 Chapter 8 Applying Percent

25. **SALES** Use the graphic at the right to find the percent of change in ketchup sales from 2000 to 2001.

26. **EDUCATION** A person with a high school diploma earns an average of $16,053 per year. A person with some college earns an average of $25,686 per year. What is the percent of increase?

27. **POPULATION** On July 1, 2003, the U.S. had an estimated 290,809,777 residents. This is 9,386,546 more than on April 1, 2000. To the nearest tenth, what was the percent of change in population?

28. **ART** In 2001, the sales of fine art were $40.8 billion. Art sales are projected to increase by 25% from 2001 to 2005. What is the projected amount of art sales in 2005?

Source: Heinz North America

ALLOWANCES For Exercises 29–31, refer to the table at the right.

29. Find the percent of increase of allowance from age 13 to 14. Round to the nearest whole percent.

30. State two consecutive ages where the change in allowance is a percent of decrease. Then find the percent of decrease. Round to the nearest tenth.

31. Between which two consecutive years is the percent of increase the greatest? What is the percent of increase? Round to the nearest whole percent.

Age	Average Weekly Allowance
12	$9.58
13	$9.52
14	$13.47
15	$15.57
16	$17.84
17	$30.66

Source: www.kidsmoney.org

32. **CRITICAL THINKING** If a quantity increases by 10% and then decreases by 10%, will the result be the original quantity? Explain.

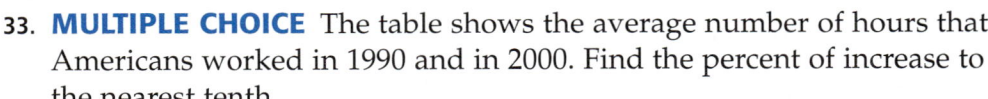

Spiral Review with Standardized Test Practice

33. **MULTIPLE CHOICE** The table shows the average number of hours that Americans worked in 1990 and in 2000. Find the percent of increase to the nearest tenth.

    Ⓐ 19.0%   Ⓑ 2.0%   Ⓒ 1.9%   Ⓓ 0.2%

Year	Time (h)
1990	1,943
2000	1,979

Source: International Labor Organization

34. **SHORT RESPONSE** Find the percent of decrease from 85 to 68.

35. **FOOD** In a survey of 150 students at Kennedy Middle School, 48% said that their favorite type of pizza crust is thick crust. Predict how many of the 1,375 students in the school prefer thick crust pizza. (Lesson 8-3)

Write an equation for each problem. Then solve. Round to the nearest tenth if necessary. (Lesson 8-2)

36. 30% of what number is 17?

37. What is 21% of 62?

GETTING READY FOR THE NEXT LESSON

PREREQUISITE SKILL Write each percent as a decimal. (Lesson 7-5)

38. 6.5%   39. $5\frac{1}{2}\%$ 40. $8\frac{1}{4}\%$ 41. $6\frac{3}{4}\%$

8-5 Sales Tax and Discount

What You'll LEARN
Solve problems involving sales tax and discount.

NEW Vocabulary
sales tax
discount

WHEN am I ever going to use this?

COMPUTERS Julie Ann plans to buy a new computer with a flat screen that costs $1,299. She lives in Florida where there is a sales tax of 6%.

1. Calculate the sales tax by finding 6% of $1,299.
2. What will be the total cost including the sales tax?
3. Use a calculator to multiply 1.06 and 1,299. How does the result compare to your answer in Exercise 2?

One everyday use of percent is sales tax. **Sales tax** is an additional amount of money charged on items that people buy. The local, state, or federal government receives this money. The total cost of an item is the regular price plus the sales tax.

EXAMPLE   Find the Total Cost

1 **TECHNOLOGY** A graphing calculator costs $90, and the sales tax is 4.25%. What is the total cost of the calculator?

First, find the sales tax.

4.25% of $90 = 0.0425 · 90

$\approx$ 3.83   The sales tax is $3.83.

Next, add the sales tax to the regular price.

3.83 + 90 = 93.83

The total cost of the calculator is $93.83.

Another way to find the cost of an item with sales tax is to add the percent of tax to 100%.

EXAMPLE   Find the Total Cost

2 **CLOTHES** What is the total cost of a sweatshirt if the regular price is $42 and the sales tax is $5\frac{1}{2}$%?

$100\% + 5\frac{1}{2}\% = 105\frac{1}{2}\%$ Add the percent of tax to 100%.

The total cost is $105\frac{1}{2}$% of the regular price.

$105\frac{1}{2}$% of $42 = 1.055 · 42   Use a calculator.

= 44.31

The total cost of the sweatshirt is $44.31.

Discount is the amount by which the regular price of an item is reduced. The sale price is the regular price minus the discount.

EXAMPLE Find the Sale Price

Standardized Test Practice

3 **MULTIPLE-CHOICE TEST ITEM** Alan wants to buy a snowboard that has a regular price of $169. This week, the snowboard is on sale at a 35% discount. What is the sale price of the snowboard?

   Ⓐ $59.15   Ⓑ $109.85   Ⓒ $134.00   Ⓓ $228.15

Read the Test Item
The sale price is $169 minus the discount.

Solve the Test Item

Method 1 First, find the amount of the discount d.

$$\underbrace{part}_{d} = \underbrace{percent}_{0.35} \cdot \underbrace{base}_{169}$$ Use the percent equation.

$d = 59.15$ The discount is $59.15.

So, the sale price is $169 - $59.15 or $109.85.

Method 2 First, subtract the percent of discount from 100%.

$100\% - 35\% = 65\%$

So, the sale price s is 65% of the regular price.

$s = 0.65 \cdot 169$ Use the percent equation.

$s = 109.85$ The sale price is $109.85.

So, the sale price of the snowboard is $109.85. The answer is B.

Test-Taking Tip

Elimination The sale price of an item is less than the regular price. Eliminate any choices in which the sale price given is greater than the regular price.

EXAMPLE Find the Percent of Discount

4 **MUSIC** An electric guitar is on sale as shown at the right. What is the percent of discount?

First, find the *amount* of discount.

$299.95 - $179.99 = $119.96

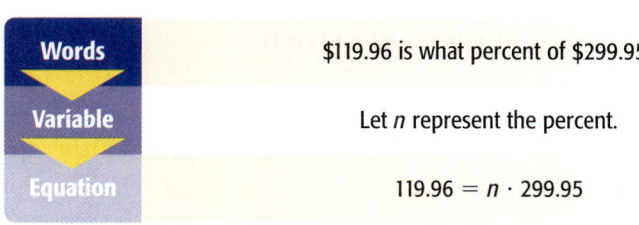
Original Price: ~~$299.95~~
Now: $179.99

Next, use the percent equation to find the percent discount.

Words	$119.96 is what percent of $299.95?
Variable	Let n represent the percent.
Equation	$119.96 = n \cdot 299.95$

$119.96 = n \cdot 299.95$ Write the equation.

$0.40 \approx n$ Divide each side by 299.95 and simplify.

The percent of discount is about 40%.

STUDY TIP

Discount You also could have used the percent proportion to find what percent $119.96 is of $299.95.

Your Turn

a. Find the percent of discount if the sale price of the guitar is $224.96.

Skill and Concept Check

1. **Find** the sales tax of a $98 chair if the tax rate is 7%.

2. **Writing Math** Describe two methods for finding the sale price of an item that is discounted 30%. Which method do you prefer? Explain.

3. **OPEN ENDED** Give an example of the regular price of an item and the total cost including sales tax if the tax rate is 5.75%.

GUIDED PRACTICE

Find the total cost or sale price to the nearest cent.

4. $2.95 notebook; 5% tax
5. $1,575 computer; 15% discount
6. $119.50 skateboard; 20% off
7. $46 shoes; 2.9% tax

Find the percent of discount to the nearest percent.

8. lotion: regular price, $4.50
   sale price, $2.25
9. in-line skates: regular price, $99
   sale price, $90

10. **VIDEO GAMES** What is the sales tax of a $178.90 video game system if the tax rate is 3.75%?

Practice and Applications

Find the total cost or sale price to the nearest cent.

11. $58 ski lift ticket; 20% discount
12. $1,500 computer; 7% tax
13. $99 CD player; 5% tax
14. $12.25 pen set; 60% discount
15. $4.30 notebook; 40% discount
16. $7.50 meal; 6.5% tax
17. $39.60 sweater; 33% discount
18. $89.75 scooter; $7\frac{1}{4}$% sales tax

19. **BOOKS** A book about candle making is $24.95. Find the total cost of the book including the 4.5% sales tax.

20. **TICKETS** A local movie theater is selling movie tickets for $5.25 during the first week of May. The regular price of a ticket is $6.75. What is the percent of discount to the nearest percent?

Find the percent of discount to the nearest percent.

21. calendar: regular price, $9
    sale price, $2.25
22. telescope: regular price, $180
    sale price, $126
23. concert tickets: regular price, $44
    sale price, $34
24. TV: regular price, $625
    sale price, $562.50

25. **MULTI STEP** A sound system has a regular price of $249. Find the total cost if it is on sale for 50% off and the sales tax is 5.75%.

26. **MULTI STEP** Suppose your restaurant bill comes to $28.35. Find your total cost if the tax is 6.25% and you leave a 20% tip on the amount before tax.

HOMEWORK HELP

For Exercises	See Examples
11–19, 25–31	1–3
20–24	4

Extra Practice
See pages 584, 603.

MOUNTAIN BIKES For Exercises 27–30, use the information below and at the right.

A mountain bike has a regular price of $575.

State	2004 Sales Tax Rate
Arkansas	6%
Illinois	6.25%
Mississippi	7%
New York	4.25%

Source: www.salestaxinstitute.com

27. Suppose Antonio lives in Mississippi. What is the total cost of the bike including tax?

28. If the mountain bike is on sale with a 25% discount, how much will Antonio pay for the bike, including tax?

29. Before 8 A.M., the bike will be discounted an additional 15% off the already discounted price. What will be the sale price, not including tax?

30. **RESEARCH** Use the Internet or another source to find the current tax rates of the states listed in the table or of other states. Find the cost of the bike including tax in one of these states.

31. **PROFIT** To make a profit, stores sell items for more than they paid. The increase in price is called the *markup*. Suppose Sports Galore purchases tennis racquets for $45 each. Find the markup price if the racquets are sold for 28% over the price paid for them.

32. **CRITICAL THINKING** Find the total percent of change on the price of an item if it is 15% off and the sales tax is 5%. Does it matter in which order the discount and the sales tax are applied? Explain.

Spiral Review with Standardized Test Practice

33. **MULTIPLE CHOICE** A T-shirt at the mall costs $14.95. It is on sale for 30% off. What is the sale price to the nearest cent?

    A $4.49   **B** $10.47   **C** $10.50   **D** $10.65

34. **SHORT RESPONSE** All of a department store's jackets are 20% off. To the nearest cent, what is the total cost of a jacket if the original price is $74.99 and the sales tax is $8\frac{1}{2}$%?

Find each percent of change. Round to the nearest whole percent if necessary. State whether the percent of change is an *increase* or *decrease*. (Lesson 8-4)

35. original: 4
    new: 6

36. original: $556
    new: $500

37. original: 20.5
    new: 35.5

38. **FOOD** The graph shows the results of a survey in which magazine readers were asked to name their favorite take-out foods. Predict how many out of 5,000 readers would choose Chinese food as one of their favorite take-out foods. (Lesson 8-3)

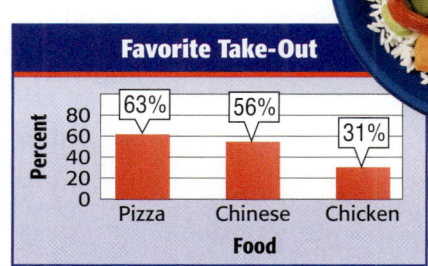

Favorite Take-Out
Pizza 63%, Chinese 56%, Chicken 31%
Source: Bon Appetit

GETTING READY FOR THE NEXT LESSON

PREREQUISITE SKILL Multiply. (Lesson 6-4)

39. $790 \cdot 0.185 \cdot 2$

40. $45 \cdot 0.04 \cdot 9$

41. $600 \cdot 0.03 \cdot 5\frac{1}{2}$

42. $162 \cdot 0.25 \cdot 3\frac{1}{2}$

msmath2.net/self_check_quiz

8-6 Simple Interest

What You'll LEARN
Solve problems involving simple interest.

NEW Vocabulary
simple interest
principal

WHEN am I ever going to use this?

INVESTING Brooke plans to invest $1,000 in a certificate of deposit (CD). The graph shows CD rates for one year at various banks.

1. Calculate 2.25% of $1,000 to find the amount of money that Brooke can earn in one year for a CD at State Credit Union.

Certificate of Deposit Rates

2. Find the amount of money that she can earn in one year at the other three banks.

When you deposit money in a CD, the amount that you earn is called interest. **Simple interest** is the amount paid or earned for the use of money. To find simple interest I, use the following formula.

Principal is the amount of money deposited or invested.

annual interest rate written as a decimal $I = prt$ amount of time that money is invested or borrowed (years)

EXAMPLES   Find Interest Earned

SAVINGS Raini has $750 in a savings account that pays 3% simple interest. How much interest will he earn in each amount of time?

① 4 years

$I = prt$ Formula for simple interest

$I = 750 \cdot 0.03 \cdot 4$ Replace p with $750, r with 0.03, and t with 4.

$I = 90$ Simplify.

Raini will earn $90 in interest in 4 years.

② 9 months

9 months = $\frac{9}{12}$ or 0.75 year   Write the time as years.

$I = prt$ Formula for simple interest

$I = 750 \cdot 0.03 \cdot 0.75$ $p = \$750, r = 0.03, t = 0.75$

$I \approx 16.88$ Simplify.

Raini will earn $16.88 in interest in 9 months.

358 Chapter 8 Applying Percent

READING Math

Formulas Read $I = prt$ as *Interest is equal to principal times rate times time.*

The formula $I = prt$ can also be used to find the interest owed when you borrow money. In this case, p is the amount of money borrowed, and t is the amount of time the money is borrowed.

EXAMPLE   Find Interest Paid on a Loan

3 LOANS Emilio's father borrows $1,200 from the bank for a riding lawn mower. The interest rate is 8% per year. How much simple interest will he pay if he takes 2 years to repay the loan?

$I = prt$	Formula for simple interest
$I = 1{,}200 \cdot 0.08 \cdot 2$	Replace p with $1,200, r with 0.08, and t with 2.
$I = 192$	Simplify.

Emilio's father will pay $192 in interest in 2 years.

EXAMPLE   Find Total Paid on a Credit Card

4 CREDIT CARDS Cory charged a $600 TV on his credit card with an interest rate of 21%. If he has no other charges on the card, how much money will he owe after one month?

$I = prt$	Formula for simple interest
$I = 600 \cdot 0.21 \cdot \frac{1}{12}$	Replace p with $600, r with 0.19, and t with $\frac{1}{12}$.
$I = 10.5$	Simplify.

The interest owed after one month is $10.50. So, the total amount owed would be $600 + $10.50 or $610.50.

Skill and Concept Check

1. **List** the steps you would use to find the simple interest on a $500 loan at 6% interest rate for 18 months.

2. **OPEN ENDED** Suppose you earn 3% on a $1,200 deposit for 5 years. Investigate how the interest is affected if the rate or the time is increased.

GUIDED PRACTICE

Find the interest earned to the nearest cent for each principal, interest rate, and time.

3. $640, 3%, 2 years
4. $1,500, 4.25%, 4 years

Find the interest paid to the nearest cent for each loan balance, interest rate, and time.

5. $4,500, 9%, 3.5 years
6. $290, 12.5%, 6 months

7. **HISTORY** In 2002, a $5 bank note from 1886 was sold to a collector for $103,500. Suppose a person had deposited the $5 in a bank in 1886 with an interest rate of 4%. After 116 years, how much simple interest would have been earned on the account?

Practice and Applications

Find the interest earned to the nearest cent for each principal, interest rate, and time.

8. $1,050, 4.6%, 2 years
9. $250, 2.85%, 3 years
10. $500, 3.75%, 1 year
11. $3,000, 5.5%, 2 years
12. $875, 6%, 4 months
13. $98.50, $6\frac{1}{2}$%, 16 months

HOMEWORK HELP

For Exercises	See Examples
8–13	1, 2
14–19	3
20–22	4

Extra Practice
See pages 584, 603.

Find the interest paid to the nearest cent for each loan balance, interest rate, and time.

14. $1,000, 7%, 2 years
15. $725, 6.25%, 1 year
16. $2,700, 8.2%, 3 years
17. $175.80, 12%, 1.25 years
18. $925, $19\frac{1}{2}$%, 3 months
19. $800, 10.5%, 30 months

INVESTING For Exercises 20–22, use the following information.
Marcus has $1,800 from his summer job to invest.

20. If he invests in a CD for 3 years at a rate of 5.25%, how much will the CD be worth after 3 years?

21. Suppose he invests the $1,800 for 2 years and earns $144. What was the rate of interest?

22. Marcus would like to have $2,340 altogether. If he invests his money at 5% interest, in how many years will he have $2,340?

23. **CRITICAL THINKING** Mrs. Williams deposits $600 in an account that pays 4.5% annually. At the end of the year, the interest earned is added to the principal. Find the total amount in her account each year for 3 years.

Spiral Review with Standardized Test Practice

24. **MULTIPLE CHOICE** Antonia opened a savings account that pays 6.5% simple interest. How much money will be in Antonia's account after 3 years if she deposited $250 at the beginning and never made any more deposits?

    Ⓐ $48.75   Ⓑ $248.75   Ⓒ $298.75   Ⓓ $300.00

25. **MULTIPLE CHOICE** Mr. McMahon bought a $562 freezer using a credit card that charges 18% annual interest. If he does not make any payments or any additional charges, how much will he owe after 1 month?

    Ⓕ $553.57   Ⓖ $570.43   Ⓗ $578.86   Ⓘ $663.16

26. Find the total cost of a $13.99 music CD if the tax rate is 7%. (Lesson 8-5)

Find each percent of change. Round to the nearest whole percent if necessary. State whether the percent of change is an *increase* or *decrease*. (Lesson 8-4)

27. original: 35
    new: 45

28. original: 60
    new: 38

29. original: $2.75
    new: $1.80

msmath2.net/self_check_quiz

8-6b Spreadsheet Investigation
A Follow-Up of Lesson 8-6

What You'll LEARN

Use a spreadsheet to calculate simple interest.

Simple Interest

A computer spreadsheet is a useful tool for quickly calculating simple interest for different values of principal, rate, and time.

ACTIVITY

Max plans on opening a "Young Savers" account at his bank. The current rate on the account is 4%. He wants to see how different starting balances, rates, and times will affect his account balance. To find the balance at the end of 2 years for different principal amounts, he enters the values B2 = 4 and C2 = 2 into the spreadsheet below.

For each principal given in column A, simple interest is calculated for any values of rate and time entered in B2 and C2, respectively.

The spreadsheet adds simple interest to the principal.

EXERCISES

1. Why is the rate in column B divided by 100?
2. What is the balance in Max's account after 2 years if the principal is $1,500 and the simple interest rate is 4%?
3. How much interest does Max earn in 2 years if his account has a principal of $2,000 and an interest rate of 4%?
4. Suppose you wanted to add a new row to the spreadsheet that represents a principal of $3,000. List each of the cell entries (A8, B8, C8, D8, and E8) that you would enter.
5. What entries for cells B2 and C2 would you use to calculate the simple interest on a principal of $1,500 at a rate of 7% for a 9-month period? What is the balance of this account at the end of the 9 months?
6. Explain how a spreadsheet is more useful than a pencil and paper when finding simple interest.

Lesson 8-6b Spreadsheet Investigation: Simple Interest **361**

Chapter 8 Study Guide and Review

Vocabulary and Concept Check

discount (p. 355)
percent equation (p. 340)
percent of change (p. 350)
percent of decrease (p. 350)
percent of increase (p. 350)
population (p. 345)
principal (p. 358)
random sample (p. 345)
sales tax (p. 354)
simple interest (p. 358)
survey (p. 345)

State whether each sentence is *true* or *false*. If *false*, replace the underlined word, number, or equation to make a true sentence.

1. The sale price of an item is the regular price <u>minus</u> the discount.
2. When taking a survey, the total group of people that the survey is interested in is called the <u>sample</u>.
3. To find a percent of increase, compare the amount of the increase to the <u>new</u> amount.
4. The formula for simple interest is <u>$I = prt$</u>.
5. A sample is representative of the population if it is <u>random</u>.
6. A method for estimating the percent of a number is to find <u>21%</u> of the number and then multiply.
7. The percent equation is <u>part = percent · base</u>.
8. A <u>tax</u> is the amount by which the regular price of an item is reduced.
9. The <u>principal</u> is the amount of money deposited or borrowed.
10. A <u>sample</u> is a question or set of questions designed to collect data about a specific group of people.

Lesson-by-Lesson Exercises and Examples

8-1 Percent and Estimation (pp. 334–337)

Estimate by using fractions.

11. 25% of 81
12. 33% of 122
13. 77% of 38
14. 19.5% of 96

Estimate by using 10%.

15. 12% of 77
16. 88% of 400
17. 52% of 1,000
18. 21% of 53

19. **PETS** About 12% of 291 households in a neighborhood have fish. Estimate how many households have fish.

Example 1 Estimate 52% of 495.

52% is about 50% or $\frac{1}{2}$, and 495 is about 500.

52% of 495 ≈ $\frac{1}{2}$ · 500 or 250

So, 52% of 495 is about 250.

Example 2 Estimate 68% of 80.

10% of 80 = 0.1 · 80 or 8 Find 10% of 80.
68% is about 70%.
7 · 8 = 56 70% of 80 ≈ 7 · (10% of 80)

So, 68% of 80 is about 56.

msmath2.net/vocabulary_review

8-2 Algebra: The Percent Equation (pp. 340–343)

Write an equation for each problem. Then solve. Round to the nearest tenth if necessary.

20. 32 is what percent of 50?
21. 65% of what number is 39?
22. Find 42% of 300.
23. 7% of 92 is what number?
24. 12% of what number is 108?

Example 3 27 is what percent of 90?

27 is the part and 90 is the base.
Let n represent the percent.

part = percent · base

$27 = n \cdot 90$ Write an equation.

$\dfrac{27}{90} = \dfrac{90n}{90}$ Divide each side by 90.

$0.3 = n$ The percent is 30%.

So, 27 is 30% of 90.

8-3 Statistics: Using Statistics to Predict (pp. 345–347)

CAREERS In a university survey, 5.5% of the incoming freshmen chose elementary teacher as a career goal and 6.4% chose engineer.

25. Predict how many of the 3,775 freshmen would choose a career as an elementary teacher.
26. How many of the 3,775 freshmen would you expect to choose a career as an engineer?

Example 4 In a survey of 150 students at McAuliffe Middle School, 12% said they have after-school jobs. Predict how many of the 644 students at the school have after-school jobs.

Find 12% of 644.

$n = 0.12 \cdot 644$ Write an equation.

$ = 77.28$ Multiply.

So, you could predict that about 77 students at McAuliffe Middle School have after-school jobs.

8-4 Percent of Change (pp. 350–353)

Find each percent of change. Round to the nearest whole percent if necessary. State whether the percent of change is an *increase* or *decrease*.

27. original: 172
    new: 254
28. original: $200
    new: $386
29. original: 75
    new: 60
30. original: $49.95
    new: $54.95

31. **GAMES** A computer game that sold for $24.95 last year is now priced at $27.95. Find the percent of change.

Example 5 A magazine that originally cost $2.75 is now $3.55. Find the percent of change. Round to the nearest whole percent.

The new price is greater than the original price, so this is a percent of increase.

amount of increase = 3.55 − 2.75 or 0.80

percent of increase = $\dfrac{\text{amount of increase}}{\text{original amount}}$

$= \dfrac{0.80}{2.75}$ Substitution

≈ 0.29 Simplify.

The percent of increase is about 29%.

Study Guide and Review continued

Mixed Problem Solving
For mixed problem-solving practice, see page 603.

8-5 Sales Tax and Discount (pp. 354–357)

Find the total cost or sale price to the nearest cent.

32. $25 backpack; 7% tax
33. $210 bicycle; 15% discount
34. $8,000 car; $5\frac{1}{2}$% tax
35. $40 sweater; 33% discount
36. $6.25 address book, 40% discount

Find the percent of discount to the nearest percent.

37. shirt: regular price: $42
    sale price: $36
38. snack: regular price, $2.50
    sale price: $1
39. boots: regular price, $78
    sale price: $70
40. DVD: regular price, $24.99
    sale price: $19.99

Example 6 A new computer system is priced at $2,499. Find the total cost if the sales tax is 6.5%.

First, find the sales tax.
6.5% of $2,499 = 0.065 · 2,499
             ≈ 162.44

Next, add the sales tax to the original price. The total cost is 162.44 + 2,499 or $2,661.44.

Example 7 A pass at a water park is $58. At the end of the season, the same pass costs $46.40. What is the percent of discount?

58 − 46.40 = 11.60   Find the amount of discount.

Next, find what percent of 58 is 11.60.
11.60 = n · 58   Write an equation.
0.2 = n   Divide each side by 58.

The percent of discount is 20%.

8-6 Simple Interest (pp. 358–360)

Find the interest earned to the nearest cent for each principal, interest rate, and time.

41. $475, 5%, 2 years
42. $5,000, 10%, 3 years
43. $2,500, 11%, $1\frac{1}{2}$ years

Find the interest paid to the nearest cent for each loan balance, interest rate, and time.

44. $3,200, 8%, 4 years
45. $450, 13.5%, 2 years
46. $1,980, 21%, 9 months
47. **LOANS** Brian has a loan balance of $1,000. If he pays off the balance over 2 years at an annual simple interest rate of 18%, what is the total amount that he will pay?

Example 8 Find the interest earned on $400 at 9% for 3 years.

$I = prt$ Formula for simple interest
$I = 400 · 0.09 · 3$ $p = \$400, r = 0.09, t = 3$
$I = 108$ Simplify.

The interest earned is $108.

Example 9 Elisa has a loan for $1,300. The interest rate is 7%. If she pays it off in 6 months, how much interest will she pay?

$I = prt$ Formula for simple interest
$I = 1,300 · 0.07 · 0.5$ $p = \$1,300, r = 0.07, t = 0.5$
$I = 45.5$ Simplify.

The interest she will pay after 6 months is $45.50.

Chapter 8 Practice Test

Vocabulary and Concepts

1. **Describe** percent of change.
2. **State** the formula used to compute simple interest.

Skills and Applications

Estimate.

3. 18% of 246
4. 145% of 81
5. 71% of 324

Write an equation for each problem. Then solve. Round to the nearest tenth if necessary.

6. Find 14% of 65.
7. What number is 36% of 294?
8. 82% of what number is 73.8?
9. 75 is what percent of 50?

BOOKS For Exercises 10 and 11, refer to the table. It shows the results of a survey in which students at Haskell Middle School were asked to name their favorite types of fiction.

Type of Fiction	Percent
historical fiction	8%
mystery	24%
science fiction	38%
sports	30%

10. Predict how many of the 845 students at Haskell Middle School would choose science fiction as their favorite type.
11. How many of the 845 students would you expect to select mystery as their favorite type of fiction?

Find each percent of change. Round to the nearest whole percent if necessary. State whether the percent of change is an *increase* or *decrease*.

12. original: $60
    new: $75
13. original: 145
    new: 216
14. original: 48
    new: 40

Find the total cost or sale price to the nearest cent.

15. $1,730 treadmill, $6\frac{1}{2}$% sales tax
16. $16 hat, 55% discount

Find the interest earned for each principal, interest rate, and time.

17. $3,000, 5.5%, 5 years
18. $2,600, 4%, 3 months

19. **LOANS** Leah borrows $2,200 to buy new furniture. Her loan has an annual interest rate of 16%. Find the simple interest that Leah will owe after 1 year.

Standardized Test Practice

20. **MULTIPLE CHOICE** James earned $38 last week from mowing lawns in his neighborhood. This week, he earned $52. What was the percent of change?

    A 37% decrease   B 27% decrease   C 27% increase   D 37% increase

CHAPTER 8 Standardized Test Practice

PART 1 Multiple Choice

Record your answers on the answer sheet provided by your teacher or on a sheet of paper.

1. Students at Karlon School are collecting canned foods for the local food pantry. The graph shows how many cans students in four of the grades collected in the first week.

   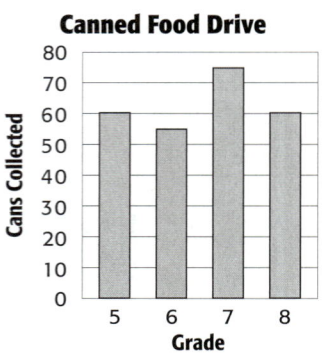

   How many total cans did students in the 5th, 6th, 7th, and 8th grades collect during the first week? (Lesson 2-7)

   Ⓐ 75 cans Ⓑ 120 cans
   Ⓒ 240 cans Ⓓ 250 cans

2. Greenapple Books has 25 copies of this week's national best-selling book. On Friday, 12 copies of the book were sold. On Saturday, 7 copies of the book were returned. Which represents the number of copies of this book that are available? (Lesson 3-5)

   Ⓕ −5 Ⓖ 5 Ⓗ 19 Ⓘ 20

3. Find the slope of the line graphed below. (Lesson 4-7)

   Ⓐ −2
   Ⓑ −$\frac{1}{2}$
   Ⓒ $\frac{1}{2}$
   Ⓓ 2

   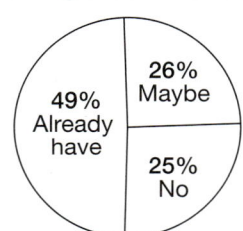

4. How far will you travel after 10 complete turns of the bicycle wheel shown at the right? Use 3.14 for π. (Lesson 6-9)

   Ⓕ 84.8 in.
   Ⓖ 270 in.
   Ⓗ 847.8 in.
   Ⓘ 1,695.6 in.

5. The graph shows the results of an online poll to which 1,721 high school students responded.

   Getting a Summer Job
   - 49% Already have
   - 26% Maybe
   - 25% No

   Source: Alloy online poll

   Estimate how many of the students already had a summer job when they responded to the poll. (Lesson 8-1)

   Ⓐ 50 Ⓑ 500 Ⓒ 850 Ⓓ 1,850

6. Zoe finished reading 85% of her book. If the book is 280 pages long, how many pages has she read? (Lesson 8-2)

   Ⓕ 185 Ⓖ 208 Ⓗ 238 Ⓘ 265

7. A survey showed that 85% of youths who went to summer camp when they were 11 years old also attended summer camp when they were 12 years old. Out of 300 11-year-olds who went to summer camp this year, predict how many will go to summer camp next year. (Lesson 8-3)

   Ⓐ 385 Ⓑ 255 Ⓒ 245 Ⓓ 45

PART 2 Short Response/Grid In

Record your answers on the answer sheet provided by your teacher or on a sheet of paper.

8. Jamal bought 2.4 meters of blue ribbon and 110 centimeters of white ribbon. How many centimeters longer is the blue ribbon than the white ribbon? (Lesson 1-8)

9. Write a percent to represent the shaded area of the model. (Lesson 5-5)

10. A 5-pound bag of potatoes costs $3.75. At that rate, how much does a 4-pound bag of the same potatoes cost? (Lesson 7-3)

11. A survey showed that 9 out of 10 teens expect to make a charitable contribution or volunteer during the holidays. Based on that survey, how many teenagers in a class of 400 expect to give money or volunteer the next holiday season? (Lesson 8-3)

12. What was the percent of increase in the price of first class stamps from 2001 to 2002? Round to the nearest whole percent. (Lesson 8-4)

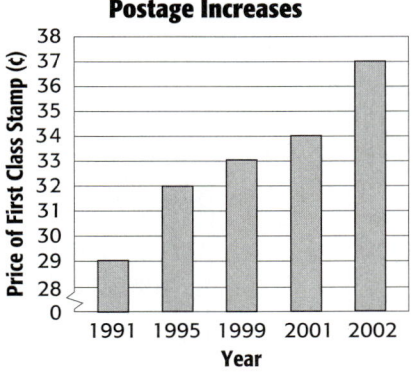

Postage Increases

Source: U.S. Postal Service

13. Midtown Veterinarians cared for 125 animals on Monday and 100 animals on Tuesday. What is the percent of decrease in number of animals? (Lesson 8-4)

14. Mr. Martinez wants to purchase a sound system that costs $245. If 6.2% sales tax is added, what will be the total cost of the system? (Lesson 8-5)

15. Kaitlyn and Camilia each have money in savings accounts, as shown below.

Money in the Bank		
Name	Principal	Rate
Kaitlyn	$400	3%
Camilia	$300	6%

Whose savings account will earn more interest after 1 year? (Lesson 8-6)

16. If a principal amount of $600 earned $75 in $2\frac{1}{2}$ years, find the interest rate. (Lesson 8-6)

PART 3 Extended Response

Record your answers on a sheet of paper. Show your work.

17. The Sybil family is working on its budget.
    a. Their rent increased from $750 to $907.50 a month. What is the percent of increase? (Lesson 8-4)
    b. Their landlord said that she would decrease the $907.50 monthly rent payment by 6% if the family swept the stairwell of the apartment building each week. If the Sybil family takes the landlord's offer, describe two ways to find how much rent they would pay. (Lesson 8-5)

TEST-TAKING TIP

Question 17 Remember to show all of your work. You may be able to get partial credit for your answers, even if they are not entirely correct.

CHAPTER 9 Probability

"What do games have to do with math?"

Rolling number cubes allows you to advance forward or backward in many board games. **Rolling doubles on a pair of number cubes happens only about 17% of the time.** In mathematics, you can use probability to help you determine how likely it is for events to take place.

You will solve problems about games in Lesson 9-1.

GETTING STARTED

▶ Diagnose Readiness

Take this quiz to see if you are ready to begin Chapter 9. Refer to the lesson number in parentheses for review.

Vocabulary Review
Choose the correct term or number to complete each sentence.

1. A fraction is in (factored, simplest) form when the GCF of the numerator and denominator is 1. (Lesson 5-3)

2. The fraction $\frac{12}{38}$ in simplest form is $\left(\frac{6}{19}, \frac{3}{17}\right)$. (Lesson 5-3)

Prerequisite Skills
Multiply.

3. 7×15
4. 24×6
5. 13×4
6. 8×21
7. 5×32
8. 30×8
9. $6 \cdot 5 \cdot 4 \cdot 3$
10. $7 \cdot 6 \cdot 5$
11. $8 \cdot 7 \cdot 6$
12. $4 \cdot 3 \cdot 2 \cdot 1$
13. $10 \cdot 9 \cdot 8 \cdot 7$
14. $11 \cdot 10 \cdot 9$

Write each fraction in simplest form. Write *simplified* if the fraction is already in simplest form. (Lesson 5-3)

15. $\frac{8}{12}$
16. $\frac{3}{18}$
17. $\frac{4}{9}$
18. $\frac{5}{15}$

Find each value. (Lesson 5-3)

19. $\frac{6 \cdot 5}{3 \cdot 2}$
20. $\frac{9 \cdot 8 \cdot 7}{5 \cdot 4 \cdot 3}$
21. $\frac{4 \cdot 3 \cdot 2}{3 \cdot 2 \cdot 1}$
22. $\frac{7 \cdot 6 \cdot 5 \cdot 4}{4 \cdot 3 \cdot 2 \cdot 1}$

Probability Make this Foldable to organize topics in this chapter. Begin with four sheets of $8\frac{1}{2}''$ by $11''$ paper.

STEP 1 Stack Pages
Place 4 sheets of paper $\frac{3}{4}$ inch apart.

STEP 2 Roll Up Bottom Edges
All tabs should be the same size.

STEP 3 Crease and Staple
Staple along fold.

STEP 4 Label
Write the chapter title on the front. Label each tab with a lesson number and title.

 Chapter Notes Each time you find this logo throughout the chapter, use your *Noteables™: Interactive Study Notebook with Foldables™* or your own notebook to take notes. Begin your chapter notes with this Foldable activity.

 Readiness To prepare yourself for this chapter with another quiz, visit **msmath2.net/chapter_readiness**

9-1 Simple Events

What You'll LEARN

Find the probability of a simple event.

NEW Vocabulary

outcome
simple event
probability
random
complementary event

WHEN am I ever going to use this?

CANDY A box of saltwater taffy contains 48 pieces, six pieces of each flavor shown at the right.

peppermint	chocolate
grape	raspberry
root beer	orange creme
cherry	vanilla

1. What fraction of the taffy is vanilla? Write in simplest form.

2. Suppose you take one piece of taffy from the box without looking. Are your chances of picking vanilla the same as picking root beer? Explain.

The 48 pieces of taffy in the box above are called **outcomes**. A **simple event** is one outcome or a collection of outcomes. For example, picking a piece of vanilla taffy is a simple event. The chance of that event happening is called **probability**.

Noteables™ — Key Concept: Probability

Words The probability of an event is a ratio that compares the number of favorable outcomes to the number of possible outcomes.

Symbols $P(\text{event}) = \dfrac{\text{number of favorable outcomes}}{\text{number of possible outcomes}}$

Outcomes occur at **random** if each outcome occurs by chance. A piece of vanilla taffy was selected at random in the activity above.

Read P(vanilla) as the probability of picking a piece of vanilla taffy.

$P(\text{vanilla}) = \dfrac{6}{48}$ ← number of pieces of vanilla taffy / total number of pieces of taffy

$= \dfrac{1}{8}$, or 12.5%   Simplify.

EXAMPLE Find Probability

1 What is the probability of rolling an even number on a number cube marked with 1, 2, 3, 4, 5, and 6 on its faces?

$P(\text{even number}) = \dfrac{\text{even numbers possible}}{\text{total numbers possible}}$

$= \dfrac{3}{6}$ Three numbers are even: 2, 4, and 6.

$= \dfrac{1}{2}$ Simplify.

The probability of rolling an even number is $\dfrac{1}{2}$ or 50%.

370 Chapter 9 Probability

EXAMPLE Find Probability

2 **PARCHEESI** Jewel rolls two number cubes. She can move her game piece if a total of 5 is shown on the number cubes, or if a 5 is shown on at least one number cube. What is the probability that she can move her game piece on one roll of the number cubes?

List all the possible outcomes. Then, find the pairs that total 5, or if a 5 is shown on at least one of the number cubes.

1, 1	1, 2	1, 3	1, 4	1, 5	1, 6
2, 1	2, 2	2, 3	2, 4	2, 5	2, 6
3, 1	3, 2	3, 3	3, 4	3, 5	3, 6
4, 1	4, 2	4, 3	4, 4	4, 5	4, 6
5, 1	5, 2	5, 3	5, 4	5, 5	5, 6
6, 1	6, 2	6, 3	6, 4	6, 5	6, 6

There are 36 possible outcomes and 15 of them are favorable. So, the probability that Jewel moves her game piece in one roll is $\frac{15}{36}$, or $\frac{5}{12}$.

REAL-LIFE MATH

PARCHEESI Parcheesi is derived from the Indian word *pacis*, which means 25.
Source: www.yesterdayland.com

The probability that an event will happen is somewhere between 0 and 1. It can be shown on a number line.

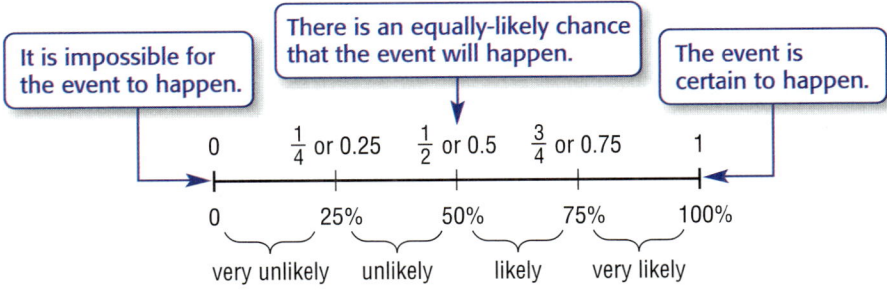

Either Jewel will be able to move her game piece or she will *not* be able to move it. Two events that are the only ones that can possibly happen are examples of **complementary events**.

EXAMPLE Find a Complementary Event

3 **PARCHEESI** Refer to Example 2. Find the probability that Jewel *cannot* move her game piece on one roll of the number cubes.

$P(A) + P(\text{not } A) = 1$

$\frac{5}{12} + P(\text{not } A) = 1$ Substitute $\frac{5}{12}$ for $P(A)$.

$-\frac{5}{12} \qquad\qquad -\frac{5}{12}$ Subtract $\frac{5}{12}$ from each side.

$P(\text{not } A) = \frac{7}{12}$ Simplify.

So, the probability that Jewel *cannot* move her game piece is $\frac{7}{12}$.

STUDY TIP

Check Your Answer You can check the answer by counting the number of outcomes that do not total 5 and that do not have a 5 on at least one number cube.

msmath2.net/extra_examples

Lesson 9-1 Simple Events **371**

Skill and Concept Check

1. **Explain** why an event with a probability of 0.7 is likely to happen.

2. **OPEN ENDED** Describe an event that has a probability of $\frac{1}{5}$.

3. **Which One Doesn't Belong?** Identify the pair of probabilities that do not represent probabilities of complementary events. Explain.

| $\frac{5}{8}, \frac{3}{8}$ | 0.65, 0.55 | $\frac{4}{6}, \frac{1}{3}$ | 0.875, $\frac{1}{8}$ |

GUIDED PRACTICE

Use the spinner at the right to find each probability. Write as a fraction in simplest form.

4. $P(J)$
5. $P(\text{vowel})$

A bag contains 7 blue, 5 purple, 12 red, and 6 orange marbles. Find each probability if you draw one marble at random from the bag. Write as a fraction in simplest form.

6. $P(\text{purple})$
7. $P(\text{red or orange})$
8. $P(\text{not blue})$

9. **CANDY** Refer to the activity on the top of page 370. What is the probability of picking a piece of taffy that is *not* peppermint nor root beer? Write the probability as a percent.

Practice and Applications

A set of 20 cards is numbered 1, 2, 3, …, 20. Suppose you pick a card at random without looking. Find the probability of each event. Write as a fraction in simplest form.

10. $P(1)$
11. $P(\text{not a factor of 10})$
12. $P(\text{multiple of 3})$
13. $P(\text{even number})$
14. $P(\text{less than or equal to 20})$
15. $P(3 \text{ or } 13)$

16. How likely is it that an event with a probability of 0.28 will occur?

17. The forecast for tomorrow says that there is a 37% chance of rain. Describe the complementary event and its probability.

HOMEWORK HELP

For Exercises	See Examples
10–15	1
18, 20–23, 25–27	2
17, 19, 24, 28–29	3

Extra Practice
See pages 584, 604.

STUDENT COUNCIL The table shows the members of the Student Council. Suppose one student is randomly selected as the president. Find the probability of each event. Write as a fraction in simplest form.

18. $P(\text{girl})$
19. $P(\text{not 7th grader})$
20. $P(\text{boy})$
21. $P(\text{8th grader})$
22. $P(\text{boy or girl})$
23. $P(\text{6th or 8th grader})$
24. $P(\text{not 6th grader})$
25. $P(\text{5th grader})$

Student Council	
girls	30
boys	20
8th graders	25
7th graders	15
6th graders	10

26. Which event has a greater chance of happening: picking a president who is a girl or a president who is not an 8th grader? Explain.

27. **TECHNOLOGY** The graph shows the cost of 22 digital cameras. If one of the 22 cameras is chosen at random, what is the probability that it costs between $111 and $160?

28. **MOVIES** The probability of buying a defective DVD is 0.002. What is the probability of buying a DVD that is *not* defective?

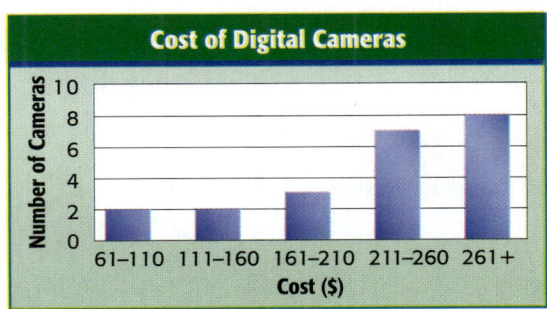

29. **MULTI STEP** The Jefferson Middle School Booster Club is selling raffle tickets for a new computer system. They sold 1,000 tickets at $2 each. Emilia's parents spent $200 on tickets. What is the probability that they will *not* win?

30. **CRITICAL THINKING** Melissa and Hakan are playing a game by rolling two number cubes. Hakan gets a point each time the sum of the number cubes is 2, 3, 4, 9, 10, 11, or 12. Melissa gets a point when the sum is 5, 6, 7, or 8. Does each player have an equal chance to win? Explain.

Spiral Review with Standardized Test Practice

31. **MULTIPLE CHOICE** A bookshelf contains the books at the right. If you randomly choose a book from the shelf, what is the probability that it is science fiction?

    A $\frac{1}{10}$ **B** $\frac{3}{20}$ **C** $\frac{9}{40}$ **D** $\frac{1}{4}$

32. **MULTIPLE CHOICE** In a school raffle, one ticket will be drawn out of a total of 500 tickets. If the Hawkins family has 12 tickets, what is the probability that they will win?

    F 0.002   **G** 2%   **H** $\frac{3}{125}$ **I** $0.08\overline{3}$

Find the interest earned to the nearest cent for each principal, interest rate, and time. (Lesson 8-6)

33. $300, 10%, 2 years

34. $900, 5.5%, 4.5 years

35. **FOOD** The United States produced almost 11 billion pounds of apples in a recent year. Use the information in the graph to find how many pounds of apples were used to make juice and cider. (Lesson 7-8)

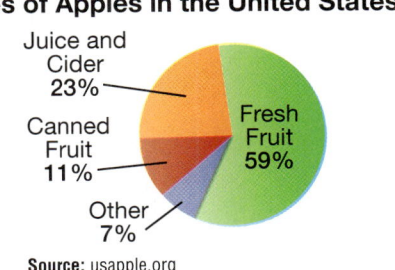

Write each fraction or mixed number as a decimal. Use bar notation if necessary. (Lesson 5-4)

36. $\frac{3}{5}$ 37. $\frac{9}{20}$ 38. $6\frac{1}{8}$ 39. $1\frac{7}{9}$

GETTING READY FOR THE NEXT LESSON

PREREQUISITE SKILL Write each fraction in simplest form. Write *simplified* if the fraction is already in simplest form. (Lesson 5-3)

40. $\frac{2}{6}$ 41. $\frac{3}{8}$ 42. $\frac{15}{30}$ 43. $\frac{6}{16}$ 44. $\frac{18}{32}$

9-2 Tree Diagrams

What You'll LEARN
Use tree diagrams to count outcomes and find probabilities.

NEW Vocabulary
fair game
tree diagram
sample space

HANDS-ON Mini Lab

Materials
- 2 counters
- marker

Work with a partner.

Here is a probability game that you can play with two counters.

- Mark one side of the first counter A. Mark the other side B. Mark both sides of the second counter A.
- Player 1 tosses the counters. If both sides shown are the same, Player 1 wins a point. If the sides are different, Player 2 wins a point. Record your results.
- Player 2 then tosses the counters and the results are recorded. Continue alternating the tosses until each player has tossed the counters ten times. The player with the most points wins.

1. Before you play, **make a conjecture**. Do you think that each player has an equal chance of winning? Explain.
2. Now, play the game. Who won? What was the final score?
3. Collect the data from the entire class. What is the combined score for Player 1 versus Player 2?
4. Do you want to change the conjecture you made in Exercise 1? Explain.

A game in which players of equal skill have an equal chance of winning is a **fair game**. One way you can analyze whether games are fair is by drawing a **tree diagram**. A tree diagram is used to show all of the possible outcomes, or **sample space**, in a probability experiment.

EXAMPLE Draw a Tree Diagram

1 GAMES Refer to the Mini Lab above. Draw a tree diagram to show the sample space. Then determine whether the game is fair.

There are four equally-likely outcomes with two favoring each player. So, the probability that each player can win is $\frac{1}{2}$. Thus, the game is fair.

EXAMPLE Find the Number of Outcomes

REAL-LIFE MATH

SCOOTERS Kickboard scooters can travel approximately 4 miles per hour.
Source: www.factmonster.com

2 SCOOTERS A certain type of kickboard scooter comes in silver, red, or purple with wheel sizes of 125 millimeters or 180 millimeters. Find the total number of color-wheel size combinations.

Make a tree diagram to show the sample space.

Scooter Color	Wheel Size	Sample Space
silver	125 mm	silver, 125 mm
silver	180 mm	silver, 180 mm
red	125 mm	red, 125 mm
red	180 mm	red, 180 mm
purple	125 mm	purple, 125 mm
purple	180 mm	purple, 180 mm

There are six different color and wheel size combinations.

Your Turn

a. How many outcomes are there if the manufacturer added an orange color and a 150-millimeter wheel size?

You can also use tree diagrams to help you find the probability of events.

EXAMPLE Find Probability Using Tree Diagrams

3 COINS Suppose Pablo tosses three pennies. Find the probability that all three will show heads.

Make a tree diagram to show the sample space. Then, find the probability of the three pennies showing heads.

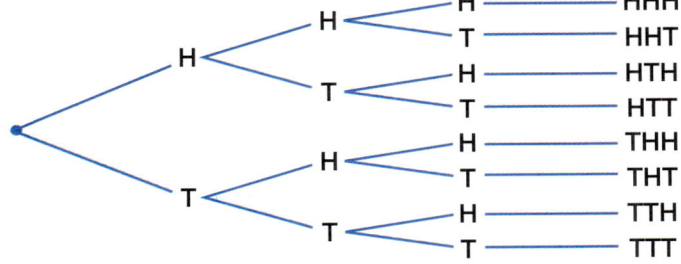

The sample space contains 8 possible outcomes. Only 1 outcome has all pennies showing heads. So, the probability of three pennies showing heads is $\frac{1}{8}$, or 12.5%.

Your Turn Find each probability.

b. P(3 tails)   c. P(exactly 2 heads)   d. P(at least 1 tail)

msmath2.net/extra_examples   Lesson 9-2 Tree Diagrams

Skill and Concept Check

1. **Writing Math** Describe a fair game between two players using one coin.

2. **OPEN ENDED** Give an example of a situation that has 8 outcomes.

GUIDED PRACTICE

The spinner is spun twice.

3. Draw a tree diagram to represent the situation.
4. How many outcomes are possible?
5. What is the probability of spinning two Cs?

6. **DELICATESSEN** A neighborhood deli sells sandwiches that can be made with ham, turkey, roast beef, salami, or bologna on rye, white, pumpernickel, or sourdough breads. Make a tree diagram to show all of the possible meat-bread choices.

Practice and Applications

For each situation, make a tree diagram to show the sample space. Then give the total number of outcomes.

HOMEWORK HELP
For Exercises	See Examples
7–15, 18, 21	1, 2
16–17, 19	3

Extra Practice
See pages 585, 604.

7. tossing a coin and rolling a number cube
8. choosing black, blue, or brown socks with boots, gym shoes, or dress shoes
9. picking a number from 1 to 5 and choosing the color red, white, or blue
10. choosing a card with a shape and spinning the spinner from the choices at the right

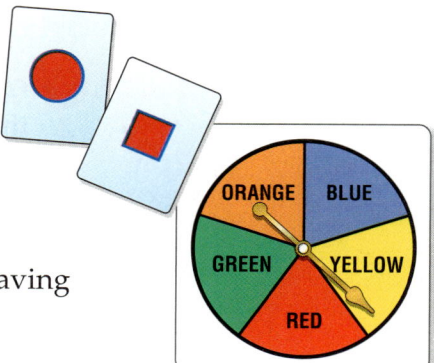

11. choosing a letter from the word SPACE and choosing a consonant from the word MATH
12. choosing a purple, green, black, or silver mountain bike having 10, 18, 21, or 24 speeds
13. tossing a quarter, a dime, and a nickel
14. rolling a number cube, tossing a coin, and choosing a card from the cards marked A and B

FAMILY For Exercises 15–17, use the information below.
Mr. and Mrs. Chen have three children. Suppose the chance of having either a boy or a girl is 50%.

15. Draw a tree diagram showing the possible arrangements of gender of the three children.
16. What is the probability of having 2 boys and 1 girl?
17. What is the probability of having the first two children boys and the last child a girl?

376 Chapter 9 Probability

UNIFORMS For Exercises 18 and 19, use the information below and at the right.

	Home Jersey	Road Jersey	Practice Jersey	Cap
White	2	0	2	0
Black	1	0	0	1
Gray	0	1	0	0
Orange	1	1	0	2

Hunter is a big fan of the Houston Astros baseball team and wears a different jersey and cap every time he goes to a game. The table shows the number of different jerseys and caps Hunter owns.

18. How many jersey/cap combinations can Hunter wear when he goes to the baseball game to cheer on the Astros?

19. If Hunter picks a jersey/cap combination at random, what is the probability that he will wear a home jersey with a black cap?

20. **RESEARCH** Use the Internet or another resource to find the number of jerseys and caps your favorite major league baseball team has as part of its uniform. How many jersey/cap combinations are there for the team you picked?

21. **GAMES** Determine whether the following game for two players is fair. Explain.

	Side 1	Side 2
Counter 1	A	B
Counter 2	A	C
Counter 3	B	C

- Three counters are labeled according to the table at the right.
- Toss the three counters.
- If exactly 2 counters match, Player 1 scores a point. Otherwise, Player 2 scores a point.

22. **CRITICAL THINKING** Refer to Exercise 21. Adjust the scoring of the game so that it is fair.

Spiral Review with Standardized Test Practice

23. **MULTIPLE CHOICE** How many three-course dinners are possible if Yolanda chooses from an appetizer of soup or salad; an entree of steak, chicken, or fish; and a dessert of cake or pie?

    Ⓐ 7   Ⓑ 10   Ⓒ 12   Ⓓ 15

24. **SHORT RESPONSE** Find the sample space of the tree diagram.

PROBABILITY A spinner is equally likely to stop on each of its regions numbered 1 to 20. Find each probability as a fraction in simplest form. (Lesson 9-1)

25. a prime number   26. GCF(12, 18)   27. multiple of 2 or 3

28. **INTEREST** If Carlota invests $2,100 in a CD for 5 years at a simple interest rate of 4.75%, how much will the CD be worth after 5 years? (Lesson 8-7)

GETTING READY FOR THE NEXT LESSON

BASIC SKILL Multiply.

29. $7 \cdot 22$ 30. $5 \cdot 36$ 31. $11 \cdot 16$ 32. $23 \cdot 20$ 33. $131 \cdot 4$

9-3 The Fundamental Counting Principle

What You'll LEARN

Use multiplication to count outcomes.

NEW Vocabulary

Fundamental Counting Principle

WHEN am I ever going to use this?

RETAIL SALES The Jean Factory sells juniors' jeans in different sizes and lengths. The table shows what they have available.

Sizes	Lengths
3	petite
5	regular
7	tall
9	
11	
13	
15	

1. According to the table, how many sizes of juniors' jeans are there?
2. How many lengths are there?
3. Find the product of the two numbers you found in Exercises 1 and 2.
4. Draw a tree diagram to help you find the number of different size and length combinations. How does the number of outcomes compare to the product you found above?

In the activity above, you discovered that multiplication, instead of a tree diagram, can be used to find the number of possible outcomes in a sample space. This is called the **Fundamental Counting Principle**.

> **Noteables** — **Key Concept: Fundamental Counting Principle**
>
> If event M can occur in m ways and is followed by event N that can occur in n ways, then the event M followed by N can occur in $m \times n$ ways.

EXAMPLE Use the Fundamental Counting Principle

1 FOOD A famous steak house allows customers to create their own steak dinners. The choices are shown at the right. How many different steak dinners are possible?

451 Steakhouse

Steak	How Steaks Are Cooked	Potatoes
New York Strip	rare	mashed
Ribeye	medium	baked
Filet	well	twice baked
Porterhouse		au gratin
T-Bone		

$$\underbrace{5}_{\text{types of steaks}} \cdot \underbrace{3}_{\substack{\text{number of ways} \\ \text{steaks can be} \\ \text{cooked}}} \cdot \underbrace{4}_{\text{types of potatoes}} = \underbrace{60}_{\substack{\text{total number of} \\ \text{steak dinners}}}$$

There are 60 different ways of choosing a steak dinner.

Check You can check your work by drawing a tree diagram and listing the 60 outcomes.

READING in the Content Area

For strategies in reading this lesson, visit **msmath2.net/reading**.

Standardized Test Practice

EXAMPLE Find Outcomes

2 MULTIPLE-CHOICE TEST ITEM There are 2 roads connecting Eastland and Harping, 3 roads connecting Harping and Sinclair, and 2 roads connecting Sinclair and Johnstown. Find the number of ways to drive from Eastland to Johnstown.

Ⓐ 8 Ⓑ 12 Ⓒ 16 Ⓓ 20

Read the Test Item
To find the number of ways to drive from Eastland to Johnstown, multiply the number of roads from Eastland to Harping, from Harping to Sinclair, and from Sinclair to Johnstown.

Solve the Test Item
There are 2 roads from Eastland to Harping, 3 roads from Harping to Sinclair, and 2 roads from Sinclair to Johnstown. So, there are $2 \cdot 3 \cdot 2$, or 12 ways to drive from Eastland to Johnstown. So the answer is B.

Test-Taking Tip

Educated Guess Find out if there is a penalty for incorrect answers. If there is no penalty, making an educated guess can only increase your score, or at worst, leave your score the same.

Skill and Concept Check

1. **Explain** how to use the Fundamental Counting Principle to count outcomes.

2. **OPEN ENDED** Give an example of a situation that has 48 outcomes using three events.

3. **NUMBER SENSE** Can a pair of events have a number of outcomes that is prime? Explain.

4. **Which One Doesn't Belong?** Identify the pair of events that do not have the same number of outcomes as the other three. Explain your reasoning.

| 12 meats, 8 breads | 16 T-shirts, 6 colors |
| 18 hats, 8 sizes | 24 teams, 4 sports |

GUIDED PRACTICE

Use the Fundamental Counting Principle to find the total number of outcomes in each situation.

5. tossing a quarter, a dime, and a nickel

6. choosing scrambled, sunny-side up, or poached eggs with bacon or sausage and milk, orange juice, or apple juice

7. choosing a card from twenty different cards and spinning the spinner at the right

8. **STUDENT GOVERNMENT** If Opal, Jacob, Luanda, Tyrone, and Erica have an equal chance of being an officer of student council, what is the probability that Luanda will be an officer?

Practice and Applications

Use the Fundamental Counting Principle to find the total number of outcomes in each situation.

9. rolling a number cube and tossing two coins
10. choosing a number from 1 to 20 and a color from 7 colors
11. choosing a plain, blueberry, garlic, or cinnamon-and-raisin bagel, with plain, chive, or sun-dried tomato cream cheese
12. choosing iced tea in regular, raspberry, lemon, or peach flavors; sweetened or unsweetened; and in a glass or in a plastic container
13. picking a date in the month of May and a day of the week
14. choosing a 4-digit Personal Identification Number (PIN) if the digits cannot be repeated

15. **ADVERTISING** The Wake-Up Restaurant advertises that you can have a different pancake breakfast every day of the year. It offers 25 different kinds of pancakes and 14 flavored syrups. If the restaurant is open every day of the year, is its claim valid? Explain.

16. **GAMES** A player rolls five number cubes to score the maximum number of points. Find the number of outcomes possible in one roll.

CRITICAL THINKING How many outcomes are possible if you toss each of the following?

17. one coin
18. two coins
19. three coins
20. n coins

HOMEWORK HELP

For Exercises	See Examples
9–14	1
15–16	2

Extra Practice
See pages 585, 604.

Spiral Review with Standardized Test Practice

21. **MULTIPLE CHOICE** Elizabeth has 3 sweaters, 5 blouses, and 6 skirts that coordinate. How many different outfits can Elizabeth make?
    - Ⓐ 105
    - Ⓑ 90
    - Ⓒ 45
    - Ⓓ 30

22. **GRID IN** WritePen makes 8 different styles of pens in several colors with 2 types of grips. If the company makes 112 kinds of pens, how many different colors do they make?

PROBABILITY What is the probability that the spinner shown at the right will stop on each of the following numbers? Write as a fraction in simplest form. (Lesson 9-1)

23. an even number
24. a multiple of 4
25. a number less than 10

26. **SCHOOL** Gustavo must choose from two geography classes, three history classes, and two statistics classes for next year. Make a tree diagram to show all of the possible schedules he can arrange. (Lesson 9-2)

GETTING READY FOR THE NEXT LESSON

BASIC SKILL Multiply.

27. $3 \cdot 2 \cdot 1$
28. $9 \cdot 8 \cdot 7$
29. $5 \cdot 4 \cdot 3 \cdot 2$
30. $7 \cdot 6 \cdot 5 \cdot 4$

msmath2.net/self_check_quiz

9-4 Permutations

What You'll LEARN
Find the number of permutations of a set of objects.

NEW Vocabulary
permutation
factorial

MATH Symbols
4! four factorial

HANDS-ON Mini Lab

Materials
- 3 index cards

Work with a partner.

How many different ways are there to arrange your first three classes if they are math, science, and language arts?

STEP 1 Write math, science, and language arts on the index cards.

STEP 2 Choose one of the subjects as the first class of the day. Choose one of the remaining two subjects for the second class. The third class is the card that remains.

| MATH | SCIENCE | LANGUAGE ARTS |

STEP 3 Record this arrangement of classes.

STEP 4 Change the order of the classes. Record this arrangement. Continue rearranging the cards until you have found all of the possible arrangements.

1. When you first started to make your list, how many choices did you have for your first class?
2. Once your first class was selected, how many choices did you have for the second class? Then, the third class?
3. Explain how you can use the Fundamental Counting Principle to find the number of arrangements.

A **permutation** is an arrangement, or listing, of objects in which order is important. You can use the Fundamental Counting Principle to find the number of possible arrangements.

There are **3** choices for the 1st class.
There are **2** choices that remain for the 2nd class.
There is **1** choice that remains for the 3rd class.

$3 \cdot 2 \cdot 1 = 6$ ← The number of arrangements of 3 classes.

The expression $3 \cdot 2 \cdot 1$ can be written as 3!, which is read *three factorial*.

Noteables™ Key Concept: Factorial

The expression n **factorial** (n!) is the product of all counting numbers beginning with n and counting backward to 1.

STUDY TIP

Technology You can find the factorial of a number on a calculator. To find 3!, press 3 [PRB] ▶ ▶ [ENTER] [ENTER].

EXAMPLES — Evaluate Factorials

Find the value of each expression.

1) 7!

$7! = 7 \cdot 6 \cdot 5 \cdot 4 \cdot 3 \cdot 2 \cdot 1$ Definition of factorial
$ = 5{,}040$ Simplify.

2) 2! · 3!

$2! \cdot 3! = 2 \cdot 1 \cdot 3 \cdot 2 \cdot 1$ Definition of factorial
$ = 12$ Simplify.

EXAMPLE — Find a Permutation

3) VOLLEYBALL In how many ways can the starting six players of a volleyball team stand in a row for a picture?

This is a permutation that can be written as 6!.

$6! = 6 \cdot 5 \cdot 4 \cdot 3 \cdot 2 \cdot 1$ Definition of factorial
$ = 720$ Simplify.

So, there are 720 ways the six starting players can stand in a row.

Arrangements can include all or only part of a group.

EXAMPLE — Find a Permutation

4) SWIMMING The finals of the Middle School Appalachian League features 8 swimmers. In how many ways can the swimmers finish in first or second place?

There are 8 choices for first place and 7 choices that remain for second place. So, there are 8 · 7 or 56 choices for first and second place.

Skill and Concept Check

1. **Writing Math** Explain the difference between 5! and 5 · 4 · 3.

2. **OPEN ENDED** Describe a real-life situation that has 6 permutations.

GUIDED PRACTICE

Find the value of each expression.

3. 3!

4. 6! · 2!

5. In how many ways can you arrange the letters in the word *equal*?

6. **TRANSPORTATION** There are 7 students waiting at the bus stop. In how many ways can the students board the bus when it arrives?

Practice and Applications

Find the value of each expression.

7. 5!
8. 9!
9. 4! · 3!
10. four factorial
11. 3! · 6!
12. 10 · 9 · 8
13. 5! · 4!
14. 8! · 2!

HOMEWORK HELP

For Exercises	See Examples
7–14	1, 2
16	3
15, 17–18	4

Extra Practice
See pages 585, 604.

15. In how many ways can a softball manager arrange the first four batters in a lineup of nine players?

16. How many different 5-digit zip codes are there if no digit is repeated?

17. **MUSIC** The chromatic scale has 12 notes. In how many ways can a song start with 4 different notes from that scale?

DOGS For Exercises 18 and 19, use the information below and at the right.

During the annual Westminster Dog Show, the best dog in each breed competes to win one of four top ribbons in the group.

18. In how many ways can a ribbon be awarded to a breed of dog in the Working group?

19. The top dog in each group competes against the other six group winners for Best of Show. If each dog has an equally-likely chance of winning Best of Show, what is the probability that a terrier will win?

2002 Westminster Dog Show

Group	Number of Breeds
Herding	19
Hounds	25
Non-sporting	18
Sporting	27
Terriers	27
Toy	22
Working	21

Source: westminsterkennelclub.org

 Data Update Which breed of dog has won the most Best of Shows? Visit msmath2.net/data_update to learn more.

20. **CRITICAL THINKING** There are 1,320 ways for three students to win first, second, and third place during a debate match. How many students are there on the debate team?

Spiral Review with Standardized Test Practice

21. **SHORT RESPONSE** In a Battle of the Bands contest, how many ways can the four participating bands be ordered?

22. **GRID IN** How many different three-digit security codes can be made from the digits 1, 2, 3, 4, and 5 if no digit is repeated in a code?

23. **BREAKFAST** Find the total number of outcomes if you can choose from 8 kinds of muffins, 3 sizes, and 4 beverages. (Lesson 9-3)

24. **LUNCH** Make a tree diagram showing different ways to make a sandwich with turkey, ham, or salami and either cheddar or Swiss cheese. (Lesson 9-2)

GETTING READY FOR THE NEXT LESSON

PREREQUISITE SKILL Find each value. (Lesson 5-3)

25. $\dfrac{5 \cdot 4}{2 \cdot 1}$
26. $\dfrac{8 \cdot 7 \cdot 6}{3 \cdot 2 \cdot 1}$
27. $\dfrac{5 \cdot 4 \cdot 3}{4 \cdot 3 \cdot 2}$
28. $\dfrac{10 \cdot 9 \cdot 8 \cdot 7}{8 \cdot 7 \cdot 6 \cdot 5}$

msmath2.net/self_check_quiz

Lesson 9-4 Permutations 383

CHAPTER 9 Mid-Chapter Practice Test

Vocabulary and Concepts

1. **Explain** what it means for an event to occur at random. (Lesson 9-1)
2. **State** the Fundamental Counting Principle. (Lesson 9-3)
3. **Define** *permutation.* (Lesson 9-4)

Skills and Applications

Two number cubes are rolled. Find each probability. (Lesson 9-1)

4. P(sum of 6)
5. P(sum less than 7)
6. P(sum of 7 or 11)

For each situation, make a tree diagram to show the sample space. Then give the total number of outcomes. (Lesson 9-2)

7. tossing a penny and tossing a dime
8. choosing cereal, French toast, or pancakes and choosing orange, apple, cranberry or grapefruit juice

Use the Fundamental Counting Principle to find the total number of outcomes in each situation. (Lesson 9-3)

9. choosing a dinner with one entrée, one salad, and one dessert
10. rolling a number cube and tossing three coins

Entrée	Salad	Dessert
ham	potato	ice cream
beef	tossed	pie
turkey	cole slaw	cake

Find the value of each expression. (Lesson 9-4)

11. $5!$
12. $8!$
13. $2! \cdot 7!$

14. **STUDENT COUNCIL** In how many ways can a president, treasurer, and a secretary be chosen from among 8 candidates? (Lesson 9-4)

Standardized Test Practice

15. **GRID IN** Jeffrey has 2 action, 3 comedy, and 4 drama DVDs. If he randomly picks one DVD to watch, what is the probability that it will be a comedy? (Lesson 9-1)

16. **SHORT RESPONSE** Molly has three windows to display three of five best-selling books. How many different displays can she make if she puts one book in each window? (Lesson 9-4)

The Game Zone

A Place To Practice Your Math Skills

Math Skill: Probability

Cherokee Butterbean Game

GET READY!
Players: two, three, or four
Materials: 6 dry lima beans, bowl, marker

GET SET!
- This is a variation of a traditional Cherokee game.
- Color one side of each bean with the marker.

GO!
- The first player places the beans in the bowl, gently tosses the beans into the air, and catches them in the bowl. Points are scored as follows.
  - If all of the beans land with the marked or unmarked sides up, score 6 points.
  - If exactly one bean lands with the marked or unmarked side up, score 3 points.
  - If three beans are marked and three beans are unmarked, score 1 point.
- If a toss scores points, the player takes another turn. If a toss does not score any points, it is the next player's turn.
- **Who Wins?** The person with the most points after a given number of rounds is the winner.

9-5a Hands-On Lab

A Preview of Lesson 9-5

Exploring Combinations

INVESTIGATE *Work with a partner.*

The student safety club at Mahomet Middle School is planning to sell sundaes with two different toppings at the summer carnival. The choices are shown at the right.

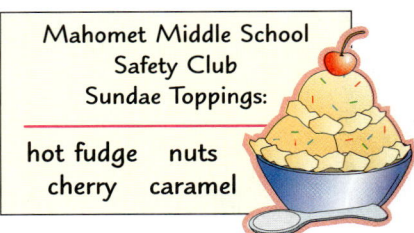

Mahomet Middle School Safety Club Sundae Toppings:
hot fudge   nuts
cherry   caramel

What You'll LEARN
Find combinations.

Materials
- 5 index cards

STEP 1 Write the name of the four sundae toppings on the index cards.

STEP 2 To make a sundae, select any pair of cards. Make a list of all the different combinations that are possible. Note that the order of the toppings is not important.

[Cherry] and [Nuts]

is considered the same as

[Nuts] and [Cherry]

Writing Math

Work with a partner.

1. How many different combinations are possible with two toppings?
2. How many different sundaes could be made with two toppings if the order of the toppings *was* important? What is this type of arrangement called?
3. **Write** a fifth topping on an index card. Now find all two-topping sundae combinations. How many are there? How many sundaes can be made with two toppings if the order of the toppings was important?
4. **Compare and contrast** the way you can find a permutation with the way you find a combination.
5. **Make a conjecture** about how to find a combination given a permutation.

9-5 Combinations

What You'll LEARN
Find the number of combinations of a set of objects.

NEW Vocabulary
combination

Link to READING
Everyday Meaning of combination: the result of putting together objects, as in a combination of ingredients

WHEN am I ever going to use this?

BASKETBALL Coach Chávez wants to select co-captains for her basketball team. She will select two girls from the four oldest members on the team: Alita, Bailey, Charmaine, and Danielle.

1. Use the first letter of each name to list all of the permutations of co-captains. How many are there?
2. Cross out any arrangement that contains the same letters as another one in the list. How many are there now?
3. Explain the difference between the two lists above.

An arrangement, or listing, of objects in which order is *not* important is called a **combination**. For example, in the activity above, choosing Alita and Bailey is the same as choosing Bailey and Alita.

Permutations and combinations are related. You can find the number of combinations of objects by dividing the number of permutations of the entire set by the number of ways each smaller set can be arranged.

A permutation of 4 players, taken 2 at a time.

$$\frac{4 \cdot 3}{2!} = \frac{4 \cdot 3}{2 \cdot 1} = \frac{12}{2} = 6$$

There are 2! ways to arrange 2 players.

EXAMPLE Find the Number of Combinations

1 **FOOD** Paul's Pizza Parlor is offering a large two-topping pizza for $14.99. There are five toppings from which to choose. How many different two-topping pizzas are possible?

Method 1 Make a list.
The five toppings are labeled pepperoni (p), sausage (s), onions (o), mushrooms (m), and green pepper (g).

| p, s | p, m | p, o | p, g | s, m |
| s, o | s, g | m, o | m, g | o, g |

Method 2 Use a permutation.
There are 5 · 4 permutations of two toppings chosen from five.

There are 2! ways to arrange the two toppings.

$$\frac{5 \cdot 4}{2!} = \frac{20}{2} = 10$$

So, there are 10 different two-topping pizzas.

EXAMPLE **Use a Combination to Solve a Problem**

② CHECKERS A checkers tournament features each of the top 8 regional players playing every opponent one time. The 2 players with the best records will then play in a final round to determine the champion. How many matches will be played if there are no ties?

Find the number of ways 2 players can be chosen from a group of 8.

There are 8 · 7 ways to choose 2 people. → $\dfrac{8 \cdot 7}{2!} = \dfrac{56}{2} = 28$
There are 2! ways to arrange 2 people. →

There are 28 matches plus 1 final match to determine the champion. So, there will be 29 matches played.

Check Make a diagram in which each person is represented by a point. Draw line segments between two points to represent the games. There are 28 line segments. Then add the final-round match to make a total of 29 matches.

Your Turn

a. How many matches will be played if the top 16 players were invited to play?

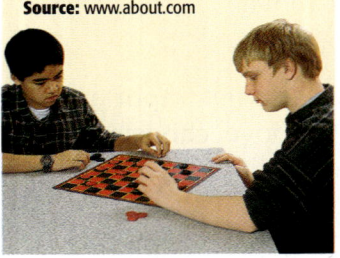

REAL-LIFE MATH

CHECKERS Checkers is played on an 8-by-8 board in the U.S., England, Australia, and Ireland. People in eastern Europe play on a 10-by-10 board.

Source: www.about.com

The difference between permutations and combinations is that order is important in permutations, while order is *not* important in combinations.

EXAMPLES **Identify Permutations and Combinations**

Tell whether each situation represents a *permutation* or *combination*. Then solve the problem.

③ STUDENT GOVERNMENT The six students listed at the right are members of Student Council. How many ways can you choose a president, vice president, and treasurer from this group?

This is a permutation because the order of president, vice president, and treasurer is important. So, the number of ways you can choose the three officers is 6 · 5 · 4, or 120 ways.

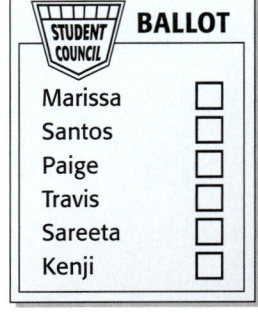

STUDENT COUNCIL **BALLOT**
Marissa ☐
Santos ☐
Paige ☐
Travis ☐
Sareeta ☐
Kenji ☐

④ In how many ways can you choose a committee of three students from the six members in student council shown above?

This is a combination because the order of the students in the committee is not important.

There are 6 · 5 · 4 ways to choose 3 people. → $\dfrac{6 \cdot 5 \cdot 4}{3!} = \dfrac{120}{6} = 20$
There are 3! ways to arrange 3 people. →

So, there are 20 ways to choose the committee.

388 Chapter 9 Probability

Skill and Concept Check

1. **Writing Math** Explain why a combination lock should be called a permutation lock.

2. **OPEN ENDED** Give an example of a permutation and combination.

3. **FIND THE ERROR** Allison and Francisca are calculating the number of ways that a 3-member committee can be chosen from a 7-member club. Who is correct? Explain.

   Allison
   $7 \cdot 6 \cdot 5 = 210$ ways

   Francisca
   $\dfrac{7 \cdot 6 \cdot 5}{3!} = \dfrac{210}{6} = 35$ ways

GUIDED PRACTICE

4. **VOLLEYBALL** Coach Malone has an 8-member volleyball team. He told his team that he would start six different players every game. How many games would it take to do this?

Tell whether each problem represents a *permutation* or *combination*. Then solve the problem.

5. How many ways can 10 students finish first, second, or third at the science fair?

6. How many ways can you pick 2 puppies from a litter of 7 puppies?

Practice and Applications

7. **FOOD** At a hot dog stand, customers can select three toppings from among chili, onions, cheese, mustard, or relish. How many combinations of three-topping hot dogs are there?

CIVICS For Exercises 8 and 9, use the information below and at the right.
If five of the nine Justices on the United States Supreme Court agree on a decision, they can issue a majority opinion.

8. How many different combinations of five Supreme Court Justices are there?

9. Before they take the Bench each day, the Justices engage in the "Conference handshake." Each Justice shakes hands with each of the other eight. How many handshakes take place?

10. **SOCCER** There are 21 players trying out for 15 spots on the soccer team. How many ways does the coach have to create her team?

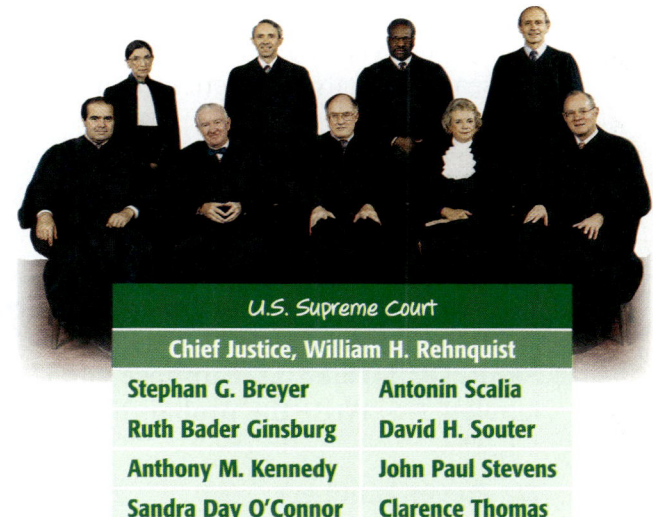

HOMEWORK HELP

For Exercises	See Examples
7–10	1, 2
11–16	3, 4

Extra Practice
See pages 586, 604.

U.S. Supreme Court

Chief Justice, William H. Rehnquist	
Stephan G. Breyer	Antonin Scalia
Ruth Bader Ginsburg	David H. Souter
Anthony M. Kennedy	John Paul Stevens
Sandra Day O'Connor	Clarence Thomas

Source: www.washingtonpost.com

msmath2.net/self_check_quiz

Tell whether each problem represents a *permutation* or *combination*. Then solve the problem.

11. How many ways can you select four essay questions out of a total of 10 on the exam?

12. Six children remain in a game of musical chairs. If two chairs are removed, how many different groups of four students can remain?

13. How many ways can three flute players be seated in the first, second, or third seats in the orchestra?

14. In how many ways can four paintings be displayed from a collection of 15?

15. How many ways can seven students line up to buy concert tickets?

16. Given 12 Web sites, how many ways can you visit half of them?

17. **WRITE A PROBLEM** Write about a real-life situation that can be solved using a permutation and one that can be solved using a combination. Then solve both problems.

18. **CRITICAL THINKING** At a party, there were 105 handshakes. If each person shook hands exactly once with every other person, how many people were at the party?

Spiral Review with Standardized Test Practice

19. **MULTIPLE CHOICE** In how many ways can three gymnasts from a team of 10 be chosen to compete in a meet?
    - Ⓐ 120
    - Ⓑ 180
    - Ⓒ 240
    - Ⓓ 720

20. **SHORT RESPONSE** Find $\dfrac{(n+6)(n+5)(n+4)}{n!}$ if $n = 3$.

21. **SAILING** When Mr. Elms purchased his sailboat, it came with six different-colored flags to be used for sending signals. A specific signal depended on the order of the flags. How many different three-flag signals can he send? (Lesson 9-4)

22. **CARS** A certain brand of car has a choice of a 2.5, 3.1, or 4.0 liter engine; a radio with a cassette-player, CD-player, or CD-changer; cloth or leather seats; and silver, green, red, or white. How many different cars are possible? (Lesson 9-3)

Estimate. (Lesson 6-1)

23. $\dfrac{1}{10} + \dfrac{7}{8}$

24. $\dfrac{5}{12} - \dfrac{1}{9}$

25. $\dfrac{4}{9} \cdot 20$

26. $15\dfrac{7}{9} \div 3\dfrac{3}{5}$

GETTING READY FOR THE NEXT LESSON

PREREQUISITE SKILL Suppose you choose a card from a set of eight cards labeled 1–8. Write the probability of each event as a fraction in simplest form. (Lesson 9-1)

27. P(even number)
28. P(number greater than 6)
29. P(multiple of 8)
30. P(number less than 9)
31. P(prime number)
32. P(not 2 or 5)

9-6a Problem-Solving Strategy
A Preview of Lesson 9-6

Act It Out

What You'll LEARN

Solve problems by acting them out.

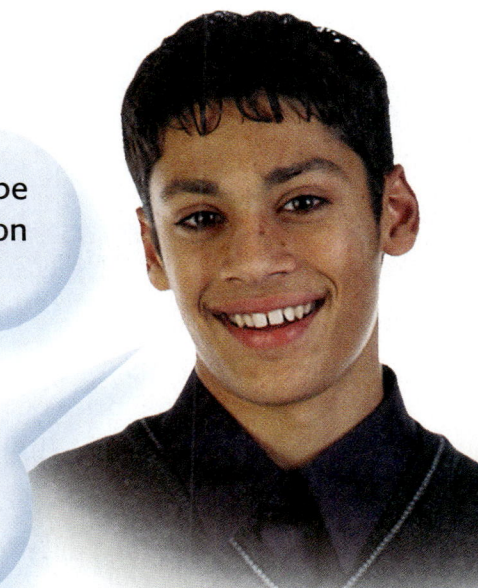

I wonder if tossing a coin would be a good way to answer a 5-question true-false quiz?

I'm not so sure, Whitney. Let's do an experiment with a coin and **act it out**!

Explore	We know that there are five true-false questions on the quiz. We can carry out an experiment to test whether tossing a coin would be a good way to answer the questions and get a good grade.
Plan	Let's toss a coin five times. If the coin shows tails, we will answer T. If the coin shows heads, we will answer F. Let's do four trials.
Solve	Suppose the correct answers are T, F, F, T, F. Let's circle them.

Answers	T	F	F	T	F	Number Correct
Trial 1	ⓣ	T	Ⓕ	F	T	2
Trial 2	F	Ⓕ	T	Ⓣ	Ⓕ	3
Trial 3	Ⓣ	Ⓕ	T	F	T	2
Trial 4	F	Ⓕ	T	F	Ⓕ	2

The experiment produces 2 or 3 correct answers.

Examine	There is an equally likely chance that tossing a coin will produce a correct answer or a wrong answer for each question. Since the experiment produced about 2–3 correct answers on a 5-question quiz, it shows that tossing a coin to answer a true-false quiz is not the way to get a good grade.

Analyze the Strategy

1. **Explain** whether the results of the experiment would be the same if it were repeated.
2. **Explain** an advantage of using the act it out strategy to solve a problem.

Apply the Strategy

Solve. Use the act it out strategy.

3. **POP QUIZ** Determine whether using a spinner with four equal sections is a good way to answer a 5-question multiple-choice quiz. Each question has choices A, B, C, and D. Explain.

4. **PHOTOGRAPHS** Samuel is taking a picture of the Spanish Club's five officers. The club president will always stand on the left, and the vice president will always stand on the right. How many different ways can he arrange the officers for the picture?

Mixed Problem Solving

Solve. Use any strategy.

5. **MONEY MATTERS** Lola purchased a $35 book bag at a sale price of $27.50. What was the percent of decrease from the original price to the sales price?

6. **PICNIC** Examine the bar graph.

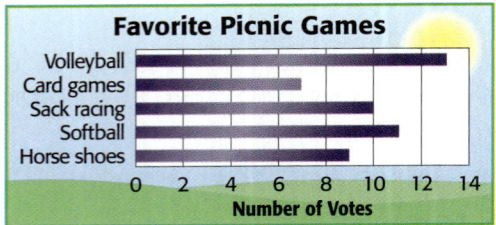

What is the probability that a student's favorite picnic game is sack racing?

7. **CROQUET** Sixteen teams are playing in a croquet tournament. If a team loses one game, it is eliminated. How many total games will be played in the tournament?

8. **EARTH SCIENCE** The Hawaiian mountain Mauna Kea is about 1.38×10^4 feet tall. Mt. Everest is about 29,000 feet. Which mountain is taller?

9. **BASEBALL** Each package of baseball cards contains a puzzle piece. If you collect all 6 different pieces, you win two tickets to a major league game. There is an equally-likely chance of getting a different puzzle piece each time. Explain how you would model this problem to find the answer.

10. **PATTERNS** The pattern below is known as Pascal's Triangle. Find the pattern and complete the 6th and 7th rows.

1st Row				1				
2nd Row			1		1			
3rd Row		1		2		1		
4th Row	1		3		3		1	
5th Row	1	4		6		4		1

11. **SPACE SCIENCE** A 21-kilogram sample of rocks from the moon is composed of about 40% oxygen and about 19.2% silicon. How much more is the oxygen mass of the rocks than the silicon mass?

12. **TRANSPORTATION** A taxi charges $2.35 for the first 0.4 mile and $0.75 for each additional 0.4 mile. Find the cost of a 4-mile taxi ride.

13. **STANDARDIZED TEST PRACTICE**
    In a middle school of 900 students, about how many prefer the season of autumn?

    A 81   **B** 207   **C** 279   **D** 333

392 Chapter 9 Probability

9-6 Theoretical and Experimental Probability

What You'll LEARN

Find and compare experimental and theoretical probabilities.

NEW Vocabulary

experimental probability
theoretical probability

REVIEW Vocabulary

sample space: a set of all possible outcomes (Lesson 9-2)

HANDS-ON Mini Lab

Materials
- 2 number cubes

Work with a partner.

STEP 1 Roll two number cubes 36 times. Record the number of times the sum of the number cubes is 7.

STEP 2 Use an addition table from 1 to 6 to help you find the expected number of times the sum of 7 should come up after rolling two number cubes 36 times. The top row represents one number cube, and the left column represents the other number cube. The table above has been started for you.

+	1	2	3
1	2	3	4
2	3	4	5

1. How many times did you roll a sum of 7? What is the probability of rolling a sum of 7?
2. How does your result compare to the results of other groups? Explain.
3. What is the expected probability of rolling a sum of 7?
4. How does your result compare to the expected probability of rolling a sum of 7? Explain any differences.

In the activity above, you found the experimental probability of rolling a sum of 7 on two number cubes. **Experimental probability** is found using frequencies obtained in an experiment or game.

The expected probability of an event occurring is called **theoretical probability**. This is the probability that you have been using since Lesson 9-1. The theoretical probability of rolling a sum of 7 on two number cubes is $\frac{6}{36}$, or $\frac{1}{6}$.

EXAMPLE   Experimental Probability

1 Two number cubes are rolled seventy-five times and a sum of 9 is rolled ten times.

What is the experimental probability of rolling a sum of 9?

$$P(9) = \frac{\text{number of times a sum of 9 occurs}}{\text{number of possible outcomes}}$$

$$= \frac{10}{75} \text{ or } \frac{2}{15}$$

The experimental probability of rolling a sum of 9 is $\frac{2}{15}$.

EXAMPLES — Experimental and Theoretical Probability

2. The graph shows the results of an experiment in which a coin was tossed thirty times. Find the experimental probability of tossing tails for this experiment.

$P(\text{tails}) = \dfrac{\text{number of times tails occurs}}{\text{number of possible outcomes}}$

$= \dfrac{18}{30}$ or $\dfrac{3}{5}$

The experimental probability of tossing tails is $\dfrac{3}{5}$.

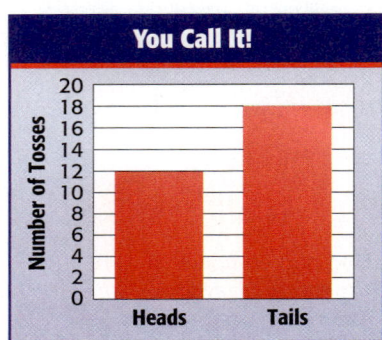

3. Compare the experimental probability you found in Example 2 to its theoretical probability.

The theoretical probability of tossing tails on a coin is $\dfrac{1}{2}$. So, the experimental probability is close to the theoretical probability.

Experimental probability can also be based on past performances and be used to make predictions on future events.

EXAMPLES — Predict Future Events

4. **FOOD** In a survey, 100 people were asked to name their favorite Independence Day side dishes. What is the experimental probability of macaroni salad being someone's favorite dish?

There were 100 people surveyed and 12 chose macaroni salad. So, the experimental probability is $\dfrac{12}{100}$, or $\dfrac{3}{25}$.

Side Dish	Number of People
potato salad	55
green salad or vegetables	25
macaroni salad	12
coleslaw	8

5. Suppose 250 people attend the city's Independence Day barbecue. How many can be expected to choose macaroni salad as their favorite side dish?

$\dfrac{3}{25} = \dfrac{x}{250}$ Write a proportion.

$3 \cdot 250 = 25x$ Find the cross products.

$30 = x$

About 30 will choose macaroni salad.

Your Turn

a. What is the experimental probability of potato salad being someone's favorite dish?

b. About how many people can be expected to choose potato salad as their favorite dish if 250 attend the barbecue?

REAL-LIFE CAREERS

How Does a Chef Use Math?
In addition to cooking, chefs order enough food and supplies based on customers' preferences and recent sales.

Research For more information about a career as a chef, visit: msmath2.net/careers

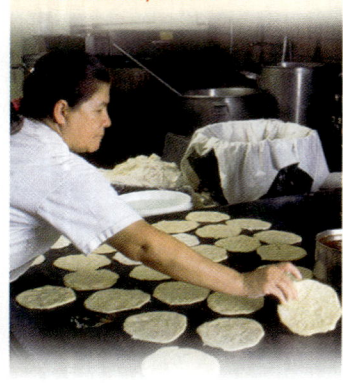

Skill and Concept Check

1. **Writing Math** Compare and contrast experimental probability and theoretical probability.

2. **OPEN ENDED** Give an example of a situation that has a theoretical probability of $\frac{1}{10}$.

GUIDED PRACTICE

For Exercises 3 and 4, a coin is tossed 50 times, and it lands heads 28 times.

3. Find the experimental probability of the coin landing heads.
4. Find the theoretical probability of the coin landing heads.

5. **TRAFFIC** Between 3:00 P.M. and 4:00 P.M., 11 sedans, 15 trucks, and 6 sports cars drove through an intersection. Based on this information, what is the probability that a vehicle that drives through the intersection is a sports car?

Practice and Applications

For Exercises 6–9, a number cube is tossed 20 times and lands on 1 two times and on 5 four times.

6. Find the experimental probability of landing on 5.
7. Find the theoretical probability of *not* landing on 5.
8. Find the theoretical probability of landing on 1.
9. Find the experimental probability of *not* landing on 1.

HOMEWORK HELP

For Exercises	See Examples
6–9	1–3
10, 15	4
11–14, 16–17	5

Extra Practice
See pages 586, 604.

X GAMES For Exercises 10–12, use the graph shown. It shows the results of a survey in which 50 students were asked to name their favorite X Game sport.

10. What is the probability of inline being someone's favorite sport?
11. Suppose 500 people attend the X Games. How many can be expected to choose inline as their favorite sport?
12. Suppose 500 people attend the X Games. How many can be expected to choose speed climbing as their favorite sport?

13. **REFRESHMENTS** In a survey taken at the beach, 47 people preferred cola, 28 preferred root beer, and 25 preferred ginger ale. If the manager of the Beach Hut is going to buy 50 cases of soda for the next day, about how many cases should be root beer?

14. **SPINNERS** A spinner marked with three sections A, B, and C was spun 100 times. The results are shown in the table. Make a drawing of the spinner based on its experimental probabilities.

Section	Frequency
A	24
B	50
C	26

GIFTS For Exercises 15–17, use the results of the survey at the right.

15. What is the probability that a mother will receive a gift of flowers? Write the probability as a fraction.

16. Out of 750 mothers who receive gifts, how many would you expect to receive flowers?

17. Out of 750 mothers who receive gifts, how many would you expect to receive jewelry?

18. **CRITICAL THINKING** *Hot numbers* in a lottery are numbers that keep coming up. *Cold numbers* are numbers that have not "hit" in awhile. Some people select hot numbers or cold numbers when buying a ticket. Using probability, explain whether this makes sense.

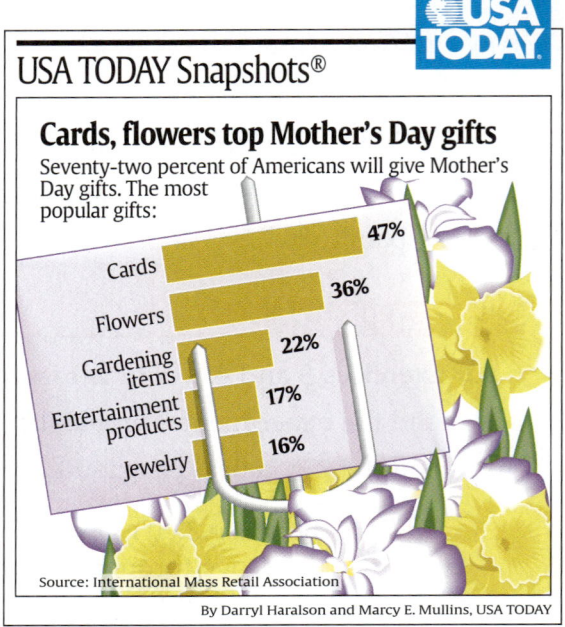

EXTENDING THE LESSON

The *odds* of an event occurring is the ratio that compares the number of ways an event can occur (success) to the number of ways it cannot occur (failure).

Example Find the odds of rolling a number less than 5 on a number cube.

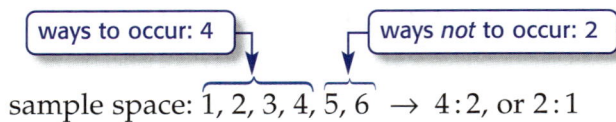

sample space: 1, 2, 3, 4, 5, 6 → 4 : 2, or 2 : 1

Find the odds of each outcome if a number cube is rolled.

19. 6

20. odd number

21. number greater than 4

Spiral Review with Standardized Test Practice

22. **MULTIPLE CHOICE** Myron has four dimes in his wallet with dates 1998, 1995, 2005, and 2000. If he randomly picks one dime from his wallet, what is the probability that it will have a date in the 1990s?

    Ⓐ $\frac{1}{4}$ Ⓑ $\frac{1}{2}$ Ⓒ $\frac{2}{3}$ Ⓓ $\frac{3}{4}$

23. **GRID IN** Rachel spins the spinner at the right 50 times, and it lands on red 15 times. What is the theoretical probability of spinning red?

24. **BASEBALL** How many ways can a baseball coach select four starting pitchers from a pitching staff of eight? (Lesson 9-5)

25. **POLE VAULT** A pole vault competition has 10 people in it. In how many ways can first-, second- and third-place ribbons be awarded? (Lesson 9-4)

GETTING READY FOR THE NEXT LESSON

PREREQUISITE SKILL Three coins are tossed. Find each probability. (Lesson 9-1)

26. P(3 tails)

27. P(2 heads)

28. P(at least 2 heads)

29. P(at least 1 tail)

9-6b HANDS-ON LAB
A Follow-Up of Lesson 9-6

Experimental Probability

INVESTIGATE *Work with a partner.*

Shannon is on the basketball team. She makes 75% of her free throws. In this lab, you will model, or *simulate* Shannon shooting free throws with a spinner and investigate the experimental probabilities of her being able to make a free throw.

- Create and cut out the spinner at the right.

- Spin the spinner 25 times to model Shannon shooting 25 free throws. Record your results in the table below. If the spinner lands on a line, spin again.

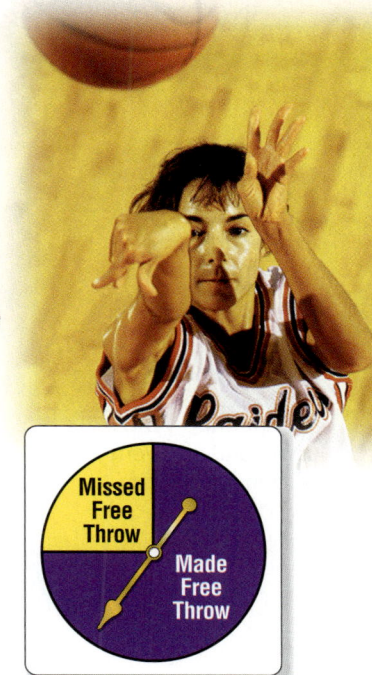

Outcome	Tally	Total
Made Free Throw		
Missed Free Throw		

- Combine your data with four other groups so you have 100 data points.

What You'll LEARN
Investigate experimental probability.

Materials
- paper
- scissors
- spinner arrows

Writing Math

Work with a partner.

1. **Explain** why you combined your data with four other groups.

2. From your table, what is the experimental probability of making a free throw?

3. How does the experimental probability compare with Shannon's past performance?

4. What is the theoretical probability of making two free throws in a row? **Explain** how you found your answer.

5. Suppose Shannon did shoot 100 free throws in a row. What factors would influence her making *or* not making 75 free throws? **Explain** your reasoning.

6. **Describe** the size of both sections of the spinner.

7. **Describe** another way to simulate Shannon shooting free throws.

8. **Draw** a spinner that simulates a free-throw shooter making 60% of her free throws.

9-7 Independent and Dependent Events

What You'll LEARN

Find the probability of independent and dependent events.

NEW Vocabulary

compound event
independent event
dependent event

WHEN am I ever going to use this?

TRACK AND FIELD The 100-meter dash features 20 runners competing in a preliminary round of 4 heats. The winner of each heat advances to the final race. Before the race, each runner chooses a number from jar 1 to determine the heat in which he runs and a number from jar 2 to determine one of five lanes he occupies. Omar is the first runner to choose from the jars.

1. What is the probability of Omar being in the second heat?
2. What is the probability of Omar being in lane 3?
3. Multiply your answers in Exercises 1 and 2 above. What does this number mean? Explain.

In the Mini Lab, choosing the heat and the lane is a compound event. A **compound event** consists of two or more simple events. Since choosing the heat number does not affect choosing the lane number, both events are called **independent events**.

Noteables™ Key Concept: Probability of Independent Events

Words The probability of two independent events can be found by multiplying the probability of the first event by the probability of the second event.

Symbols $P(A \text{ and } B) = P(A) \cdot P(B)$

EXAMPLES Independent Events

READING Math

Probability Notation
$P(A \text{ and } B)$ is read *the probability of A followed by B*.

1. A number cube is rolled, and the spinner at the right is spun. Find the probability of rolling a 2 and spinning a vowel.

$P(2) = \frac{1}{6}$ $P(\text{vowel}) = \frac{2}{5}$

$P(2 \text{ and vowel}) = \frac{1}{\underset{3}{\cancel{6}}} \cdot \frac{\overset{1}{\cancel{2}}}{5}$ or $\frac{1}{15}$

So, the probability of rolling a 2 and spinning a vowel is $\frac{1}{15}$.

Check You can make a tree diagram to check your answer.

398 Chapter 9 Probability

If the outcome of one event affects the outcome of a second event, the events are called **dependent events**.

EXAMPLE — Dependent Events

2 SNACK BARS A box contains 2 oatmeal, 3 strawberry, and 6 cinnamon snack bars. Ruby reaches in the box and randomly takes two snack bars, one after the other. Find the probability that she will choose a cinnamon bar and then a strawberry bar.

$P(\text{cinnamon}) = \dfrac{6}{11}$ ← 11 snack bars, 6 are cinnamon

$P(\text{strawberry}) = \dfrac{3}{10}$ ← 10 snack bars after 1 cinnamon snack bar has been removed, 3 are strawberry

$P(\text{cinnamon, then strawberry}) = \dfrac{\cancel{6}^{\,3}}{11} \cdot \dfrac{3}{\cancel{10}_{\,5}}$ or $\dfrac{9}{55}$

So, the probability that Ruby will choose a cinnamon snack bar and then a strawberry snack bar is $\dfrac{9}{55}$, or about 16%.

STUDY TIP

Dependent Events The probability of B following A is the same as the probability of A following B.

Noteables™ — Key Concept: Probability of Dependent Events

Words The probability of two dependent events is the probability of the first event times the probability that the second event occurs after the first.

Symbols $P(A \text{ and } B) = P(A) \cdot P(B \text{ following } A)$

Skill and Concept Check

1. **Writing Math** Explain the difference between independent and dependent events.

2. **OPEN ENDED** Describe two dependent events.

3. **FIND THE ERROR** Kimi and Shane are finding the probability of spinning two even numbers in a row. Who is correct? Explain.

Kimi
$\dfrac{2}{5} \cdot \dfrac{1}{4} = \dfrac{2}{20} = \dfrac{1}{10}$

Shane
$\dfrac{2}{5} \cdot \dfrac{2}{5} = \dfrac{4}{25}$

GUIDED PRACTICE

4. A coin is tossed, and a number cube is rolled. What is the probability of tossing heads and rolling a 3 or a 5?

5. A bag contains 5 red apples and 3 yellow apples. What is the probability of picking 2 red apples without the first being replaced?

msmath2.net/extra_examples   Lesson 9-7 Independent and Dependent Events **399**

Practice and Applications

HOMEWORK HELP
For Exercises	See Examples
6–7, 17–19	1
8–9, 11–14	2

Extra Practice
See pages 586, 604.

6. A red and a blue number cube are rolled. Find the probability that an odd number is rolled on the red cube and a number greater than 1 is rolled on the blue cube.

7. Find the probability of heads on three consecutive tosses of a coin.

8. A cooler is filled with 12 colas and 9 diet colas. If Victor randomly chooses two without replacing the first, what is the probability that he will choose a cola and then a diet cola?

9. A deck of 30 cards is made up of the numbers 1–10 in three colors: red, purple, and green. Two cards are selected without either being replaced. Find the probability of choosing a purple 5 and then a red or green card.

10. Draw a Venn diagram to show the probability of two independent events A and B.

11. **CIVICS** In the 108th Congress, Tennessee had 4 Republicans and 5 Democrats serving in the House of Representatives. If a subcommittee of 2 representatives was formed to study Internet usage among middle school students, what is the probability that both would be Republicans?

CARDS For Exercises 12–14, use the information below.
A standard deck of playing cards contains 52 cards in four suits of 13 cards each. Two suits are red and two suits are black. Two cards are chosen from the deck one after another. Find each probability.

12. $P(\text{2 hearts})$
13. $P(\text{red, black})$
14. $P(\text{Ace, King})$

Determine whether each event is *independent* or *dependent*.

15. choosing a student in the 7th grade and a student from the 8th grade

16. choosing a pair of shoes to try on, then choosing a smaller pair to try on

FASHION For Exercises 17 and 18, use the graphic at the right. Write your answers as decimals to the nearest hundredth.

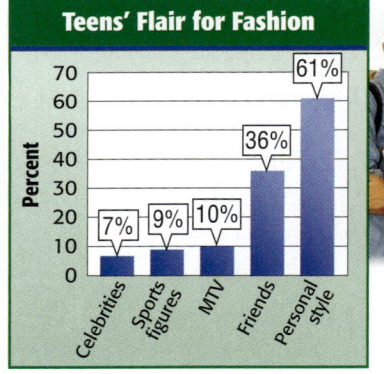

17. What is the probability that a teen chosen at random said that friends and sports figures influenced her fashion choices?

18. What is the probability that a teen chosen at random said that personal style and celebrities influenced his fashion choices?

19. **TRAFFIC SIGNALS** There are three consecutive traffic signals on a street that operate independently of each other. The first is green 45% of the time, the second is green 60% of the time, and the third is green 50% of the time. What is the probability of a person driving down the street making all three green lights? Write as a percent to the nearest tenth.

CRITICAL THINKING For Exercises 20 and 21, use the spinner.

20. Make a tree diagram of all the possible outcomes of three successive spins of the spinner shown. How many paths in the tree diagram represent two red spins and one blue?

21. Suppose the spinner is designed so that for each spin there is a 40% probability of spinning red and a 20% chance of spinning blue. What is the probability of spinning two reds and one blue?

EXTENDING THE LESSON Events that cannot happen at the same time are *mutually exclusive*. The two events are connected by the word *or* and the probability is found by *adding* their individual probabilities.

Example Suppose a number cube is rolled. Find P(even or 5).

$$P(\text{even or } 5) = \frac{3}{6} + \frac{1}{6} = \frac{4}{6} \text{ or } \frac{2}{3}$$

Find each probability as a fraction in simplest form.

22. P(1 or 3) 23. P(5 or multiple of 3) 24. P(6 or less than 5)

Spiral Review with Standardized Test Practice

25. **MULTIPLE CHOICE** You guess the answers on a two-question multiple-choice quiz with five choices: a, b, c, d, and e. What is the probability that you will get both correct?

    A $\frac{1}{50}$ **B** $\frac{1}{25}$ **C** $\frac{2}{25}$ **D** $\frac{1}{5}$

26. **GRID IN** Find the probability of rolling an even number on a number cube, tossing heads on a coin, and choosing an ace in a standard deck of cards.

27. **PROBABILITY** Paz performed a probability experiment by spinning a spinner 20 times. The results are shown in the table. If the spinner is divided into four equal sections, how many sections would you expect to be blue? (Lesson 9-6)

Color	Frequency
red	|||||
green	||||| |||||
blue	|||||

28. **LOTTERY** Balls numbered 1 to 51 are dropped into a machine and mixed together. Six balls are then selected in any order to make up the winning numbers. How many different ways can the six numbers be chosen? (Lesson 9-5)

Evaluate each expression if $a = 6$, $b = -4$, and $c = -3$. (Lesson 3-6)

29. $9c$ 30. $-8a$ 31. $2bc$ 32. $5b^2$

INTERDISCIPLINARY PROJECT

Step Right Up and Win a Prize

Math and Recreation It's time to complete your project. Use the information and data you have gathered about carnival games to prepare a Web page or poster. Be sure to include a scale drawing of the game you design with your project.

WebQuest msmath2.net/webquest

Study Guide and Review

Vocabulary and Concept Check

combination (p. 387)
complementary events (p. 371)
compound event (p. 398)
dependent events (p. 399)
experimental probability (p. 393)
factorial (p. 381)
fair game (p. 374)
Fundamental Counting Principle (p. 378)
independent events (p. 398)
outcome (p. 370)
permutation (p. 381)
probability (p. 370)
random (p. 370)
sample space (p. 374)
simple event (p. 370)
theoretical probability (p. 393)
tree diagram (p. 374)

Choose the correct term to complete the sentence.

1. The set of all possible outcomes for an experiment is called the (sample space, probability).
2. The Fundamental Counting Principle counts the number of possible outcomes using the operation of (addition, multiplication).
3. The ratio of the number of times an event occurs to the number of trials completed is called the (theoretical, experimental) probability.
4. When the outcome of one event influences the outcome of a second event, the events are called (independent, dependent).
5. A (permutation, combination) is a listing of objects in which order is important.
6. A (complementary, compound) event consists of two or more simple events.
7. A (simple, random) event occurs by chance.
8. The probability of an event is the (product, ratio) of the number of ways an event can occur to the number of possible outcomes.
9. When using a combination, the order of the arrangement (is, is not) important.
10. The expression $n!$ is the (sum, product) of all counting numbers beginning with n and counting backward to 1.

Lesson-by-Lesson Exercises and Examples

9-1 Simple Events (pp. 370–373)

A bag contains 6 red, 3 pink, and 3 white bows. Suppose you draw a bow at random. Find the probability of each event. Write each fraction in simplest form.

11. $P(\text{red})$
12. $P(\text{pink})$
13. $P(\text{white})$
14. $P(\text{red or white})$
15. $P(\text{pink or white})$
16. $P(not\text{ white})$

Example 1 What is the probability of rolling an odd number on a number cube?

$P(\text{odd}) = \dfrac{\text{odd numbers possible}}{\text{total numbers possible}}$

$= \dfrac{3}{6}$ Three numbers are odd: 1, 3, and 5.

$= \dfrac{1}{2}$ Simplify.

Therefore, $P(\text{odd}) = \dfrac{1}{2}$.

msmath2.net/vocabulary_review

9-2 Tree Diagrams (pp. 374–377)

For each situation, make a tree diagram to show the sample space. Then give the total number of outcomes.

17. rolling a number cube and then tossing a coin
18. choosing a red, blue, or white shirt with either black or gray lettering
19. tossing a coin and choosing a card from seven cards numbered from 1 to 7
20. choosing from white, wheat, or rye bread and turkey, ham, or salami to make a sandwich

Example 2 Make a tree diagram to describe the possible outcomes for a family with two children.

If a family has two children, there are 4 possible outcomes.

9-3 The Fundamental Counting Principle (pp. 378–380)

Use the Fundamental Counting Principle to find the total number of outcomes in each situation.

21. rolling two number cubes
22. selecting a car from 3 styles, 3 interior colors, and 3 exterior colors
23. making an ice cream sundae selecting from 5 flavors of ice cream and 4 different toppings
24. **SHOPPING** One catalog offers a jogging suit in two colors, gray and black. It comes in sizes S, M, L, XL, and XXL. How many possible jogging suits can be ordered?

Example 3 Use the Fundamental Counting Principle to find the total number of outcomes for a family that has three children.

There are 2 possible outcomes each time a family has a child, a boy or a girl. If a family has 3 children, there are $2 \cdot 2 \cdot 2$, or 8 outcomes.

9-4 Permutations (pp. 381–383)

Find the value of each expression.

25. $4!$
26. $9!$
27. $8 \cdot 7 \cdot 6 \cdot 5$
28. $3! \cdot 4!$
29. $5! \cdot 3!$
30. $7! \cdot 4!$

31. **BASKETBALL** In how many ways can five basketball players be placed in three positions?

Example 4 In how many ways can a director and assistant director be chosen from among 4 candidates?

There are 4 choices for director and then 3 choices remain for assistant director. So, there are $4 \cdot 3$, or 12, ways the director and assistant director can be chosen.

Chapter 9 Study Guide and Review 403

Study Guide and Review continued

Mixed Problem Solving
For mixed problem-solving practice, see page 604.

9-5 Combinations (pp. 387–390)

32. How many three-topping pizzas are possible given eight different toppings?
33. How many groups of five people are there from a committee of nine?
34. How many groups of three kittens are possible from a litter of nine?
35. In how many ways can Rondell select two board games from the ten that his family has?

Example 5 In how many ways can you choose two items from a menu with four items?

There are $4 \cdot 3$ permutations of two menu items chosen from four.

There are 2! ways to arrange the two menu items.

$\frac{4 \cdot 3}{2!} = \frac{12}{2}$ or 6

So, there are 6 ways to choose two items.

9-6 Theoretical and Experimental Probability (pp. 393–396)

The results of spinning a spinner labeled A–E fifty times are given. Find the experimental probability of each event.

Letter	Frequency
A	8
B	17
C	9
D	6
E	10

36. $P(A)$ 37. $P(D)$ 38. $P(E)$
39. If the spinner is equally likely to land on each section, what is the theoretical probability of landing on B?

Example 6 A coin is tossed 75 times, and it lands on tails 55 times. What is the experimental probability of the coin landing on heads?

The coin landed on heads 20 times.

$P(\text{heads}) = \frac{\text{number of times heads occurs}}{\text{number of possible outcomes}}$

$= \frac{20}{75}$ or $\frac{4}{15}$

So, the experimental probability of the coin landing on heads is $\frac{4}{15}$.

9-7 Independent and Dependent Events (pp. 398–401)

For Exercises 40–43, a bag contains 6 green, 8 white, and 2 blue counters. Two counters are randomly drawn. Find each probability if the first counter is replaced before the second counter is drawn.

40. $P(\text{green, blue})$ 41. $P(2 \text{ white})$

Find each probability if the first counter is *not* replaced before the second counter is drawn.

42. $P(2 \text{ green})$ 43. $P(\text{blue, white})$

Example 7 A box contains 12 solid, 14 striped, and 10 spotted marbles. Suppose you reach in and grab two marbles. Find the probability of choosing a striped marble, replacing it, and then choosing a spotted marble.

$P(\text{striped}) = \frac{14}{36}$

$P(\text{spotted, after replacing the striped}) = \frac{10}{36}$

$\frac{14}{36} \cdot \frac{10}{36} = \frac{140}{1,296}$ or $\frac{35}{324}$

So, the probability of choosing a striped marble, replacing it, and then choosing a spotted marble is $\frac{35}{324}$, or about 11%.

Chapter 9 Practice Test

Vocabulary and Concepts

1. **Explain** the purpose of a tree diagram.
2. **Define** *combination*.

Skills and Applications

A spinner with sections labeled 1–8 has an equal chance of landing on each number. Find each probability.

3. P(odd number)
4. P(number greater than 1)
5. P(1 or 7)

For each situation, make a tree diagram to show the sample space. Then give the total number of outcomes.

6. tossing a coin three times
7. choosing a letter from the word MATH and then a digit from the number 123

Use the Fundamental Counting Principle to find the total number of outcomes in each situation.

8. choosing a 3-digit security code
9. rolling four number cubes

Find the value of each expression.

10. 8!
11. $10 \cdot 9 \cdot 8 \cdot 7$
12. $5! \cdot 4!$

13. **PARADES** If there are 50 floats in a parade, how many ways can a first place and a second place trophy be awarded?

14. **SURVEY** Two hundred fifty teenagers were asked what type of pet they owned. The results of the survey are in the table. What is the experimental probability that a teenager owns a pet? Write as a percent.

Pet	Number of Teenage Pet Owners
fish	26
cat	65
dog	86
bird	20
other	38
no pet	15

15. **YOGURT** A brand of yogurt has 15 different flavors. In how many ways can you choose three flavors?

Standardized Test Practice

16. **GRID IN** A variety bag of popcorn contains 10 pieces of regular popcorn, 6 pieces of cheese popcorn, and 4 pieces of caramel popcorn. Katie pulls out one piece of popcorn and eats it. Then she pulls out a second piece and eats it. What is the probability that her first piece was caramel and her second piece was regular?

msmath2.net/chapter_test

Chapter 9 Standardized Test Practice

PART 1 Multiple Choice

Record your answers on the answer sheet provided by your teacher or on a sheet of paper.

1. The ornithologist said there is a 0.6754 chance of the bird returning next year to its nest in a tree in the Glowinski family's backyard. What is the value of the 5 in this number? (Prerequisite Skill, p. 555)
   - Ⓐ five tenths
   - Ⓑ five hundredths
   - Ⓒ five thousandths
   - Ⓓ five ten thousandths

2. The line plot below shows the weight in grams of a single serving of different brands of cookies. Between which weights is there a cluster of data? (Lesson 2-3)

   - Ⓕ between 30 and 50
   - Ⓖ between 60 and 80
   - Ⓗ between 90 and 110
   - Ⓘ between 120 and 140

3. Park rangers measured the depth of a nearby river. It is 12 inches shallower than it was 4 years ago. Which formula best expresses the yearly change in the river? (Lesson 3-7)
   - Ⓐ $-12 \div 4$
   - Ⓑ 12×4
   - Ⓒ -12×4
   - Ⓓ $-12 \div -4$

4. Which of the following is the greatest common factor of 84 and 49? (Lesson 5-2)
   - Ⓕ 2
   - Ⓖ 6
   - Ⓗ 7
   - Ⓘ 12

5. If $792 was made in the sale of pens, what was the total amount of all sales to the nearest dollar? (Lesson 8-2)

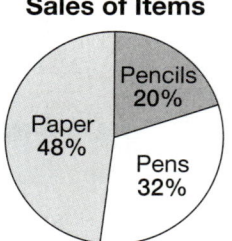

Sales of Items
Pencils 20%
Paper 48%
Pens 32%

   - Ⓐ $2,534
   - Ⓑ $2,475
   - Ⓒ $253
   - Ⓓ $248

6. The tree diagram shows the number of ways a customer can order lunch in a diner. If all choices are equally likely, what is the probability that a customer will order meat, rice, and salad? (Lesson 9-2)

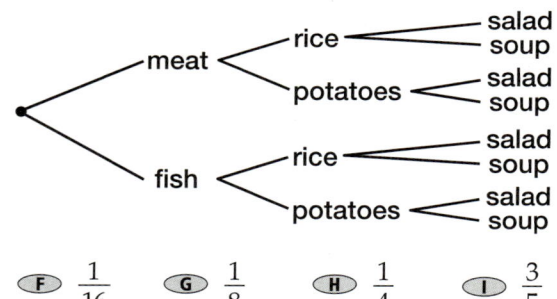

   - Ⓕ $\frac{1}{16}$
   - Ⓖ $\frac{1}{8}$
   - Ⓗ $\frac{1}{4}$
   - Ⓘ $\frac{3}{5}$

7. Fernando has to wash carrots, celery, onions, and potatoes for dinner. Assuming that he washes each vegetable only once, in how many different orders can Fernando wash all of the vegetables? (Lesson 9-4)
   - Ⓐ 4
   - Ⓑ 16
   - Ⓒ 20
   - Ⓓ 24

8. What is the theoretical probability of choosing a vowel from the word MATHEMATICS? (Lesson 9-6)
   - Ⓕ $\frac{4}{11}$
   - Ⓖ $\frac{4}{9}$
   - Ⓗ $\frac{4}{7}$
   - Ⓘ $\frac{7}{11}$

TEST-TAKING TIP

Question 7 You may want to find your own answer before looking at the answer choices. Doing so keeps you from being tempted by wrong answer choices that look correct, but are still wrong.

PART 2 Short Response/Grid In

Record your answers on the answer sheet provided by your teacher or on a sheet of paper.

9. Ebony shot −4, −2, 1, and −3 in four rounds of golf. The final score is the sum of the scores. What was her final score? (Lesson 3-4)

10. How much greater is the area of a professional basketball court than a college basketball court? (Lesson 6-8)

Basketball Court	Length (ft)	Width (ft)
college	84	50
professional	94	50

11. What is the percent of increase in two-vehicle households from 1990 to 2000? Write as a percent to the nearest tenth. (Lesson 8-4)

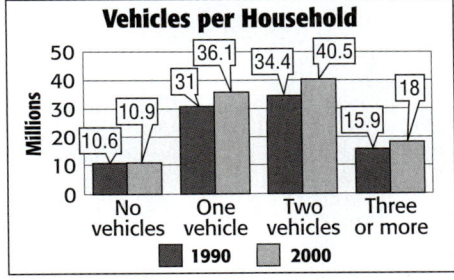

Source: Census Bureau

12. Two number cubes marked with 1, 2, 3, 4, 5, and 6 on their faces are rolled. What is the probability that their sum is 5 or 9? (Lesson 9-1)

13. Suppose you want to spend some time on the water. In how many ways can you choose a type of watercraft, the length of time, and the river? (Lesson 9-3)

Type of Watercraft	Time (hours)	River
paddleboat	1	Hamlin
canoe	2	Dustin
kayak	3	
water tricycle		

14. In how many ways can the Geiger family sit in a car if the two parents are in the front seat and the three children are in the back seat? (Lesson 9-4)

15. Twelve points are drawn on a circle. How many different line segments can be drawn to connect all pairs of points? (Lesson 9-5)

For Questions 16 and 17, use the information below and at the right.

A survey asked 400 students what type of materials they prefer to read. The table shows the results. (Lesson 9-6)

Type of Reading Materials	Number of Students
book	225
newspaper	35
magazine	140

16. What is the probability that a student prefers to read a magazine?

17. Suppose 1,000 students attend Franklin Middle School. About how many would prefer to read a book?

PART 3 Extended Response

Record your answers on a sheet of paper. Show your work.

18. Paula has a 5-disc CD player. She has a jazz, country, rap, pop, and R&B CD in the player. She listens to the CDs on RANDOM mode on both Friday and Saturday night.

   a. Make a tree diagram that shows all of the possible outcomes. (Lesson 9-2)

   b. What is the probability that Paula will hear a country song first on Friday night? (Lesson 9-6)

   c. What is the probability that Paula will hear a rap song first on Friday night and a jazz song first on Saturday night? (Lesson 9-7)

UNIT 5
Geometry and Measurement

Chapter 10
Geometry

Chapter 11
Measuring Two-Dimensional Figures

Chapter 12
Measuring Three-Dimensional Figures

In this unit, you will explore line and angle relationships, two- and three-dimensional figures, and the areas and volumes of these figures.

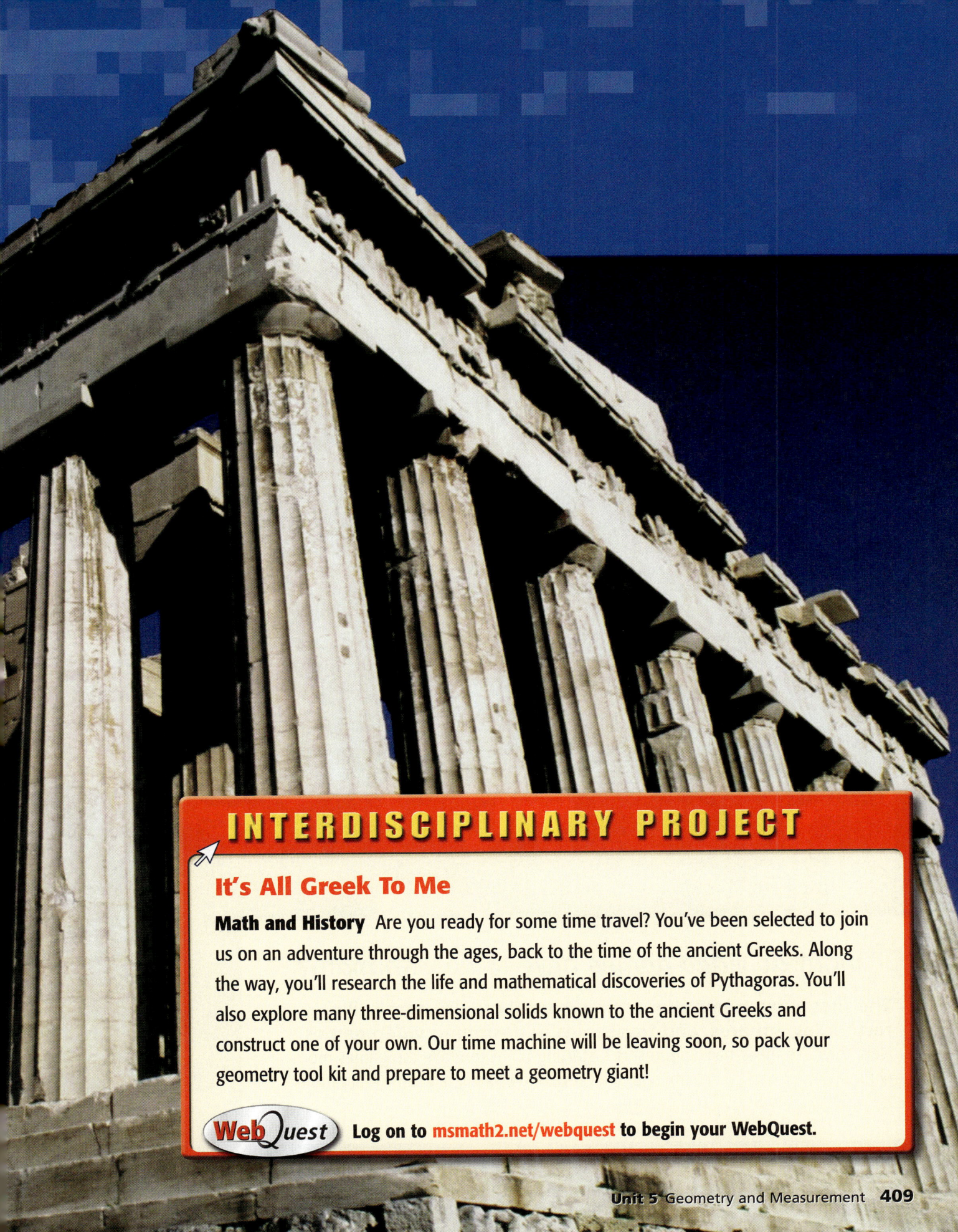

INTERDISCIPLINARY PROJECT

It's All Greek To Me

Math and History Are you ready for some time travel? You've been selected to join us on an adventure through the ages, back to the time of the ancient Greeks. Along the way, you'll research the life and mathematical discoveries of Pythagoras. You'll also explore many three-dimensional solids known to the ancient Greeks and construct one of your own. Our time machine will be leaving soon, so pack your geometry tool kit and prepare to meet a geometry giant!

WebQuest Log on to **msmath2.net/webquest** to begin your WebQuest.

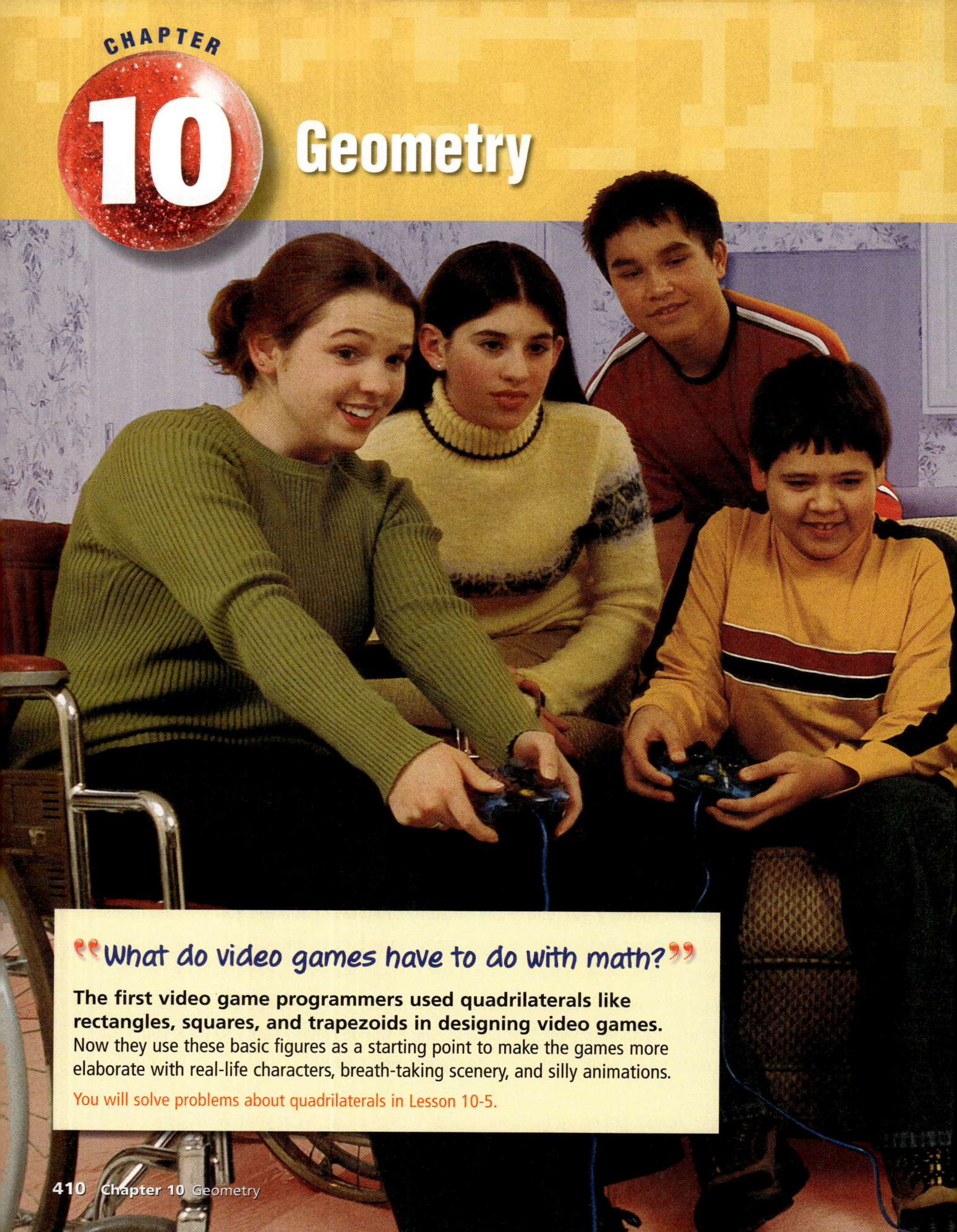

CHAPTER 10 Geometry

"What do video games have to do with math?"

The first video game programmers used quadrilaterals like rectangles, squares, and trapezoids in designing video games. Now they use these basic figures as a starting point to make the games more elaborate with real-life characters, breath-taking scenery, and silly animations.

You will solve problems about quadrilaterals in Lesson 10-5.

GETTING STARTED

▶ Diagnose Readiness

Take this quiz to see if you are ready to begin Chapter 10. Refer to the lesson or page number in parentheses for review.

Vocabulary Review

Complete each sentence.

1. A comparison of two numbers by division is called a __?__. (Lesson 7-1)
2. The __?__ is written as part = percent · base. (Lesson 8-2)

Prerequisite Skills

Multiply or divide. Round to the nearest hundredth if necessary. (Pages 560 and 562)

3. $360 \cdot 0.85$
4. $48 \div 191$
5. $24 \div 156$
6. $0.37 \cdot 360$
7. $33 \div 307$
8. $0.69 \cdot 360$

Solve each equation. (Lessons 4-2 and 4-3)

9. $b + 36 = 89$
10. $74 = 22 + s$
11. $15 + r = 146$
12. $m + 78 = 93$
13. $153 = d + 61$
14. $6x = 360$
15. $180 = 2f$
16. $120 = 3c$
17. $5n = 270$
18. $15z = 90$

Solve each proportion. (Lesson 7-3)

19. $\frac{4}{a} = \frac{3}{9}$
20. $\frac{7}{16} = \frac{h}{32}$
21. $\frac{5}{8} = \frac{15}{y}$
22. $\frac{t}{42} = \frac{6}{7}$
23. $\frac{s}{12} = \frac{2}{3}$
24. $\frac{24}{p} = \frac{2}{3}$

FOLDABLES™ Study Organizer

Geometry Make this Foldable to help you organize your notes. Begin with a sheet of 11" × 17" paper and six index cards.

STEP 1 Fold
Fold lengthwise about 3" from the bottom.
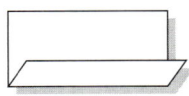

STEP 2 Fold Again
Fold the paper in thirds.

STEP 3 Open and Staple
Staple the edges on either side to form three pockets.
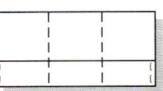

STEP 4 Label
Label the pockets as shown. Place two index cards in each pocket.

Noteables™ Chapter Notes Each time you find this logo throughout the chapter, use your *Noteables™: Interactive Study Notebook with Foldables™* or your own notebook to take notes. Begin your chapter notes with this Foldable activity.

Readiness To prepare yourself for this chapter with another quiz, visit **msmath2.net/chapter_readiness**

10-1a HANDS-ON LAB

A Preview of Lesson 10-1

Measuring Angles

What You'll LEARN

Measure angles.

Materials
- protractor

To measure lines, you use a ruler. To measure angles, you use a *protractor*. Angles are measured in units called **degrees**. In this lab, you will learn how to measure angles using a protractor.

The outer scale goes from 0 to 180 from left to right.

The inner scale goes from 0 to 180 from right to left.

ACTIVITY Work with a partner.

Find the measure of the angle below.

STEP 1 Place the protractor on the angle so that the center is on the vertex of the angle and one side goes through 0° on the protractor.

STEP 2 In this case, 0° is on the inner scale. So, follow the inner scale to the point where the other side of the angle meets the protractor. The inner number is the angle's degree measure.

So, the measure of this angle is 135°.

Your Turn Find the measure of each angle.

a.   b. c.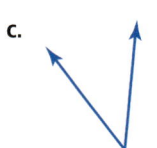

Writing Math

1. **Explain** why there are two scales on the protractor.
2. Can you place either side of an angle through 0° and get the same angle measure? **Explain** your reasoning.
3. What angle has the same measure on both the inner and outer scales?
4. **Explain** how the two scales on the protractor are related.

10-1 Angles

What You'll LEARN
Classify and draw angles.

NEW Vocabulary
angle
degrees
vertex
acute angle
right angle
obtuse angle
straight angle

Link to READING
Everyday Meaning of Acute: characterized by sharpness, as in acute pain

WHEN am I ever going to use this?

CLOCKS The hour and minute hands of a clock form angles of different sizes.

3:10
less than 90°

3:00
90°

3:40
greater than 90°

1. Name other times in which the hands of a clock form an angle less than 90°, equal to 90°, and greater than 90°.

2. How many degrees is the angle that is formed by clock hands at 6:00?

An **angle** is made up of two rays with a common endpoint and is measured in units called **degrees**. If a circle were divided into 360 equal-sized parts, each part would have an angle measure of 1 degree.

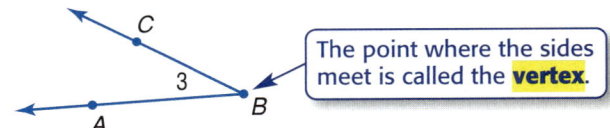

The point where the sides meet is called the **vertex**.

You can use the symbol for angle, ∠, to name an angle.

Use the vertex as the middle letter and a point from each side.	∠ABC or ∠CBA
Use the vertex only.	∠B
Use a number.	∠3

Angles are classified according to their measure.

Noteables™   Key Concept: Types of Angles

Acute Angle — less than 90°

Right Angle — exactly 90° (This symbol indicates a right angle.)

Obtuse Angle — between 90° and 180°

Straight Angle — exactly 180°

Lesson 10-1 Angles 413

EXAMPLES Classify Angles

Classify each angle as *acute*, *obtuse*, *right*, or *straight*.

1.

The angle is less than 90°, so it is an acute angle.

2.

The angle is greater than 90°, so it is an obtuse angle.

EXAMPLE Draw an Angle

3. Draw $\angle A$ with a measure of 75°.

- Draw a ray with endpoint A.
- Place the center point of a protractor on A. Align the ray with 0°.
- Use the scale that begins with 0°. Locate the mark labeled 75. Then draw the other side of the angle.

Skill and Concept Check

1. **Writing Math** Describe a right angle.

2. **OPEN ENDED** Draw and label acute angle XYZ. Then find its measure.

GUIDED PRACTICE

Classify each angle as *acute*, *obtuse*, *right*, or *straight*.

3.
4.

Draw an angle having each measurement. Then classify the angle as *acute*, *obtuse*, *right*, or *straight*.

5. 90°
6. 170°
7. 11°

8. **BICYCLE SAFETY** A cyclist should be familiar with arm and hand signals indicating his or her movements. Describe the angles formed by the arm of the cyclist below.

Practice and Applications

HOMEWORK HELP
For Exercises	See Examples
9–11, 18, 21	1, 2
12–17	3

Extra Practice
See pages 587, 605.

Classify each angle as *acute, obtuse, right,* **or** *straight.*

9.
10.
11.

Draw an angle having each measurement. Then classify each angle as *acute, obtuse, right,* **or** *straight.*

12. 56° 13. 147° 14. 180° 15. 99° 16. 8° 17. 90°

For Exercises 18–20, use the figure at the right.

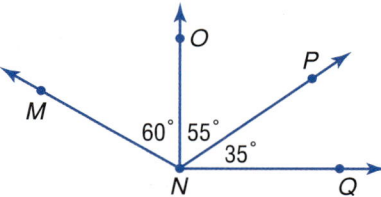

18. Name the angles that are obtuse.

19. The symbol $m\angle MNQ$ means the *measure of angle MNQ*. Find $m\angle MNQ$.

20. Angles are said to be *adjacent* if they have a common vertex and a common side between them. So, $\angle MNO$ is adjacent to $\angle ONP$. Name another angle adjacent to $\angle ONP$.

21. **SKI JUMPING** When skiers jump off a hill, they want to make the angle between their bodies and the front of their skis as small as possible. Describe where the ski jumper's legs and skis form acute and obtuse angles in the photograph.

22. **EARTH SCIENCE** Earth rotates 360° degrees in one day. Through how many degrees does it rotate in one hour?

23. **CRITICAL THINKING** How many times do the hands of a clock make a right angle in a 24-hour time period?

Spiral Review with Standardized Test Practice

24. **SHORT RESPONSE** Name the obtuse angle in the figure at the right.

25. **MULTIPLE CHOICE** Which is the measure of an acute angle?
    Ⓐ 87°   Ⓑ 95°   Ⓒ 120°   Ⓓ 180°

PROBABILITY A number cube is rolled, and a coin is tossed. Find each probability. (Lesson 9-7)

26. P(5 and heads)   27. P(odd and tails)   28. P(7 and heads or tails)

29. Anica spins a spinner fifty times, and it lands on 3 fifteen times. What is the experimental probability of *not* landing on 3? (Lesson 9-6)

GETTING READY FOR THE NEXT LESSON

PREREQUISITE SKILL Multiply or divide. Round to the nearest hundredth if necessary. (Pages 560 and 562)

30. 0.62 · 360   31. 360 · 0.25   32. 17 ÷ 146   33. 63 ÷ 199

10-1b Hands-On Lab

A Follow-Up of Lesson 10-1

Constructing and Bisecting Angles

Two angles that have the same measure are **congruent angles**. To **bisect** an angle means to divide it into two congruent angles. In this lab, you will learn how to construct congruent angles and bisect angles.

What You'll LEARN

Construct and bisect angles.

Materials
- straightedge
- compass

STUDY TIP

Symbols $\overrightarrow{LK}$ is read ray LK. A ray is a path that extends infinitely from one point in a certain direction.

ACTIVITY   Work with a partner.

1 Construct an angle congruent to ∠ABC.

STEP 1 Use a straightedge to draw $\overrightarrow{LK}$.

STEP 2 With the compass at point B, draw an arc that intersects both sides of ∠ABC. Label the two points of intersection as X and Y.

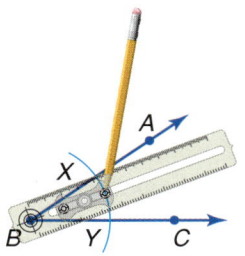

STEP 3 With the same setting on your compass, place your compass at point L. Draw another arc. Label the intersection R.

STEP 4 Open your compass to the same width as the distance between points X and Y. Then place the compass at point R. Draw an arc that intersects the arc you drew in Step 3. Label this point of intersection S.

STEP 5 Draw $\overrightarrow{LM}$ through point S. Angle MLK is congruent to ∠ABC.

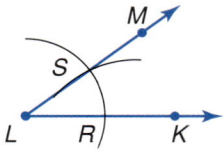

Your Turn

a. Draw an obtuse angle. Then construct an angle congruent to it.

ACTIVITY *Work with a partner.*

2 Bisect ∠DEF.

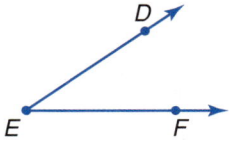

STEP 1 Place the compass at point *E* and draw an arc that intersects both sides of ∠DEF. Mark the two points of intersection as *H* and *I*.

STEP 2 Place your compass at point *H*. Draw an arc inside ∠DEF. Using the same setting on your compass, place your compass at point *I*. Draw another arc inside ∠DEF intersecting the first arc. Label the intersection point *G*.

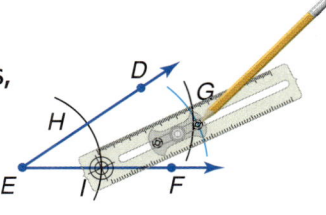

STEP 3 Draw $\overrightarrow{EG}$. $\overrightarrow{EG}$ is the bisector of ∠DEF.

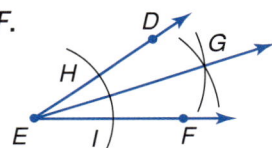

Your Turn

b. Draw a right angle. Then bisect it.

Writing Math

Work with a partner.

1. **Explain** how you could verify that ∠ABC and ∠MLK in Activity 1 are congruent.

2. **Explain** how you could verify that ∠DEF in Activity 2 is bisected.

3. A straight angle has a measure of 180°. What kind of angle is formed if you bisect a straight angle?

4. **Perpendicular lines** are lines that meet to form right angles. In the figure at the right, $\overline{PS}$ is perpendicular to $\overline{QR}$. Use a compass and a straightedge to construct perpendicular lines. (*Hint*: Use the procedure for bisecting angles.)

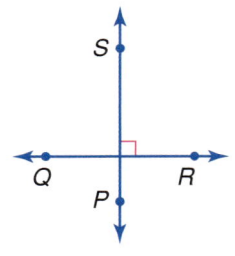

Lesson 10-1b Hands-On Lab: Constructing and Bisecting Angles **417**

10-2 Statistics: Making Circle Graphs

What You'll LEARN
Construct and interpret circle graphs.

NEW Vocabulary
circle graph

REVIEW Vocabulary
ratio: a comparison of two numbers by division (Lesson 7-1)

percent equation: part = percent · base (Lesson 8-2)

WHEN am I ever going to use this?

COLORS In a recent survey, people ages 13–20 were asked to choose their favorite shade of blue. The results are shown in the table.

1. Explain how you know that each person surveyed chose only one shade of blue.
2. If 500 people took part in the survey, how many preferred aquamarine?

Favorite Shades of Blue for People Ages 13–20

Shade	Percent
navy	35%
sky/light blue	30%
aquamarine	17%
other	18%

Source: *American Demographics*

A graph used to compare parts of a whole is called a **circle graph**. In a circle graph, the percents add up to 100.

EXAMPLE Construct a Circle Graph

1 COLORS Make a circle graph of the data in the table above.

- Find the degrees for each part. Round to the nearest whole degree.

  35% of 360° = 0.35 · 360° or 126°

  30% of 360° = 0.30 · 360° or 108°

  17% of 360° = 0.17 · 360° or about 61°

  18% of 360° = 0.18 · 360° or about 65°

- Use a compass to draw a circle with a radius as shown. Then use a protractor to draw the first angle, in this case 126°. Repeat this step for each section or *sector*.

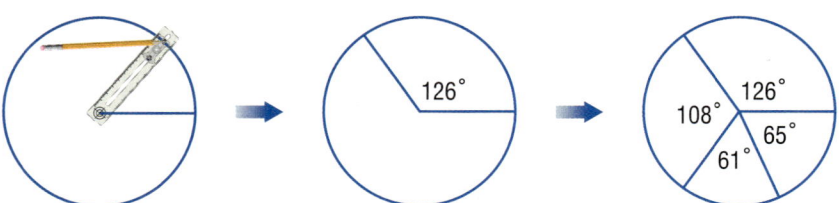

- Label each section of the graph with the category and percent. Give the graph a title.

Check To draw an accurate circle graph, make sure the sum of the angle measures equals 360°.

Favorite Shades of Blue for People Ages 13–20

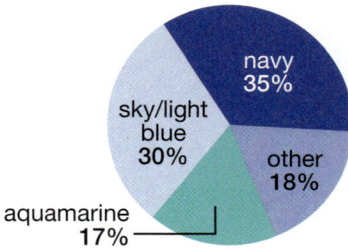

418 Chapter 10 Geometry

When constructing a circle graph, it is sometimes necessary to first convert the data to percents or decimals, and then to degrees.

EXAMPLE Construct a Circle Graph

2 OLYMPICS The table shows the number of each type of medal won by the United States during the Summer Olympics from 1896 to 2000. Make a circle graph of the data.

U.S. Summer Olympic Medals

Type	Number
gold	872
silver	658
bronze	587

Source: infoplease.com

REAL-LIFE MATH

OLYMPICS The United States has won 2,117 Summer Olympic medals through 2000, the most by any country.

Source: infoplease.com

- Find the total number of medals: 872 + 658 + 587 or 2,117.

- Find the ratio that compares each number with the total. Write the ratio as a decimal rounded to the nearest hundredth.

  gold: $\frac{872}{2,117} \approx 0.41$ silver: $\frac{658}{2,117} \approx 0.31$ bronze: $\frac{587}{2,117} \approx 0.28$

- Find the number of degrees for each section of the graph.

  gold: $0.41 \cdot 360° \approx 148°$
  silver: $0.31 \cdot 360° \approx 112°$
  bronze: $0.28 \cdot 360° \approx 101°$

  Because of rounding, the sum of the degrees is 361°.

- Draw the circle graph.

U.S. Summer Olympic Medals

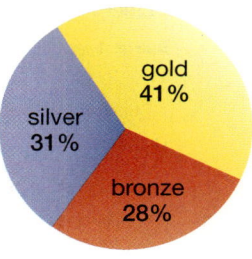

gold 41%
silver 31%
bronze 28%

Check After drawing the first two sections, you can measure the last section of a circle graph to verify that the angles have the correct measures.

EXAMPLE Interpret a Circle Graph

3 MONEY The circle graph shows the percent of Americans who favor, oppose, or don't know how they feel about a common currency for North America. Use the graph to describe the opinion of most Americans.

Do Americans Favor Common North American Currency?

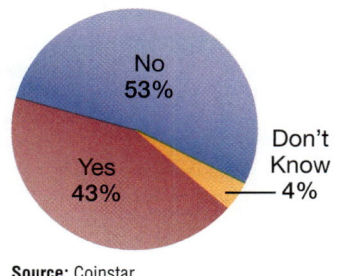

No 53%
Yes 43%
Don't Know 4%

Source: Coinstar

The greatest percent of the circle graph is the section representing the "No" response.

So, most Americans do not favor a common North American currency.

Skill and Concept Check

1. **OPEN ENDED** Draw a circle graph that has three parts. Label the parts and describe what they represent.

2. **DATA SENSE** The table shows the percent of people who use the sweeteners listed. Can the data be represented in a circle graph? Explain.

Type of Sweetener	Percent
sugar	74%
honey, molasses, syrup	46%
low calorie sweeteners	36%

   Source: Yankelovich Partners

GUIDED PRACTICE

Make a circle graph of the data in each table.

3.
Favorite Shades of Blue for People Ages 21–34	
Shade	Percent
navy	48%
sky/light blue	23%
aquamarine	12%
other	18%

Source: American Demographics

4.
Speed Limit (mph)	Number of States
55	1
65	20
70	18
75	11

Source: The World Almanac

5. **EDUCATION** The circle graph shows the percent of students by grade level in U.S. schools. In which grades are most students?

6. Refer to the circle graph in Example 1 and the graph you drew in Exercise 3. What can you conclude about the favorite shades of blue for people in different age groups?

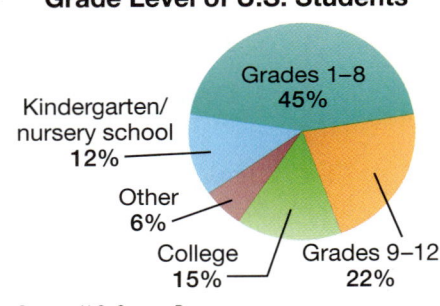

Grade Level of U.S. Students

- Kindergarten/nursery school 12%
- Other 6%
- College 15%
- Grades 9–12 22%
- Grades 1–8 45%

Source: U.S. Census Bureau

Practice and Applications

Make a circle graph of the data in each table.

7.
New York City Commuters	
Transportation	Percent
driving alone	24%
carpool	9%
public transit	53%
other	14%

Source: Time Almanac

8.
Los Angeles Commuters	
Transportation	Percent
driving alone	65%
carpool	15%
public transit	11%
other	9%

Source: Time Almanac

9. Compare and contrast the data from the circle graphs you drew in Exercises 7 and 8.

HOMEWORK HELP

For Exercises	See Examples
7–8, 12–13	1
10–11	2
9, 14	3

Extra Practice
See pages 587, 605.

Chapter 10 Geometry

Make a circle graph of the data in each table.

10.
Endangered Species in U.S.	Number of Species
mammals	63
birds	78
reptiles	14
amphibians	10

Source: U.S. Fish and Wildlife Service

11.
U.S. Regions	Population (millions)
Northeast	54
Midwest	64
South	100
West	63

Source: *Time Almanac*

POLITICS For Exercises 12–14, refer to the graphic at the right.

12. How many 13- to 15-year olds participated in the survey?

13. Calculate the angle measure for each section.

14. Measure the angles for each section on the graph. Explain any differences.

15. **RESEARCH** Use the Internet or another source to find data that add up to 100%. Make a circle graph of the data.

16. **CRITICAL THINKING** Line graphs are usually best for data that show change over time. When might it be more appropriate to display data in a circle graph? Give an example.

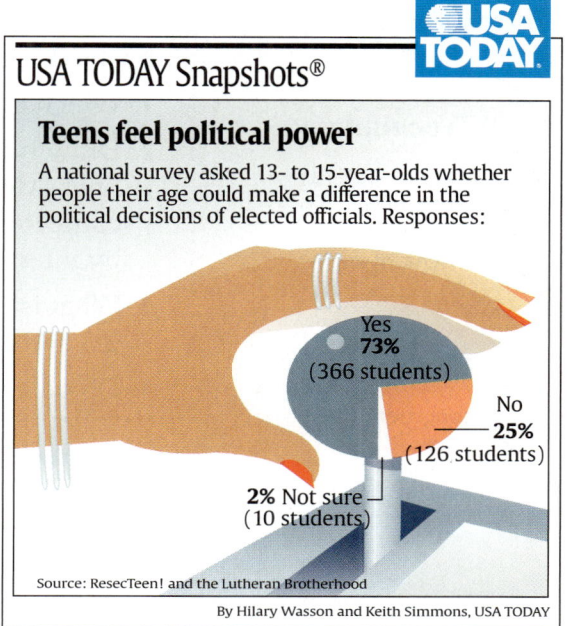

Spiral Review with Standardized Test Practice

17. **MULTIPLE CHOICE** On a circle graph of the data in the table, which section would have an angle measure of about 38°?

    Ⓐ Lake Erie   Ⓑ Lake Huron
    Ⓒ Lake Michigan   Ⓓ Lake Ontario

Lake	Area (sq mi)
Erie	9,930
Huron	23,010
Michigan	22,400
Ontario	7,520
Superior	31,820

Source: www.infoplease.com

18. **SHORT RESPONSE** Find the angle measure of the Lake Michigan section in the circle graph referred to in Exercise 17. Round to the nearest whole degree.

Classify each angle as *acute, obtuse, right,* or *straight.* (Lesson 10-1)

19. 65°   20. 102°   21. 90°

22. **PROBABILITY** Corey has an after-school activity 15% of the time. Li-Cheng has one 20% of the time. Find the probability that they will both have an after-school activity on the same day. (Lesson 9-7)

GETTING READY FOR THE NEXT LESSON

PREREQUISITE SKILL Solve each equation. (Lesson 4-2)

23. $a + 18 = 90$ 24. $44 + x = 90$ 25. $180 = 39 + n$ 26. $123 + t = 180$

msmath2.net/self_check_quiz

10-3 Angle Relationships

What You'll LEARN
Identify and apply angle relationships.

NEW Vocabulary
vertical angles
congruent angles
supplementary angles
complementary angles

MATH Symbols
≅ is congruent to

HANDS-ON Mini Lab

Materials
- protractor

Work with a partner.

Draw two intersecting lines and label the angles as shown. Then measure each angle with your protractor and record the measurements.

1. Which angles have the same measure?
2. Draw two other pairs of intersecting lines. Measure their angles. **Make a conjecture** involving four angles created by intersecting lines.
3. What is the relationship between the measures of ∠1 and ∠2? ∠3 and ∠4?
4. Do other pairs of angles share the same relationship? Explain.

When two lines intersect, they form two pairs of opposite angles called **vertical angles**. In the Mini Lab, you found that vertical angles have the same measure. Angles with the same measure are **congruent angles**.

∠1 ≅ ∠4
∠2 ≅ ∠3

The symbol ≅ is used to show that the angles are congruent.

Pairs of angles can also have other relationships. In the Mini Lab, you found pairs of angles whose sum is 180°. Two angles are **supplementary** if the sum of their measures is 180°.

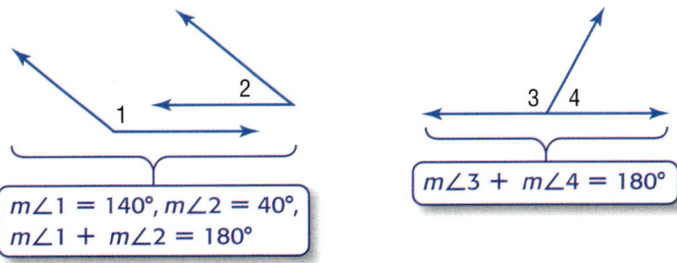

m∠1 = 140°, m∠2 = 40°, m∠1 + m∠2 = 180°

m∠3 + m∠4 = 180°

Two angles are **complementary** if the sum of their measures is 90°.

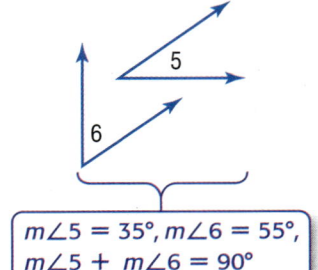

m∠5 = 35°, m∠6 = 55°, m∠5 + m∠6 = 90°

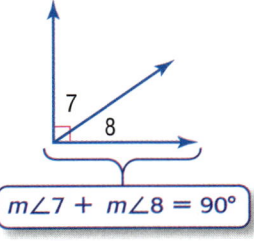

m∠7 + m∠8 = 90°

READING in the Content Area
For strategies in reading this lesson, visit **msmath2.net/reading**.

422 Chapter 10 Geometry

EXAMPLES Classify Angles

Classify each pair of angles as *complementary*, *supplementary*, or *neither*.

1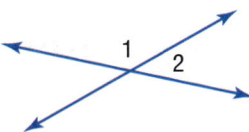

∠1 and ∠2 form a straight line. So, the angles are supplementary.

2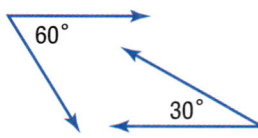

60° + 30° = 90°
The angles are complementary.

You can use angle relationships to find missing measures.

EXAMPLE Find a Missing Angle Measure

3 Angles *ABC* and *CBD* are complementary. If $m\angle ABC = 28°$, find $m\angle CBD$.

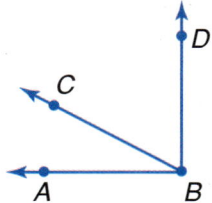

Since ∠*ABC* and ∠*CBD* are complementary, $m\angle ABC + m\angle CBD = 90°$.

$m\angle ABC + m\angle CBD =$	90	Write the equation.
$28 + m\angle CBD =$	90	Replace $m\angle ABC$ with 28.
-28 -28		Subtract 28 from each side.
$m\angle CBD =$	62	$90 - 28 = 62$

The measure of ∠*CBD* is 62°.

Your Turn Find the value of *x* in each figure.

a.

b.

REAL-LIFE MATH

BASKETBALL Canadian Dr. James Naismith invented the game of basketball in 1891 using a ball the size of a soccer ball and two peach baskets.

Source: www.allsands.com

EXAMPLE Use Angles to Solve A Problem

4 **BASKETBALL** Erin wants to make a bounce pass to Mackenzie. Find the value of *x* so that Erin's pass hits Mackenzie in the hands.

Since the sum of the three angles is 180°, $48° + x° + 48° = 180°$.

$48 + x + 48 =$	180	Write the equation.
$x + 96 =$	180	Simplify.
-96 -96		Subtract 96 from each side.
$x =$	84	$180 - 96 = 84$

So, the angle must be 84° for the ball to hit Mackenzie in the hands.

msmath2.net/extra_examples   Lesson 10-3 Angle Relationships

Skill and Concept Check

1. **OPEN ENDED** Draw two angles that are complementary.
2. **Writing Math** Can a pair of angles be vertical and supplementary? Give an example or nonexample, with angle measurements, to support your answer.

GUIDED PRACTICE

Classify each pair of angles as *complementary*, *supplementary*, or *neither*.

3.
4.
5.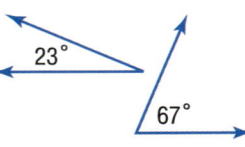

Find the value of x in each figure.

6.
7.
8.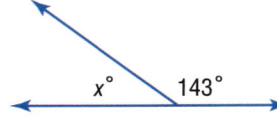

9. Suppose $\angle 1$ and $\angle 2$ are supplementary. If $m\angle 1 = 84°$, find $m\angle 2$.

Practice and Applications

Classify each pair of angles as *complementary*, *supplementary*, or *neither*.

10.
11.
12.

13.
14.
15.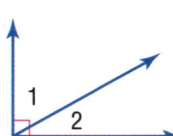

HOMEWORK HELP

For Exercises	See Examples
10–15, 22–23	1, 2
16–21	3, 4

Extra Practice
See pages 587, 605.

Find the value of x in each figure.

16.
17.
18.

19.
20.
21.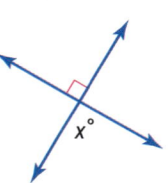

HEALTH For Exercises 22 and 23, use the graphic at the right.

22. Which two sections have angles that are complementary?

23. Which two sections have angles that are supplementary?

Determine whether each statement is *sometimes*, *always*, or *never* true. Explain or give an example to support your answer.

24. Adjacent angles are complementary.

25. Two straight angles are supplementary.

26. Vertical angles have the same angle measure.

27. Adjacent angles are congruent.

28. **WRITE A PROBLEM** Write about a real-life object that has angles. Draw a diagram of it, measure all angles, and identify any complementary, supplementary, and vertical angles.

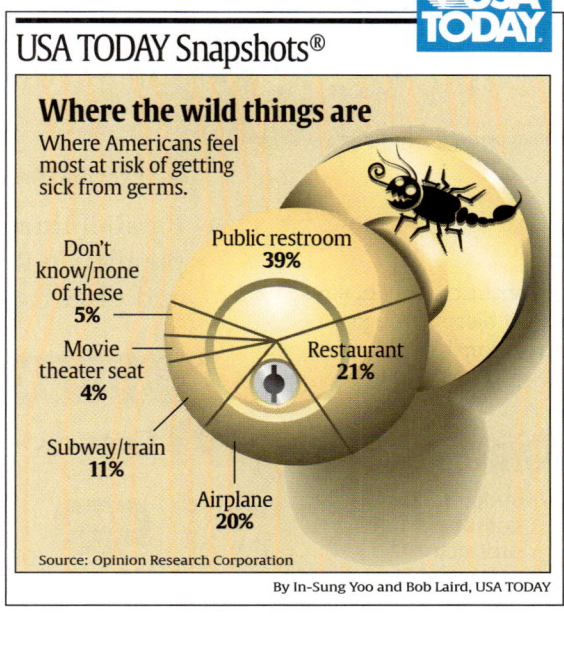

29. **CRITICAL THINKING** Angles E and F are complementary. If $m\angle E = x - 10$ and $m\angle F = x + 2$, find the measure of each angle.

Spiral Review with Standardized Test Practice

30. **SHORT RESPONSE** Find the measure of two angles that are vertical and complementary.

31. **MULTIPLE CHOICE** Find the value of a in the figure.
    - Ⓐ 90
    - Ⓑ 75
    - Ⓒ 55
    - Ⓓ 45

32. **STATISTICS** A company surveyed people about the type of crust they preferred on their pizza. Make a circle graph of the results shown at the right. (Lesson 10-2)

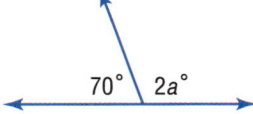

33. **TIME** Classify the angle the hands of a clock make at 3:30. (Lesson 10-1)

34. **REAL ESTATE** A house for sale has a rectangular lot with a length of 250 feet and a width of 120 feet. What is the area of the lot? (Lesson 6-8)

Write each percent as a decimal. (Lessons 5-5 and 7-6)

35. 53%
36. 78.5%
37. 431%
38. 0.23%

GETTING READY FOR THE NEXT LESSON

PREREQUISITE SKILL Solve each equation. (Lesson 4-2)

39. $x + 112 = 180$
40. $50 + t = 180$
41. $180 = 79 + y$
42. $180 = h + 125$

10-3b HANDS-ON LAB

A Follow-Up of Lesson 10-3

Constructing Parallel Lines

Parallel lines are lines that do not intersect. The symbol ∥ means parallel. In this lab, you will learn how to construct parallel lines and discover angle relationships that are created from parallel lines.

What You'll LEARN

Construct parallel lines, and discover angle relationships.

Materials
- straightedge
- compass
- protractor
- colored pencils
- notebook paper

ACTIVITY *Work with a partner.*

1 Construct a line parallel to line ℓ.

STEP 1 Draw line ℓ. Choose a point A not on the line. Then draw a line through point A that intersects ℓ. Label the point of intersection B.

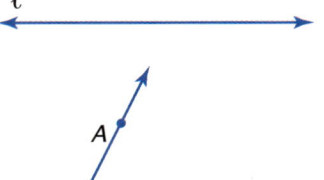

STEP 2 Place the compass at vertex B and draw an arc. Label the points of intersection C and D. With the same compass setting, place the compass at point A and draw an arc. Label the point of intersection E.

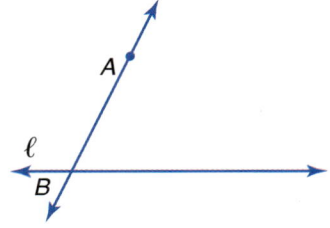

STEP 3 Open your compass to the same width as the distance between C and D. Then place the compass at point E. Draw an arc that intersects the arc you drew in Step 2. Label this point of intersection F.

STEP 4 Draw a line through points A and F. Label the line m. Line m is parallel to line ℓ.

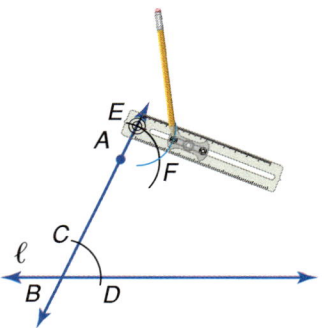

Your Turn

a. Draw a line. Then construct a line parallel to it.

426 Chapter 10 Geometry

A line that intersects parallel lines is called a **transversal**. Line t is a transversal for parallel lines ℓ and m. When a transversal cuts two parallel lines, there are several angle relationships formed.

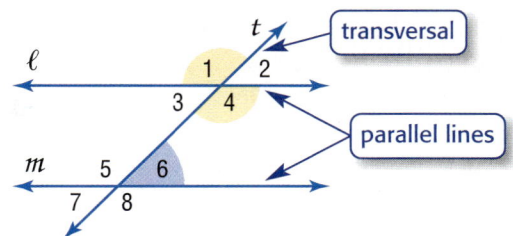

ACTIVITY   *Work with a partner.*

2 Find angle relationships in the figure above.

STEP 1 Draw parallel lines and a transversal on notebook paper. Label the angles as shown above.

STEP 2 Measure each angle with your protractor. Record each measure.

STEP 3 Using colored pencils, shade any angles with the same measure one color. If there are angles with a different measure, shade them a second color. For example, ∠1 and ∠4 shown above are both shaded yellow because they have the same measure. Angle 6 is shaded blue because it has a different measure.

In the figure above, certain pairs of angles have special names. The angles that make up each pair have the same measure.

alternate interior angles	∠3 and ∠6, ∠4 and ∠5
alternate exterior angles	∠1 and ∠8, ∠2 and ∠7
corresponding angles	∠1 and ∠5, ∠2 and ∠6, ∠3 and ∠7, ∠4 and ∠8
vertical angles	∠1 and ∠4, ∠2 and ∠3, ∠5 and ∠8, ∠6 and ∠7

Writing Math

1. **Explain** why you think ∠1, ∠2, ∠7, and ∠8 are called exterior angles.
2. **Explain** why you think ∠3, ∠4, ∠5, and ∠6 are called interior angles.
3. What are ∠2 and ∠4 called?
4. If you know only one angle measure in the figure, **explain** how you can find the measures of the other angles without measuring.
5. Predict the measure of the other angles in the figure at the right using the 45° angle. Then copy the figure onto notebook paper and check by using a protractor.

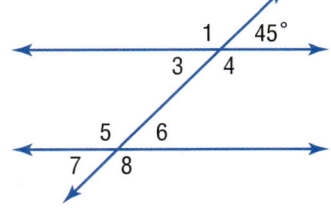

Lesson 10-3b Hands-On Lab: Constructing Parallel Lines **427**

10-4 Triangles

What You'll LEARN
Identify and classify triangles.

NEW Vocabulary
triangle
acute triangle
right triangle
obtuse triangle
congruent segments
scalene triangle
isosceles triangle
equilateral triangle

MATH Symbols
$m\angle 1$ measure of angle 1

HANDS-ON Mini Lab

Materials
- paper
- straightedge
- scissors

Work with a partner.

STEP 1 Use a straightedge to draw a triangle with three acute angles. Label the angles *A*, *B*, and *C*. Cut out the triangle.

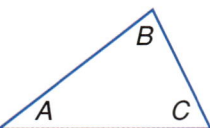

STEP 2 Fold $\angle B$ so that its vertex touches the line between angles *A* and *C*. The fold should be parallel to the base of the triangle.

STEP 3 Fold $\angle A$ and $\angle C$ so the vertices meet.

1. What kind of angle is formed where the three vertices meet?
2. Repeat the activity with another triangle. **Make a conjecture** about the sum of the measures of three angles of any triangle.

A **triangle** is a figure with three sides and three angles. In the Mini Lab, you discovered a relationship among the three angles in a triangle.

Noteables™   Key Concept: Angles of a Triangle

Words   The sum of the measures of the angles of a triangle is 180°.

Symbols $m\angle 1 + m\angle 2 + m\angle 3 = 180°$

Model

EXAMPLE   Find Angle Measures of Triangles

1 **KITES** A kite is constructed with two triangles. Find the missing measure.

$x + 68 + 37 = 180$ The sum of the measures is 180.

$x + 105 = 180$ Simplify.

$\underline{-105 \quad -105}$ Subtract 105 from each side.

$x = 75$

The missing measure is 75°.

428 Chapter 10 Geometry

STUDY TIP

Check the Reasonableness of Results You can check whether you found the correct measure by adding 43 + 119 + 18. The sum should be 180.

EXAMPLE — Find a Missing Measure

2 **ALGEBRA** Find $m\angle Z$ in $\triangle XYZ$ if $m\angle X = 43°$ and $m\angle Y = 119°$.

Draw the triangle. Then write and solve an equation to find $m\angle Z$.

$m\angle X + m\angle Y + m\angle Z =$	180	The sum of the measures is 180.
$43 + 119 + m\angle Z =$	180	Substitute.
$162 + m\angle Z =$	180	Simplify.
-162	-162	Subtract 162 from each side.
$m\angle Z =$	18	

So, the measure of $\angle Z$ is 18°.

Since every triangle has at least two acute angles, one way you can classify a triangle is by using the third angle.

Noteables™ — **Key Concept: Classify Triangles Using Angles**

Acute Triangle	Right Triangle	Obtuse Triangle
all acute angles	1 right angle	1 obtuse angle

Another way to classify triangles is by their sides. Sides with the same length are **congruent segments**.

STUDY TIP

Congruent Segments The marks on the sides of the triangles indicate that those sides are congruent.

Noteables™ — **Key Concept: Classify Triangles Using Sides**

Scalene Triangle	Isosceles Triangle	Equilateral Triangle
no congruent sides	at least 2 congruent sides	3 congruent sides

EXAMPLES — Classify Triangles

Classify each triangle by its angles and by its sides.

3

The triangle has a right angle and two congruent sides. So, it is a right, isosceles triangle.

4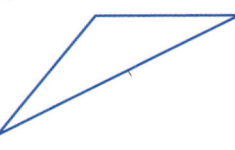

The triangle has one obtuse angle and no congruent sides. So, it is an obtuse, scalene triangle.

msmath2.net/extra_examples

Lesson 10-4 Triangles **429**

Skill and Concept Check

1. **Writing Math** Describe the angles in an obtuse triangle.

2. **OPEN ENDED** Draw an acute scalene triangle.

GUIDED PRACTICE

Find the missing measure in each triangle. Then classify the triangle as *acute*, *right*, or *obtuse*.

3.

4.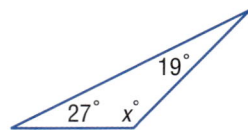

Classify the marked triangle in each flag by its angles and by its sides.

5. Puerto Rico

6. Seychelles Islands

7. Bosnia-Herzegovina

Practice and Applications

Find the missing measure in each triangle. Then classify the triangle as *acute*, *right*, or *obtuse*.

8.

9.

10.

11.

12.

13.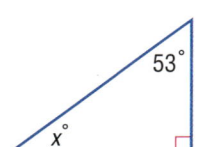

HOMEWORK HELP

For Exercises	See Examples
8–14, 17, 24	1, 2
8–13, 15–16, 18–23	3, 4

Extra Practice
See pages 588, 605.

14. Find the third angle of a right triangle if the measure of one of the angles is 10°.

15. Three angles of a triangle measure 30°, 60°, and 90°. Classify the triangle by its angles.

16. Three sides of a triangle measure 5 meters, 8 meters, and 8 meters. Classify the triangle by its sides.

17. **ALGEBRA** Find $m\angle T$ in $\triangle RST$ if $m\angle R = 88°$ and $m\angle S = 75°$.

Classify each triangle by its angles and by its sides.

18.
19.
20.
21.
22.
23.

24. **GEOGRAPHY** A triangular area in the southern Atlantic Ocean where airplanes and boats have disappeared is known as the *Bermuda Triangle*. Find the missing angle measurement at Bermuda in the figure at the right.

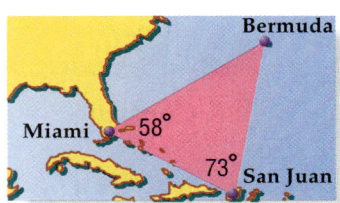

25. **WRITE A PROBLEM** Write about a real-life situation or object involving a triangle and its measures. Classify the triangle by its sides and its angles.

26. **CRITICAL THINKING** Find the missing angle measures in the figure.

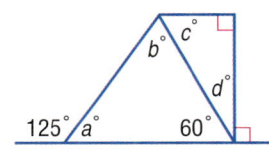

Spiral Review with Standardized Test Practice

27. **MULTIPLE CHOICE** How would you find $m\angle R$?

    Ⓐ Add 30° to 180°.　　Ⓑ Subtract 60° from 180°.
    Ⓒ Subtract 30° from 90°.　Ⓓ Subtract 180° from 60°.

28. **SHORT RESPONSE** Find the measure of two congruent angles of a triangle if the third angle measure is 54°.

Classify each pair of angles as *complementary*, *supplementary*, or *neither*.
(Lesson 10-3)

29.
30.
31.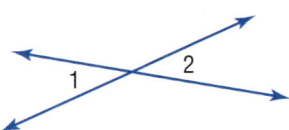

32. **BIRDS** A circle graph shows that 41% of people who bird watch live in the Northeast region of the United States. What is the measure of the angle of the Northeast section of the graph? (Lesson 10-2)

GETTING READY FOR THE NEXT LESSON

PREREQUISITE SKILL Solve each equation. (Lesson 4-2)

33. $x + 120 + 120 + 60 = 360$
34. $73 + 119 + x + 50 = 360$

msmath2.net/self_check_quiz

Lesson 10-4 Triangles **431**

10-4b Hands-On Lab

A Follow-Up of Lesson 10-4

Constructing Triangles

Equilateral triangles are triangles with three congruent sides. Isosceles triangles are triangles with at least two congruent sides. In this lab, you will learn how to construct both types of triangles.

What You'll LEARN

Construct equilateral and isosceles triangles.

Materials
- straightedge
- compass
- protractor
- ruler

ACTIVITY *Work with a partner.*

1 Construct an equilateral triangle.

STEP 1 Use your straightedge to draw line segment $\overline{AB}$.

STEP 2 Open the compass to the same length as $\overline{AB}$. With the compass at point A, draw an arc above the line.

STEP 3 With the same compass setting, place the compass at point B. Draw an arc that intersects the arc you drew in Step 2. Label the intersection C.

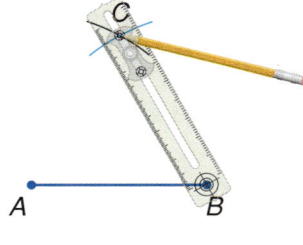

STEP 4 Connect the points to complete equilateral △ABC.

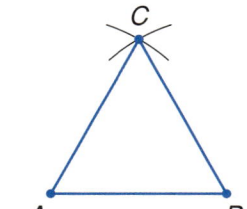

STUDY TIP

Check You can measure each side of the triangle to see if the sides are congruent.

Your Turn

a. Construct equilateral triangle *TUV* with sides measuring 3 inches.

In order to construct equilateral triangles in the activity above, you copied line segments using a compass. You constructed line segments with the same length by copying $\overline{AB}$ and making $\overline{AC}$ and $\overline{BC}$ congruent. So, a compass is also a useful tool to check whether line segments have the same length.

432 Chapter 10 Geometry

ACTIVITY *Work with a partner.*

2 Construct an isosceles triangle.

STEP 1 Use your straightedge to draw line segment $\overline{DE}$.

STEP 2 Open the compass to a length greater than $\overline{DE}$. With the compass at point D, draw an arc above the line.

STEP 3 With the same compass setting, place the compass at point E. Draw an arc that intersects the arc you drew in Step 2. Label the intersection F.

STEP 4 Connect the points to complete isosceles △DEF.

Check You can measure $\overline{DF}$ and $\overline{EF}$ to check if they are congruent.

Your Turn

b. Construct isosceles triangle XYZ with sides measuring 3 centimeters, 5 centimeters, and 5 centimeters.

Writing Math

Work with a partner.

1. **Measure** the angles in the equilateral triangle you constructed with a protractor. Compare your measurements with other groups.

2. **Make a conjecture** about the measure of the angles of an equilateral triangle.

3. **Measure** the angles in the isosceles triangle you constructed with a protractor. Compare your measurements with other groups.

4. **Make a conjecture** about the measure of the angles of an isosceles triangle.

Lesson 10-4b Hands-On Lab: Constructing Triangles **433**

10-5 Quadrilaterals

What You'll LEARN
Identify and classify quadrilaterals.

NEW Vocabulary
quadrilateral
parallelogram
trapezoid
rhombus

REVIEW Vocabulary
parallel: lines that do not intersect (Lesson 10-3b)

WHEN am I ever going to use this?

BASKETBALL The photograph shows the free throw lane used in international basketball.

1. Describe the angles inside the 4-sided figure.
2. Which sides of the figure appear to be parallel?
3. Which sides of the figure appear to be congruent?

The shape of the free-throw lane above is called a trapezoid. Squares, rectangles, and trapezoids are examples of quadrilaterals. A **quadrilateral** is a closed figure with four sides and four angles. Quadrilaterals are named based on their sides and angles.

The diagram shows how quadrilaterals are related. Notice how it goes from the most general to the most specific.

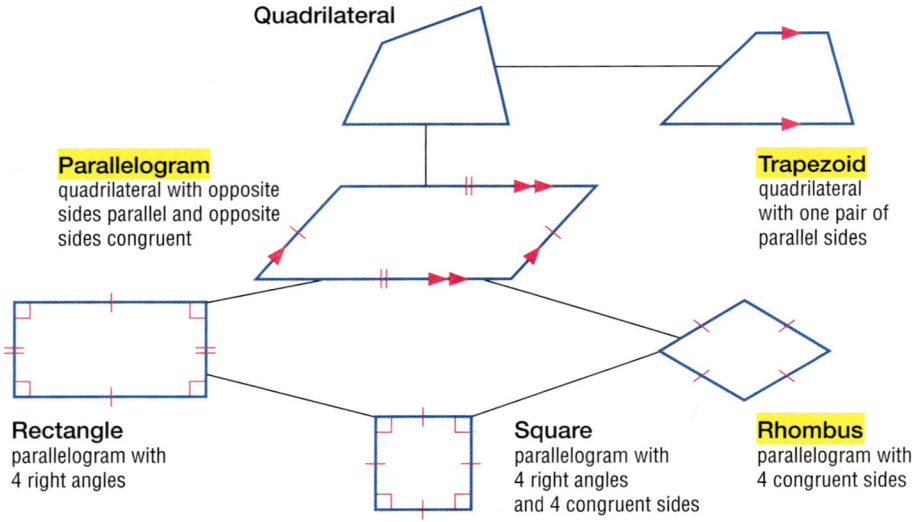

Quadrilateral

Parallelogram
quadrilateral with opposite sides parallel and opposite sides congruent

Trapezoid
quadrilateral with one pair of parallel sides

Rectangle
parallelogram with 4 right angles

Square
parallelogram with 4 right angles and 4 congruent sides

Rhombus
parallelogram with 4 congruent sides

EXAMPLES Classify Quadrilaterals

Classify the quadrilateral using the name that *best* describes it.

READING Math
Parallel Lines The sides with matching arrows are parallel.

1 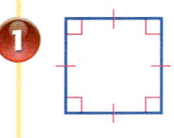 The quadrilateral has 4 right angles and 4 congruent sides. It is a square.

2 The quadrilateral has opposite sides parallel. It is a parallelogram.

A quadrilateral can be separated into two triangles, *A* and *B*. Since the sum of the angle measures of each triangle is 180°, the sum of the angle measures of both triangles is 2 · 180, or 360°.

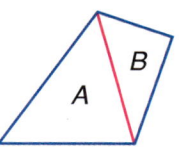

Noteables™   Key Concept: Angles of a Quadrilateral

Words   The sum of the measures of the angles of a quadrilateral is 360°.

Symbols $m\angle 1 + m\angle 2 + m\angle 3 + m\angle 4 = 360°$

Model

EXAMPLE   Find a Missing Measure

3 Find the missing angle measure in the quadrilateral.

$$85 + 73 + 59 + x = 360 \quad \text{The sum of the measures is 360°.}$$
$$217 + x = 360 \quad \text{Simplify.}$$
$$\underline{-217 -217} \quad \text{Subtract 217 from each side.}$$
$$x = 143$$

So, the missing angle measure is 143°.

Skill and Concept Check

1. **Writing Math** Explain why all squares are rectangles but not all rectangles are squares.

2. **OPEN ENDED** Describe a real-life example of a rhombus.

3. **FIND THE ERROR** Venus and Justin are describing a square. Who is correct? Explain.

> Venus
> a parallelogram with
> 4 right angles

> Justin
> a rhombus with 4
> right angles

GUIDED PRACTICE

Classify the quadrilateral using the name that *best* describes it.

4.
5.
6.

Find the missing angle measure in each quadrilateral.

7.
8.
9.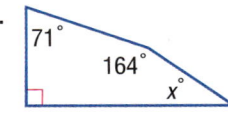

Practice and Applications

HOMEWORK HELP

For Exercises	See Examples
10–15, 25, 27, 29–31	1, 2
19–24, 28	3

Extra Practice
See pages 588, 605.

Classify the quadrilateral using the name that *best* describes it.

10.
11.
12.

13.
14.
15.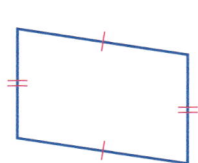

Determine whether each statement is *sometimes*, *always*, or *never* true. Explain.

16. A quadrilateral is a trapezoid.
17. A parallelogram is a rectangle.
18. A trapezoid is a parallelogram.

Find the missing angle measure in each quadrilateral.

19.
20.
21.

22.
23.
24.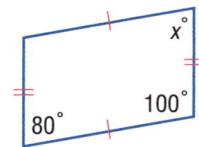

25. **VIDEO GAMES** The first video games used basic shapes like rectangles, squares, and circles in their interface. Name the quadrilaterals used in the video game shown at the right.

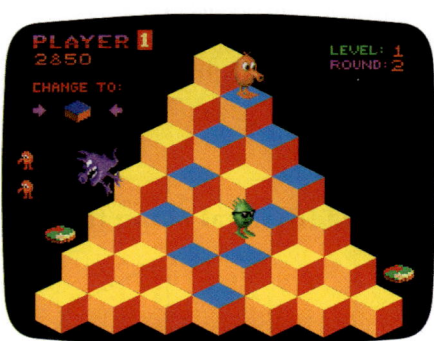

26. **RESEARCH** Use the Internet or another source to find other early video games that used different types of quadrilaterals as part of their interface. Classify the quadrilaterals in the games you found.

27. **ART** Design and draw a stained glass window that contains the following types of quadrilaterals: trapezoid, parallelogram, and rhombus.

28. **ALGEBRA** Find $m\angle B$ in quadrilateral $ABCD$ if $m\angle A = 87°$, $m\angle C = 135°$, and $m\angle D = 22°$.

Determine whether each figure described below can be drawn. Explain.

29. a quadrilateral that is both a rhombus and a rectangle

30. a trapezoid with 3 right angles

31. a trapezoid with two congruent sides

32. **CRITICAL THINKING** Find the value of x in the quadrilateral. Then find the measure of each angle.

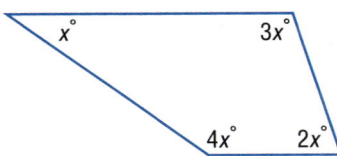

Spiral Review with Standardized Test Practice

33. **MULTIPLE CHOICE** Which property is *not* characteristic of a rhombus?
    - Ⓐ opposite sides parallel
    - Ⓑ 4 right angles
    - Ⓒ 4 congruent sides
    - Ⓓ opposite sides congruent

34. **MULTIPLE CHOICE** Which quadrilateral does *not* have opposite sides congruent?
    - Ⓕ parallelogram
    - Ⓖ square
    - Ⓗ trapezoid
    - Ⓘ rectangle

Classify each triangle by its angles and by its sides. (Lesson 10-4)

35.   36.   37.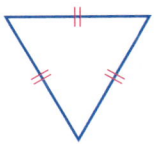

38. Suppose $\angle RST$ and $\angle TSU$ are supplementary angles. Find $m\angle TSU$ if $m\angle RST$ is 76°. (Lesson 10-3)

Find the sales tax or discount to the nearest cent. (Lesson 8-5)

39. $54 jacket; 7% sales tax

40. $23 hat; 15% discount

Find each number. Round to the nearest tenth if necessary. (Lesson 7-7)

41. 20% of what number is 17?

42. What number is 45% of 160?

43. 5 is 12% of what number?

44. 15 is what percent of 24?

45. **TRAVEL** On his summer vacation, Timothy drove 250 miles in 5 hours on the first day. He continued driving at the same rate the second day and drove for 8 hours. How many miles did Timothy drive the second day of his vacation? (Lesson 7-2)

GETTING READY FOR THE NEXT LESSON

PREREQUISITE SKILL Solve each proportion. (Lesson 7-3)

46. $\dfrac{3}{5} = \dfrac{x}{75}$
47. $\dfrac{a}{7} = \dfrac{18}{42}$
48. $\dfrac{7}{9} = \dfrac{28}{m}$
49. $\dfrac{3.5}{t} = \dfrac{16}{32}$
50. $\dfrac{3}{6} = \dfrac{c}{5}$

Chapter 10 Mid-Chapter Practice Test

Vocabulary and Concepts

1. **Define** *supplementary angles.* (Lesson 10-3)
2. **Describe** the difference between a square and a rhombus. (Lesson 10-5)

Skills and Applications

Draw an angle having each measurement. Then classify each angle as *acute, obtuse, right,* or *straight*. (Lesson 10-1)

3. 134°   4. 90°   5. 180°   6. 17°

7. **SOCCER** Make a circle graph of total injuries of high school girls soccer players by position. (Lesson 10-2)

Injuries of High School Girls Soccer Players	
Position	Percent
halfbacks	37%
fullbacks	23%
forward line	28%
goalkeepers	12%

Source: National Athletic Trainers' Association

Find the value of x in each figure. (Lesson 10-3)

8.   9.   10.

11. **ALGEBRA** Find $m\angle B$ in $\triangle ABC$ if $m\angle A = 62°$ and $m\angle C = 44°$. (Lesson 10-4)

Classify the quadrilateral using the name that *best* describes it. (Lesson 10-5)

12.   13.   14.

Standardized Test Practice

15. **MULTIPLE CHOICE** Which angle is complementary to $\angle CBD$? (Lesson 10-1)

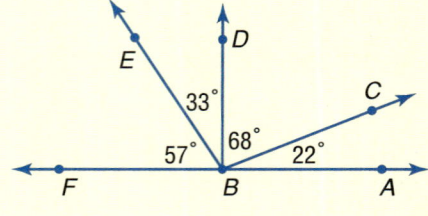

- Ⓐ $\angle ABC$
- Ⓑ $\angle DBE$
- Ⓒ $\angle FBC$
- Ⓓ $\angle EBF$

16. **SHORT RESPONSE** Which section of the circle graph has an angle measure of 63°? (Lesson 10-2)

What Do You Drink With Dinner?

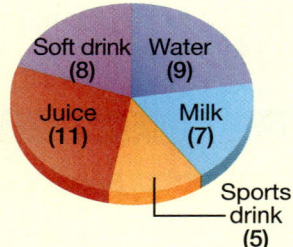

The Game Zone

A Place To Practice Your Math Skills

Math Skill: Finding Squares

Squares Everywhere!

• GET READY!

Players: two
Materials: 8 red counters, 8 yellow counters, dot paper, straight edge

• GET SET!

- Copy the game board onto dot paper.

• GO!

- The first player covers any dot with a counter. Then, players alternate turns.
- The object of the game is to cover the four vertices of a square with one set of counters. Note that the sides of a square don't have to be vertical or horizontal. The square can be "tilted" to one side.
- **Who Wins?** The first player to cover the four vertices of a square with his or her counters wins.
- Once you have played a game covering the four vertices of a square, change the game so you cover the four vertices of a rectangle or a parallelogram.

10-6 Similar Figures

What You'll LEARN
Determine whether figures are similar and find a missing length in a pair of similar figures.

NEW Vocabulary
similar figures
indirect measurement

REVIEW Vocabulary
proportion: an equation that shows that two ratios are equivalent (Lesson 7-3)

MATH Symbols
~ is similar to

HANDS-ON Mini Lab

Materials
- dot paper
- protractor

Work with a partner.

Copy each pair of rectangles and triangles onto dot paper. Then find the measure of each angle and the length of each side. Letters such as *AB* refer to the measure of the segment with those endpoints.

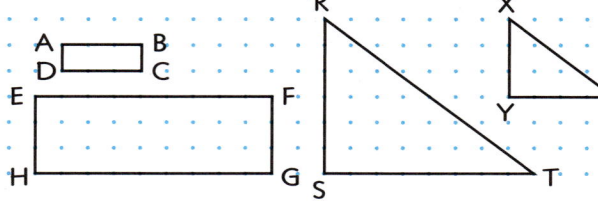

- Each pair of figures has the same shape but different sizes.
- Side $\overline{AB}$ "goes with" side $\overline{EF}$. So, they are *corresponding sides*.

1. Write each fraction in simplest form.
   a. $\dfrac{AB}{EF}, \dfrac{BC}{FG}, \dfrac{DC}{HG}, \dfrac{AD}{EH}$
   b. $\dfrac{RS}{XY}, \dfrac{ST}{YZ}, \dfrac{RT}{XZ}$
2. What do you notice about the ratios of corresponding sides?
3. Measure the corresponding angles in the figures above. What do you notice about the measure of these angles?
4. The rectangles are similar, and the triangles are similar. **Make a conjecture** about similar figures.

Figures that have the same shape but not necessarily the same size are **similar figures**. The Mini Lab illustrates the following definition.

Noteables — Key Concept: Similar Figures

Words If two figures are similar, then
- the corresponding sides are proportional, and
- the corresponding angles are congruent.

Models

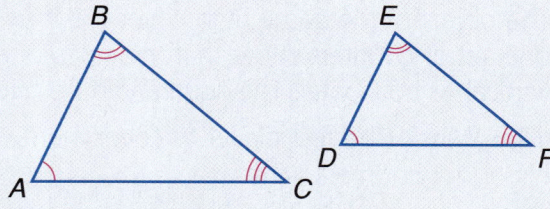

Symbols △ABC ~ △DEF   The symbol ~ means *is similar to*.
corresponding sides: $\dfrac{AB}{DE} = \dfrac{BC}{EF} = \dfrac{AC}{DF}$
corresponding angles: $\angle A \cong \angle D;\ \angle B \cong \angle E;\ \angle C \cong \angle F$

440 Chapter 10 Geometry

READING Math

Line Segments The symbol $\overline{XY}$ means the line segment $\overline{XY}$. XY means the measure of $\overline{XY}$.

EXAMPLE Find Side Measures of Similar Triangles

① If $\triangle RST \sim \triangle XYZ$, find the length of $\overline{XY}$.

Since the two triangles are similar, the ratios of their corresponding sides are equal. So, you can write and solve a proportion to find XY.

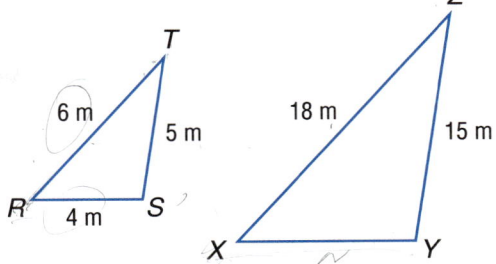

$\dfrac{RT}{XZ} = \dfrac{RS}{XY}$ Write a proportion.

$\dfrac{6}{18} = \dfrac{4}{n}$ Let n represent the length of $\overline{XY}$. Then substitute.

$6n = 18(4)$ Find the cross products.

$6n = 72$ Simplify.

$n = 12$ Divide each side by 6.

The length of $\overline{XY}$ is 12 meters.

You can use similar figures to find the length, width, or height of objects that are too difficult to measure directly. This type of measurement is called **indirect measurement**.

Standardized Test Practice

EXAMPLE Use Indirect Measurement

② **GRID-IN TEST ITEM** Natalie wants to resize a 4-inch wide by 5-inch long photograph for the school newspaper. It is to fit in a space that is 2 inches wide. What is the length of the resized photograph in inches?

Read the Test Item
To find the length of the resized photograph, draw a picture and write a proportion.

Solve the Test Item

 5 in. 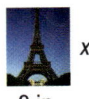 x

4 in.   2 in.

$\dfrac{4}{2} = \dfrac{5}{x}$ Write a proportion.

$4x = 10$ Find the cross products.

$x = 2.5$ Divide each side by 4.

So, the length of the photograph is 2.5 inches.

Fill in the Grid

Test-Taking Tip

Grid In
When filling in the response grid, you can print your answer with the first digit in the left answer box, or with the last digit in the right answer box.

msmath2.net/extra_examples

Lesson 10-6 Similar Figures **441**

Skill and Concept Check

1. **Writing Math** Describe characteristics common to two similar triangles.

2. **OPEN ENDED** Draw two similar quadrilaterals and label the vertices. Then write equivalent ratios comparing all corresponding sides.

GUIDED PRACTICE

Find the value of x in each pair of similar figures.

3.

4.

5. **MEASUREMENT** The height of an object and its shadow is proportional to the height of another object and its shadow. Suppose you are 6 feet tall and you cast a shadow 5 feet long. Find the height of the tree if it casts a shadow 40 feet long.

Practice and Applications

Find the value of x in each pair of similar figures.

6.

7.

8.

9.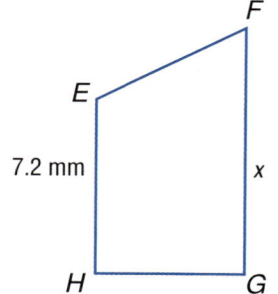

HOMEWORK HELP

For Exercises	See Examples
6–9, 11–12	1
10, 13–15	2

Extra Practice
See pages 588, 605.

10. **FURNITURE** Kidco Furniture produces children's furniture that is similar in shape to full-sized furniture. The top of a full-sized desk measures 54 inches long by 36 inches wide. If the top of a child's desk is 24 inches wide, what is the length?

442 Chapter 10 Geometry

Determine whether each pair of figures is similar. Justify your answer.

11.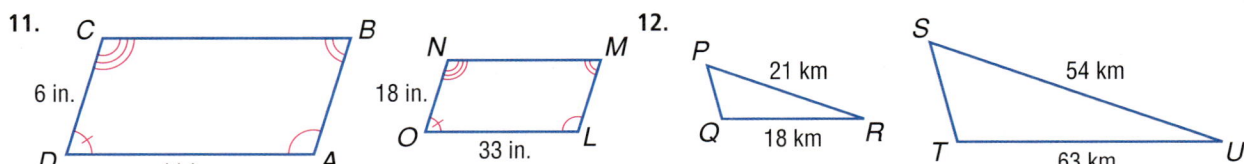
12.

STATUES For Exercises 13 and 14, use the information below and at the right.
Alyssa bought a miniature replica of the Statue of Liberty. The replica is 9 inches tall, and the length of the statue's right arm holding the torch is $1\frac{1}{4}$ inches.

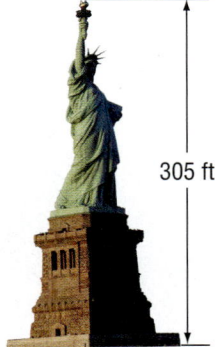

305 ft

13. **MULTI STEP** About how long is the Statue of Liberty's right arm?

14. **MULTI STEP** Alyssa's friend Garcia brought home a smaller replica. The length of the statue's right arm is $\frac{3}{4}$ inch. How tall is Garcia's statue?

15. **GEOMETRY** The ratio of square H's length to square I's length is $3:5$. If the length of square H is 18 meters, what is the perimeter of square I?

CRITICAL THINKING Two rectangles are similar. The ratio of their corresponding sides is $1:4$.

16. Find the ratio of their perimeters.
17. Find the ratio of their areas.

Spiral Review with Standardized Test Practice

18. **MULTIPLE CHOICE** Which of the following is *not* a true statement?
    - Ⓐ All squares are similar.
    - Ⓑ All rhombi are similar.
    - Ⓒ Some trapezoids are similar.
    - Ⓓ All equilateral triangles are similar.

19. **SHORT RESPONSE** Old Faithful in Yellowstone National Park shoots water 60 feet into the air and casts a shadow of 42 feet. What is the height of a nearby tree if it casts a shadow 63 feet long?

Classify the quadrilateral using the name that *best* describes it. (Lesson 10-5)

20.
21.
22.

23. **SAILING** A triangular-shaped sail has angle measures of 44° and 67°. Find the measure of the third angle. (Lesson 10-4)

GETTING READY FOR THE NEXT LESSON

PREREQUISITE SKILL Solve each equation. (Lesson 4-3)

24. $5a = 120$
25. $360 = 4a$
26. $940 = 8n$
27. $6t = 720$

10-7a Problem-Solving Strategy
A Preview of Lesson 10-7

Use Logical Reasoning

What You'll LEARN

Solve problems using logical reasoning.

Dion, how can we check that the garden we dug is in the shape of an equilateral triangle?

*There is a relationship between the angles of an equilateral triangle that can help us. Nathaniel, let's **use logical reasoning** to find out!*

Explore	Equilateral triangles have sides that are congruent. We need to find a relationship between the angles.
Plan	Let's draw several equilateral triangles and measure the angles.
Solve	sides = 1 cm sides = 2 cm sides = 3 cm sides = 4 cm Each angle of the triangles is 60°. So, if the angles of our triangular garden are 60°, then the garden is in the shape of an equilateral triangle.
Examine	Any triangle with angle measures of 60° is equilateral. If we try to draw a triangle with angle measures of 60° and different side lengths, the drawing would not be in the shape of a triangle.

Analyze the Strategy

1. When you use *inductive reasoning*, you make a rule after seeing several examples. When you use *deductive reasoning*, you use a rule to make a decision. What type of reasoning did Nathaniel and Dion use to solve the problem? **Explain** your reasoning.

2. **Explain** how Nathaniel and Dion could have solved the problem using deductive reasoning.

3. **Explain** how the *look for a pattern* problem-solving strategy is similar to inductive reasoning.

444 Chapter 10 Geometry

Apply the Strategy

Solve. Use logical reasoning.

4. **GEOMETRY** Draw several isosceles triangles and measure their angles. What can you conclude about the measures of the angles of an isosceles triangle? See Exercise 1. Did you use inductive or deductive reasoning?

5. **TRAVEL** Use $d = rt$ where d is distance, r is rate, and t is time to find how far Mrs. Petricca drives if she drives 55 miles per hour for seven hours. See Exercise 1. Did you use inductive or deductive reasoning?

Mixed Problem Solving

Solve. Use any strategy.

6. **PHOTOCOPYING** Suppose you enlarge a drawing to 120% of its original size on the copy machine. If the drawing is 2 inches long and 3 inches wide, what are the dimensions of the copy?

7. **TOYS** Hannah was finding the relationship between the time it took a yo-yo to swing back and forth and its length. Predict the length of a yo-yo if it takes 5 seconds to swing back and forth.

Time (seconds)	Length (units)
1	1
2	4
3	9
4	16

GEOMETRY For Exercises 8 and 9, draw several rectangles and measure their diagonals.

8. Find a relationship between the diagonals of a rectangle.

9. See Exercise 1. Did you use inductive or deductive reasoning in Exercise 8?

10. **BASKETBALL** Placido, Dexter, and Scott play guard, forward, and center on a team, but not necessarily in that order. Placido and the center drove Scott to practice on Saturday. Placido does not play guard. Who is the guard?

11. **GARDENING** Shelby wants to fence in her garden with the dimensions shown. How much fencing will she need?

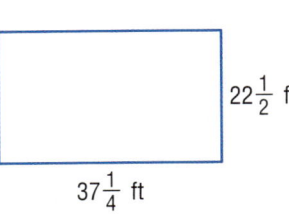

12. **CLOTHES** Careta put a $5 bill into one of seven pockets of her jacket. If she randomly reaches into one pocket, what is the probability that she will find the $5 bill?

13. **GEOMETRY** The large square has been divided into 9 squares. The lengths of the sides of two squares are given. Find the area of the entire square.

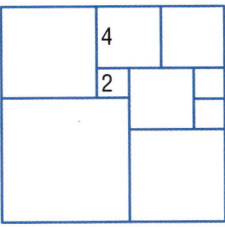

14. **STANDARDIZED TEST PRACTICE**
What can you conclude from the survey about favorite athletic shoes?

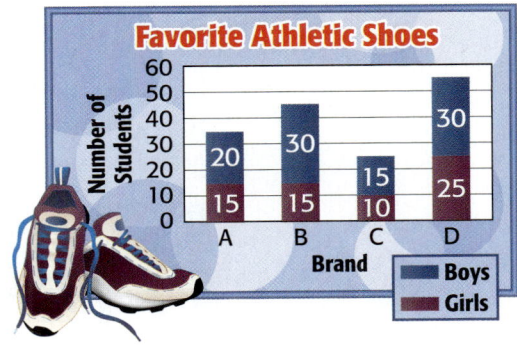

A. Brand A is the most popular among both boys and girls.

B. Brand B is two times more popular among boys than girls.

C. Brand C cost more than Brand D.

D. Brand A is less popular than Brand C.

Lesson 10-7a Problem-Solving Strategy: Use Logical Reasoning **445**

10-7 Polygons and Tessellations

What You'll LEARN

Classify polygons and determine which polygons can form a tessellation.

NEW Vocabulary

polygon
pentagon
hexagon
heptagon
octagon
nonagon
decagon
regular polygon
tessellation

WHEN am I ever going to use this?

GEOGRAPHY The size and shape of each state in the United States is different. Analyze the shapes of the states in both groups at the right.

1. Find the difference between the shapes of the states in Group 1 and the shapes of the states in Group 2.

2. Why do most states have boundaries that are not straight line segments?

Group 1	Group 2
Utah	Georgia
Wyoming	Michigan
New Mexico	Mississippi
Colorado	New York

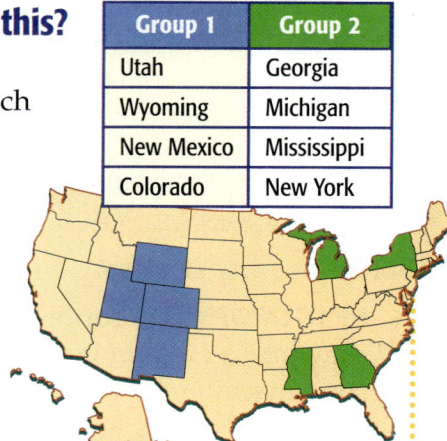

The states listed in Group 1 above are examples of polygons. A **polygon** is a simple, closed figure formed by three or more straight lines. A simple figure does not have lines that cross each other. You have drawn a closed figure when your pencil ends up where it started.

Polygons	Not Polygons
• Line segments are called sides. • Sides meet only at their endpoints. • Points of intersection are called vertices.	• Figures whose sides overlap. • Figures that are open. • Figures that have curved sides.

Just as a triangle has three sides and a quadrilateral has four sides, other polygons can be classified by the number of sides they have.

Words	pentagon	hexagon	heptagon	octagon	nonagon	decagon
Number of Sides	5	6	7	8	9	10
Models						

A polygon that has all sides congruent and all angles congruent is called a **regular polygon**. Equilateral triangles and squares are regular polygons.

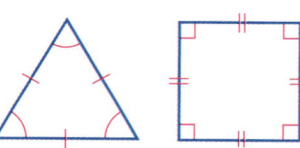

446 Chapter 10 Geometry

READING Math

Regular Polygons Since regular polygons have *equal-sized angles*, they are also called *equiangular*.

EXAMPLES Classify Polygons

Determine whether each figure is a polygon. If it is, classify the polygon and state whether it is regular. If it is *not* a polygon, explain why.

①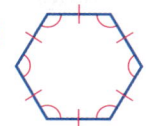

The figure has 6 congruent sides and 6 congruent angles. It is a regular hexagon.

②

The figure is not a polygon since it has a curved side.

Your Turn Determine whether each figure is a polygon. If it is, classify the polygon and state whether it is regular. If it is *not* a polygon, explain why.

a.   b.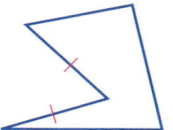

The sum of the measures of the angles of a triangle is 180°. You can use this relationship to find the measures of the angles of regular polygons.

STUDY TIP

Diagonal A diagonal of a polygon is a line segment that connects two vertices that are not next to each other.

EXAMPLE Angle Measures of a Polygon

③ **ALGEBRA** Find the measure of each angle of a regular pentagon.

- Draw all of the diagonals from one vertex as shown and count the number of triangles formed.

- Find the sum of the angle measures in the polygon.

  number of triangles formed × 180° = sum of angle measures in polygon

  $$3 \times 180° = 540°$$

- Find the measure of each angle of the polygon. Let n represent the measure of one angle in the pentagon.

  $5n = 540$ There are five congruent angles.

  $n = 108$ Divide each side by 5.

The measure of each angle in a regular pentagon is 108°.

Your Turn Find the measure of an angle in each polygon.

c. regular octagon   d. equilateral triangle

A repetitive pattern of polygons that fit together with no overlaps or holes is called a **tessellation**. The surface of a chessboard is an example of a tessellation of squares.

msmath2.net/extra_examples

The sum of the measures of the angles where the vertices meet in a tessellation is 360°. The diagrams below show tessellations of equilateral triangles and squares.

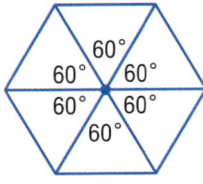

$6 \times 60° = 360°$

$4 \times 90° = 360°$

REAL-LIFE CAREERS

How Does a Landscape Architect Use Math?
Landscape architects multiply the length times the width of a garden to find the amount of mulch needed to cover the garden.

Math Online — Research
For information about a career as a landscape architect, visit: msmath2.net/careers

EXAMPLE Tessellations

4 LANDSCAPING Mr. Brooks bought hexagonal-shaped stones to pave his patio. The stones are regular hexagons. Can Mr. Brooks tessellate his patio with the stones?

The measure of each angle in a regular hexagon is 120°.

The sum of the measures of the angles where the vertices meet must be 360°. So, solve $120n = 360$.

Since the solution, $n = 3$, is a whole number, a regular hexagon makes a tessellation.

Check You can check if your answer is correct by drawing a tessellation of regular hexagons.

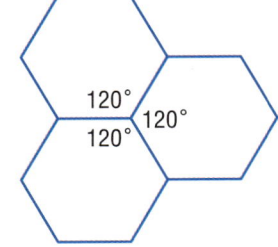

Skill and Concept Check

1. **Explain** how you know the parallelogram at the right can be used by itself to make a tessellation.

2. **OPEN ENDED** Draw examples of a pentagon, hexagon, heptagon, octagon, nonagon, and decagon.

3. *Writing Math* Explain why a rhombus is not a regular polygon.

GUIDED PRACTICE

Determine whether each figure is a polygon. If it is, classify the polygon and state whether it is regular. If it is *not* a polygon, explain why.

4.   5.

6. Find the measure of each angle of a regular hexagon.

7. Can a regular polygon with an angle measure of 140° be used by itself to make a tessellation? Explain.

448 Chapter 10 Geometry

Practice and Applications

Determine whether each figure is a polygon. If it is, classify the polygon and state whether it is regular. If it is *not* a polygon, explain why.

8.
9.
10.
11.
12.
13.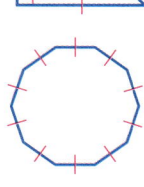

HOMEWORK HELP

For Exercises	See Examples
8–13, 18, 20, 24–25, 27	1, 2
14–17, 19, 26	3
21–23, 28	4

Extra Practice
See pages 589, 605.

Find the measure of an angle in each polygon if the polygon is regular. Round to the nearest tenth of a degree if necessary.

14. decagon
15. nonagon
16. 13-gon
17. 20-gon

MANUFACTURING For Exercises 18–20, use the information below and the figure at the right.

A company designs cafeteria trays so that four people can place their trays around a square table without bumping corners. The trays are similar to the one at the right. The top and bottom sides of the tray are parallel.

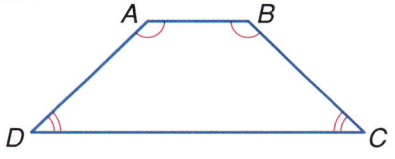

18. Classify the shape of the tray.
19. If $\angle A \cong \angle B$, $\angle C \cong \angle D$, and $m\angle A = 135°$, find $m\angle B$, $m\angle C$, and $m\angle D$.
20. Name the polygon formed by the outside of four trays when they are placed around the table with their sides touching.

Identify the polygons that are used to create each tessellation.

21.
22.
23.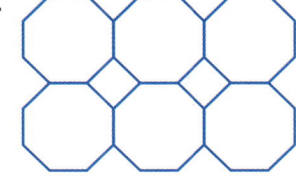

24. **ART** The mosaic at the right is decorated with handmade tiles from Pakistan. Name the polygons used in the tessellation.

25. What is the perimeter of a nonagon with sides 4.8 centimeters long?

26. Name the regular polygon if the sum of the measures of the polygon is 2,880° and the measure of one angle is 160°.

Lesson 10-7 Polygons and Tessellations

27. Find the perimeter of a pentagon having sides $7\frac{1}{2}$ yards long.

28. **SIGNS** Stop signs are made from large sheets of steel. Suppose nine stop signs can be cut from one sheet of steel. Can all nine signs be arranged on the sheet so that none of the steel goes to waste? Explain your reasoning.

29. **RESEARCH** Use the Internet or another source to find the shape of other traffic signs. Name the type of sign, its shape, and whether it is regular.

30. **WRITE A PROBLEM** Write about a real-life object or situation that contains a tessellation. Name the polygons used and draw a picture of the tessellation.

31. **CRITICAL THINKING** You can make a tessellation with equilateral triangles. Can you make a tessellation with any isosceles or scalene triangle? If so, explain your reasoning and make a drawing.

Spiral Review with Standardized Test Practice

32. **MULTIPLE CHOICE** Which of the following flowchart symbols is *not* a polygon?

    Ⓐ process box   Ⓑ decision box   Ⓒ stop box   Ⓓ preparation box

33. **MULTIPLE CHOICE** Which of the following regular shapes *cannot* be used by itself to make a tessellation?

    Ⓕ hexagon   Ⓖ triangle   Ⓗ square   Ⓘ pentagon

34. **SHORT RESPONSE** What is the sum of the measures of the angles of a quadrilateral?

For Exercises 35 and 36, use the figures at the right.

35. Classify figure *ABCD*. (Lesson 10-5)
36. The quadrilaterals are similar. Find the value of *x*. (Lesson 10-6)
37. **MEASUREMENT** How many $\frac{1}{2}$-cup servings of ice cream are there in a gallon of chocolate ice cream? (Lesson 6-5)

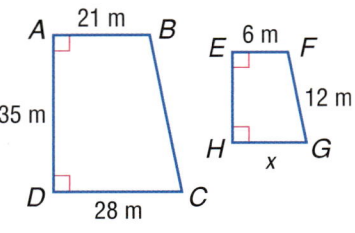

Add or subtract. Write each sum or difference in simplest form. (Lesson 6-3)

38. $3\frac{2}{9} + 5\frac{4}{9}$ 39. $5\frac{1}{3} - 2\frac{1}{6}$ 40. $1\frac{3}{7} + 6\frac{1}{4}$ 41. $9\frac{4}{5} - 4\frac{7}{8}$

GETTING READY FOR THE NEXT LESSON

PREREQUISITE SKILL On graph paper, draw a coordinate plane. Then graph and label each point. Connect the points in order. (Lesson 3-3)

42. $A(-2, 3)$ 43. $B(4, 3)$ 44. $C(2, -1)$ 45. $D(-4, -1)$

10-8 Translations

What You'll LEARN

Graph translations of polygons on a coordinate plane.

NEW Vocabulary

transformation
translation

Link to READING

Everyday Meaning of Translation: a change from one language to another, as in an English-to-Spanish translation

HANDS-ON Mini Lab

Materials
- paper
- scissors

Work with a partner.

You can make changes in the polygons that tessellate to create new shapes that also tessellate.

STEP 1 Draw a square and cut a shape from it as shown.

STEP 2 Slide the shape to the opposite side without turning it.

STEP 3 The new shape will make a tessellation.

1. Make your own tessellation. Use different colors and shapes to make an interesting design.

Anytime you move a geometric figure, it is called a **transformation**. In the Mini Lab, you slid the polygon to a new position without turning it. This sliding motion is called a **translation**.

When translating a figure, every point of the original figure is moved the same distance and in the same direction.

EXAMPLE Graph a Translation

1 Translate trapezoid *HIJK* 3 units left and 5 units up.

- Move each vertex of the figure 3 units left and 5 units up. Label the new vertices H', I', J', and K'.
- Connect the vertices to draw the trapezoid. The coordinates of the vertices of the new figure are $H'(-2, 1)$, $I'(1, 1)$, $J'(1, 5)$, and $K'(-2, 4)$.

STUDY TIP

Transformations Whenever a figure is translated, use prime symbols for the vertices in the transformed image.
$A \rightarrow A'$
$B \rightarrow B'$
$C \rightarrow C'$

msmath2.net/extra_examples

Lesson 10-8 Translations **451**

When a figure has been translated, the original figure and the translated figure, or *image*, are congruent. In Example 1, trapezoid *HIJK* is congruent to trapezoid *H'I'J'K'*. In congruent figures, the corresponding sides and angles have equal measures.

You can also add to or subtract from the coordinates of the vertices of a figure to find the translated vertices.

EXAMPLE Find Coordinates of a Translation

2 Triangle *LMN* has vertices $L(-1, -2)$, $M(6, -3)$, and $N(2, -5)$. Find the vertices of $\triangle L'M'N'$ after a translation of 6 units left and 4 units up. Then graph the figure and its translated image.

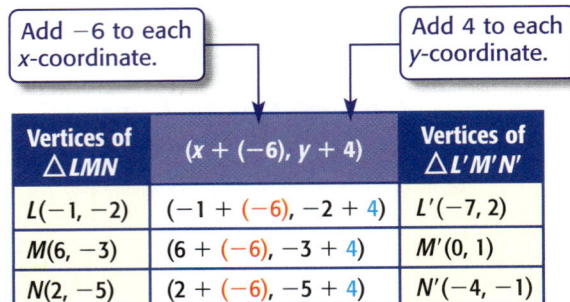

Vertices of △LMN	$(x + (-6), y + 4)$	Vertices of △L'M'N'
$L(-1, -2)$	$(-1 + (-6), -2 + 4)$	$L'(-7, 2)$
$M(6, -3)$	$(6 + (-6), -3 + 4)$	$M'(0, 1)$
$N(2, -5)$	$(2 + (-6), -5 + 4)$	$N'(-4, -1)$

The coordinates of the vertices of $\triangle L'M'N'$ are $L'(-7, 2)$, $M'(0, 1)$, and $N'(-4, -1)$.

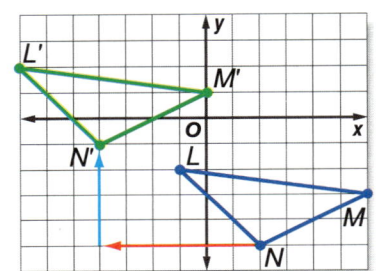

STUDY TIP

Translation
A *positive* integer describes a translation right or up on a coordinate plane. A *negative* integer describes a translation left or down.

Your Turn Triangle *TUV* has vertices $T(6, -3)$, $U(-2, 0)$, and $V(-1, 2)$. Find the vertices of $\triangle T'U'V'$ after each translation. Then graph the figure and its translated image.

a. 3 units right and 4 units down   b. 1 unit left and 6 units up

In Example 2, $\triangle LMN$ was translated 6 units left and 4 units up. This translation can be described using the ordered pair $(-6, 4)$.

EXAMPLE Naming Translations with Ordered Pairs

3 **GAMES** When playing chess, the rook can only move vertically or horizontally across a chessboard. The chessboard at the right shows the movement of a rook after two turns. Describe this translation as an ordered pair.

The rook moved 5 places left and 3 places up. The translation can be written as $(-5, 3)$.

452 Chapter 10 Geometry

Skill and Concept Check

1. **OPEN ENDED** On a coordinate plane, draw a triangle and its translation 1 unit right and 4 units down.

2. **Which One Doesn't Belong?** Identify the translation that is not the same as the other three. Explain your reasoning.

   | (-5, -4) | (1, -3) | (6, -6) | (4, -2) |

GUIDED PRACTICE

3. Translate $\triangle ABC$ 3 units left and 3 units down. Graph $\triangle A'B'C'$.

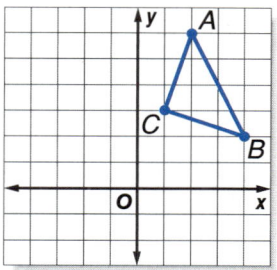

Quadrilateral $DEFG$ has vertices $D(1, 0)$, $E(-2, -2)$, $F(2, 4)$, and $G(6, -3)$. Find the vertices of $D'E'F'G'$ after each translation. Then graph the figure and its translated image.

4. 4 units right, 5 units down
5. 6 units right

Practice and Applications

6. Translate $\triangle HIJ$ 2 units right and 6 units down. Graph $\triangle H'I'J'$.

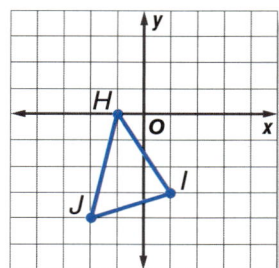

7. Translate rectangle $KLMN$ 1 unit left and 3 units up. Graph rectangle $K'L'M'N'$.

HOMEWORK HELP

For Exercises	See Examples
6–7	1
8–11, 15	2
12, 16	3

Extra Practice
See pages 589, 605.

Triangle PQR has vertices $P(0, 0)$, $Q(5, -2)$, and $R(-3, 6)$. Find the vertices of $P'Q'R'$ after each translation. Then graph the figure and its translated image.

8. 6 units right, 5 units up
9. 8 units left, 1 unit down
10. 3 units left
11. 9 units down

12. **MAPS** Payat lives at the corner of Wabash and Ohio. The school he attends is located at Huron and Dearborn. Describe Payat's walk from school to home as an ordered pair of the number of blocks.

msmath2.net/self_check_quiz

Lesson 10-8 Translations 453

13. **ART** Explain how translations and tessellations were used in *Horsemen*, created by M.C. Escher at the right.

14. **RESEARCH** Use the Internet or another source to find other pieces of art that contain tessellations of translations. Describe how the artists incorporated both ideas into their work.

Horsemen by M.C. Escher. © Cordon Art-Baarn-Holland. All rights reserved.

MULTI STEP For Exercises 15 and 16, use the following information.

△XYZ has vertices X(7, 6), Y(−3, 4), and Z(1, −5). The translated figure △X'Y'Z' has vertices X'(2, 4), Y'(−8, 2), and Z'(−4, −7).

15. Describe the translation.

16. Write the translation as an ordered pair.

17. **TESSELLATIONS** Use the figure at the right and patty paper to make a tessellation using translations.

18. **CRITICAL THINKING** Is it possible to make a tessellation with translations of equilateral triangles? Explain your reasoning.

Spiral Review with Standardized Test Practice

19. **MULTIPLE CHOICE** Which graph shows a translation of the letter Z?

20. **SHORT RESPONSE** Triangle *ABC* with vertices A(−5, 10), B(5, 10), and C(−10, 5) is translated by (−10, 5). What are the coordinates of A'?

21. **GEOMETRY** What is the name of a polygon with eight sides? (Lesson 10-7)

22. **TOWERS** Horatio is 167 centimeters tall and casts a shadow 93 centimeters long. What is the approximate height of a cellphone tower if it casts a shadow 763 centimeters long? (Lesson 10-6)

GETTING READY FOR THE NEXT LESSON

BASIC SKILL Determine whether each figure can be folded in half so that one side matches the other. Write *yes* or *no*.

23. ▭   24. ⌐⌐   25. △   26. ⬡

10-8b Spreadsheet Investigation
A Follow-Up of Lesson 10-8

What You'll LEARN

Use a spreadsheet to enlarge and reduce polygons.

Dilations

A computer spreadsheet is a useful tool for calculating the vertices of polygons. You can enlarge or reduce the polygons by using a spreadsheet to automatically calculate the new coordinates of the vertices. Enlarging or reducing a figure is a transformation called a **dilation**.

ACTIVITY

Emma has plotted a pentagon on graph paper. The coordinates of the vertices of the pentagon are (2, 2), (4, 2), (5, 4), (3, 6), and (1, 4). She wants to multiply the coordinates by 3 to enlarge the pentagon. She enters the coordinates on a spreadsheet as shown below.

Set up the spreadsheet like the one shown below.

	A	B	C	D
1	original		dilation	
2	x-coordinate	y-coordinate	x-coordinate	y-coordinate
3	2	2	=A3*3	=B3*3
4	4	2	=A4*3	=B4*3
5	5	4		
6	3	6		
7	1	4		

Continue entering the formulas in columns C and D to complete the dilation.

EXERCISE

1. How will the formulas in columns C and D change the original pentagon? How do you know?
2. Graph the original pentagon and its dilation on graph paper.
3. What is the percent of increase of the original pentagon to its dilation?
4. Find the coordinates of the pentagon enlarged 5 times.
5. Find the coordinates of the pentagon reduced by one-half.
6. What type of dilation had occurred if the new coordinates of the pentagon are (5, 5), (10, 5), (12.5, 10), (7.5, 15), and (2.5, 10)? What is the scale factor?
7. Select another geometric figure and plot its points on graph paper. Set up a spreadsheet to find two dilations, one enlargement and one reduction of the same figure.

Lesson 10-8b Spreadsheet Investigation: Dilations **455**

10-9 Reflections

HANDS-ON Mini Lab

Materials
- geomirror
- paper

Work with a partner.

- Write your first name in capital letters on a sheet of paper.
- Use the geomirror to trace the reflection of the letters in your name.
- Write your last name. Draw the reflection of the letters without using the geomirror.

1. Describe how you drew the reflection of your last name.
2. List the capital letters that look the same as their reflections.
3. Explain why the line where the geomirror and paper meet is called the *line of symmetry*.

What You'll LEARN

Identify figures with line symmetry and graph reflections on a coordinate plane.

NEW Vocabulary

line symmetry
line of symmetry
reflection

Figures that match exactly when folded in half have **line symmetry**. The figures below have line symmetry.

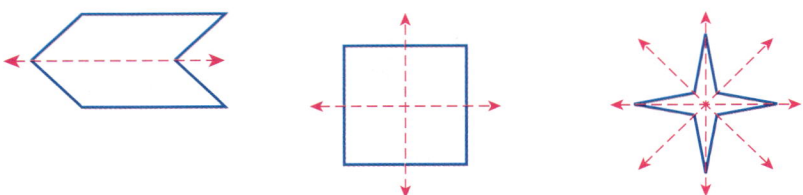

Each fold line is called a **line of symmetry**. Some figures can be folded in more than one way to show symmetry, so they have more than one line of symmetry.

EXAMPLES Identify Lines of Symmetry

Determine whether each figure has line symmetry. If so, copy the figure and draw all lines of symmetry.

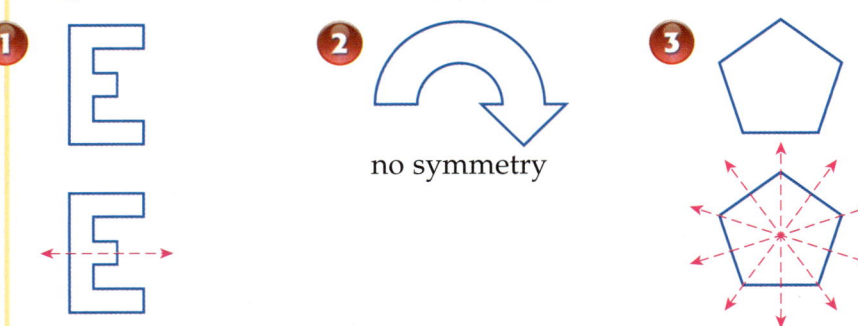

② no symmetry

456 Chapter 10 Geometry

A type of transformation where a figure is flipped over a line of symmetry is a **reflection**. As with translations, the original figure and the reflected image are congruent.

STUDY TIP

x-axis When reflecting over the x-axis, change the y-coordinates to their opposites.

EXAMPLE — Reflect a Figure Over the x-axis

4 Triangle ABC has vertices A(5, 2), B(1, 3), and C(−1, 1). Find the coordinates of ABC after a reflection over the x-axis. Then graph the figure and its reflected image.

Vertices of △ABC	Distance from x-axis	Vertices of △A'B'C'
A(5, 2)	2	A'(5, −2)
B(1, 3)	3	B'(1, −3)
C(−1, 1)	1	C'(−1, −1)

Plot the vertices and connect to form △ABC. The x-axis is the line of symmetry. So, the distance from each point on △ABC to the line of symmetry is the same as the distance from the line of symmetry to △A'B'C'.

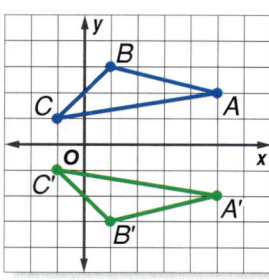

STUDY TIP

y-axis When reflecting over the y-axis, change the x-coordinates to their opposites.

EXAMPLE — Reflect a Figure Over the y-axis

5 Quadrilateral KLMN has vertices K(2, 3), L(5, 1), M(4, −2), and N(1, −1). Find the coordinates of KLMN after a reflection over the y-axis. Then graph the figure and its reflected image.

Vertices of quad KLMN	Distance from y-axis	Vertices of quad K'L'M'N'
K(2, 3)	2	K'(−2, 3)
L(5, 1)	5	L'(−5, 1)
M(4, −2)	4	M'(−4, −2)
N(1, −1)	1	N'(−1, −1)

Plot the vertices and connect to form quadrilateral KLMN. The y-axis is the line of symmetry. So, the distance from each point on quadrilateral KLMN to the line of symmetry is the same as the distance from the line of symmetry to quadrilateral K'L'M'N'.

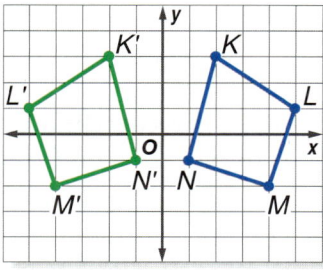

Your Turn

a. Find the coordinates of rectangle GHIJ with vertices G(3, −4), H(3, −1), I(−2, −1), and J(−2, −4) after a reflection over the x-axis. Then graph the figure and its reflected image.

b. Find the coordinates of triangle PQR with vertices P(1, 5), Q(3, 7), and R(5, −1) after a reflection over the y-axis. Then graph the figure and its reflected image.

Skill and Concept Check

1. **OPEN ENDED** Draw a figure on a coordinate plane and its reflection over the y-axis.

2. **Which One Doesn't Belong?** Identify the transformation that is not the same as the other three. Explain your reasoning.

GUIDED PRACTICE

Determine which figures have line symmetry. Write *yes* or *no*. If *yes*, draw all lines of symmetry.

3.   4.

Find the coordinates of each figure after a reflection over the given axis. Then graph the figure and its reflected image.

5. $\triangle ABC$ with vertices $A(5, 8)$, $B(1, 2)$, and $C(6, 4)$; x-axis

6. parallelogram $WXYZ$ with vertices $W(-4, -2)$, $X(-4, 3)$, $Y(-2, 4)$, and $Z(-2, -1)$; y-axis

Practice and Applications

Determine which figures have line symmetry. Write *yes* or *no*. If *yes*, draw all lines of symmetry.

7.   8.   9.   10.

HOMEWORK HELP

For Exercises	See Examples
7–10, 15–17	1–3
11–12	4
13–14	5

Extra Practice
See pages 589, 605.

Find the coordinates of each figure after a reflection over the x-axis. Then graph the figure and its reflected image.

11. quadrilateral $DEFG$ with vertices $D(-3, 6)$, $E(-2, -3)$, $F(2, 2)$, and $G(4, 9)$

12. $\triangle TUV$ with vertices $T(-6, 1)$, $U(-2, -3)$, and $V(5, -4)$

Find the coordinates of each figure after a reflection over the y-axis. Then graph the figure and its reflected image.

13. $\triangle QRS$ with vertices $Q(2, -5)$, $R(4, -5)$, and $S(2, 3)$

14. parallelogram $HIJK$ with vertices $H(-1, 3)$, $I(-1, -1)$, $J(2, -2)$, and $K(2, 2)$

15. **BUILDINGS** Describe the location of the line(s) of symmetry in the photograph of the Taj Mahal.

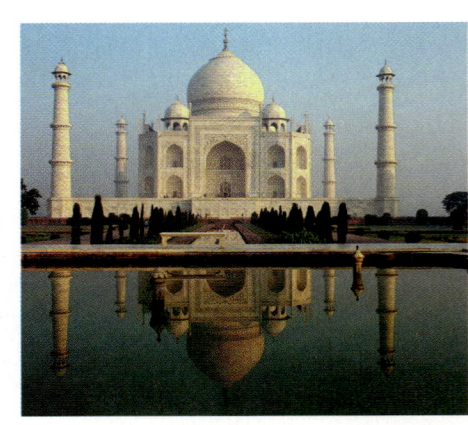

458 Chapter 10 Geometry

16. **AMBULANCE** Explain why the word "AMBULANCE" is written backward and from right to left on the front of the emergency vehicle.

17. **SIGNAL FLAGS** International Marine Signal Flags are used by sailors to send messages at sea. A flag represents each letter of the alphabet. The flags below spell out the word *MATH*. Which flags have line symmetry? Draw all lines of symmetry.

M   A   T   H

 Data Update What other maritime signal flags have line symmetry? Visit msmath2.net/data_update for more information.

18. **TESSELLATIONS** Make a tessellation using a combination of translations and reflections of polygons. Explain how you created your tessellation.

19. **CRITICAL THINKING** Triangle *JKL* has vertices $J(-7, 4)$, $K(7, 1)$, and $L(2, -2)$. Without graphing, reflect the triangle over the *x*-axis and then over the *y*-axis. What are the new coordinates of the triangle after the double reflection?

Spiral Review with Standardized Test Practice

20. **MULTIPLE CHOICE** $\triangle ABC$ with vertices $A(-2, -5)$, $B(4, 1)$, and $C(3, -2)$ is reflected over the *y*-axis. Find the coordinates of the new figure.

    A $A'(2, -5), B'(-4, 1), C'(-3, -2)$ **B** $A'(-2, 5), B'(4, -1), C'(3, 2)$

    C $A'(2, 5), B'(-4, -1), C'(-3, 2)$ **D** $A'(2, 5), B'(4, 1), C'(3, 2)$

21. **SHORT RESPONSE** How many lines of symmetry, if any, are in the rhombus at the right?

22. **GEOMETRY** Graph $\triangle FGH$ with vertices $F(-3, 7)$, $G(-1, 5)$, and $H(-2, 2)$ and its translation 4 units right and 1 unit down. Write the ordered pairs for the vertices of the new figure. (Lesson 10-8)

23. **ART** Aisha wishes to construct a tessellation for a wall hanging made only from regular decagons. Is this possible? Explain. (Lesson 10-7)

Estimate. (Lesson 6-1)

24. $\frac{4}{9} + 8\frac{1}{8}$ 25. $\frac{1}{9} \times \frac{2}{5}$ 26. $12\frac{1}{4} \div 5\frac{6}{7}$

27. Write an inequality for *six times a number is less than or equal to 18*. Then solve the inequality. (Lesson 4-5)

10-9b Hands-On Lab
A Follow-Up of Lesson 10-9

Rotations

Another type of transformation is a rotation. A **rotation** moves a figure around a central point. Another name for a rotation is a *turn*. In this lab, you will learn how to rotate a figure on a coordinate plane.

What You'll LEARN

Graph rotations on a coordinate plane.

Materials
- graph paper
- protractor
- ruler

ACTIVITY Work with a partner.

1 Rotate △ABC 90° counterclockwise about the origin. Then graph △A′B′C′.

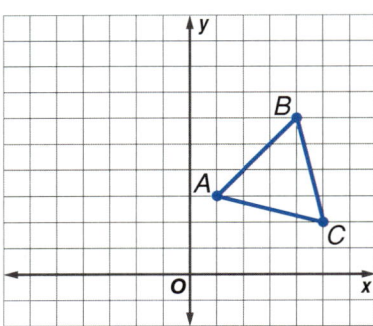

STEP 1 Draw a ray $\overrightarrow{OA}$ from the origin through point A.

STEP 2 Draw a 90° angle from $\overrightarrow{OA}$ rotated counterclockwise. Label the ray $\overrightarrow{OS}$.

STEP 3 Measure $\overline{OA}$. Then measure the same distance from O on $\overrightarrow{OS}$. This is the location of rotated vertex A′.

STEP 4 Rotate the other vertices in the same way by drawing rotated rays $\overrightarrow{OV}$ and $\overrightarrow{OT}$ to obtain vertices B′ and C′.

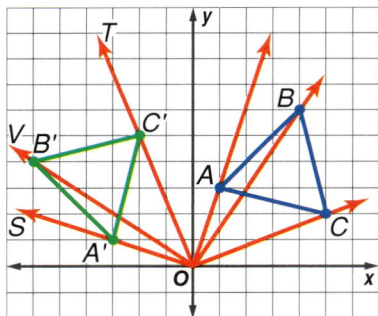

STEP 5 Connect the rotated points to make △A′B′C′.

Rotated figures are congruent because the side lengths remain the same. So, △ABC ≅ △A′B′C′.

READING Math

Rotations If you rotate an image once, the vertices contain a prime symbol (′). If you rotate a rotated image, the vertices contain a double prime symbol (″).

Your Turn

a. Rotate △DEF 90° counterclockwise about the origin. Then graph △D′E′F′.

b. Rotate △D′E′F′ from Part a above 90° counterclockwise about the origin. Then graph △D″E″F″.

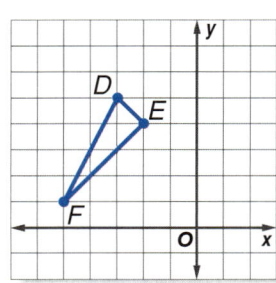

460 Chapter 10 Geometry

ACTIVITY Work with a partner.

2 Identify the transformation as a *translation, reflection,* or *rotation*. The original figure is blue and the transformation is green.

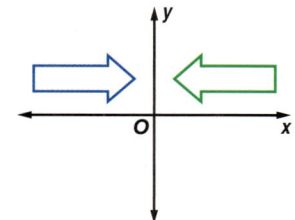

The distance from the point of the blue arrow to the line of symmetry is the same as the distance from the point of the green arrow to the line of symmetry. The *y*-axis is a line of symmetry. Therefore, the transformation is a reflection.

Your Turn Identify the transformation as a *translation, reflection,* or *rotation*. The original figure is blue and the transformation is green.

c.

d.

e.

f.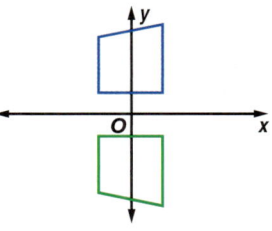

Writing Math

Work with a partner.

1. Rotate △*DEF* from Your Turn a 180° counterclockwise. Compare your rotated figure with your answer from Your Turn b.
2. **Make a conjecture** about rotating figures 180°.
3. Without rotating △*DEF* 360°, **make a conjecture** about rotating figures 360°. Describe the figure after a 360° rotation.
4. A 60° *counterclockwise* rotation is the same as a −300° *clockwise* rotation. Find an equivalent rotation of a 90° counterclockwise rotation.
5. Complete the following sentence. Since you can rotate figures in clockwise or counterclockwise directions, a 270° counterclockwise rotation produces the same result as a __?__ clockwise rotation.

Chapter 10 Study Guide and Review

Vocabulary and Concept Check

acute angle (p. 413)
acute triangle (p. 429)
angle (p. 413)
circle graph (p. 418)
complementary angles (p. 422)
congruent angles (p. 422)
congruent segments (p. 429)
decagon (p. 446)
degrees (p. 413)
equilateral triangle (p. 429)
heptagon (p. 446)
hexagon (p. 446)
indirect measurement (p. 441)
isosceles triangle (p. 429)

line of symmetry (p. 456)
line symmetry (p. 456)
nonagon (p. 446)
obtuse angle (p. 413)
obtuse triangle (p. 429)
octagon (p. 446)
parallelogram (p. 434)
pentagon (p. 446)
polygon (p. 446)
quadrilateral (p. 434)
reflection (p. 457)
regular polygon (p. 446)
rhombus (p. 434)
right angle (p. 413)

right triangle (p. 429)
scalene triangle (p. 429)
similar figures (p. 440)
straight angle (p. 413)
supplementary angles (p. 422)
tessellation (p. 447)
transformation (p. 451)
translation (p. 451)
trapezoid (p. 434)
triangle (p. 428)
vertex (p. 413)
vertical angles (p. 422)

Choose the letter of the term that best matches each phrase.

1. the point where the sides of an angle meet
2. an angle whose measure is less than 90°
3. two angles whose measures add to 180°
4. a polygon with six sides
5. a triangle with no congruent sides

a. scalene triangle
b. complementary angles
c. vertex
d. hexagon
e. acute angle
f. supplementary angles

Lesson-by-Lesson Exercises and Examples

10-1 Angles (pp. 413–415)

Classify each angle as *acute, obtuse, right,* or *straight.*

6.
7.
8.
9.

10. Draw ∠PQR with a measure of 128°.

Example 1 Classify the angle as *acute, obtuse, right,* or *straight.*

The angle is an acute angle because its measure is less than 90°.

462 Chapter 10 Geometry

msmath2.net/vocabulary_review

10-2 Statistics: Making Circle Graphs (pp. 418–421)

11. Make a circle graph of the data.

Favorite Soft Drink	Percent
Cola	36%
Diet Cola	28%
Root Beer	15%
Lemon Lime	7%
Other	14%

Example 2 Make a circle graph of the data of favorite season: 40% spring, 26% summer, 22% fall, and 12% winter.

First find the degrees for each part. Then construct the circle graph.

Favorite Season: Winter (43.2°), Spring (144°), Fall (79.2°), Summer (93.6°)

10-3 Angle Relationships (pp. 422–425)

Find the value of x in each figure.

12.   13.

14. If $\angle Y$ and $\angle Z$ are complementary angles and $m\angle Z = 35°$, find $m\angle Y$.

Example 3 Find the value of x.

Since the angles are complementary, $x + 27 = 90$.

$x + 27 = 90$ Write the equation.
$\underline{-27 \quad -27}$ Subtract 27 from each side.
$x \quad = 63$

So, the value of x is 63.

10-4 Triangles (pp. 428–431)

Classify each triangle by its angles and by its sides.

15.   16.

17. Find $m\angle S$ in $\triangle RST$ if $m\angle R = 28°$ and $m\angle T = 13°$.

Example 4 Classify the triangle by its angles and by its sides.

The triangle is acute since all three angles are acute. It is also isosceles because it has two congruent sides.

10-5 Quadrilaterals (pp. 434–437)

Classify the quadrilateral using the name that *best* describes it.

18.   19.

20. What type of quadrilateral may *not* have opposite sides congruent?

Example 5 Classify the quadrilateral using the name that *best* describes it.

The quadrilateral is a parallelogram with 4 right angles and 4 congruent sides. It is a square.

Study Guide and Review continued

Mixed Problem Solving
For mixed problem-solving practice, see page 605.

10-6 Similar Figures (pp. 440–443)

Find the value of x in each pair of similar figures.

21.

22.

Example 6 Find the value of x in the pair of similar figures.

$\frac{7}{28} = \frac{x}{16}$ Write a proportion.

$28x = 112$ Multiply.

$x = 4$ Divide.

10-7 Polygons and Tessellations (pp. 446–450)

Classify each polygon below and state whether it is regular.

23.   24.

25. **ALGEBRA** Find the measure of each angle of a regular 15-gon.

Example 7 Classify the polygon and state whether it is regular.

Since the polygon has 5 congruent sides and 5 congruent angles, it is a regular pentagon.

10-8 Translations (pp. 451–454)

Triangle PQR has coordinates $P(4, -2)$, $Q(-2, -3)$, and $R(-1, 6)$. Find the coordinates of $P'Q'R'$ after each translation. Then graph each translation.

26. 6 units left, 3 units up
27. 4 units right, 1 unit down
28. 3 units left
29. 7 units down

Example 8
Find the coordinates of $\triangle G'H'I'$ after a translation of 2 units left and 4 units up.

The vertices of $\triangle G'H'I'$ are $G'(-2, 5)$, $H'(0, 3)$, and $I'(-4, 1)$.

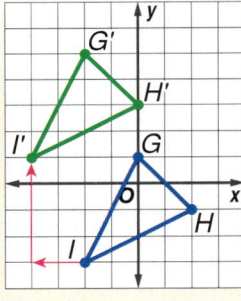

10-9 Reflections (pp. 456–459)

Find the coordinates of each figure after a reflection over the given axis. Then graph the figure and its reflected image.

30. $\triangle RST$ with coordinates $R(-1, 3)$, $S(2, 6)$, and $T(6, 1)$; x-axis
31. parallelogram $ABCD$ with coordinates $A(1, 3)$, $B(2, -1)$, $C(5, -1)$, and $D(4, 3)$; y-axis

Example 9
Find the coordinates of $\triangle C'D'E'$ after a reflection over the y-axis. Then graph its reflected image.

The vertices of $\triangle C'D'E'$ are $C'(-2, 4)$, $D'(-1, 1)$, and $E'(-4, 3)$.

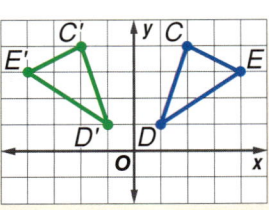

CHAPTER 10 Practice Test

Vocabulary and Concepts

1. **List** the number of sides and number of angles in an octagon.
2. **Explain** how to translate a geometric figure.

Skills and Applications

Classify each angle as *acute, obtuse, right,* **or** *straight.*

3. 53°   4. 97°   5. 180°

Classify each pair of angles as *complementary, supplementary,* **or** *neither.*

6.   7.   8.

9. **FOOD** Angelo's Pizza Parlor makes square pizzas. After baking, the pizzas are cut along one diagonal into two triangles. Describe the triangles.

Find the missing measure in each quadrilateral.

10.   11.   12.

13. **GEOGRAPHY** Luz is drawing a map similar to one found in the atlas. The map in the atlas is 7 inches wide and 10 inches long. If she draws the width 17.5 inches, how long should she draw the map?

14. Can a regular heptagon, whose angle measures total 900°, be used by itself to make a tessellation? Explain your reasoning.

15. Draw a figure with exactly one line of symmetry.

Standardized Test Practice

16. **MULTIPLE CHOICE** The table shows the results of a student survey of favorite type of book. Choose the statement that is *false.*

Favorite Type of Books	
Mystery	24
Science Fiction	8
Sports	26
Romance	30

    Ⓐ The science fiction section on a circle graph of the data at the right has an angle measure of about 33°.

    Ⓑ Romance books were the most favorite types of books.

    Ⓒ About 30% of students chose sports books as their favorite.

    Ⓓ The mystery and sports sections on a circle graph of the data at the right have supplementary angles.

msmath2.net/chapter_test

CHAPTER 10 Standardized Test Practice

PART 1 Multiple Choice

Record your answers on the answer sheet provided by your teacher or on a sheet of paper.

1. Which inequality is false? (Prerequisite Skill, p. 556)
   - A $0.0059 > 0.0005$
   - B $6.1530 < 6.1532$
   - C $89.13 > 89.10$
   - D $5.06 < 5.006$

2. On Spring Street, there is an apartment building that houses four families, each with four members. One of the families moves away. Express this change in residents as an integer. (Lesson 3-1)
   - F 12
   - G 4
   - H -4
   - I -1

3. Find the algebraic expression for the following statement. *Blanca has five more than three times as many CDs as Garrett.* (Lesson 4-1)
   - A $B = \dfrac{3g}{5}$
   - B $B = 3g - 5$
   - C $B = 3(g + 5)$
   - D $B = 5 + 3g$

4. Which is the equation for the graphed line? (Lesson 4-7)

   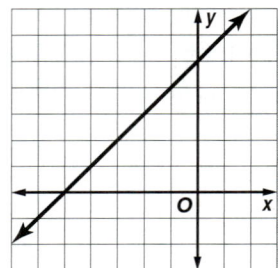

   - F $y = x + 5$
   - G $y = x - 5$
   - H $y = 5x$
   - I $y = 5x + 5$

5. The tree diagram shows the types of notebooks Esther can buy for school. If Esther selects one at random, what are the chances that she will select an orange notebook without lines? (Lesson 9-2)

   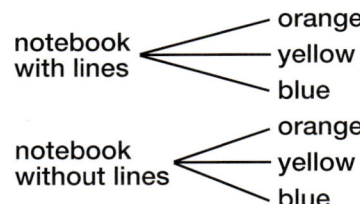

   - A $\dfrac{1}{12}$
   - B $\dfrac{1}{8}$
   - C $\dfrac{1}{6}$
   - D $\dfrac{1}{5}$

6. Which two of the figures below appear to be similar? (Lesson 10-6)

   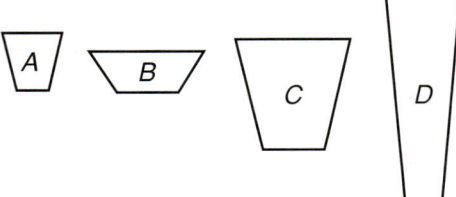

   - F figure A and figure C
   - G figure B and figure C
   - H figure A and figure D
   - I figure A and figure B

7. Kendrick chose a frame for a painting that looks like the polygon. Is this a regular polygon? (Lesson 10-7)

   - A Yes, because the angles are congruent.
   - B No, because the sides are not congruent.
   - C Yes, because the polygon is made up of line segments.
   - D Yes, because the angles and sides are congruent.

8. Which letter has two lines of symmetry? (Lesson 10-9)
   - F B
   - G E
   - H L
   - I X

PART 2 Short Response/Grid In

Record your answers on the answer sheet provided by your teacher or on a sheet of paper.

9. According to the box-and-whisker plot, what is the interquartile range? (Lesson 2-6)

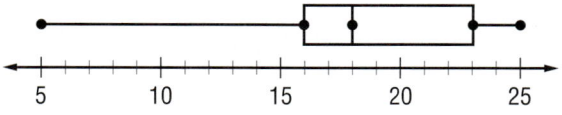

10. The table shows the values of p and q, where the values of p and q form a proportion. What are the values of Y and Z? (Lesson 7-3)

p	4	16	Z
q	7	Y	63

11. The veterinary hospital treated 162 animals last week. It treated 204 animals this week. What was the percent of increase? (Lesson 8-4)

12. Rebecca can choose seven different classes from five different class periods offered during the school day. How many possible ways can Rebecca arrange her class schedule? (Lesson 9-3)

13. How many degrees do the hands of a clock make at 9:00? (Lesson 10-1)

14. How long should the missing sides of the triangle be to make it isosceles with a perimeter of 15 centimeters? (Lesson 10-4)

3 cm

15. Find the sum of the angle measures of a hexagon. (Lesson 10-7)

16. How many lines of symmetry does the figure at the right have? (Lesson 10-9)

TEST-TAKING TIP

Question 17 Most standardized tests allow you to write in the test booklet. Use it to work out problems. For example, plot a figure and its translation on the coordinate graph given with the question in the booklet. Doing so will help keep you from making careless errors.

17. Triangle ABC with vertices $A(3, -7)$, $B(2, -3)$, and $C(8, -3)$ was translated to $A'B'C'$ with vertices $A'(-3, -3)$, $B'(-4, 1)$, and $C'(2, 1)$. Represent the translation as an ordered pair. (Lesson 10-8)

PART 3 Extended Response

Record your answers on a sheet of paper. Show your work.

18. Edmundo plotted polygon $JKLM$ on the coordinate plane below.

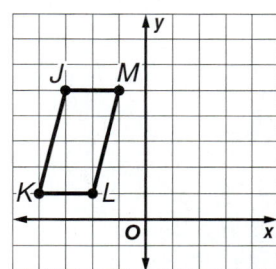

a. Classify $\angle J$. (Lesson 10-1)

b. Classify $\angle M$. (Lesson 10-1)

c. Classify polygon $JKLM$ using the name that *best* describes it. Be as specific as possible. (Lesson 10-5)

d. Can polygon $JKLM$ be used by itself to make a tessellation? Explain your reasoning. (Lesson 10-7)

e. If polygon $JKLM$ is translated 2 units right and 5 units down, what are the coordinates of the new figure? (Lesson 10-8)

f. If polygon $JKLM$ is reflected over the x-axis, what are the coordinates of the new figure? (Lesson 10-9)

CHAPTER 11

Geometry: Measuring Two-Dimensional Figures

"What does landscape design have to do with math?"

In designing a circular path, pool, or fountain, landscape architects calculate the area of the region. **They use the formula $A = \pi r^2$, where r is the radius of the circle.** There are many formulas involving two-dimensional figures that are useful in real life.

You will solve problems involving areas of circles in Lesson 11-6.

GETTING STARTED

▶ Diagnose Readiness

Take this quiz to see if you are ready to begin Chapter 11. Refer to the lesson or page number in parentheses for review.

Vocabulary Review

State whether each sentence is *true* or *false*. If *false*, replace the underlined word or number to make a true sentence.

1. 0.5 is a <u>rational</u> number. (Lesson 5-4)
2. π is approximately equal to <u>3.14</u>. (Lesson 6-9)
3. A triangle with angle measures of 60°, 90°, and 30° is an <u>acute</u> triangle. (Lesson 10-4)

Prerequisite Skills

Replace each ● with <, >, or = to make a true sentence. (Page 556)

4. 67 ● 8.2 · 8.2
5. 11.1 × 11.1 ● 123
6. 5.9(5.9) ● 34.9
7. 12.25 ● 3.5 · 3.5

Evaluate each expression. (Lesson 1-2)

8. 3^2
9. 8 squared
10. 5 to the third power
11. 6 to the second power

Find each value. (Lesson 1-3)

12. $\frac{1}{2}(5)(6)$
13. $\frac{1}{2}(4)(12 + 18)$
14. 9(3 + 3)
15. 7(2)(8)

Find the probability of rolling each number on a number cube. (Lesson 9-1)

16. P(3)
17. P(6 or 2)
18. P(odd)
19. P(greater than 4)

Measuring Figures Make this Foldable to help you organize your notes. Begin with a piece of 11" by 17" paper.

STEP 1 Fold
Fold a 2" tab along the long side of the paper.

STEP 2 Open and Fold
Unfold the paper and fold in thirds widthwise.

STEP 3 Open and Label
Draw lines along the folds and label the head of each column as shown. Label the front of the folded table with the chapter title.

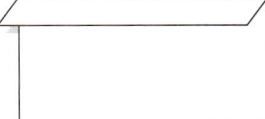

Chapter Notes Each time you find this logo throughout the chapter, use your *Noteables™: Interactive Study Notebook with Foldables™* or your own notebook to take notes. Begin your chapter notes with this Foldable activity.

Readiness To prepare yourself for this chapter with another quiz, visit **msmath2.net/chapter_readiness**

Study Skill

Use a Web

Studying Math Vocabulary

If you've surfed the World Wide Web, you know it is a collection of documents linked together to form a huge electronic library. In mathematics, a web helps you understand how concepts are linked together.

A *web* can help you understand how math concepts are related to each other. To make a web, write the major topic in a box in the center of a sheet of paper. Then, draw "arms" from the center for as many categories as you need. You can label the arms to indicate the type of information that you are listing.

Here is a partial web for the major topic of *polygons*.

SKILL PRACTICE

1. Continue the web above by adding an arm for *quadrilaterals*.

2. In Lesson 5-8, you learned that the set of rational numbers contains fractions, terminating and repeating decimals, and integers. You also know that there are proper and improper fractions and that integers are whole numbers and their opposites. Complete the web below for *rational numbers*.

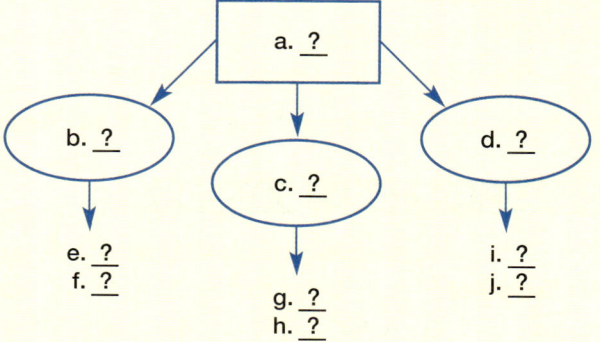

11-3 The Pythagorean Theorem

WHEN am I ever going to use this?

MOVING A square mirror 7 feet on each side must be delivered through the doorway.

1. Can the mirror fit through the doorway? Explain.
2. Make a scale drawing on grid paper to solve the problem.

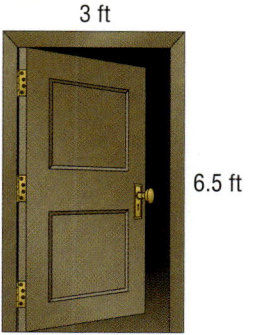

What You'll LEARN

Find length using the Pythagorean Theorem.

NEW Vocabulary

leg
hypotenuse
Pythagorean Theorem

REVIEW Vocabulary

right triangle: a triangle with exactly one angle that measures 90° (Lesson 10-4)

The sides of a right triangle have special names, as shown below.

The two sides adjacent to the right angle are the **legs**.

The side opposite the right angle is the **hypotenuse**.

The **Pythagorean Theorem** describes the relationship between the length of the hypotenuse and the lengths of the legs.

Noteables™ — Key Concept: Pythagorean Theorem

Words In a right triangle, the square of the length of the hypotenuse equals the sum of the squares of the lengths of the legs.

Symbols $c^2 = a^2 + b^2$

Model

(Right triangle with hypotenuse c, legs a and b.)

You can use the Pythagorean Theorem to find the length of the hypotenuse of a right triangle if the measures of both legs are known.

EXAMPLE Find the Length of the Hypotenuse

1 MOVING Determine whether a 7-foot square mirror will fit diagonally through the doorway shown at the right.

To solve, find the length of the hypotenuse c.

$c^2 = a^2 + b^2$ Pythagorean Theorem
$c^2 = 3^2 + 6.5^2$ Replace a with 3 and b with 6.5.
$c^2 = 9 + 42.25$ Evaluate 3^2 and 6.5^2.
$c^2 = 51.25$ Add.
$\sqrt{c^2} = \sqrt{51.25}$ Take the square root of each side.
$c \approx 7.2$ Simplify.

The length of the diagonal is about 7.2 feet. So, the mirror will fit through the doorway if it is turned diagonally.

msmath2.net/extra_examples   Lesson 11-3 The Pythagorean Theorem 479

READING Math

Theorem A *theorem* is a statement in mathematics that can be justified by logical reasoning.

You can also use the Pythagorean Theorem to find the measure of a leg if the measure of the other leg and the hypotenuse are known.

EXAMPLE Find the Length of a Leg

2 Find the missing measure of the triangle at the right.

The missing measure is of a leg of the triangle.

$$c^2 = a^2 + b^2$$ Pythagorean Theorem
$$13^2 = 5^2 + b^2$$ Replace a with 5 and c with 13.
$$169 = 25 + b^2$$ Evaluate 13^2 and 5^2.
$$169 - 25 = 25 + b^2 - 25$$ Subtract 25 from each side.
$$144 = b^2$$ Simplify.
$$\sqrt{144} = \sqrt{b^2}$$ Take the square root of each side.
$$12 = b$$ Simplify.

The length of the leg is 12 centimeters.

Your Turn Find the missing measure of each right triangle. Round to the nearest tenth if necessary.

a. b. c. $b = 7$ in., $c = 25$ in.

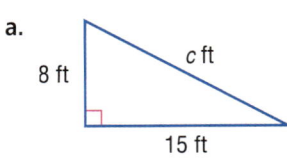

REAL-LIFE CAREERS

How Does an Archaeologist Use Math?
Before digging, archaeologists use the Pythagorean Theorem to calculate the diagonal of an excavation site to be sure that the area is a rectangle.

Research
For information about a career as an archaeologist, visit: **msmath2.net/careers**

EXAMPLE Solve a Real-Life Problem

3 **ARCHAEOLOGY** Archaeologists placed corner stakes to mark a rectangular excavation site, as shown at the right. If their stakes are placed correctly, what is the measure of the diagonal?

The diagonal of the rectangle is the hypotenuse of a right triangle. Write an equation to solve for x.

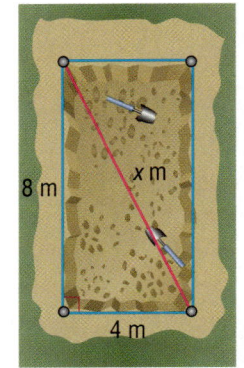

$$c^2 = a^2 + b^2$$ Pythagorean Theorem
$$x^2 = 8^2 + 4^2$$ Replace a with 8, b with 4, and c with x.
$$x^2 = 64 + 16$$ Evaluate 8^2 and 4^2.
$$x^2 = 80$$ Simplify.
$$\sqrt{x^2} = \sqrt{80}$$ Take the square root of each side.
$$x \approx 8.9$$ Simplify.

The length of the diagonal is about 8.9 meters.

480 Chapter 11 Geometry: Measuring Two-Dimensional Figures

You can determine whether a triangle is a right triangle by applying the Pythagorean Theorem.

EXAMPLES Identify Right Triangles

Determine whether a triangle with the given lengths is a right triangle.

STUDY TIP

Hypotenuse
Remember that the hypotenuse is always the longest side.

4 1.5 mm, 2 mm, 2.5 mm

$c^2 = a^2 + b^2$

$2.5^2 \stackrel{?}{=} 1.5^2 + 2^2$

$6.25 \stackrel{?}{=} 2.25 + 4$

$6.25 = 6.25$ ✓

The triangle is a right triangle.

5 4 ft, 6 ft, 8 ft

$c^2 = a^2 + b^2$

$8^2 \stackrel{?}{=} 4^2 + 6^2$

$64 \stackrel{?}{=} 16 + 36$

$64 \neq 52$

The triangle is *not* a right triangle.

Your Turn Determine whether each triangle with the given lengths is a right triangle. Write *yes* or *no*.

d. 7.5 cm, 8 cm, 12 cm   e. 9 in., 40 in., 41 in.

Skill and Concept Check

1. **Writing Math** Describe the information that you need in order to find the missing measure of a right triangle.

2. **OPEN ENDED** Draw and label a right triangle that has one side measuring 14 units. Write the length of another side. Then find the length of the third side to the nearest tenth.

3. **FIND THE ERROR** Devin and Jamie are writing an equation to find the missing measure of the triangle at the right. Who is correct? Explain.

Devin
$16^2 = 5^2 + x^2$

Jamie
$x^2 = 16^2 + 5^2$

16 cm
x cm
5 cm

GUIDED PRACTICE

Find the missing measure of each right triangle. Round to the nearest tenth if necessary.

4.
   c mm, 10 mm, 24 mm

5.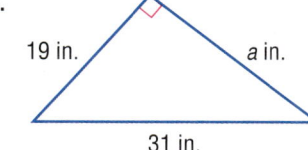
   19 in., a in., 31 in.

6. $b = 21$ cm, $c = 28$ cm

Determine whether a triangle with the given side lengths is a right triangle. Write *yes* or *no*.

7. 1.4 m, 4.8 m, 5 m   8. 21 ft, 24 ft, 30 ft

Lesson 11-3 The Pythagorean Theorem **481**

Practice and Applications

Find the missing measure of each right triangle. Round to the nearest tenth if necessary.

9.

10.

11.

12.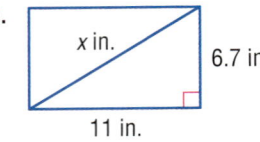

13. $a = 7$ in., $b = 24$ in.

14. $a = 13.5$ mm, $b = 18$ mm

15. $b = 13$ m, $c = 27$ m

16. $a = 2.4$ yd, $c = 3$ yd

Determine whether a triangle with the given side lengths is a right triangle. Write *yes* or *no*.

17. 12 cm, 16 cm, 20 cm

18. 8 m, 15 m, 17 m

19. 11 ft, 14 ft, 17 ft

20. 18 in., 18 in., 36 in.

21. **SAFETY** To the nearest tenth of a foot, how far up the wall *x* does the ladder shown at the right reach?

22. **TRAVEL** You drive 80 miles east, then 50 miles north, then 140 miles west, and finally 95 miles south. Make a drawing to find how far you are from your starting point.

23. **CRITICAL THINKING** What is the length of the diagonal of the cube shown at the right?

HOMEWORK HELP

For Exercises	See Examples
9–16	1, 2
17–20	4, 5
21–22	3

Extra Practice
See pages 590, 606.

Spiral Review with Standardized Test Practice

24. **MULTIPLE CHOICE** Find the missing measure of a right triangle if $a = 20$ meters and $c = 52$ meters.

    Ⓐ 24 m   Ⓑ 32 m   Ⓒ 48 m   Ⓓ 55.7 m

25. **SHORT RESPONSE** An isosceles right triangle has legs that are 8 inches long. Find the length of the hypotenuse to the nearest tenth.

Estimate each square root to the nearest whole number. (Lesson 11-2)

26. $\sqrt{61}$

27. $\sqrt{147}$

28. $\sqrt{40}$

29. $\sqrt{277}$

30. Find $\sqrt{256}$. (Lesson 11-1)

GETTING READY FOR THE NEXT LESSON

PREREQUISITE SKILL Multiply. (Lesson 6-4)

31. $17.8 \cdot 12$

32. $21.5 \cdot 27.1$

33. $3\frac{1}{2} \cdot 8$

34. $15\frac{1}{4} \cdot 18$

msmath2.net/self_check_quiz

11-4 Area of Parallelograms

What You'll LEARN
Find the areas of parallelograms.

NEW Vocabulary
base
height

REVIEW Vocabulary
parallelogram: quadrilateral with opposite sides parallel and opposite sides congruent (Lesson 10-5)

HANDS-ON Mini Lab

Materials
- grid paper
- straightedge

Work with a partner.

1. What is the value of *x* and *y* for each parallelogram?
2. Count the grid squares to find the area of each parallelogram.
3. On grid paper, draw three different parallelograms in which $x = 5$ units and $y = 4$ units. Find the area of each.
4. **Make a conjecture** about how to find the area of a parallelogram if you know the values of *x* and *y*.

In Lesson 10-5, you learned that a parallelogram is a special kind of quadrilateral. You can find the area of a parallelogram by using the values for the base and height, as described below.

The **base** is any side of a parallelogram.

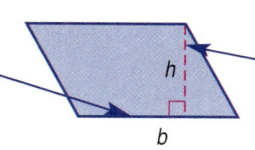

The **height** is the length of the segment perpendicular to the base with endpoints on opposite sides.

Noteables™ Key Concept: Area of a Parallelogram

Words	The area *A* of a parallelogram equals the product of its base *b* and height *h*.	**Model**
Symbols	$A = bh$	

EXAMPLE Find the Area of a Parallelogram

1 Find the area of the parallelogram.

Estimate $A = 13 \cdot 6$ or 78 cm²

$A = bh$ Area of a parallelogram

$A = 13 \cdot 5.8$ Replace *b* with 13 and *h* with 5.8.

$A = 75.4$ Multiply.

The area of the parallelogram is 75.4 square centimeters. **This is close to the estimate.**

msmath2.net/extra_examples

Lesson 11-4 Area of Parallelograms 483

STUDY TIP

Common Error
When finding the area of a parallelogram, be sure to use the base and the height, not the base and a side.

EXAMPLE Find the Area of a Parallelogram

2 Find the area of the parallelogram at the right.

The base is 11 inches, and the height is 9 inches.

Estimate $A = 10 \cdot 10$ or 100 in^2

$A = bh$ Area of a parallelogram

$A = 11 \cdot 9$ Replace b with 11 and h with 9.

$A = 99$ Multiply.

The area of the parallelogram is 99 square inches. This is close to the estimate.

Your Turn Find the area of each parallelogram.

a. base = 5 cm
   height = 8 cm

b.

Skill and Concept Check

1. **Describe** what values you would substitute in the formula $A = bh$ to find the area of the parallelogram at the right.

2. **OPEN ENDED** Draw three different parallelograms, each with an area of 24 square units.

3. **Writing Math** *True* or *False*? The area of a parallelogram doubles if you double the base and the height. Explain or give a counterexample to support your answer.

GUIDED PRACTICE

Find the area of each parallelogram. Round to the nearest tenth if necessary.

4.

5.

6.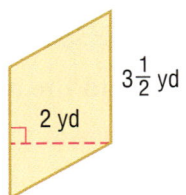

7. base = 16 in.
   height = 4 in.

8. base = 3.5 m
   height = 5 m

9. What is the area of a parallelogram with a base of 25 millimeters and a height that is half the base?

484 Chapter 11 Geometry: Measuring Two-Dimensional Figures

Practice and Applications

Find the area of each parallelogram. Round to the nearest tenth if necessary.

10.
11.
12.

13.
14.
15.

16. base = 13 mm
    height = 6 mm
17. base = 45 yd
    height = 35 yd
18. base = 8 in.
    height = 12.5 in.
19. base = 7.9 cm
    height = 7.2 cm

HOMEWORK HELP
For Exercises 10–21, See Examples 1, 2
Extra Practice See pages 591, 606.

20. **MULTI STEP** A quilted block uses eight parallelogram-shaped pieces of cloth, each with a height of $3\frac{1}{3}$ inches and a base of $3\frac{3}{4}$ inches. How much fabric is needed to make the parallelogram pieces for 24 blocks? Write in square feet. (*Hint*: 144 in^2 = 1 ft^2)

21. What is the height of a parallelogram if the base is 24 inches and the area is 360 square inches?

22. **CRITICAL THINKING** Identify two possible measures of base and height for a parallelogram that has an area of 320 square inches.

Spiral Review with Standardized Test Practice

23. **MULTIPLE CHOICE** Find the area of the parallelogram.
    - Ⓐ 75 cm^2
    - Ⓑ 150 cm^2
    - Ⓒ 200 cm^2
    - Ⓓ 300 cm^2

24. **SHORT RESPONSE** What is the base of a parallelogram if the height is 18.6 inches and the area is 279 square inches?

Determine whether a triangle with the given side lengths is a right triangle. Write *yes* or *no*. (Lesson 11-3)

25. 8 in., 10 in., 12 in.
26. 12 ft, 16 ft, 20 ft
27. 5 cm, 12 cm, 14 cm

28. Which is closer to $\sqrt{55}$, 7 or 8? (Lesson 11-2)

GETTING READY FOR THE NEXT LESSON

PREREQUISITE SKILL Find each value. (Lesson 1-3)

29. 6(4 + 10)
30. $\frac{1}{2}(8)(8)$
31. $\frac{1}{2}(24 + 15)$
32. $\frac{1}{2}(5)(13 + 22)$

CHAPTER 11 Mid-Chapter Practice Test

Vocabulary and Concepts

1. **Define** *square root*. (Lesson 11-1)

2. **State,** in words, the Pythagorean Theorem. (Lesson 11-3)

3. *True* or *False*? The area of any parallelogram equals the length times the width. Explain. (Lesson 11-4)

Skills and Applications

Find the square of each number. (Lesson 11-1)

4. 4

5. 12

Find each square root. (Lesson 11-1)

6. $\sqrt{64}$

7. $\sqrt{289}$

8. **LANDSCAPING** A bag of lawn fertilizer covers 2,500 square feet. Describe the largest square that one bag of fertilizer could cover. (Lesson 11-1)

Estimate each square root to the nearest whole number. (Lesson 11-2)

9. $\sqrt{32}$

10. $\sqrt{55}$

Find the missing measure of each right triangle. Round to the nearest tenth if necessary. (Lesson 11-3)

11.

12. $a = 8.2$ m
    $b = 15.6$ m

Find the area of each parallelogram. (Lesson 11-4)

13. base = 4.3 in.
    height = 9 in.

14.

Standardized Test Practice

15. **MULTIPLE CHOICE** Which is the best estimate for $\sqrt{120}$? (Lesson 11-2)

    Ⓐ 10   Ⓑ 11   Ⓒ 12   Ⓓ 15

16. **SHORT RESPONSE** Miranda jogs 5 kilometers north and 5 kilometers west. To the nearest kilometer, how far is she from her starting point? (Lesson 11-3)

The Game Zone

A Place To Practice Your Math Skills

Math Skill: Finding Square Roots

Tic Tac Root

GET READY!
Players: two to four
Materials: index cards, construction paper

GET SET!

- Use 20 index cards. On each card, write one of the following square roots.

$\sqrt{1}$ $\sqrt{4}$ $\sqrt{9}$ $\sqrt{16}$
$\sqrt{25}$ $\sqrt{36}$ $\sqrt{49}$ $\sqrt{64}$
$\sqrt{81}$ $\sqrt{100}$ $\sqrt{121}$ $\sqrt{144}$
$\sqrt{169}$ $\sqrt{196}$ $\sqrt{225}$ $\sqrt{256}$
$\sqrt{289}$ $\sqrt{324}$ $\sqrt{361}$ $\sqrt{400}$

- Each player should draw a tic-tac-toe board on construction paper. In each square, place a number from 1 to 20, but do not use any number more than once. See the sample board at the right.

GO!

- The dealer shuffles the index cards and places them facedown on the table.
- The player to the left of the dealer chooses the top index card and places it faceup. Any player with the matching square root on his or her board places an X on the appropriate square.
- The next player chooses the top index card and places it faceup on the last card chosen. Players mark their boards accordingly.
- **Who Wins?** The first player to get three Xs in a row wins the game.

The Game Zone: Finding Square Roots

11-5a HANDS-ON LAB

A Preview of Lesson 11-5

Triangles and Trapezoids

What You'll LEARN

Find the areas of triangles and trapezoids using models.

Materials
- centimeter grid paper
- straightedge
- scissors
- tape

INVESTIGATE *Work as a class.*

STEP 1 On grid paper, draw a triangle with a base of 6 units and a height of 3 units. Label the base b and the height h as shown.

STEP 2 Fold the grid paper in half and cut out the triangle through both sheets so that you have two congruent triangles.

STEP 3 Turn the second triangle upside down and tape it to the first triangle.

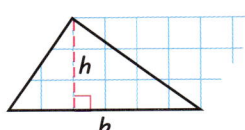

Writing Math

Work with a partner.

1. What figure is formed by the two triangles?
2. Write the formula for the area of the figure. Then find the area.
3. What is the area of each of the triangles? How do you know?
4. Repeat the activity above, drawing a different triangle in Step 1. Then find the area of each triangle.
5. Compare the area of a triangle to the area of a parallelogram with the same base and height.
6. **MAKE A CONJECTURE** Write a formula for the area of a triangle with base b and height h.

For Exercises 7–9, refer to the information below.

On grid paper, cut out two identical trapezoids. Label the bases b_1 and b_2, respectively, and label the heights h. Then turn one trapezoid upside down and tape it to the other trapezoid as shown.

7. Write an expression to represent the base of the parallelogram.
8. Write a formula for the area A of the parallelogram using b_1, b_2, and h.
9. **MAKE A CONJECTURE** Write a formula for the area A of a trapezoid with bases b_1 and b_2, and height h.

11-5 Area of Triangles and Trapezoids

What You'll LEARN

Find the areas of triangles and trapezoids.

REVIEW Vocabulary

trapezoid: quadrilateral with one pair of parallel sides (Lesson 10-5)

HANDS-ON Mini Lab

Materials
- grid paper
- straightedge
- scissors

Work with a partner.

- Draw a parallelogram with a base of 6 units and a height of 4 units.
- Draw a diagonal as shown.
- Cut out the parallelogram.

1. What is the area of the parallelogram?
2. Cut along the diagonal. What is true about the triangles formed?
3. What is the area of each triangle?
4. If the area of a parallelogram is *bh*, then write an expression for the area *A* of each of the two congruent triangles that form the parallelogram.

Like parallelograms, you can find the area of a triangle by using the base and height.

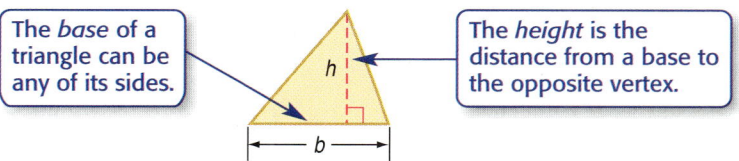

The *base* of a triangle can be any of its sides.

The *height* is the distance from a base to the opposite vertex.

Noteables™   Key Concept: Area of a Triangle

Words   The area A of a triangle equals half the product of its base b and height h.

Symbols $A = \frac{1}{2}bh$

Model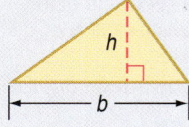

EXAMPLE   Find the Area of a Triangle

1 Find the area of the triangle below.   **Estimate** $\frac{1}{2}(10)(7) = 35$

$A = \frac{1}{2}bh$ Area of a triangle

$A = \frac{1}{2}(10)(6.5)$ Replace b with 10 and h with 6.5.

$A = 32.5$ Multiply.

6.5 m, 10 m

The area of the triangle is 32.5 square meters.   This is close to the estimate.

A trapezoid has two bases, b_1 and b_2. The height of a trapezoid is the distance between the bases.

> **Noteables™** — **Key Concept: Area of a Trapezoid**
>
> **Words** The area A of a trapezoid equals half the product of the height h and the sum of the bases b_1 and b_2.
>
> **Symbols** $A = \frac{1}{2}h(b_1 + b_2)$
>
> **Model**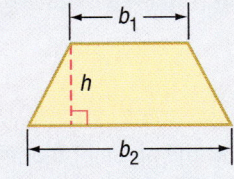

READING Math

Subscripts Read b_1 as b sub 1. Read b_2 as b sub 2. The subscripts mean that b_1 and b_2 represent different variables.

EXAMPLE — Find the Area of a Trapezoid

2 Find the area of the trapezoid at the right.

The bases are 5 inches and 12 inches. The height is 7 inches.

$A = \frac{1}{2}h(b_1 + b_2)$ Area of a trapezoid

$A = \frac{1}{2}(7)(5 + 12)$ Replace h with 7, b_1 with 5, and b_2 with 12.

$A = \frac{1}{2}(7)(17)$ Add 5 and 12.

$A = 59.5$ Multiply.

The area of the trapezoid is 59.5 square inches.

Your Turn Find the area of each triangle or trapezoid. Round to the nearest tenth if necessary.

a.   b.   c.

REAL-LIFE MATH

GEOGRAPHY The actual area of Arkansas is 52,068 square miles.

Source: Merriam Webster's Collegiate Dictionary

EXAMPLE — Use a Formula to Estimate Area

3 **GEOGRAPHY** The shape of the state of Arkansas resembles a trapezoid. Estimate its area in square miles.

$A = \frac{1}{2}h(b_1 + b_2)$

$A = \frac{1}{2}(235)(280 + 210)$ Replace h with 235, b_1 with 280, and b_2 with 210.

$A = \frac{1}{2}(235)(490)$ Add 280 and 210.

$A = 57{,}575$ Multiply.

The area of Arkansas is about 57,575 square miles.

490   Chapter 11 Geometry: Measuring Two-Dimensional Figures

Skill and Concept Check

1. **Estimate** the area of the trapezoid at the right.

2. **OPEN ENDED** Draw a trapezoid and label the bases and the height. In your own words, explain how to find the area of the trapezoid.

3. **Writing Math** Describe the relationship between the area of a parallelogram and the area of a triangle with the same height and base.

GUIDED PRACTICE

Find the area of each figure. Round to the nearest tenth if necessary.

4.
5.
6.

7. **MULTI STEP** The blueprints for a patio are shown at the right. If the cost of the patio is $4.50 per square foot, what will be the total cost of the patio?

Practice and Applications

Find the area of each figure. Round to the nearest tenth if necessary.

8.
9.
10.

11.
12.
13.

14. triangle: base = 4 cm, height = 7.5 cm
15. trapezoid: bases 13 in. and $1\frac{1}{4}$ ft, height 1 ft

Draw and label each figure on grid paper. Then find the area.

16. a triangle with no right angles
17. an isosceles triangle with a height greater than 6 units
18. a trapezoid with a right angle and an area of 40 square units
19. a trapezoid with no right angles and an area less than 25 square units

HOMEWORK HELP

For Exercises	See Examples
8–9, 12, 14	1
10–11, 13, 15	2
20–21	3

Extra Practice
See pages 591, 606.

msmath2.net/self_check_quiz   **Lesson 11-5** Area of Triangles and Trapezoids **491**

20. **GEOGRAPHY** Nevada has a shape that looks like a trapezoid, as shown at the right. Find the approximate area of the state.

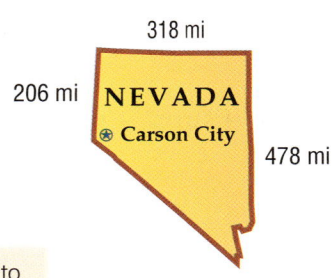

21. **GEOGRAPHY** Delaware has a shape that is roughly triangular with a base of 39 miles and a height of 96 miles. Find the approximate area of the state.

 Data Update How do the actual areas of Nevada and Delaware compare to your estimates? Visit **msmath2.net/data_update** to learn more.

22. **CRITICAL THINKING** A triangle has height h. Its base is 4. Find the area of the triangle. (*Hint*: Express your answer in terms of h.)

Spiral Review with Standardized Test Practice

23. **SHORT RESPONSE** Find the area of the triangle at the right to the nearest tenth.

24. **MULTIPLE CHOICE** A trapezoid has bases of 15 meters and 18 meters and a height of 10 meters. What is the area of the trapezoid?

    Ⓐ 30 m²   Ⓑ 60 m²   Ⓒ 165 m²   Ⓓ 330 m²

25. **GEOMETRY** Find the area of a parallelogram having a base of 2.3 inches and a height of 1.6 inches. Round to the nearest tenth. (Lesson 11-4)

Find the missing measure of each right triangle. Round to the nearest tenth if necessary. (Lesson 11-3)

26. $a = 10$ m, $b = 14$ m

27. $a = 13$ ft, $c = 18$ ft

For Exercises 28–31, refer to the graphic at the right. Classify the angle that represents each category as *acute*, *obtuse*, *right*, or *straight*. (Lesson 10-1)

28. 30–39 hours
29. 1–29 hours
30. 40 hours
31. 41–50 hours

32. **MUSIC** Use the Fundamental Counting Principle to find the number of piano instruction books in a series if there are Levels 1, 2, 3, and 4 and each level contains five different books. (Lesson 9-3)

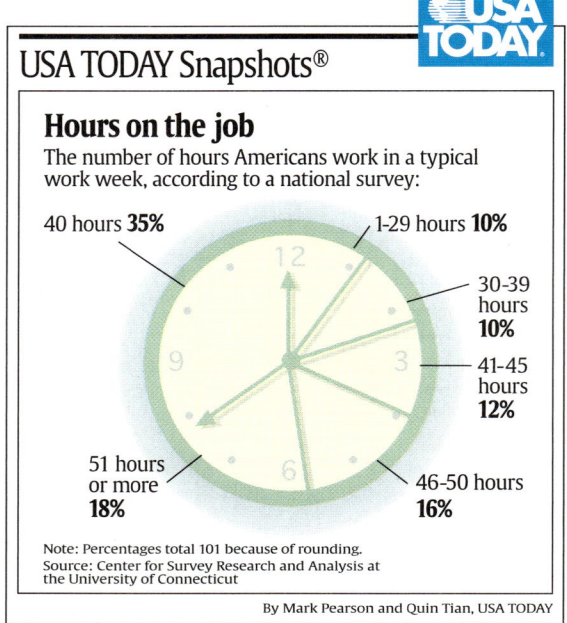

GETTING READY FOR THE NEXT LESSON

BASIC SKILL Use a calculator to find each product to the nearest tenth.

33. $\pi \cdot 13$ 34. $\pi \cdot 29$ 35. $\pi \cdot 16^2$ 36. $\pi \cdot 4.8^2$

11-6 Area of Circles

What You'll LEARN

Find the areas of circles.

REVIEW Vocabulary

pi (π): the Greek letter that represents an irrational number approximately equal to 3.14 (Lesson 6-9)

HANDS-ON Mini Lab

Materials
- large paper plate
- scissors

Work with a partner.

- Fold a paper plate in half four times to divide it into 16 equal-sized sections.
- Label the radius *r* as shown. Let *C* represent the circumference of the circle.
- Cut out each section; reassemble to form a parallelogram-shaped figure.

1. What is the measurement of the base and the height?
2. Substitute these values into the formula for the area of a parallelogram.
3. Replace *C* with the expression for the circumference of a circle, $2\pi r$. Simplify the equation and describe what it represents.

In the Mini Lab, the formula for the area of a parallelogram was used to develop a formula for the area of a circle.

Noteables™ — Key Concept: Area of a Circle

Words The area *A* of a circle equals the product of pi (π) and the square of its radius *r*.

Symbols $A = \pi r^2$

Model

EXAMPLES — Find the Areas of Circles

1 Find the area of the circle at the right.

$A = \pi r^2$ Area of a circle

$A = \pi \cdot 2^2$ Replace *r* with 2.

[π] [×] 2 [x²] [ENTER =] 12.56637061

The area of the circle is approximately 12.6 square inches.

2 Find the area of a circle with a diameter of 15.2 centimeters.

$A = \pi r^2$ Area of a circle

$A = \pi \cdot 7.6^2$ Replace *r* with 15.2 ÷ 2 or 7.6.

$A \approx 181.5$ Use a calculator.

The area of the circle is approximately 181.5 square centimeters.

Skill and Concept Check

1. **OPEN ENDED** Draw and label the radius of a circle that has an area less than 10 square units.

2. **FIND THE ERROR** Carlos and Sean are finding the area of a circle that has a diameter of 12 centimeters. Who is correct? Explain.

   Carlos
   $A = \pi(12)^2$
   $\approx 452\ cm^2$

   Sean
   $A = \pi(6)^2$
   $\approx 113\ cm^2$

3. **NUMBER SENSE** Without using a calculator, determine which has the greatest value: 2π, $\sqrt{7}$, or 1.5^2. Explain.

GUIDED PRACTICE

Find the area of each circle. Round to the nearest tenth.

4.   5 cm

5.   9 in.

6.   16 m

7. radius = 4.2 ft

8. diameter = 13 ft

9. diameter = 24 mm

10. **HISTORY** The Roman Pantheon is a circular-shaped structure that was completed about 126 A.D. Find the area of the floor if the diameter is 43 meters.

Practice and Applications

Find the area of each circle. Round to the nearest tenth.

11.   8 cm

12.   3 in.

13.   11 ft

14.   2.4 m

15.   17 cm

16.   6.5 m

17. radius = 6 ft
18. diameter = 7 ft
19. diameter = 3 cm
20. radius = 10.5 mm
21. radius = $4\frac{1}{2}$ in.
22. diameter = $20\frac{3}{4}$ yd

23. A *semicircle* is half a circle. Find the area of the semicircle at the right to the nearest tenth.

      8.6 m

24. **MONEY** Find the area of the face of a Sacagawea $1 coin if the diameter is 26.5 millimeters. Round to the nearest tenth.

HOMEWORK HELP

For Exercises	See Examples
11–16, 25	1
17–22, 24	2

Extra Practice
See pages 591, 606.

494 Chapter 11 Geometry: Measuring Two-Dimensional Figures

25. **LANDSCAPE DESIGN** A circular stone path is to be installed around a birdbath with radius 1.5 feet, as shown at the right. What is the area of the path? (*Hint*: Find the area of the large circle minus the area of the small circle.)

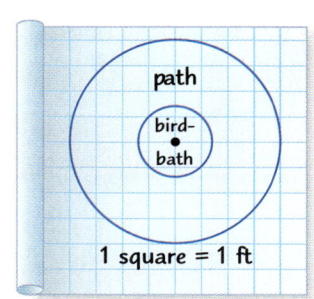

For Exercises 26 and 27, refer to the information below.
Let the *x*-axis of a coordinate plane represent the radius of a circle and the *y*-axis represent the area of a circle.

26. Graph the points that represent the circles with radii 0, 1, 2, and 3 units long. Draw a line or curve that goes through each point.

27. Consider a circle with radius of 1 unit and a circle with a radius of 2 units. Write a ratio comparing the radii. Write a ratio comparing the areas. Do these ratios form a proportion? Explain.

Find the area of the shaded region in each figure. Round to the nearest tenth.

28.

29.

30.

31. **CRITICAL THINKING** Determine whether the area of a circle is *sometimes*, *always*, or *never* doubled when the radius is doubled. Explain.

Spiral Review with Standardized Test Practice

32. **MULTIPLE CHOICE** A CD has a diameter of 12 centimeters. The hole in the middle of the CD has a diameter of 1.5 centimeters. Find the area of one side of the CD to the nearest tenth. Use 3.14 for π.

    Ⓐ 111.3 cm²   Ⓑ 113.0 cm²   Ⓒ 349.4 cm²   Ⓓ 445.1 cm²

33. **SHORT RESPONSE** Find the radius of a circle that has an area of 42 square centimeters. Use 3.14 for π and round to the nearest tenth.

34. **GEOMETRY** Find the area of a triangle with a base of 21 meters and a height of 27 meters. (Lesson 11-5)

Find the area of each parallelogram. Round to the nearest tenth if necessary. (Lesson 11-4)

35.

36.

37.

GETTING READY FOR THE NEXT LESSON

BASIC SKILL Simplify each expression. (Lessons 1-2 and 1-3)

38. 8.5^2 39. $3.14 \cdot 6^2$ 40. $\frac{1}{2} \cdot 5.4^2 + 11$ 41. $\frac{1}{2} \cdot 7^2 + (9)(14)$

11-7a Problem-Solving Strategy
A Preview of Lesson 11-7

Solve a Simpler Problem

What You'll LEARN

Solve problems by solving a simpler problem.

The diagram shows the backdrop for our fall play. How much wallpaper will we need to cover the entire front?

*We need to find the total area of the backdrop. Let's **solve a simpler problem** by breaking it down into separate geometric shapes.*

Explore	We know that the backdrop is made of one large rectangle and two semicircles, which equal an entire circle.
Plan	We can find the areas of the rectangle and the circle, and then add.
Solve	area of rectangle: $A = \ell w$ $\qquad A = (8 + 8)7$ or 112 area of circle: $A = \pi r^2$ $\qquad A = \pi \cdot 4^2$ or about 50.3 total area: $\qquad 112 + 50.3$ or 162.3 square feet So, we need at least 162.3 square feet of wallpaper.
Examine	Use estimation to check. The backdrop is 16 feet long and 11 feet high. However, it is less than a complete rectangle, so the area should be *less than* $16 \cdot 11$ or 176 feet. The area, 162.3 square feet, is less than 176 feet, so the answer is reasonable.

Analyze the Strategy

1. **Explain** why simplifying this problem is a good strategy to solve this problem.
2. **Describe** another way that the problem could have been solved.
3. **Write** a problem that can be solved by breaking it down into a simpler problem. Solve the problem and explain your answer.

Apply the Strategy

Solve. Use the solve a simpler problem strategy.

4. **LANDSCAPING** James is helping his father pour a circular sidewalk around a flower bed, as shown below. What is the area, in square feet, of the sidewalk? Use 3.14 for π.

5. **COMMUNICATION** According to a recent report, one city has 2,945,000 phone lines assigned to three different area codes. How many of the phone lines are assigned to each area code?

Area Code	Percent
888	44.3%
777	23.7%
555	31.5%

Mixed Problem Solving

Solve. Use any strategy.

6. **EARTH SCIENCE** Earth's atmosphere exerts a pressure of 14.7 pounds per square inch at the ocean's surface. The pressure increases by 12.7 pounds per square inch for every 6 feet that you descend. Find the pressure at 18 feet below the surface.

7. **SALES** Deirdre is trying to sell $3,000 in ads for the school newspaper. The prices of the ads and the number of ads that she has sold are shown in the table. Which is the smallest ad she could sell in order to meet her quota?

Ad Size	Cost Per Ad	Number Sold
quarter-page	$75	15
half-page	$125	8
full-page	$175	4

8. **THEATER** Mr. Marquez is purchasing fabric for curtains for a theatrical company. The front of the stage is 15 yards wide and 5 yards high. The fabric is sold on bolts that are 60 inches wide and 20 yards long. How many bolts are needed to make the curtains?

9. **TELEVISION** The graph shows the results of a survey in which 365,750 people were asked to name their favorite television programs. Estimate how many people chose sitcoms as their favorite.

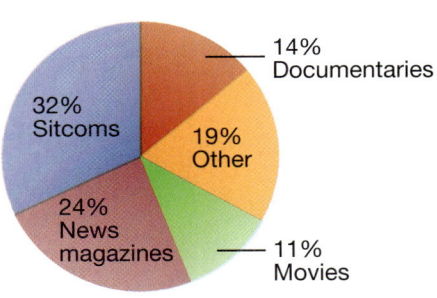

10. **STANDARDIZED TEST PRACTICE** Kara is painting one wall in her room, as shown by the shaded region below. What is the area that she is painting?

- **A** 92 ft²
- **B** 94 ft²
- **C** 96 ft²
- **D** 100 ft²

11-7 Area of Complex Figures

What You'll LEARN
Find the areas of complex figures.

NEW Vocabulary
complex figure

Link to READING
Everyday Meaning of Complex: a whole made up of interrelated parts, as in a retail complex that is made up of many different stores

WHEN am I ever going to use this?

ARCHITECTURE Rooms in a house are not always square or rectangular, as shown in the diagram at the right.

1. Describe the shape of the kitchen.
2. How could you determine the area of the kitchen?
3. How could you determine the total square footage of a house with rooms shaped like these?

A **complex figure** is made of circles, rectangles, squares, and other two-dimensional figures. To find the area of a complex figure, separate it into figures whose areas you know how to find, and then add the areas.

EXAMPLE Find the Area of an Irregular Room

1 ARCHITECTURE Refer to the diagram of the house above. The kitchen is 28 feet by 15 feet, as shown at the right. Find the area of the kitchen. Round to the nearest tenth.

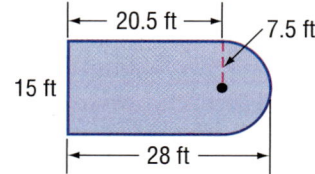

The figure can be separated into a rectangle and a semicircle.

Area of Rectangle

$A = \ell w$ Area of a rectangle
$A = 20.5 \cdot 15$ Replace ℓ with 20.5 and w with 15.
$A = 307.5$ Multiply.

Area of Semicircle

$A = \frac{1}{2}\pi r^2$ Area of a semicircle
$A = \frac{1}{2}\pi (7.5)^2$ Replace r with 7.5.
$A \approx 88.4$ Simplify.

The area of the kitchen is approximately $307.5 + 88.4$ or 395.9 square feet.

498 Chapter 11 Geometry: Measuring Two-Dimensional Figures

Standardized Test Practice

EXAMPLE Find the Area of a Complex Figure

② **GRID-IN TEST ITEM** Find the area of the figure at the right in square inches.

Read the Test Item
The figure can be separated into a rectangle and a triangle. Find the area of each.

Solve the Test Item

Area of Rectangle
$A = \ell w$ Area of a rectangle
$A = 10 \cdot 6$ Replace ℓ with 10 and w with 6.
$A = 60$ Multiply.

Area of Triangle
$A = \frac{1}{2}bh$ Area of a triangle
$A = \frac{1}{2}(4)(4)$ $b = 10 - 6$ or 4, $h = 4$
$A = 8$ Multiply.

The area is 60 + 8 or 68 square inches.

Fill in the Grid

Test-Taking Tip

Drawings Be sure to add the areas of all the separate figures and not stop once you find the area of part of the figure.

⬤ **Your Turn** Find the area of each figure. Round to the nearest tenth if necessary.

a.

b.

Skill and Concept Check

1. **Writing Math** Describe how you would find the area of the figure at the right.

2. **OPEN ENDED** Sketch a complex figure and describe how you could find the area.

GUIDED PRACTICE

Find the area of each figure. Round to the nearest tenth if necessary.

3.

4.

5.

Practice and Applications

Find the area of each figure. Round to the nearest tenth if necessary.

HOMEWORK HELP
For Exercises	See Examples
6–12	1, 2

Extra Practice
See pages 592, 606.

6.

7.

8.

9.

10.

11.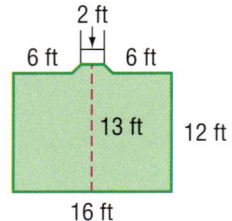

12. **INTERIOR DESIGN** The Eppicks' living room, shown at the right, has a bay window. They are planning to have the hardwood floors in the room refinished. What is the total area that needs to be refinished?

CRITICAL THINKING Describe how you could estimate the area of each state.

13.

14. (North Carolina map)

Spiral Review with Standardized Test Practice

15. **SHORT RESPONSE** Find the area of the figure if each triangle has a height of 3.5 inches and the square has side lengths of 4 inches.

16. **MULTIPLE CHOICE** A rectangular room 14 feet by 12 feet has a semicircular sitting area attached with a diameter of 12 feet. What is the total area of the room and the sitting area?
    - Ⓐ 168 ft²
    - Ⓑ 224.5 ft²
    - Ⓒ 281.1 ft²
    - Ⓓ 620.4 ft²

Find the area of each circle. Round to the nearest tenth. (Lesson 11-6)

17. radius = 4.7 cm

18. radius = 12 in.

19. diameter = 15 in.

20. Find the area of a triangle that has a base of 3.8 meters and a height of 9 meters. (Lesson 11-5)

GETTING READY FOR THE NEXT LESSON

PREREQUISITE SKILL Find the probability of rolling each number on a number cube. (Lesson 9-1)

21. P(2)

22. P(even)

23. P(3 or 4)

24. P(less than 5)

msmath2.net/self_check_quiz

11-8 Probability: Area Models

What You'll LEARN
Find probability using area models.

REVIEW Vocabulary
probability: the ratio of the number of ways an event can occur to the number of possible outcomes (Lesson 9-1)

HANDS-ON Mini Lab

Materials
- two number cubes

Work with a partner.

- Roll the number cubes and find the product of the numbers rolled.
- Repeat nine more times.
- Collect the data for the entire class. Organize the outcomes in a table.

1. Do certain products occur more often?
2. Make and complete a table like the one at the right to find all the possible outcomes.

×	1	2	3	4	5	6
1	1	2	3			
2	2	4	6			
3						

The grid at the right shows the possible products when two number cubes are rolled. The area of the grid is 36 square units. Notice that 6 and 12 make up $\frac{8}{36}$ of the area. So, the probability of rolling two numbers whose product is 6 or 12 is $\frac{8}{36}$.

1	2	3	4	5	6
2	4	6	8	10	12
3	6	9	12	15	18
4	8	12	16	20	24
5	10	15	20	25	30
6	12	18	24	30	36

The area of geometric shapes can be used to find probabilities.

EXAMPLE Use Area Models to Find Probability

1 PROBABILITY A randomly-dropped counter falls somewhere in the squares. Find the probability that it falls on the shaded squares.

$$\text{probability} = \frac{\text{number of ways to land in shaded squares}}{\text{number of ways to land on squares}}$$

$$= \frac{\text{area of shaded squares}}{\text{area of all squares}}$$

Area of Shaded Squares

$A = \frac{1}{2}bh$ Area of a triangle

$A = \frac{1}{2}(2)(3)$ $b = 2$ and $h = 3$

$A = 3$ Simplify.

Area of All Squares

$A = \frac{1}{2}h(b_1 + b_2)$ Area of a trapezoid

$A = \frac{1}{2}(5)(8 + 3)$ $h = 5, b_1 = 8, b_2 = 3$

$A = 27.5$ Simplify.

So, the probability of a counter falling in the shaded squares is $\frac{3}{27.5}$ or about 10.9%.

EXAMPLE   Find the Probability of Winning a Game

2 **GAMES** Suppose a dart is equally likely to hit any point on the board. What is the probability that it hits the white section?

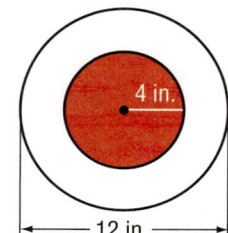

First, find the area of the white section. It equals the area of the large circle minus the area of the small circle.

REAL-LIFE MATH

GAMES The arrangement of numbers around a standard dartboard dates back over 100 years.
Source: mathpages.com

Area of Large Circle		Area of Small Circle	
$A = \pi r^2$	Area of a circle	$A = \pi r^2$	Area of a circle
$A = \pi(6)^2$	Replace r with 6.	$A = \pi(4)^2$	Replace r with 4.
$A \approx 113.1$	Simplify.	$A \approx 50.3$	Simplify.

Area of White Section = 113.1 − 50.3   large circle − small circle
$\phantom{\text{Area of White Section }}= 62.8$ Subtract.

$P(\text{white}) = \dfrac{62.8}{113.1}$ ← area of white section
← area of entire model

$\phantom{P(\text{white})} \approx 0.5553$ Use a calculator.

So, the probability of hitting the white section is about 55.6%.

Skill and Concept Check

1. **Writing Math** Explain how area models are used to solve probability problems.

2. **OPEN ENDED** Draw a spinner in which the probability of spinning and landing on a blue region is $\frac{1}{6}$. Explain your reasoning.

GUIDED PRACTICE

A randomly-dropped counter falls in the squares. Find the probability that it falls in the shaded regions. Write as a percent. Round to the nearest tenth if necessary.

3.

4.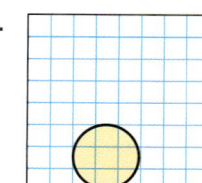

5. **GAMES** Suppose a dart is equally likely to hit any point on the dartboard at the right. What is the probability that it hits the red section?

502 Chapter 11 Geometry: Measuring Two-Dimensional Figures

Practice and Applications

A randomly-dropped counter falls in the squares. Find the probability that it falls in the shaded regions. Write as a percent. Round to the nearest tenth if necessary.

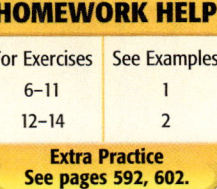

HOMEWORK HELP

For Exercises	See Examples
6–11	1
12–14	2

Extra Practice
See pages 592, 602.

6.   7.   8.

9.   10.   11.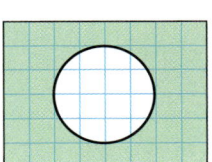

GAMES Each figure represents a dartboard. If it is equally likely that a thrown dart will land anywhere on the dartboard, find the probability that it lands in the shaded region.

12.   13.   14.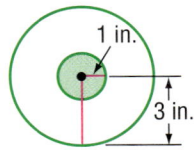

15. **CRITICAL THINKING** A quarter is randomly tossed on the grid board at the right. If a quarter has a radius of 12 millimeters, what is the probability that it does *not* touch a line when it lands? (*Hint*: Find the area where the center of the coin could land so that the edges do not touch a line.)

Spiral Review with Standardized Test Practice

16. **MULTIPLE CHOICE** At a carnival, a person wins a prize if their dart pops a balloon on the rectangular wall. If the radius of each circular balloon is 0.4 foot, approximately what is the probability that a person will win? Use 3.14 for π.

    Ⓐ 25%   Ⓑ 29%   Ⓒ 30%   Ⓓ 63%

17. **GRID IN** Find the probability that a randomly dropped counter will fall in the shaded region at the right. Write as a fraction.

18. Find the area of a figure that is a 6-inch square with a semicircle attached to each side. Each semicircle has a diameter of 6 inches. Round to the nearest tenth. (Lesson 11-7)

19. Find the area of a circle with a radius of 5.7 meters. Round to the nearest tenth. (Lesson 11-6)

msmath2.net/self_check_quiz   Lesson 11-8 Probability: Area Models **503**

CHAPTER 11 Study Guide and Review

Vocabulary and Concept Check

base (p. 483)
complex figure (p. 498)
height (p. 483)
hypotenuse (p. 479)
irrational number (p. 476)
leg (p. 479)
perfect square (p. 471)
Pythagorean Theorem (p. 479)
radical sign (p. 471)
square (p. 470)
square roots (p. 471)

Choose the correct term or number to complete each sentence.

1. In a right triangle, the square of the length of the hypotenuse is (equal to, less than) the sum of the squares of the lengths of the legs.
2. $A = \frac{1}{2}h(a + b)$ is the formula for the area of a (triangle, trapezoid).
3. The square of 25 is (625, 5).
4. A (square, square root) of 49 is 7.
5. The longest side of a right triangle is called the (leg, hypotenuse).
6. The $\sqrt{}$ symbol is called a (radical, perfect square) sign.

Lesson-by-Lesson Exercises and Examples

11-1 Squares and Square Roots (pp. 470–473)

Find the square of each number.

7. 6
8. 14
9. 23

Find each square root.

10. $\sqrt{16}$
11. $\sqrt{256}$
12. $\sqrt{900}$

Example 1 Find the square of 9.
$9^2 = 9 \cdot 9$ or 81

Example 2 Find $\sqrt{121}$.
Since $11 \cdot 11 = 121$, $\sqrt{121} = 11$.

11-2 Estimating Square Roots (pp. 475-477)

Estimate each square root to the nearest whole number.

13. $\sqrt{6}$
14. $\sqrt{99}$
15. $\sqrt{48}$
16. $\sqrt{76}$
17. $\sqrt{19}$
18. $\sqrt{52}$

Use a calculator to find each square root to the nearest tenth.

19. $\sqrt{61}$
20. $\sqrt{132}$
21. $\sqrt{444}$
22. $\sqrt{12}$

Example 3 Estimate $\sqrt{29}$ to the nearest whole number.

29 is between the perfect squares 25 and 36.

$25 < 29 < 36$

$\sqrt{25} < \sqrt{29} < \sqrt{36}$ Find the square root of each number.

$5 < \sqrt{29} < 6$ $\sqrt{25} = 5$ and $\sqrt{36} = 6$

So, $\sqrt{29}$ is between 5 and 6. Since 29 is closer to 25 than to 36, the best whole number estimate is 5.

msmath2.net/vocabulary_review

11-3 The Pythagorean Theorem (pp. 479–482)

Find the missing measure of each right triangle. Round to the nearest tenth if necessary.

23.
24.

25. $a = 5$ ft, $b = 6$ ft
26. $b = 10$ yd, $c = 12$ yd
27. $a = 7$ m, $c = 15$ m
28. $a = 12$ in., $b = 4$ in.

Example 4 Find the missing measure of the triangle. Round to the nearest tenth if necessary.

$c^2 = a^2 + b^2$ Pythagorean Theorem
$c^2 = 4^2 + 12^2$ $a = 4, b = 12$
$c^2 = 16 + 144$ Evaluate.
$c^2 = 160$ Add.
$c \approx 12.6$ Take the square root of each side.

The length of the hypotenuse is about 12.6 centimeters.

11-4 Area of Parallelograms (pp. 483–485)

Find the area of each parallelogram. Round to the nearest tenth if necessary.

29.
30.

31. base = 9 cm, height = 15 cm
32. base = 24 m, height = 16.2 m

Example 5 Find the area of a parallelogram if the base is 15 inches and the height is 8 inches.

$A = bh$ Area of a parallelogram
$A = 15 \cdot 8$ Replace b with 15 and h with 8.
$A = 120$ in^2 Multiply.

11-5 Area of Triangles and Trapezoids (pp. 489–492)

Find the area of each figure. Round to the nearest tenth if necessary.

33.
34.

35. triangle: base = 24.7 cm, height = 15.2 cm
36. trapezoid: bases = 22 yd and 35 yd, height = 18.5 yd

Example 6 Find the area of a triangle with a base of 8 meters and a height of 11.2 meters.

$A = \frac{1}{2}bh$ Area of a triangle
$A = \frac{1}{2}(8)(11.2)$ or 44.8 m^2 $b = 8, h = 11.2$

Example 7 Find the area of the trapezoid.

$A = \frac{1}{2}h(b_1 + b_2)$
$A = \frac{1}{2}(3)(2 + 10)$ $h = 3, b_1 = 2, b_2 = 10$
$A = \frac{1}{2}(3)(12)$ or 18 in^2 Simplify.

Study Guide and Review continued

Mixed Problem Solving
For mixed problem-solving practice, see page 606.

11-6 Area of Circles (pp. 493–495)

Find the area of each circle. Round to the nearest tenth.

37. radius = 11.4 in.
38. diameter = 44 cm
39. **GARDENING** A lawn sprinkler can water a circular area with a radius of 20 feet. Find the area that can be watered. Round to the nearest tenth.

Example 8 Find the area of a circle with a radius of 5 inches.

$A = \pi r^2$ Area of a circle
$A = \pi(5)^2$ Replace r with 5.
$A \approx 78.5$ Multiply.

The area of the circle is about 78.5 square inches.

11-7 Area of Complex Figures (pp. 498–500)

Find the area of each figure. Round to the nearest tenth if necessary.

40.
41.
42.
43.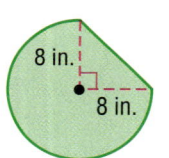

Example 9 Find the area of the figure.

The figure can be separated into a parallelogram and a trapezoid.

parallelogram: $A = bh = (12)(7)$ or 84
trapezoid: $A = \frac{1}{2}h(b_1 + b_2)$
$= \frac{1}{2}(12)(16 + 5)$ or 126

The area of the figure is 84 + 126 or 210 square centimeters.

11-8 Probability: Area Models (pp. 501–503)

A randomly-dropped counter falls in the squares. Find the probability that it falls in the shaded regions. Write as a percent. Round to the nearest tenth if necessary.

44.
45.

Example 10 Find the probability that a randomly-dropped counter will land on a white square.

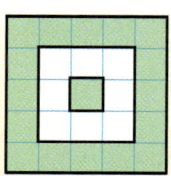

Probability of landing on a white square

$= \dfrac{\text{number of ways to land on white squares}}{\text{number of ways to land on squares}}$

$= \dfrac{\text{area of white squares}}{\text{total area}}$

$= \dfrac{8}{25}$ The probability is $\dfrac{8}{25}$ or 32%.

Chapter 11 Practice Test

Vocabulary and Concepts

1. **State** the measurements that you need in order to find the area of a parallelogram.
2. **Describe** how to find the area of a complex figure.

Skills and Applications

3. Find the square of 9.

4. Find $\sqrt{400}$.

5. **PHYSICAL FITNESS** This morning, Elisa walked 1 mile north, 0.5 mile west, and then walked straight back to her starting point. How far did Elisa walk? Round to the nearest tenth.

6. Estimate $\sqrt{23}$ to the nearest whole number.

7. Use a calculator to find $\sqrt{133}$ to the nearest tenth.

Find the missing measure of each right triangle. Round to the nearest tenth if necessary.

8. $a = 5$ m, $b = 4$ m

9. $b = 12$ in., $c = 14$ in.

Find the area of each figure. Round to the nearest tenth if necessary.

10.

11.

12.

13.

Find the area of each circle. Round to the nearest tenth.

14. radius = 9 ft

15. diameter = 5.2 cm

Standardized Test Practice

16. **MULTIPLE CHOICE** A randomly dropped counter falls in the squares. Find the probability that it falls in the shaded squares. Write as a percent. Round to the nearest tenth if necessary.

Ⓐ 9%   Ⓑ 14.1%   Ⓒ 22.5%   Ⓓ 40%

Chapter 11 Standardized Test Practice

PART 1 Multiple Choice

Record your answers on the answer sheet provided by your teacher or on a sheet of paper.

1. Victoria bought 3 notebooks and 2 gel pens. Which expression represents her total cost if n is the cost of each notebook and p is the cost of each gel pen? (Lesson 1-4)

   A. $n + p$
   B. $5(n + p)$
   C. $5n \cdot 2p$
   D. $3n + 2p$

2. The table shows the weights of 13 dogs at a dog adoption center. Which measure of central tendency for these data is the least number? (Lesson 2-4)

Weight (lb)	Number of Dogs
20	III
40	II
60	IHH II
80	I

   F. median
   G. mean
   H. mode
   I. range

3. To make muffins, Desiree used $4\frac{1}{8}$ cups of flour, $4\frac{2}{3}$ cups of water, $4\frac{1}{4}$ cups of sugar, and $4\frac{1}{3}$ cups of milk. Desiree used the least amount of which ingredient? (Lesson 5-8)

   A. flour
   B. sugar
   C. water
   D. milk

4. Find $\frac{1}{7} \times \frac{2}{9}$. (Lesson 6-4)

   F. $\frac{1}{63}$
   G. $\frac{2}{63}$
   H. $\frac{3}{16}$
   I. $\frac{9}{14}$

5. For his job, Marc drives 15,000 miles every 60 days. What is the average number of miles that Marc drives every day? (Lesson 7-2)

   A. 250 mi
   B. 300 mi
   C. 900 mi
   D. 1,500 mi

6. Trevor drives 45 miles per hour. Robin drives 54 miles per hour. What is the percent of increase from 45 miles per hour to 54 miles per hour? (Lesson 8-4)

   F. 9%
   G. 20%
   H. 109%
   I. 120%

7. Chi can take 3 different routes and 4 different modes of transportation to get to school, as shown below.

Route	Transportation
scenic	bike
quick	car
convenient	bus
	walking

   How many possible choices are there for Chi to use to get to school? (Lesson 9-3)

   A. 7
   B. 9
   C. 12
   D. 16

8. Which is the square root of 441? (Lesson 11-1)

   F. 21
   G. 22
   H. 23
   I. 24

9. Which is a reasonable *estimate* for the square root of 66? (Lesson 11-2)

   A. 7.4
   B. 7.8
   C. 8.1
   D. 8.9

10. What is the value of x in the triangle? (Lesson 11-3)

    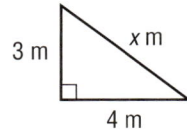

    F. 2
    G. 5
    H. 7
    I. 125

TEST-TAKING TIP

Question 8 You can use the answer choices to help you answer a multiple-choice question. For example, if you are asked to find the square root of a number, you can find the correct answer by calculating the square of the number in each answer choice.

508 Chapter 11 Geometry: Measuring Two-Dimensional Figures

PART 2 Short Response/Grid In

Record your answers on the answer sheet provided by your teacher or on a sheet of paper.

11. What value is represented by $(1 \times 10^5) + (2 \times 10^4) + (1 \times 10^3) + (1 \times 10^0)$? (Prerequisite Skill, page 555)

12. In the coordinate system, coordinates with a positive x value and a negative y value appear in what quadrant? (Lesson 3-3)

13. Write an equation to represent the following statement. (Lesson 4-1)

    The Palmas have 8 less than 2 times the number of trees in their yard as the Kandinskis have.

14. If you translate hexagon LMNOPQ 4 units to the right, what are the new coordinates of point Q? (Lesson 10-8)

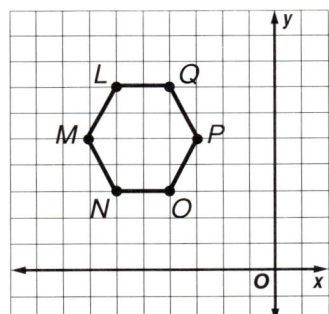

15. A window shade is being custom made to cover the triangular window shown at the right. What is the minimum area of the shade? (Lesson 11-5)

16. A kitchen chair has a circular seat that measures 14 inches across. What is the area of the seat on the kitchen chair? Use 3.14 for π and round to the nearest tenth. (Lesson 11-6)

17. The Connaught Centre building in Hong Kong has 1,748 circular windows. The diameter of each window is 2.4 meters. Find the total area of the glass in the windows. Round to the nearest tenth. (Lesson 11-6)

18. Find the area of the figure shown at the right. Use 3.14 for π. (Lesson 11-7)

19. Amy plays hopscotch by throwing a small stone onto a numbered triangle. What is the probability that Amy's stone will land on a triangle numbered 5? Write as a fraction. (Lesson 11-8)

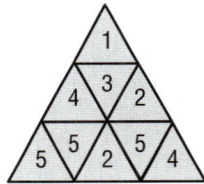

PART 3 Extended Response

Record your answers on a sheet of paper. Show your work.

20. Suppose you bought a new tent with the dimensions shown below.

   a. Is the area of the parallelogram-shaped side of the tent greater than or less than the area of the floor? Explain. (Lessons 11-3 and 11-4)

   b. The front and back triangular regions are covered with screens. What is the total area of the screens? (Lesson 11-5)

CHAPTER 12
Geometry: Measuring Three-Dimensional Figures

"What do paint cans have to do with math?"

Paint cans come in many different sizes, but they are all shaped like cylinders. To find the volume V of a paint can, you can use the formula $V = \pi r^2 h$, where r is the radius of the lid and h is the height of the can. A different formula can be used to find the surface area of a paint can.

You will solve problems involving volumes and surface areas of cylinders in Lessons 12-3 and 12-5.

GETTING STARTED

▶ Diagnose Readiness

Take this quiz to see if you are ready to begin Chapter 12. Refer to the lesson or page number in parentheses for review.

Vocabulary Review

Choose the correct term to complete each sentence.

1. The distance around a circle is called (perimeter, circumference). (Lesson 6-9)
2. The expression πr^2 is used to find the (circumference, area) of a circle. (Lesson 11-6)

Prerequisite Skills

Estimate each product. (Page 558)

3. $9 \cdot 10.4$
4. $6.25 \cdot 3.8$
5. $7.03 \cdot 5.3$

Evaluate each expression. Round to the nearest tenth if necessary. (Pages 559, 560)

6. $14.45 + 9.62$
7. $8.3 \cdot 6.4$
8. 36×5.2
9. $26.45 - 7.918$

Simplify. (Lesson 1-3)

10. $2 \cdot 6 + 4 \cdot 1 \cdot 3$
11. $1.2 \cdot 4 \cdot 4 + 3 \cdot 1.5$
12. $8 + 1 + 9 \cdot 2 + 5$
13. $7 + 3 \cdot 5 + 2 \cdot 6$

Multiply. (Lesson 6-4)

14. $4\frac{1}{2} \cdot 6$
15. $1\frac{1}{4} \cdot 5\frac{3}{5}$
16. $\frac{2}{7} \cdot 3\frac{3}{4}$

Find the area of each circle. Round to the nearest tenth. (Lesson 11-6)

17. diameter = 33 cm
18. radius = 3.8 yd
19. radius = 6 ft
20. diameter = 18 m

Surface Area and Volume Make this Foldable to help you organize information about solids. Begin with a piece of 11″ by 17″ paper.

STEP 1 **Fold** Fold the paper in fourths lengthwise.

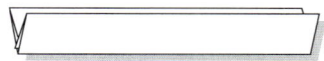

STEP 2 **Open and Fold** Fold a 2″ tab along the short side. Then fold the rest in half.

STEP 3 **Label** Draw lines along folds and label as shown.

Chapter Notes Each time you find this logo throughout the chapter, use your *Noteables™: Interactive Study Notebook with Foldables™* or your own notebook to take notes. Begin your chapter notes with this Foldable activity.

Readiness To prepare yourself for this chapter with another quiz, visit **msmath2.net/chapter_readiness**

12-1a HANDS-ON LAB
A Preview of Lesson 12-1

Building Three-Dimensional Figures

Cubes are examples of three-dimensional figures because they have length, width, and depth. In this lab, you will use centimeter cubes to build other three-dimensional figures.

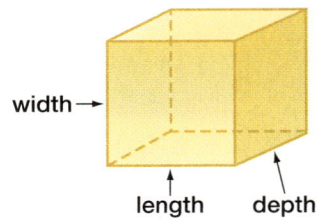

What You'll LEARN

Build three-dimensional figures given the top, side, and front views.

Materials
- centimeter cubes

STUDY TIP

Plane A *plane* is a flat surface that extends in all directions. Each face of a cube represents a different plane.

ACTIVITY Work with a partner.

The top view, side view, and front view of a three-dimensional figure are shown below. Use centimeter cubes to build the figure.

STEP 1 Use the top view to build the base of the figure. It is a 3-by-2 rectangle.

STEP 2 Use the side view to complete the figure. It is a 2-by-3 rectangle.

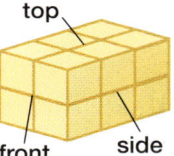

STEP 3 Use the front view to check the figure. It is a 2-by-2 square. So, the model is correct.

● **Your Turn** The top view, side view, and front view of each three-dimensional figure are shown. Use centimeter cubes to build the figure. Then make a sketch of the figure.

a. top   side   front

b. top   side   front

512 Chapter 12 Geometry: Measuring Three-Dimensional Figures

The top view, side view, and front view of each three-dimensional figure are shown. Use centimeter cubes to build the figure. Then make a sketch of the figure.

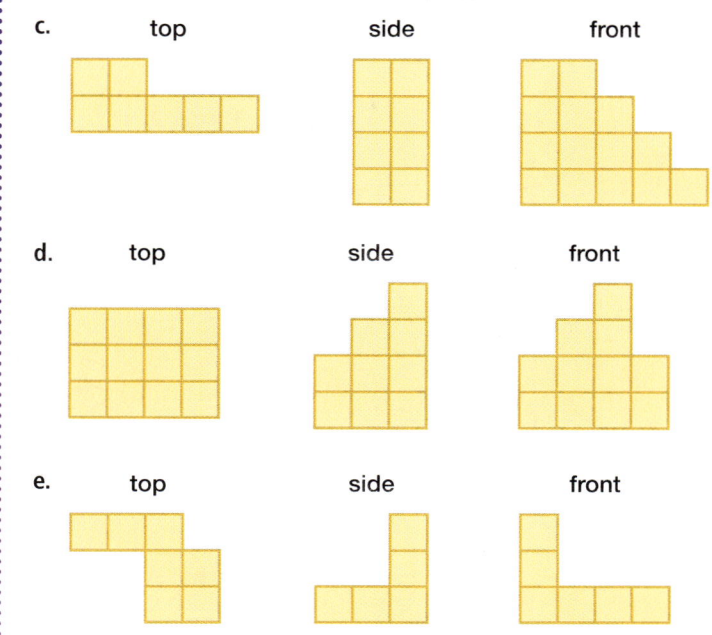

Writing Math

Work with a partner.

1. **Build** a model with cubes and draw the top, side, and front views. Give the drawing of the views to your partner and have him or her build the figure with cubes. Trade roles with your partner and repeat making the drawing and building the figure.

2. **Explain** how you began building the figures.

3. **Determine** whether there is more than one way to build each model. Explain your reasoning.

4. The figure at the right represents a building with a section that is 15 stories tall and another section that is 20 stories tall. Which view would you use to show the difference in height of each section?

5. **Build** two different models that would look the same from two views, but not the third view. Draw a top view, side view, and front view of each model.

6. **Describe** a real-life situation where it might be necessary to draw a top, side, and front view of a three-dimensional figure.

Lesson 12-1a Hands-On Lab: Building Three-Dimensional Figures **513**

12-1 Drawing Three-Dimensional Figures

What You'll LEARN

Draw a three-dimensional figure given the top, side, and front views.

NEW Vocabulary

solid

WHEN am I ever going to use this?

COMICS For Exercises 1 and 2, refer to the comic below.

1. Which view of the Washington Monument is shown in the comic?
2. Find a photograph of the Washington Monument and draw a side view.

Study Tip

Plane Figures In geometry, three-dimensional figures are *solids* and two-dimensional figures such as triangles, circles, and squares are *plane figures*.

A **solid** is a three-dimensional figure because it has length, width, and depth. You can draw different views of solids.

EXAMPLE Draw Different Views of a Solid

1. Draw a top, a side, and a front view of the figure at the right.

   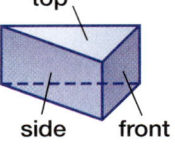

   The top view is a triangle.
   The side and front views are rectangles.

Your Turn Draw a top, a side, and a front view of each solid.

a.   b.

514 Chapter 12 Geometry: Measuring Three-Dimensional Figures

EXAMPLE Draw a Three-Dimensional Figure

2 Draw the solid using the top, side, and front views shown at the right.

top   side   front
 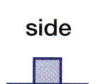

Step 1 Use the top view to draw the base of the figure, a 1-by-3 rectangle.

Step 2 Add edges to make the base a solid figure.

Step 3 Use the side and front views to complete the figure.

STUDY TIP

Paper Use isometric dot paper for the drawings in this lesson, as shown at the right. This is different from rectangular dot paper.

Your Turn Draw each solid using the top, side, and front views shown. Use isometric dot paper.

c.   top   side   front

d.   top   side   front

Skill and Concept Check

1. **OPEN ENDED** Draw the top, side, and front view of a solid in your home.

2. **Which One Doesn't Belong?** Identify the figure that does not have the same characteristic as the other three. Explain your reasoning.

GUIDED PRACTICE

Draw a top, a side, and a front view of each solid.

3.

4.

Draw each solid using the top, side, and front views shown. Use isometric dot paper.

5.   top   side   front

6.   top   side   front

Practice and Applications

HOMEWORK HELP

For Exercises	See Examples
7–12, 18	1
13–17, 19, 20	2

Extra Practice
See pages 593, 607.

Draw a top, a side, and a front view of each solid.

7.

8.

9.

10.

11.

12.

Draw each solid using the top, side, and front views shown. Use isometric dot paper.

13. top   side   front

14. top   side   front

15. top   side   front

16. top   side   front

HISTORY For Exercises 17–20, use the information below.

The Lighthouse of Alexandria was the last of the seven Wonders of the Ancient World that disappeared. It contained three different-shaped levels.

top   side   front

17. The bottom level is the largest. Make a drawing of this level using the top, side, and front views shown at the right.

18. The middle level is a prism with a base that is a regular octagon. If the height and width of this level is approximately one third the height and width of the bottom level, draw the top, side, and front views.

19. The top level has the views shown at the right. What kind of solid is this?

top   side   front

20. Make a sketch of the Lighthouse of Alexandria that shows all three levels.

Data Update How does your drawing compare to the actual Lighthouse of Alexandria? Visit msmath2.net/data_update to learn more.

516 Chapter 12 Geometry: Measuring Three-Dimensional Figures

21. **RESEARCH** Use the Internet or another source to find a photograph of the only Wonder of the Ancient World existing today, the Great Pyramid of Giza. Draw a top view, a side view, and a front view of the pyramid.

22. **ARCHITECTURE** When a building is being designed, an architect provides a set of elevation drawings. These drawings show how the building appears from each side. Draw a set of elevation drawings for your home or school.

23. **CRITICAL THINKING** Draw a three-dimensional figure in which the front and top views each have a line of symmetry but the side view does not. (*Hint*: Refer to Lesson 10-9 to review lines of symmetry.)

Spiral Review with Standardized Test Practice

24. **MULTIPLE CHOICE** Which is the top view of the cylinder at the right?

    Ⓐ triangle Ⓑ rectangle
    Ⓒ square   Ⓓ circle

25. **MULTIPLE CHOICE** Which three-dimensional figure has the top, side, and front views shown at the right?

    Ⓕ   Ⓖ

    Ⓗ   Ⓘ

26. Find the probability that a randomly dropped counter will fall in the shaded region. Write as a percent. (Lesson 11-8)

    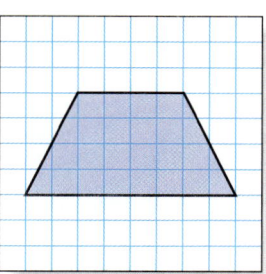

Find the area of each figure. Round to the nearest tenth if necessary.
(Lesson 11-7)

27.   28.   29.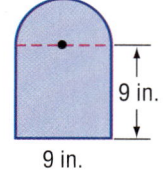

GETTING READY FOR THE NEXT LESSON

PREREQUISITE SKILL Multiply. (Lesson 6-4)

30. $7\frac{1}{2} \cdot 6$ 31. $8 \cdot 2\frac{3}{4}$ 32. $\frac{5}{6} \cdot 1\frac{4}{5}$ 33. $10\frac{1}{5} \cdot 6\frac{2}{3}$

Problem-Solving Strategy
A Follow-Up of Lesson 12-1

Make a Model

What You'll LEARN

Solve problems by making a model.

I'm going to help my mom make 3-inch soft alphabet blocks for the children at her daycare center. About how much fabric do we need for one cube if there is a $\frac{1}{2}$-inch seam on each side?

*We could **make a model** to find out.*

Explore	We know that each cube is 3 inches long with $\frac{1}{2}$-inch seams.
Plan	We can make a cardboard model of a cube with sides 3 inches long. We could then cut the model into six squares and add $\frac{1}{2}$-inch paper extensions to each side as seams.
Solve	Make the cardboard model and unfold the cube. Cut the cube into six squares and tape $\frac{1}{2}$-inch seams on each side. Now each pattern piece is 4 inches by 4 inches or *about* 16 square inches. 16 in² × 6 squares = 96 in² So, each cube needs about 96 square inches of fabric.
Examine	If the 4-inch-by-4-inch pattern pieces were laid as two columns and three rows, it would be 8 inches wide and 12 inches long or 96 square inches.

Analyze the Strategy

1. **Explain** when making a model is a better strategy than drawing a picture.
2. **Explain** why you think the students started with the three-dimensional model to make their pattern.
3. **Write** a problem that can be solved by making a model. Then solve the problem.

518 Chapter 12 Geometry: Measuring Three-Dimensional Figures

Apply the Strategy

Solve. Make a model.

4. **ART** Dominic is creating a layout of his bedroom for art class. The room measures 15 foot by 12 feet. If he uses a scale of 1 foot = $\frac{3}{4}$ inch, what are the dimensions of his bedroom on the model?

5. **BICYCLES** Eight customers lined up outside The Bike Shop with either a bicycle or a tricycle that needed repair. When the owner looked out the window, she counted 21 wheels outside the shop. How many tricycles and bicycles were there?

Mixed Problem Solving

Solve. Use any strategy.

6. **COMMUNITY SERVICE** There are four drop-off centers for the community food drive. Their total collections are shown in the table.

Center	Number of Cans
A	3,298
B	2,629
C	4,429
D	2,892

   A newsletter reported that over 13,000 cans of food were collected. Is this estimate reasonable? Explain.

7. **SWIMMING** Yeti can swim one 20-meter lap in 1.25 minutes. How long will it take her to swim 100 meters at the same rate?

8. **TRAFFIC** At the four-way intersection shown below, the traffic lights change every 90 seconds. About 8 cars in one lane travel through the light in this amount of time. Determine the number of cars that travel through the intersection in 3 minutes.

9. **MONEY** A top film actor made 16 films, which grossed over 5.09×10^9. On average, how much did each film make? Write in scientific notation. Round to the nearest tenth.

10. **DISPLAYS** Identical boxes are stacked in the corner of a store as shown below. How many boxes are *not* visible?

    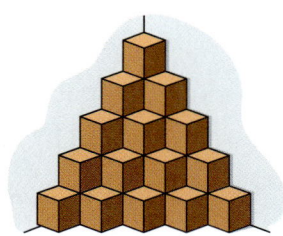

11. **REMODELING** How many square feet of wallpaper are needed to cover a wall that measures 9 feet by 16 feet?

12. **STANDARDIZED TEST PRACTICE**
    Yolanda deposited $450 in a new savings account in January, withdrew $175 in February, and then began monthly deposits of $75 from March through December. Which equation shows her balance b after her last deposit in December?

    Ⓐ $b = \$275 + 9(\$75)$
    Ⓑ $b = \$275 + 10(\$75)$
    Ⓒ $b = \$450 + 10(\$75)$
    Ⓓ $b = \$275 - 10(\$75)$

12-2 Volume of Rectangular Prisms

What You'll LEARN

Find the volumes of rectangular prisms.

NEW Vocabulary

volume
rectangular prism

Link to READING

Everyday Meaning of Volume: bulk or mass, as in shipping a large volume of merchandise

HANDS-ON Mini Lab

Materials
- centimeter grid paper
- scissors
- tape
- centimeter cubes

Work with a partner.

- On a piece of grid paper, cut out a square that is 10 centimeters on each side.
- Cut a 1-centimeter square from each corner. Fold the paper and tape the corners together to make a box.

1. What is the area of the base, or bottom, of the box? What is the height of the box?
2. How many centimeter cubes fit in the box?
3. What do you notice about the product of the base area and the height of the box?

The **volume** of a solid is the measure of space occupied by it. It is measured in cubic units such as cubic centimeters (cm^3) or cubic inches (in^3). The volume of the figure at the right can be shown using cubes.

The bottom layer, or base, has 6 · 6 or 36 cubes.

There are two layers.

It takes 36 · 2 or 72 cubes to fill the box. So, the volume of the box is 72 cubic centimeters.

A **rectangular prism** is a solid figure that has two parallel and congruent sides, or bases, that are rectangles.

Noteables™   Key Concept: Volume of a Rectangular Prism

Words The volume V of a rectangular prism is the area of the base B times the height h. It is also the product of the length ℓ, the width w, and the height h.

Model
$B = \ell w$

Symbols $V = Bh$ or $V = \ell wh$

READING in the Content Area

For strategies in reading this lesson, visit **msmath2.net/reading**.

520 Chapter 12 Geometry: Measuring Three-Dimensional Figures

EXAMPLES Find the Volumes of Prisms

1 Find the volume of the rectangular prism.

$V = \ell wh$ Volume of a rectangular prism

$V = 5 \cdot 4 \cdot 3$ Replace ℓ with 5, w with 4, and h with 3.

$V = 60$ Multiply.

The volume is 60 cubic centimeters.

STUDY TIP

Estimation It is helpful to first estimate the volume of a rectangular prism to determine whether your answer is reasonable.

2 **CEREAL** Find the volume of the cereal box.

Estimate $V \approx 8 \cdot 3 \cdot 10$ or 240

$V = \ell wh$ Volume of a rectangular prism

$V = 7.5 \cdot 2.5 \cdot 12$ Replace ℓ with 7.5, w with 2.5, and h with 12.

$V = 225$ Multiply.

The volume is 225 cubic inches.
Compare to the estimate.

Skill and Concept Check

1. **Write** the abbreviation for cubic yards using an exponent.

2. **OPEN ENDED** Draw and label a rectangular prism that has a volume between 400 and 600 cubic centimeters. State the volume of the prism.

3. **FIND THE ERROR** Cassandra and Ling are comparing the volumes of the two prisms at the right. Who is correct? Explain.

Cassandra
Prism B has a greater volume because it is twice as tall as Prism A.

Ling
Both prisms have the same volume.

GUIDED PRACTICE

Find the volume of each rectangular prism.

4.

5.

6.

7. **STORAGE** A cabinet measures 20 inches by 30 inches by 60 inches. What is its volume?

msmath2.net/extra_examples

Practice and Applications

Find the volume of each rectangular prism. Round to the nearest tenth if necessary.

8.
9.
10.
11.
12.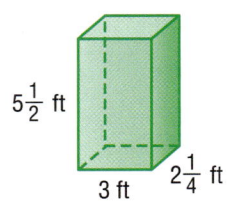

HOMEWORK HELP
For Exercises	See Examples
8–15	1, 2

Extra Practice See pages 593, 607.

13. A cube has 3-centimeter edges. Find its volume.

14. The base of a rectangular prism has an area of 19.4 square meters. Find the height if the volume is 306.52 cubic meters.

15. **TRANSPORTATION** Find the approximate volume of the California trolley car shown at the right.

16. **MEASUREMENT** How many cubic inches are in a cubic foot?

17. **MEASUREMENT** How many cubic centimeters are in a cubic meter?

18. **CRITICAL THINKING** A *triangular prism* is a prism that has bases that are triangles. Use $V = Bh$ to find the volume of the triangular prism at the right.

Spiral Review with Standardized Test Practice

19. **MULTIPLE CHOICE** An office is 20 feet long, 15 feet wide, and 12 feet high. It costs about 9¢ per year to air condition one cubic foot of space. On average, how much does it cost to air condition this office for one month?
    - Ⓐ $324
    - Ⓑ $300
    - Ⓒ $27
    - Ⓓ $2.25

20. **SHORT RESPONSE** A landscaper wants to cover a 40-foot-by-12-foot rectangular area with small stones. If she uses 120 cubic feet of stones, how deep will they be?

21. Draw a top, a side, and a front view of the solid at the right.
    (Lesson 12-1)

22. Draw an area model so that the probability of a randomly dropped counter falling in a triangular region is 25%. (Lesson 11-8)

GETTING READY FOR THE NEXT LESSON

PREREQUISITE SKILL Estimate. (Page 558)

23. $3.14 \cdot 6$
24. $5 \cdot 2.7^2$
25. $9.1 \cdot 8.3$
26. $3.1 \cdot 1.75^2 \cdot 2$

522 Chapter 12 Measuring Three-Dimensional Figures

12-2b Spreadsheet Investigation
A Follow-Up of Lesson 12-2

What You'll LEARN
Use a spreadsheet to investigate the volumes of similar solids.

Similar Solids

A computer spreadsheet can help you calculate the sizes and volumes of similar rectangular prisms. You can enlarge or reduce dimensions of the prisms and the spreadsheet will automatically calculate the new volumes.

ACTIVITY

Mrs. Sanchez owns a box-making factory. She wishes to produce a set of similar boxes so that they will nest inside each other when assembled. She will need to show her customers the dimensions and volume of each type of box she plans to sell. This information can be placed into a spreadsheet.

Set up the spreadsheet like the one shown below.

	A	B	C	D	E
1		Dimensions (in.)			
2	Box	L	W	H	Volume (in^3)
3	small	2.5	2.5	2.5	
4	medium	5	5	5	
5	large	7.5	7.5	7.5	
6					

EXERCISES

1. State the spreadsheet commands you used to find the volumes of the three boxes. Then find the volumes to the nearest tenth.

2. Expand the spreadsheet to calculate the volumes of the boxes when each of the dimensions is doubled. What happens to the volume of the small box when all of the dimensions of the box are doubled?

3. What happens to the volume of the small box when all of the dimensions of the box are tripled?

4. Suppose another box has a volume 216 times greater than the small box. What are the dimensions of this box? Use the spreadsheet to check your answer.

5. Extend the pattern in the volume column. Then express the pattern in exponential form.

Lesson 12-2b Spreadsheet Investigation: Similar Solids 523

12-3 Volume of Cylinders

HANDS-ON Mini Lab

Materials
- soup can
- centimeter grid paper
- scissors
- tape

What You'll LEARN

Find the volumes of cylinders.

NEW Vocabulary

cylinder

Work with a partner.

Set a soup can on a piece of grid paper and trace around the base, as shown at the right.

1. Estimate the number of centimeter cubes that would fit at the bottom of the can. Include parts of cubes.

2. How many layers would it take to fill the cylinder?

3. **Make a conjecture** about how you could find the volume of the soup can.

A **cylinder** is a solid figure that has two congruent, parallel circles as its bases. As with prisms, the area of the base tells the number of cubic units in one layer. The height tells how many layers there are.

Noteables™ Key Concept: Volume of a Cylinder

Words The volume V of a cylinder with radius r is the area of the base B times the height h.

Symbols $V = Bh$ or $V = \pi r^2 h$, where $B = \pi r^2$

Model
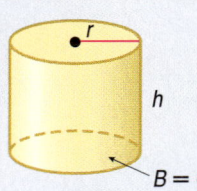
$B = \pi r^2$

EXAMPLE Find the Volume of a Cylinder

1 Find the volume of the cylinder. Round to the nearest tenth.

$V = \pi r^2 h$ Volume of a cylinder

$V = \pi (5)^2 (8.3)$ Replace r with 5 and h with 8.3.

  5 × 8.3 651.8804756

The volume is about 651.9 cubic centimeters.   **Check by using estimation.**

524 Chapter 12 Geometry: Measuring Three-Dimensional Figures

Standardized Test Practice

> **EXAMPLE** Find the Volume of a Real-Life Object
>
> **2 MULTIPLE-CHOICE TEST ITEM** Find the volume of a cylinder-shaped paint can that has a diameter of 4 inches and a height of 5 inches.
>
> (A) 20.0 in³   (B) 62.8 in³   (C) 125.7 in³   (D) 251.3 in³
>
> **Read the Test Item**
> To find the volume, use the formula $V = \pi r^2 h$.
>
> **Solve the Test Item**
> $V = \pi r^2 h$ Volume of a cylinder
> $V = \pi(2)^2(5)$ Replace r with 2 and h with 5.
> $V \approx 62.8$ Simplify.
>
> The answer is B.

Test-Taking Tip

Reading
The radius of the can is 4 ÷ 2 or 2 inches. Be sure to substitute that value into the formula. If you substitute 4 for r into the formula, you would get the incorrect answer D.

Skill and Concept Check

1. **OPEN ENDED** Draw and label a cylinder that has a larger radius, but less volume than the one shown at the right.

2. **Writing Math** Explain how the formula for the volume of a cylinder is similar to the formula for the volume of a rectangular prism.

GUIDED PRACTICE

Find the volume of each cylinder. Round to the nearest tenth.

3.
4.
5.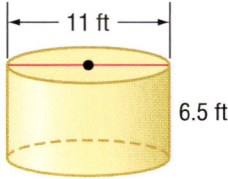

6. radius = 3 in.
   height = $5\frac{1}{2}$ in.

7. diameter = 12.4 m
   height = 2 m

8. **BAKING** Which will hold more cake batter, the rectangular pan, or the two round pans? Explain.

Lesson 12-3 Volume of Cylinders **525**

Practice and Applications

Find the volume of each cylinder. Round to the nearest tenth.

9.   4 in., 8 in.

10.   9 ft, 16 ft

11.   13.3 cm, 2 cm

12. $3\frac{1}{2}$, $7\frac{1}{2}$ in.

13.   1 m, 1.7 m

14. 15 mm, 4.8 mm

15. diameter = 21 mm
    height = 8 mm

16. diameter = 4.5 m
    height = 6.5 m

17. radius = 6 ft
    height = $5\frac{1}{3}$ ft

18. radius = 12 ft
    height = 11 yd

19. a soup can with a radius of 4 centimeters and a height of 12 centimeters

20. a hockey puck with a diameter of 3 inches and a height of 1 inch

21. a can of potato chips with a radius of $1\frac{1}{2}$ inches and a height of 8 inches

22. **SPACE SCIENCE** The Hubble Space Telescope is cylinder-shaped, approximately the size of a school bus, as shown at the right. What is the volume to the nearest tenth?

23. **WRITE A PROBLEM** Write a problem in which you find the volume of a cylinder.

24. **NUMBER SENSE** What is the ratio of the volume of a cylinder to the volume of a cylinder having twice the height but the same radius?

25. **NUMBER SENSE** What is the ratio of the volume of a cylinder to the volume of a cylinder having the same height but twice the radius?

26. **CONTAINERS** The two cans at the right have the same volume. Find h.

27. **MEASUREMENT** Firewood is usually sold by a unit of measure called a *cord*. A cord is a stack of wood that is 8 feet long, 4 feet wide, and 4 feet high. Suppose a tree has a diameter of 2 feet. Find the height of the tree trunk that would produce about 1 cord of firewood.

HOMEWORK HELP

For Exercises	See Examples
9–18	1
19–22, 26–27	2

Extra Practice
See pages 594, 607.

28. **CRITICAL THINKING** Two equal-sized sheets of paper are rolled along the length and along the width, as shown at the right. Which cylinder do you think has the greater volume? Explain.

Spiral Review with Standardized Test Practice

29. **MULTIPLE CHOICE** Which statement is true about the volumes of cylinders 1 and 2 shown below?

Cylinder 1

Cylinder 2

- Ⓐ The volume of cylinder 1 is greater.
- Ⓑ The volume of cylinder 2 is greater.
- Ⓒ The volumes are equal.
- Ⓓ cannot tell from the diagrams

30. **MULTIPLE CHOICE** A cylinder-shaped popcorn tin has a height of 1.5 feet and a diameter of 10 inches. Find the volume to the nearest cubic inch. Use 3.14 for π.

- Ⓕ 118 in^3
- Ⓖ 565 in^3
- Ⓗ 1,413 in^3
- Ⓘ 5,652 in^3

31. Find the volume of a rectangular prism with a length of 6 meters, a width of 4.9 meters, and a height of 5.2 meters. (Lesson 12-2)

Draw each solid using the top, side, and front views shown. Use isometric dot paper. (Lesson 12-1)

32. top side front

33. top side front

Find the area of each trapezoid. Round to the nearest tenth if necessary. (Lesson 11-5)

34.
35.
36.

GETTING READY FOR THE NEXT LESSON

PREREQUISITE SKILL Simplify. (Lesson 1-3)

37. $3 \cdot 5 \cdot 8 + 2 \cdot 9 \cdot 3$

38. $7.6 \cdot 11 \cdot 2 + 2 \cdot 7.6 \cdot 3$

39. $2 \cdot 7 \cdot 1.5 + 2 \cdot 4 \cdot 1.5 + 1.5 \cdot 7$

40. $2 \cdot 2\frac{1}{4} \cdot 6 + 2 \cdot 9 \cdot 6 + 2 \cdot 2\frac{1}{4} \cdot 9$

Chapter 12 Mid-Chapter Practice Test

Vocabulary and Concepts

1. Define *volume*. (Lesson 12-2)
2. State, in words, the formula for the volume of a cylinder. (Lesson 12-3)

Skills and Applications

Draw a top, a side, and a front view of each solid. (Lesson 12-1)

3.
4.
5.

6. Draw the solid using the top, side, and front views shown at the right. Use isometric dot paper. (Lesson 12-1)

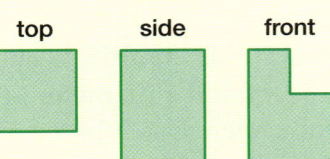

7. **GIFTS** A jewelry box measures 7 centimeters by 12 centimeters by 14 centimeters. What is its volume? (Lesson 12-2)

8. **TRUCKS** How tall is the trailer of a truck if it is 9 meters long, 7.2 meters wide and has a volume of 226.8 cubic meters? (Lesson 12-2)

9. **GARDENING** Jocelyn is buying potting soil to fill the window box. If one bag of potting soil contains 576 cubic inches, how many bags should she buy? (Lesson 12-2)

Find the volume of each cylinder. Round to the nearest tenth. (Lesson 12-3)

10. diameter = 25 ft
    height = 24 ft

11. radius = 2.2 cm
    height = 5 cm

12. radius = $1\frac{1}{2}$ in.
    height = $4\frac{1}{2}$ in.

Standardized Test Practice

13. **GRID IN** A brick is 4 inches wide, 8 inches long, and 3 inches tall. What is the volume in cubic inches of a stack of 25 bricks? (Lesson 12-2)

14. **SHORT RESPONSE** A circular pond has a radius of 12 meters and it is 2.5 meters deep. What is its volume? Use 3.14 for π. (Lesson 12-3)

The Game Zone

A Place To Practice Your Math Skills

Math Skill: 3-Dimensional Figures

Shape-Tac-Toe

GET READY!
Players: two
Materials: 1 index card, 2 large cubes, 10 two-color counters

GET SET!
One player draws a game board on the index card like the one shown. The second player labels the six faces of each cube with the following words.

First Cube	**Second Cube**
prism	circular base
pyramid	triangular base
cylinder	square base
cone	2 parallel bases
flat surfaces	1 base
curved surface	congruent faces

GO!
- The first player rolls both cubes and places a counter on any one shape that matches the conditions on the cubes. If it is impossible to cover a shape, the player loses his or her turn.
- Players alternate turns.
- **Who Wins?** The first player to cover three shapes in a row is the winner.

The Game Zone: Three-Dimensional Figures

12-4a Hands-On Lab

A Preview of Lesson 12-4

Nets and Surface Area

What You'll LEARN

Use nets to find the surface area of rectangular prisms.

Materials
• rectangular dot paper
• scissors

Suppose you cut a cardboard box along its edges, open it up, and lay it flat.

The result is a two-dimensional figure called a **net**. Nets can help you see the regions or faces that make up the surface of the figure.

ACTIVITY *Work with a partner.*

Make a net of the rectangular prism shown at the right.

5 units   4 units   3 units

STEP 1 Begin by drawing the base of the prism. On the dot paper, draw a rectangle that is 5 units long and 4 units wide.

STEP 2 Visualize unfolding the prism along its edges. Draw the rectangles that represent the front, back, and sides of the prism.

STEP 3 Finally, draw the top of the prism. *This is only one of several possible nets that you could draw.*

530 Chapter 12 Geometry: Measuring Three-Dimensional Figures

Your Turn

Draw a net for each figure. Find the area of the net. Then cut out the net, fold it, and tape it together to form a three-dimensional figure.

a.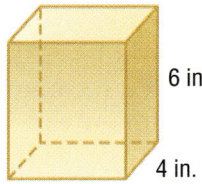
6 in.
4 in.
5 in.

b.
8 ft
3 ft
2 ft

c.
4.5 cm
4.5 cm
12 cm

Writing Math

Work in groups of three.

1. The net shown on page 530 is made of rectangles. How many rectangles are in the net?

2. Explain how you can find the total area of the rectangles.

3. The *surface area* of a prism is the total area of its net. Write an equation that shows how to find the surface area of the prism at the right using the length ℓ, width w, and height h.

4. Find the surface areas of cubes whose edges are 1 unit, 2 units, and 3 units and graph the ordered pairs (side length, surface area) on a coordinate plane. Describe the graph.

5. **Describe** what happens to the surface area of a cube as its dimensions are doubled? tripled?

6. **Describe** how you would find the surface area of a square-based pyramid.

Draw a net for each figure.

7.
tetrahedron

8.
square-based pyramid

9. **Explain** how the formula for the surface area of a tetrahedron differs from the formula for the surface area of a square-based pyramid.

Lesson 12-4a Hands-On Lab: Nets and Surface Area **531**

12-4 Surface Area of Rectangular Prisms

What You'll LEARN

Find the surface areas of rectangular prisms.

NEW Vocabulary

surface area

HANDS-ON Mini Lab

Materials
- 8 centimeter cubes

Work with a partner.

- Use the cubes to build a rectangular prism with a length of 8 centimeters.
- Count the number of squares on each face. The sum is the *surface area*.

1. Record the dimensions, volume, and surface area in a table.
2. Build two more prisms using all of the cubes. For each, record the dimensions, volume, and surface area.
3. Describe the rectangular prisms with the greatest and least surface areas.

The sum of the areas of all of the surfaces, or faces, of a three-dimensional figure is the **surface area**.

EXAMPLE Use a Net to Find Surface Area

1 Find the surface area of the rectangular prism.

You can use a net of the rectangular prism to find its surface area. There are three pairs of congruent faces.

- top and bottom
- front and back
- two sides

Faces	Area
top and bottom	$2(5 \cdot 4) = 40$
front and back	$2(5 \cdot 3) = 30$
two sides	$2(3 \cdot 4) = 24$
Sum of the areas	$40 + 30 + 24 = 94$

The surface area is 94 square centimeters.

Noteables™ Key Concept: Surface Area of a Rectangular Prism

Words The surface area S of a rectangular prism with length ℓ, width w, and height h is the sum of the areas of the faces.

Model

Symbols $S = 2\ell w + 2\ell h + 2wh$

EXAMPLE Use a Formula to Find Surface Area

2. Find the surface area of the rectangular prism.

Replace ℓ with 9, w with 7, and h with 13.

surface area $= 2\ell w + 2\ell h + 2wh$
$= 2 \cdot 9 \cdot 7 + 2 \cdot 9 \cdot 13 + 2 \cdot 7 \cdot 13$
$= 126 + 234 + 182$ Multiply first. Then add.
$= 542$

The surface area of the prism is 542 square inches.

REAL-LIFE MATH

GIFTS Gifts have been wrapped since the invention of paper in 105 A.D. in China.
Source: www.hallmark.com

EXAMPLE Use Surface Area to Solve a Problem

3. **GIFTS** Mario is wrapping a package 8 inches long, 2 inches wide, and 11 inches high. He bought a roll of wrapping paper that is 1 foot wide and 2 feet long. Did he buy enough to wrap the package?

Step 1 Find the surface area of the package.

Replace ℓ with 8, w with 2, and h with 11.
surface area $= 2 \cdot 8 \cdot 2 + 2 \cdot 8 \cdot 11 + 2 \cdot 2 \cdot 11$
$= 252$ square inches

Step 2 Find the area of the wrapping paper.

area $= 12$ in. $\cdot 24$ in. or 288 in^2

Since 288 > 252, Mario bought enough wrapping paper.

Skill and Concept Check

1. **Explain** how to find the surface area of a rectangular prism.
2. **OPEN ENDED** Draw a rectangular prism and its net with the dimensions labeled on both figures. Then find the surface area.
3. **Writing Math** Explain why surface area is measured in square units even though the figure is three-dimensional.

GUIDED PRACTICE

Find the surface area of each rectangular prism. Round to the nearest tenth if necessary.

4.

5.

6. Find the surface area of a rectangular prism that has a length of 7 inches, a width of 11 inches, and a height of 9 inches.

Practice and Applications

Find the surface area of each rectangular prism. Round to the nearest tenth if necessary.

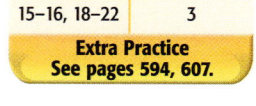

HOMEWORK HELP

For Exercises	See Examples
7–14	1, 2
15–16, 18–22	3

Extra Practice
See pages 594, 607.

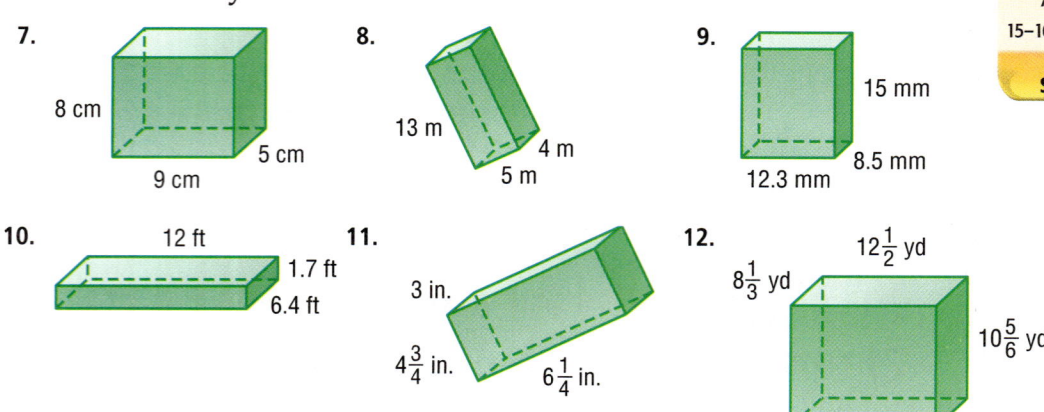

7. 8 cm, 9 cm, 5 cm
8. 13 m, 5 m, 4 m
9. 15 mm, 12.3 mm, 8.5 mm
10. 12 ft, 6.4 ft, 1.7 ft
11. 3 in., $4\frac{3}{4}$ in., $6\frac{1}{4}$ in.
12. $12\frac{1}{2}$ yd, $8\frac{1}{3}$ yd, $10\frac{5}{6}$ yd

13. A cube has a surface area of 42 square meters. What is the area of one face?

14. Find the surface area of a rectangular prism that has a length of 5 feet, a width of 3 feet, and a height of 9 inches.

BOXES For Exercises 15 and 16, use the information below.
The largest corrugated cardboard box ever constructed was in Helmond, Netherlands. It measured 22.9 feet long, 8.5 feet high, and 7.87 feet wide.

15. Find the volume of the cardboard box to the nearest tenth.
16. Find the surface area of the cardboard box to the nearest tenth.

17. **ALGEBRA** Write a formula for the surface area of a cube in which each side measures x units.

PHYSICAL SCIENCE For Exercises 18–20, use the information below.
Granulated sugar dissolves faster in water than a sugar cube.

18. Suppose the length of each edge of a sugar cube is 1 centimeter. Find the surface area of the cube.

19. Imagine cutting the cube once in half horizontally and twice vertically. Find the total surface area of the eight cubes.

20. Make a conjecture as to why granulated sugar dissolves faster than a sugar cube.

CEREAL For the Exercises 21 and 22, use the following information.
Suppose you are designing a trial size cereal box that holds 100 cubic centimeters of cereal.

21. **MULTI STEP** Find the whole number dimensions of the box that would use the least amount of cardboard.

22. **MULTI STEP** If cardboard costs $0.05 per 100 square centimeters, how much would it cost to make 100 boxes?

534 Chapter 12 Geometry: Measuring Three-Dimensional Figures

23. **WRITE A PROBLEM** Write about a real-life situation that involves finding the surface area of a rectangular prism. Then solve the problem.

24. **CRITICAL THINKING** A model is made by placing a cube with 12-centimeter sides on top of another cube with 15-centimeter sides. Find the surface area. (*Hint*: Do not include the area where the smaller cube covers the larger cube.)

Spiral Review with Standardized Test Practice

25. **MULTIPLE CHOICE** Find the surface area of the rectangular prism to the nearest tenth.

    Ⓐ 101.5 cm²   Ⓑ 165.0 cm²
    Ⓒ 198.6 cm²   Ⓓ 203.0 cm²

26. **MULTIPLE CHOICE** Find the amount of glass used for an aquarium that is 2.5 feet long, 1.6 feet wide, and 2 feet tall. (*Hint*: The top of the aquarium is open.)

    Ⓕ 19.4 ft²   Ⓖ 20.4 ft²   Ⓗ 21.2 ft²   Ⓘ 24.4 ft²

27. **GRID IN** The surface area of a cube is 294 square millimeters. What is the length of one side in millimeters?

Find the volume of each cylinder. Round to the nearest tenth. (Lesson 12-3)

28.

29.

30.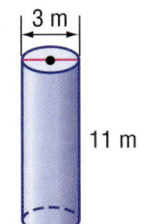

31. A rectangular prism is 14 inches long, 4.5 inches wide, and 1 inch high. Find its volume. (Lesson 12-2)

32. **CARPENTRY** The deck on a house is $25\frac{3}{4}$ feet long and $12\frac{1}{2}$ feet wide. The longer side of the deck is against the house. How many feet of wood does Jack need to buy to build a railing around the deck? (Lesson 6-6)

GETTING READY FOR THE NEXT LESSON

PREREQUISITE SKILL Find the area of each circle. Round to the nearest tenth. (Lesson 11-6)

33.

34.

35. diameter = 13.6 yd   36. radius = 23 km

12-4b Hands-On Lab

A Follow-Up of Lesson 12-4

Changes in Volume and Surface Area

What You'll LEARN

Investigate changes in volume and surface area.

Materials
- isometric dot paper

Suppose you have a model of a rectangular prism and you are asked to create a similar model whose dimensions are twice as large. In this lab, you will investigate how changing the dimensions of a three-dimensional figure affects the surface area and volume.

ACTIVITY Work with a partner.

STEP 1 Draw a cube on dot paper that measures 1 unit on each side. Calculate the volume and the surface area of the cube. Then record the data in a table like the one shown below.

1 unit

STEP 2 Double the side lengths of the cube. Calculate the volume and the surface area of this cube. Record the data in your table.

2 units

STEP 3 Triple the side lengths of the original cube. Now each side measures 3 units long. Calculate the volume and the surface area of the cube and record the data.

3 units

STEP 4 For each cube, write a ratio comparing the side length and the volume. Then write a ratio comparing the side length and the surface area. The first one is done for you.

Side Length (units)	Volume (units³)	Surface Area (units²)	Ratio of Side Length to Volume	Ratio of Side Length to Surface Area
1	$1^3 = 1$	$6(1^2) = 6$	1:1	1:6
2				
3				
4				
5				
s				

Your Turn

a. Complete the table above.

536 Chapter 12 Geometry: Measuring Three-Dimensional Figures

ACTIVITY *Work with a partner.*

2

STEP 1 Draw a cube on dot paper that measures 8 units on each side. Calculate the volume and the surface area of the cube. Record the data in a table like the one shown below.

STEP 2 Halve the side lengths of the cube in Step 1. Calculate the volume and the surface area of this cube and record the data.

STEP 3 Halve the side lengths of the cube in Step 2. Calculate the volume and the surface area of the cube and record the data.

STEP 4 For each cube, write a ratio comparing the side length and the volume and a ratio comparing the side length and the surface area. The first one is done for you.

Side Length (units)	Volume (units³)	Surface Area (units²)	Ratio of Side Length to Volume	Ratio of Side Length to Surface Area
8	$8^3 = 512$	$6(8^2) = 384$	8 : 512	8 : 384
4				
2				
s				

Your Turn

b. Complete the table above.

Writing Math

Work with a partner.

1. **Write** a formula for the volume V of a cube with side length s.
2. **Write** a formula for the surface area A of a cube with side length s.

Complete each sentence.

3. If the side length of a cube is doubled, the volume is __?__ times greater.
4. If the side length of a cube is doubled, the surface area is __?__ times greater.
5. If the side length of a cube is tripled, the volume increases by __?__ times and the surface area increases by __?__ times.
6. If the side length of a cube decreases by $\frac{1}{2}$, the surface area decreases by __?__.

Lesson 12-4b Hands-On Lab: Changes in Volume and Surface Area

12-5 Surface Area of Cylinders

What You'll LEARN
Find the surface areas of cylinders.

REVIEW Vocabulary

circumference: the distance around a circle (Lesson 6-9)

HANDS-ON Mini Lab

Materials
- soup can
- grid paper
- scissors

Work with a partner.

- Trace the top and bottom of the can on grid paper. Then cut out the shapes.
- Cut a long rectangle from the grid paper. The width of the rectangle should be the same as the height of the can. Wrap the rectangle around the side of the can. Cut off the excess paper so that the edges just meet.

1. Make a net of the cylinder.
2. Name the shapes in the net.
3. How is the length of the rectangle related to the circles?
4. Explain how to find the surface area of the cylinder.

The diagram below shows how you can put two circles and a rectangle together to make a cylinder.

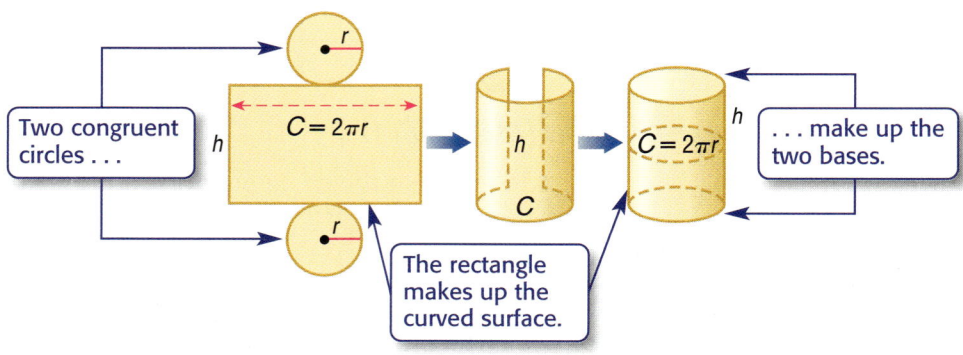

In the diagram above, the length of the rectangle is the same as the circumference of the circle. Also, the width of the rectangle is the same as the height of the cylinder.

The surface area of a cylinder	equals	the area of two bases	plus	the area of the curved surface.
S	$=$	$2(\pi r^2)$	$+$	$(2\pi r)h$

Noteables™ Key Concept: Surface Area of a Cylinder

Words The surface area S of a cylinder with height h and radius r is the sum of the areas of the circular bases and the area of the curved surface.

Symbols $S = 2\pi r^2 + 2\pi rh$

Model

538 Chapter 12 Geometry: Measuring Three-Dimensional Figures

REAL-LIFE CAREERS

How Does a Package Designer Use Math?
A package designer must know the size of the package before creating a design. This involves calculating the surface areas of the package.

Math Online Research
For information about a career as a package designer, visit: msmath2.net/careers

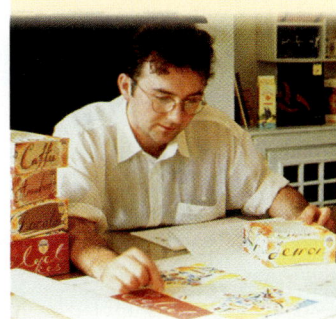

EXAMPLE Find the Surface Area of a Cylinder

1. Find the surface area of the cylinder. Round to the nearest tenth.

   $S = 2\pi r^2 + 2\pi rh$ Surface area of a cylinder
   $= 2\pi(2)^2 + 2\pi(2)(7)$ Replace r with 2 and h with 7.
   ≈ 113.1 Simplify.

   The surface area is about 113.1 square meters.

Your Turn

a. Find the surface area of the cylinder. Round to the nearest tenth.

EXAMPLE Use Surface Area to Solve a Problem

2. **DESIGN** A can of soup is 5 inches high, and its base has a diameter of 4 inches. How much paper is needed to make the label on the can?

   Since only the curved side of the can has a label, you do not need to include the areas of the top and bottom of the can.

   $S = 2\pi rh$ Curved surface of a cylinder
   $= 2\pi(2)(5)$ Replace r with 2 and h with 5.
   ≈ 62.8 Simplify.

   So, about 62.8 square inches of paper is needed to make the label.

Skill and Concept Check

1. **OPEN ENDED** Find the surface area of a cylinder found in your home.

2. **FIND THE ERROR** Matthew and Latonya have each drawn a cylinder with the dimensions shown below and insist that theirs has more surface area. Who is correct? Explain.

   Matthew
   $r = 6$ cm, $h = 3$ cm

   Latonya
   $r = 3$ cm, $h = 6$ cm

GUIDED PRACTICE

Find the surface area of each cylinder. Round to the nearest tenth.

3.

4.

5. The radius of a cylinder is 8 meters, and its height is 5 meters. Find the surface area to the nearest tenth.

Practice and Applications

Find the surface area of each cylinder. Round to the nearest tenth.

6.

7.

8.

9.

10.

11.

HOMEWORK HELP

For Exercises	See Examples
6–11	1
12–19	2

Extra Practice
See pages 595, 607.

12. Find the area of the label on a can of tuna with a radius of 5.1 centimeters and a height of 2.9 centimeters.

13. The height of a water tank is 10 meters and has a diameter of 10 meters. What is the surface area of the tank?

14. An underground oil tank has a surface area of 2,915 square feet and a radius of 6 feet. What is the height of the oil tank?

BAKING For Exercises 15 and 16, use the following information.
A *smash cake* is a smaller version of a birthday cake given to a child on his or her first birthday. Mrs. Jones baked a smash cake 3 inches high and 10 inches in diameter for her daughter when she turned 1 year old.

15. Mrs. Jones covers the top and sides of the cake with frosting. Find the area that the frosting covers to the nearest tenth.

16. Suppose Mrs. Jones first cuts the cake in two layers, frosts the top and sides of the bottom layer, puts the top layer on, and then frosts the rest of the cake. How much of the cake is covered with frosting now?

17. **MOVIES** The Student Council at Southwest Middle School is planning to sell popcorn in one of two open-top containers below at its "Movie Night." The cost depends on the amount of cardboard used to make each container. Which container should Student Council buy? Use volume and surface area measurements to explain your choice.

540 Chapter 12 Geometry: Measuring Three-Dimensional Figures

18. **CANDY** The largest piece of candy ever created was shaped like a cylinder 4.49 meters long and had a circumference of 1.14 meters. What was the surface area of the candy to the nearest tenth?

19. **MODELING** Suppose you took an $8\frac{1}{2}$-×-11-inch piece of paper and made two cylinders; one using the $8\frac{1}{2}$-inch side as the height (cylinder A) and the other using the 11-inch side as the height (cylinder B). Without calculating, which cylinder has the greater surface area? Explain.

20. **CRITICAL THINKING** If you double the height of a cylinder, will its surface area also double? Explain your reasoning.

21. **EXTENDING THE LESSON** In Lesson 12-1, you saw another solid with a curved surface, called a *cone*, as shown at the right. The curved surface is difficult to draw accurately in two dimensions since it is not a polygon. Make a cone with a piece of paper. Then draw the net for the cone using it as a pattern.

Spiral Review with Standardized Test Practice

22. **MULTIPLE CHOICE** Gilberto completely covers a cylindrical can with construction paper. The can has a height of 15 inches and a radius of 4 inches. About how much construction paper does he use?

    Ⓐ 477.5 in²   Ⓑ 375.2 in²   Ⓑ 104.5 in²   Ⓓ 60.0 in²

23. **MULTIPLE CHOICE** The three metal containers below each hold about 1 liter of liquid. Which container has the greatest surface area?

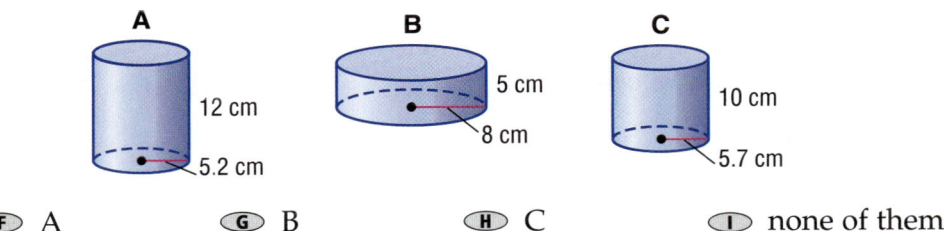

    Ⓕ A   Ⓖ B   Ⓗ C   Ⓘ none of them

Find the surface area of each rectangular prism. (Lesson 12-4)

24.   25.   26.

27. Find the volume of a cylinder having a radius of 4 inches and a height of 6.5 inches. (Lesson 12-3)

GETTING READY FOR THE NEXT LESSON

PREREQUISITE SKILL Find the value of each expression. Round to the nearest tenth. (Pages 559, 560)

28. $13.2 + 15.378$ 29. $23.7 - 9.691$ 30. $23 \cdot 7.1$ 31. $1.6(8.5)0.4$

12-6 Measurement: Precision

What You'll LEARN
Determine and apply significant digits in a real-life context.

NEW Vocabulary
precision
precision unit
significant digits

WHEN am I ever going to use this?

WORLD RECORDS In 2000, Joyce Samuels blew a bubble gum bubble that had a diameter of 27.94 centimeters.

1. What is the smallest unit of measure?
2. Explain whether a ruler whose smallest increment is 0.5 centimeter could have been used to measure the bubble.

The **precision** or *exactness* of a measurement depends on the unit of measure. The **precision unit** is the smallest unit on a measuring tool. As the units get smaller, the measurement gets "more precise."

EXAMPLES Identify Precision Units

Identify the precision unit of each measuring tool.

1

The smallest unit is a half centimeter. So, the precision unit is 0.5 centimeter.

2

The smallest unit is an eighth of an inch. So, the precision unit is $\frac{1}{8}$ inch.

Your Turn

a. Identify the precision unit of the scale at the right.

All measurements are approximate. You could estimate the measure on the scale below as 8 grams. A more precise method is to use significant digits. **Significant digits** include all of the digits of a measurement that you know for sure, plus one estimated digit.

8.3 g ← 2 significant digits

estimated digit

digit known for certain

EXAMPLE **Apply Significant Digits**

3 State the measure of the pencil using significant digits.

The precision unit is 0.1 centimeter. You know for certain that the length is between 12.7 and 12.8 centimeters. One estimate is 12.75 centimeters.

Your Turn

b. State the measure of the paper clip using significant digits.

Skill and Concept Check

1. Choose the most precise unit of measurement: *inches*, *feet*, or *yards*. Explain.

2. **OPEN ENDED** Describe some items or situations in which an approximate measure is sufficient. Then describe some items or situations in which exactness in measurement is important and necessary.

3. **Which One Doesn't Belong?** Identify the measurement that does not have the same number of significant digits as the other three. Explain your reasoning.

GUIDED PRACTICE

4. **Identify** the precision unit of the ruler at the right.

State each measure using significant digits.

5.

6.

7. **OLYMPICS** Tom Dolan broke the Olympic Swimming World Record in 2000 for the 400-meter race. His time was 4 minutes, 11.76 seconds. Describe the precision of this measure.

 msmath2.net/extra_examples

Lesson 12-6 Measurement: Precision 543

Practice and Applications

Identify the precision unit of each measuring tool.

8.

9.

10.

11.

HOMEWORK HELP

For Exercises	See Examples
8–11, 19–20	1, 2
12–18	3

Extra Practice
See pages 595, 607.

State each measure using significant digits.

12.

13.

14.

15.

16.

17.

TRIATHLONS For Exercises 18–20, use the graphic at the right.

18. How many significant digits are in the measurement describing the length of the cycling portion of the triathlon?

19. Which is a more precise measure: the number of miles cycled in a triathlon, or the number of miles run? Explain.

20. The running portion of the women's triathlon is equivalent to 5 kilometers. Which unit is more precise: a mile or a kilometer? Explain.

21. **RESEARCH** The silica tiles on space shuttles are cut accurately to 10,000th of an inch. Use the Internet or another source to investigate the technology that is used to make such precise measurements.

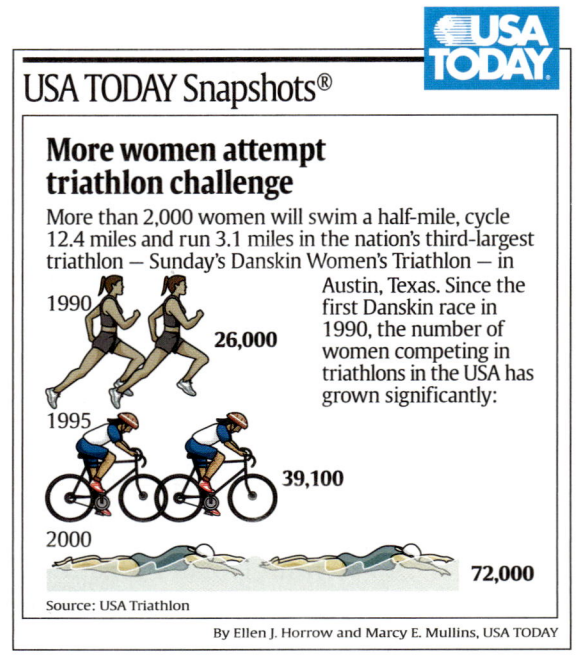

USA TODAY Snapshots®

More women attempt triathlon challenge

More than 2,000 women will swim a half-mile, cycle 12.4 miles and run 3.1 miles in the nation's third-largest triathlon — Sunday's Danskin Women's Triathlon — in Austin, Texas. Since the first Danskin race in 1990, the number of women competing in triathlons in the USA has grown significantly:

1990 — 26,000
1995 — 39,100
2000 — 72,000

Source: USA Triathlon

By Ellen J. Horrow and Marcy E. Mullins, USA TODAY

544 Chapter 12 Geometry: Measuring Three-Dimensional Figures

22. If you use a ruler with a precision unit of 1 millimeter, will your measurements *sometimes*, *always*, or *never* be exact? Explain.

23. **WRITE A PROBLEM** Write a measurement problem in which you use significant digits.

24. **CRITICAL THINKING** Choose the most precise measures to complete the sentence: The yarn at the right is between __?__ and __?__ inches long.

Spiral Review with Standardized Test Practice

25. **MULTIPLE CHOICE** Choose the measuring tool that you would use to estimate the volume of a shoebox.

    Ⓐ scale Ⓑ ruler Ⓒ protractor Ⓓ measuring cup

26. **MULTIPLE CHOICE** Choose the best precision unit for estimating the length of a gymnasium.

    Ⓕ 0.1 cm Ⓖ 0.5 in. Ⓗ 0.1 mi Ⓘ 0.5 ft

Find the surface area of each cylinder. Round to the nearest tenth. (Lesson 12-5)

27.
    5 yd
    11 yd

28.
    1.4 mm
    2 mm

29.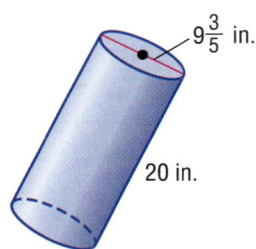
    $9\frac{3}{5}$ in.
    20 in.

30. Find the surface area of a rectangular prism that has a length of 19 centimeters, a width of 8.8 centimeters, and a height of 13 centimeters. (Lesson 12-4)

ALGEBRA Solve each equation. Check your solution. (Lesson 6-5)

31. $\frac{n}{15} = 3$

32. $-26 = \frac{x}{2.4}$

33. $\frac{3}{8}y = 9$

INTERDISCIPLINARY PROJECT

It's All Greek to Me

Math and History It's time to complete your project. Use the information and data you have gathered about Pythagoras to prepare a Web page or poster. Be sure to include the three-dimensional solid you created with your project.

WebQuest msmath2.net/webquest

Chapter 12 Study Guide and Review

Vocabulary and Concept Check

cylinder (p. 524) precision unit (p. 542) significant digits (p. 542) surface area (p. 532)
precision (p. 542) rectangular prism (p. 520) solid (p. 514) volume (p. 520)

Choose the correct term or number to complete each sentence.

1. A (rectangular prism, rectangle) is a solid figure that has three sets of parallel congruent sides.
2. The (volume, surface area) of a solid figure is the measure of the space occupied by it.
3. Volume is measured in (square, cubic) units.
4. The volume of a rectangular prism is found by (adding, multiplying) the length, the width, and the height.
5. A (cylinder, prism) is a solid that has two congruent, parallel circles as its bases.
6. Surface area is the sum of the (areas, volumes) of all of the outside surfaces of a three-dimensional figure.
7. The (smaller, larger) the precision unit, the more precise is the measurement.
8. Significant digits include all of the digits of a measurement that you know for sure, plus (one, two) estimated digit(s).

Lesson-by-Lesson Exercises and Examples

12-1 Drawing Three-Dimensional Figures (pp. 514–517)

Draw a top, a side, and a front view of each solid.

9.   10.

11. Draw a solid using the top, side, and front views shown. Use isometric dot paper.

top   side   front
  (triangle)

Example 1 Draw the solid using the top, side, and front views shown.

top   side   front

The side view is a square. The top and front views are rectangles.

546 Chapter 12 Measuring Three-Dimensional Figures

msmath2.net/vocabulary_review

12-2 Volume of Rectangular Prisms (pp. 520–522)

Find the volume of each rectangular prism. Round to the nearest tenth.

12.

13.

14. **POOLS** A swimming pool 25 yards long has 8 lanes that are each 3 yards wide. The water is 6 feet deep. Find the volume of water in the pool.

Example 2 Find the volume of the rectangular prism.

$V = \ell wh$ Volume of a rectangular prism
$V = (10)(4)(9)$ $\ell = 10$, $w = 4$, and $h = 9$.
$V = 360$ Multiply.

The volume is 360 cubic centimeters.

12-3 Volume of Cylinders (pp. 524–527)

Find the volume of each cylinder. Round to the nearest tenth.

15.

16.

17. **POTTERY** In art class, Arturo made a vase in the shape of a cylinder. The diameter is 5 inches, and the height is 10 inches. Find the maximum volume of water the vase can hold.

Example 3 Find the volume of the cylinder.

$V = \pi r^2 h$ Volume of a cylinder
$ = \pi(3)^2(7)$ Replace r with 3 and h with 7.
$ \approx 197.9$ Multiply.

The volume is about 197.9 cubic feet.

12-4 Surface Area of Rectangular Prisms (pp. 532–535)

Find the surface area of each rectangular prism. Round to the nearest tenth if necessary.

18. 19.

20. **PETS** A plastic pet carrier box is 2.5 feet long, 1 foot high, and 1.25 feet wide. How much plastic is used to make this carrier?

Example 4 Find the surface area of the rectangular prism.

surface area $= 2\ell w + 2\ell h + 2wh$
$ = 2(10)(3) + 2(10)(8) + 2(3)(8)$
$ = 268$

The surface area is 268 square centimeters.

Study Guide and Review continued

Mixed Problem Solving
For mixed problem-solving practice, see page 607.

12-5 Surface Area of Cylinders (pp. 538–541)

Find the surface area of each cylinder. Round to the nearest tenth.

21.

22.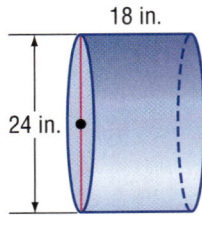

23. **DESIGN** A can of chicken noodle soup is $5\frac{1}{2}$ inches high, and its base has a radius of 2 inches. How much paper is needed to make the label on the can?

Example 5 Find the surface area of the cylinder.

surface area = $2\pi r^2 + 2\pi rh$
$= 2\pi(2)^2 + 2\pi(2)(8)$
≈ 125.7 ft^2

The surface area is about 125.7 square feet.

12-6 Measurement: Precision (pp. 542–545)

Identify the precision unit of each measuring tool.

24.

25.

26.

State each measure using significant digits.

27.

28.

Example 6 Identify the precision unit of the measuring tool.

The smallest unit is one eighth of an inch. So, the precision unit is $\frac{1}{8}$ inch.

Example 7 State the measure of the graduated cylinder using significant digits.

The precision unit is 1 milliliter. You know for certain that the volume is between 2 and 3 milliliters. One estimate is 2.4 milliliters.

548 Chapter 12 Measuring Three-Dimensional Figures

CHAPTER 12 Practice Test

Vocabulary and Concepts

1. **Define** *surface area*.
2. **Explain** what a precision unit is.

Skills and Applications

Draw a top, a side, and a front view of each solid.

3.
4.
5.

Find the volume and surface area of each rectangular prism and cylinder. Round to the nearest tenth if necessary.

6.

7. rectangular prism;
   length = 19.6 m
   width = 14 m
   height = 26.1 m

8.

9. cylinder;
   radius = 6 ft
   height = 12 ft

10.

11. cylinder;
    diameter = 11.5 mm
    height = 20.7 mm

12. **DRINKING STRAWS** A drinking straw has a radius of $\frac{1}{8}$ inch and a height of $7\frac{3}{4}$ inches. How much liquid is in a straw that is half full?

13. **PACKAGING** What is the least amount of paper needed to wrap a box that is 9 inches by 18 inches by 4 inches?

Identify the precision unit of each measuring tool.

14.
15.

Standardized Test Practice

16. **MULTIPLE CHOICE** Find the best estimate of the measure of the paper clip using significant digits.

    Ⓐ 2 cm
    Ⓑ 2.25 cm
    Ⓒ 2.4 cm
    Ⓓ 2.75 cm

CHAPTER 12 Standardized Test Practice

PART 1 Multiple Choice

Record your answers on the answer sheet provided by your teacher or on a sheet of paper.

1. Which city in the table is the closest to sea level? (Lesson 3-2)

City	Altitude Compared to Sea Level (ft)
Ocean City	−25
Redmond	1,376
Reston	24
West Orange	−13

   Ⓐ West Orange Ⓑ Reston
   Ⓒ Ocean City   Ⓓ Redmond

2. What is the value of w in the equation $6w + 12 = 30$? (Lesson 4-4)
   Ⓕ 3   Ⓖ 12   Ⓗ 18   Ⓘ 30

3. Each inch of Tyrell's train set represents $4\frac{1}{2}$ feet of an actual train system. If the engine is 6 inches long, how long is an actual engine? (Lesson 7-4)
   Ⓐ $1\frac{1}{3}$ ft   Ⓑ 27 in.
   Ⓒ $10\frac{1}{2}$ ft   Ⓓ 27 ft

4. The tree diagram shows the choices of flowers and pots sold at Ivitz Gardens. How many different flower-pot choices are possible? (Lesson 9-2)
   Ⓕ 3
   Ⓖ 6
   Ⓗ 9
   Ⓘ 12

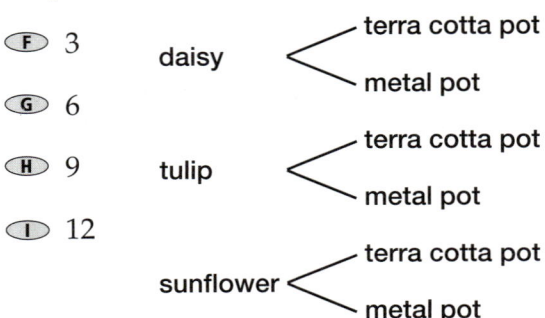

5. Which is the side view of the solid shown at the right? (Lesson 12-1)

   Ⓐ   Ⓑ

   Ⓒ   Ⓓ

6. Gillian has a window box the shape of a rectangular prism. It is 36 inches wide and 6 inches long. How much soil should she buy in order to fill the flower box with a 4-inch-deep layer of soil? (Lesson 12-2)
   Ⓕ 46 in³   Ⓖ 576 in³
   Ⓗ 864 in³   Ⓘ 1,296 in³

7. Which is the volume of the cylinder? (Lesson 12-3)
   Ⓐ 94.2 m³
   Ⓑ 1,963.5 m³
   Ⓒ 3,125.0 m³
   Ⓓ 9,812.5 m³

8. What is the surface area of the rectangular prism? (Lesson 12-4)
   Ⓕ 84 in²
   Ⓖ 122 in²
   Ⓗ 141 in²
   Ⓘ 588 in²

TEST-TAKING TIP

Questions 6–8 Most standardized tests include any necessary formulas in the test booklet. It helps to be familiar with formulas such as the volume or surface area of a cylinder, but use any formulas that are given to you.

Preparing for Standardized Tests

For test-taking strategies and more practice, see pages 608–625.

PART 2 Short Response/Grid In

Record your answers on the answer sheet provided by your teacher or on a sheet of paper.

9. To find the perimeter of the rectangular tray shown below, Oi Ying wrote the equation $2 \times 8 + 2 \times 14$. What operation should Oi Ying do first? (Lesson 1-3)

10. Marcos worked at a car wash during the summer. He worked with other employees to wipe the cars dry. Typically, Marcos wiped down 75% of the cars each day. What are two other ways that 75% can be expressed? (Lesson 5-6)

11. Find $5\frac{1}{2} \div \frac{3}{2}$. (Lesson 6-6)

12. At the Gilmour family reunion, 28 of the 70 family members ate potato salad. What percent of the Gilmours ate potato salad at the reunion? (Lesson 7-5)

13. At a sale, Shay finds a $125 coat marked down to $87.50. What percent of decrease is this? (Lesson 8-4)

14. Alisa wants a bicycle that usually costs $186. During a sale, the bicycle sells for 20% less than that. What is the price of the bicycle when it is on sale? (Lesson 8-5)

15. Triangle TUV is reflected over the x-axis. What are the new coordinates of point T? (Lesson 10-9)

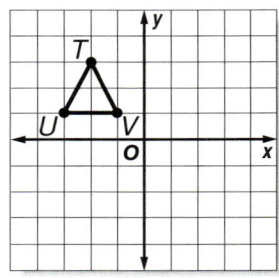

16. Find the total volume of the solid. (Lesson 12-2)

17. What is the surface area of a rectangular prism that is 14.2 centimeters long, 10.5 centimeters wide, and 3 centimeters high? (Lesson 12-4)

18. The cylindrical gasoline storage tank on Mrs. Simon's farm needs to be painted. The dimensions are shown below.

If one quart of paint covers 85 square feet, how many quarts does Mrs. Simon need for this job? (Lesson 12-5)

PART 3 Extended Response

Record your answers on a sheet of paper. Show your work.

19. Draw a top view, a front view, and a side view of the *tetrahedron* shown at the right. (Lesson 12-1)

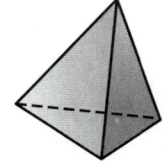

20. Determine whether the given precision unit is appropriate to use in estimating the length of each object. Explain your reasoning. If it is not, describe an appropriate precision unit. (Lesson 12-6)

   a. driveway: 1 inch
   b. compact disc: 1 meter
   c. calculator key: 0.1 centimeter

Student Handbook

Built-In Workbooks

Prerequisite Skills **554**

Extra Practice **564**

Mixed Problem Solving **596**

Preparing for Standardized Tests **608**

Reference

English-Spanish Glossary **626**

Selected Answers **645**

Photo Credits **663**

Index **664**

Formulas and Symbols .... **Inside Back Cover**

How To Cover Your Book .... **Inside Back Cover**

How To...
Use the Student Handbook

A Student Handbook is the additional skill and reference material found at the end of books. The Student Handbook can help answer these questions.

What If I Forget What I Learned Last Year?
Use the **Prerequisite Skills** section to refresh your memory about things you have learned in other math classes. Here's a list of the topics covered in your book.

1. Divisibility Patterns
2. Place Value and Decimals
3. Comparing and Ordering Decimals
4. Rounding Decimals
5. Estimating with Decimals
6. Adding and Subtracting Decimals
7. Multiplying Decimals
8. Powers of Ten
9. Dividing Decimals
10. Mixed Numbers and Improper Fractions

What If I Need More Practice?
You, or your teacher, may decide that working through some additional problems would be helpful. The **Extra Practice** section provides these problems for each lesson so you have ample opportunity to practice new skills.

What If I Have Trouble with Word Problems?
The **Mixed Problem Solving** portion of the book provides additional word problems that use the skills presented in each chapter. These problems give you real-life situations where the math can be applied.

What If I Need Help on Taking Tests?
The **Preparing for Standardized Tests** section gives you tips and practice on how to answer different types of questions that appear on tests.

What If I Forget a Vocabulary Word?
The **English-Spanish Glossary** provides a list of important, or difficult, words used throughout the textbook. It provides a definition in English and Spanish as well as the page number(s) where the word can be found.

What If I Need to Check a Homework Answer?
The answers to the odd-numbered problems are included in **Selected Answers**. Check your answers to make sure you understand how to solve all of the assigned problems.

What If I Need to Find Something Quickly?
The **Index** alphabetically lists the subjects covered throughout the entire textbook and the pages on which each subject can be found.

What If I Forget a Formula?
Inside the back cover of your math book is a list of **Formulas and Symbols** that are used in the book.

Prerequisite Skills

1 Divisibility Patterns

In 54 ÷ 6 = 9, the quotient, 9, is a whole number. So, we say that 54 is **divisible** by 6. You can use the following rules to determine whether a number is divisible by 2, 3, 4, 5, 6, 9, and 10.

A number is divisible by:
- 2 if the ones digit is divisible by 2.
- 3 if the sum of the digits is divisible by 3.
- 4 if the number formed by the last two digits is divisible by 4.
- 5 if the ones digit is 0 or 5.
- 6 if the number is divisible by both 2 and 3.
- 9 if the sum of the digits is divisible by 9.
- 10 if the ones digit is 0.

EXAMPLE Use Divisibility Rules

Determine whether 972 is divisible by 2, 3, 4, 5, 6, 9, or 10.

2: Yes; the ones digit, 2, is divisible by 2.
3: Yes; the sum of the digits, 9 + 7 + 2 = 18, is divisible by 3.
4: Yes; the number formed by the last two digits, 72, is divisible by 4.
5: No; the ones digit is not 0 or 5.
6: Yes; the number is divisible by 2 and 3.
9: Yes; the sum of the digits, 18, is divisible by 9.
10: No; the ones digit is not 0.

Exercises

Use divisibility rules to determine whether the first number is divisible by the second number.

1. 447; 3
2. 135; 6
3. 240; 4
4. 419; 3
5. 831; 3
6. 4,408; 4
7. 7,110; 5
8. 1,287; 9
9. 2,984; 9
10. 7,026; 6
11. 1,260; 10
12. 8,903; 6

Determine whether each number is divisible by 2, 3, 4, 5, 6, 9, or 10.

13. 712
14. 1,035
15. 8,901
16. 462
17. 270
18. 1,005
19. 32,221
20. 8,340
21. 920
22. 50,319
23. 64,042
24. 3,498

25. **MEASUREMENT** Jordan has 5,280 feet of rope. Can he cut the rope into 9-foot pieces and use all of the rope? Explain.

2 Place Value and Decimals

Our number system is based on units of ten. Numbers like 1.35, 0.8, and 25.09 are called decimals.

You can use a place-value chart like the one at the right to help you read decimals and write them in words. Use the word *and* to represent the decimal point.

EXAMPLES Write Decimals in Words

Write each decimal in words.

1 20.09

twenty and nine hundredths

2 6.738

six and seven hundred thirty-eight thousandths

You can also write decimals in **expanded notation** using place value and decimals or their fraction equivalents as shown at the right.

Decimal	0.1	0.01	0.001	0.0001
Fraction	$\frac{1}{10}$	$\frac{1}{10^2}$	$\frac{1}{10^3}$	$\frac{1}{10^4}$

EXAMPLE Write a Decimal in Expanded Notation

3 Write 2.814 in expanded notation using decimals and using fractions.

Write the product of each digit and its place value.

$2.814 = (2 \times 1) + (8 \times 0.1) + (1 \times 0.01) + (4 \times 0.001)$

$2.814 = (2 \times 1) + \left(8 \times \frac{1}{10}\right) + \left(1 \times \frac{1}{10^2}\right) + \left(4 \times \frac{1}{10^3}\right)$ Replace the decimals with their fraction equivalents.

Exercises

Write each decimal in words.

1. 6.37
2. 13.7
3. 145.66
4. 92.03
5. 4,280.899
6. 0.6
7. 0.69
8. 0.00012
9. 7.005
10. 2000.0002

Write each decimal in expanded notation using decimals and using fractions.

11. 0.8
12. 5.3
13. 6.79
14. 9.132
15. 65.002
16. 0.625
17. 0.0072
18. 100.001
19. 548.845
20. 4,260.705

21. **BATS** The northern blossom bat is one of the world's smallest bats. It weighs just 0.53 ounce. Write its weight in expanded notation using fractions.

3 Comparing and Ordering Decimals

To determine which of two decimals is greater, you can compare the digits in each place-value position, or you can use a number line.

EXAMPLES Order Decimals

1 Which is greater, 7.4 or 7.63?

Method 1 Use place value.

   7.4   Line up the decimal points.

   7.63   Starting at the left, compare the digits in each place-value position.
   ↑

The digits in the tenths place are not the same.

Since 6 tenths > 4 tenths, 7.63 > 7.4.

Method 2 Use a number line.

Graph each decimal on a number line and compare.

On a number line, numbers to the right are greater than numbers to the left.

7.63 is to the right of 7.4, so 7.63 > 7.4.

2 Order 12.15, 12.08, and 12.103 from least to greatest.

12.15   Line up the decimal points.

12.08   12.08 is less than 12.15 and 12.103 since 0 < 1 in the tenths place.

12.103   12.103 is less than 12.15 since 0 < 5 in the hundredths place.

So, the numbers from least to greatest are 12.08, 12.103, and 12.15.

Exercises

Replace each ● with <, >, or = to make a true sentence.

1. 1.22 ● 1.02
2. 0.97 ● 1.06
3. 7.90 ● 7.9
4. 1.3 ● 1.31
5. 4.03 ● 4.01
6. 0.77 ● 0.69
7. 0.8 ● 0.08
8. 0.68 ● 0.680
9. 3.28 ● 3.279
10. 25.23 ● 25.32
11. 77.55 ● 77.65
12. 1.29 ● 1.43
13. 310.36 ● 310.3600
14. 0.0034 ● 0.034
15. 9.09 ● 9
16. 1.67 ● 0.48
17. 738 ● 738.99
18. 1,000.01 ● 1,000

Order each group of decimals from least to greatest.

19. 5.13, 5.07, 5.009
20. 7.9, 7.088, 8.02, 7.98
21. 0.087, 0.901, 2, 1.001
22. 12.3, 12.008, 12.54, 12
23. 60.5, 60.05, 60.55, 60.505
24. 505.9, 505.09, 505.91, 505.99
25. 0.02, 0.2, 0.022, 0.002
26. 0.99, 0.90, 0.09, 0.999, 0.099

27. Place decimal points in 999, 463, 208, and 175 so that the resulting decimals are in order from least to greatest.

28. Order 3, 0.3, 3.33, 0.33, and 3.03 from least to greatest.

4 Rounding Decimals

When a number has more decimal places than you want or need, you can **round** it. Use the rules below to round a number to any place value.

Rounding Decimals	Look at the digit to the right of the place being rounded. • If the digit is 4 or less, the digit being rounded remains the same. • If the digit is 5 or greater, the digit being rounded is rounded up.

EXAMPLES   Round Decimals

1 Round 3.92 to the nearest tenth.

tenths place
↓

3.9**2**   Look at the digit to the right of the tenths place.

Since 2 is less than 5, the digit in the tenths place stays the same.

3.92 rounded to the nearest tenth is 3.9.

2 Round 46.297 to the nearest hundredth.

hundredths place
↓

46.29**7**   Look at the digit to the right of the hundredths place.

Since 7 is greater than 5, round 9 up.

46.297 rounded to the nearest hundredth is 46.30.

Exercises

Round each number to the given place value.

1. 0.315; tenth
2. 0.2456; hundredth
3. 17.499; tenth
4. 43.219; hundredth
5. 15.522; tenth
6. 9.6; ones
7. 7.0375; thousandth
8. 16.399; tenth
9. 6.95; tenth
10. 99.1283; thousandth
11. 1,000.37; tenth
12. 750.523; ones
13. 0.445; hundredth
14. 490.299; ones
15. 999.99; tenth
16. 476.835; hundredth
17. 682.596; tenth
18. 1,000.562; ones

Round each number to the underlined place-value position.

19. 0.7_8_9
20. 0._9_6
21. 1.5_7_246
22. 2_3_.48
23. 1.7_0_4
24. 0._1_63
25. 15._4_51
26. 4.52_9_88
27. 0._7_87
28. 3_8_.56
29. 5_9_.61
30. _0_.555
31. 0._5_55
32. 0.5_5_5
33. 10._9_29
34. 10.9_2_9
35. 42_5_.599
36. 0.92_9_29

37. **HORSES** A trainer recorded a racehorse running at 47.54 miles per hour. Round this speed to the nearest mile per hour.

38. **POPULATION** In 2000, the population of Texas was 20.85 million. Round to the nearest million.

5 Estimating with Decimals

Estimation can be used to provide quick answers when an exact answer is not necessary. It is also an excellent way to check whether your answer is reasonable. One method of estimating is to use rounding. Round numbers to any place value that makes estimation easier.

EXAMPLES Estimate by Rounding

Estimate by rounding.

1) $23.485 - 9.757$

$23.485 \rightarrow 23$ Round to the nearest
$-9.757 \rightarrow -10$ whole numbers.
$ 13$

The difference is about 13.

2) 43.9×37.5

$43.9 \rightarrow 40$ Round to the
$\times 37.5 \rightarrow \times 40$ nearest ten.
$ 1{,}600$

The product is about 1,600.

3) $6.43 + 2.17 + 9.1 + 4.87$

$6.43 \rightarrow 6$
$2.17 \rightarrow 2$ Round to the nearest
$9.1 \rightarrow 9$ whole numbers.
$+4.87 \rightarrow +5$
$ 22$ The sum is about 22.

4) $432.87 \div 8.9$

$8.9 \rightarrow 9$ Round the divisor.
$432.87 \rightarrow 450$ Round the dividend to a multiple of 9.

$432.87 \div 8.9 \rightarrow 450 \div 9 = 50$

The quotient is about 50.

Another way to estimate sums is to use **clustering**. This strategy is used when all the numbers are close to a common value.

EXAMPLE Estimate by Clustering

5) Estimate $9.775 + 9.862 + 9.475 + 9.724$ by clustering.

All of the numbers are clustered around 10. There are four numbers.
So, the sum is about 4×10 or 40.

Exercises

Estimate by rounding.

1. $8.56 + 5.34$
2. $34.84 - 17.69$
3. 6.8×2.4
4. $40.79 \div 6.8$
5. $6.9 + 5.2$
6. $23.84 + 12.13$
7. $34.3 - 18.9$
8. 7.5×8.4
9. $65.48 \div 9.3$
10. 26.3×9.7
11. $33.21 - 8.23$
12. $67.86 - 24.35$
13. $8.99 \div 2.6$
14. $121.5 + 487.8$
15. 32.5×81.4

Estimate by clustering.

16. $18.4 + 22.5 + 20.7$
17. $56.9 + 63.2 + 59.3 + 61.1$
18. $42.3 + 41.5 + 39.8 + 40.4$
19. $77.8 + 75.6 + 81.2 + 79.9$
20. $239.8 + 242.43 + 236.20 + 240.77$
21. $9.9 + 10.0 + 10.3 + 11.1 + 9.8 + 11.2$
22. $50.4 + 51.1 + 48.9 + 49.5 + 50.8$
23. $100.5 + 97.8 + 101.6 + 100.2 + 99.3$

6 Adding and Subtracting Decimals

To add or subtract decimals, write the numbers in a column and line up the decimal points. In some cases, you may want to *annex*, or place zeros at the end of the decimals, to help align the columns. Then add or subtract as with whole numbers and bring down the decimal point. Always estimate first to see whether your answer is reasonable.

EXAMPLES Add and Subtract Decimals

Find each sum or difference.

1 12.6 + 4.5   **Estimate** 13 + 5 = 18

   12.6   Line up the decimal points.
 + 4.5   Add.
  17.1   The sum is close to the estimate.

2 42.17 − 15.85   **Estimate** 42 − 16 = 26

  42.17   Line up the decimal points.
− 15.85   Subtract.
  26.32   The difference is close to the estimate.

3 57.125 + 7.63   **Estimate** 57 + 8 = 65

  57.125
+ 7.630   Annex a zero to align the columns.
  64.755   Compare to the estimate.

4 25 − 15.25   **Estimate** 25 − 15 = 10

  25.00   Annex two zeros to align the columns.
− 15.25
   9.75   Compare to the estimate.

Exercises

Find each sum or difference.

1. 0.132 − 0.021
2. 3.78 + 0.21
3. 13.2 + 12.8
4. 5.86 − 1.51
5. 42.07 − 38.78
6. 14.7 + 351.82
7. 42.3 + 0.81
8. 12.3 − 0.847
9. 342.9 − 0.18
10. 282.45 − 111.3
11. 100 − 0.48
12. 82.23 + 0.88
13. 128.01 − 39.117
14. 80.05 − 79.06
15. 104.98 − 0.12

16. 0.42 + 0.68
17. 0.48 + 2.901
18. 5.8 + 3.92
19. 38.63 + 38.63
20. 8 − 2.54
21. 16.354 − 0.2
22. 0.125 + 0.78
23. 8.2 − 6.9
24. 1.245 + 3.842
25. 3.2 + 1.23
26. 0.889 − 0.3
27. 22.22 + 1.475
28. 10 − 0.25
29. 33.16 − 0.08
30. 1.254 + 0.5
31. 44.698 − 14.903
32. 10 − 0.005
33. 722.86 + 0.024
34. 100.211 + 8.004
35. 86.124 + 32.822
36. 6.9 + 1.1
37. 75 − 0.24
38. 13 − 0.324
39. 0.8 + 1.2

40. What is the sum of 35.009 and 3.6?

41. Find 9.1 minus 5.625.

42. **AIR TRAVEL** The numbers of passengers, in millions, passing through four airports in a recent year were 80.2, 72.1, 68.5, and 60.7. Find the total number of passengers for the four airports.

7 Multiplying Decimals

To multiply decimals, multiply as with whole numbers. The product has the same number of decimal places as the sum of the decimal places of the factors. Use estimation to determine whether your answers are reasonable.

EXAMPLES Multiply Decimals

Multiply.

1 1.3×0.9 **Estimate** $1 \times 1 = 1$

$$\begin{array}{r} 1.3 \\ \times\, 0.9 \\ \hline 1.17 \end{array} \begin{array}{l} \leftarrow \text{1 decimal place} \\ \leftarrow \text{1 decimal place} \\ \leftarrow \text{2 decimal places} \end{array}$$

The product is reasonable.

2 0.054×1.6 **Estimate** $0 \times 2 = 0$

$$\begin{array}{r} 0.054 \\ \times\, 1.6 \\ \hline 324 \\ 540 \\ \hline 0.0864 \end{array} \begin{array}{l} \leftarrow \text{3 decimal places} \\ \leftarrow \text{1 decimal place} \\ \\ \\ \end{array}$$

0.0864 Annex a zero on the left so the answer has four decimal places. Compare to the estimate.

Exercises

Place the decimal point in each product. Add zeros if necessary.

1. $1.32 \times 4 = 528$
2. $0.07 \times 1.1 = 77$
3. $0.4 \times 0.7 = 28$
4. $1.9 \times 0.6 = 114$
5. $1.4 \times 0.09 = 126$
6. $5.48 \times 3.6 = 19728$
7. $4.5 \times 0.34 = 153$
8. $0.45 \times 0.02 = 9$
9. $150.2 \times 32.75 = 4919050$

Multiply.

10. 0.2×6
11. 0.3×0.9
12. 0.45×0.12
13. 0.0023×32
14. 1.5×2.7
15. 10.1×9
16. 2×0.3
17. 6.78×1.3
18. 200×0.004
19. 0.0023×0.35
20. 15.8×11
21. 88×2.5
22. 33×0.03
23. 36×0.46
24. 0.003×482
25. 1.88×1.11
26. 0.6×2
27. 38.3×29.1
28. 0.7×18
29. 8×0.3
30. 12.2×12.4
31. 380×1.25
32. 42×0.17
33. 0.4×16
34. 0.23×0.2
35. 0.44×0.5
36. 0.44×55
37. 44×0.55

38. **JOBS** Antonia earns $10.75 per hour. What are her total weekly earnings if she works 34.5 hours? Round to the nearest cent.

8 Powers of Ten

You can use a pattern to mentally find the product of any number and a power of 10. Count the number of zeros in the power of 10 or use the exponent. Then move the decimal point that number of places *to the right*.

Decimal Power of 10	Product
19.7×10^1 (or 10)	= 197
19.7×10^2 (or 100)	= 1,970
19.7×10^3 (or 1,000)	= 19,700
19.7×10^4 (or 10,000)	= 197,000

EXAMPLES Use Mental Math to Multiply

Multiply mentally.

1 12.562×100

$12.562 \times 100 = 12.562$ Move the decimal point two places to the right, since 100 has two zeros.
$= 1,256.2$

2 0.59×10^4

$0.59 \times 10^4 = 0.5900$ Move the decimal point four places to the right, since the exponent is 4.
$= 5,900$

To mentally multiply by a power of ten that is less than 1, count the number of decimal places. Or, if the power is written as a fraction, use the exponent in the denominator. Then move the decimal point that number of places *to the left*.

Decimal Power of 10	Product
19.7×0.1 (or $\frac{1}{10^1}$)	= 1.97
19.7×0.01 (or $\frac{1}{10^2}$)	= 0.197
19.7×0.001 (or $\frac{1}{10^3}$)	= 0.0197

EXAMPLES Use Mental Math to Multiply

Multiply mentally.

3 10.5×0.01

$10.5 \times 0.01 = 10.5$ Move the decimal point two places to the left.
$= 0.105$

4 $5,284 \times 0.00001$

$5,284 \times 0.00001 = 05284$ Move the decimal point five places to the left.
$= 0.05284$

Exercises

Multiply mentally.

1. 12.53×10
2. 4.6×10^3
3. 78.4×0.01
4. 0.05×100
5. 4.527×10^0
6. $2.78 \times 1,000$
7. 13.58×0.01
8. 5.49×10^3
9. 0.1×0.8
10. 0.925×10
11. 99.44×10^2
12. 0.01×16
13. 1.32×10^3
14. $0.56 \times 10,000$
15. 1.4×0.001
16. 11.23×10^5
17. 68.94×0.01
18. 0.8×10^4
19. 28.1×0.01
20. 9.3×10^7
21. $625,799 \times 0.0001$

9 Dividing Decimals

To divide two decimals, use the following steps.

- If necessary, change the divisor to a whole number by moving the decimal point to the right. **You are multiplying the divisor by a power of ten.**
- Move the decimal point in the dividend the same number of places to the right. **You are multiplying the dividend by the same power of ten.**
- Divide as with whole numbers.

EXAMPLES Divide Decimals

Divide.

1) 25.8 ÷ 2   **Estimate** 26 ÷ 2 = 13

```
       12.9
    2)25.8     The divisor, 2, is already a whole number, so
    − 2         you do not need to move the decimal point.
       5       Divide as with whole numbers. Then place the decimal
     − 4        point directly above the decimal point in the dividend.
      18
    − 18
       0       Compared to the estimate, the quotient, 12.9, is reasonable.
```

2) 199.68 ÷ 9.6   **Estimate** 200 ÷ 10 = 20

```
         20.8
   9.6)199.6.8   Move each decimal point one place to the right.
    − 192
       7 68
     − 7 68
         0      Compare the answer to the estimate.
```

Exercises

Divide.

1. 0.3)9.81
2. 12)0.12
3. 3.2)5.76
4. 0.22)0.0132
5. 0.04)0.008
6. 3.18)0.636
7. 0.2)8.24
8. 82.3)823
9. 12.02)24.04
10. 0.5)85
11. 74.9)5.992
12. 19.2)4.416
13. 1.9)38.57
14. 13.8)131.1
15. 6.48)259.2
16. 812 ÷ 0.4
17. 0.34 ÷ 0.2
18. 14.4 ÷ 0.12
19. 90.175 ÷ 2.5
20. 39.95 ÷ 799
21. 88.8 ÷ 444
22. 613.8 ÷ 66
23. 2,445.3 ÷ 33
24. 20.24 ÷ 2.3
25. 45 ÷ 0.09
26. 2.475 ÷ 0.03
27. 4.6848 ÷ 0.366
28. 180 ÷ 0.36
29. 97.812 ÷ 1.1
30. 23 ÷ 0.023
31. 1,680.042 ÷ 44.2

32. **OLYMPICS** In the 2000 Olympics, Michael Johnson of the U.S. ran the 400-meter run in 43.84 seconds. To the nearest hundredth, find his speed in meters per second.

33. **PLANETS** It takes Pluto 247.69 Earth years to revolve once around the Sun. It takes Jupiter 11.86 Earth years to revolve once around the Sun. About how many times longer does it take Pluto than Jupiter to revolve once around the Sun?

10 Mixed Numbers and Improper Fractions

A **mixed number** is the sum of a whole number and a fraction. An **improper fraction** is a fraction with a numerator that is greater than or equal to the denominator.

$$\text{mixed number: } 8\frac{1}{2} \quad \text{improper fraction: } \frac{17}{2}$$

To write a mixed number as an improper fraction, use the following steps.
- Multiply the whole number by the denominator.
- Add the numerator.
- Write the sum as the numerator of the improper fraction.

EXAMPLES — Write Mixed Numbers as Improper Fractions

Write each mixed number as an improper fraction.

1 $6\frac{1}{2}$

$6\frac{1}{2} = \frac{(6 \cdot 2) + 1}{2}$ ← Multiply 6 by 2. Add 1.
← The denominator is 2.

$= \frac{12 + 1}{2}$ or $\frac{13}{2}$ Simplify.

2 $3\frac{4}{5}$

$3\frac{4}{5} = \frac{(3 \cdot 5) + 4}{5}$ ← Multiply 3 by 5. Add 4.
← The denominator is 5.

$= \frac{15 + 4}{5}$ or $\frac{19}{5}$ Simplify.

To write an improper fraction as a mixed number, divide the numerator by the denominator. Write the remainder as the numerator of the fraction.

EXAMPLE — Write Improper Fractions as Mixed Numbers

3 Write $\frac{7}{4}$ as a mixed number.

$7 \div 4 = 1 \text{ R } 3$ Divide the numerator by the denominator.

$= 1\frac{3}{4}$ Write the remainder as the numerator of the fraction.

So, $\frac{7}{4} = 1\frac{3}{4}$.

Exercises

Write each mixed number as an improper fraction.

1. $2\frac{1}{3}$
2. $1\frac{5}{8}$
3. $4\frac{5}{7}$
4. $3\frac{3}{4}$
5. $9\frac{1}{3}$
6. $1\frac{1}{10}$
7. $10\frac{2}{3}$
8. $12\frac{1}{4}$
9. $6\frac{3}{8}$
10. $11\frac{1}{2}$
11. $2\frac{3}{8}$
12. $20\frac{3}{5}$
13. $5\frac{7}{8}$
14. $27\frac{1}{3}$
15. $5\frac{2}{3}$

Write each improper fraction as a mixed number.

16. $\frac{15}{2}$
17. $\frac{9}{7}$
18. $\frac{15}{8}$
19. $\frac{100}{3}$
20. $\frac{28}{5}$
21. $\frac{3}{3}$
22. $\frac{19}{8}$
23. $\frac{17}{10}$
24. $\frac{51}{3}$
25. $\frac{17}{6}$
26. $\frac{5}{3}$
27. $\frac{99}{3}$
28. $\frac{25}{7}$
29. $\frac{46}{9}$
30. $\frac{27}{4}$

Extra Practice

Lesson 1-1 (Pages 6–9)

Use the four-step plan to solve each problem.

1. The Reyes family rode their bicycles for 9 miles to the park. The ride back was along a different route for 14 miles. How many miles did they ride in all?
2. A farmer planted 389 acres of land with 78,967 corn plants. How many plants were planted per acre?
3. A group of 251 people is eating dinner at a school fund-raiser. If each person pays $8.00 for their meal, how much money is raised?
4. When Tamika calls home from college, she talks ten minutes per call for 3 calls per week. How many minutes does she call in a 15-week semester?
5. Darren runs at 6 feet per second and Kim runs at 7 feet per second. If they both start a race at the same time, how far apart are they after one minute?

Lesson 1-2 (Pages 10–13)

Write each power as a product of the same factor.

1. 13^4
2. 9^6
3. 1^7
4. 12^2
5. 5^8
6. 15^4

Evaluate each expression.

7. 5^6
8. 17^3
9. 2^{12}
10. 3^5
11. 1^4
12. 5^3
13. 10^2
14. 2^8
15. 8^2
16. 7^4
17. 20^3
18. 42^3

Write each product in exponential form.

19. $2 \cdot 2 \cdot 2 \cdot 2 \cdot 2$
20. $3 \cdot 3$
21. $1 \cdot 1 \cdot 1 \cdot 1 \cdot 1 \cdot 1$
22. $18 \cdot 18 \cdot 18 \cdot 18$
23. $9 \cdot 9 \cdot 9 \cdot 9 \cdot 9 \cdot 9 \cdot 9 \cdot 9$
24. $10 \cdot 10 \cdot 10 \cdot 10 \cdot 10 \cdot 10$

Lesson 1-3 (Pages 14–17)

Evaluate each expression.

1. $14 - (5 + 7)$
2. $(32 + 10) - 5 \times 6$
3. $(50 - 6) + (12 + 4)$
4. $12 - 2 \cdot 3$
5. $16 + 4 \times 5$
6. $(5 + 3) \times 4 - 7$
7. $2 \times 3 + 9 \times 2$
8. $6 \cdot (8 + 4) \div 2$
9. $7 \times 6 - 14$
10. $8 + (12 \times 4) \div 8$
11. $13 - 6 \cdot 2 + 1$
12. $(80 \div 10) \times 8$
13. $14 - 2 \cdot 7 + 0$
14. $156 - 6 \times 0$
15. $30 - 14 \cdot 2 + 8$
16. $3 \times 4 - 3^2$
17. $10^2 - 5$
18. $3 + (10 - 5 + 1)^2$
19. $(4 + 3)^2 \div 7$
20. 8×10^3
21. $10^4 \times 6$
22. 4.5×10^3
23. 1.8×10^2
24. $3 + 5(1.7 + 2.3)$
25. $4(3.6 + 5.4) - 9$
26. $10 + 3(6.1 + 3.7)$
27. $6(7.5 + 2.1) - 2.3$

Lesson 1-4
(Pages 18–21)

Evaluate each expression if $a = 3$, $b = 4$, $c = 12$, and $d = 1$.

1. $a + b$
2. $c - d$
3. $a + b + c$
4. $b - a$
5. $c - ab$
6. $a + 2d$
7. $b + 2c$
8. ab
9. $a + 3b$
10. $6a + c$
11. $\dfrac{c}{d}$
12. abc
13. $2(a + b)$
14. $\dfrac{2c}{b}$
15. $144 - abc$
16. $2ab$
17. $\dfrac{b}{2}$
18. a^2
19. $c^2 - 100$
20. $a^3 + 3$
21. $2b^2$
22. $b^3 + c$
23. $\dfrac{a^2}{d}$
24. $5a^2 + 2d^2$
25. $\dfrac{4d^2}{b}$
26. $\dfrac{15}{a}$
27. $3a^2$
28. $10d^3$
29. $\dfrac{ab}{c}$
30. $\dfrac{(a + b)}{d}$
31. $2.5b + c$
32. $\dfrac{10}{d}$
33. $\dfrac{(2c + b)}{b}$
34. $\dfrac{(b^2 + 2d)}{a}$
35. $\dfrac{(2c + ab)}{c}$
36. $\dfrac{(3.5c + 2)}{11}$

Lesson 1-5
(Pages 24–27)

Solve each equation mentally.

1. $b + 7 = 12$
2. $a + 3 = 15$
3. $s + 10 = 23$
4. $9 + n = 13$
5. $20 = 24 - n$
6. $4x = 36$
7. $2y = 10$
8. $15 = 5h$
9. $j \div 3 = 2$
10. $14 = w - 4$
11. $24 \div k = 6$
12. $b - 3 = 12$
13. $c \div 10 = 8$
14. $6 = t \div 5$
15. $14 + m = 24$
16. $3y = 39$
17. $\dfrac{f}{2} = 12$
18. $16 = 4v$
19. $81 = 80 + a$
20. $9 = \dfrac{72}{x}$
21. $66 = 22m$
22. $77 - 12 = a$
23. $9k = 81$
24. $95 + d = 100$
25. $b = \dfrac{72}{6}$
26. $z = 15 + 22$
27. $15b = 225$
28. $43 + s = 57$
29. $4w = 52$
30. $e - 10 = 0$
31. $62 - d = 12$
32. $14f = 14$
33. $48 \div n = 8$
34. $a - 82 = 95$
35. $\dfrac{x}{2} = 36$
36. $99 = c \div 2$

Lesson 1-6
(Pages 30–33)

Use the Distributive Property to write each expression as an equivalent expression. Then evaluate the expression.

1. $3(4 + 5)$
2. $(2 + 8)6$
3. $4(9 - 6)$
4. $8(6 - 3)$
5. $5(200 - 50)$
6. $20(3 + 6)$
7. $(20 - 5)8$
8. $50(8 + 2)$
9. $15(1{,}000 - 200)$
10. $3(2{,}000 + 400)$
11. $12(1{,}000 + 10)$
12. $7(1{,}000 - 50)$

Name the property shown by each statement.

13. $7(2 + 3) = 7(2) + 7(3)$
14. $7 \times 8 = 8 \times 7$
15. $(4 + 2) + 8 = 4 + (2 + 8)$
16. $92 + 3 = 3 + 92$
17. $4(12) + 4(3) = 4(12 + 3)$
18. $z \cdot 1 = z$
19. $(1 \times 5) \times 2 = 1 \times (5 \times 2)$
20. $72 + (8 + 6) = (72 + 8) + 6$
21. $4 \cdot (x \cdot y) = (4 \cdot x) \cdot y$
22. $a + 0 = a$
23. $a \times b = b \times a$
24. $43 + z = z + 43$

Lesson 1-7 (Pages 34–36)

Describe the pattern in each sequence and identify the sequence as *arithmetic*, *geometric*, or *neither*.

1. 5, 9, 13, 17, …
2. 3, 6, 12, 24, …
3. 10, 15, 25, 40, …
4. 90, 91, 94, 99, …
5. 8, 24, 72, 216, …
6. 4.5, 5.4, 6.3, 7.2, …
7. 0.3, 0.4, 0.5, …
8. 2.3, 3.4, 4.5, 5.6, …
9. 9.1, 8.4, 7.7, 6.0, …

Write the next three terms of each sequence.

10. 3, 11, 19, 27, …
11. 350, 300, 250, 200, …
12. 1,600, 800, 400, 200, …
13. 2, 6, 18, 54, …
14. 10, 17, 24, 31, …
15. 0, 7, 14, 21, …
16. 1, 5, 25, 125, …
17. 113, 107, 101, 95, …
18. 9, 81, 729, 6,561, …
19. 8,748, 2,916, 972, 324, …
20. 400, 1,000, 2,500, 6,250, …
21. 9.9, 8.8, 7.7, 6.6, …
22. 24, 22.5, 21, 19.5, …
23. 4,000, 2,000, 1,000, 500, …
24. 0.1, 1, 10, 100, …

Lesson 1-8 (Pages 38–41)

Complete.

1. 400 mm = __?__ cm
2. 4 km = __?__ m
3. 660 cm = __?__ m
4. 0.3 km = __?__ m
5. 30 mm = __?__ cm
6. 84.5 m = __?__ km
7. __?__ m = 54 cm
8. 18 km = __?__ cm
9. __?__ mm = 45 cm
10. 4 kg = __?__ g
11. 632 mg = __?__ g
12. 4,497 g = __?__ kg
13. __?__ mg = 21 g
14. 61.2 mg = __?__ g
15. 61 g = __?__ mg
16. __?__ mg = 0.51 kg
17. 0.63 kg = __?__ g
18. __?__ kg = 563 g
19. 662 m = __?__ km
20. 5,283 mL = __?__ L
21. 0.24 cm = __?__ mm
22. 380 kL = __?__ L
23. 10.8 g = __?__ mg
24. 83,000 mL = __?__ L
25. 17.8 m = __?__ cm
26. 2 km = __?__ m
27. 0.75 m = __?__ mm
28. 125 L = __?__ kL
29. 3 km = __?__ cm
30. 99 g = __?__ mg

Lesson 1-9 (Pages 43–45)

Write each number in standard form.

1. 7.6×10^2
2. 2.6×10^3
3. 4.7×10^5
4. 1.0×10^2
5. 6.4×10^6
6. 8.3×10^4
7. 2.76×10^3
8. 6.17×10^5
9. 8.70×10^5
10. 1.28×10^7
11. 4.32×10^9
12. 5.01×10^2
13. 1.89×10^6
14. 7.43×10^4
15. 3.03×10^9

Write each number in scientific notation.

16. 720
17. 7,560
18. 892
19. 1,400
20. 91,200
21. 51,000
22. 145,000
23. 90,100
24. 123,000,000,000
25. 4,500
26. 1,700,000
27. 10,000,000
28. 820,000,000
29. 680,000
30. 52,000
31. 60,600,000
32. 3,970,000
33. 574,000,000

Lesson 2-1 (Pages 54–57)

Make a frequency table of each set of data.

1. **Length of Time Walking (min)**

15	30	15	45
45	30	30	60
30	60	15	30
45	45	60	15

2. **Number of Raisins Eaten in Preschool**

40	49	45	49
42	41	45	41
45	41	41	40
41	43	40	41

3. **Favorite Type of TV Show**

M	S	N	S	M
M	S	S	S	S
G	M	N	S	M
S	G	G	G	G
N	S	M	G	G
S	S	N	N	N

M = movie, S = sitcom, N = news show, G = game show

4. **What Type of Transportation Did You Use on Your Last Family Vacation?**

C	A	C	T	C	A
C	C	C	C	S	A
B	B	C	T	A	A
T	S	A	C	A	A
A	A	A	A	S	

C = car, T = train, S = ship, A = airplane, B = bus

Lesson 2-2 (Pages 60–63)

1. Rachel's quiz scores in science have been steadily going up since her parents hired a tutor for her. If the trend continues, predict Rachel's score on the next quiz.

Quiz Scores

2. The table shows the average price paid to farmers per 100 pounds of sheep they sold.
   a. Make a scatter plot of the data. Use the year on the horizontal axis and the price on the vertical axis.
   b. Describe the relationship, if any, between the two sets of data.
   c. Predict the price per 100 pounds for 2010. Explain.

Year	Price Per 100 Pounds ($)
1940	4
1950	12
1960	6
1970	8
1980	21
1990	23
2000	34

Source: *The World Almanac and Book of Facts*

Lesson 2-3 (Pages 64–68)

Make a line plot for each set of data. Identify any clusters, gaps, or outliers.

1. **Number of Pets in the Home**

0	1	3	4	0
2	1	0	1	1
10	0	1	5	2

2. **High Temperatures for 18 Days (°F)**

75	81	75	65	76	81
77	80	65	65	80	80
76	85	66	75	80	75

3. **Number of Stories for Buildings in Denver**

56	43	36	42	29
54	42	32	34	
52	40	32	32	

Source: *The World Almanac and Book of Facts*

4. **Ages of Children at Sunny Day Care (years)**

4	1	6	4	5	3
4	5	1	2	5	4
3	2	4	1	3	3

Lesson 2-4
(Pages 69–72)

Find the mean, median, and mode for each set of data.

1. 1, 5, 9, 1, 2, 6, 8, 2
2. 2, 5, 8, 9, 7, 6, 3, 5, 1, 4
3. 1, 2, 1, 2, 2, 1, 2, 1
4. 12, 13, 15, 12, 12, 11
5. 82, 79, 93, 91, 95, 95, 81
6. 117, 103, 108, 120
7. 256, 265, 247, 256
8. 957, 562, 462, 847, 721
9. 47, 54, 66, 54, 46, 66
10. 81, 82, 83, 84, 85, 86, 87

11.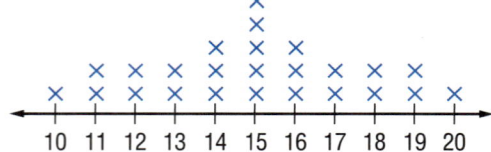

12.
Number of Absences	Tally	Frequency
0	IIII	4
1	JHT IIII	9
2	JHT I	6
3	JHT	5

Lesson 2-5
(Pages 76–79)

Make a stem-and-leaf plot for each set of data.

1. 23, 15, 39, 68, 57, 42, 51, 52, 41, 18, 29
2. 5, 14, 39, 28, 14, 6, 7, 18, 13, 28, 9, 14
3. 189, 182, 196, 184, 197, 183, 196, 194, 184
4. 71, 82, 84, 95, 76, 92, 83, 74, 81, 75, 96

5.
Average Monthly High Temperatures in Albany, NY (°F)			
21	46	72	50
24	58	70	40
34	67	61	27

Source: *The World Almanac and Book of Facts*

6.
Super Bowl Winning Scores 1987–2004					
39	55	52	27	34	20
42	20	30	35	23	48
20	37	49	31	34	32

Source: *The World Almanac and Book of Facts*

Lesson 2-6
(Pages 80–83)

For each set of data:
a. Find the lower extreme, LQ, median, UQ, and upper extreme.
b. Draw a box-and-whisker plot of the data.

1.
Price of Chicken Sandwiches at Various Restaurants			
$1.00	$3.70	$3.50	$3.90
$2.50	$1.70	$2.20	$2.10
$1.20	$1.90	$4.20	

2.
Lengths of Principal Rivers in South America (mi)					
4,000	1,750	1,400	2,485	1,000	910
1,100	2,013	1,600	1,000	1,988	1,000
808	956	1,584	2,100	1,677	1,300

Source: *The World Almanac and Book of Facts*

3.
Height of World's Tallest Free-Standing Towers (ft)			
1,831	1,772	1,535	1,379
1,214	1,230	1,411	
1,815	1,362	1,369	

Source: *The World Almanac and Book of Facts*

4.
Mathematics Test Scores in Ms. Ruff's Class				
65	100	84	80	68	69
92	69	75	80	99	90
78	87	100	72	85	

Lesson 2-7
(Pages 85–89)

For each set of data, make a bar graph.

1.
Longest Snakes	
Snake Name	Length (ft)
Royal python	35
Anaconda	28
Indian python	25
Diamond python	21
King cobra	19
Boa constrictor	16

Source: *The Top 10 of Everything*

2.
Least Densely Populated States	
State	People per Square Mile
Alaska	1
Wyoming	5
Montana	6
North Dakota	9
South Dakota	10
New Mexico	15

Source: *The Top 10 of Everything*

For each set of data, make a histogram.

3.
Cost of a Movie Ticket at Selected Theaters			
$5.25	$6.50	$3.50	$3.75
$7.50	$9.25	$10.40	$4.75
$10.00	$4.50	$8.75	$7.25
$3.50	$6.75	$4.20	$7.50

4.
Highest Recorded Wind Speeds for Selected U.S. Cities (mph)					
52	55	81	46	73	57
75	54	58	76	46	58
60	91	53	53	51	56
80	60	73	46	49	47

Source: *The World Almanac and Book of Facts*

Lesson 2-8
(Pages 92–95)

Which graph could be misleading? Why?

1. Both graphs show pounds of grapes sold to Westview School in one week.

2. Both graphs show commissions made by Mr. Turner for a four-week pay period.

Lesson 3-1
(Pages 106–108)

Write an integer for each situation.

1. seven degrees below zero
2. a loss of 3 pounds
3. a loss of 20 yards
4. a profit of $25
5. 112°F above 0
6. 2,830 feet above sea level

Evaluate each expression.

7. $|1|$
8. $|-8|$
9. $|0|$
10. $|-82|$
11. $|64|$
12. $|-128|$
13. $|-22| + 5$
14. $|-40| - 8$
15. $|-18| + |10|$
16. $|-7| + |-1|$
17. $|98| - |-5|$
18. $|-49| - |-10|$

Graph each set of integers on a number line.

19. $\{-2, 0, 2\}$
20. $\{1, 3, 5\}$
21. $\{-2, -5, 3\}$
22. $\{7, -1, 4\}$

Extra Practice **569**

Lesson 3-2
(Pages 109–111)

Replace each ● with < or > to make a true sentence.

1. 7 ● −7
2. −8 ● 4
3. −4 ● −9
4. −3 ● 0
5. 8 ● 10
6. −5 ● −4
7. 6 ● −7
8. −12 ● −13
9. 3 ● 1
10. −2 ● 2
11. 7 ● −1
12. −15 ● −20
13. −40 ● 30
14. 0 ● −3
15. −5 ● 0
16. 85 ● −17

Order the integers from least to greatest.

17. −2, −8, 4, 10, −6, −12
18. 19, −19, −21, 32, −14, 18
19. 18, 23, 95, −95, −18, −23, 2
20. 46, −48, −47, −52, −18, 12
21. 0, −10, −6, −8, 12
22. −15, 18, −1, 0, 14, −20

Lesson 3-3
(Pages 112–115)

Name the ordered pair for each point graphed at the right. Then identify the quadrant in which each point lies.

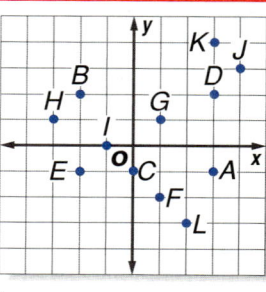

1. A
2. B
3. C
4. D
5. E
6. F
7. G
8. H
9. I
10. J
11. K
12. L

On graph paper, draw a coordinate plane. Then graph and label each point.

13. $N(-4, 3)$
14. $K(2, 5)$
15. $W(-6, -2)$
16. $X(5, 0)$
17. $Y(4, -4)$
18. $M(0, -3)$
19. $Z(-2, 0.5)$
20. $S(-1, -3)$
21. $A(0, 2)$
22. $C(-2, -2)$
23. $E(0, 1)$
24. $G(1, -1)$

Lesson 3-4
(Pages 120–124)

Add.

1. $-4 + 8$
2. $14 + 16$
3. $-7 + (-7)$
4. $-9 + (-6)$
5. $-18 + 11$
6. $-36 + 40$
7. $42 + (-18)$
8. $-42 + 29$
9. $18 + (-32)$
10. $12 + (-9)$
11. $-24 + 9$
12. $-7 + (-1)$

Evaluate each expression if $a = 6$, $b = -2$, $c = -6$, and $d = 3$.

13. $-96 + a$
14. $b + (-5)$
15. $c + (-32)$
16. $d + 98$
17. $-120 + b$
18. $-120 + c$
19. $5 + b$
20. $a + d$
21. $c + a$
22. $d + (-9)$
23. $b + c$
24. $d + c$

Simplify.

25. $x + 3 + (-8)$
26. $-7 + b + 3$
27. $9 + (-7) + f$
28. $10 + y + (-6)$
29. $m + (-11) + 11$
30. $-14 + a + (-19)$

570 Extra Practice

Lesson 3-5 (Pages 128–131)

Subtract.

1. $3 - 7$
2. $-5 - 4$
3. $-6 - 2$
4. $8 - 13$
5. $6 - (-4)$
6. $12 - 9$
7. $-2 - 23$
8. $63 - 78$
9. $0 - (-14)$
10. $15 - 6$
11. $18 - 20$
12. $-5 - 8$
13. $21 - (-37)$
14. $-60 - 32$
15. $57 - 63$

Evaluate each expression if $k = -3$, $p = 6$, $n = 1$, and $d = -8$.

16. $55 - k$
17. $p - 7$
18. $d - 15$
19. $n - 12$
20. $-51 - d$
21. $k - 21$
22. $n - k$
23. $-99 - k$
24. $p - k$
25. $d - (-1)$
26. $k - d$
27. $n - d$

Lesson 3-6 (Pages 134–137)

Multiply.

1. $5(-2)$
2. $6(-4)$
3. $4(21)$
4. $-11(-5)$
5. $-6(5)$
6. $-50(0)$
7. $-5(-5)$
8. $-4(8)$
9. $3(-13)$
10. $12(-5)$
11. $-9(-12)$
12. $15(-8)$

Evaluate each expression if $a = -5$, $b = 2$, $c = -3$, and $d = 4$.

13. $-2d$
14. $6a$
15. $3ab$
16. $-12d$
17. $-4b^2$
18. $-5cd$
19. a^2
20. $13ab$

Simplify each expression.

21. $6(7b)$
22. $-8(2a)$
23. $-3(-5n)$
24. $5(-6x)$
25. $-9(8c)$
26. $14(-2m)$
27. $25(3d)$
28. $-10(17z)$

Lesson 3-7 (Pages 138–141)

Divide.

1. $4 \div (-2)$
2. $16 \div (-8)$
3. $-14 \div (-2)$
4. $\frac{32}{8}$
5. $18 \div (-3)$
6. $-18 \div 3$
7. $8 \div (-8)$
8. $0 \div (-1)$
9. $-25 \div 5$
10. $\frac{-14}{-7}$
11. $-32 \div 8$
12. $-56 \div (-8)$
13. $-81 \div 9$
14. $-42 \div (-7)$
15. $121 \div (-11)$
16. $-81 \div (-9)$
17. $18 \div (-2)$
18. $\frac{-55}{11}$
19. $\frac{25}{-5}$
20. $-21 \div 3$

Evaluate each expression if $a = -2$, $b = -7$, $x = 8$, and $y = -4$.

21. $-64 \div x$
22. $\frac{16}{y}$
23. $x \div 2$
24. $\frac{a}{2}$
25. $ax \div y$
26. $\frac{bx}{y}$
27. $2y \div 1$
28. $\frac{x}{ay}$
29. $-y \div a$
30. $x^2 \div y$
31. $\frac{ab}{1}$
32. $\frac{xy}{a}$

Lesson 4-1
(Pages 150–152)

Write each phrase as an algebraic expression.

1. six less than p
2. twenty more than c
3. the quotient of a and b
4. Juana's age plus 6
5. x increased by twelve
6. $1,000 divided by z
7. 3 divided into y
8. the product of 7 and m
9. the difference of f and 9
10. twenty-six less q
11. 19 decreased by z
12. two less than x
13. thirty-five increased by a
14. four hundred divided into n
15. Joe's height minus nineteen
16. twice as many lemons
17. eight more than x
18. the product of m and n

Write each sentence as an algebraic equation.

19. Three times a number less four is 17.
20. The sum of a number and 6 is 5.
21. Twenty more than twice a number is -30.
22. The quotient of a number and -2 is -42.
23. Four plus three times a number is 18.
24. Five times a number minus 15 is 92.
25. Eight times a number plus twelve is 36.
26. The difference of a number and 24 is -30.
27. 42 more than five times a number is 156.
28. Thirteen and a number is 62.
29. The quotient of a number and -4 is -324.
30. The sum of a number and eleven is 41.

Lesson 4-2
(Pages 156–159)

Solve each equation. Check your solution.

1. $r - 3 = 14$
2. $t + 3 = 21$
3. $s + 10 = 23$
4. $7 + a = -10$
5. $14 + m = 24$
6. $-9 + n = 13$
7. $s - 2 = -6$
8. $6 + f = 71$
9. $x + 27 = 30$
10. $a - 7 = 23$
11. $-4 + b = -5$
12. $w + 18 = -4$
13. $k - 9 = -3$
14. $j + 12 = 11$
15. $-42 + v = -42$
16. $s + 1.3 = 18$
17. $t + 3.43 = 7.4$
18. $x + 7.4 = 23.5$
19. $p + 3.1 = 18$
20. $q - 2.17 = 21$
21. $w - 3.7 = 4.63$
22. $m - 4.8 = 7.4$
23. $x - 1.3 = 12$
24. $y + 3.4 = 18$
25. $0.013 + h = 4.0$
26. $7.2 + g = 9.1$
27. $z - 12.1 = 14$
28. $v - 18 = 13.7$
29. $w - 0.1 = 0.32$
30. $r + 6.7 = 1.2$

Lesson 4-3
(Pages 160–163)

Solve each equation. Check your solution.

1. $2m = 18$
2. $-42 = 6n$
3. $72 = 8k$
4. $-20r = 20$
5. $420 = 5s$
6. $325 = 25t$
7. $-14 = -2p$
8. $18q = 36$
9. $40 = 10a$
10. $100 = 20b$
11. $416 = 4c$
12. $45 = 9d$
13. $0.5m = 3.5$
14. $1.8 = 0.6x$
15. $0.4y = 2$
16. $1.86 = 6.2z$
17. $-8x = 24$
18. $8.34 = 2r$
19. $1.67t = 10.02$
20. $243 = 27a$
21. $0.9x = 4.5$
22. $4.08 = 1.2y$
23. $8d = 112$
24. $5f = 180.5$
25. $59.66 = 3.14m$
26. $98.4 = 8p$
27. $208 = 26k$

Lesson 4-4
(Pages 166–169)

Solve each equation. Check your solution.

1. $3x + 6 = 6$
2. $2r - 7 = -1$
3. $-10 + 2d = 8$
4. $2b + 4 = -8$
5. $5w - 12 = 3$
6. $5t - 4 = 6$
7. $2q - 6 = 4$
8. $2g - 3 = -9$
9. $15 = 6y + 3$
10. $3s - 4 = 8$
11. $18 - 7f = 4$
12. $13 + 3p = 7$
13. $7.5r + 2 = -28$
14. $4.2 + 7z = 2.8$
15. $-9m - 9 = 9$
16. $32 + 0.2c = 1$
17. $5t - 14 = -14$
18. $-0.25x + 0.5 = 4$
19. $5w - 4 = 8$
20. $4d - 3 = 9$
21. $2g - 16 = -9$
22. $4k + 13 = 20$
23. $7 = 5 - 2x$
24. $8z + 15 = -1$
25. $92 - 16b = 12$
26. $14e + 14 = 28$
27. $1.1j + 2 = 7.5$

Lesson 4-5
(Pages 172–175)

Solve each inequality. Graph the solution on a number line.

1. $x + 2 > -3$
2. $x + 2.9 \leq 9.1$
3. $8t \geq 24$
4. $v - 3 < -3$
5. $6y \geq -12$
6. $a + 3 \leq -2$
7. $k - 5 < -2$
8. $q - 3 \leq 14$
9. $c - 4 \leq -2$
10. $n + 2 > -5$
11. $j + 1.2 > 4.8$
12. $4x < 40$
13. $2y \leq 10$
14. $g + 8 < 10$
15. $2 + b < 4$

Write an inequality for each sentence. Then solve the inequality.

16. The product of a number and four is at least 32.
17. The sum of a number and −6 is at most 27.
18. Six times a number is greater than 48.
19. The difference between a number and 7 is less than 18.

Lesson 4-6
(Pages 177–181)

Copy and complete each function table. Identify the domain and range.

1.

x	2x	y
2		
1		
0		
−1		

2.

x	3x + 1	y
1		
0		
−1		
−2		

Graph each equation.

3. $y = 3x$
4. $y = 2x + 3$
5. $y = -x$
6. $y = 4x + 2$
7. $y = 0.5x + 2$
8. $y = -x + 3$
9. $y = 0.25x + 6$
10. $y = -3x + 6$
11. $y = 2x + 7$
12. $y = -5x + 1$
13. $y = 13 + x$
14. $y = 5 - 0.5x$

Make a function table for each sentence. Then write an equation using x to represent the first number and y to represent the second number.

15. The second number is four less than the first number.
16. The second number is twelve times the first number.
17. The second number is seven more than the first number.

Lesson 4-7
(Pages 182–185)

Find the slope of the line that passes through each pair of points.

1.
2.
3.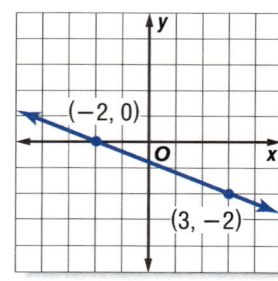

4. $(3, 4), (-2, 0)$
5. $(9, -8), (-4, 1)$
6. $(-2, -3), (-3, -6)$
7. $(0, 0), (-2, -5)$
8. $(5, 2), (-1, 3)$
9. $(1, 5), (-2, 4)$
10. $(7, 2), (-4, 3)$
11. $(2, -6), (9, -5)$
12. $(4, -1), (0, -5)$

Lesson 5-1
(Pages 197–200)

Determine whether each number is *prime* or *composite*.

1. 32
2. 41
3. 52
4. 21
5. 71
6. 102
7. 239
8. 93
9. 123
10. 251
11. 321
12. 423

Find the prime factorization of each number.

13. 81
14. 525
15. 245
16. 1,120
17. 750
18. 2,400
19. 914
20. 975
21. 423
22. 972
23. 144
24. 72

Write each expression as a product of its factors.

25. $35xy$
26. $14a^2$
27. $19bc$
28. $30n$
29. $27cd^2$
30. $37c^2$
31. $4s^2t^2$
32. $72abc$
33. $60p^2qr$

Lesson 5-2
(Pages 203–206)

Find the GCF of each set of numbers.

1. 12, 16
2. 63, 81
3. 225, 500
4. 37, 100
5. 32, 240
6. 412, 640
7. 36, 81
8. 140, 350
9. 72, 170
10. 51, 255
11. 48, 72
12. 86, 200
13. 12, 18, 42
14. 24, 56, 120
15. 48, 60, 84
16. 32, 80, 96
17. 14, 49, 70
18. 8, 10, 20

Find the GCF of each set of algebraic expressions.

19. $18b, 24b$
20. $2a, 3a$
21. $5n, 5mn$
22. $12cd, 24c$
23. $30x, 50x^2$
24. $15az, 25az$
25. $2c, 4ac, 8a$
26. $d, 6c^2d, 12d$
27. $10ab, 15bc, 20b^2$

Lesson 5-3
(Pages 207–209)

Write each fraction in simplest form.

1. $\frac{14}{28}$
2. $\frac{15}{25}$
3. $\frac{100}{300}$
4. $\frac{14}{35}$
5. $\frac{9}{51}$
6. $\frac{54}{56}$
7. $\frac{75}{90}$
8. $\frac{24}{40}$
9. $\frac{180}{270}$
10. $\frac{312}{390}$
11. $\frac{240}{448}$
12. $\frac{71}{82}$
13. $\frac{333}{900}$
14. $\frac{85}{255}$
15. $\frac{84}{128}$
16. $\frac{64}{96}$
17. $\frac{99}{99}$
18. $\frac{3}{99}$
19. $\frac{44}{55}$
20. $\frac{57}{69}$
21. $\frac{15}{37}$
22. $\frac{204}{408}$
23. $\frac{5}{125}$
24. $\frac{144}{216}$
25. $\frac{15}{75}$

Write two fractions that are equivalent to each fraction.

26. $\frac{2}{3}$
27. $\frac{1}{5}$
28. $\frac{2}{7}$
29. $\frac{3}{10}$
30. $\frac{7}{8}$
31. $\frac{5}{12}$
32. $\frac{3}{8}$
33. $\frac{8}{9}$
34. $\frac{1}{4}$
35. $\frac{2}{11}$

Lesson 5-4
(Pages 210–213)

Write each repeating decimal using bar notation.

1. 0.3333333…
2. 0.121212…
3. 4.3151515…
4. 7.023023023…
5. 0.544444…
6. 18.75484848…

Write each fraction or mixed number as a decimal. Use bar notation if the decimal is a repeating decimal.

7. $\frac{16}{20}$
8. $\frac{25}{100}$
9. $1\frac{7}{8}$
10. $\frac{1}{6}$
11. $\frac{11}{40}$
12. $5\frac{13}{50}$
13. $\frac{55}{300}$
14. $1\frac{1}{2}$
15. $\frac{5}{9}$
16. $2\frac{3}{4}$
17. $\frac{9}{11}$
18. $4\frac{1}{9}$

Write each decimal as a fraction in simplest form.

19. 0.26
20. 0.75
21. 0.4
22. 0.1
23. 4.48
24. 9.8
25. 0.91
26. 11.15

Lesson 5-5
(Pages 216–218)

Write each ratio as a percent.

1. 39 out of 100
2. $\frac{23}{100}$
3. 17 : 100
4. 72 per 100
5. 4 to 100
6. 98 in 100

Write each fraction as a percent.

7. $\frac{1}{2}$
8. $\frac{2}{5}$
9. $\frac{60}{100}$
10. $\frac{17}{20}$
11. $\frac{7}{25}$
12. $\frac{1}{20}$
13. $\frac{8}{100}$
14. $\frac{7}{7}$
15. $\frac{9}{10}$
16. $\frac{1}{100}$
17. $\frac{50}{50}$
18. $\frac{49}{50}$

Write each percent as a fraction in simplest form.

19. 12%
20. 23%
21. 1%
22. 94%
23. 36%
24. 4%
25. 72%
26. 100%
27. 65%
28. 47%
29. 15%
30. 48%

Extra Practice

Lesson 5-6 (Pages 220–223)

Write each percent as a decimal.

1. 42%
2. 100%
3. 8%
4. 20%
5. 35%
6. 3%
7. 62%
8. 50%
9. 28%
10. 87%
11. 7.5%
12. 87.5%
13. 1.8%
14. 99.9%
15. $85\frac{1}{4}$%
16. $24\frac{1}{2}$%
17. $64\frac{4}{5}$%
18. $36\frac{3}{4}$%
19. $1\frac{1}{5}$%
20. $2\frac{1}{2}$%

Write each decimal as a percent.

21. 0.16
22. 0.1
23. 0.5
24. 0.98
25. 0.31
26. 0.76
27. 0.07
28. 0.8
29. 0.07
30. 0.10
31. 0.90
32. 1.00
33. 0.666
34. 0.725
35. 0.138
36. 0.899
37. 0.256
38. 0.038
39. 0.0525
40. 0.017

Lesson 5-7 (Pages 224–226)

Find the LCM of each set of numbers.

1. 4, 9
2. 6, 16
3. 24, 36
4. 48, 84
5. 8, 9
6. 49, 56
7. 42, 66
8. 15, 39
9. 56, 64
10. 24, 42
11. 80, 250
12. 16, 24
13. 13, 14
14. 36, 48
15. 10, 100
16. 25, 200
17. 1, 2, 5
18. 2, 3, 7
19. 1, 9, 27
20. 2, 24, 36
21. 7, 21, 35
22. 12, 18, 28
23. 32, 80, 96
24. 5, 18, 45
25. 11, 22, 33
26. 35, 70, 140
27. 25, 200, 400
28. 100, 200, 300

Lesson 5-8 (Pages 227–231)

Find the LCD for each pair of fractions.

1. $\frac{3}{8}, \frac{2}{3}$
2. $\frac{5}{9}, \frac{7}{12}$
3. $\frac{4}{9}, \frac{8}{15}$
4. $\frac{11}{24}, \frac{17}{42}$
5. $\frac{12}{36}, \frac{15}{42}$
6. $\frac{25}{27}, \frac{43}{81}$
7. $\frac{32}{64}, \frac{15}{48}$
8. $\frac{2}{6}, \frac{14}{15}$

Replace each ● with <, >, or = to make a true sentence.

9. $\frac{7}{9}$ ● $\frac{3}{5}$
10. $\frac{14}{25}$ ● $\frac{3}{4}$
11. $\frac{8}{24}$ ● $\frac{20}{60}$
12. $\frac{5}{12}$ ● $\frac{4}{9}$
13. $\frac{18}{24}$ ● $\frac{10}{18}$
14. $\frac{4}{6}$ ● $\frac{5}{9}$
15. $\frac{11}{49}$ ● $\frac{12}{42}$
16. $\frac{5}{14}$ ● $\frac{2}{6}$

Order each set of ratios from least to greatest.

17. 70%, 0.6, $\frac{2}{3}$
18. 0.8, $\frac{17}{20}$, 17%
19. $\frac{61}{100}$, 0.65, 61.5%
20. $0.\overline{42}$, $\frac{3}{7}$, 42%
21. 0.15, 10.5%, $\frac{7}{50}$
22. $\frac{1}{8}$, 0.81, 18%

Determine whether each number is rational. Write *yes* or *no*.

23. $4.\overline{28}$
24. $-\frac{5}{4}$
25. $\frac{0.5}{4}$
26. $\frac{1}{7}$
27. 0.414114111…
28. 10.987…
29. 7.22…
30. $3\frac{2}{5}$

Lesson 6-1 (Pages 240–243)

Estimate.

1. $\frac{3}{7} + \frac{6}{8}$
2. $\frac{3}{9} + \frac{7}{8}$
3. $\frac{1}{8} + \frac{8}{9}$
4. $3\frac{1}{8} + 7\frac{6}{7}$
5. $4\frac{2}{3} + 6\frac{7}{8}$
6. $3\frac{2}{3} \times 2\frac{1}{3}$
7. $\frac{4}{5} \cdot 3$
8. $9\frac{7}{8} - 6\frac{2}{3}$
9. $\frac{3}{7} - \frac{1}{15}$
10. $\frac{3}{4} \cdot \frac{7}{8}$
11. $7\frac{1}{4} \div \frac{2}{3}$
12. $\frac{5}{6} \div \frac{2}{3}$
13. $9\frac{3}{5} + 3\frac{1}{8}$
14. $5\frac{1}{3} - 2\frac{3}{4}$
15. $13\frac{7}{8} - 2\frac{1}{3}$
16. $\frac{13}{15} \cdot \frac{3}{8}$
17. $\frac{1}{9} \div 2$
18. $\frac{5}{8} - \frac{1}{16}$
19. $9\frac{2}{3} + 4\frac{7}{8}$
20. $\frac{1}{2} \cdot 25$
21. $35\frac{1}{3} \div 6\frac{3}{4}$
22. $100 \div 3\frac{3}{4}$
23. $\frac{1}{3} \cdot 38$
24. $\frac{15}{16} - \frac{3}{4}$
25. $\frac{1}{5} \cdot 16$
26. $47\frac{1}{2} \div 6\frac{5}{8}$
27. $\frac{1}{9} \cdot 79$

Lesson 6-2 (Pages 244–247)

Add or subtract. Write in simplest form.

1. $\frac{5}{11} + \frac{9}{11}$
2. $\frac{5}{8} - \frac{1}{8}$
3. $\frac{7}{10} + \frac{7}{10}$
4. $\frac{9}{12} - \frac{5}{12}$
5. $\frac{2}{9} + \frac{1}{3}$
6. $\frac{1}{2} + \frac{3}{4}$
7. $\frac{1}{4} - \frac{3}{12}$
8. $\frac{3}{7} + \frac{6}{14}$
9. $\frac{1}{4} + \left(-\frac{3}{5}\right)$
10. $\frac{4}{9} + \frac{1}{2}$
11. $\frac{5}{7} - \frac{4}{6}$
12. $\frac{3}{4} - \frac{1}{6}$
13. $-\frac{3}{5} + \frac{3}{4}$
14. $-\frac{2}{3} - \frac{1}{8}$
15. $\frac{9}{10} + \frac{1}{3}$
16. $\frac{8}{15} + \frac{2}{9}$
17. $\frac{6}{7} + \frac{6}{9}$
18. $-\frac{3}{7} - \frac{3}{4}$

Evaluate each expression if $a = \frac{2}{3}$ and $b = \frac{7}{12}$.

19. $\frac{1}{5} + a$
20. $a - \frac{1}{2}$
21. $b + \frac{7}{8}$
22. $\frac{7}{8} - a$
23. $a + b$
24. $a - b$

Lesson 6-3 (Pages 248–251)

Add or subtract. Write in simplest form.

1. $2\frac{1}{3} + 1\frac{1}{3}$
2. $5\frac{2}{7} - 2\frac{3}{7}$
3. $6\frac{3}{8} + 7\frac{1}{8}$
4. $2\frac{3}{4} - 1\frac{1}{4}$
5. $5\frac{1}{2} - 3\frac{1}{4}$
6. $2\frac{2}{3} + 4\frac{1}{9}$
7. $7\frac{4}{5} + 9\frac{3}{10}$
8. $3\frac{3}{4} + 5\frac{5}{8}$
9. $10\frac{2}{3} + 5\frac{6}{7}$
10. $17\frac{2}{9} - 12\frac{1}{3}$
11. $6\frac{5}{12} + 12\frac{5}{12}$
12. $7\frac{1}{4} + 15\frac{5}{6}$
13. $6\frac{1}{8} + 4\frac{2}{3}$
14. $7 - 6\frac{4}{9}$
15. $8\frac{1}{12} + 12\frac{6}{11}$
16. $7\frac{2}{3} + 8\frac{1}{4}$
17. $12\frac{3}{11} + 14\frac{3}{13}$
18. $21\frac{1}{3} + 15\frac{3}{8}$
19. $19\frac{1}{7} + 6\frac{1}{4}$
20. $9\frac{2}{5} - 8\frac{1}{3}$
21. $18\frac{1}{4} - 3\frac{3}{8}$
22. $1\frac{1}{8} + 2\frac{1}{12}$
23. $2\frac{1}{12} - 1\frac{1}{8}$
24. $10 - \frac{2}{3}$

Lesson 6-4 (Pages 254–257)

Multiply. Write in simplest form.

1. $\frac{2}{3} \times \frac{3}{5}$
2. $\frac{1}{6} \times \frac{2}{5}$
3. $\frac{4}{9} \times \frac{3}{7}$
4. $\frac{5}{12} \times \frac{6}{11}$
5. $\frac{3}{8} \times \frac{8}{9}$
6. $\frac{3}{5} \times \frac{1}{12}$
7. $\frac{2}{5} \times \frac{5}{8}$
8. $\frac{7}{15} \times \frac{3}{21}$
9. $\frac{5}{6} \times \frac{15}{16}$
10. $\frac{6}{14} \times \frac{12}{18}$
11. $\frac{2}{3} \times \frac{3}{13}$
12. $\frac{4}{9} \times \frac{1}{6}$
13. $3 \times \frac{1}{9}$
14. $5 \times \frac{6}{7}$
15. $\frac{3}{5} \times 15$
16. $3\frac{1}{2} \times 4\frac{1}{3}$
17. $-3\frac{5}{8} \times 4\frac{1}{2}$
18. $-\frac{4}{5} \times 2\frac{3}{4}$
19. $6\frac{1}{8} \times 5\frac{1}{7}$
20. $2\frac{2}{3} \times 2\frac{1}{4}$
21. $\frac{7}{8} \times 16$
22. $5\frac{1}{5} \times 2\frac{1}{2}$
23. $-7 \times \frac{1}{14}$
24. $-22 \times \frac{3}{11}$
25. $8\frac{2}{3} \times 1\frac{1}{2}$
26. $4 \times 6\frac{1}{2}$
27. $\frac{1}{2} \times 10\frac{2}{3}$
28. $\frac{2}{3} \times 21\frac{1}{3}$
29. $-\frac{7}{8} \times \left(-\frac{8}{7}\right)$
30. $-1\frac{2}{3} \times \left(-1\frac{2}{3}\right)$
31. $21 \times \frac{1}{2}$
32. $11 \times \frac{1}{4}$

Lesson 6-5 (Pages 258–261)

Find the multiplicative inverse of each number.

1. $\frac{2}{3}$
2. $-\frac{5}{4}$
3. 1
4. 10
5. -23
6. $-\frac{1}{7}$
7. $\frac{9}{16}$
8. 15
9. $1\frac{1}{3}$
10. $-3\frac{3}{4}$
11. $7\frac{3}{8}$
12. $-\frac{19}{10}$
13. $-\frac{3}{49}$
14. $6\frac{2}{5}$
15. $33\frac{1}{3}$
16. $-66\frac{2}{3}$

Solve each equation. Check your solution.

17. $\frac{a}{13} = 2$
18. $\frac{8}{9}x = 24$
19. $\frac{3}{8}r = 36$
20. $\frac{3}{4}t = \frac{1}{2}$
21. $16 = \frac{h}{4}$
22. $\frac{m}{8} = -12$
23. $\frac{5}{8}n = -45$
24. $10 = \frac{b}{10}$
25. $\frac{1}{7}x = 7$
26. $-5 = \frac{1}{5}y$
27. $\frac{4}{3}m = -28$
28. $\frac{2}{3}z = 20$
29. $\frac{c}{9} = 81$
30. $\frac{m}{9} = 9$
31. $-16 = \frac{4}{9}f$
32. $\frac{15}{8}x = 225$

Lesson 6-6 (Pages 264–266)

Divide. Write in simplest form.

1. $\frac{2}{3} \div \frac{3}{2}$
2. $\frac{3}{5} \div \frac{2}{5}$
3. $\frac{7}{10} \div \frac{3}{8}$
4. $\frac{5}{9} \div \frac{2}{5}$
5. $4 \div \frac{2}{3}$
6. $8 \div \frac{4}{5}$
7. $9 \div \frac{5}{9}$
8. $\frac{2}{7} \div 2$
9. $\frac{1}{14} \div 7$
10. $\frac{2}{13} \div \frac{5}{26}$
11. $\frac{4}{7} \div \frac{6}{7}$
12. $\frac{7}{8} \div \frac{1}{3}$
13. $15 \div \frac{3}{5}$
14. $\frac{9}{14} \div \frac{3}{4}$
15. $-\frac{8}{9} \div \frac{5}{6}$
16. $-2\frac{1}{2} \div 5$
17. $\frac{7}{8} \div 10$
18. $16 \div \frac{3}{4}$
19. $-22 \div 3\frac{2}{3}$
20. $40\frac{5}{8} \div (-4)$
21. $\frac{3}{8} \div 2\frac{1}{2}$
22. $5\frac{1}{2} \div 2\frac{1}{2}$
23. $3\frac{1}{4} \div 5\frac{1}{2}$
24. $12\frac{5}{6} \div 2\frac{1}{6}$
25. $7\frac{1}{2} \div 3\frac{1}{2}$
26. $-3\frac{1}{2} \div \left(-7\frac{1}{2}\right)$
27. $-4\frac{2}{3} \div \left(-2\frac{1}{3}\right)$

Lesson 6-7 (Pages 267–269)

Complete.

1. 4,000 lb = __?__ T
2. 5 T = __?__ lb
3. 5 lb = __?__ oz
4. 12,000 lb = __?__ T
5. $\frac{1}{4}$ lb = __?__ oz
6. 12 pt = __?__ c
7. 3 gal = __?__ pt
8. 24 fl oz = __?__ c
9. 8 pt = __?__ c
10. 10 pt = __?__ qt
11. $2\frac{1}{4}$ c = __?__ fl oz
12. 6 lb = __?__ oz
13. 10 gal = __?__ qt
14. 4 qt = __?__ fl oz
15. 4 pt = __?__ c
16. 13,200 ft = __?__ mi
17. 120 oz = __?__ lb
18. $9\frac{1}{4}$ gal = __?__ qt
19. 7,480 yd = __?__ mi
20. $12\frac{1}{2}$ lb = __?__ oz
21. $7\frac{1}{2}$ qt = __?__ pt
22. $3\frac{1}{8}$ c = __?__ fl oz
23. $2\frac{1}{4}$ mi = __?__ ft
24. $3\frac{2}{3}$ T = __?__ lb

Lesson 6-8 (Pages 270–273)

Find the perimeter and area of each rectangle.

1. 8 yd, 3 yd
2. 15.5 cm, 12.2 cm
3. 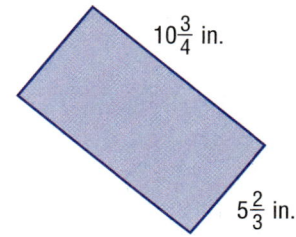 $10\frac{3}{4}$ in., $5\frac{2}{3}$ in.

4. $\ell = 80$ yd, $w = 20$ yd
5. $\ell = 75$ cm, $w = 25$ cm
6. $\ell = 5.25$ km, $w = 1.5$ km
7. $\ell = 8.6$ cm, $w = 2.5$ cm
8. $\ell = 20.25$ m, $w = 4.75$ m
9. $\ell = 12$ ft, $w = 3$ ft
10. $\ell = 5\frac{1}{4}$ mi, $w = 2\frac{1}{2}$ mi
11. $\ell = 10\frac{2}{3}$ ft, $w = 5\frac{5}{6}$ ft
12. $\ell = 12\frac{1}{3}$ yd, $w = 5\frac{2}{3}$ yd
13. $\ell = 3\frac{7}{8}$ in., $w = 1\frac{1}{4}$ in.

Lesson 6-9 (Pages 275–277)

Find the circumference of each circle. Use 3.14 or $\frac{22}{7}$ for π. Round to the nearest tenth if necessary.

1. 8 ft
2. 3 in.
3. 0.5 m
4. 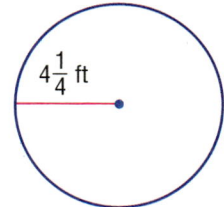 $4\frac{1}{4}$ ft

5. $r = 1$ m
6. $d = 2$ yd
7. $d = 5,280$ ft
8. $r = 350$ mi
9. $r = 1.5$ in.
10. $d = 1.5$ in.
11. $r = 0.5$ cm
12. $d = 6.4$ m
13. $r = 10.7$ km
14. $r = \frac{5}{8}$ in.
15. $d = 8\frac{1}{4}$ in.
16. $r = 2\frac{1}{3}$ yd
17. $d = \frac{3}{16}$ in.
18. $r = 5\frac{1}{2}$ mi
19. $d = 42\frac{3}{4}$ ft

Lesson 7-1 (Pages 288–291)

Write each ratio as a fraction in simplest form.

1. 45 to 15
2. 64 : 128
3. 14 to 49
4. 125 : 25
5. 18 to 81
6. 16 : 40
7. 120 to 180
8. 32 : 64
9. 90 to 100
10. 12 weeks out of 15
11. 33 min : 60 min
12. 10 ft to 8 yd
13. 25 pounds to 80 pounds
14. 15 minutes : 4 hours
15. 10 yards to 12 feet

Determine whether the ratios are equivalent. Explain.

16. 14 to 77 and 8 to 44
17. $\frac{48}{16}$ and $\frac{1}{3}$
18. 65 : 13 and 500 : 100
19. 72 to 90 and 20 to 16
20. 250 : 100 and 5 : 2
21. $\frac{32}{2}$ and $\frac{3}{48}$
22. 3 lb to 60 oz and 5 lb to 100 oz
23. 12 : 25 and 84 : 150
24. 28 in. : 5 ft and 40 in. : 8 ft
25. 8 hours to 5 days and 24 hours to 15 days

Lesson 7-2 (Pages 292–295)

Find each unit rate. Round to the nearest hundredth if necessary.

1. $240 for 4 days
2. 250 people in 5 buses
3. 500 miles in 10 hours
4. 18 cups for 24 pounds
5. 32 people in 8 cars
6. $4.50 for 3 dozen
7. 245 tickets in 5 days
8. 12 classes in 4 semesters
9. 60 people in 4 rows
10. 48 ounces in 3 pounds
11. 20 people in 4 groups
12. 1.5 pounds for $3.00
13. 45 miles in 60 minutes
14. $5.50 for 10 disks
15. 360 miles for 12 gallons
16. $8.50 for 5 yards
17. 24 cups for $1.20
18. 160 words in 4 minutes
19. $60 for 5 books
20. $24 for 6 hours

Lesson 7-3 (Pages 297–300)

Determine whether each pair of ratios forms a proportion.

1. $\frac{2}{5}$ and $\frac{5}{8}$
2. $\frac{125}{275}$ and $\frac{150}{300}$
3. $\frac{28}{42}$ and $\frac{70}{105}$
4. $\frac{55}{22}$ and $\frac{35}{14}$
5. $\frac{99}{100}$ and $\frac{100}{99}$
6. $\frac{85}{17}$ and $\frac{5}{1}$
7. $\frac{100 \text{ ft}}{8 \text{ s}}$ and $\frac{250 \text{ ft}}{20 \text{ s}}$
8. $\frac{\$15}{45 \text{ lb}}$ and $\frac{\$20}{65 \text{ lb}}$
9. $\frac{32 \text{ mi}}{6 \text{ h}}$ and $\frac{44 \text{ mi}}{8.25 \text{ h}}$

Solve each proportion.

10. $\frac{u}{72} = \frac{2}{4}$
11. $\frac{12}{m} = \frac{15}{10}$
12. $\frac{36}{90} = \frac{16}{t}$
13. $\frac{g}{32} = \frac{8}{64}$
14. $\frac{5}{14} = \frac{10}{a}$
15. $\frac{k}{18} = \frac{5}{3}$
16. $\frac{120}{150} = \frac{p}{20}$
17. $\frac{15}{w} = \frac{60}{4}$
18. $\frac{81}{90} = \frac{y}{20}$
19. $\frac{14}{s} = \frac{8}{4}$
20. $\frac{h}{3} = \frac{36}{9}$
21. $\frac{50}{8} = \frac{150}{t}$
22. $\frac{45}{8} = \frac{36}{d}$
23. $\frac{125}{v} = \frac{20}{5}$
24. $\frac{4}{5} = \frac{x}{3}$
25. $\frac{45}{75} = \frac{j}{3}$
26. $\frac{3}{7} = \frac{21}{d}$
27. $\frac{3}{10} = \frac{z}{36}$

Lesson 7-4
(Pages 304–308)

On a map, the scale is 1 inch = 50 miles. For each map distance, find the actual distance.

1. 5 inches
2. 12 inches
3. $3\frac{1}{2}$ inches
4. $2\frac{3}{8}$ inches
5. $\frac{4}{5}$ inch
6. $6\frac{3}{4}$ inches
7. $2\frac{5}{6}$ inches
8. 8 inches
9. 1.5 inches
10. 3.25 inches
11. 4.75 inches
12. 5.25 inches

On a scale drawing, the scale is $\frac{1}{2}$ inch = 2 feet. Find the dimensions of each room in the scale drawing.

13. 14 feet by 18 feet
14. 32 feet by 6 feet
15. 3 feet by 5 feet
16. 20 feet by 30 feet
17. 8 feet by 15 feet
18. 25 feet by 80 feet

Lesson 7-5
(Pages 312–315)

Write each percent as a fraction in simplest form.

1. 32%
2. 89%
3. 72%
4. 11%
5. 1%
6. 28%
7. 55%
8. 18.5%
9. 22.75%
10. 25.2%
11. 75.5%
12. 48.25%
13. 6.5%
14. 1.25%
15. 88.9%
16. $52\frac{1}{4}$%

Write each fraction as a percent. Round to the nearest hundredth if necessary.

17. $\frac{14}{25}$
18. $\frac{28}{50}$
19. $\frac{14}{20}$
20. $\frac{7}{10}$
21. $\frac{17}{17}$
22. $\frac{80}{125}$
23. $\frac{9}{12}$
24. $\frac{4}{6}$
25. $\frac{11}{12}$
26. $\frac{9}{16}$
27. $\frac{8}{9}$
28. $\frac{3}{16}$
29. $\frac{5}{32}$
30. $\frac{1}{16}$
31. $\frac{8}{15}$
32. $\frac{9}{11}$

Lesson 7-6
(Pages 316–318)

Write each percent as a decimal and as a mixed number or fraction in simplest form.

1. 895%
2. 555%
3. 480%
4. 920%
5. 122%
6. 825%
7. 0.3%
8. 0.42%
9. 0.78%
10. 765%
11. 0.99%
12. 1,000%

Write each decimal as a percent.

13. 3.5
14. 12
15. 0.002
16. 6.78
17. 0.0056
18. 1.95
19. 0.0077
20. 0.0102
21. 14.0
22. 0.0064
23. 16.2
24. 44.3

Write each number as a percent.

25. $\frac{1}{250}$
26. $\frac{7}{400}$
27. $1\frac{1}{4}$
28. $7\frac{9}{10}$
29. $\frac{9}{10}$
30. $\frac{1}{500}$
31. $12\frac{1}{2}$
32. $\frac{1}{1,000}$
33. $4\frac{3}{4}$
34. 25
35. 900
36. $18\frac{2}{5}$

Lesson 7-7
(Pages 319–321)

Find each number. Round to the nearest tenth if necessary.

1. 5% of 40 is what number?
2. What number is 10% of 120?
3. Find 12% of 150.
4. Find 12.5% of 40.
5. What number is 75% of 200?
6. Find 13% of 25.3.
7. 250% of 44 is what number?
8. What number is 0.5% of 13.7?
9. Find 600% of 7.
10. Find 1.5% of $25.
11. Find 81% of 134.
12. What number is 43% of 110?
13. What number is 61% of 524?
14. Find 100% of 3.5.
15. 20% of 58.5 is what number?
16. Find 45% of 125.5.
17. What number is 23% of 500?
18. Find 80% of 8.
19. 90% of 72 is what number?
20. What number is 32% of 54?

Lesson 7-8
(Pages 323–325)

Find each number. Round to the nearest tenth if necessary.

1. What number is 25% of 280?
2. 38 is what percent of 50?
3. 54 is 25% of what number?
4. 24.5% of what number is 15?
5. What number is 80% of 500?
6. 12% of 120 is what number?
7. Find 68% of 50.
8. What percent of 240 is 32?
9. 99 is what percent of 150?
10. Find 75% of 1.
11. What number is $33\frac{1}{3}$% of 66?
12. 50% of 350 is what number?
13. What percent of 450 is 50?
14. What number is $37\frac{1}{2}$% of 32?
15. 95% of 40 is what number?
16. Find 30% of 26.
17. 9 is what percent of 30?
18. 52% of what number is 109.2?
19. What number is 65% of 200?
20. What number is 15.5% of 45?

Lesson 8-1
(Pages 334–337)

Estimate by using fractions.

1. 28% of 48
2. 99% of 65
3. 445% of 20
4. 9% of 81
5. 73% of 240
6. 65.5% of 75
7. 48.2% of 93
8. 39.45% of 51
9. 287% of 122
10. 53% of 80
11. 414% of 72
12. 59% of 105

Estimate by using 10%.

13. 30% of 42
14. 70% of 104
15. 90% of 152
16. 67% of 70
17. 78% of 92
18. 12% of 183
19. 51% of 221
20. 23% of 504
21. 81% of 390
22. 41% of 60
23. 59% of 178
24. 22% of 450

Estimate.

25. 50% of 37
26. 18% of 90
27. 300% of 245
28. 1% of 48
29. 70% of 300
30. 35% of 35
31. 60.5% of 60
32. $5\frac{1}{2}$% of 100
33. 40.01% of 16
34. 80% of 62
35. 45% of 119
36. 14.81% of 986

Lesson 8-2

(Pages 340–343)

Write an equation for each problem. Then solve. Round to the nearest tenth if necessary.

1. Find 45% of 50.
2. 75 is what percent of 300?
3. 16% of what number is 2?
4. 75% of 80 is what number?
5. 5% of what number is 12?
6. Find 60% of 45.
7. Find 22% of 22.
8. 12% of what number is 50?
9. 38 is what percent of 62?
10. 80 is what percent of 90?
11. 90 is what percent of 95?
12. $28\frac{1}{2}$% of 64 is what number?
13. Find 46.5% of 75.
14. What number is 55.5% of 70?
15. 80.5% of what number is 80.5?
16. $66\frac{2}{3}$% of what number is 40?
17. Find 122.5% of 80.
18. 250% of what number is 75?

Lesson 8-3

(Pages 345–347)

1. The table shows the results of a survey of students' favorite cookies. Predict how many of the 424 students at Scobey High School prefer chocolate chip cookies.

Cookie	Number
chocolate chip	49
peanut butter	12
oatmeal	10
sugar	8
raisin	3

2. The circle graph shows the results of a survey of teens and where they would prefer to spend a family vacation. Predict how many of the 4,000 teens in the Central School District would prefer to go to an amusement park.

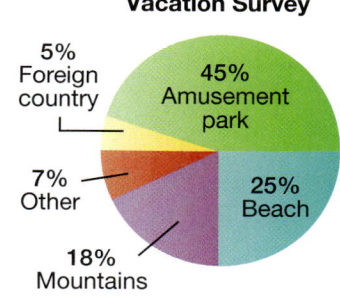

Vacation Survey
- 5% Foreign country
- 45% Amusement park
- 7% Other
- 25% Beach
- 18% Mountains

3. A survey showed that 7 out of 10 first graders received an allowance. Based on that survey, how many of the 800 first graders in the Midtown School District receive an allowance?
4. In 2000, about 29% of the foreign visitors to the U.S. were from Canada. If a particular hotel had 150,000 foreign guests in one year, how many would you predict were from Canada?

Lesson 8-4

(Pages 350–353)

Find each percent of change. Round to the nearest whole percent if necessary. State whether the percent of change is an *increase* or *decrease*.

1. original: 450
   new: 675
2. original: $1,500
   new: $1,200
3. original: 750
   new: 700
4. original: 350
   new: 420
5. original: 500
   new: 100
6. original: $75
   new: $50
7. original: 3.25
   new: 2.95
8. original: 32.5
   new: 44
9. original: 180
   new: 160
10. original: $65
    new: $75
11. original: 22
    new: 88
12. original: 450
    new: 445.5

Lesson 8-5
(Pages 354–357)

Find the total cost or sale price to the nearest cent.

1. $45 sweater; 6% tax
2. $18.99 CD; 15% discount
3. $199 ring; 10% discount
4. $29 shirt; 7% tax
5. $55 plant; 20% discount
6. $19 purse; 25% discount
7. $150 clock; 5% tax
8. $89 radio; 30% discount
9. $39 shoes; 5.5% tax
10. $145 coat; 6.25% tax
11. $300 table; 30% discount
12. $12 meal; 4.5% tax
13. $899 computer; 20% discount
14. $105 skateboard; $7\frac{1}{2}$% tax
15. $599 TV; 12% discount
16. $425 skis; 15% discount
17. $12,500 car; $3\frac{3}{4}$% tax
18. $49.95 gloves; $5\frac{1}{4}$% tax

Find the percent of discount to the nearest percent.

19. regular price: $72
    sale price: $60
20. regular price: $125
    sale price: $120
21. regular price: $360
    sale price: $280
22. regular price: $90
    sale price: $22.50
23. regular price: $25,000
    sale price: $22,000
24. regular price: $0.99
    sale price: $0.82

Lesson 8-6
(Pages 358–360)

Find the interest earned to the nearest cent for each principal, interest rate, and time.

1. $2,000, 8%, 5 years
2. $500, 10%, 8 months
3. $750, 5%, 1 year
4. $175.50, $6\frac{1}{2}$%, 18 months
5. $236.20, 9%, 16 months
6. $89, $7\frac{1}{2}$%, 6 months
7. $800, 5.75%, 3 years
8. $5,500, 7.2%, 4 years
9. $245, 6%, 13 months
10. $1,200, 3%, 45 months
11. $225, $1\frac{1}{2}$%, 2 years
12. $12,000, $4\frac{1}{2}$%, 40 months

Find the interest paid to the nearest cent for each loan balance, interest rate, and time.

13. $750, 18%, 2 years
14. $1,500, 19%, 16 months
15. $300, 9%, 1 year
16. $4,750, 19.5%, 30 months
17. $2,345, 17%, 9 months
18. $689, 12%, 2 years
19. $390, 18.75%, 15 months
20. $1,250, 22%, 8 months
21. $3,240, 18%, 14 months
22. $675, 15%, 2 years
23. $899, $10\frac{1}{2}$%, 18 months
24. $1,000, 8%, 1 year

Lesson 9-1
(Pages 370–373)

Use the spinner at the right to find each probability. Write as a fraction in simplest form.

1. P(even number)
2. P(prime number)
3. P(factor of 12)
4. P(composite number)
5. P(greater than 10)
6. P(neither prime nor composite)

A package of balloons contains 5 green, 3 yellow, 4 red, and 8 pink balloons. Suppose you reach in the package and choose one balloon at random. Find the probability of each event. Write as a fraction in simplest form.

7. P(red balloon)
8. P(yellow balloon)
9. P(pink balloon)
10. P(orange balloon)
11. P(red or yellow balloon)
12. P(*not* green balloon)

Lesson 9-2 (Pages 374–377)

For each situation, make a tree diagram to show the sample space. Then give the total number of outcomes.

1. rolling 2 number cubes
2. choosing an ice cream cone from waffle, plain, or sugar and a flavor of ice cream from chocolate, vanilla, or strawberry
3. making a sandwich from white, wheat, or rye bread, cheddar or Swiss cheese and ham, turkey, or roast beef
4. flipping a penny twice
5. choosing one math class from algebra and geometry and one foreign language class from French, Spanish, or Latin

Lesson 9-3 (Pages 378–380)

Use the Fundamental Counting Principle to find the total number of outcomes in each situation.

1. choosing a local phone number if the exchange is 398 and each of the four remaining digits is different
2. choosing a way to drive from Millville to Westwood if there are 5 roads that lead from Millville to Miamisburg, 3 roads that connect Miamisburg to Hathaway, and 4 highways that connect Hathaway to Westwood
3. tossing a quarter, rolling a number cube, and tossing a dime
4. spinning the spinners shown below

  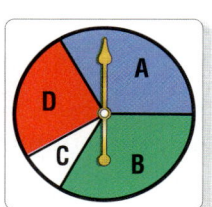

Lesson 9-4 (Pages 381–383)

Find the value of each expression.

1. 2!
2. 7!
3. 1!
4. 8!
5. 3! · 5!
6. 2! · 5!
7. 10 factorial
8. 1! · 9!
9. 9 · 8 · 7
10. 6! · 5!
11. 7 · 6 · 5 · 4
12. 8! · 3!

Solve.

13. Eight runners are competing in a 100-meter sprint. In how many ways can the gold, silver, and bronze medals be awarded?
14. In a lottery for which 30 tickets were sold (all to different people), in how many ways can the grand prize, second prize, and third prizes be awarded?
15. Five-digit locker combinations are assigned using the digits 1–9. In how many ways can the combinations be formed if no digit can be repeated?
16. How many ways can you arrange the letters in the word *factor*?

Lesson 9-5
(Pages 387–390)

Tell whether each problem represents a *permutation* or *combination*. Then solve the problem.

1. How many ways can four students be placed in a line?
2. In how many ways can six books be displayed on a rack from a collection of 12?
3. In how many ways can you select three report topics from a total of 8 topics?
4. How many ways can nine students finish first, second, third, or fourth at a speech competition?
5. In how many ways can you ride five out of nine roller coasters if you don't care in what order you ride them?
6. In how many ways can you ride five out of nine roller coasters if you care in what order you ride them?

Lesson 9-6
(Pages 393–396)

The frequency table shows the results of a fair number cube rolled 40 times.

Face	Frequency
1	5
2	9
3	2
4	8
5	12
6	4

1. Find the experimental probability of rolling a 4.
2. Find the theoretical probability of *not* rolling a 4.
3. Find the theoretical probability of rolling a 2.
4. Find the experimental probability of *not* rolling a 6.
5. Suppose the number cube was rolled 500 times. About how many times would it land on 5?
6. Suppose the number cube was rolled 500 times. About how many times would it land on a prime number?

Lesson 9-7
(Pages 398–401)

Determine whether each event is *independent* or *dependent*.

1. selecting and eating a candy from a dish and then selecting another candy
2. tossing a coin and then rolling a number cube
3. choosing a card from deck 1 and then choosing a card from deck 2

Find each probability.

4. Two evenly balanced nickels are flipped. Find the probability that one head and one tail result.
5. A wallet contains four $5 bills, two $10 bills, and eight $1 bills. Two bills are selected without the first selection being replaced. Find $P(\$5, \text{then } \$5)$.
6. Two chips are selected from a box containing 6 blue chips, 4 red chips, and 3 green chips. The first chip selected is not replaced before the second is drawn. Find $P(\text{red, then green})$.
7. A bag contains 7 blue, 4 orange, 8 red, and 5 purple marbles. Suppose one marble is chosen and then replaced. A second marble is then chosen. Find $P(\text{purple, then red})$.

Lesson 10-1 (Pages 413–415)

Classify each angle as *acute, obtuse, right,* or *straight.*

1.
2.
3.

4. 65°   5. 24°   6. 110°   7. 95°   8. 90°

Draw an angle having each measurement. Then classify each angle as *acute, obtuse, right,* or *straight.*

9. 150°   10. 45°   11. 15°   12. 180°   13. 175°
14. 88°   15. 52°   16. 160°   17. 80°   18. 55°

Lesson 10-2 (Pages 418–421)

Make a circle graph of the data in each table.

1.
Car Sales	
Style	Percent
sedan	45%
SUV	22%
pickup truck	9%
sports car	13%
compact car	11%

2.
Favorite Flavor of Ice Cream	
Flavor	Number
vanilla	11
chocolate	15
strawberry	8
mint chip	5
cookie dough	3

Lesson 10-3 (Pages 422–425)

Classify each pair of angles as *complementary, supplementary,* or *neither.*

1.
2.
3.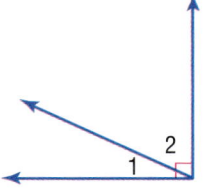

Find the value of x in each figure.

4.
5.
6.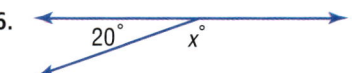

Lesson 10-4
(Pages 428–431)

Classify each triangle by its angles and by its sides.

1.
2.
3.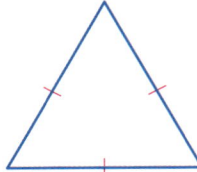

Find the missing measure in each triangle. Then classify the triangle as *acute*, *right*, or *obtuse*.

4.
5.
6.

Lesson 10-5
(Pages 434–437)

Classify the quadrilateral using the name that *best* describes it.

1.
2.
3.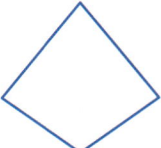

Find the missing measure in each quadrilateral.

4.
5.
6.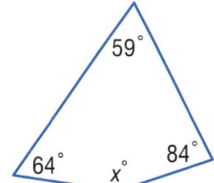

Lesson 10-6
(Pages 440–443)

Find the value of x in each pair of similar figures.

1.
2.
3.

4.
5.
6.

588 Extra Practice

Lesson 10-7 (Pages 446–450)

Determine whether each figure is a polygon. If it is, classify the polygon and state whether it is regular. If it is *not* a polygon, explain why.

1.
2.
3.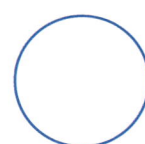

Find the measure of an angle in each regular polygon. Round to the nearest tenth degree, if necessary.

4. triangle
5. 30-gon
6. 18-gon
7. 14-gon

Lesson 10-8 (Pages 451–454)

1. Translate △ABC 2 units right and 1 unit down.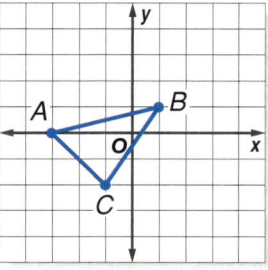
2. Translate quadrilateral RSTU 4 units left and 3 units down.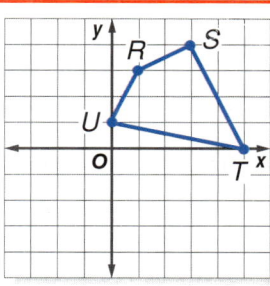

Triangle *TRI* has vertices T(1, 1), R(4, −2), and I(−2, −1). Find the vertices of T′R′I′ after each translation. Then graph the figure and its translated image.

3. 2 units right, 1 unit down
4. 5 units left, 1 unit up
5. 3 units right
6. 2 units up

Lesson 10-9 (Pages 456–459)

Determine which figures have line symmetry. Write *yes* or *no*. If *yes*, draw all lines of symmetry.

1.
2.
3.

Find the coordinates of each figure after a reflection over the *x*-axis. Then graph the figure and its reflected image.

4. quadrilateral *QUAD* with vertices Q(−1, 4), U(2, 2), A(1, 1), and D(−2, 2)
5. triangle *ABC* with vertices A(0, −1), B(4, −3), and C(−4, −5)

Find the coordinates of each figure after a reflection over the *y*-axis. Then graph the figure and its reflected image.

6. parallelogram *PARL* with vertices P(3, 5), A(5, 4), R(5, 1), and L(3, 2)
7. pentagon *PENTA* with vertices P(−1, 3), E(1, 1), N(0, −2), T(−2, −2), and A(−3, 1)

Lesson 11-1 (Pages 470–473)

Find the square of each number.

1. 4
2. 19
3. 13
4. 25
5. 9
6. 2
7. 14
8. 24
9. 40
10. 50
11. 100
12. 250

Find each square root.

13. $\sqrt{324}$
14. $\sqrt{900}$
15. $\sqrt{2,500}$
16. $\sqrt{576}$
17. $\sqrt{8,100}$
18. $\sqrt{676}$
19. $\sqrt{100}$
20. $\sqrt{784}$
21. $\sqrt{1,024}$
22. $\sqrt{841}$
23. $\sqrt{2,304}$
24. $\sqrt{3,025}$

Lesson 11-2 (Pages 475–477)

Estimate each square root to the nearest whole number.

1. $\sqrt{27}$
2. $\sqrt{112}$
3. $\sqrt{249}$
4. $\sqrt{88}$
5. $\sqrt{1,500}$
6. $\sqrt{612}$
7. $\sqrt{340}$
8. $\sqrt{495}$
9. $\sqrt{264}$
10. $\sqrt{350}$
11. $\sqrt{834}$
12. $\sqrt{3,700}$
13. $\sqrt{298}$
14. $\sqrt{101}$
15. $\sqrt{800}$

Use a calculator to find each square root to the nearest tenth.

16. $\sqrt{58}$
17. $\sqrt{750}$
18. $\sqrt{1,200}$
19. $\sqrt{1,000}$
20. $\sqrt{5,900}$
21. $\sqrt{999}$
22. $\sqrt{374}$
23. $\sqrt{512}$
24. $\sqrt{3,750}$
25. $\sqrt{255}$
26. $\sqrt{83}$
27. $\sqrt{845}$
28. $\sqrt{200}$
29. $\sqrt{500}$
30. $\sqrt{10,001}$

Lesson 11-3 (Pages 479–483)

Find the missing measure of each right triangle. Round to the nearest tenth if necessary.

1.
2.
3.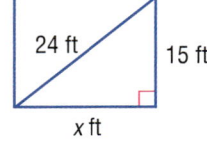

4. $a = 12$ cm, $b = 25$ cm
5. $a = 5$ yd, $c = 10$ yd
6. $b = 12$ mi, $c = 20$ mi
7. $a = 15$ yd, $b = 24$ yd
8. $a = 4$ m, $c = 12$ m
9. $a = 8$ mm, $b = 11$ mm
10. $a = 1$ mi, $c = 3$ mi
11. $a = 5$ yd, $b = 8$ yd
12. $b = 7$ in., $c = 19$ in.
13. $a = 50$ km, $c = 75$ km
14. $b = 82$ ft, $c = 100$ ft
15. $a = 100$ m, $b = 200$ m

Determine whether each triangle with the given side lengths is a right triangle. Write *yes* or *no*.

16. 6 mm, 8 mm, 10 mm
17. 12 ft, 15 ft, 20 ft
18. 300 m, 400 m, 500 m

Lesson 11-4
(Pages 483–485)

Find the area of each parallelogram. Round to the nearest tenth if necessary.

1.
2.
3.

4. base = 19 m
   height = 6 m
5. base = 15 ft
   height = 12 ft
6. base = 25 yd
   height = 53 yd
7. base = 135 in.
   height = 15 in.
8. base = 8.2 cm
   height = 5.5 cm
9. base = 29.3 m
   height = 10.1 m

Lesson 11-5
(Pages 489–492)

Find the area of each figure. Round to the nearest tenth if necessary.

1.
2.
3.

4. triangle: base = 5 in., height = 9 in.
5. trapezoid: bases = 3 cm and 8 cm, height = 12 cm
6. trapezoid: bases = 10 ft and 15 ft, height = 12 ft
7. triangle: base = 12 cm, height = 8 cm
8. trapezoid: bases = 82.6 cm and 72.2 cm, height = 44.5 cm
9. triangle: base = 500.5 ft, height = 254.5 ft

Lesson 11-6
(Pages 493–495)

Find the area of each circle. Round to the nearest tenth.

1.
2.
3.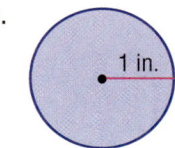

4. radius = 8 in.
5. diameter = 5 ft
6. radius = 24 cm
7. diameter = 2.3 m
8. diameter = 82 ft
9. radius = 68 cm
10. radius = 9.8 mi
11. diameter = 25.6 m
12. diameter = 6.75 in.
13. radius = $1\frac{1}{4}$ ft
14. diameter = $5\frac{2}{3}$ yd
15. diameter = $45\frac{1}{2}$ mi

Lesson 11-7
(Pages 498–500)

Find the area of each figure. Round to the nearest tenth if necessary.

1.

2.

3.

4.

5.

6.

7.

8.

9.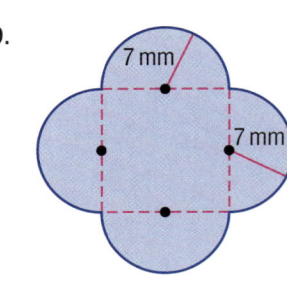

Lesson 11-8
(Pages 501–503)

A randomly dropped counter falls in the squares. Find the probability that it falls in the shaded squares. Write as a percent. Round to the nearest tenth if necessary.

1.
2.
3.
4.

5.
6.
7.
8.

9.
10.
11.
12.

Lesson 12-1
(Pages 514–517)

Draw a top, a side, and a front view of each solid.

1.
2.
3.

Draw each solid using the top, side, and front views shown. Use isometric dot paper.

4. top side front

5. top side front

6. top side front

Lesson 12-2
(Pages 520–522)

Find the volume of each rectangular prism. Round to the nearest tenth if necessary.

1.
2.
3.
4.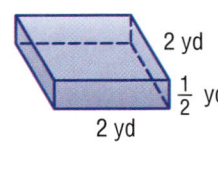

5. length = 3 ft
   width = 10 ft
   height = 2 ft

6. length = 18 cm
   width = 23 cm
   height = 15 cm

7. length = 25 mm
   width = 32 mm
   height = 10 mm

8. length = 1.5 in.
   width = 3 in.
   height = 6 in.

9. length = 4.5 cm
   width = 6.75 cm
   height = 2 cm

10. length = 16 mm
    width = 0.7 mm
    height = 12 mm

11. length = $3\frac{1}{2}$ ft
    width = 10 ft
    height = 6 ft

12. length = $5\frac{1}{2}$ in.
    width = 12 in.
    height = $3\frac{3}{8}$ in.

13. Find the volume of a rectangular prism with a length of 3 yards, a width of 5 feet, and a height of 12 feet.

Lesson 12-3
(Pages 524–527)

Find the volume of each cylinder. Round to the nearest tenth.

1.
2 cm, 4 cm

2.
3 yd, 6.5 yd

3.
7.5 mm, 16 mm

4.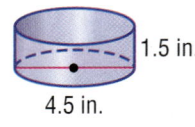
1.5 in., 4.5 in.

5. radius = 6 in.
   height = 3 in.
6. radius = 8 ft
   height = 10 ft
7. radius = 6 km
   height = 12 km
8. radius = 8.5 cm
   height = 3 cm

9. diameter = 16 yd
   height = 4.5 yd
10. diameter = 3.5 mm
    height = 2.5 mm
11. diameter = 12 m
    height = 4.75 m
12. diameter = $\frac{5}{8}$ in.
    height = 4 in.

13. diameter = 100 ft
    height = 35 ft
14. radius = 40.5 m
    height = 65.1 m
15. radius = 0.5 cm
    height = 1.6 cm
16. diameter = $8\frac{3}{4}$ in.
    height = $5\frac{1}{2}$ in.

17. Find the volume of a cylinder whose diameter is 6 inches and height is 2 feet.

18. How tall is a cylinder that has a volume of 2,123 cubic meters and a radius of 13 meters?

Lesson 12-4
(Pages 532–535)

Find the surface area of each rectangular prism. Round to the nearest tenth if necessary.

1.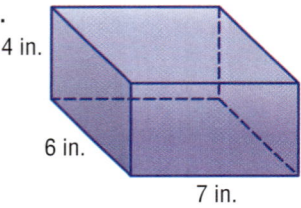
4 in., 6 in., 7 in.

2.
15 cm, 4 cm, 4 cm

3.
18 in., 10 in., 32 in.

4.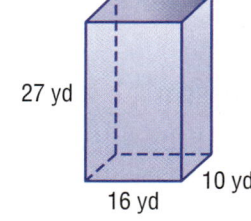
27 yd, 16 yd, 10 yd

5. length = 10 m
   width = 6 m
   height = 7 m
6. length = 20 mm
   width = 15 mm
   height = 25 mm
7. length = 16 ft
   width = 20 ft
   height = 12 ft
8. length = 52 cm
   width = 48 cm
   height = 45 cm

9. length = 8 ft
   width = 6.5 ft
   height = 7 ft
10. length = 9.4 m
    width = 2 m
    height = 5.2 m
11. length = 20.4 cm
    width = 15.5 cm
    height = 8.8 cm
12. length = 8.5 mi
    width = 3 mi
    height = 5.8 mi

13. length = $7\frac{1}{4}$ ft
    width = 5 ft
    height = $6\frac{1}{2}$ ft
14. length = $15\frac{2}{3}$ yd
    width = $7\frac{1}{3}$ yd
    height = 9 yd
15. length = $4\frac{1}{2}$ in.
    width = 10 in.
    height = $8\frac{3}{4}$ in.
16. length = 12.2 mm
    width = 7.4 mm
    height = 7.4 mm

17. Find the surface area of an open-top box with a length of 18 yards, a width of 11 yards, and a height of 14 yards.

18. Find the surface area of a rectangular prism with a length of 1 yard, a width of 7 feet, and a height of 2 yards.

Lesson 12-5 (Pages 538–541)

Find the surface area of each cylinder. Round to the nearest tenth.

1.
2.
3.
4.

5. height = 6 cm
   radius = 3.5 cm
6. height = 16.5 mm
   diameter = 18 mm
7. height = 22 yd
   radius = 10.5 yd
8. height = 6 ft
   radius = 18.5 ft

9. height = 10.2 mi
   diameter = 4 mi
10. height = 8.6 cm
    diameter = 8.2 cm
11. height = 5.8 km
    diameter = 3.6 km
12. height = 32.7 m
    radius = 21.5 m

13. height = $2\frac{2}{3}$ yd
    diameter = 6 yd
14. height = $12\frac{3}{4}$ ft
    radius = $7\frac{1}{4}$ ft
15. height = $5\frac{1}{5}$ mi
    radius = $18\frac{1}{3}$ mi
16. height = $5\frac{1}{2}$ in
    diameter = 3 in

Lesson 12-6 (Pages 542–545)

Identify the precision unit of each measuring tool.

1.
2.
3.
4.

State each measure using significant digits.

5.
6.
7.
8.

Mixed Problem Solving

Chapter 1 Decimal Patterns and Algebra (pages 4–51)

HISTORY For Exercises 1 and 2, use the following information.
In 1932, Amelia Earhart flew 2,026 miles in 14 hours 56 minutes. (Lesson 1-1)

1. How many minutes did she fly?

2. To the nearest mile, what was her speed in miles per minute?

3. **TRANSPORTATION** An MD-80 aircraft burns 950 gallons of fuel per hour. How much fuel burns during a three-hour flight? (Lesson 1-1)

4. **LIGHT** The speed of light is about 67^3 kilometers per second. About how many kilometers per second is this? (Lesson 1-2)

SALES For Exercises 5 and 6, use the following information.
A department store is having a back-to-school sale. The table shows the prices of three popular items.

Item	Price
jeans	$37.99
sweatshirt	$19.88
polo shirt	$22.50

Latonia wants to buy 2 pairs of jeans, 3 sweatshirts, and 1 polo shirt. (Lesson 1-3)

5. Write a numerical expression to represent the total cost of the six items.

6. Evaluate the expression to find the total cost.

7. **FITNESS** You can estimate how fast you walk in miles per hour by evaluating the expression $\frac{n}{30}$, where n is the number of steps you take in one minute. Find your speed if you take 96 steps in one minute. (Lesson 1-4)

8. **BASEBALL** Last year, Scott attended 13 Minnesota Twins baseball games. This year, he attended 24. Solve $13 + n = 24$ to find how many more games he attended this year than last. (Lesson 1-5)

9. **HOT AIR BALLOONS** Miyoki paid $140 for a four-hour hot air balloon ride over the Bridger Mountains. Solve $4h = 140$ to find the cost per hour of the ride. (Lesson 1-5)

ENTERTAINMENT For Exercises 10 and 11, use the following information.
The five members of the Wolff family went to an amusement park. They each purchased an all-day ride pass and a water park pass, as shown below. (Lesson 1-6)

Item	Price
all-day ride pass	$14.95
water park pass	$6.50

10. Use the Distributive Property to write two different expressions that represent the total cost for the family.

11. Find the total cost of the passes.

12. **NUMBER THEORY** Numbers that can be represented by a square arrangement of dots are called *square numbers*. The first four square numbers are shown below. (Lesson 1-7)

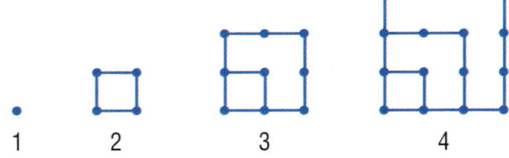

Write a sequence formed by the first eight square numbers.

13. **BIRDS** The world's smallest bird, the bee hummingbird, weighs only about 1.6 grams. How many milligrams is this? (Lesson 1-8)

14. **BRIDGES** The Akashi Kaikyo bridge in Japan is 2,003 meters long. What is the length of the bridge in kilometers? (Lesson 1-8)

15. **SPACE** The Andromeda Spiral is a galaxy 2,200,000 light years from Earth. Write this distance in scientific notation. (Lesson 1-9)

Chapter 2 Statistics: Analyzing Data (pages 52–101)

TRAVEL For Exercises 1 and 2, refer to the table below. It shows the number of passengers in twenty U.S. airports in one year. (Lesson 2-1)

Airport Passengers (millions)						
80	41	36	34	28	25	37
72	39	36	33	27	61	34
68	37	35	31	25	31	

Source: *The World Almanac*

1. Make a frequency table using the intervals 25–34, 35–44, 45–54, and so on.
2. Which interval has the greatest frequency? Explain what this means.
3. **DRIVING** The frequency table below shows the rural interstate speed limits for the fifty states. How many states have a speed limit of 75? (Lesson 2-1)

Speed Limit	Tally	Frequency
55	I	1
60		0
65	IIII IIII IIII IIII	20
70	IIII IIII IIII III	18
75	?	?

Source: *The World Almanac*

SWIMMING For Exercises 4 and 5, refer to the table at the right. It shows the winning Olympic times for the Women's 4 × 100-meter Freestyle Relay in swimming. (Lesson 2-2)

Year	Time (s)
1976	225
1980	223
1984	224
1988	221
1992	220
1996	219
2000	217

Source: *ESPN Sports Almanac*

4. Make a line graph of the data.
5. Predict the winning time in this event in 2008.

NUTRITION For Exercises 6–9, use the data below. They are the grams of Carbohydrates in fifteen different energy bars. 16, 15, 20, 24, 16, 16, 16, 2, 20, 26, 14, 20, 20, 16, 16

6. Make a line plot of the data. (Lesson 2-3)
7. What is the range of the data? (Lesson 2-3)
8. Identify any clusters, gaps, or outliers and explain what they represent. (Lesson 2-3)
9. Find the mean, median, and mode. (Lesson 2-4)

BASKETBALL For Exercises 10–12, refer to the table below. It shows the number of games played by Michael Jordan each year from 1986–1987 to 2001–2002.

Number of Games Played							
82	82	81	82	82	80	78	0
17	82	82	82	0	0	0	60

Source: www.nba.com

10. Find the mean, median, and mode of the data. (Lesson 2-4)
11. Make a stem-and-leaf plot of the data. (Lesson 2-5)
12. Find the lower extreme, LQ, median, UQ, and upper extreme. (Lesson 2-6)

VACATION For Exercises 13–15, refer to the box-and-whisker plot below. It shows the number of vacation days for employees at a company. (Lesson 2-6)

13. What fraction of the employees receive 6 to 14 vacation days?
14. What fraction of the employees receive 14 or more vacation days?
15. What is the range of the data?
16. **TOURISTS** The table shows the countries from which the most tourists in the United States came. Make a bar graph of the data. (Lesson 2-7)

Country	Number of Visitors (millions)
Canada	14.6
Mexico	10.3
Japan	5.0
United Kingdom	4.7

Source: Department of Commerce

TRAVEL For Exercises 17 and 18, refer to the data in Exercises 1 and 2.

17. Make a histogram of the data. (Lesson 2-7)
18. Would the mean or median best describe the data? Explain your reasoning. (Lesson 2-8)

Chapter 3 Algebra: Integers

(pages 104–147)

1. **AIR CONDITIONING** Jacob turned on the air conditioning, and the temperature in his apartment decreased 8 degrees. Write an integer to represent the drop in temperature. (Lesson 3-1)

EARTH SCIENCE For Exercises 2 and 3, use the table below. It describes the deepest land depressions in the world in feet below sea level.

Depth (ft)				
220	436	511	433	282
383	505	235	1,312	230

Source: *The Top 10 of Everything*

2. Write an integer to represent each depth. (Lesson 3-1)

3. Order the integers from greatest depth to least depth. (Lesson 3-2)

ENTERTAINMENT For Exercises 4–8, use the diagram below. It shows the locations of several rides at the Outlook Amusement Park. (Lesson 3-3)

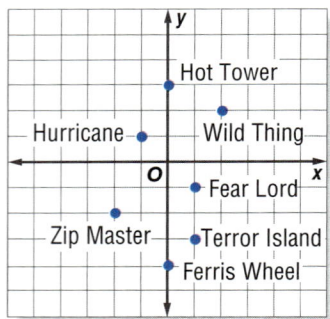

4. Which ride(s) is located in quadrant III?
5. Which ride(s) is located on the *y*-axis? Name the coordinates.
6. Which ride(s) has coordinates in which the *x*- and *y*-coordinates are equal?
7. In which quadrant is the Hurricane located?
8. A new ride is built with a location on the *x*-axis and 5 units left of the origin. Name the coordinates of this point.

9. **CAVERNS** Adriana is 52 feet underground touring the Lewis and Clark Caverns. She climbs a ladder up 15 feet. What is her new location? (Lesson 3-4)

10. **RECORDS** The lowest temperature recorded in Verkhoyansk, Russia, was about −90°F. The highest temperature was about 99°F. What is the difference between these temperatures? (Lesson 3-5)

11. **EARTH SCIENCE** The highest and lowest points in California are shown in the table. What is the difference in elevations? (Lesson 3-5)

Location	Elevation
Mount Whitney	14,494 ft above sea level
Death Valley	282 ft below sea level

Source: *The World Almanac of the U.S.A.*

12. **TEMPERATURE** Refer to the table below. How much colder is outer space than liquid helium? (Lesson 3-5)

Item	Temperature (°C)
dry ice	−108
outer space	−457
liquid helium	−452
bird's body	108

Source: *The Sizesaurus*

13. **RIDES** A glider ride over the Crazy Mountains has a maximum altitude of 12,000 feet, and it descends about 300 feet per minute. How long will it take the glider to reach an altitude of 6,000 feet? (Lesson 3-6)

14. **POOLS** A swimming pool holds 5,000 gallons of water. A hose can drain the pool at the rate of 225 gallons per hour. How much water will still be in the pool after it has been draining for 5 hours? (Lesson 3-6)

15. **TEMPERATURE** A temperature of −89°C was recorded in Antarctica. Use the expression $\frac{9C}{5} + 32$, where C is the temperature in degrees Celsius, to find the temperature in degrees Fahrenheit. (Lesson 3-7)

16. **DIVING** In two minutes, a scuba diver ascended from 40 feet below sea level to 8 feet below sea level. By how many feet did she ascend per minute? (Lesson 3-7)

Chapter 4 Algebra: Linear Equations and Functions (pages 148–191)

1. **TOURISM** The Statue of Liberty in New York, New York, and the Eiffel Tower in Paris, France, were designed by the same person. The Statue of Liberty is 152 feet tall. It is 732 feet shorter than the Eiffel Tower, x. Write an equation that models this situation. (Lesson 4-1)

ELECTIONS For Exercises 2 and 3, use the table at the right and the following information.
New York has one more electoral vote than Texas. Pennsylvania has 9 fewer electoral votes than Texas. (Lesson 4-2)

Number of Electoral Votes 2000	
California	54
New York	33
Texas	?
Florida	25
Pennsylvania	23

Source: *The World Almanac*

2. Write two different equations to find the number of electoral votes in Texas, n.
3. Find the number of electoral votes.

4. **ROLLER COASTERS** The track length of a popular roller coaster is 5,106 feet. The roller coaster has an average speed of about 2,000 feet per minute. At that speed, how long will it take to travel its length of 5,106 feet? Use the formula $d = rt$. (Lesson 4-3)

5. **BUSINESS** Carla's Catering charges a $25 fee to serve 15 or fewer people. In addition to that fee, they charge $10 per appetizer. You are having a party for 12 people and can spend a total of $85. How many appetizers can you order from Carla's Catering? (Lesson 4-4)

6. **MONEY** Koto works in a hardware store and earns $6 per hour. She plans on saving her money to buy a DVD player that costs $189. Write an inequality for the least number of hours x Koto needs to work in order to be able to purchase the DVD player. Then solve the inequality. (Lesson 4-5)

7. **CIVICS** The 26th amendment to the United States Constitution guarantees the right to vote to citizens who are eighteen years of age or older. Write an inequality showing the ages a of all voters. (Lesson 4-5)

8. **GEOMETRY** The formula for the perimeter of a square is $P = 4s$, where P is the perimeter and s is the length of a side. Graph the equation. (Lesson 4-6)

AGES For Exercises 9–12, use the table below. It shows how Jared's age and his sister Emily's age are related. (Lesson 4-6)

Jared's age (yr)	1	2	3	4	5
Emily's age (yr)	7	8	9	10	11

9. Write a verbal expression to describe how the ages are related.
10. Write an equation for the verbal expression. Let x represent Jared's age and y represent Emily's age.
11. Predict how old Emily will be when Jared is 10 years old.
12. Graph the equation.

13. **RAMPS** A ramp used to load cargo onto ships is modeled below. Is the slope of the ramp greater than or less than 1? Explain. (Lesson 4-7)

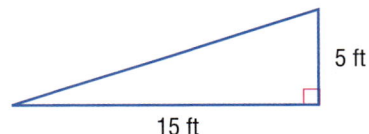

CYCLING For Exercises 14–16, use the table below. It shows the distances traveled by Jamal and Jasmine. (Lesson 4-7)

Hours Traveled	Cycling Distance (miles)	
	Jamal	Jasmine
1	8	11
2	16	22
3	24	33
4	32	44

14. Suppose each of these functions was graphed on a coordinate plane. Without calculating, which line is steeper? Explain.
15. Find the slope of each line.
16. What does the slope of each line represent?

Chapter 5 Fractions, Decimals, and Percents (pages 194–237)

LAND For Exercises 1–3, use the information below.
A section of land is one mile long and one mile wide. (Lesson 5-1)

1. Write the prime factorization of 5,280.
2. Find the area of the section of land in square feet. (*Hint*: 1 mile = 5,280 feet)
3. Write the prime factorization of the area that you found in Exercise 2.

DECORATIONS For Exercises 4 and 5, use the information below.
Benito is cutting streamers from crepe paper to decorate for a party. He has a red roll of crepe paper 144 inches long, a white roll 192 inches long, and a blue roll 360 inches long. (Lesson 5-2)

4. If he wants to have all colors of streamers the same length, what is the longest length that he can cut?
5. How many total streamers can he cut?

OLYMPICS For Exercises 6 and 7, refer to the table below. It shows the medals won by the top three countries in the 2000 Summer Olympics.

Country	Medals		
	Gold	Silver	Bronze
United States	40	24	33
Russia	32	28	28
China	28	16	15

Source: *The World Almanac*

6. Write the number of gold medals that Russia won as a fraction of the total number that Russia won in simplest form. (Lesson 5-3)
7. Write the fraction that you wrote in Exercise 6 as a decimal. (Lesson 5-4)
8. **GEOGRAPHY** Alaska makes up about 0.16 of the total land area of the United States. Write this decimal as a fraction in simplest form. (Lesson 5-4)
9. **SPORTS** At Belgrade Intermediate School, 75 out of every 100 students participate in sports. What percent of students do *not* participate in sports? (Lesson 5-5)

ADVERTISING For Exercises 10–13, use the table below. It shows the results of a survey in which teens 13 to 17 years old were asked which types of advertising they pay attention to.

Type of Advertising	Percent of Teens
television	80%
magazine	62%
product in a movie	48%
ad in an E-mail	24%

Source: E-Poll

Write each percent as a fraction in simplest form. (Lesson 5-5)

10. television
11. magazine
12. product in a movie
13. ad in an E-mail

14. **POPULATION** In 2000, about 5.4% of the people in Wyoming lived in Laramie. Write this percent as a decimal. (Lesson 5-6)

GEOMETRY For Exercises 15–17, refer to the grid at the right. (Lesson 5-6)

15. Write a decimal and a percent to represent the "T" shaded area.
16. Write a decimal and a percent to represent the area shaded pink.
17. What percent of the grid is *not* shaded?

18. **FLOWERS** Roses can be ordered in bunches of 6 and carnations, in bunches of 15. If Ingrid wants to have the same number of roses as carnations for parent night, what is the least number of each flower that she must order? (Lesson 5-7)

19. **WATER** The table at the right shows the fraction of each state that is water. Order the states from least to greatest fraction of water. (Lesson 5-8)

What Part is Water?	
State	Fraction
Alaska	$\frac{3}{41}$
Michigan	$\frac{40}{97}$
Wisconsin	$\frac{1}{6}$

Source: *The World Almanac of the U.S.A.*

Chapter 6 Applying Fractions

(pages 238–283)

1. **MEALS** A box of instant potatoes contains 20 cups of flakes. A family-sized bowl of potatoes uses $3\frac{2}{3}$ cups of the flakes. Estimate how many family-sized bowls can be made from one box. (Lesson 6-1)

RAINFALL For Exercises 2 and 3, use the table. It shows the average annual precipitation for three of the driest locations on Earth. (Lesson 6-2)

Location	Precipitation (in.)
Arica, Chile	$\frac{3}{100}$
Iquique, Chile	$\frac{1}{5}$
Callao, Peru	$\frac{12}{25}$

Source: *The Top 10 of Everything*

2. How much more rain does Iquique get per year than Arica?
3. How much more annual rain does Callao get than Iquique?
4. **CRAFTS** Kyle uses $2\frac{3}{4}$ yards of fabric for each craft item that he makes. If he uses a 25-yard fabric bolt and sells each item for $35, what is the greatest amount of money that he can earn? (Lesson 6-3)
5. **STARS** The star Sirius is about $8\frac{7}{10}$ light years from Earth. Alpha Centauri is half this distance from Earth. How far is Alpha Centauri from Earth? (Lesson 6-4)

LIFE SCIENCE For Exercises 6 and 7, use the table below. It shows the average growth per month of hair and fingernails. (Lesson 6-5)

Average Monthly Growth	
hair	$\frac{1}{2}$ in.
fingernails	$\frac{2}{25}$ in.

6. Solve $3 = \frac{1}{2}t$ to find how long it takes hair to grow 3 inches.
7. Solve $\frac{2}{25} = \frac{d}{12}$ to find how much fingernails grow in 1 year.

8. **SEWING** Jocelyn has nine yards of cloth to make table napkins for the senior citizens' center. She needs $\frac{3}{8}$ yard for each napkin. Use $\frac{3}{8}c = 9$ to find the number of napkins that she can make. (Lesson 6-5)
9. **WHALES** During the first year, a baby whale gains about $27\frac{3}{5}$ tons. What is the average weight gain per month? (Lesson 6-6)
10. **LIFE SCIENCE** An adult has about 5 quarts of blood. If a person donates 1 pint of blood, how many pints are left? (Lesson 6-7)
11. **COFFEE** In Switzerland, the average amount of coffee consumed per year is 1,089 cups per person. How many pints is this? (Lesson 6-7)

CHESS For Exercises 12–15, use the chess board below. (Lesson 6-8)

12 in.
12 in.

12. What is the perimeter of the chess board?
13. What is the area of the chess board?
14. What is the area of each small square?
15. A travel chess board has half the length and width of the board shown. What is the perimeter and area?
16. **LIFE SCIENCE** A nest built by bald eagles had a diameter of $9\frac{1}{2}$ feet. What was the circumference of the nest? (Lesson 6-9)
17. **EARTH SCIENCE** Earth has a diameter of 7,926 miles. Use the formula for the circumference of a circle to approximate the circumference of Earth at its equator. (Lesson 6-9)

Chapter 7 Ratios and Proportions

(pages 286–331)

1. **SCHOOLS** In a recent year, Oregon had 924 public elementary schools and 264 public high schools. Write a ratio in simplest form comparing the number of public high schools to elementary schools. (Lesson 7-1)

2. **TEACHERS** The table shows the ratio of students to teachers for three states. Are any of the ratios equivalent? Explain. (Lesson 7-1)

State	Students : Teachers
Alabama	152 : 10
Kentucky	231 : 15
Montana	76 : 5

   Source: *The World Almanac*

3. **EXERCISE** A person jumps rope 14 times in 10 seconds. What is the unit rate in jumps per second? (Lesson 7-2)

4. **FOOD** A 16-ounce box of cereal costs $3.95. Find the unit price to the nearest cent. (Lesson 7-2)

5. **PHOTOGRAPHS** Mandy is enlarging a photograph that is 3 inches wide and 4.5 inches long. If she wants the width of the enlargement to be 10 inches, what will be the length? (Lesson 7-3)

6. **GEOMETRY** If the ratio of length to width of a rectangle is about 1.6 to 1, then the rectangle is said to be a *golden rectangle*. Determine whether either figure below is golden rectangle. (Lesson 7-3)

   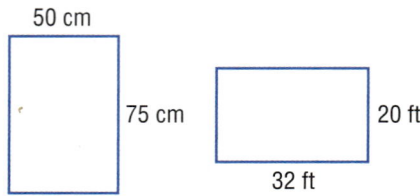

7. **MAPS** Washington, D.C., and Baltimore, Maryland, are $2\frac{7}{8}$ inches apart on a map. If the scale is $\frac{1}{2}$ inch : 6 miles, what is the actual distance between the cities? (Lesson 7-4)

8. **MODELS** Ian is making a miniature bed for his daughter's doll house. The actual bed is $6\frac{3}{4}$ feet long. If he uses the scale $\frac{1}{2}$ inch = $1\frac{1}{2}$ feet, what will be the length of the miniature bed? (Lesson 7-4)

9. **POPULATION** According to the U.S. Census Bureau, 6.6% of all people living in Florida are 10–14 years old. What fraction is this? (Lesson 7-5)

10. **INTERNET** In 2001, 50.5% of U.S. households had access to the Internet. Write this percent as a fraction in simplest form. (Lesson 7-5)

COINS For Exercises 11 and 12, use the table below. It shows the fraction of a quarter that is made up of the metals nickel and copper. Write each fraction as a percent. Round to the nearest hundredth if necessary. (Lesson 7-5)

Metal	Fraction of Quarter
nickel	$\frac{1}{12}$
copper	$\frac{11}{12}$

Source: *The Top 10 of Everything*

11. nickel

12. copper

13. **FARMS** Approximately 0.0009 of the land in Montana is farmland. Write this as a percent. (Lesson 7-6)

14. **POPULATION** In 2003, 1.6% of the U.S. population was 85 years and older. Write this percent as a fraction in simplest form. (Lesson 7-6)

15. **SEEDS** A packet of beans guarantees that 95% of its 200 seeds will germinate. How many seeds are expected to germinate? (Lesson 7-7)

16. **SKIS** Toshiro spent $520 on new twin-tip skis. This was 40% of the money he earned at his summer job. How much did he earn at his summer job? (Lesson 7-8)

Chapter 8 Applying Percent

1. **GEOGRAPHY** In Washington, about 5.7% of the total area is water. If the total area of Washington is 70,637 square miles, estimate the number of square miles of water by using 10%. (Lesson 8-1)

2. **POPULATION** According to the U.S. Census Bureau, 9.4% of the population in the Seattle-Bellevue metropolitan area is Asian. If this area has a population of 2,414,616, estimate the number of people who are Asian. (Lesson 8-1)

3. **FORESTS** In the United States, forests cover 24.7% of the land. If the total land area of the U.S. is 3,536,338 square miles, how many square miles are forests? Round to the nearest square mile. (Lesson 8-2)

4. **GOVERNMENT** Of the 435 members in the U.S. House of Representatives, 53 are from California and 13 are from North Carolina. To the nearest whole percent, what percent of the representatives are from California? from North Carolina? (Lesson 8-2)

FOOD For Exercises 5 and 6, use the graph below. It shows the results of a survey in which people were asked how they determine how long food has been in their freezer. (Lesson 8-3)

Frozen Foods
- Guess 44%
- Arrange by date 12%
- Other 9%
- Label the containers 35%

Source: Opinion Research Corp. for Tupperware

5. In a city of 56,000 people, predict how many of them guess to determine how long food has been in their freezer.

6. Suppose 420 people surveyed said they label their freezer containers. How many people took part in the survey?

SPORTS For Exercises 7–9, use the table below. It shows the number of participants ages 7 to 17 in the sports listed. (Lesson 8-4)

Sport	Number (millions)	
	1990	2000
in-line skating	3.6	21.8
snowboarding	1.5	4.3
roller hockey	1.5	2.2
golf	23.0	26.4

Source: National Sporting Goods Association

7. What is the percent of change in in-line skaters 7 to 17 years old from 1990 to 2000? Round to the nearest percent and state whether the percent of change is an *increase* or *decrease*.

8. Find the percent of change from 1990 to 2000 in the number of children and teens who played roller hockey. Round to the nearest percent.

9. Which sport had a greater percent of increase in participants from 1990 to 2000: golf or snowboarding? Explain.

10. **MUSIC** Find the total cost of a $15.50 CD if the sales tax is 8%. (Lesson 8-5)

COMPUTERS For Exercises 11 and 12, use the following information.
The Wares want to buy a new computer with a regular price of $1,049. (Lesson 8-5)

11. If the store is offering a 20% discount, what will be the sale price of the computer?

12. If the sales tax on the computer is 5.25%, what will be the total cost with the discount?

BANKING For Exercises 13–16, complete the table below. The interest earned is simple interest. (Lesson 8-6)

	Principal	Rate	Time (yr)	Interest Earned
13.	$1,525.00	5%	$2\frac{1}{2}$	?
14.	$2,250.00	4%	?	$337.50
15.	?	3.5%	4	$498.40
16.	$5,080.00	?	3	$952.50

Chapter 9 Probability (pages 368–407)

1. **DENTISTS** A dental hygienist randomly chooses a toothbrush in a drawer containing 17 white, 12 green, and 5 blue toothbrushes. What is the probability that she chooses a green toothbrush? Write as a fraction in simplest form. (Lesson 9-1)

2. **SURVEYS** For Exercises 2 and 3, use the table below. It shows the results of a survey in which adults were asked how proud they were to be an American. (Lesson 9-1)

How Proud Are You?	
Response	Number
extremely	650
very	250
moderately	60
little/not at all	30
no opinion	10

   Source: Gallup Poll

2. If one person participating in the survey is chosen at random, what is the probability that the person is extremely patriotic? Write as a fraction in simplest form.

3. If one person participating in the survey is chosen at random, what is the probability that he is *not* moderately patriotic? Write as a fraction in simplest form.

RANCHING For Exercises 4–6, use the following information.
For Roger to reach his cattle pasture, he must pass through three consecutive gates. Any of the three gates can be either *open* or *closed*. (Lesson 9-2)

4. Make a tree diagram to show all of the possible positions of the gates.

5. What is the probability that all three gates will be closed when Roger visits this pasture? Write as a fraction.

6. What is the probability that the first two gates are open? Write as a fraction in simplest form.

7. **HIKING** To *run the ridge* in the Bridger Mountains, you can choose two paths up the face of the first peak, four paths across the ridge, and three routes down the final peak. Find the number of ways you can *run the ridge*. (Lesson 9-3)

8. **SKATEBOARDS** World Sports makes skateboards with several different patterns on the deck. You can choose one of four deck lengths and one of six types of wheels. If they advertise that they have 120 different skateboards, how many deck patterns are there? (Lesson 9-3)

9. **READING** Mr. Steadman plans to read eight children's novels to his second graders during the school year. In how many ways can he arrange the books to be read? (Lesson 9-4)

10. **CRAFTS** Marina has print fabric in pink, blue, magenta, green, yellow, and tan. How many different stuffed bears can she make if each bear has only four different fabrics, and the order of the fabrics is not important? (Lesson 9-5)

11. **FOOD** The graph shows the results of a survey in which 7th graders at Plentywood Middle School were asked to name their favorite fruit.

    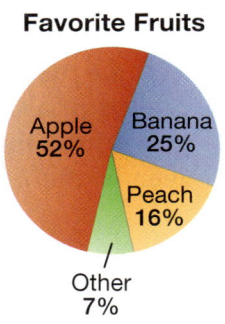
    Favorite Fruits
    Apple 52%, Banana 25%, Peach 16%, Other 7%

    If a 7th grader at the school is randomly selected, what is the probability that they chose bananas as their favorite? Write as a fraction in simplest form. (Lesson 9-6)

12. **CANDY** A bag has 32 candies as shown in the table. What is the probability of randomly selecting one yellow candy and then one purple candy? Assume that the first candy is not replaced. (Lesson 9-7)

Color	Number
red	10
blue	6
purple	10
yellow	4
green	2

Chapter 10 Geometry

(pages 410–467)

ART For Exercises 1 and 2, use the diagram of the Native American artifact. (Lesson 10-1)

1. Name a right angle and a straight angle.

2. Suppose one side of an angle is $\overline{OA}$. Describe the location of point E so that $\angle AOE$ is acute.

TELEVISION For Exercises 3 and 4, use the table shown below. It shows the results of a survey in which families were asked how many T.V. channels they watched. (Lesson 10-2)

| Channels that Families Watch ||
Number	Percent
5 or fewer	30%
6–12	33%
13–25	19%
26 or more	14%

3. The fifth category in the survey is *No T.V. or no opinion*. What percent of the people surveyed were in this category?

4. Make a circle graph of the data.

PHYSICAL SCIENCE For Exercises 5 and 6, use the following information.
In physics, the *law of reflection* states that the angle of incidence of incoming light is equal in measure to the angle of reflection of outgoing light when light strikes a mirror, as shown in the diagram below. (Lesson 10-3)

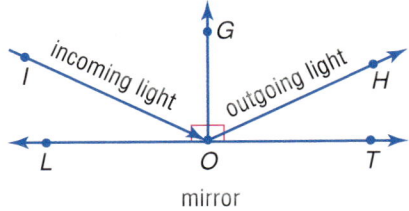

5. The angle of incidence is $\angle IOG$ and the angle of reflection is $\angle HOG$. If $m\angle IOG = 65°$, find $m\angle HOG$.

6. Find $m\angle LOI$. Name another angle with this measure.

For Exercises 7 and 8, use the figure below.

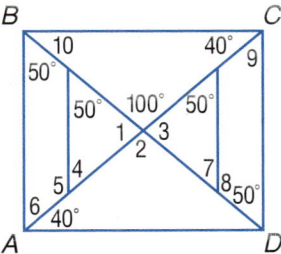

7. Find the measure of each angle numbered from 1–10. (Lesson 10-4)

8. Find the *best* name to classify quadrilateral *ABCD*. Explain your reasoning. (Lesson 10-5)

9. **CRAFTS** Priscilla makes porcelain dolls that are proportional to a real child. If Jody is $4\frac{2}{3}$ feet tall with a 23-inch waist, what should be the waist measure of a doll that is 13 inches tall? Round to the nearest inch. (Lesson 10-6)

10. **ART** Draw a tessellation using two of the polygons listed at the right. Identify the polygons and explain why the tessellation works. (Lesson 10-7)

> regular triangles
> quadrilaterals
> pentagons
> hexagons
> octagons

For Exercises 11 and 12, use the quadrilateral *MOVE* shown below.

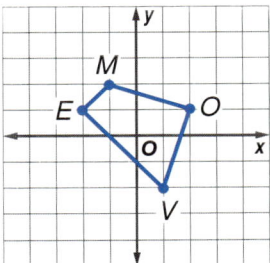

11. Describe the translation that will move M to the point at $(2, -2)$. Then graph quadrilateral $M'O'V'E'$ using this translation. (Lesson 10-8)

12. Find the coordinates of the vertices of quadrilateral *MOVE* after a reflection over the *y*-axis. Then graph the reflection. (Lesson 10-9)

Chapter 11 Geometry: Measuring Two-Dimensional Figures (pages 468–509)

1. **PROPERTY** If a square section of land contains 1,742,400 square feet, what are the side dimensions? (Lesson 11-1)

2. **RECORDS** In Japan, a gigantic omelet was made with an area of 1,383 square feet. If the omelet was a square, what would be the side lengths? Round to the nearest tenth. (Lesson 11-2)

3. **SOFTBALL** A softball diamond is a square measuring 60 feet on each side.

   How far does a player on second base throw when she throws from second base to home? Round to the nearest tenth. (Lesson 11-3)

4. **BANDS** Mr. Garcia is planning a formation for the middle school band at a football game. The diagram shows the dimensions of the field.

   To the nearest foot, what is the distance from A to B? (Lesson 11-3)

5. **CRAFTS** A quilt pattern uses 25 parallelogram-shaped pieces of fabric, each with a base of 4 inches and a height of $2\frac{1}{2}$ inches. How much fabric is used to make the 25 pieces? (Lesson 11-4)

6. **FURNITURE** A corner table is in the shape of a right triangle. If the side lengths of the tabletop are 3.5 feet, 3.5 feet, and 4.9 feet, what is the area? Round to the nearest tenth if necessary. (Lesson 11-5)

7. **PUZZLES** A *tangram* is a square puzzle formed by the seven geometric figures shown at the right. Find the area of each small and large shaded triangle. (Lesson 11-5)

   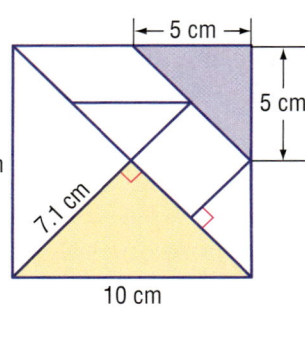

8. **COOKIES** In New Zealand, a giant circular chocolate chip cookie was baked with a diameter of 81 feet 8 inches. To the nearest square foot, what was the area of the cookie? (Lesson 11-6)

9. **LANDSCAPING** Find the area of the flower garden shown in the diagram at the right. Round to the nearest square foot. (Lesson 11-7)

   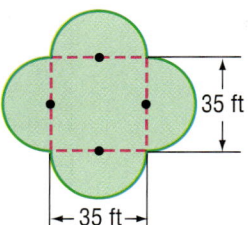

10. **CONSTRUCTION** A design for a deck is shown below. Find the area of the deck. (Lesson 11-7)

11. **GAMES** At the Moore Middle School carnival, one game has a dartboard as shown below.

    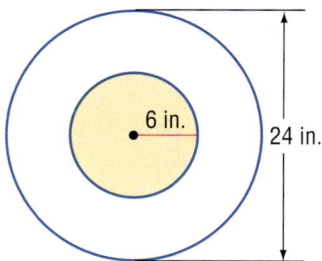

    If it is equally likely that a thrown dart will land anywhere on the dartboard, find the probability that a dart lands in the shaded region. Write as a percent. (Lesson 11-8)

Chapter 12 Geometry: Measuring Three-Dimensional Figures (pages 510–551)

1. **RECORDS** According to the *Guinness Book of World Records*, the tallest hotel in the world is the 1,053-foot sail-shaped Burj Al Arab in Dubai, United Arab Emirates.

   Draw possible sketches of the top, side, and front views of the hotel. (Lesson 12-1)

OCEANS For Exercises 2 and 3, use the following information.
The Atlantic Ocean has an area of about 33,420,000 square miles. Its average depth is 11,730 feet. (Lesson 12-2)

2. To the nearest hundredth, what is the average depth of the Atlantic Ocean in miles? (*Hint*: 1 mi = 5,280 ft)

3. What is the approximate volume of the Atlantic Ocean in cubic miles?

WATER For Exercises 4–6, use the cylinder-shaped water tank. (Lesson 12-3)

4. Find the volume of the tank. Round to the nearest cubic foot.

5. One gallon of water equals 231 cubic inches of water. Find the volume of the water tank in gallons.

6. A cylinder-shaped water tank holds 200,000 gallons of water. The height of the tank is 48 feet. To the nearest tenth of a foot, what is the radius of the tank?

For Exercises 7 and 8, use the following information.
Liz is designing some gift boxes. The small box is 6 inches long, 4 inches wide, and 2.5 inches high. The medium box has dimensions that are each 3 times the dimensions of the small box. (Lesson 12-4)

7. Find the surface area of the small box.

8. What are the dimensions of the medium box? Find the surface area of the medium box.

STORAGE For Exercises 9–11, use the following information.
The two canisters shown below each have a volume of about 628.3 cubic inches. (Lesson 12-5)

9. What is the radius of the blue canister? Round to the nearest tenth.

10. What is the height of the yellow canister? Round to the nearest tenth.

11. What is the difference between the surface area of the two canisters?

WATER For Exercises 12 and 13, refer to the water tank shown in Exercise 4. (Lesson 12-5)

12. What is the surface area of the water tank, including the bottom? Round to the nearest tenth.

13. A gallon of paint covers 400 square feet of surface. How many gallons are needed to cover the top and curved surface of the tank? Round to the nearest gallon.

For Exercises 14 and 15, use the following information.
In science class, Yvette and Tyrell use different rulers to measure the length of an earthworm. Using significant digits, Yvette records 13.2 centimeters, and Tyrell records 13.25 centimeters. (Lesson 12-6)

14. What is the most precise unit on Yvette's ruler? Explain.

15. What is the most precise unit on Tyrell's ruler? Explain.

Mixed Problem Solving **607**

Preparing For Standardized Tests

Becoming a Better Test-Taker

At some time in your life, you will probably have to take a standardized test. Sometimes this test may determine if you go on to the next grade level or course, or even if you will graduate from high school. This section of your textbook is dedicated to making you a better test-taker.

TYPES OF TEST QUESTIONS In the following pages, you will see examples of four types of questions commonly seen on standardized tests. A description of each type is shown in the table below.

Type of Question	Description	See Pages
multiple choice	Four or five possible answer choices are given from which you choose the best answer.	610–613
gridded response	You solve the problem. Then you enter the answer in a special grid and shade in the corresponding circles.	614–617
short response	You solve the problem, showing your work and/or explaining your reasoning.	618–621
extended response	You solve a multi-part problem, showing your work and/or explaining your reasoning.	622–625

For each type of question, worked-out examples are provided that show you step-by-step solutions. Strategies that are helpful for solving the problems are also provided.

PRACTICE After being introduced to each type of question, you can practice that type of question. Each set of practice questions is divided into five sections that represent the concepts most commonly assessed on standardized tests.

- Number and Operations
- Algebra
- Geometry
- Measurement
- Data Analysis and Probability

USING A CALCULATOR On some tests, you are permitted to use a calculator. You should check with your teacher to determine if calculator use is permitted on the test you will be taking, and if so, what type of calculator can be used.

If you *are* allowed to use a calculator, make sure you are familiar with how it works so that you won't waste time trying to figure out the calculator when taking the test.

TEST-TAKING TIPS In addition to the Test-Taking Tips like the one shown at the right, here are some additional thoughts that might help you.

> **Test-Taking Tip**
> The units of measure asked for in the answer may be different than the units given in the question. Check that your answer is in the correct units.

- Get a good night's rest before the test. Cramming the night before does not improve your results.

- Watch for key words like NOT and EXCEPT. Also look for order words like LEAST, GREATEST, FIRST, and LAST.

- For multiple-choice questions that ask for the answer choice that is *not* true, check each answer choice, labeling it with the letter T or F to show whether it is true or false.

- Cross out information that is not important.

- Underline key words, and circle numbers in a question.

- Label your answers for open-ended questions.

- Rephrase the question you are being asked.

- Become familiar with common formulas and when they should be used.

- When you read a chart, table, or graph, pay attention to the words, numbers, and patterns of the data.

- Budget your time when taking a test. Don't dwell on problems that you cannot solve. Just make sure to leave that question blank on your answer sheet.

YOUR TEXTBOOK Your textbook contains many opportunities for you to get ready for standardized tests. Take advantage of these so you don't need to cram before the test.

- Each lesson contains two standardized test practice problems to provide you with ongoing opportunities to sharpen your test-taking skills.

- Every chapter contains a completely worked-out Standardized Test Practice Example, along with a Test-Taking Tip to help you solve problems that are similar.

- Each chapter contains two full pages of Standardized Test Practice with Test-Taking Tips. These two pages contain practice questions in the various formats that can be found on the most frequently given standardized tests.

HELP ON THE INTERNET There are many online resources to help you prepare for standardized tests.

- Glencoe's Web site contains Online Study Tools that include Standardized Test Practice. For hundreds of multiple choice practice problems, visit:

    msmath2.net/standardized_test

- Some states provide online help for students preparing to take standardized tests. For more information, visit your state's Board of Education Web site.

Multiple-Choice Questions

Multiple-choice questions are the most common type of question on standardized tests. These questions are sometimes called *selected-response questions*. You are asked to choose the best answer from four or five possible answers.

To record a multiple-choice answer, you may be asked to shade in a bubble that is a circle or an oval or just to write the letter of your choice. Always make sure that your shading is dark enough and completely covers the bubble.

The answer to a multiple-choice question may not stand out from the choices. However, you may be able to eliminate some of the choices. Another answer choice might be that the correct answer is not given.

Incomplete shading
Ⓐ Ⓑ Ⓒ Ⓓ

Too light shading
Ⓐ Ⓑ Ⓒ Ⓓ

Correct shading
Ⓐ Ⓑ ● Ⓓ

EXAMPLE

1 Mrs. Hon's seventh grade students are purchasing stuffed animals to donate to a charity. They bought 3 boxes containing eight animals each and 5 boxes containing twelve animals each. Which expression *cannot* be used to find the total number of animals they bought to give to the charity?

Notice that the problem asks for the expression that cannot represent the situation.

Ⓐ $8 + 8 + 8 + 12 + 12 + 12 + 12 + 12$
Ⓑ $3 \times 8 + 5 \times 12$
Ⓒ $3(8) + 5(12)$
Ⓓ $8 \times (8 + 12)$

Read the problem carefully and locate the important information. There are 3 boxes that have eight animals, so that is 3×8, or 24 animals. There are 5 boxes of twelve animals, so that is 5×12, or 60 animals. The total number of animals is $24 + 60$, or 84.

You know from reading the problem that you are looking for the expression that *does not* simplify to 84. Simplify each expression to find the answer.

Ⓐ $8 + 8 + 8 + 12 + 12 + 12 + 12 + 12 = (8 + 8 + 8) + (12 + 12 + 12 + 12 + 12)$
$= 24 + 60$
$= 84$

Ⓑ $3 \times 8 + 5 \times 12 = 24 + 60$
$= 84$

Ⓒ $3(8) + 5(12) = 24 + 60$
$= 84$

Ⓓ $8 \times (8 + 12) = 8 \times 20$
$= 160$

The only expression that *does not* simplify to 84 is D. The correct choice is D.

Some problems are easier to solve if you draw a diagram. If you cannot write in the test booklet, draw a diagram on scratch paper.

EXAMPLE

STRATEGY

Diagrams
Draw a diagram for the situation.

2 On a hiking trip, Grace and Alicia traveled 10 miles south and 4 miles west. If they take the shortest return route, how far will the hike be back to their starting point? Round to the nearest tenth of a mile.

   Ⓐ 6.0 mi   Ⓑ 9.2 mi   Ⓒ 10.8 mi   Ⓓ 14.0 mi

To solve this problem, you need to draw a diagram of the situation. Label the directions and the important information from the problem.

Use the Pythagorean Theorem to find the distance that they will hike back to their starting point.

$c^2 = a^2 + b^2$ Pythagorean Theorem
$c^2 = 4^2 + 10^2$ Replace a with 4 and b with 10.
$c^2 = 16 + 100$ Simplify.
$c^2 = 116$ Add.
$\sqrt{c^2} = \sqrt{116}$ Take the square root of each side.
$c \approx 10.8$ Use a calculator to simplify.

Round the answer to the correct decimal place.

The hike back will be about 10.8 miles. The correct choice is C.

Some problems give you more information than you need to solve the problem. Read the question carefully to determine the information you need.

EXAMPLE

STRATEGY

Formulas
Use the reference sheet to find the correct formula.

3 One of the biggest pieces of cheese ever produced was made in 1866 in Ingersoll, Canada. It weighed 7,300 pounds. It was shaped as a cylinder with a diameter of 7 feet and a height of 3 feet. To the nearest cubic foot, what was the volume of the cheese? Use 3.14 for π.

   Ⓐ 462 ft³   Ⓑ 143 ft³   Ⓒ 115 ft³   Ⓓ 63 ft³

You need to use the formula for the volume of a cylinder. The diameter is 7 feet, so the radius is $\frac{7}{2}$ or 3.5 feet. The height is 3 feet.

$V = \pi r^2 h$ Volume of a cylinder
$V \approx (3.14)(3.5)^2(3)$ Replace π with 3.14, r with 3.5, and h with 3.
$V \approx 115.395$ Simplify.

The volume of the cheese is about 115 cubic feet. The correct choice is C.

Multiple-Choice Practice

Choose the best answer.

Number and Operations

1. The world's smallest fruit is the fruit of a wolffia plant, which measures about 0.01 inch in length. Another small fruit is the eye of a sewing needle, with a length of 0.20 inch. How many times longer is the eye of a sewing needle fruit than a wolffia fruit?

   A 0.002
   B 0.005
   C 2
   D 20

2. The table shows what types of trash fill landfills in the United States. What fraction of the trash in landfills is plastic?

Type of Trash	Percent in Landfills
metal	8%
plastic	24%
food, yard waste	11%
rubber, leather	6%
paper	21%
other trash	30%

   Source: *The World Almanac for Kids*

   A $\frac{1}{100}$
   B $\frac{1}{24}$
   C $\frac{1}{6}$
   D $\frac{6}{25}$

3. A recent movie earned 317 million dollars in ticket sales. What is this value in scientific notation?

   A 3.17
   B 317×10^6
   C 3.17×10^8
   D 3.17×10^{11}

4. Mercury orbits the sun at 29.75 miles per second. Earth orbits the sun at 18.51 miles per second. How many more miles does Mercury travel in one minute than Earth?

   A 11.24 mi
   B 269.76 mi
   C 674.4 mi
   D 40,464 mi

Algebra

5. The table shows the population growth of a certain bacteria. How many bacteria will there be after 5 hours?

Hours	0	1	2	3	4	5
Number of Bacteria	32	48	72	104	144	?

   A 243
   B 200
   C 192
   D 178

6. Which function rule describes the relationship between distance from home y and hours traveled x?

Time (h) x	Distance from Home (mi) y
0	0
1	65
2	130
3	195

   A $65y = x$
   B $y = 65 \div x$
   C $y = x + 65$
   D $y = 65x$

7. For a family portrait, a photographer charges a sitting fee and an amount of money per portrait ordered. Which function rule describes the relationship between the total cost y and number of portraits x?

   Family Portrait

   A $y = 10 + 40x$
   B $y = 40x + 10x$
   C $y = 40 + 10x$
   D $y = 10x$

Geometry

8. The three towns on the map form a triangle. Which term *best* describes the angle with vertex at Worthington?

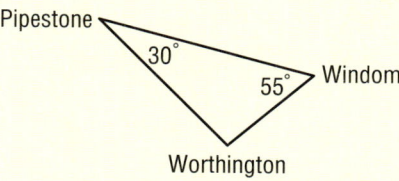

   A obtuse
   B right
   C acute
   D straight

9. Pedro is using a triangle in a computer graphics design. On a coordinate plane, the vertices of the triangle are $A(-1, 1)$, $B(0, 2)$, and $C(5, -1)$. If Pedro translates the triangle 4 units left and 3 units down, what will be the coordinates of B'?

   A $(-1, -4)$
   B $(4, -1)$
   C $(4, 5)$
   D $(-4, -1)$

10. Jasmine is using this polygon on a poster she is making for the basketball team. Which term *best* describes the polygon?

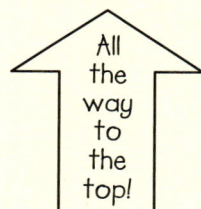

   A quadrilateral
   B decagon
   C hexagon
   D heptagon

Measurement

11. Crispy Crackers are packaged in a box that measures 6 inches by 2.5 inches by 10 inches. Which dimensions are of a prism that has the same volume as the Crispy Crackers box?

   A 6.5 in. by 2 in. by 10 in.
   B 6.25 in. by 3 in. by 8 in.
   C 7.25 in. by 2 in. by 9.5 in.
   D 4 in. by 7.5 in. by 6 in.

Test-Taking Tip

Question 11 Most standardized tests will include any commonly used formulas at the front of the test booklet. Quickly review the list before you begin so that you know what formulas are available.

12. The Crab nebula is a cloud of gas and dust particles in space that is expanding at a rate of 930 miles per second. What is its rate of expansion in miles per hour?

   A 22,320 mph
   B 55,800 mph
   C 1,339,200 mph
   D 3,348,000 mph

13. Super Toys makes two sizes of building blocks shaped as cubes. The large block has side length four times the length of the small block. What is the ratio of the surface area of the small block to the surface area of the large block?

   A 1 to 4
   B 1 to 6
   C 1 to 16
   D 1 to 32

Data Analysis and Probability

14. The table shows the number of students playing each sport at Wilson Junior High. Find the mean of the data.

Sport	Number of Students
baseball/softball	49
basketball	74
soccer	82
swimming	21
track and field	115
volleyball	25

   A 23
   B 49
   C 61
   D 94

15. The spinner is divided into four equal-sized sections. If you spin the spinner 62 times, which is the *best* estimate for the number of times you will land on 2?

   A 15
   B 30
   C 40
   D 50

Gridded-Response Questions

Gridded-response questions are another type of question on standardized tests. These questions are sometimes called *student-produced response* or *grid in*.

For gridded response, you must mark your answer on a grid printed on an answer sheet. The grid contains a row of four or five boxes at the top, two rows of ovals or circles with decimal and fraction symbols, and four or five columns of ovals, numbered 0–9. An example of a grid from an answer sheet is shown.

EXAMPLE

1. Mr. Byrd builds and sells storage buildings. The dimensions of his most popular model are shown in the diagram. What is the volume of the building in cubic feet?

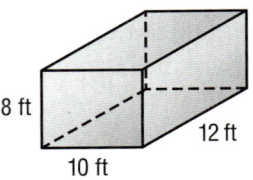

What do you need to find?

You need to find the volume of a rectangular prism. Use the formula and the dimensions given in the diagram.

$V = \ell wh$ Volume of a rectangular prism

$V = 12 \cdot 10 \cdot 8$ Replace ℓ with 12, w with 10, and h with 8.

$V = 960$ Multiply.

The volume is 960 cubic feet.

STRATEGY

Formulas
Use the reference sheet to find the formula you need.

How do you fill in the grid for the answer?

- Write your answer in the answer boxes.
- Write only one digit or symbol in each answer box.
- Do not write any digits or symbols outside the answer boxes.
- You may write your answer with the first digit in the left answer box, or with the last digit in the right answer box. You may leave blank any boxes you do not need on the right or the left side of your answer.
- Fill in only one bubble for every answer box that you have written in. Be sure not to fill in a bubble under a blank answer box.

614 Preparing for Standardized Tests

Many gridded-response questions result in an answer that is a fraction or a decimal. These values can also be filled in on the grid.

EXAMPLE

2 Mr. Benitez has a prize box that contains 12 glitter pencils, 5 fluorescent pens, and 13 mechanical pencils. If Alex randomly selects a prize, what is the probability that he will choose a glitter pencil?

$$P(\text{glitter pencil}) = \frac{\text{number of favorable outcomes}}{\text{number of possible outcomes}}$$

$$= \frac{12}{12 + 5 + 13} = \frac{12}{30} \text{ or } \frac{2}{5}$$

You can either grid the fraction $\frac{12}{30}$ or $\frac{2}{5}$. You also can rewrite the fraction as a decimal and grid 0.4. Be sure to write the decimal point or fraction bar in the answer box. The following are acceptable answers.

Any equivalent fraction that fits the grid will be counted as correct.

Do not leave a blank answer box in the middle of an answer.

Some problems may result in an answer that is an improper fraction or mixed number. Before filling in the grid, change the mixed number to an equivalent improper fraction or decimal.

EXAMPLE

3 The bee hummingbird measures $2\frac{1}{2}$ inches in length. Another small bird is the least sandpiper, measuring $4\frac{1}{2}$ inches in length. How many times longer is the least sandpiper than the bee hummingbird?

Divide the length of the sandpiper by the length of the hummingbird.

$4\frac{1}{2} \div 2\frac{1}{2} = \frac{9}{2} \div \frac{5}{2}$ Rename the mixed numbers as improper fractions.

$= \frac{9}{2} \cdot \frac{2}{5}$ Multiply by the reciprocal of $\frac{5}{2}$, which is $\frac{2}{5}$.

$= \frac{9}{5}$ Multiply.

You can either grid the improper fraction $\frac{9}{5}$, or rewrite it as 1.8 and grid the decimal. Do not enter 14/5, as this will be interpreted as $\frac{14}{5}$.

Preparing for Standardized Tests

Gridded-Response Practice

Solve each problem. Then copy and complete a grid like the one shown on page 614.

Number and Operations

1. The seventh grade class at Willow Creek Middle School is planning a class trip. Each student will need to pay $4.50 for the bus ride, $8.00 for a ticket to the museum, and $5.25 for lunch. If there are 52 students in the class, what will be the total cost in dollars of the trip?

2. The highest point in Louisiana is Driskill Mountain at 585 feet. The lowest point is −8 feet in New Orleans. What is the difference in feet between the highest and lowest elevation points?

3. People in the United States own about 135 million cars. If the number of cars is written in scientific notation, what is the exponent of the 10 in the expression?

4. Ashlee has $5\frac{1}{4}$ cups of cocoa powder. If each batch of chocolate cookies uses $\frac{1}{2}$ cup of cocoa powder, how many batches could she make?

Algebra

5. The number of televisions per 1,000 people in France is 598. The number of televisions per 1,000 people in the United States is 208 greater than the number in France. How many televisions are there in the United States per 1,000 people?

6. The table shows the number of white beads that Carmen uses in each row for a particular pattern in a necklace that she designed. How many white beads will there be in the sixth row?

Row	1	2	3	4	5
Beads	1	2	4	8	16

7. The table shows the cost of renting a booth at the week-long Fall Festival. There is an initial charge for reserving a booth and a fee per day. What is the cost in dollars of renting a booth for the 7 days of the festival?

Days	0	1	2	3
Cost ($)	50	90	130	170

8. The graph shows the cost to rent a power paint sprayer for painting a house. Let x be the number of hours the sprayer is rented and y be the total cost of the rental. Suppose an equation of the form $y = ax$ represents the data in the graph. What is the value of a?

Renting a Power Paint Sprayer

9. The temperature on a January morning is −18°F. The temperature is expected to rise at a rate of 5° each hour for the next several hours. In how many hours will the temperature be 7°F?

Geometry

10. Triangle ABC is similar to triangle XYZ. What is the measure of $\angle Z$ in degrees?

11. Triangle DEF has vertices $D(1, 1)$, $E(4, -2)$, and $F(0, -3)$. Find the y-coordinate of point E after the triangle is reflected over the x-axis.

12. In the diagram, ∠1 and ∠2 are supplementary. Find the measure of ∠3 in degrees.

13. Rectangle B has width and length 3 times the width and length of Rectangle A. As a fraction, what is the ratio of the area of Rectangle A to the area of Rectangle B?

18. The Browns plan to paint the top, bottom, and sides of the cylinder-shaped barrel with a waterproof coating. What is the surface area in square feet of the barrel? Round to the nearest square foot.

Measurement

14. In the United States, the average amount of meat eaten per person is 261 pounds per year. If there are 365 days in a year, what is the average number of ounces of meat eaten per day? Round to the nearest ounce.

Test-Taking Tip
Question 14 The units of measure given in a question may not be the same as the units of measure asked for in the answer. Check that your solution is in the correct unit.

15. A rectangle has an area of 318 square centimeters and a width of 12 centimeters. What is the length of the rectangle in centimeters?

16. The circle graph shows the results of a survey in Ms. Chen's fifth period math class. What is the exact degree measure of the section of the circle graph representing soccer?

Favorite Sports of Ms. Chen's Students

17. One acre of land is 43,560 square feet. How many square yards are in one acre?

Data Analysis and Probability

19. The table shows the world's highest dams and the countries in which they are located. Find the range of the data.

Country	Height (ft)
Tajikistan	984
Switzerland	935
Georgia	892
Italy	859
Mexico	856

Source: *Scholastic Book of World Records*

20. The table shows the prices in dollars that Mike spent on his college textbooks. What is the mean of the data?

66.00	51.25	50.00	63.00
56.25	9.00	82.00	50.50

21. Kaya has the following scores for the first four tests in her science class: 82%, 95%, 100%, and 90%. She wants the mean of her first five tests to be 93%. What score as a percent must she earn on the fifth test?

22. In a carnival game, the probability of winning is 0.005. As a percent, what is the probability of losing?

23. A beverage cooler contains 6 regular colas, 5 orange drinks, 7 iced teas, and 7 diet colas. James reaches into the cooler and randomly takes two drinks, one after the other. Find the probability that he will choose a regular cola and then an orange drink.

Short-Response Questions

Short-response questions require you to provide a solution to the problem as well as any method, explanation, and/or justification you used to arrive at the solution. These are sometimes called *constructed-response, open-response, open-ended, free-response,* or *student-produced questions*.

The following is a sample **rubric**, or scoring guide, for scoring short-response questions.

Credit	Score	Criteria
Full	2	Full credit: The answer is correct and a full explanation is provided that shows each step in arriving at the final answer.
Partial	1	Partial credit: There are two different ways to receive partial credit. • The answer is correct, but the explanation provided is incomplete or incorrect. • The answer is incorrect, but the explanation and method of solving the problem is correct.
None	0	No credit: Either an answer is not provided or the answer does not make sense.

On some standardized tests, no credit is given for a correct answer if your work is not shown.

EXAMPLE

Hanna is buying a computer. She is deciding between a desktop and a laptop that are on sale. The desktop costs $609.00 with a 10% discount. The laptop costs $725.00 with a 25% discount. There is also a 6.75% sales tax on all purchases. Which computer is less expensive? What will be the total cost of the computer including discount and sales tax?

Full Credit Solution

STRATEGY

Reread the Problem Look for the important information in the problem.

Since there are two computers to compare, I will first find the discounted price of each computer. I will change each percent to a decimal to make the calculations.

desktop	laptop
$609.00 \times 0.10 = 60.90$	$725.00 \times 0.25 = 181.25$
$609.00 - 60.90 = 548.10$	$725.00 - 181.25 = 543.75$

The laptop is less expensive after the discount.

I still need to find the cost of the laptop computer with tax.

543.75×0.0675 $\qquad$ $6.75\% = 0.0675$
$= 36.703125$ $\qquad$ Use a calculator.
≈ 36.70

The steps, calculations, and reasoning are clearly stated.

Now I will add the sales tax to the cost of the laptop.

$543.75 + 36.70 = 580.45$

The laptop will cost Hanna $580.45, including sales tax.

Partial Credit Solution

In this sample solution, the calculations are correct and the answer is correct. However, there is no explanation for any of the calculations.

> Notice that the student multiplies the discounted price by 1.0675 since the cost is 1 and the tax is 0.0675. So, the total is then 1.0675.

$609.00 \times 0.10 = 60.90$
$609.00 - 60.90 = 548.10$
$548.10 \times 1.0675 = 585.09675$

$725.00 \times 0.25 = 181.25$
$725.00 - 181.25 = 543.75$
$543.75 \times 1.0675 = 580.453125$

The laptop is cheaper for $580.45.

Partial Credit Solution

In this sample solution, the answer is partially incorrect because the student does not add the sales tax.

I will find the discount price for each set.

Desktop: Since the current price is 100% and the discount is 10%, the sale price will be
$100 - 10 = 90\%$ or 0.9.
$609.00 \times 0.9 = 548.10$

Laptop: Since the current price is 100% and the discount is 25%, the sale price will be
$100 - 25 = 75\%$ or 0.75.
$725.00 \times 0.75 = 543.75$

Hanna should get the laptop for $543.75.

> The student does not add the cost of the tax.

No Credit Solution

In this sample solution, the student does not understand how to find discounted prices and the sales tax. There are just some calculations using the numbers in the problem.

$609.00 - 10\% = 602.91$
$725.00 - 25\% = 625.00$
$602.91 + 6.75\% = 609.66$
$625.00 + 6.75 = 631.75$

I think Hanna should buy the desktop for $609.66.

Short-Response Practice

Solve each problem. Show all your work.

Number and Operations

1. A main unit of currency in Egypt is the pound. One U.S. dollar is equal to $3\frac{4}{5}$ pounds. How many pounds are equivalent to $10.00 in the U.S.?

2. The average daytime temperature on Venus is 870°F. The average temperature on Jupiter is −160°F. What is the difference between the average temperatures on Venus and Jupiter?

3. A bag of chocolate candies has a nutrition label stating that each serving contains 20% of the recommended daily amount of fat. A serving has 13 grams of fat. Using this information, what is the total recommended daily amount of fat in grams?

4. The Montana Department of Fish, Wildlife, and Parks raised the price of a tag to catch a paddlefish from $2.50 to $5.00 for residents and from $7.50 to $15.00 for nonresidents. Which percent of increase is greater, the increase for residents or for nonresidents?

5. A recent article in the newspaper said that there were 75 cell phones for every 100 people in Finland. The number of cell phones in Finland was given to be 3,893,000. Estimate the population of Finland using this information.

Algebra

6. Florida has 8,426 miles of shoreline. Alaska has 25,478 more miles of shoreline than Florida. Write and solve an equation to find the number of miles of shoreline for Alaska.

7. Juana is saving money to buy a skateboard that costs $95. Her grandfather gave her $25 for her birthday, and she plans to save $5 per week from her allowance. In how many weeks will she have enough money for the skateboard?

8. Tyler delivers televisions for Electronics Depot. The graph shows the amount Tyler charges for delivery based on distance. Name the slope and y-intercept of the graph and describe what they mean in this situation.

9. Solve $\frac{2}{3}b = \frac{8}{7}$.

Geometry

10. The formula for the area of a trapezoid is $A = \frac{1}{2}h(b_1 + b_2)$. Find the area of the trapezoid.

11. Angles MNP and PNO are supplementary. Find $m\angle PNO$.

12. What is the value of x?

13. A regular polygon has 18 sides. What is the measure of one angle of this polygon?

14. Colby is planning to use △CDE for a design by using transformations. He plans to reflect it over the *y*-axis. Find the coordinates of △CDE after this reflection. Graph the reflected image of the triangle.

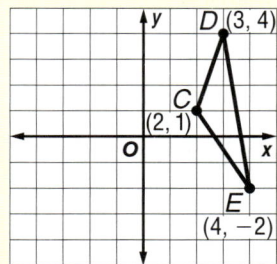

Measurement

15. Diego ran the 10K, or 10 kilometer, race at the Sweet Pea Festival. What is the distance of this race in meters?

16. What is the surface area of the box?

17. A certain aircraft uses 350 gallons of fuel per hour. If the plane used 1,925 gallons of fuel on a flight between two cities, how many hours long was the flight?

18. Curtis is painting the four columns for the set of a school play. What is the surface area that needs to be painted? Assume that the tops and bottoms of the columns do *not* need to be painted.

19. The record for women running the Boston Marathon was set in 1994 when Uta Pippig of Germany ran the 26.2-mile course in approximately 2 hours and 22 minutes. What was her average speed in miles per hour?

Test-Taking Tip

Questions 18 and 19 After finding the solution, always go back and read the problem again to make sure your solution answers what the problem is asking.

Data Analysis and Probability

20. The stem-and-leaf plot shows the scores on the last math test in Mr. Hill's class. What is the range of the data?

Stem	Leaf
9	1 2 3 5 6 9
8	0 2 5 6 6 7 8 8
7	1 2 3 5 7 9 9
6	0 5 8 9 9 8	2 = 82 points

21. The table shows the heights of the world's largest flightless birds. Make a bar graph of the data.

Bird	Height (in.)
Ostrich	96
Emu	60
Cassowary	60
Rhea	54
Emperor Penguin	45

Source: *Scholastic Book of World Records*

22. The table shows the average precipitation in inches for each month in Syracuse, New York. Make a scatter plot of the data. Use the months on the horizontal axis and the precipitation on the vertical axis. Describe the graph.

Month	Precipitation (in.)	Month	Precipitation (in.)
Jan.	2.6	July	4.0
Feb.	2.1	Aug.	3.6
Mar.	3.0	Sep.	4.1
Apr.	3.3	Oct.	3.2
May	3.4	Nov.	3.8
June	3.7	Dec.	3.1

Source: *The World Almanac*

23. Two number cubes each marked with 1, 2, 3, 4, 5, 6 on their faces are rolled. List all the possible outcomes. Then find the probability that a sum of 10 is rolled.

Preparing for Standardized Tests **621**

Extended-Response Questions

Extended-response questions are often called *open-ended* or *constructed-response questions*. Most extended-response questions have multiple parts. You must answer all parts to receive full credit.

Extended-response questions are similar to short-response questions in that you must show all of your work in solving the problem and a rubric is used to determine whether you receive full, partial, or no credit. The following is a sample rubric for scoring extended-response questions.

Credit	Score	Criteria
Full	4	Full credit: A correct solution is given that is supported by well-developed, accurate explanations.
Partial	3, 2, 1	Partial credit: A generally correct solution is given that may contain minor flaws in reasoning or computation, or an incomplete solution is given. The more correct the solution, the greater the score.
None	0	No credit: An incorrect solution is given indicating no mathematical understanding of the concept, or no solution is given.

On some standardized tests, no credit is given for a correct answer if your work is not shown.

Make sure that when the problem says to *Show your work*, you show every aspect of your solution including figures, sketches of graphing calculator screens, or the reasoning behind computations.

EXAMPLE

Each fall, the outdoor swimming pool at Franklin County Park is drained and prepared for the winter. The pool contains 100,000 gallons of water. The pool has two drains that together drain the pool at a rate of 10,000 gallons per hour.

a. Make a function table for this situation. Let x represent the time in hours the pool drains. Let y represent the gallons of water left in the pool. Use values of x starting with 0.
b. Make a graph of the data in the function table.
c. Predict how many hours it will take for the pool to drain completely.

Full Credit Solution

Part a   A complete table includes labeled columns.

Each hour there will be 10,000 gallons less water. I used every 2 hours.

Hours (x)	Water Left (y)	(x, y)
0	100,000	(0, 100,000)
2	80,000	(2, 80,000)
4	60,000	(4, 60,000)
6	40,000	(6, 40,000)
8	20,000	(8, 20,000)

Part b A complete graph includes a title for the graph, appropriate scales and labels for the axes, and correctly graphed points.

The student gives reasoning in graphing the ordered pairs.

To make a graph, I graphed the ordered pairs from my table and decided to connect them with a line.

Part c

The pool is drained when the y value is 0. That is at 10 hours. It will take 10 hours to drain the pool.

Partial Credit Solution

Part a This sample answer has an incomplete table.

1	90,000
2	80,000
3	70,000

Part b Partial credit is given because the points are graphed correctly, but the scale on the *y*-axis jumps from 0 to 70,000 with no break shown.

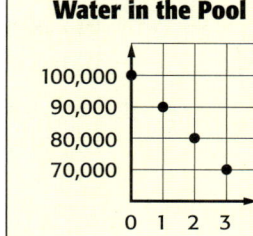

Partial credit can be given even if parts of the table and graph are missing.

Part c

I wrote a function rule to find how long it will take to drain the pool. I let $y = 0$ to find the number of hours.

$$y = 100{,}000 - 10{,}000x$$
$$0 = 100{,}000 - 10{,}000x$$
$$10{,}000x = 100{,}000$$
$$x = 10$$

It will take 10 hours to drain the pool.

Full credit is given for Part c.

No Credit Solution

If the student demonstrates no understanding of making a table or graph for a situation and using the graph to predict, then the solution receives no credit.

Extended-Response Practice

Solve each problem. Show all your work.

Number and Operations

1. The table shows the land area and water area of five states.

| Land and Water Area of Five States |||
State	Land Area (mi^2)	Water Area (mi^2)
Florida	53,937	5991
Colorado	103,729	371
Alaska	570,374	44,856
Iowa	55,875	401
Rhode Island	1,045	186

Source: USA Almanac

   a. Find the percent of each state that is water. The entire state area is the sum of the land and water areas.

   b. Order the states from the state with the least percent water to the greatest percent water.

   c. Suppose a state had land and water areas such that the water area was 20% of the total area. Give a possible land and water area, with land area greater than 1,000 square miles, such that this is true.

Algebra

2. The table shows the rental rates for a crane rental service. There is an initial fee to reserve the crane and a daily fee.

Days	Cost	Days	Cost
0	$100	4	$280
1	$145	5	$325
2	$190	6	$370
3	$235		

   a. Graph the data. Let x = number of days and y = cost. Connect the points with a line.

   b. What is the initial fee charged to reserve the crane? Where is the fee shown on the graph?

   c. What is the slope of the line? What does the slope represent?

   d. What would be the charge for renting the crane for 10 days?

3. A long-distance phone company charges 40¢ per call plus 4¢ per minute.

   a. Write and solve an equation to find the number of minutes used for a call that costs $2.80.

   b. Another long-distance company charges 6¢ per minute and no connection fee. What is the cost of a 60-minute call?

   c. Which service would charge less for a 90-minute call?

Geometry

4. Ava and Bryn are making rectangular fleece blankets. Ava is making her blanket 40 inches by 60 inches. Bryn says she wants to make her blanket "twice as big."

   a. What is the area of Ava's blanket?

   b. Bryn makes her blanket such that both the length and width are each twice the length and width of Ava's blanket. What will be the area of Bryn's blanket?

   c. What is the ratio of the area of Bryn's blanket in Part b to the area of Ava's blanket in Part a?

   d. Ava tells Bryn that a blanket is "twice as big" if the area is twice the area of the other. Give a possible length and width for a blanket twice as big as Ava's using this idea of "twice as big."

5. The map shows five towns and the angle measures for roads connecting them.

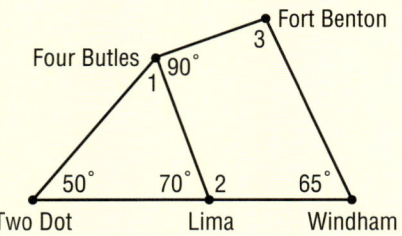

   a. Find the measure of ∠1. Explain your reasoning.

   b. Find the measures of ∠2 and ∠3. Explain your reasoning.

Measurement

6. City planners are considering enlarging the circular manholes in their streets. This would require new covers for the manholes. Currently, the diameter of a manhole cover is 24 inches. The new size being considered is 2 inches greater all the way around.

a. Find the area of a current manhole cover with diameter of 24 inches. Round to the nearest square inch.

b. Find the area of a new manhole cover. Round to the nearest square inch.

c. What would be the percent increase in the area of a manhole cover if the new covers are made?

> **Test-Taking Tip**
> **Question 6** Be sure to completely and carefully read the problem before beginning any calculations. If you read too quickly, you may miss a key piece of information.

7. A company that sells beads for craft projects has the two containers shown for packaging the beads.

a. Find the volume of each container. Round to the nearest tenth of a cubic inch.

b. Find the ratio of the volume of container A to the volume of container B.

c. The company wants to make a container shaped as a rectangular prism. Find a possible length, width, and height for the container such that its volume is twice the volume of container B.

Data Analysis and Probability

8. The table shows the normal monthly temperatures in degrees Fahrenheit for Honolulu, Hawaii, and Minneapolis, Minnesota.

Month	Honolulu	Minneapolis
Jan	73	13
Feb	73	20
Mar	74	32
Apr	76	47
May	77	59
Jun	80	68
Jul	81	73
Aug	82	71
Sep	82	61
Oct	80	49
Nov	78	33
Dec	75	19

Source: *The World Almanac*

a. Make a scatter plot of the data for Honolulu. Place the months on the x-axis and the temperatures on the y-axis. Describe the graph.

b. Make a scatter plot of the data for Minneapolis. Place the months on the x-axis and the temperatures on the y-axis. Describe the graph.

c. Compare the range for each city's temperature data. Explain a possible reason for the difference in the data.

9. One white and one black 4-sided number cubes are rolled for a game. Each number cube has faces numbered 1 through 4.

a. List the sample space for rolling the two number pieces.

b. What is the theoretical probability that a player will roll a sum of 8?

c. If you played this game 50 times, how many times would you expect to get a sum of 4?

Glossary/Glosario

Cómo usar el glosario en español:
1. Busca el término en inglés que desees encontrar.
2. El término en español, junto con la definición, se encuentran en la columna de la derecha.

 A mathematics multilingual glossary is available at www.math.glencoe.com/multilingual_glossary. The glossary includes the following languages.

Arabic	English	Korean	Tagalog
Bengali	Haitian Creole	Russian	Urdu
Cantonese	Hmong	Spanish	Vietnamese

English | Español

A

English	Español
absolute value (p. 107) The distance the number is from zero on a number line.	**valor absoluto** Distancia a la que se encuentra un número de cero en la recta numérica.
acute angle (p. 413) An angle with a measure greater than 0° and less than 90°.	**ángulo agudo** Ángulo que mide más de 0° y menos de 90°.
acute triangle (p. 429) A triangle having three acute angles.	**triángulo acutángulo** Triángulo con tres ángulos agudos.
additive inverse (p. 121) The opposite of an integer. The sum of an integer and its additive inverse is zero.	**inverso aditivo** El opuesto de un entero. La suma de un entero y su inverso aditivo es cero.
adjacent angles (p. 415) Angles that have a common vertex and a common side. ∠1 and ∠2 are adjacent angles.	**ángulos adyacentes** Ángulos que comparten un lado y poseen el mismo vértice. 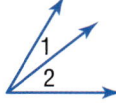 ∠1 y ∠2 son adyacentes.
algebra (p. 18) The branch of mathematics that involves expressions with variables.	**álgebra** Rama de las matemáticas que involucra expresiones con variables.
algebraic expression (p. 18) A combination of variables, numbers, and at least one operation.	**expresión algebraica** Combinación de variables, números y por lo menos una operación.
angle (p. 413) Two rays with a common endpoint form an angle. The rays and vertex are used to name the angle. 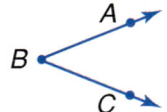 ∠ABC, ∠CBA, or ∠B	**ángulo** Dos rayos con un extremo común forman un ángulo. Los rayos y el vértice se usan para nombrar el ángulo. 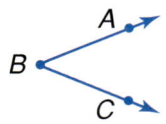 ∠ABC, ∠CBA, o ∠B
area (p. 271) The number of square units needed to cover a surface enclosed by a geometric figure.	**área** El número de unidades cuadradas necesarias para cubrir una superficie cerrada por una figura geométrica.

arithmetic sequence (p. 34) A sequence in which each term is found by adding the same number to the previous term.

Associative Property (p. 31) The way in which three numbers are grouped when they are added or multiplied does not change their sum or product.

sucesión aritmética Sucesión en que cada término se encuentra sumando el mismo número al término anterior.

propiedad asociativa La manera de agrupar tres números al sumarlos o multiplicarlos no cambia su suma o producto.

B

bar graph (p. 85) A graphic form using bars to make comparisons of statistics.

gráfica de barras Forma gráfica que usa barras para hacer comparaciones estadísticas.

bar notation (p. 211) In repeating decimals, the line or bar placed over the digits that repeat. For example, $2.\overline{63}$ indicates that the digits 63 repeat.

base (p. 10) In a power, the number used as a factor. In 10^3, the base is 10. That is, $10^3 = 10 \times 10 \times 10$.

base (p. 323) In a percent proportion, the whole quantity, or the number to which the part is being compared.

base (p. 483) The base of a parallelogram or triangle is any side of the figure. The bases of a trapezoid are the parallel sides.

bisect (p. 416) To divide an angle into two congruent angles.

box-and-whisker plot (p. 80) A diagram that summarizes data using the median, the upper and lower quartiles, and the extreme values. A box is drawn around the quartile values and whiskers extend from each quartile to the extreme data points.

notación de barra Línea o barra que se coloca sobre los dígitos que se repiten en decimales periódicos. Por ejemplo, $2.\overline{63}$ indica que los dígitos 63 se repiten.

base En una potencia, el número usado como factor. En 10^3, la base es 10. Es decir, $10^3 = 10 \times 10 \times 10$.

base En una proporción porcentual, la cantidad total o el número con el que se compara la parte.

base La base de un paralelogramo o triángulo es el lado de la figura. Las bases de un trapecio son los lados paralelos.

bisecar Dividir un ángulo en dos ángulos congruentes.

diagrama de caja y patillas Diagrama que resume información usando la mediana, los cuartiles superior e inferior y los valores extremos. Se dibuja una caja alrededor de los cuartiles y se trazan patillas que los unan a los valores extremos respectivos.

C

center (p. 275) The given point from which all points on a circle or sphere are the same distance.

circle (p. 275) The set of all points in a plane that are the same distance from a given point called the center.

centro Un punto dado del cual equidistan todos los puntos de un círculo o de una esfera.

círculo Conjunto de todos los puntos en un plano que equidistan de un punto dado llamado centro.

circle graph (p. 418) A type of statistical graph used to compare parts of a whole.

Area of Oceans

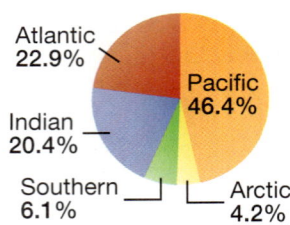

circumference (p. 275) The distance around a circle.

cluster (p. 65) Data that are grouped closely together.

coefficient (p. 19) The numerical factor of a term that contains a variable.

combination (p. 387) An arrangement, or listing, of objects in which order is not important.

common denominator (p. 227) A common multiple of the denominators of two or more fractions. 24 is a common denominator for $\frac{1}{3}$, $\frac{5}{8}$, and $\frac{3}{4}$ because 24 is the LCM of 3, 8, and 4.

Commutative Property (p. 31) The order in which two numbers are added or multiplied does not change their sum or product.

compatible numbers (p. 242) Numbers that are easy to compute mentally.

complementary angles (p. 422) Two angles are complementary if the sum of their measures is 90°.

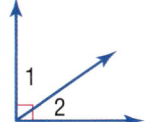

∠1 and ∠2 are complementary angles.

complementary events (p. 371) The events of one outcome happening and that outcome not happening are complementary events. The sum of the probabilities of complementary events is 1.

complex figure (p. 498) A figure made of circles, rectangles, squares, and other two-dimensional figures.

composite number (p. 197) A whole number greater than 1 that has more than two factors.

compound event (p. 398) An event consisting of two or more simple events.

compound inequality (p. 175) Two inequalities connected by the words *and* or *or*.

cone (p. 541) A three-dimensional figure with a curved surface and a circular base.

gráfica circular Tipo de gráfica estadística que se usa para comparar las partes de un todo.

Área de superficie de los océanos

circunferencia La distancia alrededor de un círculo.

agrupamiento Datos estrechamente agrupados.

coeficiente El factor numérico de un término que contiene una variable.

combinación Arreglo o lista de objetos donde el orden no es importante.

común denominador El múltiplo común de los denominadores de dos o más fracciones. 24 es un denominador común para $\frac{1}{3}$, $\frac{5}{8}$ y $\frac{3}{4}$ porque 24 es el mcm de 3, 8 y 4.

propiedad conmutativa El orden en que se suman o multiplican dos números no afecta su suma o producto.

números compatibles Números que son fáciles de computar mentalmente.

ángulos complementarios Dos ángulos son complementarios si la suma de sus medidas es 90°.

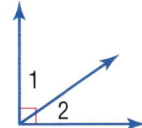

∠1 y ∠2 son complementarios.

eventos complementarios Se dice de los eventos de un resultado que ocurren y el resultado que no ocurre. La suma de las probabilidades de eventos complementarios es 1.

figura compleja Una figura compuesta por círculos, rectángulos, cuadrados y otras dos figuras bidimencionales.

número compuesto Un número entero mayor que 1 que tiene más de dos factores.

evento compuesto Un evento que consiste en dos o más eventos simples.

desigualdad compuesta Dos desigualdades conectadas por las palabras *y* u *o*.

cono Figura tridimensional con una superficie curva y una base circular.

congruent angles (p. 422) Angles that have the same measure.

∠1 and ∠2 are congruent angles.

congruent segments (p. 429) Sides of a triangle having the same length.

Side $\overline{AB}$ is congruent to side $\overline{BC}$.

constant (p. 19) A term that does not contain a variable.

coordinate grid (p. 112) A plane in which a horizontal number line and a vertical number line intersect at their zero points. Also called a coordinate plane.

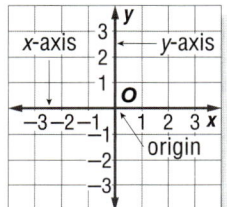

coordinate plane (p. 112) See coordinate grid.

counterexample (p. 33) An example showing that a statement is not true.

cross product (p. 297) In a proportion, a cross product is the product of the numerator of one ratio and the denominator of the other ratio.

cubed (p. 10) The product in which a number is a factor three times. Two cubed is 8 because $2 \times 2 \times 2 = 8$.

cup (p. 267) A customary unit of capacity equal to 8 fluid ounces.

cylinder (p. 524) A three-dimensional figure with two parallel congruent circular bases.

ángulos congruentes Ángulos que tienen la misma medida.

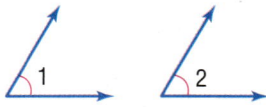

∠1 y ∠2 son congruentes.

segmentos congruentes Los lados de un triángulo que tienen la misma longitud.

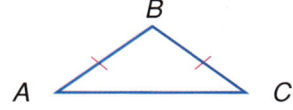

$\overline{AB}$ es congruente a $\overline{BC}$.

constante Término que no contiene una variable.

sistema de coordenadas Plano en el cual se han trazado dos rectas numéricas, una horizontal y una vertical, que se intersecan en sus puntos cero. También conocido como plano de coordenadas.

plano de coordenadas Véase sistema de coordenadas.

contraejemplo Ejemplo que demuestra que un enunciado no es verdadero.

productos cruzados En una proporción, un producto cruzado es el producto del numerador de una razón y el denominador de la otra razón.

al cubo El producto de un número por sí mismo, tres veces. Dos al cubo es 8 porque $2 \times 2 \times 2 = 8$.

taza Unidad de capacidad del sistema inglés de medidas que equivale a 8 onzas líquidas.

cilindro Figura tridimensional que tiene dos bases circulares congruentes y paralelas.

data (p. 54) Pieces of information, often numerical, which are gathered for statistical purposes.

decagon (p. 446) A polygon having ten sides.

datos Información, a menudo numérica, que se recoge con fines estadísticos.

decágono Un polígono con diez lados.

defining the variable (p. 25) Choosing a variable to represent an unknown value in a problem, and using it to write an expression or equation to solve the problem.

degrees (p. 413) The most common unit of measure for angles. If a circle were divided into 360 equal-sized parts, each part would have an angle measure of 1 degree.

dependent events (p. 399) Two or more events in which the outcome of one event affects the outcome of the other event(s).

diameter (p. 275) The distance across a circle through its center.

dilation (p. 455) A type of transformation in which a figure is reduced or enlarged.

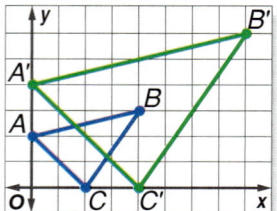

discount (p. 355) The amount by which the regular price of an item is reduced.

Distributive Property (p. 30) To multiply a sum by a number, multiply each addend of the sum by the number outside the parentheses.

domain (p. 177) The set of input values for a function.

definir una variable El elegir una variable para representar un valor desconocido en un problema y usarla para escribir una expresión o ecuación para resolver el problema.

grados La unidad más común para medir ángulos. Si un círculo se divide en 360 partes iguales, cada parte tiene una medida angular de 1 grado.

eventos dependientes Dos o más eventos en que el resultado de un evento afecta el resultado de otro u otros eventos.

diámetro La distancia a través de un círculo pasando por el centro.

dilatación Tipo de transformación en que se amplía o se reduce el tamaño de una figura.

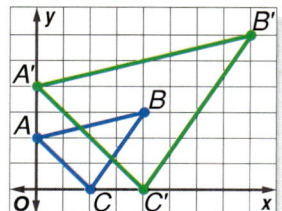

descuento La cantidad de reducción del precio normal.

propiedad distributiva Para multiplicar una suma por un número, multiplica cada sumando de la suma por el número fuera del paréntesis.

dominio El conjunto de valores de entrada de una función.

equation (p. 24) A mathematical sentence that contains an equals sign, =.

equilateral triangle (p. 429) A triangle having three congruent sides.

equivalent expressions (p. 30) Expressions that have the same value.

equivalent fractions (p. 207) Fractions that have the same value. $\frac{2}{3}$ and $\frac{4}{6}$ are equivalent fractions.

equivalent ratios (p. 289) Two ratios that have the same value.

evaluate (p. 11) To find the value of an expression.

ecuación Enunciado matemático que contiene un signo de igualdad, =.

triángulo equilátero Triángulo con tres lados congruentes.

expresiones equivalentes Expresiones que tienen el mismo valor.

fracciones equivalentes Fracciones que tienen el mismo valor. $\frac{2}{3}$ y $\frac{4}{6}$ son fracciones equivalentes.

razones equivalentes Dos razones que tienen el mismo valor.

evaluar Calcular el valor de una expresión.

experimental probability (p. 393) An estimated probability based on the relative frequency of positive outcomes occurring during an experiment.

exponent (p. 10) In a power, the number that tells how many times the base is used as a factor. In 5^3, the exponent is 3. That is, $5^3 = 5 \times 5 \times 5$.

exponential form (p. 11) Numbers written with exponents.

probabilidad experimental Estimado de una probabilidad que se basa en la frecuencia relativa de los resultados positivos que ocurren durante un experimento.

exponente En una potencia, el número que indica las veces que la base se usa como factor. En 5^3, el exponente es 3. Es decir, $5^3 = 5 \times 5 \times 5$.

forma exponencial Números escritos usando exponentes.

factorial (p. 381) The expression $n!$ is the product of all counting numbers beginning with n and counting backward to 1.

factors (p. 10) Two or more numbers that are multiplied together to form a product.

factor tree (p. 198) A diagram showing the prime factorization of a number. The factors branch out from the previous factors until all of the factors are prime numbers.

fair game (p. 374) A game in which players of equal skill have an equal chance of winning.

formula (p. 271) An equation that shows a relationship among certain quantities.

frequency table (p. 54) A table for organizing a set of data that shows the number of pieces of data that fall within given intervals or categories.

function (p. 177) A relation in which each element of the input is paired with exactly one element of the output according to a specified rule.

function table (p. 177) A table used to organize the input numbers, output numbers, and the function rule.

Fundamental Counting Principle (p. 378) Uses multiplication of the number of ways each event in an experiment can occur to find the number of possible outcomes in a sample space.

factorial La expresión $n!$ es el producto de todos los números naturales, comenzando con n y contando al revés hasta 1.

factores Dos o más números que se multiplican entre sí para formar un producto.

diagrama de árbol Diagrama que muestra la factorización prima de un número. Los factores se ramifican a partir de los factores previos hasta que todos los factores son números primos.

juego justo Juego en que jugadores con la misma habilidad tienen igual oportunidad de ganar.

fórmula Ecuación que muestra una relación entre ciertas cantidades.

tabla de frecuencias Tabla que se usa para organizar un conjunto de datos y que muestra cuántas piezas de datos caen dentro de intervalos o categorías dadas.

función Relación en que cada elemento de entrada es apareado con un único elemento de salida, según una regla específica.

tabla de funciones Tabla que organiza las entradas, la regla y las salidas de una función.

Principio Fundamental de Contar Este principio usa la multiplicación del número de veces que puede ocurrir cada evento en un experimento para calcular el número de posibles resultados en un espacio muestral.

gallon (p. 267) A customary unit of capacity equal to four quarts.

geometric sequence (p. 34) A sequence in which each term can be found by multiplying the previous term by the same number.

gram (p. 39) A unit of mass in the metric system equivalent to 0.001 kilogram.

graph (p. 106) The process of placing a point on a number line at its proper location.

galón Unidad de capacidad del sistema inglés de medidas que equivale a 4 cuartos de galón.

sucesión geométrica Sucesión en que cada término se puede calcular multiplicando el término anterior por el mismo número.

gramo Unidad de masa del sistema métrico. Un gramo equivale a 0.001 de kilogramo.

graficar Proceso de dibujar o trazar un punto en una recta numérica en su ubicación correcta.

greatest common factor (GCF) (p. 203) The greatest of the common factors of two or more numbers. The GCF of 18 and 24 is 6.

máximo común divisor (MCD) El mayor factor común de dos o más números. El MCD de 18 y 24 es 6.

height (p. 483) The length of the segment perpendicular to the base with endpoints on opposite sides. In a triangle, the distance from a base to the opposite vertex.

altura Longitud del segmento perpendicular a la base y con extremos en lados opuestos. En un triángulo, es la distancia desde una base al vértice opuesto.

heptagon (p. 446) A polygon having seven sides.

heptágono Polígono con siete lados.

hexagon (p. 446) A polygon having six sides.

hexágono Polígono con seis lados.

histogram (p. 86) A special kind of bar graph in which the bars are used to represent the frequency of numerical data that have been organized in intervals.

histograma Tipo especial de gráfica de barras que usa barras para representar la frecuencia de los datos numéricos, los cuales han sido organizados en intervalos iguales.

hypotenuse (p. 479) The side opposite the right angle in a right triangle.

hipotenusa El lado opuesto al ángulo recto en un triángulo rectángulo.

Identity Property (p. 31) The sum of an addend and 0 is the addend. The product of a factor and 1 is the factor.

propiedad de identidad La suma de un sumando y 0 es el sumando mismo. El producto de un factor y 1 es el factor mismo.

independent events (p. 398) Two or more events in which the outcome of one event does not affect the outcome of the other event(s).

eventos independientes Dos o más eventos en los cuales el resultado de uno de ellos no afecta el resultado de los otros eventos.

indirect measurement (p. 441) Finding a measurement by using similar triangles and writing a proportion.

medida indirecta Técnica que se usa para calcular una medida a partir de triángulos semejantes y proporciones.

inequality (p. 172) A mathematical sentence that contains the symbols $<$, $>$, $\leq$, or $\geq$.

integer (p. 106) The whole numbers and their opposites.
… , −3, −2, −1, 0, 1, 2, 3, …

interquartile range (p. 81) The range of the middle half of a set of numbers.
Interquartile range = UQ − LQ

interval (p. 54) On a scale, the difference between the greatest and least values in each category.

inverse operations (p. 156) Operations that "undo" each other. Addition and subtraction are inverse operations.

irrational number (p. 476) A number that cannot be written as a fraction.

isosceles triangle (p. 429) A triangle having at least two congruent sides.

desigualdad Enunciado matemático que contiene los símbolos $<$, $>$, $\leq$ o $\geq$.

entero Los números enteros y sus opuestos.
… , −3, −2, −1, 0, 1, 2, 3, …

amplitud intercuartílica El rango de la mitad central de un conjunto de datos o números. La amplitud intercuartílica = CS − CI

intervalo En una escala, la diferencia entre el valor mayor y el menor en cada categoría.

operaciones inversas Operaciones que se "anulan" mutuamente. La adición y la sustracción son operaciones inversas.

número irracional Número que no se puede escribir como fracción.

triángulo isósceles Triángulo que tiene por lo menos dos lados congruentes.

kilogram (p. 39) The base unit of mass in the metric system equivalent to 1,000 grams.

kilogramo Unidad fundamental de masa del sistema métrico. Un kilogramo equivale a mil gramos.

leaf (p. 76) The second greatest place value of data in a stem-and-leaf plot.

least common denominator (LCD) (p. 227) The least common multiple of the denominators of two or more fractions.

least common multiple (LCM) (p. 224) The least of the common multiples of two or more numbers. The LCM of 2 and 3 is 6.

leg (p. 479) Either of the two sides that form the right angle of a right triangle.

linear equation (p. 178) An equation for which the graph is a straight line.

line graph (p. 60) A type of statistical graph using lines to show how values change over a period of time.

hoja El segundo valor de posición mayor en un diagrama de tallo y hojas.

mínimo común denominador (mcd) El menor múltiplo común de los denominadores de dos o más fracciones.

mínimo común múltiplo (mcm) El menor múltiplo común de dos o más números. El mcm de 2 y 3 es 6.

cateto Cualquiera de los lados que forman el ángulo recto de un triángulo rectángulo.

ecuación lineal Ecuación cuya gráfica es una recta.

gráfica lineal Tipo de gráfica estadística que usa segmentos de recta para mostrar cómo cambian los valores durante un período de tiempo.

line of symmetry (p. 456) A line that divides a figure into two halves that are reflections of each other.

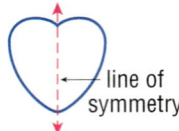

line plot (p. 64) A diagram that shows the frequency of data. An × is placed above a number on a number line each time that number occurs in a set of data.

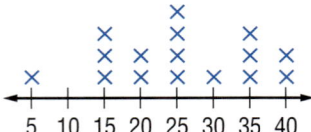

line symmetry (p. 456) Figures that match exactly when folded in half have line symmetry.

liter (p. 39) The base unit of capacity in the metric system. A liter is a little more than a quart.

lower extreme (p. 80) The least number of a set of data.

lower quartile (p. 80) The median of the lower half of a set of numbers, indicated by LQ.

eje de simetría Recta que divide una figura en dos mitades que son reflexiones entre sí.

esquema lineal Grafica que muestra la frecuencia de datos. Se coloca una × sobre la recta numérica, cada vez que el número aparece en un conjunto de datos.

simetría lineal Exhiben simetría lineal las figuras que coinciden exactamente al doblarse una sobre otra.

litro Unidad básica de capacidad del sistema métrico. Un litro es un poco más de un cuarto de galón.

extremo inferior El número menor de un conjunto de datos.

cuartil inferior La mediana de la mitad inferior de un conjunto de datos, la cual se denota por CI.

mean (p. 69) The sum of the data divided by the number of items in the data set.

measures of central tendency (p. 69) Numbers that are used to describe the center of a set of data. These measures include the mean, median, and mode.

median (p. 70) The middle number in a set of data when the data are arranged in numerical order. If the data has an even number, the median is the mean of the two middle numbers.

meter (p. 38) The base unit of length in the metric system.

metric system (p. 38) A base-ten system of measurement using the base units: meter for length, kilogram for mass, and liter for capacity.

mode (p. 70) The number or numbers that appear most often in a set of data.

multiple (p. 224) The product of a number and any whole number.

multiplicative inverse (p. 258) The product of a number and its multiplicative inverse is 1. The multiplicative inverse of $\frac{2}{3}$ is $\frac{3}{2}$.

mutually exclusive (p. 401) Events that cannot happen at the same time.

media La suma de los datos dividida entre el número total de artículos en el conjunto de datos.

medidas de tendencia central Números que se usan para describir el centro de un conjunto de datos. Estas medidas incluyen la media, la mediana y la moda.

mediana Número central de un conjunto de datos, una vez que los datos han sido ordenados numéricamente. Si hay un número par de datos, la mediana es el promedio de los dos datos centrales.

metro Unidad fundamental de longitud del sistema métrico.

sistema métrico Sistema de medidas de base diez que usa las unidades fundamentales: metro para longitud, kilogramo para masa y litro para capacidad.

moda Número o números que aparece(n) más frecuentemente en un conjunto de datos.

múltiplo El producto de un número y cualquier número entero.

inverso multiplicativo El producto de un número y su inverso multiplicativo es 1. El inverso multiplicativo de $\frac{2}{3}$ es $\frac{3}{2}$.

mutuamente exclusivo Eventos que no pueden ocurrir al mismo tiempo.

N

negative integer (p. 106) An integer that is less than zero.

net (p. 530) A two-dimensional figure that can be used to build a three-dimensional figure.

nonagon (p. 446) A polygon having nine sides.

nonlinear equation (p. 181) An equation whose graph is not a straight line.

numerical expression (p. 14) A combination of numbers and operations.

entero negativo Un entero menor que cero.

red Figura bidimensional que sirve para hacer una figura tridimensional.

enágono Polígono que tiene nueve lados.

ecuación no lineal Ecuación cuya gráfica no forma una recta.

expresión numérica Combinación de números y operaciones.

O

obtuse angle (p. 413) Any angle that measures greater than 90° but less than 180°.

obtuse triangle (p. 429) A triangle having one obtuse angle.

octagon (p. 446) A polygon having eight sides.

odds (p. 396) A ratio that compares the number of ways an event can occur to the number of ways it cannot occur.

opposites (p. 121) Two integers are opposites if they are represented on the number line by points that are the same distance from zero, but on opposite sides of zero. The sum of two opposites is zero.

order of operations (p. 14) The rules to follow when more than one operation is used in a numerical expression.
  1. Do all operations within grouping symbols first.
  2. Evaluate all powers before other operations.
  3. Multiply and divide in order from left to right.
  4. Add and subtract in order from left to right.

ángulo obtuso Cualquier ángulo que mide más de 90° pero menos de 180°.

triángulo obtusángulo Triángulo que tiene un ángulo obtuso.

octágono Polígono que tiene ocho lados.

posibilidades Proporción en que se compara el número de veces en que un evento puede ocurrir, con el número de veces en que el evento no puede ocurrir.

opuestos Dos enteros son opuestos si, en la recta numérica, están representados por puntos que equidistan de cero, pero en direcciones opuestas. La suma de dos opuestos es cero.

orden de operaciones Reglas a seguir cuando se usa más de una operación en una expresión numérica.
  1. Primero ejecuta todas las operaciones dentro de los símbolos de agrupamiento.
  2. Evalúa todas las potencias antes que las otras operaciones.
  3. Multiplica y divide en orden de izquierda a derecha.
  4. Suma y resta en orden de izquierda a derecha.

ordered pair (p. 112) A pair of numbers used to locate a point in the coordinate plane. An ordered pair is written in the form (*x*-coordinate, *y*-coordinate).

origin (p. 112) The point at which the *x*-axis and the *y*-axis intersect in a coordinate plane.

ounce (p. 267) A customary unit of weight. 16 ounces equals 1 pound.

outcome (p. 370) One possible result of a probability event. For example, 4 is an outcome when a number cube is rolled.

outlier (p. 65) A piece of data that is quite separated from the rest of the data. In a box-and-whisker plot, data that are more than 1.5 times the interquartile range from the quartiles.

par ordenado Par de números que se utiliza para ubicar un punto en un plano de coordenadas. Se escribe de la siguiente forma: (coordenada *x*, coordenada *y*).

origen Punto en que el eje *x* y el eje *y* se intersecan en un plano de coordenadas.

onza Unidad de peso del sistema inglés de medidas. 16 onzas equivalen a una libra.

resultado Uno de los resultados posibles de un evento probabilístico. Por ejemplo, 4 es un resultado posible cuando se lanza un dado.

valor atípico Dato que se encuentra muy separado del resto de los datos. En un diagrama de caja y patillas, los datos que distan de los cuartiles respectivos más de 1.5 veces la amplitud intercuartílica.

P

parallel lines (p. 426) Lines in a plane that do not intersect.

líneas paralelas Rectas situadas en un mismo plano y que no se intersecan.

parallelogram (p. 434) A quadrilateral with opposite sides parallel and opposite sides congruent.

paralelogramo Cuadrilátero cuyos lados opuestos son paralelos y congruentes.

part (p. 323) In a percent proportion, the number that is compared to the whole quantity.

pentagon (p. 446) A polygon having five sides.

parte En una proporción porcentual, el número que se compara con la cantidad total.

pentágono Polígono que tiene cinco lados.

percent (p. 216) A ratio that compares a number to 100.

percent equation (p. 340) An equation that describes the relationship between the part, base, and percent.
part = percent · base

percent of change (p. 350) A ratio that compares the change in a quantity to the original amount.

percent of decrease (p. 350) A percent of change when the original quantity decreased.

percent of increase (p. 350) A percent of change when the original quantity increased.

percent proportion (p. 323) Compares part of a quantity to the whole quantity using a percent.
$$\frac{part}{base} = \frac{percent}{100}$$

por ciento Razón que compara un número con 100.

ecuación porcentual Ecuación que describe la relación entre la parte, la base y el por ciento.
parte = por ciento · base

porcentaje de cambio Razón que compara el cambio en una cantidad, con la cantidad original.

porcentaje de disminución Porcentaje de cambio cuando disminuye la cantidad original.

porcentaje de aumento Porcentaje de cambio cuando aumenta la cantidad original.

proporción porcentual Compara parte de una cantidad con la cantidad total mediante un por ciento.
$$\frac{parte}{base} = \frac{por\ ciento}{100}$$

perfect squares (p. 471) Numbers whose square roots are whole numbers. 25 is a perfect square because the square root of 25 is 5.

perimeter (p. 270) The distance around a closed geometric figure.

permutation (p. 381) An arrangement, or listing, of objects in which order is important.

perpendicular lines (p. 417) Lines that meet to form right angles.

pi (π) (p. 275) The ratio of the circumference of a circle to its diameter. An approximation often used for π is 3.14.

pint (p. 267) A customary unit of capacity equal to two cups.

polygon (p. 446) A simple closed figure in a plane formed by three or more line segments.

population (p. 345) The entire group of items or individuals from which the samples under consideration are taken.

positive integer (p. 106) An integer that is greater than zero.

pound (p. 267) A customary unit of weight equal to 16 ounces.

powers (p. 10) Numbers expressed using exponents. The power 3^2 is read *three to the second power,* or *three squared.*

precision (p. 542) The exactness of a measurement which depends on the unit of measure.

precision unit (p. 542) The smallest unit on a measuring tool.

prime factorization (p. 198) Expressing a composite number as a product of prime numbers. For example, the prime factorization of 63 is $3 \times 3 \times 7$.

prime number (p. 197) A whole number greater than 1 that has exactly two factors, 1 and itself.

principal (p. 358) The amount of money deposited or invested.

probability (p. 370) The chance that some event will happen. It is the ratio of the number of ways a certain event can occur to the number of possible outcomes.

properties (p. 31) Statements that are true for any number or variable.

proportion (p. 297) An equation that shows that two ratios are equivalent.

protractor (p. 412) An instrument used to measure angles.

cuadrados perfectos Números cuya raíz cuadrada es un número entero. 25 es un cuadrado perfecto porque la raíz cuadrada de 25 es 5.

perímetro La distancia alrededor de una figura geométrica cerrada.

permutación Arreglo o lista en que el orden es importante.

rectas perpendiculares Rectas que al encontrarse forman ángulos rectos.

pi (π) Razón entre la circunferencia de un círculo y su diámetro. A menudo, se usa 3.14 como aproximación del valor de π.

pinta Unidad de capacidad del sistema inglés de medidas que equivale a dos tazas.

polígono Figura simple cerrada en un plano, formada por tres o más segmentos de recta.

población El grupo total de individuos o de artículos del cual se toman las muestras bajo estudio.

entero positivo Un entero mayor que cero.

libra Unidad de peso del sistema inglés de medidas que equivale a 16 onzas.

potencias Números que se expresan usando exponentes. La potencia 3^2 se lee *tres a la segunda potencia* o *tres al cuadrado.*

precisión El grado de exactitud de una medida, lo cual depende de la unidad de medida.

unidad de precisión La unidad más pequeña de un instrumento de medición.

factorización prima Escritura de un número compuesto como el producto de números primos. La factorización prima de 63 es $3 \times 3 \times 7$.

número primo Número entero mayor que 1 que sólo tiene dos factores, 1 y sí mismo.

capital La cantidad de dinero depositada o invertida.

probabilidad La posibilidad de que suceda un evento. Es la razón del número de maneras en que puede ocurrir un evento al número total de resultados posibles.

propiedades Enunciados que se cumplen para cualquier número o variable.

proporción Ecuación que afirma la igualdad de dos razones.

transportador Instrumento que sirve para medir ángulos.

Pythagorean Theorem (p. 479) In a right triangle, the square of the length of the hypotenuse is equal to the sum of the squares of the lengths of the legs. $c^2 = a^2 + b^2$

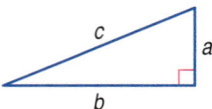

Teorema de Pitágoras En un triángulo rectángulo, el cuadrado de la longitud de la hipotenusa es igual a la suma de los cuadrados de las longitudes de los catetos. $c^2 = a^2 + b^2$

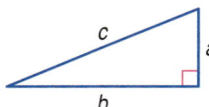

Q

quadrant (p. 113) One of the four regions into which the two perpendicular number lines of the coordinate plane separate the plane.

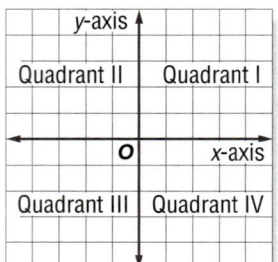

cuadrante Una de las cuatro regiones en que dos rectas numéricas perpendiculares dividen el plano de coordenadas.

quadrilateral (p. 434) A closed figure having four sides and four angles.

quart (p. 267) A customary unit of capacity equal to two pints.

cuadrilátero Figura cerrada que tiene cuatro lados y cuatro ángulos.

cuarto de galón Unidad de capacidad del sistema inglés de medidas que equivale a dos pintas.

R

radical sign (p. 471) The symbol used to indicate a nonnegative square root, $\sqrt{\ }$.

radius (p. 275) The distance from the center of a circle to any point on the circle.

signo radical Símbolo que se usa para indicar una raíz cuadrada no negativa, $\sqrt{\ }$.

radio Distancia desde el centro de un círculo hasta cualquier punto del mismo.

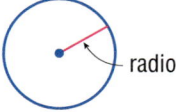

random (p. 370) Outcomes occur at random if each outcome is equally likely to occur.

random sample (p. 345) A sample in which the members are selected purely on the basis of chance.

range (p. 65) The difference between the greatest and least numbers in a data set.

range (p. 177) The set of output values for a function.

aleatorio Un resultado ocurre al azar si la posibilidad de ocurrir de cada resultado es equiprobable.

muestra aleatoria Muestra en que todos sus miembros se eligen al azar.

rango La diferencia entre el número mayor y el menor en un conjunto de datos.

rango Conjunto de los valores de salida de una función.

rate (p. 292) A ratio that compares two quantities with different kinds of units.

rate of change (p. 296) A ratio that shows a change in one quantity with respect to a change in another quantity.

ratio (p. 216) A comparison of two numbers by division. The ratio of 2 to 3 can be written as 2 out of 3, 2 to 3, 2:3, or $\frac{2}{3}$.

rational numbers (p. 229) Numbers that can be written as fractions, including terminating and repeating decimals, and integers.

reciprocal (p. 258) The multiplicative inverse of a number.

rectangle (p. 434) A parallelogram having four right angles.

rectangular prism (p. 520) A solid figure that has two parallel and congruent bases that are rectangles.

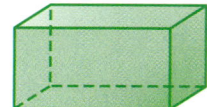

reflection (p. 457) A type of transformation in which a figure is flipped over a line of symmetry.

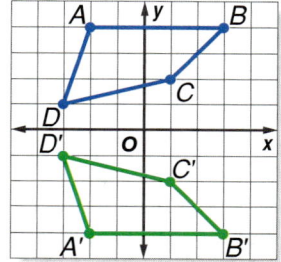

regular polygon (p. 446) A polygon that has all sides congruent and all angles congruent.

relatively prime (p. 206) Numbers that have a greatest common factor of 1.

repeating decimals (p. 211) A decimal whose digits repeat in groups of one or more. Examples are 0.181818… and 0.83333… .

rhombus (p. 434) A parallelogram having four congruent sides.

tasa Razón que compara dos cantidades que tienen distintas unidades de medida.

tasa de cambio Razón que representa el cambio en una cantidad con respecto al cambio en otra cantidad.

razón Comparación de dos números mediante división. La razón de 2 a 3 puede escribirse como 2 de cada 3, 2 a 3, 2:3 ó $\frac{2}{3}$.

números racionales Números que pueden escribirse como fracciones, incluyendo decimales terminales y periódicos y enteros.

recíproco El inverso multiplicativo de un número.

rectángulo Paralelogramo con cuatro ángulos rectos.

prisma rectangular Figura sólida con dos bases paralelas y congruentes que son rectángulos.

reflexión Tipo de transformación en el que se da vuelta a una figura sobre un eje de simetría.

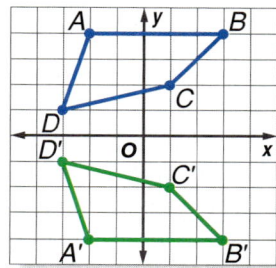

polígono regular Polígono con todos los lados y todos los ángulos congruentes.

relativamente primos Números cuyo máximo común divisor es igual a 1.

decimales periódicos Decimal cuyos dígitos se repiten en grupos de uno o más. Por ejemplo: 0.181818… y 0.83333… .

rombo Paralelogramo que tiene cuatro lados congruentes.

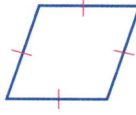

right angle (p. 413) An angle that measures 90°.

right triangle (p. 429) A triangle having one right angle.

rotation (p. 460) A type of transformation in which a figure is moved or turned around a central point.

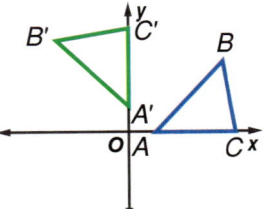

90° rotation about the origin

ángulo recto Ángulo que mide exactamente 90°.

triángulo rectángulo Triángulo que tiene un ángulo recto.

rotación Tipo de transformación en que se hace girar una figura alrededor de un punto central.

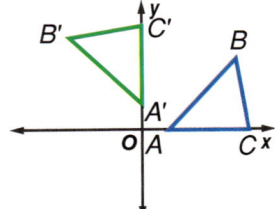

Rotación de 90° alrededor del origen

sales tax (p. 354) An additional amount of money charged on items that people buy.

sample space (p. 374) The set of all possible outcomes of a probability experiment.

sampling (p. 344) A practical method used to survey a representative group.

scale (p. 54) The set of all possible values of a given measurement, including the least and greatest numbers in the set, separated by equal intervals.

scale (p. 304) On a map, intervals used representing the ratio of distance on the map to the actual distance.

scale drawing (p. 304) A drawing that is similar but either larger or smaller than the actual object.

scale factor (p. 305) A scale written as a ratio in simplest form.

scale model (p. 306) A model used to represent something that is too large or too small for an actual-size model.

scalene triangle (p. 429) A triangle having no congruent sides.

impuesto sobre las ventas Cantidad adicional de dinero que se cobra por un artículo comprado.

espacio muestral Conjunto de todos los resultados posibles de un experimento probabilístico.

muestreo Método conveniente que facilita la elección y el estudio de un grupo representativo.

escala Conjunto de todos los posibles valores de una medida dada, el cual incluye los valores máximo y mínimo del conjunto, separados por intervalos iguales.

escala En un mapa, los intervalos que se usan para representar la razón de las distancias en el mapa a las distancias verdaderas.

dibujo a escala Dibujo que es semejante, pero más grande o más pequeño que el objeto real.

factor de escala Escala escrita como una tasa en forma reducida.

modelo a escala Réplica de un objeto real, el cual es demasiado grande o demasiado pequeño como para construirlo de tamaño natural.

triángulo escaleno Triángulo sin lados congruentes.

scatter plot (p. 61) In a scatter plot, two sets of related data are plotted as ordered pairs on the same graph.

diagrama de dispersión Diagrama en que dos conjuntos de datos relacionados aparecen graficados como pares ordenados en la misma gráfica.

scientific notation (p. 43) A number written as a product of a number that is at least 1 but less than 10, and a power of ten. For example, $687,000 = 6.87 \times 10^5$.

notación científica Números escritos como el producto de un número que es al menos igual a 1, pero menor que 10, por una potencia de diez. Por ejemplo, $687,000 = 6.87 \times 10^5$.

sequence (p. 34) A list of numbers in a certain order, such as 0, 1, 2, 3, or 2, 4, 6, 8.

sucesión Lista de números en cierto orden, tales como 0, 1, 2, 3 ó 2, 4, 6, 8.

significant digits (p. 542) All of the digits of a measurement that you know for sure, plus one estimated digit.

dígitos significativos Todos los dígitos de una medición que se sabe son exactos, más un dígito aproximado.

similar figures (p. 440) Figures that have the same shape but not necessarily the same size.

figuras semejantes Figuras que tienen la misma forma, pero no necesariamente el mismo tamaño.

simple event (p. 370) One outcome or a collection of outcomes.

eventos simples Un resultado o una colección de resultados.

simple interest (p. 358) The amount paid or earned for the use of money. The formula for simple interest is $I = prt$.

interés simple Cantidad que se paga o que se gana por el uso del dinero. La fórmula para calcular el interés simple es $I = prt$.

simplest form (p. 207) A fraction is in simplest form when the GCF of the numerator and the denominator is 1.

forma reducida Una fracción está escrita en forma reducida si el MCD de su numerador y denominador es 1.

simulate (p. 397) A way of acting out or modeling a problem situation.

simulación Manera de modelar o representar un problema.

slope (p. 182) A number that tells how steep a line is. Slope of a line is found by looking at the change in y with respect to the change in x.

pendiente Número que indica el grado de inclinación de una recta. La pendiente de una recta se calcula comparando el cambio en y con respecto al cambio en x.

solid (p. 514) A three-dimensional figure. Prisms, pyramids, cones, and cylinders are examples of solids.

sólido Una figura tridimensional. Los prismas, las pirámides, los conos y los cilindros son ejemplos de sólidos.

solution (p. 24) A value for the variable that makes an equation true. The solution of $12 = x + 7$ is 5.

solución Valor de la variable de una ecuación que hace verdadera la ecuación. La solución de $12 = x + 7$ es 5.

solving an equation (p. 24) The process of finding a solution to an equation.

resolver una ecuación Proceso de encontrar el número o números que satisfagan una ecuación.

square (p. 434) A parallelogram having four right angles and four congruent sides.

cuadrado Paralelogramo con cuatro ángulos rectos y cuatro lados congruentes.

square (p. 470) The product of a number and itself. 36 is the square of 6.

squared (p. 10) A number multiplied by itself; 4×4, or 4^2.

square root (p. 471) One of the two equal factors of a number. The square root of 9 is 3.

standard form (p. 11) Numbers written without exponents.

statistics (p. 54) The study of collecting, organizing, and interpreting data.

stem (p. 76) The greatest place value common to all the data values is used for the stem of a stem-and-leaf plot.

stem-and-leaf plot (p. 76) A system used to condense a set of data where the greatest place value of the data forms the stem and the next greatest place value forms the leaves.

straight angle (p. 413) An angle that measures exactly 180°.

supplementary angles (p. 422) Two angles are supplementary if the sum of their measures is 180°.

∠1 and ∠2 are supplementary angles.

surface area (p. 532) The sum of the areas of all the surfaces (faces) of a three-dimensional figure.

survey (p. 345) A question or set of questions designed to collect data about a specific group of people.

cuadrado El producto de un número por sí mismo. 36 es el cuadrado de 6.

al cuadrado Número multiplicado por sí mismo; 4×4, ó 4^2.

raíz cuadrada Uno de dos factores iguales de un número. La raíz cuadrada de 9 es 3.

forma estándar Números escritos sin exponentes.

estadística Estudio de la recolección, organización e interpretación de datos.

tallo El mayor valor de posición común a todos los datos es el que se usa como tallo en un diagrama de tallo y hojas.

diagrama de tallo y hojas Sistema que se usa para condensar un conjunto de datos y en el cual el mayor valor de posición de los datos forma el tallo y el segundo mayor valor de posición de los datos forma las hojas.

ángulo llano Ángulo que mide exactamente 180°.

ángulos suplementarios Dos ángulos son suplementarios si la suma de sus medidas es 180°.

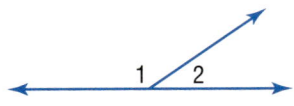

∠1 y ∠2 son suplementarios.

área de superficie La suma de las áreas de todas las superficies (caras) de una figura tridimensional.

encuesta Pregunta o conjunto de preguntas diseñadas para recoger datos sobre un grupo específico de personas.

term (p. 19) A number, a variable, or a product or quotient of numbers and variables.

term (p. 34) Each number in a sequence.

terminating decimals (p. 211) A decimal whose digits end. Every terminating decimal can be written as a fraction with a denominator of 10, 100, 1,000, and so on.

tessellation (p. 447) A repetitive pattern of polygons that fit together with no holes or gaps.

theoretical probability (p. 393) The ratio of the number of ways an event can occur to the number of possible outcomes.

ton (p. 267) A customary unit of weight equal to 2,000 pounds.

término Un número, una variable o un producto o un cociente de números y variables.

término Cada número en una sucesión.

decimales terminales Decimal cuyos dígitos terminan. Todo decimal terminal puede escribirse como una fracción con un denominador de 10, 100, 1,000, etc.

teselado Un patrón repetitivo de polígonos que coinciden perfectamente, sin dejar huecos o espacios.

probabilidad teórica La razón del número de maneras en que puede ocurrir un evento al número total de resultados posibles.

tonelada Unidad de peso del sistema inglés que equivale a 2,000 libras.

transformation (p. 451) A movement of a geometric figure.

translation (p. 451) One type of transformation where a geometric figure is slid horizontally, vertically, or both.

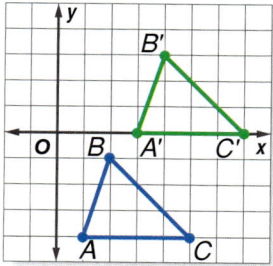

transversal (p. 427) A line that intersects parallel lines.

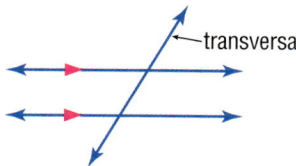

trapezoid (p. 434) A quadrilateral with one pair of parallel sides.

tree diagram (p. 374) A diagram used to show the total number of possible outcomes in a probability experiment.

triangle (p. 428) A polygon that has three sides and three angles.

triangular prism (p. 522) A prism that has bases that are triangles.

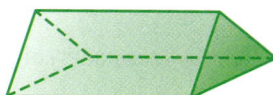

twin primes (p. 200) Primes that differ by two.

two-step equation (p. 166) An equation having two different operations.

transformación Movimientos de figuras geométricas.

traslación Tipo de transformación en que una figura se desliza horizontal o verticalmente o de ambas maneras.

transversal Recta que interseca rectas paralelas.

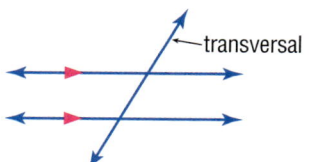

trapecio Cuadrilátero con un único par de lados paralelos.

diagrama de árbol Diagrama que se usa para mostrar el número total de resultados posibles en experimento probabilístico.

triángulo Polígono que posee tres lados y tres ángulos.

prisma triangular Prisma cuyas bases son triángulos.

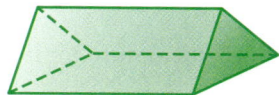

primos gemelos Números primos que difieren en dos unidades.

ecuación de dos pasos Ecuación que contiene dos operaciones distintas.

U

unbiased (p. 344) Not limited or not favoring a particular outcome.

unit rate (p. 292) A rate with denominator of 1.

upper extreme (p. 80) The greatest number of a set of data.

upper quartile (p. 80) The median of the upper half of a set of numbers, indicated by UQ.

no sesgado Que no está limitado a ningún resultado particular o que no favorece un resultado.

tasa unitaria Una tasa con un denominador de 1.

extremo superior El número máximo de un conjunto de datos o números.

cuartil superior La mediana de la mitad superior de un conjunto de números, denotada por CS.

variable (p. 18) A placeholder, usually a letter, used to represent an unspecified value in mathematical expressions or sentences. In $3 + a = 6$, a is a variable.

Venn diagram (p. 203) A diagram that uses circles to show how elements among sets of numbers or objects are related.

vertex (p. 413) A vertex of an angle is the common endpoint of the rays forming the angle.

vertical angles (p. 422) Opposite angles formed by the intersection of two lines.

∠1 and ∠2 are vertical angles.

volume (p. 520) The number of cubic units needed to fill the space occupied by a solid.

variable Marcador de posición, por lo general, una letra, que se usa para representar un valor desconocido en expresiones o enunciados matemáticos. En $3 + a = 6$, a es una variable.

diagrama de Venn Diagrama que usa círculos para mostrar la relación entre los elementos en un conjunto de números u objetos.

vértice El vértice de un ángulo es el extremo común de los rayos que lo forman.

ángulos opuestos por el vértice Ángulos opuestos que se forman de la intersección de dos rectas.

∠1 y ∠2 son ángulos opuestos por el vértice.

volumen Número de unidades cúbicas que se requieren para llenar el espacio que ocupa un sólido.

x-axis (p. 112) The horizontal number line in a coordinate plane.

x-coordinate (p. 112) The first number of an ordered pair. It corresponds to a number on the x-axis.

eje x La recta numérica horizontal en el plano de coordenadas.

coordenada x El primer número de un par ordenado. Corresponde a un número en el eje x.

y-axis (p. 112) The vertical number line in a coordinate plane.

y-coordinate (p. 112) The second number of an ordered pair. It corresponds to a number on the y-axis.

eje y La recta numérica vertical en el plano de coordenadas.

coordenada y El segundo número de un par ordenado. Corresponde a un número en el eje y.

zero pair (p. 118) The result when one positive counter is paired with one negative counter.

par nulo Resultado que se obtiene cuando una ficha positiva se aparea con una ficha negativa.

Selected Answers

Chapter 1 Decimal Patterns and Algebra

Page 5 Chapter 1 Getting Started
1. true   **3.** 105.8   **5.** 60.64   **7.** 11.3   **9.** 27.69   **11.** 30.8
13. 17.01   **15.** 8.15   **17.** 1.1   **19.** 2,450   **21.** 61

Pages 8–9 Lesson 1-1
1. Sample answer: It helps to organize your thoughts and focus on how to approach solving the problem.
5. 13 E-mails   **7.** 1,280,000
9.   **11.** 2006   **13.** Sample answer: sweater, gloves, and chocolates   **15.** 100   **17.** 625

Pages 11–13 Lesson 1-2
1. Five is used as a factor four times, or $5 \cdot 5 \cdot 5 \cdot 5 = 5^4$.
3. Sample answer: exponential form: 4^2; standard form: 125
5. $6 \cdot 6$ **7.** $8 \cdot 8 \cdot 8 \cdot 8 \cdot 8$ **9.** 3,125   **11.** 5^6 **13.** 1,331
15. $4 \cdot 4$ **17.** $9 \cdot 9 \cdot 9$ **19.** $11 \cdot 11 \cdot 11$ **21.** 64   **23.** 625
25. 10   **27.** 4,096   **29.** 20,736   **31.** 125,000   **33.** 7^4 **35.** 6^5
37. 13^5 **39.** 2^8 **41.** Sample answer: A number taken to the third power is the same as the volume of a cube, or the amount of space inside a cube.   **43.** false   **45.** false
47. false   **49.** $6^3, 15^2, 3^5, 2^8$ **51.** $8^3, 3^6, 6^4, 5^5$ **53.** $11^3, 5^5, 9^4, 6^5$ **55.** Sample answer: It is easier and less error-prone to write 10^{15} than writing out 15 zeros.   **57.** $5^4 \cdot 4^3$
59. Sample answer: $390,625 < 534,000 < 823,543$, or $5^8 < 534,000 < 7^7$ **61.** $5^1 \cdot 10^3$ **63.** C   **65.** 3   **67.** false
69. 3.4   **71.** 3.07   **73.** 33.98   **75.** 3.5

Pages 16–17 Lesson 1-3
1a. $14 - 7$ **1b.** $24 \div 3$ **3.** Cynthia; the first step is to do the division, $24 \div 6$. Yutaka incorrectly multiplied 6 and 2 first.   **5.** 5   **7.** 27   **9.** 28   **11.** 53.5   **13.** 58   **15.** 27   **17.** 13
19. 13   **21.** 14.2   **23.** 5.8   **25.** 4   **27.** 10   **29.** 2   **31.** 88
33. 600   **35.** 195   **37.** 22.8   **39.** 19   **41.** Sample answer: $2(7) + 2(9.5) + 6$; 39 cm   **43.** 450   **45.** 296,500   **47.** 57.9
49. $72 \div (9 + 27) - 2 = 0$ **51.** 16; The calculator followed the order of operations. Ruby incorrectly added $2 + 2$ first when she should have divided $12 \div 2$ first.   **53.** D
55. Sample answer: $30 - 20$, or about 10 days longer
57. 121   **59.** 1,024   **61.** $8 \cdot 8 \cdot 8 \cdot 8 \cdot 8$ **63.** $4 \cdot 4 \cdot 4 \cdot 4 \cdot 4$
65. 9   **67.** 18   **69.** 13

Pages 20–21 Lesson 1-4
1. coefficient: 2; constant: -6 **3.** $\frac{x}{y}$ is a division expression; the others are multiplication expressions.   **5.** 20
7. 21   **9.** 5   **11.** 16   **13.** 5   **15.** 7   **17.** 8   **19.** 12   **21.** 48
23. 1.6   **25.** 3   **27.** 32   **29.** 65.2   **31.** 5   **33.** 18   **35.** 3
37. Always; ab means a is multiplied by b.   **39.** $16n$ **41.** 2, 2.5, 3, 3.5   **43.** about 118   **45.** $155; $200; $245   **47.** Sample answer: $x = 5, y = 2$ **49.** August   **51.** 12   **53.** 4   **55.** true
57. false

Pages 26–27 Lesson 1-5
1. Sample answer: Find the value of the variable that makes a true sentence.   **3.** Ivan; $75 - 25 = 50$ is a true statement.
5. 17   **7.** 54   **9.** 23   **11.** 4   **13.** 23 points   **15.** 7   **17.** 38
19. 63   **21.** 5   **23.** 11   **25.** 20   **27.** 64   **29.** 20   **31.** 12

33. $x = 18$
35. z = 12

37. 40   **39.** 25   **41.** 7.5 h   **43.** 3,500 mi   **45.** $t = 51 - 18$ or $51 = t + 18$; 33 in.   **47.** Sample answer: $b = 0$; a is any number   **49.** 958   **51.** 31   **53.** 19   **55.** 28   **57.** 150

Pages 32–33 Lesson 1-6
1. $p(q + r)$ **3.** false; $(11 + 18) \times 5 = 145$; $11 + 18 \times 5 = 101$
5. $(10)2 + (8)2$; 36   **7.** Commutative Property of Addition
9. Distributive Property   **11.** $2(6) + 2(7)$; 26   **13.** $(11)8 + (3)8$; 112   **15.** $7(8) - 7(6)$; 14   **17.** Sample answer: The second expression, because it is easy to multiply $3 \cdot 400$, $3 \cdot 50$, and $3 \cdot 6$ mentally, and then add.   **19.** Associative Property of Multiplication   **21.** Commutative Property of Multiplication   **23.** Associative Property of Addition
25. Distributive Property   **27.** $239 + 531 = 531 + 239$
29. $152 + 0 = 152$ **31.** $y + 5$ **33.** $32b$ **35.** $2x + 6$
37. $5c + 6$ **39.** $6(40 + 8) = 6(40) + 6(8) = 240 + 48 = 288$; Distributive Property   **43.** A   **45.** 8   **47.** 7.1   **49.** $7y$
51. 625   **53.** 54   **55.** 20   **57.** 32.4

Pages 35–36 Lesson 1-7
1. Sample answer: Each term in an arithmetic sequence is found by adding or subtracting the same number to or from the previous term. Each term in a geometric sequence is found by multiplying or dividing the previous term by the same number.   **5.** Add 2 more than what was added to the previous term; neither.   **7.** 324, 972, 2,916   **9.** 6.6, 7, 7.4
11. 70, 7, 0.7, 0.07   **13.** $+ 12$; arith.   **15.** $\times 3$; geo.
17. Multiply by 1 more than was multiplied by the previous term; neither.   **19.** 20.6, 24.6, 28.6   **21.** 1.3, 1.6, 1.9
23. 0.2, 0.04, 0.008   **25.** 1, 3, 6, 10, 15, 21, 28, 36   **27.** 76
29. 100   **31.** Distributive   **33.** 520   **35.** 3.2

Pages 40–41 Lesson 1-8
1. There are a greater number of smaller units.   **3.** Hunter; Arturo multiplied 45.7 by 1,000; he should have divided by 1,000.   **5.** 45,000   **7.** 370   **9.** 2,340   **11.** 7.2   **13.** 64,000
15. 0.925   **17.** 8.5   **19.** 9,100   **21.** 0.043   **23.** 0.997
25. 0.0821   **27.** 57   **29.** 0.169 km   **31.** 0.02 km, 3,000 cm, 50 m   **33.** 0.06 L, 660 mL, 6.6 kL   **35.** 93,000   **37.** 0.04
39. 50 cm   **41.** 5 cobblers   **43.** 3,900,000,000 m   **45.** C
47. $+ 9$; arithmetic   **49.** $+ 1.1$; arithmetic   **51.** $5(9) + 5(7)$; 80
53. $8(7) - 8(2)$; 40   **55.** 800   **57.** 400,000   **59.** 6,500

Pages 44–45 Lesson 1-9
1. Sample answer: It is easier to keep track of the place value.   **3.** 3.0×10^6; Sample answer: Chicago has a population of about 3 million people: $3.0 \times 10^6 = 3,000,000$.
5. 100,000   **7.** 7.0×10^4 **9.** 1.75×10^7 **11.** 73,000
13. 338,000   **15.** 8,980,000   **17.** 798,000   **19.** 990,000,000
21. 80,600,000   **23.** 53,400, 5.03×10^5, 4.98×10^6, 4,980,100
25. 1.98×10^4 **27.** 9.97×10^6 **29.** 5.05×10^5 **31.** 2.36×10^7 **33.** 1.1×10^6 **35.** 6.08×10^{10} **37.** 54.3 quadrillion
39. $>$ **41.** $<$ **43.** 2.94×10^{14} mi   **45.** B   **47.** 1,010
49. 2,330

Pages 46–48 Chapter 1 Study Guide and Review
1. b 3. d 5. f 7. $89.40 9. 40,353,607 11. 324
13. 10,000 15. 100 17. 25 19. 20 21. 33 23. 0 25. 11
27. 36 29. 56 31. $965 33. 8 35. 7 37. 108 39. 67
41. Associative Property of Multiplication 43. Identity Property of Multiplication 45. arithmetic; +1.2; 6.8, 8.0, 9.2
47. $125, $62.50, $31.25 49. 29,000 51. 0.007
53. 5.9×10^4 55. 3.24×10^5 57. 1.03×10^6 59. 9.1×10^6

Chapter 2 Statistics: Analyzing Data

Page 53 Chapter 2 Getting Started
1. hundredths 3. greatest to least 5. thousands 7. ones
9. 5.062, 5.16, 5.61 11. 0.398, 3.98, 39.8 13. 10.26, 1.26, 1.026 15. 69.79, 68.99, 68.9 17. 86 19. 555

Pages 56–57 Lesson 2-1
1. An advantage of using a frequency table is that it organizes data to provide a quick summary. It makes it easier to see quickly the number of times each item of data appears. A disadvantage is that a frequency table may not show the individual pieces of data. 3. The interval in each category must be the same, 50.
5. Sample answer:

Number of Movie Rentals	Tally	Frequency
0–2	III	3
3–5	JHT	5
6–8	II	2
9–11	JHT III	8
12–14	III	3
15–17	II	2
18–20	I	1

7. cereal
9. Sample answer:

Year	Tally	Frequency
1900–1919	II	2
1920–1939	IIII	4
1940–1959	IIII	4
1960–1979	JHT IIII	9
1980–1999	JHT III	8

11. Sample answer:

Chore	Tally	Frequency
wash dishes	JHT III	8
vacuum	JHT	5
clean room	JHT JHT II	12
do laundry	JHT I	6
do yard work	IIII	4

13. Sample answer:

Year	Tally	Frequency
1970–1979	II	2
1980–1989	III	3
1990–1999	JHT III	8
2000–2009	JHT II	7

17. Sample answer: 91–110 Calories

19. Sample answer:

Price ($)	Tally	Frequency
11–20	II	2
21–30	IIII	4
31–40	IIII	4
41–50	JHT	5
51–60	JHT	5
61–70	II	2
71–80	IIII	4
81–90	III	3
91–100	JHT I	6

Most of the scooter prices are spread out evenly among the intervals, with the $91–$100 category containing the greatest number. An advantage of using this scale and interval is that you can get a more detailed distribution. A disadvantage is that it is difficult to see clusters or trends.
21. F 23. 533,000
25. 60,000,000 27. 78°F

Pages 61–63 Lesson 2-2
1. Graphs often show trends over time. If you continue the pattern, you can use it to make a prediction. 3. Tally mark; the other three are ways to display data. 5. Yes; in the 7th week he will be able to run 1 mile in about 7 minutes. 9. Sample answer: 210 nations 11. Sample answer: As the speed increases, the distance required to come to a complete stop increases. 13. The data do not appear to fall along a straight line. So, birth month and height do not seem to be related. 15. A 17. Sample answer: 2; 18–47 19. 9^2 21. 7^3 23. 6.1, 6.2, 6.2, 6.4, 6.5, 6.5, 6.7, 6.7, 6.8, 6.8, 6.8, 6.8, 7.0, 7.0

Pages 66–68 Lesson 2-3
1. Sample answer: the greatest and least values, clusters of data, gaps in data, how the data are spread out, the range, and the pieces of data that appear most often 3. Darnell; 16 is the mode and 20 is the greatest value.
5.

7. Sample answer: Most of the students in the class own 6–12 music CDs. 9. 104°F 11. The range would be 29 rather than 37.
13.

Sample answer: cluster 90–93; gap 79–85; outlier 79
15. Sometimes; clusters may appear anywhere on the line plot. 17. 5
19.

21. Sample answer: Most of the animals represented on the line plot have a life span of 10–20 years. 23. Sample answer: Bowler B is more consistent because excluding the outlier of 90, the range of scores is 4 points, compared with a range of 10 for Bowler A. Bowler A has scores evenly spread out from 90 to 100. 25. D 27. Sample answer: The more it rains, the taller the grass becomes. 29. 75 31. 29
33. 47 35. 16.1 37. 12.6 39. 36.4

Pages 71–72   Lesson 2-4

1. Sometimes; if there is an odd number of items, the median is the middle number. If there is an even number of items, the median is the mean of the two middle numbers.
3. Jared; the median is the 11th term of the ordered data, or 2. The mode is the number that appears most often, or 3.
5. 52.3; 57; 59   **7.** 8.2; 8; 7   **9.** Sample answer: Any of the three measures could be used to represent the data. The mean is slightly greater than most of the data items, and therefore is a less accurate description of the data.   **11.** 8.7; 9.0; 8.0 and 9.3   **13.** $12; $9; $6   **15.** $7.90; $8; $7.50
17. Sample answer: The mean or median best represent the data because half of the items are less than these values and half are greater. Since there are four modes, the mode does not best represent the data.
19. Sample answer: The mean would be most affected; the mode would be the least affected, since it would not change.   **21.** 12   **23.** line graph   **25.** ones   **27.** tens

Pages 78–79   Lesson 2-5

1. Both a stem-and-leaf plot and frequency table show the frequency of data occurring. However, a stem-and-leaf plot shows individual data values and a frequency table shows only the number of data items occurring in each interval.
3. 0, 1, 2, 3
5.

Stem	Leaf
1	6 9
2	1 3 5 5 9
3	1 3 4 5 9
4	2 7 8 $3\mid4 = 34$

7. 31   **9.** 32.9; 35; 40
11.

Stem	Leaf
1	3 3 5
2	4 8
4	0 1 2 2 5 6 8 8 8 $1\mid3 = 13$

13.

Stem	Leaf
40	3
41	1
42	
43	3 4 9
44	2 9
45	1 4 4 7 8 9 $43\mid9 = 439$

15. 26   **17.** $110; $115 and $150
19.

Stem	Leaf
5	1 2 2 7
6	0 1 2 3 5 7 7 8 8
7	3 3 3 4 5 6 7 8 9
8	1 2 2 4 5 9
8	1 2 2 4 5 9
9	4 4 7
10	6 8 $9\mid4 = 94$ points

21. Sample answer: Most of the game point totals are clustered from 60 to 89 points. There do not appear to be any gaps or outliers.
23. Sample answer:

Price ($)	Tally	Frequency
20–29	IIII	4
30–39	JHT JHT	10
40–49	JHT	5
50–59	I	1

Stem	Leaf
2	4 6 6 8
3	0 2 3 4 5 5 6 6 8 8
4	0 0 0 5 9
5	6 $3\mid5 = \$35$

All three representations show the frequency of data occurring. The frequency table shows intervals of data and is useful in comparing price ranges. The line plot gives a good picture of the spread of the data. The stem-and-leaf plot shows individual prices as in the line plot, as well as intervals, such as $20s, $30s, and so on.
25. 97   **27.** 4.3; 3.6; 3.6 and 2.4   **29.** 21   **31.** 13   **33.** 6, 7.9, 8, 8.4   **35.** 0.09, 0.14, 0.33, 0.7

Pages 82–83   Lesson 2-6

1. Sample answer: In a stem-and-leaf plot, you can see each data item and how the data are clustered. In a box-and-whisker plot, you can immediately see the median and how the data are dispersed.   **3.** 270; 350; 380; 588; 1,221
5.

7. 15; 30.5; 45; 51.5; 58   **9.** one-fourth   **11.** 0 and 4,600; no outliers   **13.** one half   **15.** A set in which an extreme value and a quartile are the same has no whisker on that side of the box-and-whisker plot.   **17.** 20.5   **19.** 31.6; 32; none
21. Sample answer: 20; 20–119

Pages 87–89   Lesson 2-7

1. Sample answer: Each interval represents a portion of the data set. The number of items in each interval is indicated by the frequency, typically shown along the vertical scale. By adding the frequencies for each interval, you can determine the number of values in the data set.   **3.** World Cup Soccer titles of seven countries   **5.** The horizontal axis represents the countries and the vertical axis represents the number of wins.   **7.** the number of candies found in 50 snack-size packages

9.

Candies in Snack-Size Packages

11. Sample answer: The bars are horizontal; the categories are on the vertical axis and the scale is on the horizontal axis.
13. Sample answer: 200,000

15. Sample answer: The histogram is strongly skewed toward the lower end, with a high concentration in the number of states with 10 or fewer representatives. The distribution spreads out from there, with fewer and fewer states as the number per state increases.   **17.** Sample answer: Both histograms show that many players have salaries in the lowest interval of the histogram, less than

$1 million. Both histograms show one, two, or three players in several of the other intervals. The Diamondbacks' salaries show only three players' salaries of more than $7 million, while the Yankees have seven players in that same range.   **19.** Sample answer: It is easier to compare two sets of data.   **21.** histogram
23.

Stem	Leaf
0	2 5 9
1	0 2 3 6 7
2	3 5 5
3	1 2|3 = 23

25. false

Pages 93–95   Lesson 2-8
1. Sample answer: Outliers may distort measures of central tendency; data shown in graphs may be exaggerated or minimized by manipulating scales and intervals.   **3.** The cost is increasing over time. In about a year's time, the cost has more than tripled.   **5.** Sample answer: The median, because it will not be affected by the high prices.
7. Sample answer: Although both graphs show a decrease in network viewers, in Graph B, a change in the vertical scale makes it appear that the viewers are decreasing more rapidly.   **9.** 488.3; none; 150   **11.** median   **13.** Yes; all the other test scores were lower than 96.   **15.** B   **17.** the number of in-line skaters in New York, almost 2 million
19. $7.57; $7.50; $7.50 and $8.00   **21.** 8,500   **23.** 0.643

Pages 96–98   Chapter 2   Study Guide and Review
1. true   **3.** true   **5.** false; outlier   **7.** true
9. Sample answer:

Hours of Sleep	Tally	Frequency
5	I	1
6	III	3
7	JHT	5
8	IIII	4
9	III	3

11. Sample answer: 17 million
13.
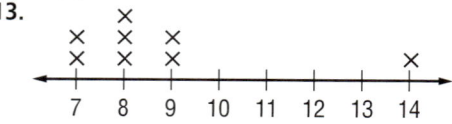
Sample answer: cluster 7–9; gap 9–14; outlier 14
15.

Sample answer: cluster 19–23; gap 23–28; outlier 28
17. 24.4; 24; 18 and 31   **19.** 132.3, 130.5, 136

21.
Stem	Leaf
5	3
6	
7	5 8 8
8	3 5 7 7 9
9	1 2 7|5 = 75

23.
Stem	Leaf
20	1 4
21	8 9
22	1 5 9
23	4 23|4 = 234

25. 11.5
27.

29. 8

Chapter 3   Algebra: Integers
Page 105   Chapter 3   Getting Started
1. false; median   **3.** <   **5.** <   **7.** 123   **9.** 1,730   **11.** 30
13. 90   **15.** 7   **17.** 25   **19.** 14; 17

Pages 107–108   Lesson 3-1
1. Sample answers: temperature, golf scores, elevation, loss of money   **3.** −3; The other expressions equal 3.   **5.** −11   **7.** 7
9. 6   **11.** 9   **13.** 0   **15.** 2008   **17.** −50   **19.** 6   **21.** 9   **23.** 4
25. 14   **27.** 0   **29.** 2 protons: 2; 5 electrons: −5
31.
33.
35. true   **37.** B   **39.** The very high salary increases the mean.   **41.** >   **43.** <   **45.** >

Pages 110–111   Lesson 3-2
1. 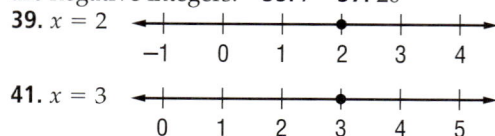   **3.** right   **5.** >   **7.** >
9. 15, −13, 10, −4, 13, −5   **11.** >   **13.** >
15. >   **17.** >   **19.** >   **21.** >   **23.** false; Sample answer: 8 > 5   **25.** true   **27.** true   **29.** Mar. 2, Mar. 6, Mar. 1, Mar. 5, Mar. 3, Mar. 4, Mar. 7   **31.** −7, −6, −4, 1, 3, 5   **33.** They are negative integers.   **35.** 7   **37.** 20
39. $x = 2$
41. $x = 3$

Pages 114–115   Lesson 3-3
1. Sample answer: Point A is 1 unit to the right and 2 units down from the origin, in quadrant IV. Point B is 2 units to the left and 1 unit up from the origin, in quadrant II.
3. (−2, −4), III   **5.** (0, 3), y-axis
6–8.
9. (−2, 2), II   **11.** (−5, 0), x-axis   **13.** (2, −2), IV
15. (3, −5), IV   **17.** (2, 2), I
19. (0, −32)

20–31.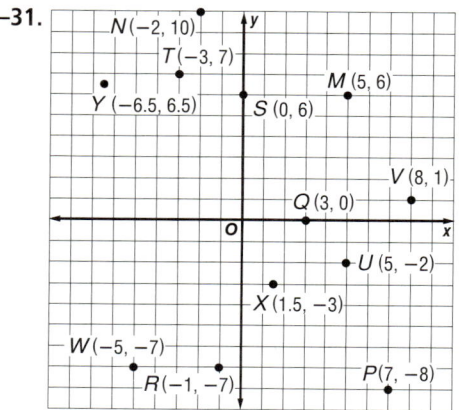

33. Sometimes; both (−2, 0) and (2, 0) lie on the x-axis.
35. China   **37.** Children's Zoo   **39.** (−3, −2)   **41.** Sample answer: On the coordinate plane, I would walk 1 unit right,

then 4 units down.   **43.** Quadrant I when both x- and y-coordinates are positive; quadrant III when both x- and y-coordinates are negative; the origin when both x- and y-coordinates are 0.   **45.** G   **47.** <   **49.** >   **51.** 101
53. 384   **55.** 7,470

Pages 122–124   Lesson 3-4
1. $2 + (-7) = -5$ Sample model:

3. Brooke; Javier added correctly but did not apply the sign of the greater absolute value.   **5.** -14 **7.** -4 **9.** less since $-\$42 + \$35 = -\$7$ **11.** 0   **13.** -6 **15.** -25 **17.** -14
19. 0   **21.** -38 **23.** 8   **25.** -30 **27.** $-3 + (-6)$; $-9°$F
29. $-6 + 14$; 8 ft   **31.** 4   **33.** -13 **35.** 12   **37.** 0   **39.** -3
41. -18 **43.** -13 **45.** Sorenstam; $-13 < -10$ **47.** $y + 2$
49. n **51.** $f + (-8)$
53. Sample answer:
$-6 + 9 + (-4) = -6 + (-4) + 9$ Comm. (+)
$ = (-6 + (-4)) + 9$ Assoc. (+)
$ = -10 + 9$ Simplify.
$ = -1$
55. Sample answer:
$8 + 10 + (-8) = 8 + (-8) + 10$ Comm. (+)
$ = (8 + (-8)) + 10$ Assoc. (+)
$ = 0 + 10$ Additive Inv.
$ = 10$
57. Sample answer:
$8 + (-9) + 9 = 8 + ((-9) + 9)$ Assoc. (+)
$ = 8 + 0$ Additive Inv.
$ = 8$ Simplify.
61. -2; -2; No, addition is associative.   **63.** -11
65. $(0, -2)$, y-axis   **67.** $(1, 1)$, I
69.   **71.** 24

Pages 130–131   Lesson 3-5
1. Sample answer: To subtract an integer, add its additive inverse.   **3.** Mitsu; Bradley did not add the additive inverse of -19.   **5.** -3 **7.** -2 **9.** -8 **11.** 265°F   **13.** -10
15. 4   **17.** -20 **19.** 23   **21.** 29   **23.** 2   **25.** -6 **27.** -3
29. 1   **31.** -4 **33.** 18   **35.** 0   **37.** -12 **39.** 11   **41.** -3
43. 8   **45.** 2,139 m   **47.** Tiberius; 79 yr   **49.** always; $-2 - 3 = -5$ **51.** sometimes; $5 - 2 = 3$, $2 - 5 = -3$
53. true   **55.** 300°C   **57.** -11 **59.** -14 **61.** 70   **63.** 192

Pages 136–137   Lesson 3-6
1. $2(-3) = -6$

3. negative; Sample answer: $-2(-3)(-4) = -24$ **5.** 45   **7.** $-10a$ **9.** $45c$ **11.** 70   **13.** -875, or 875 feet below sea level
15. 80   **17.** 170   **19.** 36   **21.** -450 **23.** -220 **25.** -64
27. 49   **29.** $-70d$ **31.** $-72f$ **33.** $24h$ **35.** $-10rs$ **37.** -40

39. -400 **41.** -300 **43.** -700 **45.** $-32, 64$; geometric: multiply the previous term by -2
47. Sample answer: new triangle $P'Q'R'$ is on the other side of the origin (in quadrant III) from original triangle PQR.

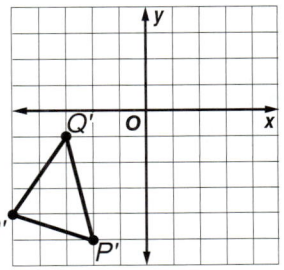

49. when n is an even number   **51.** F   **53.** 8   **55.** 0   **57.** 2   **59.** 20 and younger, 21–25, 26–30   **61.** 8   **63.** 4

Pages 140–141   Lesson 3-7
1. $-42 \div 7 = -6$, $-42 \div (-6) = 7$ **3.** $-18 \div (-9)$ has a positive quotient; the others have negative quotients.
5. -8 **7.** -1 **9.** -4 **11.** -2 **13.** 5   **15.** -7 **17.** 4
19. 15   **21.** -1 **23.** -5 **25.** -1 **27.** -3 **29.** 2   **31.** 3
33. $p = -45 \div 3$; $p = -15$ or 15-yd penalty   **35.** $-3, 1$; Divide the previous term by -3.   **37.** 16 sea otters per year   **39.** 1, 2, 4, 5, 10, 20, $-1, -2, -4, -5, -10, -20$
41. H   **43.** -28 **45.** 60   **47.** 18   **49.** 0.003 g

Pages 142–144   Chapter 3   Study Guide and Review
1. negative   **3.** 7   **5.** origin   **7.** x-coordinate   **9.** negative
11. -150 **13.** 8   **15.** 11   **17.** 5   **19.** 25   **21.** >   **23.** <
25. <   **27.** $-32, -23, -21, 14, 19, 25$ **29.** 3°, 1°, 0°, $-1°, -2°$ **31.** $(-3, -5)$, III   **33.** $(0, 3)$, y-axis
34–37.

39. -13 **41.** 0
43. -13 **45.** 7
47. -12 **49.** 35   **51.** 28
53. -35 **55.** 5   **57.** -2
59. -50 ft per h

Chapter 4   Algebra: Linear Equations and Functions

Page 149   Chapter 4   Getting Started
1. solved   **3.** 4   **5.** -11
7.

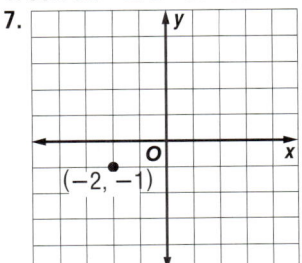

9. -5 **11.** 5   **13.** -2
15. -2 **17.** -4 **19.** 6

Pages 151–152   Lesson 4-1
1. Sample answer: 5 dollars more than my allowance is 8 dollars.   **3.** $x + 8$ **5.** $10h$ **7.** $n + 4 = -9$ **9.** $a + 2.4 = 35.3$
11. $y + 5$ **13.** $n - 6$ **15.** $2f$ **17.** $\frac{z}{-12}$ **19.** $c + 2 = 4$
21. $10k = 280$ **23.** $n - 5 = 31$ **25.** $2z + 3 = 15$
27. $y + 7$ or $4.5y$ **29.** $x + 2$; $x - 2$ **31.** H   **33.** -5 **35.** -8
37. -1 **39.** 0

Selected Answers   **649**

Pages 158–159   Lesson 4-2

1. Addition Property of Equality   3. $b + 5 = 4$; All of the other equations have solutions equal to 1. This one equals -1.   5. -2 7. 11   9. -3 11. 7   13. -3 15. -9 17. 17   19. 7   21. -4 23. 61   25. -5 27. -12 29. -5 31. 18.4   33. 6.4   35. -0.68 37. $415 = 105 + x$; 310 ft   39. The value of y must decrease by 3.   41. 3   43. -5 45. 9   47. 1.2   49. 301

Pages 162–163   Lesson 4-3

1. No; $-3(-4)$ equals 12.   3. Jesse; the variable is multiplied by -5. To solve for x, you need to divide each side of the equation by the entire coefficient, -5.   5. 2   7. -4 9. 7   11. 6   13. -7 15. 3   17. 14   19. -5 21. -133 23. 10   25. 55   27. 40   29. 2.5 h   31. 24 in.   33. 1.5   35. -5.4 37. 0.8   39. 100 s   41. *Blue Streak* 24.36 ft/s; *Corkscrew* 17.08 ft/s; *Magnum* 42.55 ft/s; *Mean Streak* 32.89 ft/s   43. C   45. -10 47. -3 49. $a + 5$ 51. $2r$ 53. -14 55. -21

Pages 168–169   Lesson 4-4

1. Sample answer: $3x + 10 = 4$ 3. 2   5. 3   7. 6   9. 3   11. -4 13. 4   15. 9   17. 8   19. 11   21. -8 23. 0   25. 3   27. 2   29. 14   31. 2.25   33. 2.1   35. 28   37. 16   39. 3 h   41. Less than; $-3°$F is about $-19.4°$C.   43. C   45. 7   47. 41   49. 14 ft   51. $>$ 53. $<$

Pages 174–175   Lesson 4-5

1.
3. Sample answer: $2x < 4$ and $x + 1 < 3$

5.
6.
7.
8. 9. $d \geq 4$
9.
10.
11.
12.
13.
14.
15.
16.
17.
18.
19.

21. $x < -6$ 23. $g > 7$ 25. $u \geq 6$ 27. $y \leq \frac{4}{3}$ 29. $p \leq -1$ 31. $d < 0$ 33. $w + 1 \geq 5$; $w \geq 4$ 35. $a \geq 16$ 37. all except baseball-softball   39. $h > 2,000,000$ 41. $j \leq 2.14$, so 1 or 2 pairs of jeans   43. B   45. 8   47. -13

49.   51.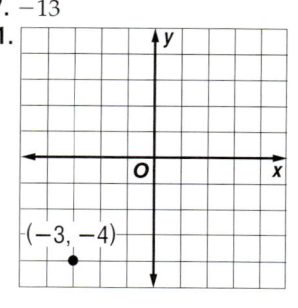

Pages 179–181   Lesson 4-6

1. Sample answer: $y = 2x$

3.
x	x − 2	y
1	1 − 2	−1
2	2 − 2	0
3	3 − 2	1
4	4 − 2	2

domain: {1, 2, 3, 4}; range: {−1, 0, 1, 2}

5.   7.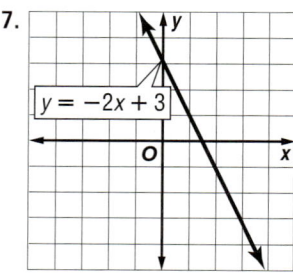

9.
x	x + 5	y
1	1 + 5	6
2	2 + 5	7
3	3 + 5	8
4	4 + 5	9

domain: {1, 2, 3, 4}; range: {6, 7, 8, 9}

11.
x	−6x	y
−1	−6(−1)	6
0	−6(0)	0
1	−6(1)	−6
2	−6(2)	−12

domain: {−1, 0, 1, 2}; range: {6, 0, −6, −12}

13.
x	−2x − 2	y
−1	−2(−1) − 2	0
0	−2(0) − 2	−2
1	−2(1) − 2	−4
2	−2(2) − 2	−6

domain: {−1, 0, 1, 2}; range: {0, −2, −4, −6}

15.   17.

19.   21.

23.   25.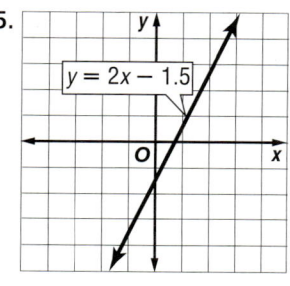

650   Selected Answers

27. $y = x - 5$

x	x − 5	y
−1	−1 − 5	−6
0	0 − 5	−5
1	1 − 5	−4
2	2 − 5	−3

29.

x	2.67x	y
1	2.67(1)	2.67
2	2.67(2)	5.34
3	2.67(3)	8.01
4	2.67(4)	10.68

27–29. Sample answers are given.

31. Average Defense Spending (per person)

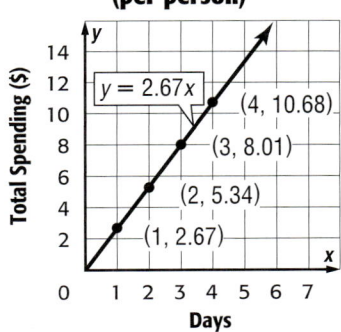

33. $y = 20x$

35. Rectangles

37. $y = x + 2$
39. $y = -4x$

41. nonlinear;

43. linear;

45. nonlinear;

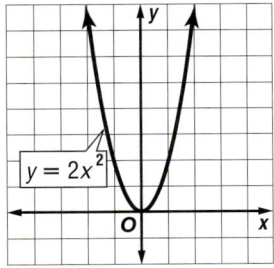

47. D   **49.** $x \geq 25$ **51.** −13
53. −2   **55.** 3

Pages 184–185   Lesson 4-7
1. the steepness of a line from left to right
3. Sample answer:

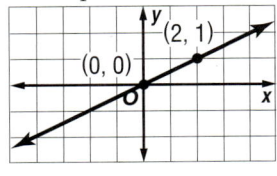

5. −1   **7.** 3   **9.** 1   **11.** −3
13. $\frac{1}{3}$ **15.** $-\frac{1}{2}$ **17.** Monica; her earnings are increasing faster.

19. The slope represents the amount of money earned with respect to the number of hours worked.   **21.** 1994–1995; The slope is steepest.   **23.** The sales remained constant; there is no change in the slope of the line. It is horizontal and the slope is equal to zero.
25. nonlinear

27.

29.

31. C

33.

35.

Pages 186–188   Chapter 4 Study Guide and Review
1. true   **3.** true   **5.** true   **7.** false; 7   **9.** false; −6
11. $n + 5$ **13.** $2a$ **15.** $a + 10 = 23$ **17.** $\frac{56}{z} = 14$ **19.** 3
21. −1   **23.** −13   **25.** 23   **27.** −2   **29.** 4   **31.** −12   **33.** −9
35. 11   **37.** 6   **39.** −1   **41.** 9.8   **43.** 3
45. $y > 3$
47. $d \leq 5$
49. $s \leq 11$
51. $m > -3$
53. $t \geq 2$

55.

57.

59.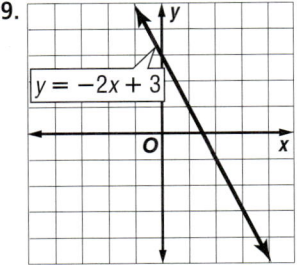

61. $y = 6x$ **63.** 4   **65.** −1

Chapter 5 Fractions, Decimals, and Percents

Page 195 Chapter 5 Getting Started
1. greater 3. 0.61 5. 0.33 7. 5 9. none 11. 0.75
13. $5 \times 5 \times 5 \times 5 \times 5$ 15. 4^3 17. 1^5

Pages 199–200 Lesson 5-1
1. Sample answer: Both prime and composite numbers are whole numbers greater than 1. A prime number has exactly two factors, 1 and itself. Composite numbers have more than two factors. 3. Sample answer: 9 is not prime.
5. prime 7. prime 9. $2 \times 3 \times 5$ 11. $2 \cdot 5 \cdot a \cdot c$
13. composite 15. prime 17. composite 19. prime
21. $2^2 \times 3^2$ 23. $3^2 \times 11$ 25. $2 \times 3 \times 5 \times 7$ 27. $2 \times 3^2 \times 7$
29. composite 31. $2 \cdot 2 \cdot 5 \cdot p \cdot q$ 33. $7 \cdot 7 \cdot y \cdot y$ 35. $2 \cdot 2 \cdot 2 \cdot 3 \cdot a \cdot a \cdot b \cdot b$ 37. 8 39. 7 41. 5^2 43. Sample answer: 5 and 7, 11 and 13, and 17 and 19 45. $2^3 \times 3 \times 5$
47. Sample answer: twenty 2-ft-by-3-ft tiles, or twelve 2-ft-by-5-ft tiles 49. 36 51. composite
53.
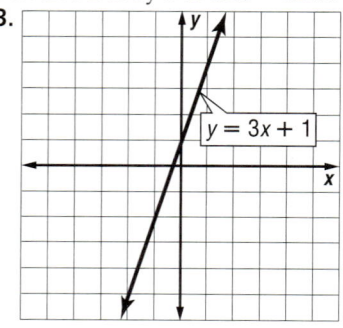
55. negative 57. zero 59. 2, 5, 10 61. 3, 9

Pages 205–206 Lesson 5-2
1. Sample answer: 32 and 48 3. Tiffany is correct; Charles included extra factors of 2 and 3. 5. 15 7. 6 9. 6 11. 5
13. 24 15. 8 17. 6 19. 7 21. 5 23. never
25. sometimes 27. $9b$ 29. $10x$ 31. Sample answer: The first building is 9 in. high, 3 in. long, and 4 in. wide. The second building is 9 in. high, 6 in. long, and 5 in. wide. The third building is 9 in. high, 5 in. long, and 5 in. wide.
33. No; Sample answer: There are a total of 16 pieces if the sandwiches are cut into 4-inch long pieces. 35. True; their only common factor is 1. 37. True; their only common factor is 1. 39. G 41. composite
43. composite 45. $\frac{1}{2}$ 47. 0.02 L 49. 24 51. 16

Pages 208–209 Lesson 5-3
1. Sample answer: A fraction is in simplest form if the GCF of the numerator and denominator is 1. 3. Seki; Luther did not divide the numerator and denominator by the GCF.
5. $\frac{3}{5}$ 7. $\frac{4}{13}$ 9. $\frac{5}{7}$ 11. $\frac{7}{10}$ 13. $\frac{4}{7}$ 15. $\frac{6}{7}$ 17. 1 19. $\frac{5}{6}$
21. $\frac{6}{10}, \frac{15}{25}$ 23. $\frac{10}{16}, \frac{20}{32}$ 25. $\frac{3}{4}$ 27. $\frac{3}{4}$ 29. No, because both the numerator and denominator can be divided by 2.
31. Sample answer:

33. $2 \times 3^2 \times 5 \times 7$
35. 0.25 37. 0.375

Pages 212–213 Lesson 5-4
1. Sample answer: $\frac{1}{2} = 0.5$; $\frac{2}{9} = 0.\overline{2}$ 3. $0.\overline{63}$ 5. $12.\overline{470}$
7. 3.625 9. $1.8\overline{3}$ 11. $\frac{1}{10}$ 13. 0.522 15. $5.92\overline{1}$ 17. $13.1\overline{46}$
19. $30.6\overline{841}$ 21. 0.8 23. $0.\overline{72}$ 25. 0.15 27. 6.425
29. $0.\overline{714285}$ 31. $4.\overline{203}$ 33. $\frac{3}{4}$ 35. $\frac{1}{5}$ 37. $5\frac{24}{25}$ 39. $15\frac{2}{5}$;
$3\frac{3}{20}$ 41. $8\frac{4}{5}$; $1\frac{4}{5}$ 43. 2.3 h 45. 3.16049; Archimedes'
47. B 49. $\frac{5}{12}$ 51. $\frac{2}{7}$ 53. 12 55. $\frac{3}{9}$ 57. $\frac{4}{12}$

Pages 218–219 Lesson 5-5
1. $\frac{9}{25}$; 36% 3. 1%; the other ratios equal 10%. 5. 29.2%
7. 25% 9. 80% 11. $\frac{9}{10}$ 13. $\frac{3}{4}$ 15. 75% 17. 12.2%
19. $11\frac{2}{7}$% 21. 99.9% 23. 20% 25. 80% 27. 74%
29. 26% 31. 95% 33. 12% 35. 76% 37. $\frac{1}{4}$ 39. $\frac{9}{20}$
41. $\frac{1}{2}$ 43. $\frac{31}{50}$ 45. $\frac{17}{25}$ 47. $\frac{11}{25}$ 49. $\frac{1}{50}$ 51. 36% 53. 17
55. D 57. $0.\overline{65}$ 59. $0.\overline{41}$ 61. 162 63. 0.144

Pages 222–223 Lesson 5-6
1. Sample answer: 0.25, $\frac{1}{4}$, 25% 3. 0.68 5. 0.276 7. 56%
9. 8% 11. 0.012 13. 0.7 15. 0.01 17. 0.022 19. 0.3025
21. 8% 23. 60% 25. 14.5% 27. 70.25% 29. 90.5%
31. 80.8% 33. 10% to 67% 35. Z 37. O 39. 76.5%
41. 99% 43. $\frac{45}{80}$ 45. G 47. 72% 49. 18% 51. 9.375
53. 23 students 55. 2×5^2 57. $2^2 \times 19$

Pages 225–226 Lesson 5-7
1. 3 and 4 are factors of 12; 12 is a multiple (and LCM) of 3 and 4. 3. Sometimes; the LCM of 3 and 9 is 9; the LCM of 3 and 5 is 15. 5. 42 7. 4 packages of hot dogs and 5 packages of buns 9. 48 11. 72 13. 315 15. 72
17. 420 19. 144 21. $45n$, where $n = 5, 6, 7, \ldots$ 23. 5 s; 3 s; 30 s 25. $2^2 \times 5^2$, or 100 27. 2012 29. $\frac{17}{25}$ 31. $\frac{3}{25}$
33. < 35. <

Pages 229–231 Lesson 5-8
1. Sample answer: $\frac{2}{15}, \frac{5}{6}$ 3. 45 5. > 7. > 9. 0.59, 60%, $\frac{31}{50}$ 11. yes 13. 20 15. 70 17. 36 19. > 21. < 23. =
25. < 27. > 29. > 31. < 33. > 35. 3 saves out of 4; 3 out of 4 has an average of 0.75; 7 out of 11 has an average of about 0.64. 37. 19%, $\frac{1}{5}$, 0.23 39. 0.615, 62%, $\frac{5}{8}$
41. $-\frac{49}{50}$, -0.5, -0.49 43. Christopher 45. no 47. no
49. yes 51. Sample answer: $2\frac{3}{4}$ is the same as 2.75. Fill the cup to halfway between 2.7 and 2.8. 53. $\frac{63}{32}$; the difference of $\frac{63}{32}$ and 2 is the least. 55. McGwire: $\frac{70}{509} = 0.138$; Sosa: $\frac{66}{643} = 0.103$; McGwire 57. Sample answer: When the numerators are the same, the fraction with the largest denominator is the least fraction. 59. 47%, $\frac{4}{7}$, $47\frac{2}{5}$, 47.41, $47.\overline{4}$ 61. 0.04 63. 0.14

Pages 232–234 Chapter 5 Study Guide and Review
1. c 3. d 5. a 7. prime 9. composite 11. 2^7 13. 5×19
15. 9 17. 14 19. $18w$ 21. $\frac{4}{5}$ 23. $\frac{1}{11}$ 25. $\frac{37}{45}$ 27. $\frac{1}{20}$

29. 0.875 **31.** 4.$\overline{3}$ **33.** 1.$\overline{857142}$ **35.** $\frac{11}{25}$ **37.** $\frac{9}{50}$ **39.** $\frac{2}{25}$
41. 44% **43.** 40% **45.** $\frac{19}{20}$ **47.** $\frac{4}{25}$ **49.** 0.48 **51.** 0.125
53. 61% **55.** 19% **57.** 0.23 **59.** 8 **61.** 24 **63.** 120
65. = **67.** > **69.** =

Chapter 6 Applying Fractions

Page 239 Chapter 6 Getting Started
1. Property **3.** 35 **5.** 30 **7.** 21.6 **9.** 0.83 **11.** $2\frac{6}{5}$
13. $5\frac{5}{4}$ **15.** $\frac{43}{4}$ **17.** $\frac{22}{5}$

Pages 242–243 Lesson 6-1
1. Sample answer: Models are useful to determine whether fractions are closer to 0, $\frac{1}{2}$, or 1, and to check estimates.
3a. Yes; the sum of $\frac{1}{2}$ and a fraction greater than $\frac{1}{2}$ is greater than 1. **3b.** Yes; both fractions are greater than $\frac{1}{2}$, so the sum is greater than $\frac{1}{2} + \frac{1}{2}$ or 1. **3c.** No; both fractions are less than $\frac{1}{2}$, so the sum is less than 1. **5–37. Sample answers are given.** **5.** 6 × 3 = 18 **7.** 1 − 0 = 1 **9.** 22 ÷ 11 = 2 **11.** 4 + 5 = 9 **13.** 5 − 3 = 2 **15.** 3 · 6 = 18
17. $1 + \frac{1}{2} = 1\frac{1}{2}$ **19.** $\frac{1}{2} - 0 = \frac{1}{2}$ **21.** 0 × 1 = 0 **23.** 1 ÷ 1 = 1 **25.** $\frac{1}{2} \times 14 = 7$ **27.** 25 ÷ 5 = 5 **29.** 12 ÷ 4 = 3
31. 24 ÷ 3 = 8 **33.** −3 × 9 = −27 **35.** $-\frac{1}{6} \times (-66) = 11$
37. 3 − 1 or 2 c **39.** It will be greater than the actual quotient. **41.** H **43.** > **45.** < **47.** 12 **49.** 24

Pages 246–247 Lesson 6-2
1. Sample answer:

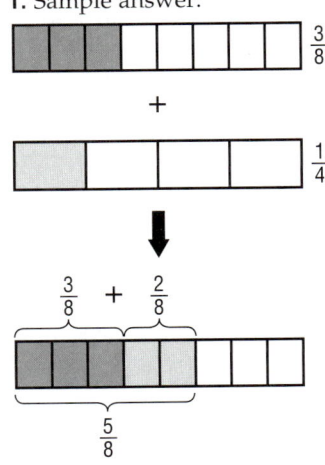

3. Jacinta; Marissa did not rename the fractions using the LCD. **5.** $1\frac{2}{7}$
7. $\frac{1}{4}$ **9.** $\frac{1}{4}$ **11.** $1\frac{5}{18}$
13. $\frac{8}{15}$ **15.** $\frac{4}{7}$ **17.** $\frac{2}{3}$
19. $\frac{8}{15}$ **21.** $\frac{19}{30}$ **23.** $\frac{1}{5}$
25. $1\frac{1}{4}$ **27.** $\frac{1}{24}$ **29.** $\frac{7}{24}$
31. $\frac{1}{20}$ **33.** $-1\frac{1}{15}$ **35.** $\frac{5}{6}$
37. $\frac{1}{12}$ **39.** Jon; $\frac{1}{18}$ of the race **41.** Sierra; she has $\frac{1}{4}$ of her allowance left over; Jacob has $\frac{1}{15}$ left over.
43. C **45.** Sample answer: $1 - \frac{1}{2} = \frac{1}{2}$ **47.** Sample answer: 16 ÷ 8 = 2 **49.** friend/family **51.** $4\frac{5}{3}$ **53.** $11\frac{6}{5}$

Pages 250–251 Lesson 6-3
1. Sample answer: A table is $8\frac{1}{2}$ feet long. A tablecloth is $2\frac{1}{3}$ feet longer. How long is the tablecloth? Answer: $8\frac{1}{2} + 2\frac{1}{3}$ or $10\frac{5}{6}$ ft **3.** less than; 7 + 1 < 2 + 7 **5.** $4\frac{2}{3}$ **7.** $9\frac{13}{20}$ **9.** $8\frac{1}{4}$
11. $17\frac{5}{24}$ **13.** $4\frac{5}{8}$ **15.** $9\frac{5}{9}$ **17.** $8\frac{1}{3}$ **19.** $7\frac{5}{12}$ **21.** $7\frac{1}{2}$

23. $8\frac{1}{4}$ **25.** $5\frac{1}{15}$ **27.** $6\frac{5}{6}$ **29.** $19\frac{2}{3}$ **31.** $6\frac{3}{5}$ **33.** $4\frac{1}{3}$
35. $31\frac{3}{16}$ **37.** $38\frac{5}{8}$ **39.** $6\frac{6}{11}$ **41.** $13\frac{7}{18}$ **43.** 20 ft **45.** H
47. Sample answer: 1 ÷ 1 = 1 **49.** Sample answer: 9 × 7 = 63 **51.** > **53.** < **55.** $\frac{19}{8}$ **57.** $\frac{51}{10}$

Pages 256–257 Lesson 6-4
1. Sample answer: $\frac{2}{3}$ and $\frac{4}{5}$ **3.** Less than; $\frac{4}{5}$ of 18 is a fraction of 18. **5.** $\frac{1}{5}$ **7.** $-\frac{2}{3}$ **9.** $6\frac{3}{5}$ **11.** $\frac{3}{32}$ **13.** $\frac{1}{6}$
15. $-\frac{1}{6}$ **17.** $\frac{1}{2}$ **19.** $\frac{5}{36}$ **21.** $4\frac{1}{2}$ **23.** 2,700 votes **25.** $2\frac{2}{3}$
27. 16 **29.** −42 **31.** $8\frac{2}{3}$ **33.** $\frac{1}{4}$ **35.** $\frac{7}{40}$ or 0.175 km
37. Always; sample answer: the numerators of the fractions are less than the denominators, so the product of the numerators will be less than the product of the denominators, resulting in a fraction that is less than 1. **39.** 7/16 **41.** $\frac{6}{7}$
43. $1\frac{7}{18}$ **45.** 4 **47.** 3 **49.** −5°F **51.** 62.7 **53.** 2.88

Pages 260–261 Lesson 6-5
1. no; $\frac{8}{3} \neq 24$ **3.** $\frac{2}{3}$, 3; The other pairs of numbers are reciprocals, but these are not. **5.** $\frac{9}{2}$ or $4\frac{1}{2}$ **7.** $\frac{5}{29}$
9. −12 **11.** $1\frac{1}{2}$ **13.** 20.5 **15.** $\frac{2}{11}$ **17.** $-\frac{8}{3}$ or $-2\frac{2}{3}$
19. $\frac{1}{3}$ **21.** $\frac{5}{22}$ **23.** 36 **25.** 30 **27.** −24 **29.** $-\frac{4}{5}$
31. 42.12 **33.** 2.88 **35.** 137.5 mi **37.** Sample answer: Use the Multiplicative Property to multiply each side by 2. Then use the Division Property to divide each side by h. So, $b = \frac{2A}{h}$. **39.** 16 **41.** 9 **43.** $2\frac{2}{3}$ **45.** $\frac{6}{17}$ **47.** $\frac{5}{6}$
49. Sample answer: 24 ÷ 12 = 2 **51.** Sample answer: 1 ÷ 1 = 1

Page 266 Lesson 6-6
1. Sample answer: How many $\frac{1}{4}$-pound hamburgers can be made with 10 pounds of ground beef? **3.** $2\frac{2}{5}$ **5.** $\frac{1}{15}$
7. 24 boxes **9.** $\frac{2}{3}$ **11.** $1\frac{1}{6}$ **13.** $\frac{2}{9}$ **15.** $3\frac{3}{4}$ **17.** $-\frac{1}{6}$
19. $2\frac{7}{15}$ **21.** greater than 1 because $7\frac{1}{6} > 3\frac{2}{3}$ **23.** $\frac{11}{16}$ lb
25. $\frac{7}{6}$ or $1\frac{1}{6}$ **27.** $\frac{1}{8}$ **29.** $\frac{1}{16}$ **31.** 14 **33.** 16

Pages 268–269 Lesson 6-7
1. Sample answer: A recipe calls for 2 pints of strawberries. How many cups is this? Answer: 4 c **3.** 3 **5.** 3 **7.** 500
9. 100 T **11.** 6 **13.** 16 **15.** 9 **17.** 10,560 **19.** $4\frac{1}{2}$ **21.** $5\frac{1}{2}$
23. $1\frac{1}{3}$ **25.** $10\frac{3}{4}$ **27.** 82.8 **29.** $\frac{1}{2}$ **31.** $43\frac{3}{4}$ gal
33.

Pounds	Ounces
1	16
2	32
3	48
4	64

The graph is a straight line. For each x-value increase of 1, the y value increases by 16. **35.** A **37.** 28 **39.** −10.8
41. 34 **43.** 19

Pages 272–273 Lesson 6-8
1. Perimeter is the sum of measures, so the units are the same. Area is the product of measures, so the units are squared. **3.** 18 yd; 20 yd² **5.** 26 cm; 42 cm² **7.** $23\frac{1}{3}$ ft **9.** 82 in.; 414 in² **11.** 65.4 m; 183.6 m² **13.** $30\frac{1}{5}$ in.; $47\frac{7}{10}$ in² **15.** 60 cm; 216 cm² **17.** $15\frac{3}{20}$ in.; 14 in² **19.** $37\frac{1}{3}$ yd or 112 ft; $81\frac{2}{3}$ yd² or 735 ft² **21.** The perimeter is doubled, the area is quadrupled. **23.** $35\frac{13}{24}$ ft **25.** 7 boxes **27.** 1.5 acres; 65,010 ÷ 43,560 **29.** The rectangles have the same perimeters; as the lengths and widths get closer, the areas increase. **31.** 36 **33.** A **35.** 10,500 **37.** $7\frac{1}{2}$ **39.** 33 **41.** 44 **43.** $4\frac{5}{7}$

Pages 276–277 Lesson 6-9
1. The circumference increases by 2π for every 1-unit increase in r. **3.** Mya; Aidan incorrectly applied the formula that uses the diameter. **5.** 44 m **7.** 73.5 cm **9.** 11 yd **11.** 50.2 ft **13.** 44 yd **15.** 47.4 m **17.** 12.6 km **19.** 285.7 m **21.** $16\frac{1}{2}$ in. **23.** 785 ft **25.** 2π or 6.28 **27.** 69.1 in. **29.** 17.6 cm; 19 cm² **31.** 8 yd

Pages 278–280 Chapter 6 Study Guide and Review
1. diameter **3.** Area **5.** reciprocal **7.** Sample answer: 3 ÷ 1 = 3 **9.** Sample answer: 1 × 0 = 0 **11.** Sample answer: $\frac{1}{2} \cdot 26 = 13$ **13.** $\frac{1}{6}$ **15.** $\frac{1}{12}$ **17.** $1\frac{3}{5}$ **19.** $\frac{5}{24}$ **21.** $1\frac{1}{10}$ **23.** $7\frac{3}{5}$ **25.** $10\frac{1}{7}$ **27.** $2\frac{4}{15}$ **29.** $15\frac{1}{8}$ **31.** $\frac{6}{35}$ **33.** $\frac{2}{7}$ **35.** $1\frac{3}{4}$ **37.** $1\frac{11}{16}$ mi **39.** $\frac{1}{5}$ **41.** 16 **43.** −22.8 **45.** $\frac{7}{10}$ **47.** $6\frac{1}{2}$ **49.** $3\frac{3}{10}$ **51.** $\frac{3}{22}$ **53.** $\frac{3}{4}$ **55.** 24 **57.** 4 **59.** 3 **61.** $6\frac{1}{4}$ ft; $1\frac{5}{16}$ ft² **63.** 11 in.; $2\frac{1}{2}$ in² **65.** $13\frac{2}{3}$ ft; $10\frac{1}{2}$ ft² **67.** 25.1 yd **69.** 26.4 ft

Chapter 7 Ratios and Proportions

Page 287 Chapter 7 Getting Started
1. ratio **3.** 48.1 **5.** 7.4 **7.** $\frac{1}{5}$ **9.** $\frac{19}{23}$ **11.** $\frac{8}{25}$ **13.** 24 **15.** $5\frac{1}{2}$ **17.** 6 **19.** $1\frac{1}{2}$

Pages 290–291 Lesson 7-1
1. $\frac{8}{15}$, 8 : 15, 8 to 15 **3.** 3 to 2; The other ratios equal $\frac{2}{3}$. **5.** $\frac{4}{15}$ **7.** $\frac{11}{4}$ **9.** $\frac{4}{3}$ **11.** yes; $\frac{9}{15} = \frac{3}{5}$ **13.** $\frac{21}{23}$ **15.** $\frac{1}{5}$ **17.** $\frac{15}{7}$ **19.** $\frac{11}{30}$ **21.** $\frac{3}{16}$ **23.** $\frac{5}{1}$ **25.** $\frac{3}{8}$ **27.** yes; $\frac{9}{15} = \frac{3}{5}$ and $\frac{6}{10} = \frac{3}{5}$ **29.** yes; 8 : 21 = $\frac{8}{21}$ and 16 : 42 = $\frac{8}{21}$ **31.** yes; 6 lb : 72 oz = $\frac{4}{3}$ and 2 lb : 24 oz = $\frac{4}{3}$ **33.** $\frac{1}{13}$ **35.** no **37.** $\frac{2}{5}$; $\frac{7}{3}$; No, the ratio is greater for the dollar bill. **39.** 4; The ratios of successive terms increase by 1. **41.** G **43.** 26 ft; 36 ft² **45.** 19 cm; 15 cm² **47.** 4.9 **49.** $0.31

Pages 294–295 Lesson 7-2
1. Sample answer: $\frac{30 \text{ mi}}{2 \text{ h}} = \frac{15 \text{ mi}}{\text{h}}$ **3.** a; Sample answer: $\frac{20 \text{ ft}}{1 \text{ min}} \rightarrow \frac{40 \text{ ft}}{1 \text{ min}}$ represents an increase in speed since 40 ft/min > 20 ft/min. If the same distance is covered in a longer period of time, then the rate is decreased. Consider $\frac{20 \text{ ft}}{1 \text{ min}} \rightarrow \frac{20 \text{ ft}}{2 \text{ min}}$. Since $\frac{20 \text{ ft}}{2 \text{ min}} = 10$ ft/min, the rate is decreased. **5.** 60 words per min **7.** $0.98 per can **9.** Music Rox; $14.95/CD < $17.99/CD **11.** 152 customers per day **13.** $1.48 per lb **15.** 270 C per serving **17.** 32 mi per gal **19.** 150 for $11.99 **21.** 4,000 gal per min **23.** about 52 people per sq mi **25.** Sometimes; a ratio that compares two measurements with different units is a rate, such as $\frac{2 \text{ mi}}{10 \text{ min}}$. A ratio that compares two numbers or two measurements with like units is not a rate, such as $\frac{2 \text{ c}}{3 \text{ c}}$. **27.** human, 72 beats/min; elephant, 35 beats/min **29.** 32 tickets per h **31.** I **33.** $\frac{6}{1}$ **35.** $\frac{10}{7}$ **37.** 56 **39.** $1\frac{1}{3}$

Pages 299–300 Lesson 7-3
1. Compare the cross products. If they are equal, the ratios are equivalent. **3.** $\frac{2}{9}$; The other ratios equal $\frac{2}{3}$. **5.** no **7.** 2 **9.** 21 **11.** 9 **13.** 29.4 **15.** $6\frac{2}{3}$ or about 6.7 oz **17.** no **19.** no **21.** no **23.** 8 kittens **25.** 6 **27.** 6 **29.** 75 **31.** 2 **33.** 63 **35.** 22.5 **37.** 1,040 **39.** 13.5 **41.** $\frac{13}{15} = \frac{x}{100}$; 87% **43.** 4 c **45.** 210 words **47.** Sample answers: 1 and 36; 2 and 18; 4 and 9 **49.** 64 **51.** $\frac{5}{9}$ **53.** $\frac{7}{40}$ **55.** 56 **57.** $1\frac{3}{5}$ **59.** 52

Pages 306–308 Lesson 7-4
3. 200 km **5.** 3.5 cm **7.** $\frac{1}{90}$ **9.** 255 km **11.** 1,205 km **13.** 6 in.; $\frac{1}{240}$ **15.** 12 cm; $\frac{1}{300}$ **17.** 4 in. **19.** $\frac{3}{4}$ in. **21.** $\frac{1}{48}$ **23.** $12\frac{3}{16}$ mi **25.** $\frac{1}{12}$ **27.** 1 in. = 70 mi **29.** D **31.** 40 mi/h **33.** $1.55 per lb **35.** $\frac{11}{40}$ **37.** $\frac{23}{75}$

Pages 314–315 Lesson 7-5
1. Sample answer: Write a proportion using the given fraction and $\frac{n}{100}$. Then solve for n and add a %. Or, write the fraction as a decimal. Then multiply by 100 and add a %. **3.** Sample answer: It is easier to compare percents than it is to compare fractions. **5.** $\frac{3}{16}$ **7.** $\frac{23}{40}$ **9.** 62.5% **11.** 36.36% **13.** $\frac{7}{40}$ **15.** $\frac{31}{500}$ **17.** $\frac{23}{80}$ **19.** $\frac{1}{3}$ **21.** $\frac{13}{16}$ **23.** $\frac{63}{80}$ **25.** 55% **27.** 37.5% **29.** 96.67% **31.** 71.43% **33.** 1.25% **35.** 41.67% **37.** 48% **39.** less than; $\frac{26}{125} = 20.8\%$ **41.** D **43.** 67.5 ft **45.** $1\frac{2}{3}$ **47.** 17.5 **49.** $\frac{9}{100}$ **51.** $\frac{42.5}{100}$

Pages 317–318 Lesson 7-6
1. 170%; 1.7 **3.** $63\frac{1}{2}$; It is the only number not equivalent to $63\frac{1}{2}$%. **5.** 2; 2 **7.** 0.0015; $\frac{3}{2,000}$ **9.** 0.15% **11.** 0.05% **13.** 3.5; $3\frac{1}{2}$ **15.** 6; 6 **17.** 0.006; $\frac{3}{500}$ **19.** 0.0055; $\frac{11}{2,000}$ **21.** 2.6; $2\frac{3}{5}$ **23.** 0.0005; $\frac{1}{2,000}$ **25.** 850% **27.** 0.9% **29.** 264% **31.** 0.34% **33.** 350% **35.** 0.4% **37.** 162.5% **39.** 0.003; $\frac{3}{1,000}$ **41.** about 0.595% **43.** A **45.** $\frac{3}{40}$ **47.** $\frac{1}{16}$ **49.** 6 in. **51.** 0.065 **53.** 0.123

Pages 320–321 Lesson 7-7
1. $\frac{x}{22} = \frac{0.5}{100}$ **3.** Sample answer: Find the number of people surveyed who use a white erasable markerboard to record information during a meeting; 525 people. **5.** 110.5 **7.** 0.1 **9.** 65 **11.** $160 **13.** 51.3 **15.** 12.5 **17.** 0.4 **19.** 0.6 **21.** $\frac{x}{1,746} = \frac{22}{100}$; about 384 students **23.** $300.96 **25.** B **27.** 750% **29.** 0.04% **31.** 73.33% **33.** 20 **35.** 4.5

Page 325 Lesson 7-8
1. b **3.** 36% **5.** 0.7 **7.** 25% **9.** 8.6 **11.** 375 **13.** 7.5% **15.** 4.1 **17.** 30 **19.** 0.04 **21.** 60% **23.** 20% of 500, 20% of 100, 5% of 100; Sample answer: If the percent is the same but the base is greater, then the part is greater. If the base is the same but the percent is greater, then the part is greater. **25.** 65% **27.** 31.5

Pages 326–328 Chapter 7 Study Guide and Review
1. e **3.** a **5.** c **7.** g **9.** d **11.** $\frac{4}{3}$ **13.** $\frac{2}{3}$ **15.** yes; $\frac{3}{5} = \frac{21}{35}$ **17.** yes; 27:15 = $\frac{9}{5}$ and 9:5 = $\frac{9}{5}$ **19.** $4.75 per lb **21.** $9.50 per h **23.** $3\frac{1}{2}$ laps per min **25.** 6 **27.** 12.5 **29.** 9 **31.** 94 cm **33.** 27 cm **35.** 5.2 cm **37.** $\frac{27}{500}$ **39.** 12.5% **41.** 17.5% **43.** 0.0075; $\frac{3}{400}$ **45.** 0.005; $\frac{1}{200}$ **47.** 475% **49.** 0.95% **51.** 34.1 **53.** 0.6 **55.** 0.3 **57.** 130.2

Chapter 8 Applying Percent

Page 333 Chapter 8 Getting Started
1. true **3.** 48 **5.** 1,512 **7.** 1.75 **9.** 0.8 **11.** 75 **13.** 38.9 **15.** 0.325 **17.** 39.5%

Pages 336–337 Lesson 8-1
1. Sample answer: Find $\frac{1}{5} \cdot 140$ or find 10% of 140 and multiply by 2; both estimates equal 28. **3.** Ian; Mandy incorrectly changed 1.5% to 1.5, which is 150%. **5–31.** Sample answers are given. **5.** $\frac{1}{2} \cdot 160 = 80$ **7.** $\frac{3}{4} \cdot 20 = 15$ **9.** $0.1 \cdot 400 = 40$ and $2 \cdot 40 = 80$ **11.** $(1 \cdot 70) + (\frac{1}{2} \cdot 70) = 105$ **13.** $0.01 \cdot 80 = 0.8$ and $\frac{1}{2} \cdot 0.8 = 0.4$ **15.** $\frac{1}{4} \cdot 400 = 100$ **17.** $\frac{3}{4} \cdot 280 = 210$ **19.** $\frac{2}{3} \cdot 15 = 10$ **21.** $0.1 \cdot 40 = 4$ and $6 \cdot 4 = 24$ **23.** $0.1 \cdot 80 = 8$ and $8 \cdot 8 = 64$ **25.** $0.1 \cdot 120 = 12$ and $3 \cdot 12 = 36$ **27.** $(1 \cdot 54) + (\frac{1}{3} \cdot 54) = 72$ **29.** $4 \cdot 12 = 48$ **31.** $0.01 \cdot 400 = 4$ and $\frac{1}{2} \cdot 4 = 2$ **33.** Sample answer: $0.01 \cdot 3,000 = 30$ lb **35.** Sample answer: Find 1% of a number, then multiply by $\frac{3}{8}$. **37.** H **39.** 64.8 **41.** 157.1 **43.** 50 **45.** 357.1

Pages 342–343 Lesson 8-2
1. missing base **3.** $n = 0.88 \cdot 300$; 264 **5.** $3 = 0.12 \cdot n$; 25 **7.** 32.4%; 0.324 **9.** $n = 0.53 \cdot 470$; 249.1 **11.** $26 = n \cdot 96$; 27.1% **13.** $17 = 0.40 \cdot n$; 42.5 **15.** $30 = n \cdot 64$; 46.9% **17.** $1.45 = 0.33 \cdot n$; 4.4 **19.** $n = 0.752 \cdot 600$; 451.2 **21.** $14 = 0.028 \cdot n$; 500 **23.** $230 = n \cdot 200$; 115% **25.** 6% **27.** 42% **29.** 28% **31.** Sample answer: If the percent is less than 100%, then the part is less than the number; if the percent equals 100%, then the part equals the number; if the percent is greater than 100%, then the part is greater than the number. **33.** I **35.** Sample answer: $\frac{3}{10} \cdot 50 = 15$ **37.** 1,312.5 mg **39.** 18 **41.** 16.7% **43.** 31.6

Pages 346–347 Lesson 8-3
1. Sample answer: Randomly select a part of the group to get a sample. Find their preferences, and use the results to find the percent of the total group. **3.** 8.96 million **5.** 1,396 students **7.** about 110 owners **9.** 6,500 teens **11.** 800,000 people **13.** $54 = 0.72 \cdot n$; 75 **15.** 0.3 **17.** 4.5

Pages 352–353 Lesson 8-4
1. Sample answer: The sale price of a T-shirt is $14. If the shirt originally cost $25, what is the percent discount? Answer: 44% decrease **3.** Jada; Miranda did not write a ratio comparing the change to the original amount. **5.** 19% inc. **7.** 25% inc. **9.** 73% inc. **11.** 20% inc. **13.** 50% dec. **15.** 71% dec. **17.** 30% dec. **19.** 18% inc. **21.** 1% inc. **23.** 100% **25.** 21% inc. **27.** 3.2% increase **29.** 41% **31.** 16 to 17; 72% **33.** C **35.** 660 students **37.** $n = 0.21 \cdot 62$; 13.0 **39.** 0.055 **41.** 0.0675

Pages 356–357 Lesson 8-5
1. $6.86 **3.** Sample answer: The regular price of a CD is $17.95. The total cost is $18.98. **5.** $1,338.75 **7.** $47.33 **9.** 9% **11.** $46.40 **13.** $103.95 **15.** $2.58 **17.** $26.53 **19.** $26.07 **21.** 75% **23.** 23% **25.** $131.66 **27.** $615.25 **29.** $366.56 **31.** $57.60 **33.** B **35.** 50% inc. **37.** 73% inc. **39.** 292.3 **41.** 99

Pages 359–360 Lesson 8-6
1. Sample answer: Write the interest rate as a decimal and write the months in years. Then multiply $500 \cdot 0.06 \cdot 1.5$. **3.** $38.40 **5.** $1,417.50 **7.** $23.20 **9.** $21.38 **11.** $330 **13.** $8.54 **15.** $45.31 **17.** $26.37 **19.** $210 **21.** 4% **23.** $627, $655.22, $684.70 **25.** G **27.** 29% inc. **29.** 35% dec.

Pages 362–364 Chapter 8 Study Guide and Review
1. true **3.** false; original **5.** true **7.** true **9.** true **11–17.** Sample answers are given. **11.** $\frac{1}{4} \cdot 80 = 20$ **13.** $\frac{3}{4} \cdot 40 = 30$ **15.** $0.1 \cdot 80 = 8$ **17.** $0.1 \cdot 1,000 = 100$ and $5 \cdot 100 = 500$ **19.** Sample answer: $\frac{1}{10} \cdot 300 = 30$ households **21.** $39 = 0.65 \cdot n$; 60 **23.** $n = 0.07 \cdot 92$; 6.4 **25.** about 208 **27.** 48% inc. **29.** 20% dec. **31.** about 12% increase **33.** $178.50 **35.** $26.80 **37.** 14% **39.** 10% **41.** $47.50 **43.** $412.50 **45.** $121.50 **47.** $1,360

Chapter 9 Probability

Page 369 Chapter 9 Getting Started
1. simplest **3.** 105 **5.** 52 **7.** 160 **9.** 360 **11.** 336 **13.** 5,040 **15.** $\frac{2}{3}$ **17.** simplified **19.** 5 **21.** 4

Pages 372–373 Lesson 9-1
1. Sample answer: Since 0.7 is greater than 0.5 and less than 0.75, the event is likely to happen. **3.** 0.65, 0.55 are probabilities that are not complementary because $0.65 + 0.55 \neq 1$. The other pairs are complementary. **5.** $\frac{1}{4}$ **7.** $\frac{3}{5}$ **9.** 75% **11.** $\frac{4}{5}$ **13.** $\frac{1}{2}$ **15.** $\frac{1}{10}$ **17.** Sample answer: The complementary event is the chance of no rain. Its probability is 63%. **19.** $\frac{7}{10}$ **21.** $\frac{1}{2}$ **23.** $\frac{7}{10}$ **25.** 0 **27.** $\frac{1}{11}$ **29.** $\frac{9}{10}$ **31.** A **33.** $60 **35.** about 2.5 billion **37.** 0.45 **39.** $1.\overline{7}$ **41.** simplified **43.** $\frac{3}{8}$

Pages 376–377 Lesson 9-2
1. Sample game: Each player tosses a coin 10 times. If it comes up heads, player 1 receives 1 point. If it comes up tails, player 2 receives 1 point. The player with the most points at the end of 10 tosses wins.

3.

7. Coin / Number Cube / Sample Space

12 outcomes

9. Number / Color / Sample Space

15 outcomes

11. Space / Math / Sample Space

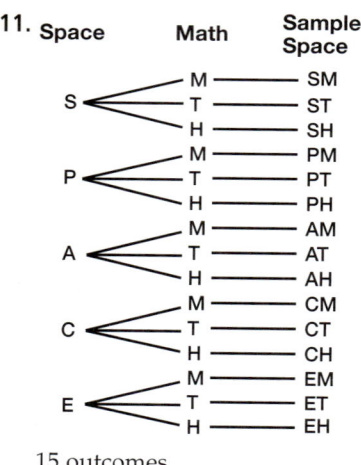

15 outcomes

5. $\frac{1}{16}$

13.

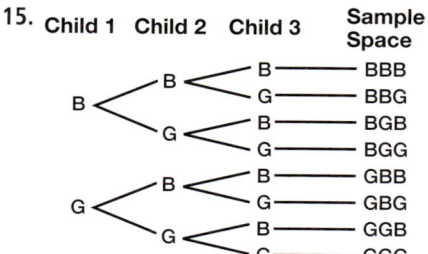

8 outcomes

15. Child 1 / Child 2 / Child 3 / Sample Space

- B — B — B — BBB
- B — B — G — BBG
- B — G — B — BGB
- B — G — G — BGG
- G — B — B — GBB
- G — B — G — GBG
- G — G — B — GGB
- G — G — G — GGG

17. $\frac{1}{8}$ 19. $\frac{4}{24}$ or $\frac{1}{6}$ 21. P(Player 1) = $\frac{6}{8}$ or $\frac{3}{4}$; P(Player 2) = $\frac{2}{8}$ or $\frac{1}{4}$; Player 2 does not have an equal chance of winning.   23. C   25. $\frac{2}{5}$ 27. $\frac{13}{20}$ 29. 154
31. 176   33. 524

Pages 379–380   Lesson 9-3
1. Sample answer: Multiply the number of ways each event can occur to find the total number of outcomes.   3. Sample answer: Yes, a pair of events can have a number of outcomes that is prime. For example, 11 different pairs of shoes can come in 1 color. 11 · 1, or 11 outcomes   5. 8
7. 60   9. 24   11. 12   13. 217   15. No; the number of selections is 350, which is less than 365.   17. 2   19. 8
21. B   23. $\frac{1}{2}$ 25. $\frac{3}{4}$ 27. 6   29. 120

Pages 382–383   Lesson 9-4
1. Sample answer: Five factorial has extra factors of 2 and 1, 5! = 5 · 4 · 3 · 2 · 1.   3. 6   5. 120   7. 120   9. 144   11. 4,320
13. 2,880   15. 3,024   17. 11,880   19. $\frac{1}{7}$ 21. 24   23. 96
25. 10   27. 2.5

Pages 389–390   Lesson 9-5
1. Sample answer: The order of the numbers is important in a lock.   3. Francisca; Allison found the number of permutations of three people chosen from a group of seven.
5. permutation; 720 ways   7. 10   9. 36   11. combination; 210   13. permutation; 6 ways   15. permutation; 5,040
19. A   21. 120 signals   23. Sample answer: 0 + 1 = 1
25. Sample answer: $\frac{1}{2}$ · 20 = 10   27. $\frac{1}{2}$ 29. $\frac{1}{8}$ 31. $\frac{1}{2}$

Pages 395–396   Lesson 9-6
1. Sample answer: Both probabilities are ratios that compare the number of favorable outcomes to the total number of possible outcomes. Experimental probability is the result of an experiment. Theoretical probability is what is expected to happen.   3. $\frac{14}{25}$ 5. $\frac{3}{16}$ 7. $\frac{5}{6}$ 9. $\frac{9}{10}$ 11. 150
13. 14   15. $\frac{9}{25}$ 17. 120   19. 1 : 5   21. 2 : 4 or 1 : 2   23. 1/5
25. 720   27. $\frac{3}{8}$ 29. $\frac{7}{8}$

Pages 399–401 **Lesson 9-7**
1. Sample answer: Events are independent when neither outcome affects the other. Events are dependent when the outcome of the first event affects the outcome of the second event. **3.** Shane; Kimi treated spinning the spinner twice as dependent events. **5.** $\frac{5}{14}$ **7.** $\frac{1}{8}$ **9.** $\frac{2}{87}$ **11.** $\frac{1}{6}$ **13.** $\frac{13}{51}$ **15.** independent **17.** 0.03 **19.** 13.5% **21.** 0.032 **23.** $\frac{1}{2}$ **25.** B **27.** 1 **29.** −27 **31.** 24

Pages 402–404 **Chapter 9 Study Guide and Review**
1. sample space **3.** experimental **5.** permutation **7.** random **9.** is not **11.** $\frac{1}{2}$ **13.** $\frac{1}{4}$ **15.** $\frac{1}{2}$

17.

12 outcomes

19.
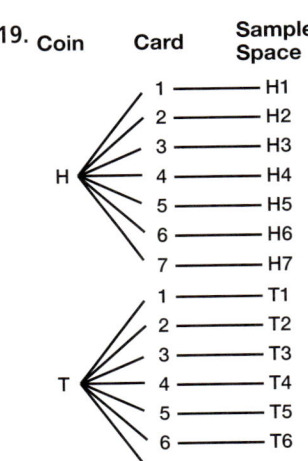
14 outcomes

21. 36 **23.** 20 **25.** 24 **27.** 1,680 **29.** 720 **31.** 60 **33.** 126 groups **35.** 45 ways **37.** $\frac{3}{25}$ **39.** $\frac{1}{5}$ **41.** $\frac{1}{4}$ **43.** $\frac{1}{15}$

Chapter 10 Geometry

Page 411 Chapter 10 Getting Started
1. ratio **3.** 306 **5.** 0.15 **7.** 0.11 **9.** 53 **11.** 131 **13.** 92 **15.** 90 **17.** 54 **19.** 12 **21.** 24 **23.** 8

Pages 414–415 **Lesson 10-1**
1. Sample answer: An angle with a measure of exactly 90°, which is indicated by a right angle symbol. **3.** obtuse
5. right **7.** acute
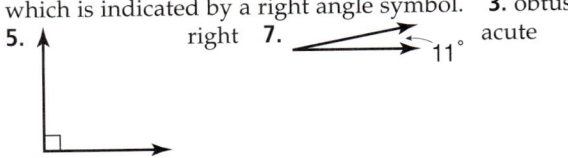
9. acute **11.** obtuse **13.** obtuse

15. obtuse **17.** right
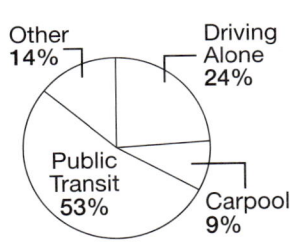
19. 150° **21.** Sample answer: acute angles: between their legs and the front of their skis; obtuse angles: between their legs and the back of their skis **23.** 48 times **25.** A **27.** $\frac{1}{4}$ **29.** $\frac{7}{10}$ **31.** 90 **33.** 0.32

Pages 420–421 **Lesson 10-2**
3. Favorite Shades of Blue **5.** grades 1–8

7. NYC Commuters

9. Sample answer: A greater percent of commuters in Los Angeles drive alone, while in New York, a greater percent use public transportation. A greater percent of people carpool in Los Angeles than in New York.

11. U.S. Regions

13. yes: 262.8°; no: 90°; not sure: 7.2° **17.** A **19.** acute **21.** right **23.** 72 **25.** 141

Pages 424–425 **Lesson 10-3**
1. Sample answer: $m\angle 1 + m\angle 2 = 90°$
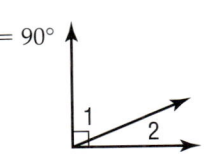

3. supplementary **5.** complementary **7.** 75 **9.** 96° **11.** supplementary **13.** supplementary **15.** complementary **17.** 69 **19.** 42 **21.** 90 **23.** Public Restroom and Subway/train **25.** never; Sample answer: One straight angle measures 180°. **27.** sometimes; Sample answer: Adjacent angles can have the same or different angle measure. **29.** $m\angle E = 39°, m\angle F = 51°$ **31.** C **33.** acute **35.** 0.53 **37.** 4.31 **39.** 68 **41.** 101

Pages 430–431 Lesson 10-4
1. An obtuse triangle has one obtuse angle and two acute angles. **3.** 44°; acute **5.** acute, equilateral **7.** right, isosceles **9.** 60°; right **11.** 65°; acute **13.** 37°; right **15.** right **17.** 17° **19.** obtuse, scalene **21.** right, scalene **23.** acute, isosceles **27.** B **29.** supplementary **31.** neither **33.** 60

Pages 435–437 Lesson 10-5
1. Sample answer: A square has all the properties of a rectangle. However, a rectangle may not have four congruent sides, which is a property of a square. **3.** Justin; Venus did not mention the 4 congruent sides of a square. **5.** quadrilateral **7.** 100° **9.** 35° **11.** trapezoid **13.** rhombus **15.** parallelogram **17.** sometimes; Sample answer: A parallelogram can also have angles with different degree measures. **19.** 56° **21.** 67° **23.** 90° **25.** rhombus, rectangle **29.** Yes; A square is a rhombus and a rectangle. **31.** Yes; One pair of opposite sides are parallel, while the other pair of opposite sides are congruent. This figure is called an isosceles trapezoid.

33. B **35.** obtuse, isosceles **37.** acute, equilateral **39.** $3.78 **41.** 85 **43.** 41.7 **45.** 400 mi **47.** 3 **49.** 7

Pages 442–443 Lesson 10-6
1. Corresponding sides are proportional and corresponding angles are congruent. **3.** 35 m **5.** 48 ft **7.** 12 m **9.** 10.2 mm **11.** yes; $\frac{6}{18} = \frac{11}{33}$ **13.** about 42 ft **15.** 120 m **17.** 1:16 **19.** 90 ft **21.** trapezoid **23.** 69° **25.** 90 **27.** 120

Pages 448–450 Lesson 10-7
1. the point where vertices meet: 45° + 45° + 135° + 135° = 360° **3.** All of the angles of a rhombus do not necessarily have the same measure. **5.** not a polygon, it is not closed **7.** No; 140 does not go into 360 evenly. **9.** regular octagon **11.** not a polygon; the figure has a curved side **13.** regular decagon **15.** 140° **17.** 162° **19.** 135°, 45°, 45° **21.** hexagon and triangle **23.** octagons and squares **25.** 43.2 cm **27.** $37\frac{1}{2}$ yd **31.** Yes; the sum of the measures of angles of any triangle is 180°.

33. I **35.** trapezoid **37.** 32 servings **39.** $3\frac{1}{6}$ **41.** $4\frac{37}{40}$

42–45.

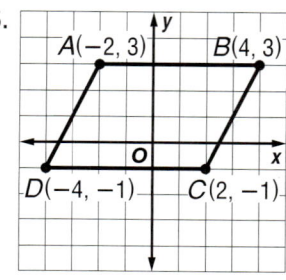

Pages 453–454 Lesson 10-8
1. Sample answer:

3.

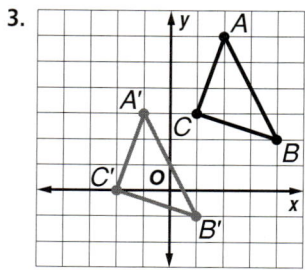

5. $D'(7, 0)$, $E'(4, -2)$, $F'(8, 4)$, $G'(12, -3)$

7.

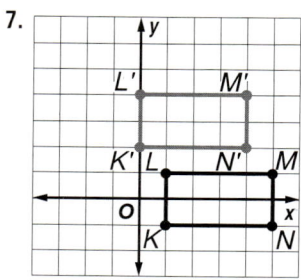

9. $P'(-8, -1)$, $Q'(-3, -3)$, and $R'(-11, 5)$

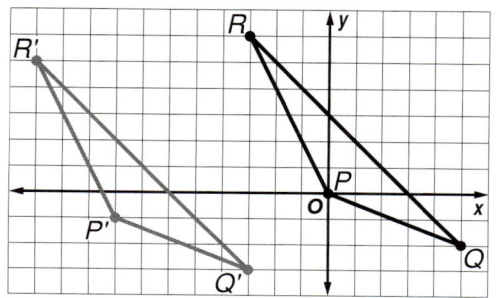

11. $P'(0, -9)$, $Q'(5, -11)$, and $R'(-3, -3)$

13. Sample answer: There are two main images, the brown horseman and the yellow horseman. Both main horsemen are translated to different parts of the picture. These translations allow for the tessellation of both horsemen. **15.** Sample answer: $\triangle X'Y'Z'$ was translated 5 units left and 2 units down.

17.

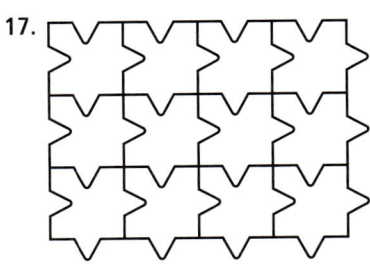

19. C **21.** octagon **23.** yes **25.** no

Pages 458–459 Lesson 10-9

1.

3. no

5. $A'(5, -8)$, $B'(1, -2)$, and $C'(6, -4)$

7. yes

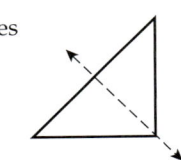

9. yes

11. $D'(-3, -6)$, $E'(-2, 3)$, $F'(2, -2)$, and $G'(4, -9)$

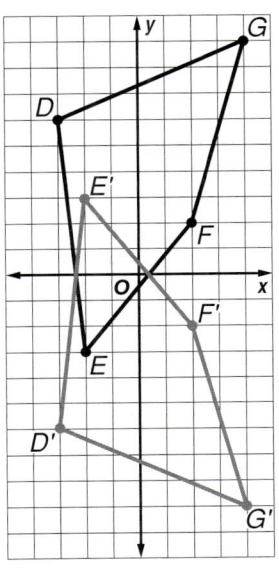

13. $Q'(-2, -5)$, $R'(-4, -5)$, and $S'(-2, 3)$

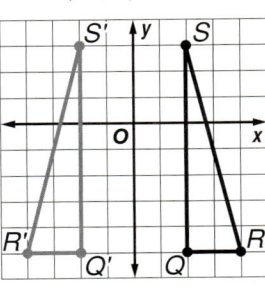

15. vertical line that divides the building in half and a line at the top of the water from the reflection in the water
17. All four flags have line symmetry.

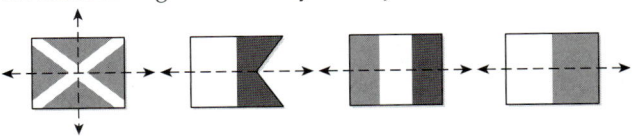

19. $J''(7, -4)$, $K''(-7, -1)$, and $L''(-2, 2)$ **21.** 2 **23.** No, the measure of one angle is 144°, which is not a factor of 360°.
25. Sample answer: $0 \times \frac{1}{2} = 0$ **27.** $6n \leq 18; n \leq 3$

Pages 462–464 Chapter 10 Study Guide and Review

1. c **3.** f **5.** a **7.** right **9.** acute
11. Favorite Soft Drinks

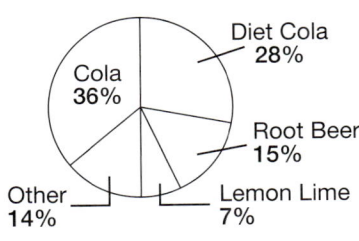

13. 79 **15.** obtuse, scalene **17.** 139° **19.** parallelogram
21. 10 cm **23.** heptagon **25.** 156°
27. $P'(8, -3)$, $Q'(2, -4)$, and $R'(3, 5)$

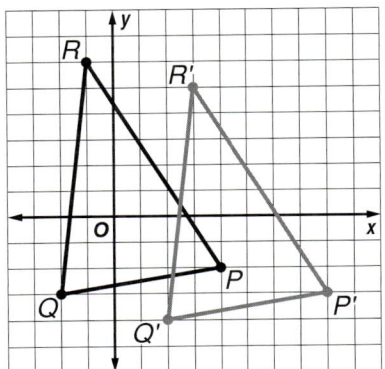

29. $P'(4, -9)$, $Q'(-2, -10)$, and $R'(-1, -1)$

31. $A'(-1, 3)$, $B'(-2, -1)$, $C'(-5, -1)$, $D'(-4, 3)$

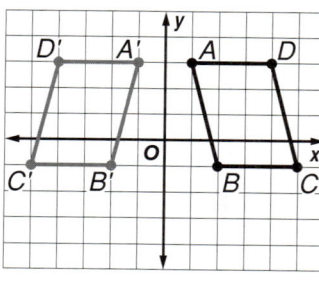

Chapter 11 Geometry: Measuring Two-Dimensional Figures

Page 469 Chapter 11 Getting Started
1. true **3.** false; right **5.** > **7.** = **9.** 64 **11.** 36 **13.** 60
15. 112 **17.** $\frac{1}{3}$ **19.** $\frac{1}{3}$

Pages 472–473 Lesson 11-1
1. Both involve multiplying a number by itself. **3.** 116; There are not two equal factors whose product is 116.
5. 100 **7.** 900 **9.** 11 **11.** 23 **13.** 25 **15.** 49 **17.** 256
19. 324 **21.** 2 **23.** 12 **25.** 27 **27.** 35 **29.** 484 **31.** 10, −10 **33.** 1,312 **35.** 202,788 mi² **37.** Yes; for example, a garden that measures 30 ft by 30 ft has the same perimeter, but its area is 900 ft², which is greater than 500 ft². **39.** $\frac{3}{5}$

40–41.

43. when the side length is between 0 and 4; when the side length is 0 or 4 **45.** 55 m² greater; The area of an 8-by-8-m square is 64 m². **47.** 1,024 cm²
49. $A'(3, 6)$, $B'(4, 3)$, $C'(1, 2)$

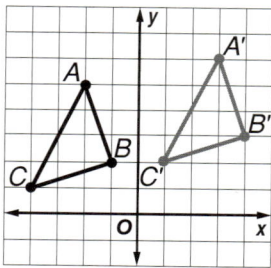

51. > **53.** >

Pages 476–477 Lesson 11-2
1. It cannot be written as a fraction. **3.** 51 is between the perfect squares 49 and 64. Since 51 is closer to 49 than to 64, $\sqrt{51}$ is closer to $\sqrt{49}$, or 7. **5.** 10 **7.** 12 **9.** 7.1 **11.** 21.5
13. 3 **15.** 6 **17.** 9 **19.** 12 **21.** 3.9 **23.** 6.6 **25.** 12.6
27. 25.4 **29.** π, $\frac{14}{3}$, $\sqrt{87}$, 10 **31.** 3.4 **35.** D **37.** 13
39. 28 **41.** 74 **43.** 16

Pages 481–482 Lesson 11-3
1. the lengths of both legs or the length of one leg and the hypotenuse **3.** Devin; the hypotenuse is 16 cm, not x cm as Jamie assumed. **5.** 24.5 in. **7.** yes **9.** 35 in.
11. 8.0 cm **13.** 25 in. **15.** 23.7 m **17.** yes **19.** no
21. 19.1 ft **23.** about 10.4 in. **25.** 11.3 in. **27.** 12 **29.** 17
31. 213.6 **33.** 28

Pages 484–485 Lesson 11-4
1. $b = 6$, $h = 3$ **3.** False; if the base and height are each doubled, then the area is $2b \cdot 2h = 4bh$, or 4 times greater.
5. 1.1 m² **7.** 64 in² **9.** 312.5 mm² **11.** 428.4 mm²
13. 207 in² **15.** 180 ft² or 20 yd² **17.** 1,575 yd²
19. 56.9 cm² **21.** 15 in. **23.** C **25.** no **27.** no **29.** 84
31. 19.5

Pages 491–492 Lesson 11-5
1. Sample answer: $\frac{1}{2}(10)(20 + 30) = 250$ in² **3.** The area of a parallelogram is twice the area of a triangle with the same base and height. **5.** 105.6 m² **7.** $1,417.50 **9.** 38.4 mm²
11. 112 m² **13.** 161.5 ft² **15.** 168 in² or 1.2 ft²
17. Sample answer: 28 in² **19.** Sample answer: 7 ft²

21. about 1,872 mi² **23.** 382.4 cm² **25.** 3.7 in² **27.** 12.4 ft
29. acute **31.** obtuse **33.** 40.8 **35.** 804.2

Pages 494–495 Lesson 11-6
1. Sample answer:

3. 2π; It is about $2 \cdot 3$, or 6; $\sqrt{7}$ is less than $\sqrt{9}$, or 3; 1.5^2 is less than 2^2, or 4. **5.** 254.5 in²
7. 55.4 ft² **9.** 452.4 mm²

11. 201.1 cm² **13.** 95.0 ft² **15.** 227.0 cm² **17.** 113.1 ft²
19. 7.1 cm² **21.** 63.6 in² **23.** 29.0 m² **25.** 43.2 ft²
27. 1:2; 1:4; No, $\frac{1}{2} \neq \frac{1}{4}$. **29.** 5.9 in² **31.** Never; the area is quadrupled because $A = \pi(2r)^2$, or $A = 4\pi r^2$. **33.** 3.7 cm
35. 120 in² **37.** 100.1 m² **39.** 113.04 **41.** 150.5

Pages 499–500   Lesson 11-7
1. Sample answer: Separate it into a rectangle and a triangle, find the area of each, and add. **3.** 112 m²
5. 145 m² **7.** 58.6 in² **9.** 257.1 mm² **11.** 510.3 ft²
13. Sample answer: Add the areas of a trapezoid and a small rectangle. **15.** 44 in² **17.** 69.4 cm² **19.** 176.7 in²
21. $\frac{1}{6}$ **23.** $\frac{1}{3}$

Pages 502–503   Lesson 11-8
1. Sample answer: The probability of an event occurring is the ratio of the shaded area (the number of ways an event can occur) to the entire area (the number of possible outcomes). **3.** 31.3% **5.** 16% **7.** 34.3% **9.** 23.1%
11. 73.8% **13.** 21.5% **15.** $\frac{4}{25}$ or 16% **17.** 3/10
19. 102.1 m²

Pages 504–506   Chapter 11   Study Guide and Review
1. equal to **3.** 625 **5.** hypotenuse **7.** 36 **9.** 529 **11.** 16
13. 2 **15.** 7 **17.** 4 **19.** 7.8 **21.** 21.1 **23.** 13 mm
25. 7.8 ft **27.** 13.3 m **29.** 99 cm² **31.** 135 cm² **33.** 36 ft²
35. 187.7 cm² **37.** 408.3 in² **39.** 1,256.6 ft² **41.** 60 m²
43. 182.8 in² **45.** 25%

Chapter 12   Geometry: Measuring Three-Dimensional Figures

Page 511   Chapter 12   Getting Started
1. circumference **3.** 90 **5.** 35 **7.** 53.1 **9.** 18.5 **11.** 23.7
13. 34 **15.** 7 **17.** 855.3 cm² **19.** 113.1 ft²

Pages 515–517   Lesson 12-1
3–17. Sample drawings are given.

11.

13.   **15.**

17.   **19.** cylinder

21. Sample answer:

23. Sample answer:
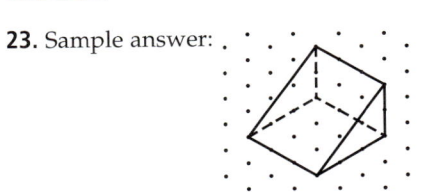

25. G **27.** 96 ft² **29.** 112.8 in² **31.** 22 **33.** 68

Pages 521–522   Lesson 12-2
1. yd³ **3.** Ling; the width of Box B is half that of Box A. All three dimensions have to be considered when comparing volumes. **5.** 9.5 cm³ **7.** 36,000 in³
9. 960 in³ **11.** 236.3 cm³ **13.** 27 cm³ **15.** 2,425.5 ft³
17. 1,000,000 cm³ **19.** C
21. Sample answer:
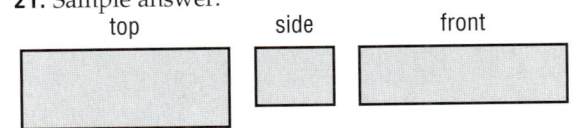

23. Sample answer: 18 **25.** Sample answer: 72

Pages 525–527   Lesson 12-3
1. Sample answer:

3. 141.4 in³ **5.** 617.7 ft³ **7.** 241.5 m³ **9.** 402.1 in³
11. 167.1 cm³ **13.** 1.3 m³ **15.** 2,770.9 mm³ **17.** 603.2 ft³
19. 603.2 cm³ **21.** 56.5 in³ **25.** 1 to 4 **27.** about 41 ft
29. B **31.** 152.88 m³

Selected Answers **661**

33. Sample answer: **35.** 572 m² **37.** 174 **39.** 43.5

Pages 533–535   Lesson 12-4
1. Find the sum of the areas of the faces.   **3.** Surface area measures the area of the faces, and area is measured in square units.   **5.** 183.4 cm²   **7.** 314 cm²   **9.** 833.1 mm²   **11.** 125.4 in² or $125\frac{3}{8}$ in²   **13.** 7 m²   **15.** 1,531.9 ft³   **17.** surface area = $6x^2$ **19.** 12 cm²   **21.** 5 cm long, 4 cm wide, 5 cm high   **25.** D   **27.** 7   **29.** 72.9 cm³   **31.** 63 in³   **33.** 804.2 ft²   **35.** 145.3 yd²

Pages 539–541   Lesson 12-5
3. 88.0 mm²   **5.** 653.5 m²   **7.** 1,215.8 m²   **9.** 183.8 ft²   **11.** 1,120.0 in² or $1,119\frac{44}{45}$ in²   **13.** 471.2 m²   **15.** 172.8 in²   **17.** Sample answer: prism: volume = 275 in³, surface area = 237.5 in²; cylinder: volume = 274.3 in³, surface area = 211.1 in²; Both containers hold about the same amount of popcorn. However, the cylinder uses less cardboard. So, the cylinder would be a more cost-effective container.   **19.** Sample answer: The areas of both curved surfaces are the same. However, the areas of the top and bottom of cylinder A are greater than the areas of the top and bottom of cylinder B because it has a greater circumference (11 in.).
21. Sample answer: 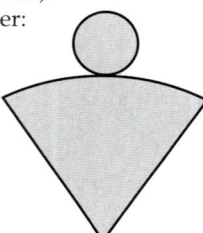   **23.** G   **25.** 112 cm²   **27.** 326.7 in³   **29.** 14.0   **31.** 5.4

Pages 543–545   Lesson 12-6
1. Inches, because it is the smallest unit of measure.   **3.** 8.6 mi, because it has 2 significant digits. The other measures have 3 significant digits.   **5.** Sample answer: 15.65 cm   **7.** Sample answer: The measure is precise to the nearest 0.01 second.   **9.** 0.1 cm   **11.** 2 lb   **13.** Sample answer: 13.6 ft   **15.** Sample answer: 9.15 cm   **17.** Sample answer: 2.45 mL   **19.** Neither; they are both given to the nearest tenth of a mile.   **25.** B   **27.** 502.7 yd²   **29.** 748.0 in²   **31.** 45   **33.** 24

Pages 546–548   Chapter 12 Study Guide and Review
1. rectangular prism   **3.** cubic   **5.** cylinder   **7.** smaller
9. Sample answer:

top   side   front

11. Sample answer:
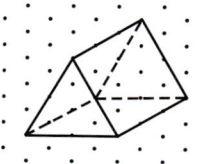

13. 639.6 in³ or $639\frac{5}{8}$ in³   **15.** 4,042.4 m³   **17.** 196.3 in³   **19.** 531.3 m²   **21.** 427.3 mm²   **23.** 69.1 in²   **25.** 5 min   **27.** Sample answer: 0.8 cm

Photo Credits

Cover (tl b)Mark Gallup, (tr)PhotoDisc; **i** Mark Gallup; **iv v x** Aaron Haupt; **xi** Robert Laberge/Getty Images; **xii** CORBIS; **xiii** NASA; **xiv** Aaron Haupt; **xv** Haslin/Sygma/CORBIS; **xvi** PhotoDisc; **xvii** NASA; **xviii** CORBIS; **xix** William Sallaz/DUOMO/CORBIS; **xx** CORBIS; **xxi** Ryan/Beyer/Getty Images; **xxiii** John Evans; **xxiv** (t)PhotoDisc, (b)John Evans; **xxv** (t bl)PhotoDisc, (br)John Evans; **xxvi** John Evans; **2–3** Nicole Duplaix/Getty Images; **4–5** Jeff Greenberg/PhotoEdit; **7** Rob Lewine/CORBIS; **8** Dimitri Lundt/CORBIS; **13** Robert Laberge/Getty Images; **15** Chuck Savage/CORBIS; **19** Aaron Haupt; **22** (l)PhotoDisc, (r)Richard Hutchings/Photo Researchers; **24** Getty Images; **27** Tim Davis/Photo Researchers; **29** John Evans; **32** Miles Ertman/Masterfile; **40** Phyllis Picardi/Stock Boston; **41** Bob Daemmrich/Stock Boston; **43** Owen Franken/Stock Boston; **52–53** Roger Ressmeyer/CORBIS; **54** Grant Symon/Getty Images; **55** Tim De Waele/TempSport/CORBIS; **57** George Clark, Hattiesburg American/AP/Wide World Photos; **58** John Evans; **64** Joe Sohm/Photo Researchers; **67** CORBIS; **70** CORBIS; **73** First Image; **75** John Evans; **77** Matt Meadows; **79** Aaron Haupt; **80** John E. Kelly/Getty Images; **82** (tl tr)White House Historical Association; **85 86** CORBIS; **87** David Cannon/Getty Images; **95** Bob Daemmrich/Stock Boston; **102–103** Mike Powell/Getty Images; **104–105** Roger Ressmeyer/CORBIS; **106** Jed Jacobsohn/Getty Images; **109** Mark Tomalty/Masterfile; **113** Geoff Butler; **117** John Evans; **120** Richard Kaylin/Getty Images; **122** Australian Picture Library/CORBIS; **123** Andy Lyons/Getty Images; **129** National Weather Service/Unisys Corp.; **131** Richard T. Nowitz/Getty Images; **132** John Evans; **137** Tim Wright/CORBIS; **139** NASA; **148–149** First Light Associated Photographers; **150** United Features Syndicate; **158** (t)Michael Zito/SportsChrome, (c)CORBIS; **161 162** PhotoDisc; **163** Cedar Point; **164** PhotoDisc; **167** Tim Courlas; **169** PhotoDisc; **171** John Evans; **172** Aaron Haupt; **173** PhotoDisc; **176** CORBIS; **177** Aaron Haupt; **179** Brett Froomer/Getty Images; **184** PhotoDisc; **192–193** ThinkStock/SuperStock; **194–195** Neil Rabinowitz/CORBIS; **199** Aaron Haupt; **201** John Evans; **206** KS Studios; **208** David Tardif/CORBIS; **210** Haslin/CORBIS SYGMA; **215** John Evans; **216** Matt Meadows; **217** Doug Martin; **221** Ralph A. Clevenger/CORBIS; **223** Bob Krist/CORBIS; **227** Aaron Haupt; **228** Getty Images; **231** Ron J. Berard/DUOMO/CORBIS; **238–239** Roy Ooms/Masterfile; **243** Doug Martin; **245** PhotoDisc; **249** Aaron Haupt; **251** KS Studios; **252** John Evans; **257** (t)Joseph Nettis/Photo Researchers, (b)Latent Image; **263** John Evans; **265** Aaron Haupt; **268** Christopher Marona; **273** Charlotte Observer; **275** Ben Wood/CORBIS; **277** Circlemakers; **284–285** James Richard/Stock Connection/PictureQuest; **286–287** Joseph Sohm/CORBIS; **289** Elaine Shay; **291** David Young-Wolff/PhotoEdit; **295** Johnny Johnson; **297** Doug Martin; **298** Project Exploration; **299** Aaron Haupt; **301** Aaron Haupt; **302** (l)Ryan McVay/PhotoDisc, (r)KS Studios; **308** Underwood & Underwood/CORBIS; **311** John Evans; **312** Richard Lord/The Image Works; **315** Aaron Haupt; **317** Johnny Johnson; **319** Lynn M. Stone; **323** NASA; **332–333** Richard Price/Getty Images; **334** CORBIS; **337** AFP/CORBIS; **338** (l)John Evans, (r)Matt Meadows; **341** Spencer Grant/PhotoEdit; **344** Mary Kate Denny/PhotoEdit; **346** Lisette Le Bon/SuperStock; **349** John Evans; **351** Charles O'Rear/CORBIS; **352** David Young-Wolff/PhotoEdit; **355** KS Studios; **356** Tony Freeman/PhotoEdit; **357** Morton & White; **359** Courtesy George Schweighofer/Currency Quest; **368–369** KS Studios; **371** Matt Meadows; **373** Laura Sifferlin; **375** Tim Fuller; **377** Aaron Haupt; **378** Geoff Butler; **385** John Evans; **388** Geoff Butler; **389** The Supreme Court Historical Society; **391** Laura Sifferlin; **394** Michael Newman/PhotoEdit; **397** William Sallaz/DUOMO/CORBIS; **398** CORBIS; **400** Digital Vision/Getty Images; **408–409** Charles O'Rear/CORBIS; **410–411** Aaron Haupt; **414** Doug Martin; **415** Wally McNamee/CORBIS; **419** Jon Feingersh/CORBIS; **423** Bettmann/CORBIS; **428** CORBIS; **434** Jed Jacobsohn/Getty Images; **439** John Evans; **441 443** PhotoDisc; **444** Laura Sifferlin; **447** Aaron Haupt; **448** (l)Tom Stack & Associates, (r)Aaron Haupt; **449** Christine Osborne/CORBIS; **450** EyeWire/Getty Images; **454** Cordon Art-Baarn-Holland; **458** CORBIS; **459** Mug Shots/CORBIS; **468–469** John & Lisa Merrill/CORBIS; **471** Mark Gibson/H. Armstrong Roberts; **477** Stephen R. Swinburne/Stock Boston; **480** Collin Reid/AP/Wide World Photos; **487** John Evans; **490** Ryan/Beyer/Getty Images; **496** (l)Laura Sifferlin, (r)Doug Martin; **502** Doug Martin; **510–511** LWA-Dann Tardif/CORBIS; **514** Tribune Media Services, Inc.; **518** Laura Sifferlin; **522** Kevin Fleming/CORBIS; **523** Geoff Butler; **526** CORBIS; **529** John Evans; **533** Tim Fuller; **539** Alan Klehr/Getty Images; **540** (t)Ariel Skelley/CORBIS, (b)Geoff Butler; **542** Claudia Kunin/CORBIS; **552** Mark Gallup.

Index

Absolute value (| |), 107

Acute angle, 413

Acute triangle, 429

Addition
  Associative Property of, 31
  Commutative Property of, 31
  decimals, 559
  fractions, 244
  Identity Property of, 31
  integers, 120–122
  Inverse Property of, 121
  mixed numbers, 248, 249
  phrases indicating, 150
  solving equations, 156

Addition Property of Equality, 157

Additive Inverse Property, 121

Additive inverses, 121

Algebra, 18
  equation, 24–27
  equivalent expressions, 30
  evaluating expressions, 18–21, 123, 129–130, 135–136, 139–140
  functions, 176–181
  graphing inequalities, 172–175
  graphing linear equations, 178, 179
  graphing solutions of equations, 25
  linear equations, 178
  solving equations, 24–27, 154–155, 156–159, 160–163, 166–169, 258–261
  solving proportions, 298
  variable, 18–21
  writing equations, 25, 151, 153, 158, 161, 167
  writing expressions, 150

Algebraic equations. *See* Equations

Algebraic expressions, 18
  evaluating, 18, 19
  factoring, 198
  order of operations, 15
  simplifying, 122, 123
  translating words into, 150

Angles, 413, 414
  acute, 413
  adjacent, 415
  alternate exterior, 427
  alternate interior, 427
  bisecting, 417
  classifying, 414, 423
  complementary, 422
  congruent, 422
  constructing, 416, 417
  corresponding, 427
  drawing, 414
  measuring, 412
  obtuse, 413
  relationships, 422, 423
  right, 413
  straight, 413
  sum in a polygon, 447
  sum in a quadrilateral, 435
  sum in a triangle, 428
  supplementary, 422
  vertex of, 413
  vertical, 422, 427

Applications. *See also* Interdisciplinary Connections; Real-Life Careers; Real-Life Math
  advertising, 380
  age, 165
  airplanes, 262
  allowances, 99, 353
  ambulance, 459
  animals, 6, 49, 67, 85, 86, 153, 161, 257, 267
    birds, 223, 308, 431
    bears, 317
    cats, 347
    dogs, 212, 383
    elephants, 253
    horses, 148
    insects, 36
    penguins, 27
    pets, 77, 247, 447
    turtles, 161, 257
    whales, 27, 295
  aquariums, 250
  architecture, 81, 498
  aspect ratio of photographs, 291
  baby-sitting, 162
  baking, 41, 230, 243, 525, 540
  balloons, 223
  behavior of dog owners, 212
  bicycles, 214, 414, 519
  blood pressure, 21, 173
  blueprints, 306, 307, 329
  books, 365
  boxes, 534
  breakfast, 172, 383
  bridges, 83, 306, 307
  buildings, 64, 458
  business, 92, 303
  cameras, 347
  candy, 370, 372, 541
  capacity, 49
  cars, 23, 46, 315, 390
  cellular phones, 46, 88
  cereal, 57, 133, 521, 534
  children's sports injuries, 175
  chores, 339
  clocks, 413
  clothes, 354, 445
  coins, 133, 217, 375
  collecting/collections, 9
    coin collecting, 32
    ornaments, 26
    souvenirs, 23
  colors, 418
  comics, 514
  commission, 320
  communication, 339, 346, 497
  community service, 519
  computers, 306, 347, 354
  congressional representatives per state, 88
  construction, 242, 249
    carpentry, 535
    flooring, 273
    painting, 235
    remodeling, 519
  containers, 526
  cooking, 9, 281, 300
  cost of gasoline, 182
  cotton production, 343
  crafts, 266
  credit cards, 359
  crops, 343
  currencies, 300
  data sense, 56, 66, 72, 420
  delicatessen, 376
  design, 436, 539, 548
  diamonds, 226
  dining, 27
  dinosaurs, 268
  discount, 355
  dog ages, 33
  drinking straws, 549
  driving, 174, 477
  earnings, 184
  earthquakes, 163
  education, 216, 353
  elections, 256
  electricity, 218
  elevators, 59, 253
  employment, 99, 162
  entertainment, 43, 94, 97, 348
  environment, 303
  family, 303, 376
  fashion, 400
  Ferris wheels, 238, 275
  fiber-optic cable, 23
  firefly flashes, 133
  flag day, 281
  flags, 257

food, 133, 151, 165, 177, 202, 206, 231, 234, 256, 266, 353, 357, 373, 378, 387, 389, 394, 465
fountains, 294
furniture, 442
games, 14, 202, 303, 368, 374, 377, 380, 452, 502, 503
    cards, 400
    checkers, 388
    chess, 78
    croquet, 392
    darts, 230
    Parcheesi, 371
    video games, 15, 356, 436
gardening, 46, 445, 506, 528
gears, 288
gifts, 396, 533
grades, 98, 253
grilling, 334
heart rates, 295
heating, 74
heights, 290
home run totals, 58
household size, 72
indirect measurement, 442
industrial technology, 205
interest, 358, 359
interior design, 500
Internet, 180, 203
investing, 358, 360
juice, 253
kites, 162, 240, 428
landscaping, 200, 448, 486, 495
lawn service, 170
light years, 47
loans, 359, 365
lottery, 401
lunch, 383
manufacturing, 449
maps, 112, 304, 306, 307, 453
measurement, 57, 206, 260, 300, 442, 450, 473, 522, 526
meteorology, 130
military, 137
models, 299, 308, 315
model trains, 223
money, 8, 17, 27, 110, 122, 123, 152, 153, 165, 168, 177, 247, 261, 291, 300, 303, 392, 419, 494, 519
monuments, 251
    Mount Rushmore, 286
    Statue of Liberty, 60
movies, 95, 189, 291, 325, 356, 373, 540
moving, 209
number names, 45
number theory, 23, 59, 165, 200, 202, 226, 342
nutrition, 69, 72, 77, 80, 294, 299, 343
Olympics, 55, 62, 218, 419, 543

packaging, 199, 256, 549
parade, 204, 405
parking, 209
parks, 270
pasta consumption, 77
patterns, 136, 140, 165, 226, 392
photography, 291, 317, 392
physical fitness, 507
picnics, 225, 392
politics, 421
ponds, 289
Pony Express, 31
pools, 305, 547
population, 13, 151, 295, 318, 336, 339, 353
pottery, 547
presidents, 82, 246
probability, 377, 380, 401, 415, 421, 473, 501
production, 13
profits, 111, 357
rainfall, 83
rain forest, 253
real estate, 425
record temperatures, 66, 168
refreshments, 395
research, 13, 41, 57, 62, 79, 206, 277, 295, 318, 357, 377, 421, 436, 450, 477, 544
reservoirs, 44
restaurants, 30
retail sales, 378
roller coasters, 159, 163
Roman history, 131
safety, 59, 482
salary, 93
sales, 98, 99, 342, 353, 497
sales tax, 354
savings, 202, 262, 358
scale drawing, 318
scaling surface area and volume, 536
school, 62, 71, 202, 380
science fair, 202
scooters, 54, 375
seat belts, 339
sewing, 243
shoes, 65
shopping, 78, 175, 189, 253, 310, 325, 339, 403
signal flags, 459
skid marks, 477
snacks, 61, 399
snowfall, 23
space, 70
speed of sound, 13
spending, 180
spinners, 395
sports, 8, 28, 59, 83, 97, 115, 175, 352, 471
    baseball, 20, 231, 342, 392, 396

    basketball, 24, 36, 27, 33, 79, 133, 213, 339, 387, 403, 423, 434, 445
    boating, 21
    bowling, 59, 167
    cycling, 214
    diving, 141, 169
    football, 106, 116, 140, 273
    golf, 123
    hockey, 312
    marathons, 16
    mountain biking, 228, 357
    racing, 9, 13, 54
    running, 62
    sailing, 390, 443
    skateboarding, 123, 179
    skating, 95
    skiing, 184, 268, 415
    snowboarding, 332
    soccer, 45, 212, 389
    softball, 210, 227
    swimming, 61, 329, 382, 519
    tennis, 222
    track & field, 41, 396, 398
    triathlons, 544
    volleyball, 271, 382, 389
static electricity, 108
statistics, 63, 108, 124, 425
statues, 443
stock market, 107, 124, 145, 251, 351
storage, 521
student governent, 372, 379, 384, 388
submarine, 123, 136
supplies, 253
surveying, 265
surveys, 312, 319, 405
teachers' salaries, 62
television, 7, 256, 345, 497
temperature, 169
tessellations, 454, 459
tests, 70, 94, 99, 223, 299, 303, 313
textbooks, 86
theater, 497
time, 165, 425
time spent preparing for testing, 17
tourism, 152
towers, 454
traffic, 395, 400, 519
transportation, 92, 382, 392, 522
travel, 32, 153, 162, 253, 437, 445, 482
trees, 291, 351
trucks, scale model, 310
uniforms, 377
utilities, 94
vacation, 260, 339
videos, 9, 165
volunteering, 136
weather, 66, 106, 108, 109, 123, 145, 168, 174, 230, 234, 245

world records, 542
X games, 395
yogurt, 405

Area, 271. *See also* Surface area
and probability, 501, 502
circles, 493, 494
complex figures, 498, 499
parallelograms, 483
rectangles, 271
trapezoids, 490
triangles, 489

Arithmetic sequences, 34

Assessment. *See also* Preparing for
Standardized Tests, Prerequisite
Skills, Standardized Test Practice
Mid-Chapter Practice Test, 28, 74,
116, 170, 214, 262, 310, 348, 384,
438, 486, 528
Practice Test, 49, 99, 145, 189, 235,
281, 329, 365, 405, 465, 507, 549

Associative Property
of Addition, 31
of Multiplication, 31

Average, 69, 73

Bar graphs, 85, 86

Bar notation, 211

Base. *See also* Percents
for power, 10
parallelogram, 483
trapezoid, 490
triangle, 489

Benchmark. *See* Estimating

Bias, 344

Box-and-whisker plots, 80, 81

Calculator, 11, 44, 120, 210, 211, 276,
313, 382, 471, 476, 493

Capacity
changing customary units,
267–268
changing metric units, 39
cup, 267
fluid ounce, 267
gallon, 267
kiloliter, 39
liter, 39
milliliter, 39
pint, 267
quart, 267

Careers. *See* Real-Life Careers

Celsius temperature, 140

Center, 275

Centimeter, 38

Central tendency. *See* Measures of
central tendency

Circle graphs, 418, 419

Circles, 275
area, 493, 494
center, 275
circumference, 274, 275
diameter, 275
radius, 275

Circumference, 275

Closure Property, 266

Clustering, 558

Clusters, 65

Coefficient, 19

Combinations, 386–388
counting, 387

Commission, 342

Common denominator, 227

Commutative Property
of Addition, 31
of Multiplication, 31

Comparing and ordering
decimals, 556
fractions, 227
integers, 109–111
irrational numbers, 477
percents, 228
rational numbers, 227
scientific notation, 45

Compass, 416

Compatible numbers, 242

Complementary angles, 422

Complementary events, 371

Complex figures, 498
area, 498, 499

Composite numbers, 197

Composite shapes. *See* Complex
figures

Compound events, 398

Compound inequality, 175

Cones, 541
net of, 541

Congruent angles, 422

Congruent figures, 452

Congruent segments, 429

Constant, 19

Constructed Response. *See*
Preparing for Standardized Tests

Constructions
bisecting an angle, 417
congruent angles, 416
equilateral triangle, 432
isosceles triangle, 433
parallel lines, 426
perpendicular lines, 417

Contractions. *See* Dilations

Coordinate grid. *See* Coordinate
plane

Coordinate plane, 112, 113
axes, 112
graphing points on, 113
ordered pair, 112
origin, 112
quadrants, 113
x-axis, 112
y-axis, 112

Coordinates, 112
x-coordinate, 112
y-coordinate, 112

Coordinate system. *See* Coordinate
plane

Counting
fundamental principle, 378, 379
permutations, 381, 382
tree diagrams, 375

Critical thinking, 9, 13, 17, 21, 27,
33, 36, 41, 45, 57, 63, 68, 72, 79,
83, 89, 95, 108, 111, 115, 124,
131, 137, 141, 152, 159, 163, 169,
175, 181, 185, 200, 206, 209, 213,
219, 223, 226, 231, 243, 247, 251,
257, 261, 266, 269, 273, 277, 291,
295, 300, 308, 315, 318, 321, 325,
337, 343, 347, 353, 357, 360, 373,
377, 380, 383, 390, 396, 401, 415,
421, 425, 431, 437, 443, 450, 454,
459, 473, 477, 482, 485, 492, 495,
500, 503, 517, 522, 527, 535, 541,
545

Cross products, 297

Cubed, 10

Cup, 267

Customary system, 267–269
capacity units, 267
conversions, 267–268
length units, 267
weight units, 267

Cylinders, 524
surface area, 538, 539
volume, 524, 525

D

Data, 54
  bar graphs, 85, 86
  box-and-whisker plots, 80, 81
  center, 69, 70
  clusters, 65
  interquartile range, 81
  line graphs, 60
  line plots, 64, 65
  mean, 69
  median, 70
  mode, 70
  outliers, 65
  quartiles, 80
  range, 65
  scatter plots, 61
  stem-and-leaf plots, 76, 77

Data Updates. *See* Internet Connections

Decagon, 446

Decimals, 210, 211
  adding, 559
  comparing and ordering, 556
  dividing, 562
  estimating, 558
  expanded notation, 555
  multiplying, 560
  place value, 555
  repeating, 211
    bar notation, 211
  rounding, 557
  subtracting, 559
  terminating, 211
  writing as percents, 221

Deductive reasoning, 444

Degrees, 413

Dependent events, 399

Diagnose Readiness. *See* Prerequisite Skills

Diameter, 275

Dilations, 455

Discount, 355

Discrete Mathematics. *See* Combinations, Counting, Permutations, Probability, Sequences, Statistics

Distance formula, 161

Distributive Property, 30

Dividend, 139

Divisibility patterns, 554

Division
  decimals, 562
  dividend, 139
  divisor, 139
  fractions, 264
  integers, 138, 139
  mixed numbers, 265
  phrases indicating, 150
  quotient, 139
  solving equations, 259

Division Property of Equality, 160

Domain, 177

E

Equality
  Addition Property, 157
  Division Property, 160
  Multiplication Property, 259
  Subtraction Property, 157

Equals sign (=), 24

Equations, 24, 151
  addition, 156
  checking solutions of, 157, 161
  division, 259
  linear, 178
  models of, 154–155, 160, 166
  multiplication, 161
  nonlinear, 181
  solution of, 24
  solving, 24, 25, 156–158, 160, 161, 259
    two-step, 166, 167
  subtraction, 157
  writing sentences as, 151

Equilateral triangle, 429

Equivalent expressions, 30

Equivalent fractions, 207

Estimating
  clustering, 558
  compatible numbers, 242
  front end, 240
  percents, 334–336
  square roots, 475, 476
  with decimals, 558
  with fractions, 240

Evaluating, 11
  algebraic expressions, 18, 19
  expressions with powers, 15
  numerical expressions, 14
  powers, 11

Events. *See* Probability

Expanded notation, 555

Experimental probability, 393, 394

Exponential form, 11

Exponents, 10, 11

Expressions, 18, 19
  equivalent, 30
  numerical, 14
  simplifying, 122, 123
  translating words into, 150

Extended Response. *See* Preparing for Standardized Tests *and* Standardized Test Practice

F

Factorial, 381
  $n!$, 381

Factoring
  algebraic expressions, 198
  whole numbers, 197, 198

Factors, 10, 196

Factor tree, 198

Fahrenheit temperature, 140

Fair game, 374

Figures
  similar, 440, 441
  three-dimensional (*See* Solids)

Find the Error, 16, 26, 40, 66, 71, 78, 122, 130, 151, 162, 168, 174, 205, 208, 222, 246, 250, 276, 290, 294, 336, 352, 389, 399, 435, 481, 494, 521, 539

Flowcharts, 125
  symbols, 125

Foldables™ Study Organizer
  decimal patterns and algebra, 5
  fractions, 239
  fractions, decimals, and percents, 195
  geometry, 411
  integers, 105
  measuring figures, 469
  percents, 333
  probability, 369
  ratios and proportions, 287
  solving equations, 149
  statistics, 53
  surface area and volume, 511

Formula, 271

Four-step problem-solving plan, 6–9, 22, 58, 132, 164, 201, 252, 302, 338, 391, 496, 518

Fractions
  adding and subtracting, 244, 245
  comparing, 227, 228
  dividing, 264
  equivalent, 207
  estimating with, 240
  like, adding and subtracting, 244

multiplying, 254, 255
ordering, 227, 228
simplest form, 207
simplifying, 207, 208
unlike, adding and subtracting, 245
writing as decimals, 211
writing as percents, 217, 313, 314

Free Response. *See* Preparing for Standardized Tests

Frequency distributions. *See* Frequency tables

Frequency tables, 54, 55
interval, 54
scale, 54
tally, 54
tally marks, 54

Functions, 177, 178
domain, 177
graphing, 176, 178
input, 177
linear, 178, 179
nonlinear, 181
ordered pairs, 176
output, 177
range, 177
real-world, 179–181
rule, 178
tables, 177–181, 182
writing expressions for, 179–181

Function table, 177

Fundamental Counting Principle, 378, 379

Gallon, 267

Game Zone
algebraic expressions, 29
describing solids, 529
factoring integers, 215
finding squares, 439
finding the mean, 75
graphing points on a coordinate plane, 117
multiplying fractions, 263
percent of a number, 349
probability, 385
solving equations, 171
solving proportions, 311
square roots, 487

GCF. *See* Greatest common factor (GCF)

Geometric sequences, 34

Geometry. *See* Angles; Area; Circles; Constructions; Lines; Perimeter; Polygons; Quadrilaterals; Transformations; Triangles

giga, 41

Gram, 39

Graphing
a line using slope, 185
linear equations, 178, 179
on coordinate plane, 113
solutions of equations, 25
solutions of inequalities, 172

Graphing Calculator Investigation, box-and-whisker plots, 84

Graphs
bar graphs, 85, 86
box-and-whisker plots, 80, 81
circle, 418, 419
coordinate plane, 112
histograms, 86
linear equations, 178
line graphs, 60
line plots, 64, 65
misleading, 92
multiple bar, 90
multiple line, 90
number line, 106
scatter plots, 61
stem-and-leaf plots, 76, 77

Greatest common factor (GCF), 203, 204
of algebraic expressions, 204

Grid In. *See* Preparing for Standardized Tests *and* Standardized Test Practice

Gridded Response. *See* Preparing for Standardized Tests

Hands-On Lab
adding integers, 118–119
average student, 73
building three-dimensional figures, 512
changes in volume and surface area, 536–537
circumference, 274
constructing and bisecting angles, 416–417
constructing parallel lines, 426
constructing triangles, 432–433
experimental probability, 397
exploring combinations, 386
exploring factors, 196
exploring sequences, 37
functions and graphs, 176
measuring angles, 412
nets and surface area, 530–531
Pythagorean Theorem, 478
rate of change, 296
rotations, 460–461
sampling, 344
solving equations using models, 154–155
subtracting integers, 126
triangles and trapezoids, 488
using a percent model, 322
wildlife sampling, 301

Hands-On Mini Lab
adding fractions on a ruler, 244
area of a circle, 493
area of a parallelogram, 483
area of a triangle, 489
converting length units, 38
finding a pattern, 18
finding the mean, 69
folding triangles, 428
making a scale drawing, 304
making predictions, 60
modeling division by a fraction, 264
modeling division of integers, 138
modeling multiples, 224
modeling multiplicative inverses, 258
modeling percent of change, 350
modeling percents, 316
modeling primes, 197
modeling subtraction of integers, 128
patterns and sequences, 34
perimeter and area, 470
probability game, 374
similar figures, 440
solving equations using models, 160
tessellations, 451

Height
parallelogram, 483
trapezoid, 490
triangle, 489

Heptagon, 446

Hexagon, 446

Histograms, 86–89

Homework Help, 9, 12, 16, 20, 26, 32, 36, 40, 45, 56, 62, 67, 72, 78, 83, 88, 94, 108, 111, 114, 123, 130, 136, 140, 152, 159, 162, 168, 174, 180, 184, 199, 205, 209, 212, 218, 222, 226, 230, 243, 246, 250, 256, 260, 266, 269, 272, 277, 290, 294, 299, 307, 315, 318, 321, 325, 337, 342, 347, 352, 356, 360, 372,

376, 380, 383, 389, 395, 400, 415, 420, 424, 430, 436, 442, 449, 453, 458, 472, 477, 482, 485, 491, 494, 500, 503, 516, 522, 526, 534, 540, 544

Hypotenuse, 479

Identity Property
of Addition, 31
of Multiplication, 31

Improper fractions, 563

Independent events, 398

Indirect measurement, 441

Inductive reasoning, 444

Inequalities, 172, 173
graphing solutions, 172
solving, 172, 173
symbols, 172
two-step, 173

Input. *See* Functions

Integers, 106, 107
absolute value, 107
adding, 120–122
different signs, 121
on a number line, 120
same sign, 120
using models, 118–119
Additive Inverse Property, 121
comparing, 109
greater than (>), 109
less than (<), 109
dividing, 138, 139
different signs, 138, 139
same sign, 139
multiplying, 134, 135
different signs, 134
same sign, 135
negative, 106
on a number line, 106, 109
ordering, 110
positive, 106
subtracting, 128, 129
negative, 129
positive, 128
using models, 126–127
zero, 106
zero pairs, 118

Interdisciplinary connections. *See also* Applications
agriculture, 208
archaeology, 480
art, 308, 449, 459, 519, 535
astronomy, 139, 248
civics, 88, 181, 389, 400

geography, 45, 94, 113, 114, 130, 165, 220, 221, 222, 308, 431, 446, 472
geology, 337
health, 19, 21, 173, 425
history, 31, 55, 76, 268, 359, 494, 516
math history, 213
music, 23, 72, 194, 208, 209, 219, 276, 294, 339, 355, 383, 492
reading, 59, 220
science, 156
Earth Science, 45, 59, 120, 129, 133, 137, 140, 202, 253, 393, 415, 497
Life Science, 33, 133, 233, 257, 295, 298, 336
Physical Science, 13, 23, 141, 303, 340, 471, 472, 534
space science, 392, 526
technology, 10, 88, 341, 354, 373

Interdisciplinary Project. *See* WebQuest

Internet Connections
msmath2.net/careers, 19, 77 113, 173, 208, 249, 298, 351, 394, 448, 480, 539
msmath2.net/chapter_readiness, 5, 53, 105, 149, 195, 239, 287, 333, 369, 411, 469, 511
msmath2.net/chapter_test, 49, 99, 145, 189, 235, 281, 329, 365, 405, 465, 507, 549
msmath2.net/data_update, 27, 66, 72, 123, 169, 174, 231, 261, 300, 342, 383, 459, 492, 516
msmath2.net/extra_examples, 7, 11, 15, 19, 25, 31, 35, 39, 44, 55, 61, 65, 69, 77, 81, 87, 93, 107, 109, 113, 121, 129, 135, 139, 151, 157, 161, 167, 173, 179, 183, 197, 203, 207, 211, 217, 221, 225, 229, 241, 245, 249, 255, 259, 265, 267, 271, 275, 289, 293, 297, 305, 313, 316, 319, 323, 335, 341, 345, 351, 355, 359, 371, 375, 379, 382, 387, 393, 399, 414, 419, 423, 429, 435, 441, 447, 451, 457, 471, 475, 479, 483, 489, 493, 499, 501, 515, 521, 525, 533, 539, 543
msmath2.net/other_calculator_ keystrokes, 84
msmath2.net/reading, 14, 64, 109, 156, 203, 244, 288, 345, 378, 422, 475, 520
msmath2.net/self_check_quiz, 9, 13, 17, 21, 27, 33, 36, 41, 45, 57, 63, 67, 72, 79, 83, 89, 95, 108, 111, 115, 123, 131, 137, 141, 152, 159, 163, 169, 175, 181, 185, 199, 205, 209, 213, 219, 223, 226, 231,

243, 247, 251, 257, 261, 266, 269, 273, 277, 291, 295, 299, 307, 315, 318, 321, 325, 337, 343, 347, 353, 357, 360, 373, 377, 380, 383, 389, 395, 401, 415, 421, 425, 431, 437, 443, 449, 453, 459, 473, 477, 482, 485, 491, 495, 500, 503, 517, 522, 527, 535, 541, 545
msmath2.net/standardized_test, 51, 101, 147, 191, 237, 283, 331, 367, 407, 467, 509, 551
msmath2.net/vocabulary_review, 46, 96, 142, 186, 232, 278, 326, 362, 402, 462, 504, 546
msmath2.net/webquest, 3, 95, 103, 185, 193, 277, 285, 401, 409, 545

Interquartile range, 81

Interval. *See* Frequency tables

Inverse operations, 156

Inverse Property
of Addition, 121
of Multiplication, 258

Irrational numbers, 476
models of, 475, 476

Irregular figures. *See* Complex figures

Isosceles triangle, 429

Kilogram, 39

Kilometer, 38

Labs. *See* Hands-On Lab and Hands-On Mini Lab

LCD. *See* Least common denominator (LCD)

LCM. *See* Least common multiple (LCM)

Leaf. *See* Stem-and-leaf plots

Least common denominator (LCD), 227

Least common multiple (LCM), 224, 225

Leg, 479

Length
changing customary units, 267
foot, 267
inch, 267
metric units, 38
mile, 267
yard, 267

Index **669**

Linear equations, 178, 179

Line graphs, 60
  multiple, 63

Line plots, 64, 65

Lines. *See also* Linear equations
  parallel, 426
  perpendicular, 417
  slope, 182, 183

Lines of symmetry, 456

Line symmetry, 456

Liter, 39

Logical reasoning, 444

Lower extreme, 80

Lower quartile (LQ), 80

LQ. *See* Lower quartile

M

Magnifications. *See* Dilations

Markup, 357

Mass
  changing units of, 39
  gram, 39
  kilogram, 39
  milligram, 39

Mean, 69
  versus median, 93

Measurement. *See also* Customary system and Metric system
  angles, 412
  area
    circles, 493
    complex figures, 498–500
    converting, 273
    parallelograms, 483
    rectangles, 271
    trapezoids, 488, 490
    triangles, 488, 489
  circumference, 274–276
  effect of changing dimensions, 271, 484, 495, 526, 536–537
  indirect, 441
  perimeter
    irregular figures, 271
    rectangles, 270
  precision, 542
  Pythagorean Theorem, 479–482
  significant digits, 542, 543
  similar solids, 523
  surface area
    cylinders, 538–539
    rectangular prisms, 532–533
  volume
    cylinders, 524–525
    rectangular prisms, 520–521

Measures of central tendency, 69
  mean, 69
  median, 70
  mode, 70
  summary, 71

Median, 70
  versus mean, 93

Mental Math. *See also* Number sense 198, 211, 221, 320

Meter, 38

Metric system, 38, 39
  capacity units, 38
  conversions
    capacity units, 38
    length units, 38
    mass units, 39
    unit conversion table, 38
  length units, 38
  mass units, 39
  unit abbreviations, 39

Mid-Chapter Practice Test, 28, 74, 116, 170, 214, 262, 310, 348, 384, 438, 486, 528

Millimeter, 38

Misleading statistics, 92, 93

Mixed numbers
  adding and subtracting, 248, 249
  dividing, 265
  improper fractions and, 563
  multiplying, 255

Mode, 70

Multiple, 224

Multiple Choice. *See* Preparing for Standardized Tests *and* Standardized Test Practice

Multiplication
  Associative Property of, 31
  Commutative Property of, 31
  decimals, 560
  fractions, 254, 255
  Identity Property of, 31
  integers, 134, 135
  Inverse Property of, 258
  mixed numbers, 255
  phrases indicating, 150
  solving equations, 259
  using dot for, 10

Multiplication Property of Equality, 259

Multiplicative Inverse Property, 258

Multiplicative inverses, 258

nano, 41

Negative integers, 106

Nets, 530, 531, 541

Nonagon, 446

Noteables™, 5, 7, 14, 30, 31, 43, 53, 69, 70, 105, 107, 109, 120, 121, 128, 134, 135, 138, 139, 149, 157, 195, 216, 220, 221, 229, 239, 244, 254, 258, 259, 264, 270, 271, 275, 287, 297, 314, 323, 333, 350, 369, 370, 378, 398, 399, 411, 413, 428, 429, 435, 440, 469, 471, 479, 483, 489, 490, 493, 511, 520, 524, 532, 538

Note taking. *See* Noteables™ *and* Study Skill

Number line
  absolute value, 107
  adding integers, 120–121
  graphing inequalities, 172–175
  graphing integers, 106
  graphing irrational numbers, 476
  graphing solutions of equations, 25

Numbers
  comparing and ordering rational numbers, 228
  compatible, 242
  composite, 197
  exponential form, 11
  integers, 106
  irrational, 476
  perfect squares, 471
  prime, 197
  rational, 229
  squares, 470, 471
  standard form, 11, 43

Number sense, 32, 44, 110, 122, 136, 225, 231, 242, 250, 256, 294, 336, 352, 379, 476, 494, 526

Numerical expressions, 14

Obtuse angle, 413

Obtuse triangle, 429

Octagon, 446

Odds, 396

Open Ended, 8, 11, 16, 20, 26, 28, 32, 35, 40, 44, 56, 61, 66, 71, 78, 82, 91, 93, 107, 110, 114, 122, 130, 136, 140, 151, 158, 162, 168, 174, 179, 184, 199, 205, 208, 212, 218, 222, 225, 229, 242, 246, 250, 256, 260, 266, 268, 272, 276, 290, 294, 299, 306, 314, 317, 320, 325, 336, 342, 346, 352, 356, 359, 372, 376, 379, 382, 389, 395, 399, 414, 420, 424, 430, 435, 442, 448, 453, 458, 472, 476, 481, 484, 491, 494, 499, 502, 515, 521, 525, 533, 539, 543

Open Response. *See* Preparing for Standardized Tests

Opposites, 121

Ordered pairs, 112
graphing, 113

Order of operations, 15

Origin, 112

Ounce, 267

Outcomes. *See* Probability

Outliers, 65
box-and-whisker plots, 81, 82
effect on mean, 72

Output. *See* Functions

Parallel lines, 426
properties of, 427

Parallelograms, 434
area, 483, 484
base, 483
height, 483

Part. *See* Percents

Pascal's Triangle, 392

Patterns, 8–9, 18, 34–36, 41, 132–133, 136, 140, 165, 226, 392

Pentagon, 446

Percent equation, 340

Percent proportion, 324

Percents, 216, 312, 313
base
  in percent equation, 340, 341
  in percent proportion, 324
commission, 342
discount, 355
estimating, 334–336
greater than 100%, 317, 318
less than 1%, 317, 318
markup, 357
models of, 216, 217, 322
part
  in percent equation, 340, 341
  in percent proportion, 324
part/whole relationship, 323
percent equation, 340, 341
percent of a number, 319, 320
percent of change, 350, 351
percent of decrease, 350
percent of increase, 350
percent proportion, 324, 325
sales tax, 354
simple interest, 358, 359, 361
writing as decimals, 220, 221
writing as fractions, 217, 312

Perfect squares. *See* Numbers

Perimeter, 270, 271
rectangles, 270

Permutations, 381, 382
counting, 382

Perpendicular lines, 417

Pi (π), 275
benchmark for, 213

Pictograph, 59

Pint, 267

Place value, 555

Plane, 512

Plane figures, 514

Point, on a coordinate plane, 112

Polygons, 446, 447
angle measurements, 447
classifying, 446, 447
decagon, 446
heptagon, 446
hexagon, 446
nonagon, 446
octagon, 446
pentagon, 446
regular, 446
tessellations, 447, 448
translations, 451, 452

Population, 345

Population density, 295

Positive integers, 106

Pound, 267

Power Notes, 42

Powers, 10, 11
base, 10
of ten, 561

Practice Test, 49, 99, 145, 189, 235, 281, 329, 365, 405, 465, 507, 549

Precision, 542

Precision unit, 542

Predictions, 60, 61
using line graphs, 60
using probability, 394
using scatter plots, 61
using statistics, 345

Preparing for Standardized Tests, 608–625
Constructed Response, 618–621, 622–625
Extended Response, 608, 622–625
Free Response, 618–621
Grid In, 614–617
Gridded Response, 608, 614–617
Multiple Choice, 608, 610–613
Open Response, 618–621
Selected Response, 610–613
Short Response, 608, 618–621
Student-Produced Questions, 618–621
Student-Produced Response, 614–617
Test-Taking Tips, 609, 613, 617, 621, 625

Prerequisite Skills
Adding and Subtracting Decimals, 559
Comparing and Ordering Decimals, 556
Diagnose Readiness, 5, 53, 105, 149, 195, 239, 287, 333, 369, 411, 469, 511
Dividing Decimals, 562
Divisibility Patterns, 554
Estimating with Decimals, 558
Getting Ready for the Next Lesson, 9, 13, 17, 21, 27, 33, 36, 41, 57, 63, 68, 72, 79, 83, 89, 108, 111, 115, 124, 131, 137, 152, 159, 163, 169, 175, 181, 200, 206, 209, 213, 219, 223, 226, 243, 247, 251, 257, 261, 266, 269, 273, 291, 295, 300, 308, 315, 318, 321, 337, 343, 347, 353, 357, 373, 377, 380, 383, 390, 396, 415, 421, 425, 431, 437, 443, 450, 454, 473, 477, 482, 485, 492, 495, 500, 517, 522, 527, 535, 541
Mixed Numbers and Improper Fractions, 563
Multiplying Decimals, 560
Place Value and Decimals, 555
Powers of Ten, 561
Rounding Decimals, 557

Prime factorization, 198

Prime numbers, 197
relatively prime, 206
twin primes, 200

Principal, 358

Prisms, rectangular
surface area, 532, 533
volume, 520, 521

Probability, 370, 371
area models, 501, 502
events, 370
  complementary, 371
  compound, 398
  dependent, 399
  independent, 398
  mutually exclusive, 401
  simple, 370
experimental, 393, 394, 397
odds, 396
outcomes, 370
  counting, 375, 378
  random, 370

sample space, 374
theoretical, 393, 394
tree diagrams, 374, 375

Problem solving, 6–8
four-step plan, 6
list of strategies, 7

Problem-Solving Strategy
act it out, 391
determine reasonable answers, 338
draw a diagram, 302
eliminate possibilities, 252
guess and check, 22
look for a pattern, 132
make a model, 518
make an organized list, 201
solve a simpler problem, 496
use a graph, 58
use logical reasoning, 444
work backward, 164

Products, integers, 134

Projects. *See* WebQuest

Properties, 30, 31
Addition Property of Equality, 157
Additive Inverse Property, 121
Associative Property, 31
Closure Property, 266
Commutative Property, 31
Distributive Property, 30
Division Property of Equality, 160
of geometric figures, 427, 434
Identity Property, 31
Multiplication Property of Equality, 259
Multiplicative Inverse Property, 258
Subtraction Property of Equality, 157

Proportional reasoning
aspect ratio, 291
circumference, 275
distance formula, 161
indirect measurement, 441
linear functions, 178–179
measurement conversions, 267–268, 38–39
percent proportion, 324
population density, 295
rate of change, 296
ratio, 288–289
scale drawings, 304–306
scale factor, 305
scaling, 304–309
similar figures, 440–443
similar solids, 523
slope, 182–185
solving proportions, 298
speed, 161
unit price, 293
unit rate, 293

Proportions, 297, 298
cross product, 297
identifying, 297
solving, 298

Protractor, 412

Pythagorean Theorem, 479–481
model of, 478

Quadrants, 113

Quadrilaterals, 434, 435
classifying, 434
parallelogram, 434
rectangle, 434
rhombus, 434
square, 434
sum of angles, 435
trapezoid, 434

Quart, 267

Quartiles, 80, 81
interquartile range, 81
lower, 80
upper, 80

Quotients, integers, 138

Radical sign, 471

Radius, 275

Random, 370

Random sample, 345

Range
of a data set, 65
of a function, 177

Rate of change, 296

Rates, 292, 293
distance formula, 161
population density, 295
speed, 161
unit, 292, 293
unit price, 293

Rational numbers, 229
comparing and ordering, 228
identifying, 229
models of, 229
percents, 217, 220, 221
repeating decimals, 211
terminating decimals, 210–211

Ratios, 216, 288, 289. *See also* Fractions
comparing, 289
equivalent, 289, 297
simplifying, 288
with units, 289
writing as percents, 216
writing in simplest form, 288

Ray, 416

Reading in the Content Area, 14, 64, 109, 156, 203, 244, 288, 345, 378, 422, 475, 520

Reading, Link to, 30, 34, 64, 69, 120, 182, 210, 258, 340, 387, 413, 451, 498, 520

Reading Math
addends and sums, 118
division sentences, 139
first power, 10
formula for simple interest, 359
line segments, 441
minuends, subtrahends, and differences, 127
names for variables, 18
parallel lines, 434
percent, 217
percents, 317
probability notation, 398
quartiles, 81
regular polygons, 447
rotations, 460
scatter plots, 61
set theory, 107
square roots, 471
squaring a number, 470
subscripts, 490
theorem, 480
unit prefixes, 39

Real-Life Careers
archaeologist, 480
athletic trainer, 19
cartographer, 113
chef, 394
composer, 208
landscape architect, 448
nurse, 173
nutritionist, 77
package designer, 539
paleontologist, 298
plumber, 249
stockbroker, 351

Real-Life Math
animal speeds, 161
basketball, 423
bowling, 167
checkers, 388
coins, 217
communicating by E-mail, 346
games, 502
geography, 221, 490
gift wrapping, 533
International Space Station, 70
mountain biking, 228

oceanography, 122
Olympic Games, 55, 419
Parcheesi, 371
physical science, 471
ponds, 289
presidents' native states, 82
scooters, 375
skateboarding, 179
skiing, 268
surveying, 265
technology use, 341
television, 7
video games, 15
weather, 245

Real numbers, 476

Reasoning, logical
deductive, 444
inductive, 444

Reciprocals, 258

Rectangles, 434
area, 271
perimeter, 270

Reflections, 457

Regular. *See* Polygons

Repeating. *See* Decimals

Rhombus, 434

Right angle, 413

Right triangle, 429
hypotenuse, 479
legs, 479
Pythagorean Theorem, 479–481

Rounding
decimals, 213, 557
fractions, 240

Rule. *See* Functions

Sales tax, 354

Sample space, 374

Sampling, 301, 344
bias, 344
random, 344, 345

Scale, 304. *See also* Frequency tables

Scale drawings, 304, 305
draw objects to scale, 306, 309
equivalent scales, 305
scale factor, 305
spreadsheets and, 309

Scale factor, 305

Scale model, 306

Scalene triangle, 429

Scaling, 304–309

Scatter plots, 61

Scientific notation, 43, 44

Sector, 418

Segments, congruent, 429

Selected Response. *See* Preparing for Standardized Tests

Sentences, as algebraic equations, 151

Sequences, 34, 35, 37
arithmetic, 34
geometric, 34
terms, 34

Sets
element, 107
intersection, 175
list by roster, 107
list by rule, 229
subset, 107
union, 175
Venn diagram, 203, 229

Short Response. *See* Preparing for Standardized Tests *and* Standardized Test Practice

Significant digits, 542, 543

Similar figures, 440, 441

Similarity
aspect ratio, 291
scaling, 304–309

Similar solids, 523
changes in volume and surface area, 536–537

Simple events, 370

Simple interest, 358, 359

Simplest form, 207
for ratios, 288

Simplifying
expressions, 122, 123
fractions, 207, 208

Slope, 182, 183
constant, 185
graph a line using, 185
negative, 182
positive, 182
rate of change, 296

Solids
drawing, 514
similar, 523

Solution, of an equation, 24

Solving
equations, 24, 25
two-step, 166, 167
equations involving addition and subtraction, 156–158
equations involving multiplication, 160, 161, 259
inequalities, one-step, 173
proportions, 298

Spreadsheet Investigation
dilations, 455
multiple line and multiple bar graphs, 90
scale drawings, 309
similar solids, 523
simple interest, 361

Square. *See also* Numbers

Squared, 10

Square roots, 471
estimating, 475, 476

Squares, 434

Standard form, 11, 43

Standardized Test Practice
Extended Response, 51, 101, 147, 191, 237, 283, 331, 367, 407, 467, 509, 551
Multiple Choice, 9, 13, 17, 21, 23, 27, 28, 33, 41, 45, 49, 50, 57, 59, 63, 68, 72, 74, 79, 83, 89, 95, 99, 100, 108, 111, 115, 116, 124, 131, 133, 137, 141, 145, 146, 152, 159, 163, 165, 169, 170, 175, 181, 185, 189, 190, 200, 202, 206, 209, 213, 214, 219, 223, 226, 231, 235, 236, 243, 247, 251, 253, 257, 261, 262, 269, 273, 277, 281, 282, 291, 295, 300, 303, 308, 315, 318, 321, 325, 329, 330, 337, 339, 343, 347, 348, 353, 357, 360, 365, 366, 373, 377, 380, 390, 392, 396, 401, 406, 415, 421, 425, 431, 437, 438, 443, 445, 450, 454, 459, 465, 466, 473, 477, 482, 485, 486, 492, 495, 497, 500, 503, 507, 508, 517, 519, 522, 527, 535, 541, 545, 549, 550
Short Response/Grid In, 9, 13, 21, 27, 28, 33, 36, 41, 45, 51, 63, 72, 74, 79, 83, 89, 95, 101, 108, 111, 115, 116, 124, 131, 137, 141, 147, 159, 169, 170, 185, 191, 200, 206, 209, 213, 214, 219, 223, 226, 231, 237, 257, 261, 266, 269, 273, 277, 283, 300, 308, 310, 315, 321, 325, 331, 347, 353, 357, 367, 377, 380, 383, 384, 390, 396, 401, 405, 407, 415, 421, 425, 431, 438, 443, 450, 454, 459, 467, 473, 477, 482, 485, 486, 492, 495, 500, 503, 509, 522, 528, 535, 551
Worked-Out Example, 25, 87, 110, 183, 229, 259, 293, 355, 379, 441, 499, 525

Statistics, 54. *See also* Graphs
  measures of central tendency (*See* Measures of central tendency)
  misleading, 93
  population, 345
  predicting using, 345, 346
  random sample, 345
  survey, 345

Stem. *See* Stem-and-leaf plots

Stem-and-leaf plots, 76, 77
  leaf, 76
  stem, 76

Straight angle, 413

Straightedge, 416

Student-Produced Questions. *See* Preparing for Standardized Tests

Student-Produced Responses. *See* Preparing for Standardized Tests

Study Guide and Review, 46–48, 96–98, 142–144, 186–188, 232–234, 278–280, 326–328, 362–364, 402–404, 462–464, 504–506, 546–548

Study Organizer. *See* Foldables™ Study Organizer

Study Skill
  reading math problems, 153
  studying math vocabulary, 474
  taking good notes, 42, 125

Study Tip
  alternate method for converting percents to decimals, 316
  area units, 271
  canceling units, 289
  check, 433
  checking solutions, 157
  checking solutions of inequalities, 173
  check the reasonableness of results, 429
  check your answer, 371
  choosing a method, 204, 225, 313
  clusters, 65
  common error, 484
  the Commutative and Associative Properties, 31
  Commutative Property, 198
  compatible numbers, 242
  congruent segments, 429
  defining the variable, 25
  dependent events, 399
  diagonals, 447
  discount, 355
  entering a negative integer on a calculator, 120
  equivalent scales, 305
  estimating percents, 335
  estimation, 293, 521
  expressions with powers, 15
  factorials on a calculator, 382
  finding circumference on a calculator, 276
  finding percents mentally, 320
  finding the base, 324
  finding the median, 70
  four-step plan, 7
  fraction-decimal equivalencies, 211
  fractions, 241
  front-end estimation, 240
  hypotenuse, 481
  input and output, 178
  irrational numbers, 476
  length of parts of a box-and-whisker plots, 81
  look back, 135, 155, 158, 259, 296, 298, 314
  mental math, 38, 198, 211, 255
  Multiplicative Property of Zero, 135
  percents and decimals, 221
  plane, 512
  plane figures, 514
  reading histograms, 86
  reflecting over the x-axis, 457
  reflecting over the y-axis, 457
  scale, 306
  scientific notation on the calculator, 44
  simplest form, 207
  simplifying, 255
  slope, 183
  solving equations, 167
  subtraction and negative numbers on a calculator, 129
  symbols, 416
  transformations, 451
  translation, 452
  unbiased samples, 344
  writing percents, 341

Subtraction
  decimals, 559
  fractions, 244
  integers, 128, 129
  mixed numbers, 248, 249
  phrases indicating, 150
  solving equations, 156

Subtraction Property of Equality, 157

Supplementary angles, 422

Surface area, 532
  cylinders, 538, 539
  nets, 530, 531
  rectangular prisms, 532, 533

Survey, 345

Symbols
  absolute value (| |), 107
  factorial (!), 381
  is congruent to (≅), 422
  is equal to (=), 24
  is greater than (>), 109
  is greater than or equal to (≥), 172
  is less than (<), 109
  is less than or equal to (≤), 172
  is not equal to (≠), 24
  is similar to (~), 440
  measure of angle 1 ($m\angle 1$), 428
  pi (π), 275

Symmetry, lines of, 456

Tally. *See* Frequency tables

Technology, *See also* Calculators, Graphing Calculator Investigation, Internet Connections, Spreadsheet Investigation, and WebQuest

Temperature, 140

Term, 19

Terminating. *See* Decimals

Terms, of a sequence, 34

Tessellations, 447, 448

Test-Taking Tip, 51, 101, 146, 191, 236, 283, 331, 367, 406, 467, 508, 550. *See also* Preparing for Standardized Tests
  backsolving, 259
  drawings, 499
  educated guess, 379
  eliminate possibilities, 229
  eliminating answer choices, 110
  elimination, 355
  grid-in answers, 293, 441
  make a drawing, 183
  read a graph, 87
  reading, 525
  solutions, 25

Theoretical probability, 393, 394

Three-dimensional figures
  building, 512–513
  cone, 541
  cylinder, 524
  drawing, 514–517
  effect of changing dimensions, 526, 536–537
  modeling, 518–519
  nets, 530–531
  pyramid, 531
  rectangular prism, 520–521
  similar solids, 523, 536–537

solids, 514
surface area of, 532–535, 538–541
triangular prism, 522
volume of, 520–527

Ton, 267

Transformations, 451
dilations, 455
identifying, 461
reflections, 456
rotations, 460, 461
translations, 451, 452

Translations, 451, 452
coordinates, 452

Transversal, 427

Trapezoid, 434
area, 490
base, 490
height, 490

Tree diagrams, 374, 375

Triangles, 428, 429
acute, 429
area, 489
base, 489
classifying, 429
equilateral, 429
height, 489
isosceles, 429
obtuse, 429
right, 429
  hypotenuse, 479
  leg, 479
scalene, 429
similar, 440, 441

Triangular numbers, 36

Two-dimensional figures. *See also*
Measurement
complex, 498
constructing triangles, 432–433
decagon, 446
heptagon, 446
hexagon, 446
octagon, 446
parallelogram, 434
pentagon, 446
polygons, 446
quadrilateral, 434–437
rectangle, 434

regular polygon, 446
similar, 440–443
square, 434
tessellation, 447–448
transformations of, 451–461
triangle, 428–431

Two-step equations, 166, 167

Unit rates, 292
best buy, 293

Units
customary, 267
  converting between, 267, 268
metric, 38
  converting between, 38, 39

Upper extreme, 80

Upper quartile (UQ), 80

UQ. *See* Upper quartile

USA TODAY Snapshots®, 6, 17, 62, 175, 180, 209, 222, 261, 320, 343, 345, 396, 421, 425, 492, 544

Variables, 18
defining, 25

Venn diagram, 203

Verbal phrases, as algebraic expressions, 150

Vertex, 413

Vertical angles, 422

Volume, 520
cylinders, 524, 525
rectangular prisms, 520, 521

WebQuest, 3, 95, 103, 185, 193, 277, 285, 401, 409, 545

Webs, 474

Weight
changing customary units, 267–268
ounce, 267
pound, 267
ton, 267

Which One Doesn't Belong?, 11, 20, 61, 107, 140, 158, 199, 218, 229, 256, 260, 268, 290, 299, 317, 372, 379, 453, 458, 472, 515, 543

Whiskers. *See* Box-and-whisker plots

Workbooks, Built-In
Extra Practice, 564–595
Mixed Problem Solving, 596–607
Preparing for Standardized Tests, 608–625
Prerequisite Skills, 554–563

Writing Math, 8, 11, 26, 35, 37, 40, 44, 56, 61, 71, 73, 78, 82, 87, 93, 107, 114, 119, 127, 130, 155, 174, 176, 179, 184, 196, 199, 208, 212, 225, 242, 272, 274, 276, 296, 299, 301, 314, 322, 346, 356, 376, 382, 386, 389, 395, 397, 399, 412, 414, 417, 424, 427, 430, 433, 435, 442, 448, 461, 472, 476, 478, 481, 484, 488, 491, 499, 502, 513, 525, 531, 533, 537, 543

x-axis, 112

x-coordinate, 112

y-axis, 112

y-coordinate, 112

Symbols

Number and Operations

$+$	plus or positive
$-$	minus or negative
$a \cdot b$, $a \times b$, ab or $a(b)$	a times b
$\div$	divided by
$\pm$	plus or minus
$=$	is equal to
$\neq$	is not equal to
$>$	is greater than
$<$	is less than
$\geq$	is greater than or equal to
$\leq$	is less than or equal to
$\approx$	is approximately equal to
$\%$	percent
$a:b$	the ratio of a to b, or $\frac{a}{b}$
$0.7\overline{5}$	repeating decimal $0.75555\ldots$

Algebra and Functions

$-a$	opposite or additive inverse of a
a^n	a to the nth power
a^{-n}	$\frac{1}{a^n}$
$|x|$	absolute value of x
$\sqrt{x}$	principal (positive) square root of x
$f(n)$	function, f of n

Geometry and Measurement

$\cong$	is congruent to
$\sim$	is similar to
$\circ$	degree(s)
$\overleftrightarrow{AB}$	line AB
$\overrightarrow{AB}$	ray AB
$\overline{AB}$	line segment AB
AB	length of $\overline{AB}$
⌐	right angle
$\perp$	is perpendicular to
$\parallel$	is parallel to
$\angle A$	angle A
$m\angle A$	measure of angle A
$\triangle ABC$	triangle ABC
(a, b)	ordered pair with x-coordinate a and y-coordinate b
O	origin
π	pi (approximately 3.14 or $\frac{22}{7}$)

Probability and Statistics

$P(A)$	probability of event A
$n!$	n factorial
$P(n, r)$	permutation of n things taken r at a time
$C(n, r)$	combination of n things taken r at a time

Formulas

Perimeter	square	$P = 4s$
	rectangle	$P = 2\ell + 2w$ or $P = 2(\ell + w)$
Circumference	circle	$C = 2\pi r$ or $C = \pi d$
Area	square	$A = s^2$
	rectangle	$A = \ell w$
	parallelogram	$A = bh$
	triangle	$A = \frac{1}{2}bh$
	trapezoid	$A = \frac{1}{2}h(b_1 + b_2)$
	circle	$A = \pi r^2$
Surface Area	cube	$S = 6s^2$
	rectangular prism	$S = 2\ell w + 2\ell h + 2wh$
	cylinder	$S = 2\pi rh + 2\pi r^2$
Volume	cube	$V = s^3$
	prism	$V = \ell wh$ or Bh
	cylinder	$V = \pi r^2 h$ or Bh
	pyramid	$V = \frac{1}{3}Bh$
	cone	$V = \frac{1}{3}\pi r^2 h$ or $\frac{1}{3}Bh$
Pythagorean Theorem	right triangle	$a^2 + b^2 = c^2$
Temperature	Fahrenheit to Celsius	$C = \frac{5}{9}(F - 32)$
	Celsius to Fahrenheit	$F = \frac{9}{5}C + 32$

Measurement Conversions

Length	1 kilometer (km) = 1,000 meters (m) 1 meter = 100 centimeters (cm) 1 centimeter = 10 millimeters (mm)	1 foot (ft) = 12 inches (in.) 1 yard (yd) = 3 feet or 36 inches 1 mile (mi) = 1,760 yards or 5,280 feet
Volume and Capacity	1 liter (L) = 1,000 milliliters (mL) 1 kiloliter (kL) = 1,000 liters	1 cup (c) = 8 fluid ounces (fl oz) 1 pint (pt) = 2 cups 1 quart (qt) = 2 pints 1 gallon (gal) = 4 quarts
Weight and Mass	1 kilogram (kg) = 1,000 grams (g) 1 gram = 1,000 milligrams (mg) 1 metric ton = 1,000 kilograms	1 pound (lb) = 16 ounces (oz) 1 ton (T) = 2,000 pounds
Time	1 minute (min) = 60 seconds (s) 1 hour (h) = 60 minutes 1 day (d) = 24 hours	1 week (wk) = 7 days 1 year (yr) = 12 months (mo) or 52 weeks or 365 days 1 leap year = 366 days
Metric to Customary	1 meter ≈ 39.37 inches 1 kilometer ≈ 0.62 mile 1 centimeter ≈ 0.39 inch	1 kilogram ≈ 2.2 pounds 1 gram ≈ 0.035 ounce 1 liter ≈ 1.057 quarts